Introduction to Modern Analysis

Oxford Graduate Texts in Mathematics

OXFORD GRADUATE TEXTS IN MATHEMATICS

---

1. Keith Hannabuss: *An Introduction to Quantum Theory*
2. Reinhold Meise and Dietmar Vogt: *Introduction to Functional Analysis*
3. James G. Oxley: *Matroid Theory*
4. N. J. Hitchin, G. B. Segal, and R. S. Ward: *Integrable Systems: Twistors, Loop Groups, and Riemann Surfaces*
5. Wulf Rossmann: *Lie Groups: An Introduction Through Linear Groups*
6. Qing. Liu: *Algebraic Geometry and Arithmetic Curves*
7. Martin R. Bridson and Simon M, Salamon (eds): *Invitations to Geometry and Topology*
8. Shmuel Kantorovitz: *Introduction to Modern Analysis*
9. Terry Lawson: *Topology: A Geometric Approach*
10. Meinolf Geck: *An Introduction to Algebraic Geometry and Algebraic Groups*
11. Alastair Fletcher and Vladimir Markovic: *Quasiconformal Maps and Teichmüller Theory*
12. Dominic Joyce: *Riemannian Holonomy Groups and Calibrated Geometry*
13. Fernando Villegas: *Experimental Number Theory*
14. Péter Medvegyev: *Stochastic Integration Theory*
15. Martin A. Guest: *From Quantum Cohomology to Integrable Systems*
16. Alan D. Rendall: *Partial Differential Equations in General Relativity*
17. Yves Félix, John Oprea, and Daniel Tanré: *Algebraic Models in Geometry*
18. Jie Xiong: *Introduction to Stochastic Filtering Theory*
19. Maciej Dunajski: *Solitons, Instantons, and Twistors*
20. Graham R. Allan: *Introduction to Banach Spaces and Algebras*
21. James Oxley: *Matroid Theory, Second Edition*
22. Simon Donaldson: *Riemann Surfaces*
23. Clifford Henry Taubes: *Differential Geometry: Bundles, Connections, Metrics and Curvature*
24. Gopinath Kallianpur and P. Sundar: *Stochastic Analysis and Diffusion Processes*
25. Selman Akbulut: *4-Manifolds*
26. Fon-Che Liu: *Real Analysis*
27. Dusa McDuff and Dietmar Salamon: *Introduction to Symplectic Topology, Third Edition*
28. Chris Heunen, Jamie Vicary: *Categories for Quantum Theory: An Introduction*
29. Shmuel Kantorovitz, Ami Viselter: *Introduction to Modern Analysis, Second Edition*

Introduction to Modern Analysis

Second Edition

Shmuel Kantorovitz

Bar Ilan University, Israel

Ami Viselter

University of Haifa, Israel

OXFORD
UNIVERSITY PRESS

Great Clarendon Street, Oxford, OX2 6DP,
United Kingdom

Oxford University Press is a department of the University of Oxford.
It furthers the University's objective of excellence in research, scholarship,
and education by publishing worldwide. Oxford is a registered trade mark of
Oxford University Press in the UK and in certain other countries

Impression: 1

Published in the United States of America by Oxford University Press
198 Madison Avenue, New York, NY 10016, United States of America

British Library Cataloguing in Publication Data
Data available

Library of Congress Control Number: 2022933188

ISBN 978–0–19–284954–0 (hbk)
ISBN 978–0–19–284955–7 (pbk)

DOI: 10.1093/oso/9780192849540.001.0001

Printed and bound by
CPI Group (UK) Ltd, Croydon, CR0 4YY

To Ita, Bracha, Pnina, Pinchas, Ruth, and Lilach

Contents

Preface to the First Edition

This book grew out of lectures given since 1964 at Yale University, the University of Illinois at Chicago, and Bar Ilan University. The material covers the usual topics of Measure Theory and Functional Analysis, with applications to Probability Theory and to the theory of linear partial differential equations. Some relatively advanced topics are included in each chapter (excluding the first two): the Riesz–Markov representation theorem and differentiability in Euclidean spaces (Chapter 3); Haar measure (Chapter 4); Marcinkiewicz's interpolation theorem (Chapter 5); the Gelfand–Naimark–Segal representation theorem (Chapter 7); the von Neumann double commutant theorem (Chapter 8); the spectral representation theorem for normal operators (Chapter 9); the extension theory for unbounded symmetric operators (Chapter 10); the Lyapounov Central Limit theorem and the Kolmogoroff "Three Series theorem" (Application I); the Hormander–Malgrange theorem, fundamental solutions of linear partial differential equations with variable coefficients, and Hormander's theory of convolution operators, with an application to integration of pure imaginary order (Application II). Some important complementary material is included in the 'Exercises' sections, with step-by-step detailed hints leading to the wanted results. Solutions to the end of chapter exercises may be found on the companion website for this text: http://www.oup.co.uk/academic/companion/mathematics/kantorovitz.

Ramat Gan
July 2002

S. K.

Preface to the Second Edition

The purpose of the second edition is to make our *Introduction to Modern Analysis* more modern. We did this mostly by broadening and deepening the presentation of operator algebras, which form a central area in functional analysis. There are three new chapters: Chapter 11 on C^*-algebras, Chapter 12 on von Neumann algebras, and Chapter 13 on constructions of C^*-algebras. They contain much more material on these subjects than the first edition. These chapters are also more advanced than the previous parts of the book and require more from the reader, occasionally in the form of guided exercises. Nevertheless, what we give here is merely a taste of operator algebras.

In addition, we made numerous corrections and added quite a lot of exercises. There are also new subjects of independent interest: fixed-point theorems (Chapter 5); the bounded *weak**-topology (Chapter 5); the Arens products (Chapter 7); tensor products of vector spaces and of Hilbert spaces (Chapter 8); and quadratic forms (Chapter 10).

Ramat Gan and Haifa S. K. and A. V.
December 2021

1

Measures

This chapter begins the study of measure theory, which spans Chapters 1–3 and most of Chapter 4. Let us explain first what necessitated this theory.

Consider the set of all Riemann integrable functions on an interval $[a, b]$. It becomes a semi-normed space with respect to the semi-norm $\|f\| := \int_a^b |f(x)|dx$. The problem is that this space is *not complete*: it admits non-convergent Cauchy sequences. As discussed later in the book, in modern analysis it is especially important for (semi-) normed spaces to be complete. Measure theory, invented by H. L. Lebesgue, introduces the concept of a *measure space*, which is a triple $(X, \mathcal{A}, \mu)$, where X is a set, $\mathcal{A}$ is the σ-algebra of all *measurable subsets* of X, and $\mu : \mathcal{A} \to [0, \infty]$ is a *measure*. To each such triple there is an associated *Lebesgue integral*. In the particular case when $X = [a, b]$ and μ is the *Lebesgue measure* (very roughly, $\mu([c, d]) = d - c$, which justifies the word "measure"), every Riemann integrable function is also Lebesgue integrable (but not conversely!) and the integrals coincide.

One big virtue of Lebesgue integration is that the space of integrable functions that comes out of it is complete. In fact, to every measure space we associate not one complete normed space, but a continuum of them—the so-called L^p-spaces ($p \in [1, \infty]$).

The chapter is structured as follows. We first introduce positive measure spaces and Lebesgue integration on them and prove several *convergence theorems* that are fundamental in the theory. We then define the L^p-spaces and prove that they are *Banach spaces*, that is, complete normed spaces. Next, we prove a few basic facts on Hilbert spaces culminating in the *"little" Riesz representation theorem* (we return to Hilbert spaces in Chapter 8). Hilbert spaces are needed in the proof of the *Lebesgue–Radon–Nikodym theorem*, a deep result about the relationship between two arbitrary measures. *Complex* measures are then introduced and studied. A few notions of *convergence* of sequences of measurable functions are defined and the relations between them are explained, including a surprising theorem of *Egoroff* saying that on *finite* measure spaces, pointwise convergence is "almost uniform". A short treatment of the *distribution function* of

Introduction to Modern Analysis. Second Edition. Shmuel Kantorovitz and Ami Viselter, Oxford University Press.
 DOI: 10.1093/oso/9780192849540.003.0001

a measurable function follows. The chapter ends with the notion of a *truncation* of a function.

1.1 Measurable sets and functions

The setting of abstract measure theory is a family $\mathcal{A}$ of so-called *measurable* subsets of a given set X, and a function

$$\mu : \mathcal{A} \to [0, \infty],$$

so that the *measure* $\mu(E)$ of the set $E \in \mathcal{A}$ has some "intuitively desirable" property, such as "countable additivity":

$$\mu\left(\bigcup_{i=1}^{\infty} E_i\right) = \sum_{i=1}^{\infty} \mu(E_i),$$

for mutually disjoint sets $E_i \in \mathcal{A}$. In order to make sense, this setting has to deal with a family $\mathcal{A}$ that is closed under countable unions. We then arrive to the concept of a *measurable space*.

Definition 1.1. Let X be a (non-empty) set. A σ-algebra of subsets of X (briefly, a σ-algebra *on* X) is a subfamily $\mathcal{A}$ of the family $\mathbb{P}(X)$ of all subsets of X, with the following properties:

(1) $X \in \mathcal{A}$;

(2) if $E \in \mathcal{A}$, then the complement E^{c} of E belongs to $\mathcal{A}$;

(3) if $\{E_i\}$ is a *sequence* of sets in $\mathcal{A}$, then its *union* belongs to $\mathcal{A}$.

The ordered pair $(X, \mathcal{A})$, with $\mathcal{A}$ a σ-algebra on X, is called a *measurable space*. The sets of the family $\mathcal{A}$ are called *measurable sets* (or $\mathcal{A}$-measurable sets) in X.

Observe that by (1) and (2), the empty set $\emptyset$ belongs to the σ-algebra $\mathcal{A}$. Taking then $E_i = 0$ for all $i > n$ in (3), we see that $\mathcal{A}$ is closed under *finite unions*; if this weaker condition replaces (3), $\mathcal{A}$ is called an *algebra* of subsets of X (briefly, an algebra *on* X).

By (2) and (3), and DeMorgan's Law, $\mathcal{A}$ is closed under countable intersections (finite intersections, in the case of an algebra). In particular, any algebra on X is closed under differences $E - F := E \cap F^{\mathrm{c}}$.

The intersection of an arbitrary family of σ-algebras on X is a σ-algebra on X. If all the σ-algebras in the family contain some fixed collection $\mathcal{E} \subset \mathbb{P}(X)$, the said intersection is the smallest σ-algebra on X (with respect to set inclusion) that contains $\mathcal{E}$; it is called *the σ-algebra generated by $\mathcal{E}$*, and is denoted by $[\mathcal{E}]$.

An important case comes up naturally when X is a topological space (for some topology τ). The σ-algebra $[\tau]$ generated by the topology is called the *Borel* (σ)-algebra [denoted $\mathcal{B}(X)$], and the sets in $\mathcal{B}(X)$ are the *Borel sets* in X. For example, the countable intersection of τ-open sets (a so-called G_δ-set) and the countable union of τ-closed sets (a so-called F_σ-set) are Borel sets.

Definition 1.2. Let $(X, \mathcal{A})$ and $(Y, \mathcal{B})$ be measurable spaces. A map $f : X \to Y$ is *measurable* if for each $B \in \mathcal{B}$, the set

$$f^{-1}(B) := \{x \in X; f(x) \in B\} := [f \in B]$$

belongs to $\mathcal{A}$.

A constant map $f(x) = p \in Y$ is trivially measurable, since $[f \in B]$ is either $\emptyset$ or X (when $p \in B^c$ and $p \in B$, respectively), and so belongs to $\mathcal{A}$.

When Y is a topological space, we shall usually take $\mathcal{B} = \mathcal{B}(Y)$, the Borel algebra on Y. In particular, for $Y = \mathbb{R}$ (the real line), $Y = [-\infty, \infty]$ (the "extended real line"), or $Y = \mathbb{C}$ (the complex plane), with their usual topologies, we shall call the measurable map a *measurable function* (more precisely, an $\mathcal{A}$-measurable function). If X is a topological space, a $\mathcal{B}(X)$-measurable map (function) is called a Borel map (function).

Given a measurable space $(X, \mathcal{A})$ and a map $f : X \to Y$, for an arbitrary set Y, the family

$$\mathcal{B}_f := \{F \in \mathbb{P}(Y); f^{-1}(F) \in \mathcal{A}\}$$

is a σ-algebra on Y (because the inverse image operation preserves the set theoretical operations: $f^{-1}(\bigcup_\alpha F_\alpha) = \bigcup_\alpha f^{-1}(F_\alpha)$, etc.), and it is the largest σ-algebra on Y for which f is measurable.

If Y is a topological space, and $f^{-1}(V) \in \mathcal{A}$ for every *open* V, then $\mathcal{B}_f$ contains the topology τ, and so contains $\mathcal{B}(Y)$; that is, f is measurable. Since $\tau \subset \mathcal{B}(Y)$, the converse is trivially true.

Lemma 1.3. *A map f from a measurable space $(X, \mathcal{A})$ to a topological space Y is measurable if and only if $f^{-1}(V) \in \mathcal{A}$ for every open $V \subset Y$.*

In particular, if X is also a topological space, and $\mathcal{A} = \mathcal{B}(X)$, it follows that every continuous map $f : X \to Y$ is a Borel map.

Lemma 1.4. *A map f from a measurable space $(X, \mathcal{A})$ to $[-\infty, \infty]$ is measurable if and only if*

$$[f > c] \in \mathcal{A}$$

for all real c.

The non-trivial direction in the lemma follows from the fact that $(c, \infty] \in \mathcal{B}_f$ by hypothesis for all real c; therefore, the σ-algebra $\mathcal{B}_f$ contains the sets

$$\bigcup_{n=1}^{\infty} (b - 1/n, \infty]^c = \bigcup_{n=1}^{\infty} [-\infty, b - 1/n] = [-\infty, b)$$

and $(a, b) = [-\infty, b) \cap (a, \infty]$ for every real $a < b$, and so contains all countable unions of "segments" of the above type, that is, all open subsets of $[-\infty, \infty]$.

The sets $[f > c]$ in the condition of Lemma 1.4 can be replaced by any of the sets $[f \geq c], [f < c]$, or $[f \leq c]$ (for all real c), respectively. The proofs are analogous.

For $f : X \to [-\infty, \infty]$ measurable and α real, the function αf (defined pointwise, with the usual arithmetics $\alpha \cdot \infty = \infty$ for $\alpha > 0, = 0$ for $\alpha = 0$, and $= -\infty$ for $\alpha < 0$, and similarly for $-\infty$) is measurable, because for all real c, $[\alpha f > c] = [f > c/\alpha]$ for $\alpha > 0, = [f < c/\alpha]$ for $\alpha < 0$, and αf is constant for $\alpha = 0$.

If $\{a_n\} \subset [-\infty, \infty]$, one denotes the superior (inferior) limit, that is, the "largest" ("smallest") limit point, of the sequence by $\limsup a_n$ ($\liminf a_n$, respectively).

Let $b_n := \sup_{k \geq n} a_k$. Then $\{b_n\}$ is a decreasing sequence, and therefore

$$\exists \lim_n b_n = \inf_n b_n.$$

Let $\alpha := \limsup a_n$ and $\beta = \lim b_n$. For any given $n \in \mathbb{N}, a_k \leq b_n$ for all $k \geq n$, and therefore $\alpha \leq b_n$. Hence $\alpha \leq \beta$.

On the other hand, for any $t > \alpha, a_k > t$ for at most *finitely many* indices k. Therefore, there exists n_0 such that $a_k \leq t$ for all $k \geq n_0$, hence $b_{n_0} \leq t$. But then $b_n \leq t$ for all $n \geq n_0$ (because $\{b_n\}$ is decreasing), and so $\beta \leq t$. Since $t > \alpha$ was arbitrary, it follows that $\beta \leq \alpha$, and the conclusion $\alpha = \beta$ follows. We showed

$$\limsup a_n = \lim_n \left(\sup_{k \geq n} a_k \right) = \inf_{n \in \mathbb{N}} \left(\sup_{k \geq n} a_k \right). \tag{1}$$

Similarly

$$\liminf a_n = \lim_n \left(\inf_{k \geq n} a_k \right) = \sup_{n \in \mathbb{N}} \left(\inf_{k \geq n} a_k \right). \tag{2}$$

Lemma 1.5. *Let $\{f_n\}$ be a sequence of measurable $[-\infty, \infty]$-valued functions on the measurable space $(X, \mathcal{A})$. Then the functions $\sup f_n, \inf f_n, \limsup f_n, \liminf f_n$, and $\lim f_n$ (when it exists), all defined pointwise, are measurable.*

Proof. Let $h = \sup f_n$. Then for all real c,

$$[h > c] = \bigcup_n [f_n > c] \in \mathcal{A},$$

so that h is measurable by Lemma 1.4.

As remarked, $-f_n = (-1)f_n$ are measurable, and therefore $\inf f_n = -\sup (-f_n)$ is measurable.

The proof is completed by the relations (1), (2), and

$$\lim f_n = \limsup f_n = \liminf f_n,$$

when the second equality holds (i.e. if and only if $\lim f_n$ exists). □

In particular, taking a sequence with $f_k = f_n$ for all $k > n$, we see that $\max\{f_1, \ldots, f_n\}$ and $\min\{f_1, \ldots, f_n\}$ are measurable, when $f_1, \ldots, f_n$ are measurable functions into $[-\infty, \infty]$. For example, the *positive* (*negative*) parts $f^+ := \max\{f, 0\}$ ($f^- := -\min\{f, 0\}$) of a measurable function $f : X \to [-\infty, \infty]$ are (non-negative) measurable functions. Note the decompositions

$$f = f^+ - f^-; \quad |f| = f^+ + f^-.$$

Lemma 1.6. *Let $(X, \mathcal{A})$, $(Y, \mathcal{B})$ and $(Z, \mathcal{C})$ be measurable spaces. If $f : X \to Y$ and $g : Y \to Z$ are measurable, then so is the composite function $h := g \circ f : X \to Z$.*

Indeed, for every $C \in \mathcal{C}$ we have $g^{-1}(C) \in \mathcal{B}$ by measurability of g, thus $h^{-1}(C) = f^{-1}(g^{-1}(C)) \in \mathcal{A}$ by measurability of f.

In particular, if Y, Z are topological spaces and $g : Y \to Z$ is continuous, then $g \circ f$ is measurable.

If

$$Y = \prod_{k=1}^{n} Y_k$$

is the product space of topological spaces Y_k, the projections $p_k : Y \to Y_k$ are continuous. Therefore, if $f : X \to Y$ is measurable, so are the "component functions" $f_k(x) := p_k(f(x)) : X \to Y_k$ $(k = 1, \ldots, n)$, by Lemma 1.6. Conversely, if the topologies on Y_k have countable bases (for all k), a countable base for the topology of Y consists of sets of the form $V = \prod_{k=1}^{n} V_k$ with V_k varying in a countable base for the topology of Y_k (for each k). Now,

$$[f \in V] = \bigcap_{k=1}^{n} [f_k \in V_k] \in \mathcal{A}$$

if all f_k are measurable. Since every open $W \subset Y$ is a countable union of sets of the above type, $[f \in W] \in \mathcal{A}$, and f is measurable. We proved:

Lemma 1.7. *Let Y be the Cartesian product of topological spaces $Y_1, \ldots, Y_n$ with countable bases to their topologies. Let $(X, \mathcal{A})$ be a measurable space. Then $f : X \to Y$ is measurable iff the components f_k are measurable for all k.*

For example, if $f_k : X \to \mathbb{C}$ are measurable for $k = 1, \ldots, n$, then $f := (f_1, \ldots, f_n) : X \to \mathbb{C}^n$ is measurable, and since $g(z_1, \ldots, z_n) := \Sigma \alpha_k z_k (\alpha_k \in \mathbb{C})$ and $h(z_1, \ldots, z_n) = z_1 \ldots z_n$ are continuous from $\mathbb{C}^n$ to $\mathbb{C}$, it follows from Lemma 1.6 that (finite) linear combinations and products of complex measurable functions are measurable. Thus, the complex measurable functions form an algebra over the complex field (similarly, the real measurable functions form an algebra over the real field), for the usual pointwise operations.

If f has values in $\mathbb{R}$, $[-\infty, \infty]$, or $\mathbb{C}$, its measurability implies that of $|f|$, by Lemma 1.6.

By Lemma 1.7, a complex function is measurable iff its real part $\Re f$ and imaginary part $\Im f$ are both measurable.

If f, g are measurable with values in $[0, \infty]$, the functions $f + g$ and fg are well-defined pointwise (with values in $[0, \infty]$) and measurable, since the functions $(s, t) \to s + t$ and $(s, t) \to st$ from $[0, \infty]^2$ to $[0, \infty]$ are Borel (cf. Lemmas 1.6 and 1.7).

The function $f : X \to \mathbb{C}$ is *simple* if its range is a *finite* set $\{c_1, \ldots, c_n\} \subset \mathbb{C}$. Let $E_k := [f = c_k]$, $k = 1, \ldots, n$. Then X is the disjoint union of the sets E_k, and

$$f = \sum_{k=1}^{n} c_k I_{E_k},$$

where I_E denotes the *indicator* of E (also called the *characteristic function* of E by non-probabilists, while probabilists reserve the later name to a different concept):

$$I_E(x) = 1 \quad \text{for } x \in E \quad \text{and} \quad = 0 \quad \text{for } x \in E^c.$$

Since a singleton $\{c\} \subset \mathbb{C}$ is closed, it is a Borel set. Suppose now that the simple (complex) function f is defined on a measurable space $(X, \mathcal{A})$. If f is measurable, then $E_k := [f = c_k]$ is measurable for all $k = 1, \ldots, n$. Conversely, if all E_k are measurable, then for each open $V \subset \mathbb{C}$,

$$[f \in V] = \bigcup_{\{k; c_k \in V\}} E_k \in \mathcal{A},$$

so that f is measurable. In particular, an indicator I_E is measurable iff $E \in \mathcal{A}$.

Let $B(X, \mathcal{A})$ denote the complex algebra of all *bounded* complex $\mathcal{A}$-measurable functions on X (for the pointwise operations), and denote

$$\|f\| = \sup_X |f| \quad (f \in B(X, \mathcal{A})).$$

The map $f \to \|f\|$ of $B(X, \mathcal{A})$ into $[0, \infty)$ has the following properties:

(1) $\|f\| = 0$ iff $f = 0$ (the zero function);

(2) $\|\alpha f\| = |\alpha| \, \|f\|$ for all $\alpha \in \mathbb{C}$ and $f \in B(X, \mathcal{A})$;

(3) $\|f + g\| \le \|f\| + \|g\|$ for all $f, g \in B(X, \mathcal{A})$;

(4) $\|fg\| \le \|f\| \, \|g\|$ for all $f, g \in B(X, \mathcal{A})$.

For example, (3) is verified by observing that for all $x \in X$,

$$|f(x) + g(x)| \le |f(x)| + |g(x)| \le \sup_X |f| + \sup_X |g|.$$

A map $\|\cdot\|$ from any (complex) vector space Z to $[0, \infty)$ with Properties (1)–(3) is called a *norm* on Z. The previous example is the *supremum norm* or *uniform norm* on the vector space $Z = B(X, \mathcal{A})$. Property (1) is the *definiteness* of the norm; Property (2) is its *homogeneity*; Property (3) is the *triangle inequality*. A vector space with a specified norm is a *normed space*. If Z is an algebra, and the specified norm satisfies Property (4) also, Z is called a *normed algebra*. Thus, $B(X, \mathcal{A})$ is a normed algebra with respect to the supremum norm. Any normed space Z is a metric space for the metric *induced by the norm*

$$d(u, v) := \|u - v\| \quad u, v \in Z.$$

Convergence in Z is convergence with respect to this metric (unless stated otherwise). Thus, convergence in the normed space $B(X, \mathcal{A})$ is precisely *uniform convergence* on X (this explains the name "uniform norm").

If $x, y \in Z$, the triangle inequality implies $\|x\| = \|(x-y)+y\| \le \|x-y\|+\|y\|$, so that $\|x\| - \|y\| \le \|x - y\|$. Since we may interchange x and y, we have

$$\big| \|x\| - \|y\| \big| \le \|x - y\|.$$

In particular, *the norm function is continuous on* Z.

The simple functions in $B(X, \mathcal{A})$ form a subalgebra $B_0(X, \mathcal{A})$; it is *dense* in $B(X, \mathcal{A})$:

Theorem 1.8 (Approximation theorem). *Let $(X, \mathcal{A})$ be a measurable space. Then:*

(1) $B_0(X, \mathcal{A})$ is dense in $B(X, \mathcal{A})$ (i.e., every bounded complex measurable function is the uniform limit of a sequence of simple measurable complex functions).

(2) If $f : X \to [0, \infty]$ is measurable, then there exists a sequence of measurable simple functions

$$0 \le \phi_1 \le \phi_2 \le \cdots \le f,$$

such that $f = \lim \phi_n$.

Proof. (1) Since any $f \in B(X, \mathcal{A})$ can be written as

$$f = u^+ - u^- + iv^+ - iv^-$$

with $u = \Re f$ and $v = \Im f$, it suffices to prove (1) for f with range in $[0, \infty)$. Let N be the first integer such that $N > \sup f$. For $n = 1, 2, \ldots$, set

$$\phi_n := \sum_{k=1}^{N2^n} \frac{k-1}{2^n} I_{E_{n,k}},$$

where

$$E_{n,k} := f^{-1}\left(\left[\frac{k-1}{2^n}, \frac{k}{2^n}\right)\right).$$

The simple functions ϕ_n are measurable,

$$0 \le \phi_1 \le \phi_2 \le \cdots \le f,$$

and

$$0 \le f - \phi_n < \frac{1}{2^n},$$

so that indeed $\|f - \phi_n\| \le (1/2^n)$, as wanted.

If f has range in $[0, \infty]$, set

$$\phi_n := \sum_{k=1}^{n2^n} \frac{k-1}{2^n} I_{E_{n,k}} + nI_{F_n},$$

where $F_n := [f \ge n]$. Again $\{\phi_n\}$ is a non-decreasing sequence of non-negative measurable simple functions $\le f$. If $f(x) = \infty$ for some $x \in X$, then $x \in F_n$

for all n, and therefore $\phi_n(x) = n$ for all n; hence $\lim_n \phi_n(x) = \infty = f(x)$. If $f(x) < \infty$ for some x, let $n > f(x)$. Then there exists a unique $k, 1 \le k \le n2^n$, such that $x \in E_{n,k}$. Then $\phi_n(x) = ((k-1)/2^n)$ while $((k-1)/2^n) \le f(x) < (k/2^n)$, so that

$$0 \le f(x) - \phi_n(x) < 1/2^n \quad (n > f(x)).$$

Hence $f(x) = \lim_n \phi_n(x)$ for all $x \in X$. □

1.2 Positive measures

Definition 1.9. Let $(X, \mathcal{A})$ be a measurable space. A (*positive*) *measure* on $\mathcal{A}$ is a function

$$\mu : \mathcal{A} \to [0, \infty]$$

such that $\mu(\emptyset) = 0$ and

$$\mu\Big(\bigcup_{k=1}^{\infty} E_k\Big) = \sum_{k=1}^{\infty} \mu(E_k) \tag{1}$$

for any sequence of mutually disjoint sets $E_k \in \mathcal{A}$. Property (1) is called *σ-additivity* of the function μ. The ordered triple $(X, \mathcal{A}, \mu)$ will be called a (*positive*) *measure space*.

Taking in particular $E_k = \emptyset$ for all $k > n$, it follows that

$$\mu\Big(\bigcup_{k=1}^{n} E_k\Big) = \sum_{k=1}^{n} \mu(E_k) \tag{2}$$

for any *finite* collection of mutually disjoint sets $E_k \in \mathcal{A}, k = 1, \dots, n$. We refer to Property (2) by saying that μ is (*finitely*) *additive*.

Any finitely additive function $\mu \ge 0$ on an algebra $\mathcal{A}$ is necessarily *monotonic*, that is, $\mu(E) \le \mu(F)$ when $E \subset F (E, F \in \mathcal{A})$; indeed

$$\mu(F) = \mu(E \cup (F - E)) = \mu(E) + \mu(F - E) \ge \mu(E).$$

If $\mu(E) < \infty$, we get

$$\mu(F - E) = \mu(F) - \mu(E).$$

Lemma 1.10. *Let $(X, \mathcal{A}, \mu)$ be a positive measure space, and let*

$$E_1 \subset E_2 \subset E_3 \subset \cdots$$

be measurable sets with union E. Then

$$\mu(E) = \lim_n \mu(E_n).$$

Proof. The sets E_n and E can be written as *disjoint* unions

$$E_n = E_1 \cup (E_2 - E_1) \cup (E_3 - E_2) \cup \cdots \cup (E_n - E_{n-1}),$$
$$E = E_1 \cup (E_2 - E_1) \cup (E_3 - E_2) \cup \cdots ,$$

where all differences belong to $\mathcal{A}$. Set $E_0 = \emptyset$. By σ-additivity,

$$\mu(E) = \sum_{k=1}^{\infty} \mu(E_k - E_{k-1})$$
$$= \lim_n \sum_{k=1}^{n} \mu(E_k - E_{k-1}) = \lim_n \mu(E_n).$$

□

In general, if E_j belong to an *algebra* $\mathcal{A}$ of subsets of X, set $A_0 = \emptyset$ and $A_n = \bigcup_{j=1}^{n} E_j$, $n = 1, 2, \ldots$. The sets $A_j - A_{j-1}$, $1 \leq j \leq n$, are disjoint $\mathcal{A}$-measurable subsets of E_j with union A_n. If μ is a non-negative *additive* set function on $\mathcal{A}$, then

$$\mu\left(\bigcup_{j=1}^{n} E_j\right) = \mu(A_n) = \sum_{j=1}^{n} \mu(A_j - A_{j-1}) \leq \sum_{j=1}^{n} \mu(E_j). \tag{*}$$

This is the *subadditivity property* of non-negative additive set functions (on algebras).

If $\mathcal{A}$ is a σ-algebra and μ is a positive measure on $\mathcal{A}$, then since $A_1 \subset A_2 \subset \cdots$ and $\bigcup_{n=1}^{\infty} A_n = \bigcup_{j=1}^{\infty} E_j$, letting $n \to \infty$ in (*), it follows from Lemma 1.10 that

$$\mu\left(\bigcup_{j=1}^{\infty} E_j\right) \leq \sum_{j=1}^{\infty} \mu(E_j).$$

This property of positive measures is called *σ-subadditivity.*

For *decreasing* sequences of measurable sets, the "dual" of Lemma 1.10 is false in general, unless we assume that the sets have *finite* measure:

Lemma 1.11. *Let $\{E_k\} \subset \mathcal{A}$ be a decreasing sequence (with respect to set-inclusion) such that $\mu(E_1) < \infty$. Let $E = \bigcap_k E_k$. Then*

$$\mu(E) = \lim_n \mu(E_n).$$

Proof. The sequence $\{E_1 - E_k\}$ is increasing, with union $E_1 - E$. By Lemma 1.10 and the *finiteness* of the measures of E and E_k (subsets of E_1!),

$$\mu(E_1) - \mu(E) = \mu\left(\bigcup_k (E_1 - E_k)\right)$$
$$= \lim \mu(E_1 - E_n) = \mu(E_1) - \lim \mu(E_n),$$

and the result follows by cancelling the *finite* number $\mu(E_1)$. □

If $\{E_k\}$ is an *arbitrary* sequence of subsets of X, set $F_n = \bigcap_{k\geq n} E_k$ and $G_n = \bigcup_{k\geq n} E_k$. Then $\{F_n\}$ ($\{G_n\}$) is increasing (decreasing, respectively), and $F_n \subset E_n \subset G_n$ for all n.

One defines

$$\liminf_n E_n := \bigcup_n F_n; \quad \limsup_n E_n := \bigcap_n G_n.$$

These sets belong to $\mathcal{A}$ if $E_k \in \mathcal{A}$ for all k. The set $\liminf E_n$ consists of all x that *belong to E_n for all but finitely many n*; the set $\limsup E_n$ consists of all x that *belong to E_n for infinitely many n*. By Lemma 1.10,

$$\mu(\liminf E_n) = \lim_n \mu(F_n) \leq \liminf \mu(E_n). \tag{3}$$

If the measure of G_1 is finite, we also have by Lemma 1.11

$$\mu(\limsup E_n) = \lim_n \mu(G_n) \geq \limsup \mu(E_n). \tag{4}$$

1.3 Integration of non-negative measurable functions

Definition 1.12. Let $(X, \mathcal{A}, \mu)$ be a positive measure space, and $\phi : X \to [0, \infty)$ a measurable simple function. The *integral over X of ϕ with respect to μ*, denoted

$$\int_X \phi \, d\mu$$

or briefly

$$\int \phi \, d\mu,$$

is the finite sum

$$\sum_k c_k \mu(E_k) \in [0, \infty],$$

where

$$\phi = \sum_k c_k I_{E_k}, \quad E_k = [\phi = c_k],$$

and c_k are the distinct values of ϕ.

Note that

$$\int I_E \, d\mu = \mu(E) \quad E \in \mathcal{A}$$

and

$$0 \leq \int \phi \, d\mu \leq \|\phi\| \, \mu([\phi \neq 0]). \tag{1}$$

For an *arbitrary* measurable function $f : X \to [0, \infty]$, consider the (non-empty) set S_f of measurable simple functions ϕ such that $0 \le \phi \le f$, and define

$$\int f\,d\mu := \sup_{\phi \in S_f} \int \phi\,d\mu. \tag{2}$$

For any $E \in \mathcal{A}$, the integral *over* E of f is defined by

$$\int_E f\,d\mu := \int f I_E\,d\mu. \tag{3}$$

Let ϕ, ψ be measurable simple functions; let c_k, d_j be the distinct values of ϕ and ψ, taken on the (mutually disjoint) sets E_k and F_j, respectively. Denote $Q := \{(k,j) \in \mathbb{N}^2; E_k \cap F_j \ne \emptyset\}$.

If $\phi \le \psi$, then $c_k \le d_j$ for $(k,j) \in Q$. Hence

$$\int \phi\,d\mu = \sum_k c_k \mu(E_k) = \sum_{(k,j)\in Q} c_k \mu(E_k \cap F_j)$$

$$\le \sum_{(k,j)\in Q} d_j \mu(E_k \cap F_j) = \sum_j d_j \mu(F_j) = \int \psi\,d\mu.$$

Thus, the integral is *monotonic on simple functions.*

If f is simple, then $\int \phi\,d\mu \le \int f\,d\mu$ for all $\phi \in S_f$ (by monotonicity of the integral on simple functions), and therefore the supremum in (2) is less than or equal to the integral of f as a simple function; since $f \in S_f$, the reverse inequality is trivial, so that the two definitions of the integral of f coincide for f simple.

Since $S_{cf} = cS_f := \{c\phi; \phi \in S_f\}$ for $0 \le c < \infty$, we have (for f as above)

$$\int cf\,d\mu = c \int f\,d\mu \quad (0 \le c < \infty). \tag{4}$$

If $f \le g$ (f, g as above), $S_f \subset S_g$, and therefore $\int f\,d\mu \le \int g\,d\mu$ (*monotonicity* of the integral with respect to the "integrand").

In particular, if $E \subset F$ (both measurable), then $fI_E \le fI_F$, and therefore $\int_E f\,d\mu \le \int_F f\,d\mu$ (monotonicity of the integral with respect to the set of integration).

If $\mu(E) = 0$, then any $\phi \in S_{fI_E}$ assumes its non-zero values c_k on the sets $E_k \cap E$, that have measure 0 (as measurable subsets of E), and therefore $\int \phi\,d\mu = 0$ for all such ϕ, hence $\int_E f\,d\mu = 0$.

If $f = 0$ on E (for some $E \in \mathcal{A}$), then fI_E is the zero function, hence has zero integral (by definition of the integral of simple functions!); this means that $\int_E f\,d\mu = 0$ when $f = 0$ on E.

Consider now the set function

$$\nu(E) := \int_E \phi\,d\mu \quad E \in \mathcal{A}, \tag{5}$$

for a *fixed* simple measurable function $\phi \geq 0$. As a special case of the preceding remark, $\nu(\emptyset) = 0$. Write $\phi = \sum c_k I_{E_k}$, and let $A_j \in \mathcal{A}$ be mutually disjoint $(j = 1, 2, \ldots)$ with union A. Then

$$\phi I_A = \sum c_k I_{E_k \cap A},$$

so that, by the σ-additivity of μ and the possibility of interchanging summation order when the summands are non-negative,

$$\begin{aligned}\nu(A) := \sum_k c_k \mu(E_k \cap A) &= \sum_k c_k \sum_j \mu(E_k \cap A_j) \\ &= \sum_j \sum_k c_k \mu(E_k \cap A_j) = \sum_j \nu(A_j).\end{aligned}$$

Thus ν is a positive measure. This is actually true for *any* measurable $\phi \geq 0$ (not necessarily simple), but this will be proved later.

If ψ, χ are *simple* functions as above (the distinct values of ψ and χ being $a_1, \ldots, a_p$ and $b_1, \ldots, b_q$, assumed on the measurable sets $F_1, \ldots, F_p$ and $G_1, \ldots, G_q$, respectively), then the simple measurable function $\phi := \psi + \chi$ assumes the constant value $a_i + b_j$ on the set $F_i \cap G_j$, and therefore, defining the measure ν as shown, we have

$$\nu(F_i \cap G_j) = (a_i + b_j)\mu(F_i \cap G_j). \tag{6}$$

But a_i and b_j are the constant values of ψ and χ on the set $F_i \cap G_j$ (respectively), so that the right-hand side of (6) equals $\nu'(F_i \cap G_j) + \nu''(F_i \cap G_j)$, where ν' and ν'' are the measures defined as ν, with the integrands ψ and χ instead of ϕ. Summing over all i, j, since X is the disjoint union of the sets $F_i \cap G_j$, the additivity of the measures ν, ν', and ν'' implies that $\nu(X) = \nu'(X) + \nu''(X)$, that is,

$$\int (\psi + \chi)\, d\mu = \int \psi\, d\mu + \int \chi\, d\mu. \tag{7}$$

Property (7) is the additivity of the integral over non-negative measurable simple functions. This property too is extended later to *arbitrary* non-negative measurable functions.

Theorem 1.13. *Let $(X, \mathcal{A}, \mu)$ be a positive measure space. Let*

$$f_1 \leq f_2 \leq f_3 \leq \cdots : X \to [0, \infty]$$

be measurable, and denote $f = \lim f_n$ (defined pointwise). Then

$$\int f\, d\mu = \lim \int f_n\, d\mu. \tag{8}$$

This is the Monotone Convergence theorem *of Lebesgue.*

Proof. By Lemma 1.5, f is measurable (with range in $[0,\infty]$). The monotonicity of the integral (and the fact that $f_n \le f_{n+1} \le f$) implies that

$$\int f_n \, d\mu \le \int f_{n+1} \, d\mu \le \int f \, d\mu,$$

and therefore the limit in (8) exists ($:= c \in [0,\infty]$) and the *inequality* $\ge$ holds in (8). It remains to show the inequality $\le$ in (8). Let $0 < t < 1$. Given $\phi \in S_f$, denote

$$A_n = [t\phi \le f_n] = [f_n - t\phi \ge 0] \quad (n = 1, 2, \ldots).$$

Then $A_n \in \mathcal{A}$ and $A_1 \subset A_2 \subset \cdots$ (because $f_1 \le f_2 \le \cdots$). If $x \in X$ is such that $\phi(x) = 0$, then $x \in A_n$ (for all n). If $x \in X$ is such that $\phi(x) > 0$, then $f(x) \ge \phi(x) > t\phi(x)$, and there exists therefore n, for which $f_n(x) \ge t\phi(x)$, that is, $x \in A_n$ (for that n). This shows that $\bigcup_n A_n = X$. Consider the measure ν defined by (5) (for the simple function $t\phi$). By Lemma 1.10,

$$t \int \phi \, d\mu = \nu(X) = \lim_n \nu(A_n) = \lim_n \int_{A_n} t\phi \, d\mu.$$

However, $t\phi \le f_n$ on A_n, so the integrals on the right are $\le \int_{A_n} f_n \, d\mu \le \int_X f_n \, d\mu$ (by the monotonicity property of integrals with respect to the set of integration). Therefore $t \int \phi \, d\mu \le c$, and so $\int \phi \, d\mu \le c$ by the arbitrariness of $t \in (0,1)$. Taking the supremum over all $\phi \in S_f$, we conclude that $\int f \, d\mu \le c$ as wanted. □

For *arbitrary* sequences of non-negative measurable functions we have the following *inequality*:

Theorem 1.14 (Fatou's lemma). *Let $f_n : X \to [0,\infty]$, $n = 1, 2, \ldots$, be measurable. Then*

$$\int \liminf_n f_n \, d\mu \le \liminf_n \int f_n \, d\mu.$$

Proof. We have

$$\liminf_n f_n := \lim_n \big(\inf_{k \ge n} f_k \big).$$

Denote the infimum on the right by g_n. Then $g_n, n = 1, 2, \ldots$, are measurable, $g_n \le f_n$,

$$0 \le g_1 \le g_2 \le \cdots,$$

and $\lim_n g_n = \liminf_n f_n$. By Theorem 1.13,

$$\int \liminf_n f_n \, d\mu = \int \lim g_n \, d\mu = \lim \int g_n \, d\mu.$$

But the integrals on the right are $\le \int f_n \, d\mu$, therefore their limit is $\le \liminf \int f_n \, d\mu$. □

Another consequence of Theorem 1.13 is the additivity of the integral of non-negative measurable functions.

Theorem 1.15. *Let $f, g : X \to [0,\infty]$ be measurable. Then*

$$\int (f+g)\,d\mu = \int f\,d\mu + \int g\,d\mu.$$

Proof. By the Approximation theorem (Theorem 1.8), there exist simple measurable functions ϕ_n, ψ_n such that

$$\begin{aligned} &0 \le \phi_1 \le \phi_2 \le \dots, \quad \lim \phi_n = f, \\ &0 \le \psi_1 \le \psi_2 \le \dots, \quad \lim \psi_n = g. \end{aligned}$$

Then the measurable simple functions $\chi_n = \phi_n + \psi_n$ satisfy

$$0 \le \chi_1 \le \chi_2 \le \dots, \quad \lim \chi_n = f + g.$$

By Theorem 1.13 and the additivity of the integral of (non-negative measurable) *simple* functions (cf. (7)), we have

$$\begin{aligned} \int (f+g)\,d\mu &= \lim \int \chi_n\,d\mu = \lim \int (\phi_n + \psi_n)\,d\mu \\ &= \lim \int \phi_n\,d\mu + \lim \int \psi_n\,d\mu = \int f\,d\mu + \int g\,d\mu. \end{aligned}$$

□

The additivity property of the integral is also true for *infinite* sums of non-negative measurable functions:

Theorem 1.16 (Beppo Levi). *Let $f_n : X \to [0,\infty], n = 1, 2, \dots,$ be measurable. Then*

$$\int \sum_{n=1}^{\infty} f_n\,d\mu = \sum_{n=1}^{\infty} \int f_n\,d\mu.$$

Proof. Let

$$g_k = \sum_{n=1}^{k} f_n; \quad g = \sum_{n=1}^{\infty} f_n.$$

The measurable functions g_k satisfy

$$0 \le g_1 \le g_2 \le \dots, \quad \lim g_k = g,$$

and by Theorem 1.15 (and induction)

$$\int g_k\,d\mu = \sum_{n=1}^{k} \int f_n\,d\mu.$$

Therefore, by Theorem 1.13

$$\int g\,d\mu = \lim_k \int g_k\,d\mu = \lim_k \sum_{n=1}^{k} \int f_n\,d\mu = \sum_{n=1}^{\infty} \int f_n\,d\mu.$$

□

We may extend now the measure property of ν, defined earlier with a *simple* integrand, to the general case of a non-negative measurable integrand.

Theorem 1.17. *Let $f : X \to [0,\infty]$ be measurable, and set*

$$\nu(E) := \int_E f\,d\mu, \quad E \in \mathcal{A}.$$

Then ν is a (positive) measure on $\mathcal{A}$, and for any measurable $g : X \to [0,\infty]$,

$$\int g\,d\nu = \int gf\,d\mu. \tag{*}$$

Proof. Let $E_j \in \mathcal{A}, j = 1,2,\dots$ be mutually disjoint, with union E. Then

$$fI_E = \sum_{j=1}^{\infty} fI_{E_j},$$

and therefore, by Theorem 1.16,

$$\nu(E) := \int fI_E\,d\mu = \sum_j \int fI_{E_j}\,d\mu = \sum_j \nu(E_j).$$

Thus, ν is a measure.

If $g = I_E$ for some $E \in \mathcal{A}$, then

$$\int g\,d\nu = \nu(E) = \int I_E f\,d\mu = \int gf\,d\mu.$$

By (4) and Theorem 1.15 (for the measures μ and ν), (*) is valid for g *simple*. Finally, for general g, the Approximation theorem (Theorem 1.8) provides a sequence of simple measurable functions

$$0 \le \phi_1 \le \phi_2 \le \cdots; \quad \lim \phi_n = g.$$

Then the measurable functions $\phi_n f$ satisfy

$$0 \le \phi_1 f \le \phi_2 f \le \cdots; \quad \lim \phi_n f = gf,$$

and Theorem 1.13 implies that

$$\int g\,d\nu = \lim_n \int \phi_n\,d\nu = \lim_n \int \phi_n f\,d\mu = \int gf\,d\mu. \qquad \square$$

Relation (*) is conveniently abbreviated as

$$d\nu = f\,d\mu.$$

Observe that if f_1 and f_2 coincide *almost everywhere* (briefly, "a.e." or μ-a.e., if the measure needs to be specified), that is, if they coincide except on a *null set* $A \in \mathcal{A}$ (more precisely, a *μ-null set*, that is, a measurable set A such that $\mu(A) = 0$), then the corresponding measures ν_i are equal, and in particular $\int f_1\,d\mu = \int f_2\,d\mu$. Indeed, for all $E \in \mathcal{A}$, $\mu(E \cap A) = 0$, and therefore

$$\nu_i(E \cap A) = \int_{E\cap A} f_i\,d\mu = 0, \quad i = 1, 2$$

by one of the observations following Definition 1.12. Hence

$$\begin{aligned}\nu_1(E) &= \nu_1(E \cap A) + \nu_1(E \cap A^c) = \nu_1(E \cap A^c)\\ &= \nu_2(E \cap A^c) = \nu_2(E).\end{aligned}$$

1.4 Integrable functions

Let $(X, \mathcal{A}, \mu)$ be a positive measure space, and let f be a measurable function with range in $[-\infty, \infty]$ or $\overline{\mathbb{C}} := \mathbb{C} \cup \{\infty\}$ (the Riemann sphere). Then $|f| : X \to [0, \infty]$ is measurable, and has therefore an integral ($\in [0, \infty]$). In case this integral is *finite*, we shall say that f is *integrable*. In that case, the measurable set $[|f| = \infty]$ has measure zero. Indeed, it is contained in $[|f| > n]$ for all $n = 1, 2, \ldots$, and

$$n\mu([|f| > n]) = \int_{[|f|>n]} n\,d\mu \le \int_{[|f|>n]} |f|\,d\mu \le \int |f|\,d\mu.$$

Hence for all n

$$0 \le \mu([|f| = \infty]) \le \frac{1}{n}\int |f|\,d\mu,$$

and since the integral on the right is finite, we must have $\mu([|f| = \infty]) = 0$.

In other words, an integrable function is *finite a.e.*

We observed previously that non-negative measurable functions that coincide a.e. have equal integrals. This property is desirable in the general case now considered. If f is measurable, and if we redefine it as the finite arbitrary constant c on a set $A \in \mathcal{A}$ of measure zero, then the new function g is also measurable. Indeed, for any open set V in the range space,

$$[g \in V] = \{[g \in V] \cap A^c\} \cup \{[g \in V] \cap A\}.$$

The second set on the right is empty if $c \in V^c$, and is A if $c \in V$, thus belongs to $\mathcal{A}$ in any case. The first set on the right is equal to $[f \in V] \cap A^c \in \mathcal{A}$, by the measurability of f. Thus $[g \in V] \in \mathcal{A}$.

If f is integrable, we can redefine it as an arbitrary *finite* constant on the set $[|f| = \infty]$ (that has measure zero) and obtain a new *finite-valued* measurable function, whose integral should be the same as the integral of f (by the "desirable" property mentioned before). This discussion shows that we may restrict ourselves to *complex* (or, as a special case, to *real*) valued measurable functions.

Definition 1.18. Let $(X, \mathcal{A}, \mu)$ be a positive measure space. The function $f : X \to \mathbb{C}$ is *integrable* if it is measurable and

$$\|f\|_1 := \int |f| \, d\mu < \infty.$$

The set of all (complex) *integrable* functions will be denoted by

$$L^1(X, \mathcal{A}, \mu),$$

or briefly by $L^1(\mu)$ or $L^1(X)$ or L^1, when the unmentioned "objects" of the measure space are understood.

Defining the operations pointwise, L^1 is a complex vector space, since the inequality

$$|\alpha f + \beta g| \leq |\alpha| \, |f| + |\beta| \, |g|$$

implies, by monotonicity, additivity, and homogeneity of the integral of non-negative measurable functions:

$$\|\alpha f + \beta g\|_1 \leq |\alpha| \, \|f\|_1 + |\beta| \, \|g\|_1 < \infty,$$

for all $f, g \in L^1$ and $\alpha, \beta \in \mathbb{C}$.

In particular, $\| \cdot \|_1$ satisfies the triangle inequality (take $\alpha = \beta = 1$), and is trivially homogeneous.

Suppose $\|f\|_1 = 0$. For any $n = 1, 2, \ldots,$

$$\begin{aligned} 0 \leq \mu([|f| > 1/n]) &= \int_{[|f|>1/n]} d\mu = n \int_{[|f|>1/n]} (1/n) \, d\mu \\ &\leq n \int_{|f|>1/n]} |f| \, d\mu \leq n \, \|f\|_1 = 0, \end{aligned}$$

so $\mu([|f| > 1/n]) = 0$. Now the set where f is *not* zero is

$$[|f| > 0] = \bigcup_{n=1}^{\infty} [|f| > 1/n],$$

and by the σ-subadditivity property of positive measures, it follows that this set has measure zero. Thus, the vanishing of $\|f\|_1$ implies that $f = 0$ a.e.

(the converse is trivially true). One verifies easily that the relation "$f = g$ a.e." is an equivalence relation for complex measurable functions (transitivity follows from the fact that the union of two sets of measure zero has measure zero, by subadditivity of positive measures). All the functions f in the same equivalence class have the same value of $\|f\|_1$ (cf. discussion following Theorem 1.17).

We use the same notation L^1 for the space of all *equivalence classes* of integrable functions, with operations performed as usual on representatives of the classes, and with the $\|\cdot\|_1$-norm of a class equal to the norm of any of its representatives; L^1 is a normed space (for the norm $\|\cdot\|_1$). It is customary, however, to think of the elements of L^1 as functions (rather than equivalence classes of functions!).

If $f \in L^1$, then $f = u + \mathrm{i}v$ with $u := \Re f$ and $v := \Im f$ real measurable functions (cf. discussion following Lemma 1.7), and since $|u|, |v| \le |f|$, we have $\|u\|_1, \|v\|_1 \le \|f\|_1 < \infty$, that is, u, v are *real* elements of L^1 (conversely, if u, v are real elements of L^1, then $f = u + \mathrm{i}v \in L^1$, since L^1 is a complex vector space).

Writing $u = u^+ - u^-$ (and similarly for v), we obtain four non-negative (finite) measurable functions (cf. remarks following Lemma 1.5), and since $u^+ \le |u| \le |f|$ (and similarly for u^-, etc.), they have *finite integrals.* It makes sense therefore to *define*

$$\int u\,d\mu := \int u^+\,d\mu - \int u^-\,d\mu$$

(on the right, one has the difference of two *finite* non-negative real numbers!).

Doing the same with v, we then let

$$\int f\,d\mu := \int u\,d\mu + \mathrm{i}\int v\,d\mu.$$

Note that according to this definition,

$$\Re\int f\,d\mu = \int \Re f\,d\mu,$$

and similarly for the imaginary part.

Theorem 1.19. *The map $f \to \int f\,d\mu \in \mathbb{C}$ is a continuous linear functional on the normed space $L^1(\mu)$.*

Proof. Consider first real-valued functions $f, g \in L^1$. Let $h = f + g$. Then

$$h^+ - h^- = (f^+ - f^-) + (g^+ - g^-),$$

and since all functions here have *finite* values,

$$h^+ + f^- + g^- = h^- + f^+ + g^+.$$

By Theorem 1.15,

$$\int h^+\,d\mu + \int f^-\,d\mu + \int g^-\,d\mu = \int h^-\,d\mu + \int f^+\,d\mu + \int g^+\,d\mu.$$

All integrals above are finite, so we may subtract $\int h^- + \int f^- + \int g^-$ from both sides of the equation. This yields:

$$\int h\,d\mu = \int f\,d\mu + \int g\,d\mu.$$

The additivity of the integral extends trivially to *complex* functions in L^1.

If $f \in L^1$ is real and $c \in [0,\infty)$, $(cf)^+ = cf^+$ and similarly for f^-. Therefore, by (4) (following Definition 1.12),

$$\int cf\,d\mu = \int cf^+\,d\mu - \int cf^-\,d\mu = c\int f^+\,d\mu - c\int f^-\,d\mu = c\int f\,d\mu.$$

If $c \in (-\infty, 0)$, $(cf)^+ = -cf^-$ and $(cf)^- = -cf^+$, and a similar calculation shows again that $\int(cf) = c\int f$. For $f \in L^1$ complex and c real, write $f = u + iv$. Then

$$\int cf = \int(cu + \mathrm{i}cv) := \int(cu) + \mathrm{i}\int(cv) = c\Big(\int u + \mathrm{i}\int v\Big) := c\int f.$$

Note next that

$$\int(\mathrm{i}f) = \int(-v + \mathrm{i}u) = -\int v + \mathrm{i}\int u = \mathrm{i}\int f.$$

Finally, if $c = a + ib$ (a, b real), then by additivity of the integral and the previous remarks,

$$\int(cf) = \int(af + \mathrm{i}bf) = \int(af) + \int \mathrm{i}bf = a\int f + \mathrm{i}b\int f = c\int f.$$

Thus

$$\int(\alpha f + \beta g)\,d\mu = \alpha\int f\,d\mu + \beta\int g\,d\mu$$

for all $f, g \in L^1$ and $\alpha, \beta \in \mathbb{C}$.

For $f \in L^1$, let $\lambda := \int f\,d\mu (\in \mathbb{C})$. Then, since the left-hand side of the following equation is *real*,

$$\begin{aligned}|\lambda| = \mathrm{e}^{\mathrm{i}\theta}\lambda = \mathrm{e}^{\mathrm{i}\theta}\int f\,d\mu = \int(\mathrm{e}^{\mathrm{i}\theta}f)\,d\mu = \Re\int(\mathrm{e}^{\mathrm{i}\theta}f)\,d\mu = \int\Re(\mathrm{e}^{\mathrm{i}\theta}f)\,d\mu \\ \leq \int|\mathrm{e}^{\mathrm{i}\theta}f|\,d\mu = \int|f|\,d\mu.\end{aligned}$$

We thus obtained the important inequality

$$\left|\int f\,d\mu\right| \leq \int |f|\,d\mu. \tag{1}$$

If $f, g \in L^1$, it follows from the linearity of the integral and (1) that

$$\left|\int f\,d\mu - \int g\,d\mu\right| = \left|\int(f-g)\,d\mu\right| \leq \|f-g\|_1. \tag{2}$$

In particular, if f and g represent the same equivalence class, then $\|f-g\|_1=0$, and therefore $\int f\,d\mu = \int g\,d\mu$. This means that the functional $f \to \int f\,d\mu$ is *well-defined* as a functional on the *normed space* $L^1(\mu)$ (of equivalence classes!), and its continuity follows trivially from (2). □

In term of sequences, continuity of the integral on the normed space L^1 means that if $\{f_n\} \subset L^1$ converges to f in the L^1-metric, then

$$\int f_n\,d\mu \to \int f\,d\mu. \tag{3}$$

A useful sufficient condition for convergence in the L^1-metric, and therefore, for the validity of (3), is contained in the *Dominated Convergence theorem* of Lebesgue:

Theorem 1.20. *Let $(X,\mathcal{A},\mu)$ be a measure space. Let $\{f_n\}$ be a sequence of complex measurable functions on X that converge pointwise to the function f. Suppose there exists $g \in L^1(\mu)$ (with values in $[0,\infty)$) such that*

$$|f_n| \le g \quad (n=1,2,\ldots). \tag{4}$$

Then $f, f_n \in L^1(\mu)$ for all n, and $f_n \to f$ in the $L^1(\mu)$-metric.

In particular, (3) is valid.

Proof. By Lemma 1.5, f is measurable. By (4) and monotonicity

$$\|f\|_1, \|f_n\|_1 \le \|g\|_1 < \infty,$$

so that $f, f_n \in L^1$.

Since $|f_n - f| \le 2g$, the measurable functions $2g - |f_n - f|$ are non-negative. By Fatou's Lemma (Theorem 1.14),

$$\int \liminf_n (2g - |f_n - f|)\,d\mu \le \liminf_n \int (2g - |f_n - f|)\,d\mu. \tag{5}$$

The left-hand side of (5) is $\int 2g\,d\mu$. The integral on the right-hand side is $\int 2g\,d\mu + (-\|f_n - f\|_1)$, and its lim inf is

$$= \int 2g\,d\mu + \liminf_n (-\|f_n - f\|_1) = \int 2g\,d\mu - \limsup_n \|f_n - f\|_1.$$

Subtracting the finite number $\int 2g\,d\mu$ from both sides of the inequality, we obtain

$$\limsup_n \|f_n - f\|_1 \le 0.$$

However, if a non-negative sequence $\{a_n\}$ satisfies $\limsup a_n \le 0$, then it converges to 0 (because $0 \le \liminf a_n \le \limsup a_n \le 0$ implies $\liminf a_n = \limsup a_n = 0$). Thus $\|f_n - f\|_1 \to 0$. □

Rather than assuming pointwise convergence of the sequence $\{f_n\}$ at every point of X, we may assume that the sequence converges *almost everywhere*, that is, $f_n \to f$ on a set $E \in \mathcal{A}$ and $\mu(E^c) = 0$. The functions f_n could be defined only a.e., and we could include the countable union of all the sets where these functions are not defined (which is a set of measure zero, by the σ-subadditivity of measures) in the "exceptional" set E^c. The limit function f is defined a.e., in any case. For such a function, measurability means that $[f \in V] \cap E \in \mathcal{A}$ for each open set V.

If f_n (defined a.e.) converge pointwise a.e. to f, then with E as mentioned, the restrictions $f_n|_E$ are $\mathcal{A}_E$-measurable, where $\mathcal{A}_E$ is the σ-algebra $\mathcal{A} \cap E$, because

$$[f_n|_E \in V] = [f_n \in V] \cap E \in \mathcal{A}_E.$$

By Lemma 1.5, $f|_E := \lim f_n|_E$ is $\mathcal{A}_E$-measurable, and therefore the a.e.-defined function f is "measurable" in the above sense. We may define f as an arbitrary constant $c \in \mathbb{C}$ on E^c; the function thus extended to X is $\mathcal{A}$-measurable, as seen by the argument preceding Definition 1.18.

Now $f_n I_E$ are $\mathcal{A}$-measurable, converge pointwise everywhere to $f I_E$, and if $|f_n| \le g \in L^1$ for all n (wherever the functions are defined), then $|f_n I_E| \le g \in L^1$ (everywhere!). By Theorem 1.20,

$$\|f_n - f\|_1 = \|f_n I_E - f I_E\|_1 \to 0.$$

We then have the following *a.e. version* of the Lebesgue Dominated Convergence theorem:

Theorem 1.21. *Let $\{f_n\}$ be a sequence of a.e.-defined measurable complex functions on X, converging a.e. to the function f. Let $g \in L^1$ be such that $|f_n| \le g$ for all n (at all points where f_n is defined). Then f and f_n are in L^1, and $f_n \to f$ in the L^1-metric (in particular, $\int f_n \, d\mu \to \int f \, d\mu$).*

A useful "almost everywhere" proposition is the following:

Proposition 1.22. *If $f \in L^1(\mu)$ satisfies $\int_E f \, d\mu = 0$ for every $E \in \mathcal{A}$, then $f = 0$ a.e.*

Proof. Let $E = [u := \Re f \ge 0]$. Then $E \in \mathcal{A}$, so

$$\|u^+\|_1 = \int_E u \, d\mu = \Re \int_E f \, d\mu = 0,$$

and therefore $u^+ = 0$ a.e. Similarly $u^- = v^+ = v^- = 0$ a.e. (where $v := \Im f$), so that $f = 0$ a.e. □

We should remark that, in general, a measurable a.e.-defined function f can be extended as a measurable function on X only by defining it as *constant* on the exceptional null set E^c. Indeed, the null set E^c could have a *non-measurable* subset A. Suppose $f : E \to \mathbb{C}$ is not onto, and let $a \in f(E)^c$. If we assign on A the (constant complex) value a, and any value $b \in f(E)$ on $E^c - A$, then the extended function is not measurable, because $[f = a] = A \notin \mathcal{A}$.

In order to be able to extend f in an arbitrary fashion and always get a measurable function, it is sufficient that subsets of null sets should be measurable (recall that a "null set" is measurable by definition!). A measure space with this property is called a *complete* measure space. Indeed, let f' be an arbitrary extension to X of an a.e.-defined measurable function f, defined on $E \in \mathcal{A}$, with E^c null. Then for any open $V \subset \mathbb{C}$,

$$[f' \in V] = ([f' \in V] \cap E) \cup ([f' \in V] \cap E^c).$$

The first set in the union is in $\mathcal{A}$, by measurability of the a.e.-defined function f; the second set is in $\mathcal{A}$ as a subset of the null set E^c (by completeness of the measure space). Hence $[f' \in V] \in \mathcal{A}$, and f' is measurable.

We say that the measure space $(X, \mathcal{M}, \nu)$ is an *extension* of the measure space $(X, \mathcal{A}, \mu)$ (both on X!) if $\mathcal{A} \subset \mathcal{M}$ and $\nu = \mu$ on $\mathcal{A}$. It is important to know that any measure space $(X, \mathcal{A}, \mu)$ has a (unique) "minimal" complete extension $(X, \mathcal{M}, \nu)$, where minimality means that if $(X, \mathcal{N}, \sigma)$ is any complete extension of $(X, \mathcal{A}, \mu)$, then it is an extension of $(X, \mathcal{M}, \nu)$. Uniqueness is of course trivial. The existence is proved below by a "canonical" construction.

Theorem 1.23. *Any measure space* $(X, \mathcal{A}, \mu)$ *has a unique minimal complete extension* $(X, \mathcal{M}, \nu)$ *(called the completion of the given measure space).*

Proof. We let $\mathcal{M}$ be the collection of all subsets E of X for which there exist $A, B \in \mathcal{A}$ such that

$$A \subset E \subset B, \quad \mu(B - A) = 0. \tag{6}$$

If $E \in \mathcal{A}$, we may take $A = B = E$ in (6), so $\mathcal{A} \subset \mathcal{M}$. In particular $X \in \mathcal{M}$.

If $E \in \mathcal{M}$ and A, B are as in (6), then $A^c, B^c \in \mathcal{A}$,

$$B^c \subset E^c \subset A^c$$

and $\mu(A^c - B^c) = \mu(B - A) = 0$, so that $E^c \in \mathcal{M}$.

If $E_j \in \mathcal{M}, j = 1, 2, \ldots$ and A_j, B_j are as in (6) (for E_j), then if E, A, B are the respective unions of E_j, A_j, B_j, we have $A, B \in \mathcal{A}$, $A \subset E \subset B$, and

$$B - A = \bigcup_j (B_j - A) \subset \bigcup_j (B_j - A_j).$$

The union on the right is a null set (as a countable union of null sets, by σ-subadditivity of measures), and therefore $B - A$ is a null set (by monotonicity of measures). This shows that $E \in \mathcal{M}$, and we conclude that $\mathcal{M}$ is a σ-algebra.

For $E \in \mathcal{M}$ and A, B as in (6), we let $\nu(E) = \mu(A)$. The function ν is *well defined* on $\mathcal{M}$, that is, the definition does not depend on the choice of A, B as in (6). Indeed, if A', B' satisfy (6) with E, then

$$A - A' \subset E - A' \subset B' - A',$$

so that $A - A'$ is a null set. Hence by additivity of μ, $\mu(A) = \mu(A \cap A') + \mu(A - A') = \mu(A \cap A')$. Interchanging the roles of A and A', we also have $\mu(A') = \mu(A \cap A')$, and therefore $\mu(A) = \mu(A')$, as wanted.

If $E \in \mathcal{A}$, we could choose $A = B = E$, and so $\nu(E) = \mu(E)$. In particular, $\nu(\emptyset) = 0$. If $\{E_j\}$ is a sequence of mutually disjoint sets in $\mathcal{M}$ with union E, and A_j, B_j are as in (6) (for E_j), we observed above that we could choose A for E (for (6)) as the union of the sets A_j. Since $A_j \subset E_j, j = 1, 2, \ldots$ and E_j are mutually disjoint, so are the sets A_j. Hence

$$\nu(E) := \mu(A) = \sum_j \mu(A_j) := \sum_j \nu(E_j),$$

and we conclude that $(X, \mathcal{M}, \nu)$ is a measure space extending $(X, \mathcal{A}, \mu)$. It is complete, because if $E \in \mathcal{M}$ is ν-null and A, B are as in (6), then for any $F \subset E$, we have

$$\emptyset \subset F \subset B,$$

and since $\mu(B - A) = 0$,

$$\mu(B - \emptyset) = \mu(B) = \mu(A) := \nu(E) = 0,$$

so that $F \in \mathcal{M}$.

Finally, suppose $(X, \mathcal{N}, \sigma)$ is any complete extension of $(X, \mathcal{A}, \mu)$, let $E \in \mathcal{M}$, and let A, B be as in (6). Write $E = A \cup (E - A)$. The set $B - A \in \mathcal{A} \subset \mathcal{N}$ is σ-null ($\sigma(B - A) = \mu(B - A) = 0$). By completeness of $(X, \mathcal{N}, \sigma)$, the subset $E - A$ of $B - A$ belongs to $\mathcal{N}$ (and is of course σ-null). Since $A \in \mathcal{A} \subset \mathcal{N}$, we conclude that $E \in \mathcal{N}$ and $\mathcal{M} \subset \mathcal{N}$. Also since $\sigma = \mu$ on $\mathcal{A}$, $\sigma(E) = \sigma(A) + \sigma(E - A) = \mu(A) := \nu(E)$, so that $\sigma = \nu$ on $\mathcal{M}$. □

1.5 L^p-spaces

Let $(X, \mathcal{A}, \mu)$ be a (positive) measure space, and let $p \in [1, \infty)$. If $f : X \to [0, \infty]$ is measurable, so is f^p by Lemma 1.6, and therefore $\int f^p \, d\mu \in [0, \infty]$ is well defined. We denote

$$\|f\|_p := \left(\int f^p \, d\mu \right)^{1/p}.$$

Theorem 1.24 (Holder's inequality). *Let $p, q \in (1, \infty)$ be conjugate exponents, that is,*

$$\frac{1}{p} + \frac{1}{q} = 1. \tag{1}$$

Then for all measurable functions $f, g : X \to [0, \infty]$,

$$\int fg \, d\mu \leq \|f\|_p \, \|g\|_q. \tag{2}$$

Proof. If $\|f\|_p = 0$, then $\|f^p\|_1 = 0$, and therefore $f = 0$ a.e.; hence $fg = 0$ a.e., and the left-hand side of (2) vanishes (as well as the right-hand side). By symmetry, the same holds true if $\|g\|_q = 0$. So we may consider only the case where $\|f\|_p$ and $\|g\|_q$ are both *positive*. Now if one of these quantities is infinite, the right-hand side of (2) is infinite, and (2) is trivially true. So we may assume that both quantities belong to $(0,\infty)$ (*positive and finite*). Denote

$$u = f/\|f\|_p, \quad v = g/\|g\|_q. \tag{3}$$

Then

$$\|u\|_p = \|v\|_q = 1. \tag{4}$$

It suffices to prove that

$$\int uv\,d\mu \le 1, \tag{5}$$

because (2) would follow by substituting (3) in (5).

The logarithmic function is concave ($(\log t)'' = -(1/t^2) < 0$). Therefore, by (1)

$$\frac{1}{p}\log s + \frac{1}{q}\log t \le \log\left(\frac{s}{p} + \frac{t}{q}\right)$$

for all $s,t \in (0,\infty)$. Equivalently,

$$s^{1/p}t^{1/q} \le \frac{s}{p} + \frac{t}{q}, \quad s,t \in (0,\infty). \tag{6}$$

When $x \in X$ is such that $u(x), v(x) \in (0,\infty)$, we substitute $s = u(x)^p$ and $t = v(x)^q$ in (6) and obtain

$$u(x)v(x) \le \frac{u(x)^p}{p} + \frac{v(x)^q}{q}, \tag{7}$$

and this inequality is trivially true when $u(x), v(x) \in \{0,\infty\}$. Thus (7) is valid on X, and integrating the inequality over X, we obtain by (4) and (1)

$$\int uv\,d\mu \le \frac{\|u\|_p^p}{p} + \frac{\|v\|_q^q}{q} = \frac{1}{p} + \frac{1}{q} = 1.$$

□

Theorem 1.25 (Minkowski's inequality). *For any measurable functions* $f, g : X \to [0,\infty]$,

$$\|f+g\|_p \le \|f\|_p + \|g\|_p \quad (1 \le p < \infty). \tag{8}$$

Proof. Since (8) is trivial for $p = 1$ (by the additivity of the integral of non-negative measurable functions, we get even an equality), we consider $p \in (1,\infty)$. The case $\|f+g\|_p = 0$ is trivial. By convexity of the function t^p (for $p > 1$), $((s+t)/2)^p \le (s^p + t^p)/2$ for $s,t \in (0,\infty)$. Therefore, if $x \in X$ is such that $f(x), g(x) \in (0,\infty)$,

$$(f(x)+g(x))^p \le 2^{p-1}[f(x)^p + g(x)^p], \tag{9}$$

and (9) is trivially true if $f(x), g(x) \in \{0, \infty\}$, and holds therefore on X. Integrating, we obtain

$$\|f+g\|_p^p \leq 2^{p-1}[\|f\|_p^p + \|g\|_p^p]. \tag{10}$$

If $\|f+g\|_p = \infty$, it follows from (10) that at least one of the quantities $\|f\|_p, \|g\|_p$ is infinite, and (8) is then valid (as the trivial equality $\infty = \infty$). This discussion shows that we may restrict our attention to the case

$$0 < \|f+g\|_p < \infty. \tag{11}$$

We write

$$(f+g)^p = f(f+g)^{p-1} + g(f+g)^{p-1}. \tag{12}$$

By Holder's inequality,

$$\int f(f+g)^{p-1}\, d\mu \leq \|f\|_p \|(f+g)^{p-1}\|_q = \|f\|_p \|f+g\|_p^{p/q},$$

since $(p-1)q = p$ for conjugate exponents p, q. A similar estimate holds for the integral of the second summand on the right hand side of (12). Adding these estimates, we obtain

$$\|f+g\|_p^p \leq (\|f\|_p + \|g\|_p)\|f+g\|_p^{p/q}.$$

By (11), we may divide this inequality by $\|f+g\|_p^{p/q}$, and (8) follows since $p - p/q = 1$. $\square$

In a manner analogous to that used for L^1, if $p \in [1, \infty)$, we consider the set

$$L^p(X, \mathcal{A}, \mu)$$

(or briefly, $L^p(\mu)$, or $L^p(X)$, or L^p, when the unmentioned parameters are understood) of all (*equivalence classes*) of measurable complex functions f on X, with

$$\|f\|_p := \| \, |f| \, \|_p < \infty.$$

Since $\|\cdot\|_p$ is trivially homogeneous, it follows from (8) that L^p is a normed space (over $\mathbb{C}$) for the pointwise operations and the norm $\|\cdot\|_p$. We can restate Holder's inequality in the form:

Theorem 1.26. *Let $p, q \in (1, \infty)$ be conjugate exponents. If $f \in L^p$ and $g \in L^q$, then $fg \in L^1$, and*

$$\|fg\|_1 \leq \|f\|_p \, \|g\|_q.$$

A sufficient condition for convergence in the L^p-metric follows at once from Theorem 1.21:

Proposition. *Let $\{f_n\}$ be a sequence of a.e.-defined measurable complex functions on X, converging a.e. to the function f. For some $p \in [1, \infty)$, suppose* *there exists $g \in L^p$ such that $|f_n| \leq g$ for all n (with the usual equivalence class ambiguity). Then $f, f_n \in L^p$, and $f_n \to f$ in the L^p-metric.*

Proof. The first statement follows from the inequalities $|f|^p, |f_n|^p \leq g^p \in L^1$. Since $|f - f_n|^p \to 0$ a.e. and $|f - f_n|^p \leq (2g)^p \in L^1$, the second statement follows from Theorem 1.21. □

The positive measure space $(X, \mathcal{A}, \mu)$ is said to be *finite* if $\mu(X) < \infty$. When this is the case, the Holder inequality implies that $L^p(\mu) \subset L^r(\mu)$ topologically (i.e., the inclusion map is continuous) when $1 \leq r < p < \infty$. Indeed, if $f \in L^p(\mu)$, then by Holder's inequality with the conjugate exponents p/r and $s := p/(p-r)$,

$$\begin{aligned} \|f\|_r^r &= \int |f|^r .1 \, d\mu \\ &\leq \left[\int (|f|^r)^{p/r} \, d\mu\right]^{r/p} \left[\int 1^s \, d\mu\right]^{1/s} = \mu(X)^{1/s} \|f\|_p^r. \end{aligned}$$

Since $1/rs = (1/r) - (1/p)$, we obtain

$$\|f\|_r \leq \mu(X)^{1/r - 1/p} \|f\|_p. \tag{13}$$

Hence, $f \in L^r(\mu)$, and (13) (with $f - g$ replacing f) shows the continuity of the inclusion map of $L^p(\mu)$ into $L^r(\mu)$.

Taking in particular $r = 1$, we get that $L^p(\mu) \subset L^1(\mu)$ (topologically) for all $p \geq 1$, and

$$\|f\|_1 \leq \mu(X)^{1/q} \|f\|_p, \tag{14}$$

where q is the conjugate exponent of p.

We formalize this discussion for future reference.

Proposition. *Let $(X, \mathcal{A}, \mu)$ be a finite positive measure space. Then $L^p(\mu) \subset L^r(\mu)$ (topologically) for $1 \leq r < p < \infty$, and the norms inequality (13) holds.*

Let $(X, \mathcal{A})$ and $(Y, \mathcal{B})$ be measurable spaces, and let $h : X \to Y$ be a measurable map (cf. Definition 1.2). If μ is a measure on $\mathcal{A}$, the function $\nu : \mathcal{B} \to [0, \infty]$ given by

$$\nu(E) = \mu(h^{-1}(E)), \quad E \in \mathcal{B} \tag{15}$$

is well defined, and is clearly a measure on $\mathcal{B}$. Since $I_{h^{-1}(E)} = I_E \circ h$, we can write (15) in the form

$$\int_Y I_E \, d\nu = \int_X I_E \circ h \, d\mu.$$

By linearity of the integral, it follows that

$$\int_Y f \, d\nu = \int_X f \circ h \, d\mu \tag{16}$$

for every $\mathcal{B}$-measurable simple function f on Y. If $f : Y \to [0, \infty]$ is $\mathcal{B}$-measurable, use the Approximation Theorem 1.8 to obtain a non-decreasing sequence $\{f_n\}$ of $\mathcal{B}$-measurable non-negative simple functions converging pointwise to f; then $\{f_n \circ h\}$ is a similar sequence converging to $f \circ h$, and the Monotone Convergence Theorem shows that (16) is true for all such f.

If $f : Y \to \mathbb{C}$ is $\mathcal{B}$-measurable, then $f \circ h$ is a (complex) $\mathcal{A}$-measurable function on X, and for any $1 \leq p < \infty$,

$$\int_Y |f|^p \, d\nu = \int_X |f|^p \circ h \, d\mu = \int_X |f \circ h|^p \, d\mu.$$

Thus, $f \in L^p(\nu)$ for some $p \in [1, \infty)$ if and only if $f \circ h \in L^p(\mu)$, and

$$\|f\|_{L^p(\nu)} = \|f \circ h\|_{L^p(\mu)}.$$

In particular (case $p = 1$), f is ν-integrable on Y if and only if $f \circ h$ is μ-integrable on X. When this is the case, writing f as a linear combination of four non-negative ν-integrable functions, we see that (16) is valid for all such f.

Proposition. *Let $(X, \mathcal{A})$ and $(Y, \mathcal{B})$ be measurable spaces, and let $h : X \to Y$ be a measurable map. For any (positive) measure μ on $\mathcal{A}$, define $\nu(E) := \mu(h^{-1}(E))$ for $E \in \mathcal{B}$. Then:*

(1) ν is a (positive) measure on $\mathcal{B}$,

(2) if $f : Y \to [0, \infty]$ is $\mathcal{B}$-measurable, then $f \circ h$ is $\mathcal{A}$-measurable and (16) is valid;

(3) if $f : Y \to \mathbb{C}$ is $\mathcal{B}$-measurable, then $f \circ h$ is $\mathcal{A}$-measurable; $f \in L^p(\nu)$ for some $p \in [1, \infty)$ if and only if $f \circ h \in L^p(\mu)$, and in that case, the map $f \to f \circ h$ is norm-preserving; *in the special case $p = 1$, the map is* integral preserving *(i.e. (16) is valid).*

If ϕ is a simple complex measurable function with distinct *non-zero* values c_j assumed on E_j, then

$$\|\phi\|_p^p = \sum_j |c_j|^p \mu(E_j)$$

is finite if and only if $\mu(E_j) < \infty$ for all j, that is, equivalently, iff

$$\mu([|\phi| > 0]) < \infty.$$

Thus, the simple functions in L^p (for *any* $p \in [1, \infty)$) are the (measurable) simple functions vanishing outside a measurable set of *finite* measure (depending on the function). These functions are *dense* in L^p. Indeed, if $0 \leq f \in L^p$ (without loss of generality, we assume that f is everywhere defined!), the Approximation Theorem provides a sequence of simple measurable functions

$$0 \leq \phi_1 \leq \phi_2 \leq \cdots \leq f$$

such that $\phi_n \to f$ pointwise. By the proposition following Theorem 1.26, $\phi_n \to f$ in the L^p-metric.

For $f \in L^p$ complex, we may write $f = \sum_{k=0}^3 i^k g_k$ with $0 \leq g_k \in L^p$ ($g_0 := u^+$, etc., where $u = \Re f$). We then obtain four sequences $\{\phi_{n,k}\}$ of simple functions in L^p converging, respectively, to $g_k, k = 0, \dots, 3$, in the L^p-metric; if $\phi_n := \sum_{k=0}^{3} \mathrm{i}^k \phi_{n,k}$, then ϕ_n are simple L^p-functions, and $\phi_n \to f$ in the L^p-metric. We proved

Theorem 1.27. *For any $p \in [1, \infty)$, the simple functions in L^p are dense in L^p.*

Actually, L^p is the *completion* of the normed space of all measurable simple functions vanishing outside a set of finite measure, with respect to the L^p-metric (induced by the L^p-norm). The meaning of this statement is made clear by the following definition.

Definition 1.28. Let Z be a metric space, with metric d. A *Cauchy sequence* in Z is a sequence $\{z_n\} \subset Z$ such that $d(z_n, z_m) \to 0$ when $n, m \to \infty$. The space Z is *complete* if every Cauchy sequence in Z converges in Z. If $Y \subset Z$ is dense in Z, and Z is complete, we also say that Z is the *completion* of Y (for the metric d). The completion of Y (for the metric d) is *unique* in a suitable sense.

A *complete normed space* is called a *Banach space.*

In order to get the conclusion preceding Definition 1.28, we still have to prove that L^p is complete:

Theorem 1.29. *L^p is a Banach space for each $p \in [1, \infty)$.*

We first prove the following.

Lemma 1.30. *Let $\{f_n\}$ be a Cauchy sequence in $L^p(\mu)$. Then it has a subsequence converging pointwise μ-a.e.*

Proof of Lemma. Since $\{f_n\}$ is Cauchy, there exists $m_k \in \mathbb{N}$ such that $\|f_n - f_m\|_p < 1/2^k$ for all $n > m > m_k$. Set

$$n_k = k + \max(m_1, \dots, m_k).$$

Then $n_{k+1} > n_k > m_k$, and therefore $\{f_{n_k}\}$ is a subsequence of $\{f_n\}$ satisfying

$$\|f_{n_{k+1}} - f_{n_k}\|_p < 1/2^k \quad k = 1, 2, \dots \tag{17}$$

Consider the series

$$g = \sum_{k=1}^{\infty} |f_{n_{k+1}} - f_{n_k}|, \tag{18}$$

and its partial sums g_m. By Theorem 1.25 and (17),

$$\|g_m\|_p \le \sum_{k=1}^{m} \|f_{n_{k+1}} - f_{n_k}\|_p < \sum_{k=1}^{\infty} 1/2^k = 1$$

for all m. By Fatou's lemma,

$$\int g^p \, d\mu \le \liminf_m \int g_m^p \, d\mu = \liminf_m \|g_m\|_p^p \le 1.$$

Therefore, $g < \infty$ a.e., that is, the series (18) converges a.e., that is, the series

$$f_{n_1} + \sum_{k=1}^{\infty} (f_{n_{k+1}} - f_{n_k}) \tag{19}$$

converges absolutely pointwise a.e. to its sum f (extended as 0 on the null set where the series does not converge). Since the partial sums of (19) are precisely f_{n_m}, the lemma is proved. □

Proof of Theorem 1.29. Let $\{f_n\} \subset L^p$ be Cauchy. Thus for any $\epsilon > 0$, there exists $n_\epsilon \in \mathbb{N}$ such that

$$\|f_n - f_m\|_p < \epsilon \tag{20}$$

for all $n, m > n_\epsilon$. By the lemma, let then $\{f_{n_k}\}$ be a subsequence converging pointwise a.e. to the (measurable) complex function f. Applying Fatou's lemma to the non-negative measurable functions $|f_{n_k} - f_m|$, we obtain

$$\|f - f_m\|_p^p = \int \lim_k |f_{n_k} - f_m|^p \, d\mu \le \liminf_k \|f_{n_k} - f_m\|_p^p \le \epsilon^p \tag{21}$$

for all $m > n_\epsilon$. In particular, $f - f_m \in L^p$, and therefore $f = (f - f_m) + f_m \in L^p$, and (21) means that $f_m \to f$ in the L^p-metric. □

Definition 1.31. Let $(X, \mathcal{A}, \mu)$ be a positive measure space, and let $f : X \to \mathbb{C}$ be a measurable function. We say that $M \in [0, \infty]$ is an a.e. upper bound for $|f|$ if $|f| \le M$ a.e. The infimum of all the a.e. upper bounds for $|f|$ is called the *essential supremum* of $|f|$, and is denoted $\|f\|_\infty$. The set of all (*equivalence classes of*) measurable complex functions f on X with $\|f\|_\infty < \infty$ will be denoted by $L^\infty(\mu)$ (or $L^\infty(X)$, or $L^\infty(X, \mathcal{A}, \mu)$, or L^∞, depending on which "parameter" we wish to stress, if at all).

By definition of the essential supremum, we have

$$|f| \le \|f\|_\infty \quad \text{a.e.} \tag{22}$$

In particular, $\|f\|_\infty = 0$ implies that $f = 0$ a.e. (that is, f is the zero class).

If $f, g \in L^\infty$, then by (22), $|f + g| \le |f| + |g| \le \|f\|_\infty + \|g\|_\infty$ a.e., and so $\|f + g\|_\infty \le \|f\|_\infty + \|g\|_\infty$.

The homogeneity $\|\alpha f\|_\infty = |\alpha| \, \|f\|_\infty$ is trivial if either $\alpha = 0$ or $\|f\|_\infty = 0$. Assume then $|\alpha|, \|f\|_\infty > 0$. For any $t \in (0, 1)$, $t\|f\|_\infty < \|f\|_\infty$, hence it is not an a.e. upper bound for $|f|$, so that $\mu([|f| > t\|f\|_\infty]) > 0$, that is, $\mu([|\alpha f| > t|\alpha| \, \|f\|_\infty]) > 0$. Therefore, $\|\alpha f\|_\infty \ge t|\alpha| \, \|f\|_\infty$ for all $t \in (0, 1)$, hence $\|\alpha f\|_\infty \ge |\alpha| \, \|f\|_\infty$. The reversed inequality follows trivially from (22), and the homogeneity of $\| \cdot \|_\infty$ follows. We conclude that L^∞ is a normed space (over $\mathbb{C}$) for the pointwise operations and the L^∞-*norm* $\| \cdot \|_\infty$.

We verify its completeness as follows. Let $\{f_n\}$ be a Cauchy sequence in L^∞. In particular, it is a bounded set in L^∞. Let then $K = \sup_n \|f_n\|_\infty$. By (22), the sets $F_k := [|f_k| > K] \quad (k \in \mathbb{N})$ and

$$E_{n,m} := [|f_n - f_m| > \|f_n - f_m\|_\infty] \quad (n, m \in \mathbb{N})$$

are μ-null, so their (countable) union E is null. For all $x \in E^c$,

$$|f_n(x) - f_m(x)| \leq \|f_n - f_m\|_\infty \to 0$$

as $n, m \to \infty$ and $|f_n(x)| \leq K$. By completeness of $\mathbb{C}$, the limit $f(x) := \lim_n f_n(x)$ exists for all $x \in E^c$ and $|f(x)| \leq K$. Defining $f(x) = 0$ for all $x \in E$, we obtain a measurable function on X such that $|f| \leq K$, that is, $f \in L^\infty$. Given $\epsilon > 0$, let $n_\epsilon \in \mathbb{N}$ be such that

$$\|f_n - f_m\|_\infty < \epsilon \quad (n, m > n_\epsilon).$$

Since $|f_n(x) - f_m(x)| < \epsilon$ for all $x \in E^c$ and $n, m > n_\epsilon$, letting $m \to \infty$, we obtain $|f_n(x) - f(x)| \leq \epsilon$ for all $x \in E^c$ and $n > n_\epsilon$, and since $\mu(E) = 0$,

$$\|f_n - f\|_\infty \leq \epsilon \quad (n > n_\epsilon),$$

that is, $f_n \to f$ in the L^∞-metric. We proved

Theorem 1.32. *L^∞ is a Banach space.*

Defining the conjugate exponent of $p = 1$ to be $q = \infty$ (so that $(1/p) + (1/q) = 1$ is formally valid in the usual sense), Holder's inequality remains true for this pair of conjugate exponents. Indeed, if $f \in L^1$ and $g \in L^\infty$, then $|fg| \leq \|g\|_\infty |f|$ a.e., and therefore $fg \in L^1$ and

$$\|fg\|_1 \leq \|g\|_\infty \|f\|_1.$$

Formally

Theorem 1.33. *Holder's inequality (Theorem 1.26) is valid for conjugate exponents $p, q \in [1, \infty]$.*

1.6 Inner product

For the conjugate pair $(p, q) = (2, 2)$, Theorem 1.26 asserts that if $f, g \in L^2$, then the product $f\bar{g}$ is integrable, so we may define

$$(f, g) := \int f\bar{g}\, d\mu \tag{1}$$

($\bar{g}$ denotes here the complex conjugate of g). The function (or *form*) $(\cdot\,,\cdot)$ has obviously the following properties on $L^2 \times L^2$:

(i) $(f, f) \geq 0$, and $(f, f) = 0$ if and only if $f = 0$ (the zero element);

(ii) $(\cdot\,, g)$ is linear for each given $g \in L^2$;

(iii) $(g, f) = \overline{(f, g)}$.

Property (i) is called *positive definiteness* of the form $(\cdot,\cdot)$; Properties (ii) and (iii) (together) are referred to as *sesquilinearity* or *Hermitianity* of the form. We may also consider the weaker condition

$$(\mathrm{i}')(f,f) \geq 0 \text{ for all } f,$$

called (positive) *semi-definiteness* of the form.

Definition 1.34. Let X be a complex vector space (with elements $x, y, \ldots$). A *(semi)-inner product on* X is a (semi)-definite sesquilinear form $(\cdot,\cdot)$ on X. The space X with a given (semi)-inner product is called a *(semi)-inner product space.*

If X is a semi-inner product space, the non-negative square root of (x,x) is denoted $\|x\|$.

Thus L^2 is an inner product space for the inner product (1) and $\|f\| := (f,f)^{1/2} = \|f\|_2$. By Theorem 1.26 with $p = q = 2$,

$$|(f,g)| \leq \|f\|_2 \, \|g\|_2 \tag{2}$$

for all $f, g \in L^2$. This special case of the Holder inequality is called the *Cauchy–Schwarz inequality*. We demonstrate below that it is valid in *any* semi-inner product space.

Observe that any sesquilinear form $(\cdot,\cdot)$ is *conjugate linear* with respect to its second variable, that is, for each given $x \in X$,

$$(x, \alpha u + \beta v) = \bar{\alpha}(x,u) + \bar{\beta}(x,v) \tag{3}$$

for all $\alpha, \beta \in \mathbb{C}$ and $u, v \in X$.

In particular

$$(x,0) = (0,y) = 0 \tag{4}$$

for all $x, y \in X$.

By (ii) and (3), for all $\lambda \in \mathbb{C}$ and $x, y \in X$,

$$(x + \lambda y, x + \lambda y) = (x,x) + \bar{\lambda}(x,y) + \lambda(y,x) + |\lambda|^2 (y,y).$$

Since $\lambda(y,x)$ is the conjugate of $\bar{\lambda}(x,y)$ by (iii), we obtain the identity (for all $\lambda \in \mathbb{C}$ and $x, y \in X$)

$$\|x + \lambda y\|^2 = \|x\|^2 + 2\Re[\bar{\lambda}(x,y)] + |\lambda|^2 \|y\|^2. \tag{5}$$

In particular, for $\lambda = 1$ and $\lambda = -1$, we have the identities

$$\|x + y\|^2 = \|x\|^2 + 2\Re(x,y) + \|y\|^2 \tag{6}$$

and

$$\|x - y\|^2 = \|x\|^2 - 2\Re(x,y) + \|y\|^2. \tag{7}$$

Adding, we obtain the so-called *parallelogram identity* for *any* s.i.p. (semi-inner product):

$$\|x + y\|^2 + \|x - y\|^2 = 2\|x\|^2 + 2\|y\|^2. \tag{8}$$

Subtracting (7) from (6), we obtain

$$4\Re(x,y) = \|x+y\|^2 - \|x-y\|^2. \tag{9}$$

If we replace y by iy in (9), we obtain

$$4\Im(x,y) = 4\Re[-\mathrm{i}(x,y)] = 4\Re(x,\mathrm{i}y) = \|x+\mathrm{i}y\|^2 - \|x-\mathrm{i}y\|^2. \tag{10}$$

By (9) and (10),

$$(x,y) = \frac{1}{4}\sum_{k=0}^{3} \mathrm{i}^k \|x+\mathrm{i}^k y\|^2, \tag{11}$$

where $\mathrm{i} = \sqrt{-1}$. This is the so-called *polarization identity* (which expresses the s.i.p. in terms of "induced norms").

By (5),

$$0 \le \|x\|^2 + 2\Re[\bar{\lambda}(x,y)] + |\lambda|^2\|y\|^2 \tag{12}$$

for all $\lambda \in \mathbb{C}$ and $x,y \in X$. If $\|y\| > 0$, take $\lambda = -(x,y)/\|y\|^2$; then $|(x,y)|^2/\|y\|^2 \le \|x\|^2$, and therefore

$$|(x,y)| \le \|x\|\,\|y\|. \tag{13}$$

If $\|y\| = 0$ but $\|x\| > 0$, interchange the roles of x and y and use (iii) to reach the same conclusion. If both $\|x\|$ and $\|y\|$ vanish, take $\lambda = -(x,y)$ in (12): we get $0 \le -2|(x,y)|^2$, hence $|(x,y)| = 0 = \|x\|\,\|y\|$, and we conclude that (13) is valid for *all* $x,y \in X$. This is the general Cauchy–Schwarz inequality for semi-inner products.

By (6) and (13),

$$\|x+y\|^2 \le \|x\|^2 + 2|(x,y)| + \|y\|^2 \le \|x\|^2 + 2\|x\|\,\|y\| + \|y\|^2 = (\|x\| + \|y\|)^2,$$

hence

$$\|x+y\| \le \|x\| + \|y\|$$

for all $x,y \in X$. Taking $x = 0$ in (5), we get $\|\lambda y\| = |\lambda|\,\|y\|$ for all $\lambda \in \mathbb{C}$ and $y \in X$. We conclude that $\|\cdot\|$ is a semi-norm on X; it is a norm iff the s.i.p. is an *inner product*, that is, iff it is *definite*. Thus, an inner product space X is a normed space for the norm $\|x\| := (x,x)^{1/2}$ *induced* by its inner product (unless stated otherwise, this will be the standard norm for such spaces). In case X is *complete*, it is called a *Hilbert space*. Thus Hilbert spaces are special cases of Banach spaces.

The norm induced by the inner product (1) on L^2 is the usual L^2-norm $\|\cdot\|_2$, so that, by Theorem 1.29, L^2 is a Hilbert space.

1.7 Hilbert space: a first look

We consider some "geometric" properties of Hilbert spaces.

Theorem 1.35 (Distance theorem). *Let X be a Hilbert space, and let $K \subset X$ be non-empty, closed, and convex (i.e., $(x+y)/2 \in K$ whenever $x, y \in K$). Then for each $x \in X$, there exists a unique $k \in K$ such that*

$$d(x, k) = d(x, K). \tag{1}$$

The notation $d(x, y)$ is used for the metric induced by the norm, $d(x, y) := \|x - y\|$. As in any metric space, $d(x, K)$ denotes the *distance from x to K*, that is,

$$d(x, K) := \inf_{y \in K} d(x, y). \tag{2}$$

Proof. Let $d = d(x, K)$. Since $d^2 = \inf_{y \in K} \|x - y\|^2$, there exist $y_n \in K$ such that

$$(d^2 \leq) \|x - y_n\|^2 < d^2 + 1/n, \quad n = 1, 2, \ldots. \tag{3}$$

By the parallelogram identity,

$$\begin{aligned} \|y_n - y_m\|^2 &= \|(x - y_m) - (x - y_n)\|^2 \\ &= 2\|x - y_m\|^2 + 2\|x - y_n\|^2 - \|(x - y_m) + (x - y_n)\|^2. \end{aligned}$$

Rewrite the last term on the right-hand side in the form

$$4\|x - (y_m + y_n)/2\|^2 \geq 4d^2,$$

since $(y_m + y_n)/2 \in K$, by hypothesis. Hence by (3)

$$\|y_n - y_m\|^2 \leq 2/m + 2/n \to 0$$

as $m, n \to \infty$. Thus, the sequence $\{y_n\}$ is Cauchy. Since X is *complete*, the sequence converges in X, and its limit k in necessarily in K because $y_n \in K$ for all n and K is closed. By continuity of the norm on X, letting $n \to \infty$ in (3), we obtain $\|x - k\| = d$, as wanted.

To prove uniqueness, suppose $k, k' \in K$ satisfy

$$\|x - k\| = \|x - k'\| = d.$$

Again by the parallelogram identity,

$$\begin{aligned} \|k - k'\|^2 &= \|(x - k') - (x - k)\|^2 \\ &= 2\|x - k'\|^2 + 2\|x - k\|^2 - \|(x - k') + (x - k)\|^2. \end{aligned}$$

As before, write the last term as $4\|x-(k+k')/2\|^2 \geq 4d^2$ (since $(k+k')/2 \in K$ by hypothesis). Hence

$$\|k-k'\|^2 \leq 2d^2+2d^2-4d^2=0,$$

and therefore $k=k'$. □

We say that the vector $y \in X$ is *orthogonal* to the vector x if $(x,y)=0$. In that case also $(y,x)=\overline{(x,y)}=0$, so that the orthogonality relation is symmetric. For x given, let $x^\perp$ denote the set of all vectors orthogonal to x. This is the kernel of the linear functional $\phi=(\cdot\,,x)$, that is, the set $\phi^{-1}(\{0\})$. As such a kernel, it is a subspace. Since $|\phi(y)-\phi(z)|=|(y-z,x)| \leq \|y-z\|\,\|x\|$ by Schwarz's inequality, ϕ is continuous, and therefore $x^\perp=\phi^{-1}(\{0\})$ is closed. Thus, $x^\perp$ is a closed subspace. More generally, for any non-empty subset A of X, define

$$A^\perp := \bigcap_{x\in A} x^\perp = \{y \in Y; (y,x)=0 \text{ for all } x \in A\}.$$

As the intersection of closed subspaces, $A^\perp$ is a closed subspace of X.

Theorem 1.36 (Orthogonal decomposition theorem). *Let Y be a closed subspace of the Hilbert space X. Then X is the direct sum of Y and $Y^\perp$, that is, each $x \in X$ has the unique orthogonal decomposition $x=y+z$ with $y \in Y$ and $z \in Y^\perp$.*

Note that the so-called *components* y and z of x (in Y and $Y^\perp$, respectively) are orthogonal.

Proof. As a closed subspace of X, Y is a non-empty, closed, convex subset of X. By the distance theorem, there exists a unique $y \in Y$ such that

$$\|x-y\| = d := d(x,Y).$$

Letting $z:=x-y$, the existence part of the theorem will follow if we show that $(z,u)=0$ for all $u \in Y$. Since Y is a subspace, and $Y \neq \{0\}$ without loss of generality, every $u \in Y$ is a scalar multiple of a unit vector in Y, so it suffices to prove that $(z,u)=0$ for *unit vectors* $u \in Y$. For all $\lambda \in \mathbb{C}$, by the identity (5) (following Definition 1.34),

$$\|z-\lambda u\|^2 = \|z\|^2 - 2\Re[\bar{\lambda}(z,u)] + |\lambda|^2.$$

The left-hand side is

$$\|x-(y+\lambda u)\|^2 \geq d^2,$$

since $y+\lambda u \in Y$. Since $\|z\|=d$, we obtain

$$0 \leq -2\Re[\bar{\lambda}(z,u)] + |\lambda|^2.$$

Choose $\lambda=(z,u)$. Then $0 \leq -|(z,u)|^2$, so that $(z,u)=0$ as claimed.

If $x = y + z = y' + z'$ are two decompositions with $y, y' \in Y$ and $z, z' \in Y^{\perp}$, then $y - y' = z' - z \in Y \cap Y^{\perp}$, so that in particular $y - y'$ is orthogonal to itself (i.e., $(y - y', y - y') = 0$), which implies that $y - y' = 0$, whence $y = y'$ and $z = z'$. □

We observed in passing that for each given $y \in X$, the function $\phi := (\cdot\,, y)$ is a continuous linear functional on the inner product space X. For *Hilbert* spaces, this is the *general* form of continuous linear functionals:

Theorem 1.37 ("Little" Riesz representation theorem). *Let $\phi : X \to \mathbb{C}$ be a continuous linear functional on the Hilbert space X. Then there exists a unique $y \in X$ such that $\phi = (\cdot\,, y)$.*

Proof. If $\phi = 0$ (the zero functional), take $y = 0$. Assume then that $\phi \neq 0$, so that its kernel Y is a closed subspace $\neq X$. Therefore $Y^{\perp} \neq \{0\}$, by Theorem 1.36. Let then $z \in Y^{\perp}$ be a unit vector. Since $Y \cap Y^{\perp} = \{0\}$, $z \notin Y$, so that $\phi(z) \neq 0$. For any given $x \in X$, we may then define

$$u := x - \frac{\phi(x)}{\phi(z)} z.$$

By linearity,

$$\phi(u) = \phi(x) - \frac{\phi(x)}{\phi(z)} \phi(z) = 0,$$

that is, $u \in Y$, and

$$x = u + \frac{\phi(x)}{\phi(z)} z \tag{4}$$

is the (unique) orthogonal decomposition of x (corresponding to the particular subspace Y, the kernel of ϕ). Define now $y = \overline{\phi(z)} z (\in Y^{\perp})$. By (4),

$$(x, y) = (u, y) + \frac{\phi(x)}{\phi(z)} \phi(z)(z, z) = \phi(x)$$

since $(u, y) = 0$ and $\|z\| = 1$. This proves the existence part of the theorem. Suppose now that $y, y' \in X$ are such that $\phi(x) = (x, y) = (x, y')$ for all $x \in X$. Then $(x, y - y') = 0$ for all x, hence in particular $(y - y', y - y') = 0$, which implies that $y = y'$. □

1.8 The Lebesgue–Radon–Nikodym theorem

We apply the Riesz representation theorem to prove the Lebesgue decomposition theorem and the Radon–Nikodym theorem for (positive) measures.

We start with a measure-theoretic lemma.

The positive measure space $(X, \mathcal{A}, \mu)$ is *σ-finite* if there exists a *sequence* of mutually disjoint measurable sets X_j with union X, such that $\mu(X_j) < \infty$ for all j.

Lemma 1.38 (The averages lemma). *Let $(X, \mathcal{A}, \sigma)$ be a σ-finite positive measure space. Let $g \in L^1(\sigma)$ be such that, for all $E \in \mathcal{A}$ with $0 < \sigma(E) < \infty$, the "averages"*

$$A_E(g) := \frac{1}{\sigma(E)} \int_E g \, d\sigma$$

are contained in some given closed *set $F \subset \mathbb{C}$. Then $g(x) \in F$ σ-a.e.*

Proof. We need to prove that $g^{-1}(F^c)$ is σ-null. Write the *open* set F^c as the countable union of the closed discs

$$\Delta_n := \{z \in \mathbb{C}; |z - a_n| \leq r_n\}, \quad n = 1, 2, \ldots .$$

Then

$$g^{-1}(F^c) = \bigcup_{n=1}^{\infty} g^{-1}(\Delta_n),$$

and it suffices to prove that $E_\Delta := g^{-1}(\Delta)$ is σ-null whenever Δ is a closed disc (with center a and radius r) contained in F^c.

Write X as the countable union of mutually disjoint measurable sets X_k with $\sigma(X_k) < \infty$. Set $E_{\Delta,k} := E_\Delta \cap X_k$, and suppose $\sigma(E_{\Delta,k}) > 0$ for some Δ as above and some k. Since $|g(x) - a| \leq r$ on $E := E_{\Delta,k}$, and $0 < \sigma(E) < \infty$, we have

$$|A_E(g) - a| = |A_E(g - a)| \leq \frac{1}{\sigma(E)} \int_E |g - a| \, d\sigma \leq r,$$

so that $A_E(g) \in \Delta \subset F^c$, contradicting the hypothesis. Hence $\sigma(E_{\Delta,k}) = 0$ for all k and therefore $\sigma(E_\Delta) = 0$ for all Δ as above. □

Lemma 1.39. *Let $0 \leq \lambda \leq \sigma$ be finite measures on the measurable space $(X, \mathcal{A})$. Then there exists a measurable function $g : X \to [0, 1]$ such that*

$$\int f \, d\lambda = \int fg \, d\sigma \tag{1}$$

for all $f \in L^2(\sigma)$.

Proof. By Definition 1.12, the relation $\lambda \leq \sigma$ between positive measures implies that $\int f \, d\lambda \leq \int f \, d\sigma$ for all non-negative measurable functions f. Hence $L^2(\sigma) \subset L^2(\lambda)(\subset L^1(\lambda)$, by the second proposition following Theorem 1.26.)

For all $f \in L^2(\sigma)$, we have then by Schwarz's inequality:

$$\left| \int f \, d\lambda \right| \leq \int |f| \, d\lambda \leq \int |f| \, d\sigma \leq \sigma(X)^{1/2} \|f\|_{L^2(\sigma)}.$$

Replacing f by $f - h$ (with $f, h \in L^2(\sigma)$), we get

$$\left| \int f \, d\lambda - \int h \, d\lambda \right| = \left| \int (f - h) d\lambda \right| \leq \sigma(X)^{1/2} \|f - h\|_{L^2(\sigma)},$$

so that the functional $f \to \int f\,d\lambda$ is a continuous linear functional on $L^2(\sigma)$. By the Riesz representation theorem for the Hilbert space $L^2(\sigma)$, there exists an element $g_1 \in L^2(\sigma)$ such that this functional is $(\cdot\,, g_1)$. Letting $g = \overline{g_1}$ $(\in L^2(\sigma))$, we get the wanted relation (1).

Since $I_E \in L^2(\sigma)$ (because σ is a finite measure), we have in particular

$$\lambda(E) = \int I_E\,d\lambda = \int_E g\,d\sigma$$

for all $E \in \mathcal{A}$. If $\sigma(E) > 0$,

$$\frac{1}{\sigma(E)} \int_E g\,d\sigma = \frac{\lambda(E)}{\sigma(E)} \in [0,1].$$

By the Averages Lemma 1.38, $g(x) \in [0,1]$ σ-a.e., and we may then choose a representative of the equivalence class g with range in $[0,1]$. □

Terminology. Let $(X, \mathcal{A}, \lambda)$ be a positive measure space. We say that the set $A \in \mathcal{A}$ *carries* the measure λ (or that λ is supported by A) if $\lambda(E) = \lambda(E \cap A)$ for all $E \in \mathcal{A}$.

This is, of course, equivalent to $\lambda(E) = 0$ for all measurable subsets E of A^c.

Two (positive) measures λ_1, λ_2 on $(X, \mathcal{A})$ are *mutually singular* (notation $\lambda_1 \perp \lambda_2$) if they are carried by *disjoint* measurable sets A_1, A_2. Equivalently, each measure is carried by a null set relative to the other measure.

On the other hand, if $\lambda_2(E) = 0$ whenever $\lambda_1(E) = 0$ (for $E \in \mathcal{A}$), we say that λ_2 is *absolutely continuous* with respect to λ_1 (notation: $\lambda_2 \ll \lambda_1$).

Equivalently, $\lambda_2 \ll \lambda_1$ if and only if any (measurable) set that carries λ_1 also carries λ_2.

Theorem 1.40 (Lebesgue–Radon–Nikodym). *Let $(X, \mathcal{A}, \mu)$ be a σ-finite positive measure space, and let λ be a finite positive measure on $(X, \mathcal{A})$. Then*

(a) λ has the unique (so-called) Lebesgue decomposition

$$\lambda = \lambda_a + \lambda_s$$

with $\lambda_a \ll \mu$ and $\lambda_s \perp \mu$;

(b) there exists a unique $h \in L^1(\mu)$ such that

$$\lambda_a(E) = \int_E h\,d\mu$$

for all $E \in \mathcal{A}$.

(part (a) is the Lebesgue decomposition theorem; part (b) is the Radon–Nikodym theorem.)

Proof. Case $\mu(X) < \infty$.

Let $\sigma := \lambda + \mu$. Then the finite positive measures λ, σ satisfy $\lambda \leq \sigma$, so that by Lemma 1.39, there exists a measurable function $g : X \to [0, 1]$ such that (1) holds, that is, after rearrangement,

$$\int f(1-g)\,d\lambda = \int fg\,d\mu \tag{2}$$

for all $f \in L^2(\sigma)$. Define

$$A := g^{-1}([0,1)); \quad B := g^{-1}(\{1\}).$$

Then A, B are disjoint measurable sets with union X.

Taking $f = I_B$ ($\in L^2(\sigma)$, since σ is a finite measure) in (2), we obtain $\mu(B) = 0$ (since $g = 1$ on B). Therefore, the measure λ_s defined on $\mathcal{A}$ by

$$\lambda_s(E) := \lambda(E \cap B)$$

satisfies $\lambda_s \perp \mu$.

Define similarly $\lambda_a(E) := \lambda(E \cap A)$; this is a positive measure on $\mathcal{A}$, mutually singular with λ_s (since it is carried by $A = B^c$), and by additivity of measures,

$$\lambda(E) = \lambda(E \cap A) + \lambda(E \cap B) = \lambda_a(E) + \lambda_s(E),$$

so that the Lebesgue decomposition will follow if we show that $\lambda_a \ll \mu$. This follows trivially from the integral representation (b), which we proceed to prove.

For each $n \in \mathbb{N}$ and $E \in \mathcal{A}$, take in (2)

$$f = f_n := (1 + g + \cdots + g^n)I_E.$$

(Since $0 \leq g \leq 1$, f is a bounded measurable function, hence $f \in L^2(\sigma)$.) We obtain

$$\int_E (1 - g^{n+1})\,d\lambda = \int_E (g + g^2 + \cdots + g^{n+1})\,d\mu. \tag{3}$$

Since $g = 1$ on B, the left-hand side equals $\int_{E\cap A}(1 - g^{n+1})\,d\lambda$. However, $0 \leq g < 1$ on A, so that the integrands form a non-decreasing sequence of non-negative measurable functions converging pointwise to 1. By the monotone convergence theorem, the left-hand side of (3) converges therefore to $\lambda(E \cap A) = \lambda_a(E)$. The integrands on the right-hand side of (3) form a non-decreasing sequence of non-negative measurable functions converging pointwise to the (measurable) function

$$h := \sum_{n=1}^{\infty} g^n.$$

Again, by monotone convergence, the right-hand side of (3) converges to $\int_E h\,d\mu$, and the representation (b) follows. Taking in particular $E = X$, we get

$$\|h\|_{L^1(\mu)} = \int_X h\,d\mu = \lambda_a(X) = \lambda(A) < \infty,$$

so that $h \in L^1(\mu)$, and the existence part of the theorem is proved in case $\mu(X) < \infty$.

General case. Let $X_j \in \mathcal{A}$ be mutually disjoint, with union X, such that $0 < \mu(X_j) < \infty$. Define

$$w = \sum_j \frac{1}{2^j \mu(X_j)} I_{X_j}.$$

This is a strictly positive μ-integrable function, with $\|w\|_1 = 1$. Consider the positive measure

$$\nu(E) = \int_E w \, d\mu.$$

Then $\nu(X) = \|w\|_1 = 1$, and $\nu \ll \mu$. On the other hand, if $\nu(E) = 0$, then $\sum_j (1/2^j \mu(X_j))\mu(E \cap X_j) = 0$, hence $\mu(E \cap X_j) = 0$ for all j, and therefore $\mu(E) = 0$. This shows that $\mu \ll \nu$ as well (one says that the measures μ and ν are *mutually absolutely continuous*, or *equivalent*).

Since ν is a finite measure, the first part of the proof gives the decomposition $\lambda = \lambda_a + \lambda_s$ with $\lambda_a \ll \nu$ (hence $\lambda_a \ll \mu$ by the trivial transitivity of the relation $\ll$), and $\lambda_s \perp \nu$ (hence $\lambda_s \perp \mu$, because λ_s is supported by a ν-null set, which is also μ-null, since $\mu \ll \nu$). The first part of the proof gives also the representation (cf. Theorem 1.17)

$$\lambda_a(E) = \int_E h \, d\nu = \int_E hw \, d\mu = \int_E \tilde{h} \, d\mu,$$

where $\tilde{h} := hw$ is non-negative, measurable, and

$$\|\tilde{h}\|_1 = \int_X \tilde{h} \, d\mu = \lambda_a(X) \le \lambda(X) < \infty.$$

This completes the proof of the "existence part" of the theorem in the general case.

To prove the *uniqueness* of the Lebesgue decomposition, suppose

$$\lambda = \lambda_a + \lambda_s = \lambda'_a + \lambda'_s,$$

with

$$\lambda_a, \lambda'_a \ll \mu \quad \text{and} \quad \lambda_s, \lambda'_s \perp \mu.$$

Let B be a μ-null set that carries both λ_s and λ'_s. Then

$$\lambda_a(B) = \lambda'_a(B) = 0 \quad \text{and} \quad \lambda_s(B^c) = \lambda'_s(B^c) = 0,$$

so that for all $E \in \mathcal{A}$,

$$\begin{aligned}\lambda_a(E) &= \lambda_a(E \cap B^c) = \lambda(E \cap B^c)\\ &= \lambda'_a(E \cap B^c) = \lambda'_a(E),\end{aligned}$$

hence also $\lambda_s(E) = \lambda'_s(E)$.

In order to prove the uniqueness of h in (b), suppose $h, h' \in L^1(\mu)$ satisfy

$$\lambda_a(E) = \int_E h\,d\mu = \int_E h'\,d\mu.$$

Then $h - h' \in L^1(\mu)$ satisfies $\int_E (h-h')\,d\mu = 0$ for all $E \in \mathcal{A}$, and it follows from Proposition 1.22 that $h - h' = 0$ μ-a.e., that is, $h = h'$ *as elements of* $L^1(\mu)$. □

If the measure λ is absolutely continuous with respect to μ, it has the trivial Lebesgue decomposition $\lambda = \lambda + 0$, with the zero measure as singular part. By uniqueness, it follows that $\lambda_a = \lambda$, and therefore Part 2 of the theorem gives the representation $\lambda(E) = \int_E h\,d\mu$ for all $E \in \mathcal{A}$. Conversely, such an integral representation of λ implies trivially that $\lambda \ll \mu$ (if $\mu(E) = 0$, the function $hI_E = 0$ μ-a.e., and therefore $\lambda(E) = \int fI_E\,d\mu = 0$). Thus

Theorem 1.41 (Radon–Nikodym). *Let $(X, \mathcal{A}, \mu)$ be a σ-finite positive measure space. A finite positive measure λ on $\mathcal{A}$ is absolutely continuous with respect to μ if and only if there exists $h \in L^1(\mu)$ such that*

$$\lambda(E) = \int_E h\,d\mu \quad (E \in \mathcal{A}). \tag{*}$$

By Theorem 1.17, Relation (*) implies that

$$\int g\,d\lambda = \int gh\,d\mu \tag{**}$$

for all non-negative measurable functions g on X. Since we may take $g = I_E$ in (**), this last relation implies (*). As mentioned after Theorem 1.17, these equivalent relations are *symbolically* written in the form $d\lambda = h\,d\mu$. It follows easily from Theorem 1.17 that, in that case, if $g \in L^1(\lambda)$, then $gh \in L^1(\mu)$ and (**) is valid for such (complex) functions g. The function h is called the *Radon–Nikodym derivative* of λ with respect to μ, and is denoted $d\lambda/d\mu$.

1.9 Complex measures

Definition 1.42. Let $(X, \mathcal{A})$ be an arbitrary measurable space. A *complex measure* on $\mathcal{A}$ is a σ-additive function $\mu : \mathcal{A} \to \mathbb{C}$, that is,

$$\mu\Big(\bigcup_n E_n\Big) = \sum_n \mu(E_n) \tag{1}$$

for any sequence of mutually disjoint sets $E_n \in \mathcal{A}$.

Since the left-hand side of (1) is independent of the order of the sets E_n and is a complex number, the right-hand side converges in $\mathbb{C}$ unconditionally, hence *absolutely*. Taking $E_n = \emptyset$ for all n, the convergence of (1) shows that $\mu(\emptyset) = 0$. It follows that μ is (finitely) additive, and since its values are complex numbers, it is "subtractive" as well (i.e., $\mu(E - F) = \mu(E) - \mu(F)$ whenever $E, F \in \mathcal{A}$, $F \subset E$).

A *partition* of $E \in \mathcal{A}$ is a sequence of mutually disjoint sets $A_k \in \mathcal{A}$ with union equal to E. We set

$$|\mu|(E) := \sup \sum_k |\mu(A_k)|, \tag{2}$$

where the supremum is taken over all partitions of E.

Theorem 1.43. *Let μ be a complex measure on $\mathcal{A}$, and define $|\mu|$ by (2). Then $|\mu|$ is a finite positive measure on $\mathcal{A}$ that dominates μ (i.e., $|\mu(E)| \le |\mu|(E)$ for all $E \in \mathcal{A}$).*

Proof. Let $E = \bigcup E_n$ with $E_n \in \mathcal{A}$ mutually disjoint ($n \in \mathbb{N}$). For any partition $\{A_k\}$ of E, $\{A_k \cap E_n\}_k$ is a partition of E_n ($n = 1, 2, \ldots$), so that

$$\sum_k |\mu(A_k \cap E_n)| \le |\mu|(E_n), \quad n = 1, 2, \ldots.$$

We sum these inequalities over all n, interchange the order of summation in the double sum (of non-negative terms!), and use the triangle inequality to obtain

$$\sum_n |\mu|(E_n) \ge \sum_k |\sum_n \mu(A_k \cap E_n)| = \sum_k |\mu(A_k)|,$$

since $\{A_k \cap E_n\}_n$ is a partition of A_k, for each $k \in \mathbb{N}$. Taking now the supremum over all partitions $\{A_k\}$ of E, it follows that

$$\sum_n |\mu|(E_n) \ge |\mu|(E). \tag{3}$$

On the other hand, given $\epsilon > 0$, there exists a partition $\{A_{n,k}\}_k$ of E_n such that

$$\sum_k |\mu(A_{n,k})| > |\mu|(E_n) - \epsilon/2^n, \quad n = 1, 2, \ldots.$$

Since $\{A_{n,k}\}_{n,k}$ is a partition of E, we obtain

$$|\mu|(E) \ge \sum_{n,k} |\mu(A_{n,k})| > \sum_n |\mu|(E_n) - \epsilon.$$

Letting $\epsilon \to 0+$ and using (3), we conclude that $|\mu|$ is σ-additive. Since $|\mu|(\emptyset) = 0$ is trivial, $|\mu|$ is indeed a positive measure on $\mathcal{A}$. □

In order to show that the measure $|\mu|$ is finite, we need the following.

Lemma. *Let $F \subset \mathbb{C}$ be a finite set. Then it contains a subset E such that*

$$|\sum_{z \in E} z| \geq \sum_{z \in F} |z|/4\sqrt{2}.$$

Proof of lemma. Let S be the sector

$$S = \{z = re^{\mathrm{i}\theta}; r \geq 0, |\theta| \leq \pi/4\}.$$

For $z \in S$, $\Re z = |z| \cos\theta \geq |z| \cos \pi/4 = |z|/\sqrt{2}$. Similarly, if $z \in -S$, then $-z \in S$, so that $-\Re z = \Re(-z) \geq |-z|/\sqrt{2} = |z|/\sqrt{2}$. If $z \in \mathrm{i}S$, then $-\mathrm{i}z \in S$, so that $\Im z = \Re(-\mathrm{i}z) \geq |-\mathrm{i}z|/\sqrt{2} = |z|/\sqrt{2}$. Similarly, if $z \in -\mathrm{i}S$, one obtains as before $-\Im z \geq |z|/\sqrt{2}$. Denote

$$a := \sum_{z \in F} |z|; \qquad a_k = \sum_{z \in F \cap (i^k S)} |z|, \quad k = 0, 1, 2, 3.$$

Since $\mathbb{C} = \bigcup_{k=0}^{3} \mathrm{i}^k S$, we have $\sum_{k=0}^{3} a_k \geq a$, and therefore there exists $k \in \{0, 1, 2, 3\}$ such that $a_k \geq a/4$. Fix such a k and define $E = F \cap (i^k S)$.

In case $k = 0$, that is, in case $a_0 \geq a/4$, we have

$$\begin{aligned} |\sum_{z \in E} z| &\geq \Re \sum_{z \in E} z = \sum_{z \in F \cap S} \Re z \\ &\geq \sum_{z \in F \cap S} |z|/\sqrt{2} = a_0/\sqrt{2} \geq a/4\sqrt{2}. \end{aligned}$$

Similarly, in case $k = 2$, replacing $\Re$ by $-\Re$ and S by $-S$, the same inequality is obtained. In cases $k = 1 (k = 3)$, we replace $\Re$ and S by $\Im$ $(-\Im)$ and $\mathrm{i}S$ $(-\mathrm{i}S)$ respectively, in the calculation. In all cases, we obtain $|\sum_{z \in E} z| \geq a/4\sqrt{2}$, as wanted. □

Returning to the proof of the finiteness of the measure $|\mu|$, suppose $|\mu|(A) = \infty$ for some $A \in \mathcal{A}$. Then there exists a partition $\{A_i\}$ of A such that $\sum_i |\mu(A_i)| > 4\sqrt{2}(1 + |\mu(A)|)$, and therefore there exists n such that

$$\sum_{i=1}^{n} |\mu(A_i)| > 4\sqrt{2}(1 + |\mu(A)|).$$

Take in the lemma

$$F = \{\mu(A_i);\ i = 1, \ldots, n\},$$

let the corresponding subset E be associated with the set of indices $J \subset \{1, \ldots, n\}$, and define

$$B := \bigcup_{i \in J} A_i \ (\subset A).$$

Then

$$|\mu(B)| = \Big|\sum_{i \in J} \mu(A_i)\Big| \geq \sum_{i=1}^{n} |\mu(A_i)|/4\sqrt{2} > 1 + |\mu(A)|.$$

If $C := A - B$, then

$$|\mu(C)| = |\mu(A) - \mu(B)| \geq |\mu(B)| - |\mu(A)| > 1.$$

Also $\infty = |\mu|(A) = |\mu|(B) + |\mu|(C)$ (since $|\mu|$ is a measure), so one of the summands at least is infinite, and for *both* subsets, the μ-measure has modulus >1.

Thus, we proved that any $A \in \mathcal{A}$ with $|\mu|(A) = \infty$ is the disjoint union of subsets $B_1, C_1 \in \mathcal{A}$, with $|\mu|(B_1) = \infty$ and $|\mu(C_1)| > 1$.

Since $|\mu|(B_1) = \infty$, B_1 is the disjoint union of subsets $B_2, C_2 \in \mathcal{A}$, with $|\mu|(B_2) = \infty$ and $|\mu(C_2)| > 1$. Continuing, we obtain two sequences $\{B_n\}, \{C_n\} \subset \mathcal{A}$ with the following properties:

$$B_n = B_{n+1} \cup C_{n+1} \quad \text{(disjoint union);}$$

$$|\mu|(B_n) = \infty; \quad |\mu(C_n)| > 1; \qquad n = 1, 2, \ldots$$

For $i > j \geq 1$, since $B_{n+1} \subset B_n$ for all n, we have $C_i \cap C_j \subset B_{i-1} \cap C_j \subset B_j \cap C_j = \emptyset$, so $C := \bigcup_n C_n$ is a disjoint union. Hence the series $\sum_n \mu(C_n) = \mu(C) \in \mathbb{C}$ converges. In particular $\mu(C_n) \to 0$, contradicting the fact that $|\mu(C_n)| > 1$ for all $n = 1, 2, \ldots$.

Finally, since $\{E, \emptyset, \emptyset, \ldots\}$ is a partition of $E \subset \mathcal{A}$, the inequality $|\mu(E)| \leq |\mu|(E)$ follows from (2). $\square$

Definition 1.44. The finite positive measure $|\mu|$ is called the *total variation measure* of μ, and $\|\mu\| := |\mu|(X)$ is called the *total variation* of μ.

Let $M(X, \mathcal{A})$ denote the complex vector space of all complex measures on $\mathcal{A}$ (with the "natural" operations $(\mu + \nu)(E) = \mu(E) + \nu(E)$ and $(c\mu)(E) = c\mu(E)$, for $\mu, \nu \in M(X, \mathcal{A})$ and $c \in \mathbb{C}$). With the total variation norm, $M(X, \mathcal{A})$ is a normed space. We verify below its completeness.

Proposition. *$M(X, \mathcal{A})$ is a Banach space.*

Proof. Let $\{\mu_n\} \subset M := M(X, \mathcal{A})$ be Cauchy. For all $E \in \mathcal{A}$,

$$|\mu_n(E) - \mu_m(E)| \leq \|\mu_n - \mu_m\| \to 0$$

as $n, m \to \infty$, so that

$$\mu(E) := \lim_n \mu_n(E)$$

exists. Clearly μ is additive. Let E be the union of the mutually disjoint sets $E_k \in \mathcal{A}$, $k = 1, 2, \ldots$, let $A_N = \bigcup_{k=1}^N E_k$, and let $\epsilon > 0$. Let $n_0 \in \mathbb{R}$ be such that

$$\|\mu_n - \mu_m\| < \epsilon \tag{*}$$

for all $n, m > n_0$. We have

$$\left|\mu(E) - \sum_{k=1}^N \mu(E_k)\right| = |\mu(E) - \mu(A_N)| = |\mu(E - A_N)|$$

$$\leq |(\mu - \mu_n)(E - A_N)| + |\mu_n(E - A_N)|.$$

Since

$$|(\mu_n - \mu_m)(E - A_N)| \leq \|\mu_n - \mu_m\| < \epsilon$$

for all $n, m > n_0$, letting $m \to \infty$, we obtain $|(\mu_n - \mu)(E - A_N)| \le \epsilon$ for all $n > n_0$ and all $N \in \mathbb{N}$. Therefore,

$$\left|\mu(E) - \sum_{k=1}^{N} \mu(E_k)\right| \le \epsilon + \left|\mu_n(E) - \sum_{k=1}^{N} \mu_n(E_k)\right|$$

for all $n > n_0$ and $N \in \mathbb{N}$. Fix $n > n_0$ and let $N \to \infty$. Since $\mu_n \in M$, we obtain

$$\limsup_N \left|\mu(E) - \sum_{k=1}^{N} \mu(E_k)\right| \le \epsilon,$$

so that $\mu(E) = \sum_{k=1}^{\infty} \mu(E_k)$. Thus $\mu \in M$.

Finally, we show that $\|\mu - \mu_n\| \to 0$. Let $\{E_k\}$ be a partition of X. By (*), for all $N \in \mathbb{N}$ and $n, m > n_0$,

$$\sum_{k=1}^{N} |(\mu_m - \mu_n)(E_k)| \le \|\mu_m - \mu_n\| < \epsilon.$$

Letting $m \to \infty$, we get

$$\sum_{k=1}^{N} |(\mu - \mu_n)(E_k)| \le \epsilon$$

for all N and $n > n_0$. Letting $N \to \infty$, we obtain $\sum_{k=1}^{\infty} |(\mu - \mu_n)(E_k)| \le \epsilon$ for all $n > n_0$, and since the partition was arbitrary, it follows that $\|\mu - \mu_n\| \le \epsilon$ for $n > n_0$. □

If $\mu \in M(X, \mathcal{A})$ has *real* range, it is called a *real measure*. For example, $\Re\mu$ (defined by $(\Re\mu)(E) = \Re[\mu(E)]$) and $\Im\mu$ are real measures for any complex measure μ, and $\mu = \Re\mu + i\Im\mu$.

If μ is a *real* measure, since $|\mu(E)| \le |\mu|(E)$ for all $E \in \mathcal{A}$, the measures

$$\mu^+ := (1/2)(|\mu| + \mu); \qquad \mu^- := (1/2)(|\mu| - \mu)$$

are *finite positive measures*, called the *positive and negative variation measures*, respectively.

Clearly

$$\mu = \mu^+ - \mu^- \tag{4}$$

and

$$|\mu| = \mu^+ + \mu^-.$$

Representation (4) of a real measure as the difference of two finite positive measures is called the *Jordan decomposition* of the real measure μ.

For a complex measure λ, write first $\lambda = \nu + i\sigma$ with $\nu := \Re\lambda$ and $\sigma = \Im\lambda$; then write the Jordan decompositions of ν and σ. It then follows that *any* complex measure λ can be written as the linear combination

$$\lambda = \sum_{k=0}^{3} i^k \lambda_k \tag{5}$$

of four finite positive measures λ_k.

If μ is a positive measure and λ is a *complex* measure (both on the σ-algebra $\mathcal{A}$), we say that λ is absolutely continuous with respect to μ (notation: $\lambda \ll \mu$) if $\lambda(E) = 0$ whenever $\mu(E) = 0, E \in \mathcal{A}$; λ is carried (or supported) by the set $A \in \mathcal{A}$ if $\lambda(E) = 0$ for all measurable subsets E of A^c; λ is singular with respect to μ if it is carried by a μ-null set. Two complex measures λ_1, λ_2 are *mutually singular* (notation $\lambda_1 \perp \lambda_2$) if they are carried by disjoint measurable sets.

It follows immediately from (5) that Theorems 1.40 and 1.41 extend verbatim to the case of a *complex* measure λ. This is stated formally for future reference.

Theorem 1.45. *Let $(X, \mathcal{A}, \mu)$ be a σ-finite positive measure space, and let λ be a complex measure on $\mathcal{A}$. Then*

(1) λ has a unique Lebesgue decomposition

$$\lambda = \lambda_a + \lambda_s$$

with $\lambda_a \ll \mu$ and $\lambda_s \perp \mu$;

(2) there exists a unique $h \in L^1(\mu)$ such that

$$\lambda_a(E) = \int_E h\, d\mu \quad (E \in \mathcal{A});$$

(3) $\lambda \ll \mu$ iff there exists $h \in L^1(\mu)$ such that $\lambda(E) = \int_E h\, d\mu$ for all $E \in \mathcal{A}$.

Another useful representation of a complex measure in terms of a finite positive measure is the following:

Theorem 1.46. *Let μ be a complex measure on the measurable space $(X, \mathcal{A})$. Then there exists a measurable function h with $|h| = 1$ on X such that $d\mu = h\, d|\mu|$.*

Proof. Since $|\mu(E)| \le |\mu|(E)$ for all $E \in \mathcal{A}$, it follows that $\mu \ll |\mu|$, and therefore, by Theorem 1.45 (since $|\mu|$ is a *finite* positive measure), there exists $h \in L^1(|\mu|)$ such that $d\mu = h\, d|\mu|$. For each $E \in \mathcal{A}$ with $|\mu|(E) > 0$,

$$\left| \frac{1}{|\mu|(E)} \int_E h\, d|\mu| \right| = \frac{|\mu(E)|}{|\mu|(E)} \le 1.$$

By Lemma 1.38, it follows that $|h| \le 1$ $|\mu|$-a.e.

We wish to show that $|\mu|([|h|<1])=0$. Since $[|h|<1]=\bigcup_n[|h|<1-1/n]$, it suffices to show that $|\mu|([|h|<r])=0$ for each $r<1$. Denote $A=[|h|<r]$ and let $\{E_k\}$ be a partition of A. Then

$$\sum_k|\mu(E_k)|=\sum_k\left|\int_{E_k}h\,d|\mu|\right|\le r\sum_k|\mu|(E_k)=r|\mu|(A).$$

Hence

$$|\mu|(A)\le r|\mu|(A),$$

so that indeed $|\mu|(A)=0$.

We conclude that $|h|=1$ $|\mu|$-a.e., and since h is only determined a.e., we may replace it by a $|\mu|$-equivalent function which satisfies $|h|=1$ everywhere. □

If μ is a complex measure on $\mathcal{A}$ and $f\in L^1(|\mu|)$, there are two natural ways to define $\int_X f\,d\mu$. One way uses decomposition (5) of μ as a linear combination of four finite positive measures μ_k, which clearly satisfy $\mu_k\le|\mu|$. Therefore $f\in L^1(\mu_k)$ for all $k=0,\dots,3$, and we define $\int_X f\,d\mu=\sum_{k=0}^3 \mathrm{i}^k\int f\,d\mu_k$. A second possible definition uses Theorem 1.46. Since $|h|=1$, $fh\in L^1(|\mu|)$, and we may define $\int_X f\,d\mu=\int_X fh\,d|\mu|$. One verifies easily that the two definitions above give the same value to the integral $\int_X f\,d\mu$. The integral thus defined is a linear functional on $L^1(|\mu|)$. As usual, $\int_E f\,d\mu:=\int_X fI_E\,d\mu$ for $E\in\mathcal{A}$.

By Theorem 1.46, every complex measure λ is of the form $d\lambda=g\,d\mu$ for some (finite) positive measure μ $(=|\lambda|)$ and a uniquely determined $g\in L^1(\mu)$.

Conversely, given a positive measure μ and $g\in L^1(\mu)$, we may *define* $d\lambda:=g\,d\mu$ (as before, the meaning of this symbolic relation is that $\lambda(E)=\int_E g\,d\mu$ for all $E\in\mathcal{A}$). If $\{E_k\}$ is a partition of E, and $F_n=\bigcup_{k=1}^n E_k$, then

$$\lambda(F_n)=\int gI_{F_n}\,d\mu=\int\sum_{k=1}^n gI_{E_k}\,d\mu$$
$$=\sum_{k=1}^n\lambda(E_k).$$

Since $gI_{F_n}\to gI_E$ pointwise as $n\to\infty$ and $|gI_{F_n}|\le|g|\in L^1(\mu)$, the dominated convergence theorem implies that the series $\sum_{k=1}^\infty\lambda(E_k)$ converges to $\int gI_E\,d\mu:=\lambda(E)$. Thus, λ is a complex measure. The following theorem gives its total variation measure.

Theorem 1.47. *Let $(X,\mathcal{A},\mu)$ be a σ-finite positive measure space, $g\in L^1(\mu)$, and let λ be the complex measure $d\lambda:=g\,d\mu$. Then $d|\lambda|=|g|\,d\mu$.*

Proof. By Theorem 1.46, $d\lambda=h\,d|\lambda|$ with h measurable and $|h|=1$ on X. For all $E\in\mathcal{A}$

$$\int_E\bar{h}g\,d\mu=\int_E\bar{h}\,d\lambda=\int_E\bar{h}h\,d|\lambda|=|\lambda|(E). \tag{*}$$

Therefore, whenever $0 < \mu(E) < \infty$,

$$\frac{1}{\mu(E)} \int_E \bar{h} g \, d\mu = \frac{|\lambda|(E)}{\mu(E)} \geq 0.$$

By Lemma 1.38, $\bar{h}g \geq 0$ μ-a.e. Hence $\bar{h}g = |\bar{h}g| = |g|$ μ-a.e., and therefore, by (*),

$$|\lambda|(E) = \int_E |g| \, d\mu \quad (E \in \mathcal{A}).$$

□

(The σ-finiteness hypothesis can be dropped: use Proposition 1.22 instead of Lemma 1.38.)

If μ is a *real* measure and $\mu = \mu^+ - \mu^-$ is its Jordan decomposition, the following theorem expresses the positive measures μ^+ and μ^- in terms of a decomposition of the space X as the disjoint union of two measurable sets A and B that carry them (respectively). In particular, $\mu^+ \perp \mu^-$.

Theorem 1.48 (Hahn decomposition). *Let μ be a real measure on the measurable space $(X, \mathcal{A})$. Then there exist disjoint sets $A, B \in \mathcal{A}$ such that $X = A \cup B$ and*

$$\mu^+(E) = \mu(E \cap A), \quad \mu^-(E) = -\mu(E \cap B)$$

for all $E \in \mathcal{A}$.

Proof. By Theorem 1.46, $d\mu = h \, d|\mu|$ with h measurable and $|h| = 1$ on X. For all $E \in \mathcal{A}$ with $|\mu|(E) > 0$,

$$\frac{1}{|\mu|(E)} \int_E h \, d|\mu| = \frac{\mu(E)}{|\mu|(E)} \in \mathbb{R}.$$

By the averages lemma, it follows that h is real $|\mu|$-a.e., and since it is only determined a.e., we may assume that h is real everywhere on X. However, $|h| = 1$; hence $h(X) = \{-1, 1\}$. Let $A := [h = 1]$ and $B = [h = -1]$. Then X is the disjoint union of these measurable sets, and for all $E \in \mathcal{A}$,

$$\begin{aligned} \mu^+(E) &:= (1/2)(|\mu|(E) + \mu(E)) \\ &= (1/2) \int_E (1 + h) \, d|\mu| \\ &= \int_{E \cap A} h \, d|\mu| = \mu(E \cap A). \end{aligned}$$

An analogous calculation shows that $\mu^-(E) = -\mu(E \cap B)$. □

1.10 Convergence

This section considers some modes of convergence and relations between them. In order to avoid repetitions, $(X, \mathcal{A}, \mu)$ will denote throughout a positive measure space; $f, f_n (n = 1, 2, \ldots)$ are complex measurable functions on X.

Definition 1.49.

(1) f_n converge to f *almost uniformly* if for any $\epsilon > 0$, there exists $E \in \mathcal{A}$ with $\mu(E) < \epsilon$, such that $f_n \to f$ uniformly on E^c.

(2) $\{f_n\}$ is almost uniformly Cauchy if for any $\epsilon > 0$, there exists $E \in \mathcal{A}$ with $\mu(E) < \epsilon$, such that $\{f_n\}$ is uniformly Cauchy on E^c.

Remark 1.50.

(1) Taking $\epsilon = 1/k$ with $k = 1, 2, \ldots$, we see that if $\{f_n\}$ is almost uniformly Cauchy, then there exist $E_k \in \mathcal{A}$ such that $\mu(E_k) < 1/k$ and $\{f_n\}$ is uniformly Cauchy on E_k^c. Let $E = \bigcap E_k$; then $E \in \mathcal{A}$ and $\mu(E) < 1/k$ for all k, so that $\mu(E) = 0$. If $x \in E^c = \bigcup E_k^c$, then $x \in E_k^c$ for some k, so that $\{f_n(x)\}$ is Cauchy, and consequently $\exists \lim_n f_n(x) := f(x)$. We may define $f(x) = 0$ on E. The function f is measurable, and $f_n \to f$ almost everywhere (since $\mu(E) = 0$). For any $\epsilon, \delta > 0$, let $F \in \mathcal{A}$, $n_0 \in \mathbb{N}$ be such that $\mu(F) < \epsilon$ and $|f_n(x) - f_m(x)| < \delta$ for all $x \in F^c$ and $n, m > n_0$. Setting $G = F^c \cap E^c$, we have $\mu(G^c) < \epsilon$, and letting $m \to \infty$ in the last inequality, we get $|f_n(x) - f(x)| \le \delta$ for all $x \in G$ and $n > n_0$. Thus $f_n \to f$ uniformly on G, and consequently $f_n \to f$ almost uniformly. This shows that *almost uniformly Cauchy sequences converge almost uniformly*; the converse follows trivially from the triangle inequality.

(2) A trivial modification of the first argument here shows that if $f_n \to f$ almost uniformly, then $f_n \to f$ almost everywhere. In particular, the almost uniform limit f is uniquely determined up to equivalence.

Definition 1.51.

(1) The sequence $\{f_n\}$ converges to f *in measure* if for any $\epsilon > 0$,

$$\lim_n \mu([|f_n - f| \ge \epsilon]) = 0.$$

(2) The sequence $\{f_n\}$ is *Cauchy in measure* if for any $\epsilon > 0$,

$$\lim_{n,m \to \infty} \mu([|f_n - f_m| \ge \epsilon]) = 0.$$

Remark 1.52.

(1) If $f_n \to f$ and $f_n \to f'$ in measure, then for any $\epsilon > 0$, the triangle inequality shows that

$$\mu([|f - f'| \ge \epsilon]) \le \mu([|f - f_n| \ge \epsilon/2]) + \mu([|f_n - f'| \ge \epsilon/2]) \to 0$$

as $n \to \infty$, that is, $[|f - f'| \ge \epsilon]$ is μ-null. Therefore, $[f \neq f'] = \bigcup_k [|f - f'| \ge 1/k]$ is μ-null, and $f = f'$ a.e. This shows that *limits in measure are uniquely determined (up to equivalence).*

(2) A similar argument based on the triangle inequality shows that if $\{f_n\}$ converges in measure, then it is Cauchy in measure.

(3) *If $f_n \to f$ almost uniformly, then $f_n \to f$ in measure.* Indeed, for any $\epsilon, \delta > 0$, there exists $E \in \mathcal{A}$ and $n_0 \in \mathbb{N}$, such that $\mu(E) < \delta$ and $|f_n - f| < \epsilon$ on E^c for all $n \geq n_0$. Hence for all $n \geq n_0$, $[|f_n - f| \geq \epsilon] \subset E$, and therefore $\mu([|f_n - f| \geq \epsilon]) < \delta$.

Theorem 1.53. *The sequence $\{f_n\}$ converges in measure iff it is Cauchy in measure.*

Proof. By Remark 1.52.2, we need only to show the "if" part of the theorem. Let $\{f_n\}$ be Cauchy in measure. As in the proof of Lemma 1.30, we obtain integers $1 \leq n_1 < n_2 < n_3, \dots$ such that $\mu(E_k) < (1/2^k)$, where

$$E_k := \left[|f_{n_{k+1}} - f_{n_k}| \geq \frac{1}{2^k} \right].$$

The set $F_m = \bigcup_{k \geq m} E_k$ has measure $< \sum_{k \geq m} 2^{-k} = (1/2^{m-1})$, and on F_m^c we have for $j > i \geq m$

$$|f_{n_j} - f_{n_i}| \leq \sum_{k=i}^{j-1} |f_{n_{k+1}} - f_{n_k}| < \sum_{k=i}^{j-1} 2^{-k} < \frac{1}{2^{i-1}}.$$

This shows that $\{f_{n_k}\}$ is almost uniformly Cauchy. By Remark 1.50, $\{f_{n_k}\}$ converges almost uniformly to a measurable function f. Hence $f_{n_k} \to f$ in measure, by Remark 1.52.3. For any $\epsilon > 0$, we have

$$[|f_n - f| \geq \epsilon] \subset [|f_n - f_{n_k}| \geq \epsilon/2] \cup [|f_{n_k} - f| \geq \epsilon/2]. \tag{1}$$

The measure of the first set on the right-hand side tends to zero when $n, k \to \infty$, since $\{f_n\}$ is Cauchy in measure. The measure of the second set on the right-hand side of (1) tends to zero when $k \to \infty$, since $f_{n_k} \to f$ in measure. Hence the measure of the set on the left-hand side of (1) tends to zero as $n \to \infty$. □

Theorem 1.54. *If $f_n \to f$ in L^p for some $p \in [1, \infty]$, then $f_n \to f$ in measure.*

Proof. For any $\epsilon > 0$ and $n \in \mathbb{N}$, set $E_n = [|f_n - f| \geq \epsilon]$.

Case $p < \infty$. We have

$$\epsilon^p \mu(E_n) \leq \int_{E_n} |f_n - f|^p \, d\mu \leq \|f_n - f\|_p^p,$$

and consequently $f_n \to f$ in L^p implies $\mu(E_n) \to 0$.

Case $p = \infty$. Let $A = \bigcup_n A_n$, where $A_n = [|f_n - f| > \|f_n - f\|_\infty]$. By definition of the L^∞-norm, each A_n is null, and therefore A is null, and $|f_n - f| \leq \|f_n - f\|_\infty$ on A^c for all n (hence $f_n \to f$ uniformly on A^c and $\mu(A) = 0$; in such a situation, one says that $f_n \to f$ *uniformly almost everywhere*). If $f_n \to f$ in L^∞, there exists n_0 such that $\|f_n - f\|_\infty < \epsilon$ for all $n \geq n_0$. Thus $|f_n - f| < \epsilon$ on A^c for all $n \geq n_0$, hence $E_n \subset A$ for all $n \geq n_0$, and consequently E_n is null for all $n \geq n_0$. □

1.11 Convergence on finite measure space

On *finite* measure spaces, there exist some additional relations between the various types of convergence. A sample of such relations is discussed in this section.

Theorem 1.55. *Let $(X, \mathcal{A}, \mu)$ be a finite measure space, and let $f_n : X \to \mathbb{C}$ be measurable functions converging almost everywhere to the (measurable) function f. Then $f_n \to f$ in measure.*

Proof. Translating the definition of *non-convergence* into set theoretic operations, we have (on any measure space!)

$$[f_n \text{ does not converge to } f] = \bigcup_{k \in \mathbb{N}} \limsup_n [|f_n - f| \geq 1/k].$$

This set is null (i.e., $f_n \to f$ a.e.) iff $\limsup_n [|f_n - f| \geq 1/k]$ is null for all k. Since $\mu(X) < \infty$, this is equivalent to

$$\lim_n \mu\Big(\bigcup_{m \geq n} [|f_m - f| \geq 1/k] \Big) = 0$$

for all k, which clearly implies that $\lim_n \mu([|f_n - f| \geq 1/k]) = 0$ for all k (i.e., $f_n \to f$ in measure). □

Remark 1.56.

(1) Conversely, if $f_n \to f$ in measure (in an arbitrary measure space), then there exists a subsequence f_{n_k} converging a.e. to f (cf. proof of Theorem 1.53).

(2) If the *bounded* sequence $\{f_n\}$ converges a.e. to f, then $f_n \to f$ in L^p for any $1 \leq p < \infty$ (by the proposition following Theorem 1.26; in an arbitrary measure space, the boundedness condition on the sequence must be replaced by its majoration by a fixed L^p-function, not necessarily constant).

(3) If $1 \leq r < p \leq \infty$, L^p-convergence implies L^r-convergence (by the second proposition following Theorem 1.26).

Theorem 1.57 (Egoroff). *Let $(X, \mathcal{A}, \mu)$ be a finite positive measure space, and let $\{f_n\}$ be measurable functions converging pointwise a.e. to the function f. Then $f_n \to f$ almost uniformly.*

Proof. We first prove the following.

Lemma (*Assumptions and notation as in theorem*). *Given $\epsilon, \delta > 0$, there exist $A \in \mathcal{A}$ with $\mu(A) < \delta$ and $N \in \mathbb{N}$ such that $|f_n - f| < \epsilon$ on A^c for all $n \geq N$.*

Proof of lemma. Denote $E_n := [|f_n - f| \geq \epsilon]$ and $A_N := \bigcup_{n \geq N} E_n$. Then $\{A_N\}$ is a decreasing sequence of measurable sets, and since μ is a finite measure, $\mu(\bigcap_N A_N) = \lim_N \mu(A_N)$. Clearly $f_n(x)$ does not converge to $f(x)$ when $x \in \bigcap_N A_N$, and since $f_n \to f$ a.e., it follows that $\mu(\bigcap_N A_N) = 0$, that is, $\lim_N \mu(A_N) = 0$. Fix then N such that $\mu(A_N) < \delta$ and choose $A := A_N$. Since $A^c = \bigcap_{n \geq N} [|f_n - f| < \epsilon]$, the set A satisfies the lemma's requirements. □

Proof of theorem. Given $\epsilon, \delta > 0$, apply the lemma with $\epsilon_m = 1/m$ and $\delta_m = (\delta/2^m)$, $m = 1, 2, \ldots$. We get measurable sets A_m with $\mu(A_m) < \delta_m$ and integers N_m, such that $|f_n - f| < 1/m$ on A_m^c for all $n \geq N_m$ $(m = 1, 2, \ldots)$. Let $A := \bigcup_m A_m$; then $\mu(A) < \delta$, and on $A^c (= \bigcap_m A_m^c)$, we have $|f_n - f| < 1/m$ for all $n \geq N_m$, $m = 1, 2, \ldots$. Fix an integer $m_0 > 1/\epsilon$, and let $N := N_{m_0}$; then $|f_n - f| < \epsilon$ on A^c for all $n \geq N$. □

1.12 Distribution function

Definition 1.58. Let $(X, \mathcal{A}, \mu)$ be a positive measure space, and let $f : X \to [0, \infty]$ be measurable. The *distribution function* of f is defined by

$$m(y) = \mu([f > y]) \quad (y > 0). \tag{1}$$

This is a non-negative non-increasing function on $\mathbb{R}^+$, so that $m(\infty) := \lim_{y \to \infty} m(y)$ exists and is ≥ 0. We shall *assume* in the sequel that m *is finite-valued and* $m(\infty) = 0$. The finiteness of m implies that

$$m(a) - m(b) = \mu([a < f \leq b]) \quad (0 < a < b < \infty).$$

Let $\{y_n\}$ be any positive sequence increasing to ∞. If $E_n := [f > y_n]$, then $E_{n+1} \subset E_n$ and $\bigcap E_n = [f = \infty]$. Since m is finite-valued, we have by Lemma 1.11

$$m(\infty) = \lim m(y_n) = \lim \mu(E_n) = \mu(\bigcap E_n) = \mu([f = \infty]).$$

Thus our second assumption here means that f is finite μ-a.e.

Both assumptions here are satisfied in particular when $\int_X f^p \, d\mu < \infty$ for some $p \in [1, \infty)$. This follows from the inequality

$$m(y) \leq \left(\frac{\|f\|_p}{y} \right)^p \quad (y > 0) \tag{2}$$

(cf. proof of Theorem 1.54).

Theorem 1.59. *Suppose the distribution function m of the non-negative measurable function f is finite and vanishes at infinity. Then:*

(1) For all $p \in [1, \infty)$

$$\int_X f^p \, d\mu = -\int_0^\infty y^p \, dm(y), \tag{3}$$

where the integral on the right-hand side is the improper Riemann–Stieltjes integral

$$\lim_{a\to 0+;b\to\infty}\int_a^b y^p\,dm(y).$$

(2) *If either one of the integrals* $-\int_0^\infty y^p\,dm(y)$ *and* $p\int_0^\infty y^{p-1}m(y)\,dy$ *is finite, then*

$$\lim_{y\to 0} y^p m(y) = \lim_{y\to\infty} y^p m(y) = 0, \tag{4}$$

and the integrals coincide.

Proof. Let $0 < a < b < \infty$ and $n \in \mathbb{N}$. Denote

$$y_j = a + j\frac{b-a}{n2^n}, \quad j = 0,\ldots,n2^n;$$

$$E_j = [y_{j-1} < f \le y_j]; \quad E_{a,b} = [a < f \le b];$$

$$s_n = \sum_{j=1}^{n2^n} y_{j-1} I_{E_j}.$$

The sequence $\{s_n^p\}$ is a non-decreasing sequence of non-negative measurable functions with limit f^p (cf. proof of Theorem 1.8). By the Monotone Convergence theorem,

$$\int_{E_{a,b}} f^p\,d\mu = \lim_n \int_{E_{a,b}} s_n^p\,d\mu = \lim_n \sum_{j=1}^{n2^n} y_{j-1}^p \mu(E_j)$$

$$= -\lim_n \sum_{j=1}^{n2^n} y_{j-1}^p [m(y_j) - m(y_{j-1})] = -\int_a^b y^p\,dm(y).$$

The first integral above converges to the (finite or infinite) limit $\int_X f^p\,d\mu$ when $a \to 0$ and $b \to \infty$. It follows that the last integral converges to the same limit, that is, the improper Riemann–Stieltjes integral $\int_0^\infty y^p\,dm(y)$ exists and (3) is valid.

Since $-dm$ is a positive measure, we have

$$0 \le a^p[m(a) - m(b)] = -\int_a^b a^p\,dm(y) \le -\int_a^b y^p\,dm(y), \tag{5}$$

hence

$$0 \le a^p m(a) \le a^p m(b) - \int_a^b y^p\,dm(y). \tag{6}$$

Let $b \to \infty$. Since $m(\infty) = 0$,

$$0 \le a^p m(a) \le -\int_a^\infty y^p\,dm(y). \tag{7}$$

In case $\int_0^\infty y^p\,dm(y)$ is finite, letting $a \to \infty$ in (7) shows that $\lim_{a\to\infty} a^p m(a) = 0$. Also letting $a \to 0$ in (6) (with b fixed arbitrary) shows that

$$0 \le \limsup_{a\to 0} a^p m(a) \le -\int_0^b y^p\,dm(y).$$

Letting $b \to 0$, we conclude that $\exists \lim_{a\to 0} a^p m(a) = 0$, and (4) is verified.

An integration by parts gives

$$-\int_a^b y^p\,dm(y) = a^p m(a) - b^p m(b) + p\int_a^b y^{p-1} m(y)\,dy. \tag{8}$$

Letting $a \to 0$ and $b \to \infty$, we obtain from (4) (in case $\int_0^\infty y^p\,dm(y)$ is finite)

$$-\int_0^\infty y^p\,dm(y) = p\int_0^\infty y^{p-1} m(y)\,dy. \tag{9}$$

Consider finally the case when $\int_0^\infty y^{p-1} m(y)\,dy < \infty$. We have

$$(1-2^{-p})b^p m(b) = [b^p - (b/2)^p]m(b) = m(b)\int_{b/2}^b p y^{p-1}\,dy$$
$$\le p\int_{b/2}^b y^{p-1} m(y)\,dy \to 0$$

as $b \to \infty$ or $b \to 0$ (by Cauchy's criterion). Thus (4) is verified, and (9) was seen to follow from (4). □

Corollary 1.60. *Let $f \in L^p(\mu)$ for some $p \in [1,\infty)$, and let m be the distribution function of $|f|$. Then*

$$\|f\|_p^p = -\int_0^\infty y^p\,dm(y) = p\int_0^\infty y^{p-1} m(y)\,dy.$$

1.13 Truncation

Technique 1.61. The technique of truncation of functions is useful in real methods of analysis. Let $(X, \mathcal{A}, \mu)$ be a positive measure space, and let $f : X \to \mathbb{C}$. For each $u > 0$, we define the *truncation at u of f* by

$$f_u := f I_{[|f|\le u]} + u(f/|f|) I_{[|f|>u]}. \tag{1}$$

Denote

$$f'_u := f - f_u = (f - u(f/|f|)) I_{[|f|>u]} = (|f| - u)(f/|f|) I_{[|f|>u]}. \tag{2}$$

We have

$$|f_u| = |f| I_{[|f|\le u]} + u I_{[|f|>u]} = \min(|f|, u), \tag{3}$$
$$|f'_u| = (|f| - u) I_{[|f|>u]}, \tag{4}$$
$$f = f_u + f'_u, \qquad |f| = |f_u| + |f'_u|. \tag{5}$$

It follows in particular that $f \in L^p(\mu)$ for some $p \in [1, \infty]$ iff *both* f_u and f'_u are in $L^p(\mu)$. In this case $f_u \in L^r(\mu)$ *for any* $r \geq p$, because $|f_u/u| \leq 1$, so that

$$u^{-r}|f_u|^r = |f_u/u|^r \leq |f_u/u|^p \leq u^{-p}|f|^p.$$

Similarly (still when $f \in L^p$), $f'_u \in L^r(\mu)$ *for any* $r \leq p$. Indeed, write

$$\int_X |f'_u|^r \, d\mu = \int_{[|f'_u|>1]} + \int_{[0<|f'_u|\leq 1]}.$$

For $r \leq p$, the first integral on the right-hand side is

$$\leq \int_{[|f'_u|>1]} |f'_u|^p \, d\mu \leq \|f'_u\|_p^p;$$

the second integral on the right-hand side is

$$\begin{aligned} &\leq \mu([0 < |f'_u| \leq 1]) = \mu([0 < |f| - u \leq 1]) \\ &= \mu([u < |f| \leq u+1]) = m(u) - m(u+1). \end{aligned}$$

Thus, for any $r \leq p$,

$$\|f'_u\|_r^r \leq \|f'_u\|_p^p + m(u) - m(u+1) < \infty,$$

as claimed.

Since $|f_u| = \min(|f|, u)$, we have $[|f_u| > y] = [|f| > y]$ whenever $0 < y < u$ and $[|f_u| > y] = \emptyset$ whenever $y \geq u$. Therefore, if m_u and m are the distribution functions of $|f_u|$ and $|f|$, respectively, we have

$$m_u(y) = m(y) \quad \text{for } 0 < y < u; \qquad m_u(y) = 0 \quad \text{for } y \geq u. \tag{6}$$

For the distribution function m'_u of $|f'_u|$, we have the relation

$$m'_u(y) = m(y+u) \quad (y > 0), \tag{7}$$

since by (4)

$$m'_u(y) := \mu([|f'_u| > y]) = \mu([|f| - u > y]) = \mu([|f| > y + u]) = m(y+u).$$

By (6), (7), and Corollary 1.60, the following formulae are valid for any $f \in L^p(\mu)$ $(1 \leq p < \infty)$ and $u > 0$:

$$\|f_u\|_r^r = r \int_0^u v^{r-1} m(v) \, dv \quad (r \geq p); \tag{8}$$

$$\|f'_u\|_r^r = r \int_u^\infty (v-u)^{r-1} m(v) \, dv \quad (r \leq p). \tag{9}$$

These formulae are used in Section 5.40.

Exercises

1. Let $(X, \mathcal{A}, \mu)$ be a positive measure space, and let f be a non-negative measurable function on X. Let $E := [f < 1]$. Prove

   (a) $\mu(E) = \lim_n \int_E \exp(-f^n)\, d\mu$.

   (b) $\sum_{n=1}^{\infty} \int_E f^n\, d\mu = \int_E (f/(1-f))\, d\mu$.

2. Let $(X, \mathcal{A}, \mu)$ be a positive measure space, and let p, q be conjugate exponents. Prove that the map $[f, g] \in L^p(\mu) \times L^q(\mu) \to fg \in L^1(\mu)$ is continuous.

3. Let $(X, \mathcal{A}, \mu)$ be a positive measure space, $f_n : X \to \mathbb{C}$ measurable functions converging pointwise to f, and $h : \mathbb{C} \to \mathbb{C}$ continuous and bounded. Prove that $\lim_n \int_E h(f_n)\, d\mu = \int_E h(f)\, d\mu$ for each $E \in \mathcal{A}$ with finite measure.

4. Let $(X, \mathcal{A}, \mu)$ be a positive measure space, and let $\mathcal{B} \subset \mathcal{A}$ be a σ-finite σ-algebra. If $f \in L^1(\mathcal{A}) := L^1(X, \mathcal{A}, \mu)$, consider the complex measure on $\mathcal{B}$ defined by

$$\lambda_f(E) := \int_E f\, d\mu \quad (E \in \mathcal{B}).$$

   Prove:

   (a) There exists a unique element $Pf \in L^1(\mathcal{B}) := L^1(X, \mathcal{B}, \mu)$ such that

$$\lambda_f(E) = \int_E (Pf)\, d\mu \quad (E \in \mathcal{B}).$$

   (b) The map $P : f \to Pf$ is a continuous linear map of $L^1(\mathcal{A})$ onto the *subspace* $L^1(\mathcal{B})$, such that $P^2 = P$ (P^2 denotes the composition of P with itself). In particular, $L^1(\mathcal{B})$ is a *closed* subspace of $L^1(\mathcal{A})$.

5. Let $(X, \mathcal{A}, \mu)$ be a finite positive measure space and $f_n \in L^p(\mu)$ for all $n \in \mathbb{N}$ (for some $p \in [1, \infty)$). Suppose there exists a measurable function $f : X \to \mathbb{C}$ such that $\sup_n \sup_X |f_n - f| < \infty$ and $f_n \to f$ in measure. Prove that $f \in L^p(\mu)$ and $f_n \to f$ in L^p-norm.

6. Let λ and μ be positive σ-finite measures on the measurable space $(X, \mathcal{A})$. State and prove a version of the Lebesgue–Radon–Nikodym theorem for this situation.

7. Let $\{\lambda_n\}$ be a sequence of complex measures on the measurable space $(X, \mathcal{A})$ such that $\sum_n \|\lambda_n\| < \infty$. Prove

   (a) For each $E \in \mathcal{A}$, the series $\sum_n \lambda_n(E)$ converges absolutely in $\mathbb{C}$ and defines a complex measure λ; the series $\sum_n |\lambda_n|(E)$ converges in $\mathbb{R}^+$, and defines a finite positive measure σ, and $\lambda \ll \sigma$.

(b)

$$\frac{d\lambda}{d\sigma} = \sum_n \frac{d\lambda_n}{d\sigma}.$$

8. Let $(X, \mathcal{A}, \mu)$ be a positive measure space, and let $M := M(\mathcal{A})$ denote the vector space (over $\mathbb{C}$) of all complex measures on $\mathcal{A}$. Set

$$M_a := \{\lambda \in M; \lambda \ll \mu\};$$
$$M_s := \{\lambda \in M; \lambda \perp \mu\}.$$

Prove:

(a) If $\lambda \in M$ is supported by $E \in \mathcal{A}$, then so is $|\lambda|$.

(b) M_a and M_s are subspaces of M and $M_a \perp M_s$ (in particular, $M_a \cap M_s = \{0\}$).

(c) If $(X, \mathcal{A}, \mu)$ is σ-finite, then $M = M_a \oplus M_s$.

(d) $\lambda \in M_a$ iff $|\lambda| \in M_a$ (and similarly for M_s).

(e) If $\lambda_k \in M$ $(k = 1, 2)$, then $\lambda_1 \perp \lambda_2$ iff $|\lambda_1| \perp |\lambda_2|$.

(f) $\lambda \ll \mu$ iff for each $\epsilon > 0$, there exists $\delta > 0$ such that $|\lambda(E)| < \epsilon$ for all $E \in \mathcal{A}$ with $\mu(E) < \delta$.

(Hint: if the ϵ, δ condition fails, there exist $E_n \in \mathcal{A}$ with $\mu(E_n) < 1/2^n$ such that $|\lambda(E_n)| \geq \epsilon$ (hence $|\lambda|(E_n) \geq \epsilon$), for some $\epsilon > 0$; consider the set $E = \limsup E_n$.)

9. Let $(X, \mathcal{A}, \mu)$ be a *probability space* (i.e., a positive measure space such that $\mu(X) = 1$). Let f, g be (complex) measurable functions. Prove that $\|f\|_1 \|g\|_1 \geq \inf_X |fg|$.

10. Let $(X, \mathcal{A}, \mu)$ be a positive measure space and f a complex measurable function on X.

(a) If $\mu(X) < \infty$, prove that

$$\lim_{p \to \infty} \|f\|_p = \|f\|_\infty. \tag{*}$$

(The cases $\|f\|_\infty = 0$ or ∞ are trivial; we may then assume that $\|f\|_\infty = 1$; given ϵ, there exists $E \in \mathcal{A}$ such that $\mu(E) > 0$ and $(1 - \epsilon)\mu(E)^{1/p} \leq \|f\|_p \leq \mu(X)^{1/p}$.)

(b) For an arbitrary positive measure space, if $\|f\|_r < \infty$ for some $r \in [1, \infty)$, then (*) is valid.

(Consider the finite positive measure $\nu(E) = \int_E |f|^r \, d\mu$. We may assume as in part (a) that $\|f\|_\infty = 1$. Verify that $\|f\|_{L^\infty(\nu)} = 1$ and $\|f\|_p = \|f\|_{L^{p-r}(\nu)}^{1-r/p}$ for all $p \geq r + 1$.)

11. Let $(X, \mathcal{A}, \mu)$ be a positive measure space, $1 \leq p < \infty$, and $\epsilon > 0$.

    (a) Suppose f_n, f are unit vectors in $L^p(\mu)$ such that $f_n \to f$ a.e. Consider the probability measure $d\nu = |f|^p \, d\mu$. Show that there exists $E \in \mathcal{A}$ such that $f_n/f \to 1$ uniformly on E and $\nu(E^c) < \epsilon$. (Hint: Egoroff's theorem.)

    (b) For E as in part (a), show that $\limsup_n \int_{E^c} |f_n|^p \, d\mu < \epsilon$.

    (c) Deduce from parts (a) and (b) that $f_n \to f$ in $L^p(\mu)$-norm.

    (d) If $g_n, g \in L^p(\mu)$ are such that $g_n \to g$ a.e. and $\|g_n\|_p \to \|g\|_p$, then $g_n \to g$ in $L^p(\mu)$-norm. (Consider $f_n = g_n/\|g_n\|_p$ and $f = g/\|g\|_p$.)

12. Let $(X, \mathcal{A}, \mu)$ be a positive measure space and $f \in L^p(\mu)$ for some $p \in [1, \infty)$. Prove that the set $[f \neq 0]$ has σ-finite measure.

13. Let $(X, \mathcal{A})$ be a measurable space, and let $f_n : X \to \mathbb{C}$, $n \in \mathbb{N}$, be measurable functions. Prove that the set of all points $x \in X$ for which the complex sequence $\{f_n(x)\}$ converges in $\mathbb{C}$ is measurable.

14. Let $(X, \mathcal{A})$ be a measurable space, and let E be a dense subset of $\mathbb{R}$. Suppose $f : X \to \mathbb{R}$ is such that $[f \geq c] \in \mathcal{A}$ for all $c \in E$. Prove that f is measurable.

15. Let $(X, \mathcal{A})$ be a measurable space, and let $f : X \to \mathbb{R}^+$ be measurable. Prove that there exist $c_k > 0$ and $E_k \in \mathcal{A}$ $(k \in \mathbb{N})$ such that $f = \sum_{k=1}^{\infty} c_k I_{E_k}$. Conclude that for any positive measure μ on $\mathcal{A}$, $\int f \, d\mu = \sum_{k=1}^{\infty} c_k \mu(E_k)$; in particular, if $f \in L^1(\mu)$, the series converges (in the strict sense) and $\mu(E_k) < \infty$ for all k. (Hint: get s_n as in the approximation theorem, and observe that $f = \sum_n (s_n - s_{n-1})$.)

16. Let $(X, \mathcal{A}, \mu)$ be a positive measure space, and let $\{E_k\} \subset \mathcal{A}$ be such that $\sum_k \mu(E_k) < \infty$. Prove that almost all $x \in X$ lie in at most finitely many of the sets E_k. (Hint: the set of all x's that lie in infinitely many E_ks is $\limsup E_k$.)

17. Let X be a (complex) normed space. Define

$$f(x, y) = \frac{\|x + y\|^2 + \|x - y\|^2}{2\|x\|^2 + 2\|y\|^2} \qquad (x, y \in X).$$

(We agree that the fraction is 1 when $x = y = 0$.) Prove:

    (a) $1/2 \leq f \leq 2$.

    (b) X is an inner product space iff $f = 1$ (identically).

2

Construction of measures

This chapter introduces Constantin Carathéodory's powerful technique for constructing positive measures from primitive objects called *semi-measures* on *semi-algebras*. In contrast to σ-algebras, which are normally "big" and intangible, semi-algebras are often "small" and concrete. The first four sections cover the development of Caratheodory's method, including a structure theorem characterizing measurability of sets. Next, we use this method to construct the *Lebesgue–Stieltjes measures*, a special case of which is the *Lebesgue measure*, which is arguably the most important example in basic mathematics of a non-trivial measure. This measure is the "natural" measure on $\mathbb{R}$ in the sense that it maps an interval to its length. Intuitively, the Lebesgue measure μ is translation invariant: $\mu(t+E) = \mu(E)$ for measurable $E \subset \mathbb{R}$ and $t \in \mathbb{R}$. This makes it the *Haar measure* of $\mathbb{R}$; see Chapter 4.

We prove that every Riemann integrable complex function on an interval $[a, b]$ is also Lebesgue integrable there with respect to the Lebesgue measure (the converse being evidently false), a result mentioned in Chapter 1.

The final section applies Caratheodory's extension theorem to construct the *product* of two positive measure spaces. The two most fundamental results are *Fubini's and Tonelli's theorems*, roughly saying that under suitable conditions, the "double integral" equals both "iterated" integrals.

2.1 Semi-algebras

The purpose of this chapter is to construct measure spaces from more primitive objects. We start with a *semi-algebra* $\mathcal{C}$ of subsets of a given set X and a *semi-measure* μ defined on it.

Definition 2.1. Let X be a (non-empty) set. A *semi-algebra* of subsets of X (briefly, a semi-algebra *on* X) is a subfamily $\mathcal{C}$ of $\mathbb{P}(X)$ with the following properties:

(1) if $A, B \in \mathcal{C}$, then $A \cap B \in \mathcal{C}$; $\emptyset \in \mathcal{C}$;

Introduction to Modern Analysis. Second Edition. Shmuel Kantorovitz and Ami Viselter, Oxford University Press.
 DOI: 10.1093/oso/9780192849540.003.0002

(2) if $A \in \mathcal{C}$, then A^c is the union of *finitely* many mutually disjoint sets in $\mathcal{C}$.

Any algebra is a semi-algebra, but not conversely. For example, the family

$$\mathcal{C} = \{(a,b]; a,b \in \mathbb{R}\} \cup \{(-\infty,b]; b \in \mathbb{R}\} \cup \{(a,\infty); a \in \mathbb{R}\} \cup \{\emptyset\}$$

is a semi-algebra on $\mathbb{R}$, but is not an algebra. Similar semi-algebras of *half-closed cells* arise naturally in the Euclidean space $\mathbb{R}^k$.

Definition 2.2. Let $\mathcal{C}$ be a semi-algebra on X. A *semi-measure* on $\mathcal{C}$ is a function

$$\mu : \mathcal{C} \to [0,\infty]$$

with the following properties:

(1) $\mu(\emptyset) = 0$;

(2) if $E_i \in \mathcal{C}$, $i = 1,\dots,n$ are mutually disjoint with union $E \in \mathcal{C}$, then $\mu(E) = \sum_i \mu(E_i)$;

(3) if $E_i \in \mathcal{C}$, $i = 1,2,\dots$ are mutually disjoint with union $E \in \mathcal{C}$, then $\mu(E) \le \sum_i \mu(E_i)$.

If $\mathcal{C}$ is a σ-algebra, any measure on $\mathcal{C}$ is a semi-measure. A simple "natural" example of a semi-measure on the semi-algebra of half-closed intervals on $\mathbb{R}$ mentioned above is given by

$$\mu((a,b]) = b - a, \quad a,b \in \mathbb{R},\ a < b;$$
$$\mu(\emptyset) = 0; \quad \mu((-\infty,b]) = \mu((a,\infty)) = \infty.$$

Let $\mathcal{C}$ be a semi-algebra on X, and let $\mathcal{A}$ be the family of all finite unions of mutually disjoint sets from $\mathcal{C}$. Then $\emptyset \in \mathcal{A}$; if $A = \bigcup E_i, B = \bigcup F_j \in \mathcal{A}$, with $E_i \in \mathcal{C}$, $i = 1,\dots,m$ disjoint and $F_j \in \mathcal{C}$, $j = 1,\dots,n$ disjoint, then $A \cap B = \bigcup_{i,j} E_i \cap F_j \in \mathcal{A}$ as a finite union of disjoint sets from $\mathcal{C}$ by Condition (1) in Definition 2.1. Also $A^c = \bigcap E_i^c \in \mathcal{A}$, since $E_i^c \in \mathcal{A}$ by Condition (2) in Definition 2.1, and we just saw that $\mathcal{A}$ is closed under finite intersections. We conclude that $\mathcal{A}$ is an *algebra* on X that includes $\mathcal{C}$, and it is obviously contained in any algebra on X that contains $\mathcal{C}$. Thus, $\mathcal{A}$ is *the algebra generated by the semi-algebra* $\mathcal{C}$ (i.e., the algebra, minimal under inclusion, that contains $\mathcal{C}$).

Definition 2.3. Let $\mathcal{A}$ be *any* algebra on the set X. A *measure on the algebra* $\mathcal{A}$ is a function $\mu : \mathcal{A} \to [0,\infty]$ such that $\mu(\emptyset) = 0$, and if $E \in \mathcal{A}$ is the countable union of mutually disjoint sets $E_i \in \mathcal{A}$, then $\mu(A) = \sum \mu(E_i)$ (i.e., μ is countably additive *whenever this makes sense*).

Theorem 2.4. *Let $\mathcal{C}$ be a semi-algebra on the set X, let μ be a semi-measure on $\mathcal{C}$, let $\mathcal{A}$ be the algebra generated by $\mathcal{C}$, and extend μ to $\mathcal{A}$ by letting*

$$\mu\Big(\bigcup E_i\Big) = \sum \mu(E_i) \tag{1}$$

for $E_i \in \mathcal{C}$, $i = 1, \dots, n$, *mutually disjoint. Then* μ *is a* measure on the algebra $\mathcal{A}$.

Proof. First, μ is well-defined by (1) on $\mathcal{A}$. Indeed, if $E_i, F_j \in \mathcal{C}$ are such that $A = \bigcup E_i = \bigcup F_j$ (finite *disjoint* unions), then each $F_j \in \mathcal{C}$ is the finite disjoint union of the sets $E_i \cap F_j \in \mathcal{C}$; by Condition (2) in Definition 2.2,

$$\mu(F_j) = \sum_i \mu(E_i \cap F_j),$$

and therefore

$$\sum \mu(F_j) = \sum_{i,j} \mu(E_i \cap F_j).$$

The symmetry of the right-hand side in E_i and F_j implies that it is also equal to $\sum \mu(E_i)$, and the definition (1) is indeed independent of the representation of $A \in \mathcal{A}$ as a finite disjoint union of sets in $\mathcal{C}$.

It is now clear that μ is *finitely* additive on $\mathcal{A}$, hence monotonic.

Let $E \in \mathcal{A}$ be the disjoint union of $E_i \in \mathcal{A}, i = 1, 2, \dots$. For each $n \in \mathbb{N}$, $\bigcup_{i=1}^n E_i \subset E$, hence

$$\sum_{i=1}^n \mu(E_i) = \mu\left(\bigcup_{i=1}^n E_i\right) \le \mu(E),$$

and therefore

$$\sum_{i=1}^\infty \mu(E_i) \le \mu(E). \tag{2}$$

Next, write $E_i = \bigcup_j F_{ij}$, a finite disjoint union of $F_{ij} \in \mathcal{C}$, for each fixed i, and similarly, since $E \in \mathcal{A}$, $E = \bigcup G_k$, a finite disjoint union of sets $G_k \in \mathcal{C}$. Then $G_k \in \mathcal{C}$ is the (countable) disjoint union of the sets $F_{ij} \cap G_k \in \mathcal{C}$ (by Condition (1) in Definition 2.1), over all i, j. By Condition (3) in Definition 2.2, it follows that for all k,

$$\mu(G_k) \le \sum_{i,j} \mu(F_{ij} \cap G_k).$$

Hence

$$\begin{aligned} \mu(E) := \sum \mu(G_k) &\le \sum_{i,j,k} \mu(F_{ij} \cap G_k) \\ &= \sum_i \sum_{j,k} \mu(F_{ij} \cap G_k) = \sum_i \mu(E_i), \end{aligned}$$

by the definition (1) of μ on $\mathcal{A}$, because $E_i = \bigcup_{j,k} F_{ij} \cap G_k$, a finite disjoint union of sets in $\mathcal{C}$. Together with (2), this proves that μ is indeed a measure on the algebra $\mathcal{A}$. □

2.2 Outer measures

A measure on an algebra can be extended to an *outer measure* on $\mathbb{P}(X)$.

Definition 2.5. An outer measure on the set X (in fact, on $\mathbb{P}(X)$) is a function

$$\mu^* : \mathbb{P}(X) \to [0, \infty]$$

with the following properties:

(1) $\mu^*(\emptyset) = 0$;

(2) μ^* is monotonic (i.e., $\mu^*(E) \leq \mu^*(F)$ whenever $E \subset F \subset X$);

(3) μ^* is countably subadditive, that is,

$$\mu^*\left(\bigcup E_i\right) \leq \sum \mu^*(E_i),$$

for any sequence $\{E_i\} \subset \mathbb{P}(X)$.

By (1) and (3), outer measures are *finitely* subadditive (i.e., (3) is valid for *finite* sequences $\{E_i\}$ as well).

Theorem 2.6. *Let μ be a measure on the algebra $\mathcal{A}$ of subsets of X. For any $E \in \mathbb{P}(X)$, let*

$$\mu^*(E) := \inf \sum \mu(E_i),$$

where the infimum is taken over all sequences $\{E_i\} \subset \mathcal{A}$ with $E \subset \bigcup E_i$ (briefly, call such sequences "$\mathcal{A}$-covers of E"). Then μ^* is an outer measure on X, called the outer measure generated by μ, *and $\mu^*|_{\mathcal{A}} = \mu$.*

Proof. We begin by showing that $\mu^*|_{\mathcal{A}} = \mu$. If $E \in \mathcal{A}$, then $\{E, \emptyset, \emptyset, \ldots\}$ is an $\mathcal{A}$-cover of E, hence $\mu^*(E) \leq \mu(E)$. Next, if $\{E_i\}$ is any $\mathcal{A}$-cover of E, then for all $n \in \mathbb{N}$,

$$F_n := E \cap E_n \cap E_{n-1}^c \cap \cdots \cap E_1^c \in \mathcal{A}$$

(since $\mathcal{A}$ is an algebra), and $E \in \mathcal{A}$ is the *disjoint* union of the sets $F_n \subset E_n$. Therefore, since μ is a measure on the algebra $\mathcal{A}$,

$$\sum_n \mu(E_n) \geq \sum_n \mu(F_n) = \mu(E).$$

Taking the infimum over all $\mathcal{A}$-covers $\{E_n\}$ of E, we obtain $\mu^*(E) \geq \mu(E)$, and the wanted equality follows.

In particular, $\mu^*(\emptyset) = \mu(\emptyset) = 0$.

If $E \subset F \subset X$, then every $\mathcal{A}$-cover of F is also an $\mathcal{A}$-cover of E; this implies that $\mu^*(E) \leq \mu^*(F)$.

Let $E_n \subset X, n \in \mathbb{N}$, and $E = \bigcup_n E_n$. For $\epsilon > 0$ given, and for each $n \in \mathbb{N}$, there exists an $\mathcal{A}$-cover $\{E_{n,i}\}_i$ of E_n such that

$$\sum_i \mu(E_{n,i}) < \mu^*(E_n) + \epsilon/2^n.$$

Since $\{E_{n,i}; n, i \in \mathbb{N}\}$ is an $\mathcal{A}$-cover of E, we have

$$\mu^*(E) \le \sum_{n,i} \mu(E_{n,i}) \le \sum_n \mu^*(E_n) + \epsilon,$$

and the arbitrariness of ϵ implies that μ^* is countably sub-additive. □

Definition 2.7 (The Caratheodory measurability condition). Let μ^* be an outer measure on X. A set $E \subset X$ is μ^*-measurable if

$$\mu^*(A) = \mu^*(A \cap E) + \mu^*(A \cap E^c) \tag{1}$$

for every $A \subset X$.

We shall denote by $\mathcal{M}$ the family of all μ^*-measurable subsets of X.

By subadditivity of outer measures, (1) is equivalent to the *inequality*

$$\mu^*(A) \ge \mu^*(A \cap E) + \mu^*(A \cap E^c) \tag{2}$$

(for every $A \subset X$). Since (2) is trivial when $\mu^*(A) = \infty$, we can use only subsets A of *finite* outer measure in the measurability test (2).

Theorem 2.8. *Let μ^* be an outer measure on X, let $\mathcal{M}$ be the family of all μ^*-measurable subsets of X, and let $\bar{\mu} := \mu^*|_{\mathcal{M}}$. Then $(X, \mathcal{M}, \bar{\mu})$ is a complete positive measure space (called the measure space* induced *by the given outer measure).*

Proof. If $\mu^*(E) = 0$, also $\mu^*(A \cap E) = 0$ by monotonicity (for all $A \subset X$), and (2) follows (again by monotonicity of μ^*). Hence $E \in \mathcal{M}$ whenever $\mu^*(E) = 0$, and in particular $\emptyset \in \mathcal{M}$. By monotonicity, this implies also that the measure space of the theorem is automatically complete.

The symmetry of the Caratheodory condition in E and E^c implies that $E^c \in \mathcal{M}$ whenever $E \in \mathcal{M}$.

Let $E, F \in \mathcal{M}$. Then for all $A \subset X$, it follows from (2) (first for F with the "test set" A, and then for E with the test set $A \cap F^c$) and the finite subadditivity of μ^*, that

$$\begin{aligned}
\mu^*(A) &\ge \mu^*(A \cap F) + \mu^*(A \cap F^c) \\
&\ge \mu^*(A \cap F) + \mu^*(A \cap F^c \cap E) + \mu^*(A \cap F^c \cap E^c) \\
&\ge \mu^*([A \cap F] \cup [A \cap (E - F)]) + \mu^*(A \cap (E \cup F)^c) \\
&= \mu^*(A \cap (E \cup F)) + \mu^*(A \cap (E \cup F)^c),
\end{aligned}$$

and we conclude that $E \cup F \in \mathcal{M}$, and so $\mathcal{M}$ is an algebra on X. It follows in particular that any countable union E of sets from $\mathcal{M}$ can be written as a *disjoint* countable union of sets $E_i \in \mathcal{M}$. Set $F_n = \bigcup_{i=1}^n E_i (\subset E), n = 1, 2, \ldots$. Then $F_n \in \mathcal{M}$, and therefore, by (2) and monotonicity, we have for all $A \subset X$

$$\begin{aligned}
\mu^*(A) &\ge \mu^*(A \cap F_n) + \mu^*(A \cap F_n^c) \\
&\ge \mu^*(A \cap F_n) + \mu^*(A \cap E^c).
\end{aligned} \tag{3}$$

By (1), since $E_n \in \mathcal{M}$, we have

$$\begin{aligned} \mu^*(A \cap F_n) &= \mu^*(A \cap F_n \cap E_n) + \mu^*(A \cap F_n \cap E_n^c) \\ &= \mu^*(A \cap E_n) + \mu^*(A \cap F_{n-1}). \end{aligned} \tag{4}$$

The recursion (4) implies that for all $n \in \mathbb{N}$,

$$\mu^*(A \cap F_n) = \sum_{i=1}^{n} \mu^*(A \cap E_i). \tag{5}$$

Substitute (5) in (3), let $n \to \infty$, and use the σ-subadditivity of μ^*; whence

$$\begin{aligned} \mu^*(A) &\geq \sum_{i=1}^{\infty} \mu^*(A \cap E_i) + \mu^*(A \cap E^c) \\ &\geq \mu^*(A \cap E) + \mu^*(A \cap E^c). \end{aligned}$$

This shows that $E \in \mathcal{M}$, and we conclude that $\mathcal{M}$ is a σ-algebra.

Choosing $A = F_n$ in (5), we obtain (by monotonicity)

$$\mu^*(E) \geq \mu^*(F_n) = \sum_{i=1}^{n} \mu^*(E_i), \quad n = 1, 2, \ldots.$$

Letting $n \to \infty$, we see that

$$\mu^*(E) \geq \sum_{i=1}^{\infty} \mu^*(E_i).$$

Together with the σ-subadditivity of μ^*, this proves the σ-additivity of μ^* restricted to $\mathcal{M}$, as wanted. □

2.3 Extension of measures on algebras

Combining Theorems 2.6 and 2.8, we obtain the *Caratheodory extension theorem.*

Theorem 2.9. *Let μ be a measure* on the algebra $\mathcal{A}$, *let μ^* be the outer measure induced by μ, and let $(X, \mathcal{M}, \bar{\mu})$ be the (complete, positive) measure space induced by μ^*. Then $\mathcal{A} \subset \mathcal{M}$, and $\bar{\mu}$ extends μ to a measure (on the σ-algebra $\mathcal{M}$), which is finite (σ-finite) if μ is finite (σ-finite, respectively).*

The measure $\bar{\mu}$ is called the Caratheodory extension of μ.

Proof. By Theorems 2.6 and 2.8, we need only to prove the inclusion $\mathcal{A} \subset \mathcal{M}$, for then

$$\bar{\mu}|_{\mathcal{A}} = (\mu^*|_{\mathcal{M}})|_{\mathcal{A}} = \mu^*|_{\mathcal{A}} = \mu.$$

Let then $E \in \mathcal{A}$, and let $A \subset X$ be such that $\mu^*(A) < \infty$. For any given $\epsilon > 0$, there exists an $\mathcal{A}$-cover $\{E_i\}$ of A such that

$$\mu^*(A) + \epsilon > \sum_i \mu(E_i). \tag{1}$$

Since $E \in \mathcal{A}$, $\{E_i \cap E\}$ and $\{E_i \cap E^c\}$ are $\mathcal{A}$-covers of $A \cap E$ and $A \cap E^c$, respectively, and therefore

$$\sum_i \mu(E_i \cap E) \geq \mu^*(A \cap E)$$

and

$$\sum_i \mu(E_i \cap E^c) \geq \mu^*(A \cap E^c).$$

Adding these relations and using the additivity of the measure μ on the algebra $\mathcal{A}$, we obtain

$$\sum_i \mu(E_i) \geq \mu^*(A \cap E) + \mu^*(A \cap E^c). \tag{2}$$

By (1), (2), and the arbitrariness of ϵ, we get

$$\mu^*(A) \geq \mu^*(A \cap E) + \mu^*(A \cap E^c)$$

for all $A \subset X$, so that $E \in \mathcal{M}$, and $\mathcal{A} \subset \mathcal{M}$.

If μ is finite, then since $X \in \mathcal{A}$ and $\mu^*|_{\mathcal{A}} = \mu$, we have $\mu^*(X) = \mu(X) < \infty$. The σ-finite case is analogous. □

If we start from a *semi-measure* μ on a *semi-algebra* $\mathcal{C}$, we first extend it to a measure (same notation) on the *algebra* $\mathcal{A}$ generated by $\mathcal{C}$ (as in Theorem 2.4). We then apply the Caratheodory extension theorem to obtain the complete positive measure space $(X, \mathcal{M}, \bar{\mu})$ with $\mathcal{A} \subset \mathcal{M}$ and $\bar{\mu}$ extending μ. Note that if μ is finite (σ-finite) on $\mathcal{C}$, then its extension $\bar{\mu}$ is finite (σ-finite, respectively).

2.4 Structure of measurable sets

Let μ be a measure on the *algebra* $\mathcal{A}$ on X. Denote by μ^* the outer measure induced by μ, and let $\mathcal{M}$ be the σ-algebra of all μ^*-measurable subsets of X. Consider the family $\mathcal{A}_\sigma \subset \mathcal{M}$ of all countable unions of sets from $\mathcal{A}$. Note that if we start from a semi-algebra $\mathcal{C}$ and $\mathcal{A}$ is the algebra generated by it, then $\mathcal{A}_\sigma = \mathcal{C}_\sigma$.

Lemma 2.10. *For any $E \subset X$ with $\mu^*(E) < \infty$ and for any $\epsilon > 0$, there exists $A \in \mathcal{A}_\sigma$ such that $E \subset A$ and*

$$\mu^*(A) \leq \mu^*(E) + \epsilon.$$

Proof. By definition of $\mu^*(E)$, there exists an $\mathcal{A}$-cover $\{E_i\}$ of E such that $\sum \mu(E_i) \le \mu^*(E) + \epsilon$. Then $A := \bigcup E_i \in \mathcal{A}_\sigma$, $E \subset A$, and

$$\mu^*(A) \le \sum \mu^*(E_i) = \sum \mu(E_i) \le \mu^*(E) + \epsilon,$$

as wanted. □

If $\mathcal{B}$ is any family of subsets of X, denote by $\mathcal{B}_\delta$ the family of all countable intersections of sets from $\mathcal{B}$. Let $\mathcal{A}_{\sigma\delta} := (\mathcal{A}_\sigma)_\delta$.

Proposition 2.11. *Let $\mu, \mathcal{A}, \mu^*$ be as before. Then for each $E \subset X$ with $\mu^*(E) < \infty$, there exists $A \in \mathcal{A}_{\sigma\delta}$ such that $E \subset A$ and $\mu^*(E) = \mu^*(A)(= \bar{\mu}(A))$.*

Proof. Let E be a subset of X with finite outer measure. For each $n \in \mathbb{N}$, there exists $A_n \in \mathcal{A}_\sigma$ such that $E \subset A_n$ and $\mu^*(A_n) < \mu^*(E) + 1/n$ (by Lemma 2.10). Therefore, $A := \bigcap A_n \in \mathcal{A}_{\sigma\delta}$, $E \subset A$, and

$$\mu^*(E) \le \mu^*(A) \le \mu^*(A_n) \le \mu^*(E) + 1/n$$

for all n, so that $\mu^*(E) = \mu^*(A)$. □

The structure of μ^*-measurable sets is described in the next theorem.

Theorem 2.12. *Let μ be a σ-finite measure on the* algebra *$\mathcal{A}$ on X, and let μ^* be the outer measure induced by it. Then $E \subset X$ is μ^*-measurable iff there exists $A \in \mathcal{A}_{\sigma\delta}$ such that $E \subset A$ and $\mu^*(A - E) = 0$.*

Proof. We observed in the proof of Theorem 2.8 that $\mathcal{M}$ contains every set of μ^*-measure zero. Thus, if $E \subset A \in \mathcal{A}_{\sigma\delta}$ and $\mu^*(A - E) = 0$, then $A - E \in \mathcal{M}$ and $A \in \mathcal{M}$ (because $\mathcal{A}_{\sigma\delta} \subset \mathcal{M}$), and therefore $E = A - (A - E) \in \mathcal{M}$.

Conversely, suppose $E \in \mathcal{M}$. By the σ-finiteness hypothesis, we may write $X = \bigcup X_i$ with $X_i \in \mathcal{A}$ mutually disjoint and $\mu(X_i) < \infty$. Let $E_i := E \cap X_i$. By Lemma 2.10, there exist $A_{ni} \in \mathcal{A}_\sigma$ such that $E_i \subset A_{ni}$ and $\mu^*(A_{ni}) \le \mu^*(E_i) + (1/n2^i)$, for all $n, i \in \mathbb{N}$. Set $A_n := \bigcup_i A_{ni}$. Then for all n, $A_n \in \mathcal{A}_\sigma$, $E \subset A_n$, and $A_n - E \subset \bigcup_i (A_{ni} - E_i)$, so that

$$\mu^*(A_n - E) \le \sum_i \mu^*(A_{ni} - E_i) \le \sum_i \frac{1}{n2^i} = 1/n.$$

Let $A := \bigcap A_n$. Then $E \subset A$, $A \in \mathcal{A}_{\sigma\delta}$, and since $A - E \subset A_n - E$ for all n, $\mu^*(A - E) = 0$. □

We can use Lemma 2.10 to prove a *uniqueness* theorem for the extension of measures on algebras.

Theorem 2.13 (Uniqueness of extension). *Let μ be a measure on the algebra $\mathcal{A}$ on X, and let $\bar{\mu}$ be the Caratheodory extension of μ (as a measure on the σ-algebra $\mathcal{M}$, cf. Theorem 2.9). Consider the σ-algebra $\mathcal{B}$ generated by $\mathcal{A}$ (of course, $\mathcal{B} \subset \mathcal{M}$). If μ_1 is any measure that extends μ to $\mathcal{B}$, then $\mu_1(E) = \bar{\mu}(E)$ for any set $E \in \mathcal{B}$ with $\bar{\mu}(E) < \infty$. If μ is σ-finite, then $\mu_1 = \bar{\mu}$ on $\mathcal{B}$.*

Proof. Since $\mu_1 = \mu = \bar{\mu}$ on $\mathcal{A}$, and each set in $\mathcal{A}_\sigma$ is a *disjoint* countable union of sets $A_i \in \mathcal{A}$, we have $\mu_1 = \bar{\mu}$ on $\mathcal{A}_\sigma$.

Let $E \in \mathcal{B}$ with $\bar{\mu}(E) < \infty$, and let $\epsilon > 0$. By Lemma 2.10, there exists $A \in \mathcal{A}_\sigma$ such that $E \subset A$ and $\bar{\mu}(A) \leq \bar{\mu}(E) + \epsilon$. Hence

$$\mu_1(E) \leq \mu_1(A) = \bar{\mu}(A) \leq \bar{\mu}(E) + \epsilon,$$

and therefore, $\mu_1(E) \leq \bar{\mu}(E)$, by the arbitrariness of ϵ.

Note in passing that $A - E \in \mathcal{B}$ with $\bar{\mu}(A - E) \leq \epsilon < \infty$ (for any A as discussed). Therefore, we have in particular $\mu_1(A - E) \leq \bar{\mu}(A - E) \leq \epsilon$. Hence

$$\bar{\mu}(E) \leq \bar{\mu}(A) = \mu_1(A) = \mu_1(E) + \mu_1(A - E) \leq \mu_1(E) + \epsilon,$$

and the reverse inequality $\bar{\mu}(E) \leq \mu_1(E)$ follows.

If μ is σ-finite, write X as the disjoint union of $X_i \in \mathcal{A}$ with $\mu(X_i) < \infty$, $i = 1, 2, \ldots$. Then, each $E \in \mathcal{B}$ is the disjoint union of $E_i := E \cap X_i$ with $\mu(E_i) < \infty$; since $\mu_1(E_i) = \bar{\mu}(E_i)$ for all i, also $\mu_1(E) = \bar{\mu}(E)$, by σ-additivity of both measures. □

2.5 Construction of Lebesgue–Stieltjes measures

Here we apply the general method of construction of measures described in the preceding sections to the special semi-algebra $\mathcal{C}$ in the example following Definition 2.1, and to the semi-measure μ *induced* by a given non-decreasing right-continuous function $F : \mathbb{R} \to \mathbb{R}$. Denote $F(\infty) := \lim_{x\to\infty} F(x)(\in (-\infty, \infty])$, and similarly $F(-\infty)(\in [-\infty, \infty))$ (both limits exist, because F is non-decreasing). We define the semi-measure μ (induced by F) by

$$\mu(\emptyset) = 0; \mu((a, b]) = F(b) - F(a) \quad (a, b \in \mathbb{R}, a < b);$$
$$\mu((-\infty, b]) = F(b) - F(-\infty); \quad \mu((a, \infty)) = F(\infty) - F(a), a, b \in \mathbb{R}.$$

The example following Definition 2.2 is the special case with $F(x) = x, x \in \mathbb{R}$.

We verify that the properties (2) and (3) of Definition 2.2 are satisfied.

Suppose $(a, b]$ is the disjoint finite union of similar intervals. Then we may index the subintervals so that

$$a = a_1 < b_1 = a_2 < b_2 = a_3 < \cdots < b_n = b.$$

Therefore,

$$\sum_{i=1}^{n} \mu((a_i, b_i]) = \sum_{i=1}^{n} [F(b_i) - F(a_i)] = \sum_{i=1}^{n-1} [F(a_{i+1}) - F(a_i)] + F(b) - F(a_n)$$
$$= F(b) - F(a) = \mu((a, b])$$

for $a, b \in \mathbb{R}, a < b$. A similar argument for the cases $(-\infty, b]$ and (a, ∞) completes the verification of Property (2). In order to verify Property (3), we show that whenever $(a, b] \subset \bigcup_{i=1}^{\infty}(a_i, b_i]$, then

$$F(b) - F(a) \le \sum_i [F(b_i) - F(a_i)]. \tag{1}$$

This surely implies Property (3) for a, b finite. If $(-\infty, b]$ is contained in such a union, then $(-n, b]$ is contained in it as well, for all $n \in \mathbb{N}$, so that $F(b) - F(-n)$ is majorized by the sum on the right-hand side of (1) for all n; letting $n \to \infty$, we deduce that this sum majorizes $\mu((-\infty, b])$. A similar argument works for $\mu((a, \infty))$.

Let $\epsilon > 0$. By the right continuity of F, there exist $c_i, i = 0, 1, 2, \dots$ such that

$$a < c_0; \quad F(c_0) < F(a) + \epsilon;$$

$$b_i < c_i; \quad F(c_i) < F(b_i) + \epsilon/2^i; \qquad i = 1, 2, \dots. \tag{2}$$

We have $[c_0, b] \subset \bigcup_{i=1}^{\infty}(a_i, c_i)$, so that, by compactness, a *finite* number n of intervals (a_i, c_i) covers $[c_0, b]$. Thus, c_0 is in one of these intervals, say (a_1, c_1) (to simplify notation), that is,

$$a_1 < c_0 < c_1.$$

Assuming we got $(a_i, c_i), 1 \le i < k$ such that

$$a_i < c_{i-1} < c_i, \tag{3}$$

and $c_{k-1} \le b$ (that is, $c_{k-1} \in [c_0, b]$), there exists one of the n intervals shown, say (a_k, c_k) to simplify notation, that contains c_{k-1}, so that (3) is valid for $i = k$ as well. This (finite) inductive process will end after at most n steps (this will exhaust our finite cover), which means that for some $k \le n$, we must get

$$b < c_k. \tag{4}$$

By (4), (3), and (2) (in this order),

$$\begin{aligned}
\mu((a, b]) := F(b) - F(a) &\le F(c_k) + \sum_{i=2}^{k}[F(c_{i-1}) - F(a_i)] - F(c_0) + \epsilon \\
&\le \sum_{i=1}^{k}[F(c_i) - F(a_i)] + \epsilon \le \sum_{i=1}^{\infty}[F(c_i) - F(a_i)] + \epsilon \\
&\le \sum_{i=1}^{\infty}[F(b_i) - F(a_i)] + 2\epsilon = \sum_i \mu((a_i, b_i]) + 2\epsilon,
\end{aligned}$$

as wanted (by the arbitrariness of ϵ). By Theorems 2.4 and 2.9, the semi-measure μ has an extension as a complete measure (also denoted μ) on a σ-algebra $\mathcal{M}$

containing the Borel algebra $\mathcal{B}$ (= the σ-algebra generated by $\mathcal{C}$, or by the algebra $\mathcal{A}$). By Theorem 2.13, the extension is uniquely determined on $\mathcal{B}$. The (complete) measure space $(\mathbb{R}, \mathcal{M}, \mu)$ is called the *Lebesgue–Stieltjes measure space induced by the given function* F (in the special case $F(x) = x$, this is the *Lebesgue measure space*). By Theorem 2.13, the "Lebesgue–Stieltjes measure μ induced by F" is the unique measure on $\mathcal{B}$ such that $\mu((a,b]) = F(b) - F(a)$. It is customary to write the integral $\int f\,d\mu$ in the form $\int f\,dF$, and to call F the "distribution of μ". Accordingly, in the special case of Lebesgue measure, the described integral is customarily written in the form $\int f\,dx$.

The Lebesgue measure μ is the unique measure on $\mathcal{B}$ such that $\mu((a,b]) = b - a$ for all real $a < b$; in particular, it is translation invariant on $\mathcal{C}$, hence on $\mathcal{A}$ (if $E \in \mathcal{A}$, write E as a finite disjoint union of intervals $(a_i, b_i]$, then $\mu(t+E) = \mu(\bigcup_i \{t + (a_i, b_i]\} = \sum_i \mu((t + a_i, t + b_i]) = \sum \mu((a_i, b_i]) = \mu(E)$ for all real t). Let μ^* be the outer measure induced by μ. Then for all $E \subset \mathbb{R}$ and $t \in \mathbb{R}$, $\{E_i\}$ is an $\mathcal{A}$-cover of E if and only if $\{t + E_i\}$ is an $\mathcal{A}$-cover of $t + E$, and therefore

$$\mu^*(t+E) := \inf \sum \mu(t + E_i) = \inf \sum \mu(E_i) = \mu^*(E).$$

In particular, if $\mu^*(E) = 0$, then $\mu^*(t+E) = 0$ for all real t. By Theorem 2.12, $E \in \mathcal{M}$ (that is, Lebesgue measurable on $\mathbb{R}$) iff there exists $A \in \mathcal{A}_{\sigma\delta}$ such that $E \subset A$ and $\mu^*(A - E) = 0$. However, translations of unions and intersections of sets are unions and intersections of the translated sets (respectively). Thus, the existence of $A \in \mathcal{A}_{\sigma\delta}$ as shown implies that $t + A \in \mathcal{A}_{\sigma\delta}$, $t + E \subset t + A$, and $\mu^*((t+A) - (t+E)) = \mu^*(t + (A - E)) = 0$, that is, $t + E \in \mathcal{M}$ (by Theorem 2.12), for all $t \in \mathbb{R}$. Since Lebesgue measure is the restriction of μ^* to $\mathcal{M}$, we conclude from this discussion that the Lebesgue measure space $(\mathbb{R}, \mathcal{M}, \mu)$ is *translation invariant*, which means that $t + E \in \mathcal{M}$ and $\mu(t+E) = \mu(E)$ for all $t \in \mathbb{R}$ and $E \in \mathcal{M}$. In the terminology of Section 1.26, the map $h(x) = x - t$ is a measurable map of $(\mathbb{R}, \mathcal{M})$ onto itself (for each given t), and the corresponding measure $\nu(E) := \mu(h^{-1}(E)) = \mu(t+E) = \mu(E)$. Therefore, by the Proposition there, for any non-negative measurable function and for any μ-integrable complex function f on $\mathbb{R}$,

$$\int f\,d\mu = \int f_t\,d\mu,$$

where $f_t(x) := f(x - t)$. This is the translation invariance of the Lebesgue integral.

Consider the quotient group $\mathbb{R}/\mathbb{Q}$ of the additive group $\mathbb{R}$ by the subgroup $\mathbb{Q}$ of rationals. Let A be an arbitrary bounded Lebesgue measurable subset of $\mathbb{R}$ of positive measure. By the Axiom of Choice, there exists a set $E \subset A$ that contains *precisely one point from each coset in* $\mathbb{R}/\mathbb{Q}$ *that meets* A. Since A is bounded, $A \subset (-a, a)$ for some $a \in (0, \infty)$. We claim that A is contained in the disjoint union $S := \bigcup_{r \in \mathbb{Q} \cap (-2a, 2a)} (r + E)$. Indeed, if $x \in A$, there exists a unique $y \in E$ such that x, y are in the same coset of $\mathbb{Q}$, that is, $x - y = r \in \mathbb{Q}$, hence $x = r + y \in r + E$, $|r| \le |x| + |y| < 2a$, so that indeed $x \in S$. If $r, s \in \mathbb{Q} \cap (-2a, 2a)$ are distinct and $x \in (r + E) \cap (s + E)$, then there exist $y, z \in E$ such that

$x = r + y = s + z$. Hence, $y - z = s - r \in \mathbb{Q} - \{0\}$, which means that y, z are distinct points of E belonging to the same coset, contrary to the definition of E. Thus, the union S is indeed a *disjoint* union. Write $\mathbb{Q} \cap (-2a, 2a) = \{r_k\}$. Suppose E is (Lebesgue) measurable. Since $r_k + E \subset r_k + A \subset (-3a, 3a)$ for all k, it follows that S is a measurable subset of $(-3a, 3a)$. Therefore, by σ-additivity and translation invariance of μ,

$$6a \geq \mu(S) = \sum_{k=1}^{\infty} \mu(r_k + E) \geq \sum_{k=1}^{n} \mu(E) = n\mu(E)$$

for all $n \in \mathbb{N}$, hence $\mu(E) = 0$ and $\mu(S) = 0$. Since $A \subset S$, also $\mu(A) = 0$, contradicting our hypothesis. This shows that E is *not* Lebesgue measurable. Since any measurable set on $\mathbb{R}$ of positive measure contains a *bounded* measurable subset of positive measure, we proved the following

Proposition. *Every (Lebesgue-) measurable subset of $\mathbb{R}$ of positive measure contains a non-measurable subset.*

2.6 Riemann vs. Lebesgue

Let $-\infty < a < b < \infty$, and let $f : [a, b] \to \mathbb{R}$ be bounded. Denote

$$m = \inf_{[a,b]} f; \quad M = \sup_{[a,b]} f.$$

Given a "partition" $P = \{x_k; k = 0, \dots, n\}$ of $[a, b]$, where $a = x_0 < x_1 < \dots < x_n = b$, we denote

$$m_k = \inf_{[x_{k-1}, x_k]} f; \quad M_k = \sup_{[x_{k-1}, x_k]} f;$$

$$L_P = \sum_{k=1}^{n} m_k (x_k - x_{k-1}); \quad U_P = \sum_{k=1}^{n} M_k (x_k - x_{k-1}).$$

Recall that the lower and upper Riemann integrals of f over $[a, b]$ are defined as the supremum and infimum of L_P and U_P (respectively) over all partitions P, and f is Riemann integrable over $[a, b]$ if these lower and upper integrals coincide (their common value is the Riemann integral, denoted $\int_a^b f(x)\,dx$). For bounded complex functions $f = u + \mathrm{i}v$ with u, v real, one says that f is Riemann integrable iff both u and v are Riemann integrable, and $\int_a^b f\,dx := \int_a^b u\,dx + \mathrm{i}\int_a^b v\,dx$.

Proposition. *If a bounded (complex) function on the real interval $[a, b]$ is Riemann integrable, then it is Lebesgue integrable on $[a, b]$, and its Lebesgue integral $\int_{[a,b]} f\,dx$ coincides with its Riemann integral $\int_a^b f\,dx$.*

Proof. It suffices to consider bounded *real* functions f. Given a partition P, consider the simple Borel functions

$$l_P = f(a)I_{\{a\}} + \sum_k m_k I_{(x_{k-1}, x_k]}; \quad u_P = f(a)I_{\{a\}} + \sum_k M_k I_{(x_{k-1}, x_k]}.$$

Then $l_P \le f \le u_P$ on $[a,b]$, and

$$\int_{[a,b]} l_P\,dx = L_P; \qquad \int_{[a,b]} u_P\,dx = U_P. \tag{1}$$

If f is Riemann integrable, there exists a sequence of partitions P_j of $[a,b]$ such that P_{j+1} is a refinement of P_j, $\|P_j\| := \max_{x_k \in P_j}(x_k - x_{k-1}) \to 0$ as $j \to \infty$, and

$$\lim_j L_{P_j} = \lim_j U_{P_j} = \int_a^b f\,dx. \tag{2}$$

The sequences $l_j := l_{P_j}$ and $u_j := u_{P_j}$ are monotonic (non-decreasing and non-increasing, respectively) and bounded. Let then $l := \lim_j l_j$ and $u := \lim_j u_j$. These are bounded Borel functions, and by (1), (2), and the Lebesgue dominated convergence theorem,

$$\int_{[a,b]} l\,dx = \lim_j \int_{[a,b]} l_j\,dx = \lim_j L_{P_j} = \int_a^b f\,dx, \tag{3}$$

and similarly $\int_{[a,b]} u\,dx = \int_a^b f\,dx$. In particular, $\int_{[a,b]}(u-l)\,dx = 0$, and since $u - l \ge 0$, it follows that $u = l$ a.e.; however $l \le f \le u$, hence $f = u = l$ a.e.; therefore, f is Lebesgue measurable (hence Lebesgue integrable, since it is bounded) and $\int_{[a,b]} f\,dx = \int_{[a,b]} l\,dx = \int_a^b f\,dx$ by (3). □

A similar proposition is valid for *absolutely convergent improper* Riemann integrals (on finite or infinite intervals). The easy proofs are omitted.

Let $Q = \bigcup_j P_j$ (a countable set, hence a Lebesgue null set). If $x \in [a,b]$ is *not* in Q, f is continuous at x iff $l(x) = u(x)$. It follows from the preceding proof that if f is Riemann integrable, then it is *continuous* at almost all points not in Q, that is, almost everywhere in $[a,b]$. Conversely, if f is continuous a.e., then $l = f = u$ a.e., hence $\int_{[a,b]} l\,dx = \int_{[a,b]} u\,dx$. Therefore, given $\epsilon > 0$, there exists j such that $\int_{[a,b]} u_j\,dx - \int_{[a,b]} l_j\,dx < \epsilon$, that is, $U_{P_j} - L_{P_j} < \epsilon$. This means that f is Riemann integrable on $[a,b]$. Formally:

Proposition. *Let f be a bounded complex function on $[a,b]$. Then f is Riemann integrable on $[a,b]$ iff it is continuous almost everywhere in $[a,b]$.*

2.7 Product measure

Let $(X, \mathcal{A}, \mu)$ and $(Y, \mathcal{B}, \nu)$ be measure spaces. A *measurable rectangle* is a cartesian product $A \times B$ with $A \in \mathcal{A}$ and $B \in \mathcal{B}$. The set $\mathcal{C}$ of all measurable rectangles is a semi-algebra, since

$$(A \times B) \cap (C \times D) = (A \cap C) \times (B \cap D)$$

and

$$(A \times B)^c = (A^c \times B) \cup (A \times B^c) \cup (A^c \times B^c),$$

where the union on the right is clearly disjoint. Define λ on $\mathcal{C}$ by

$$\lambda(A \times B) = \mu(A)\nu(B).$$

We claim that λ is a semi-measure on $\mathcal{C}$ (cf. Definition 2.2). Indeed, Property (1) is trivial, while Properties (2) and (3) follow from the stronger property:

If $A_i \times B_i \in \mathcal{C}$ are mutually disjoint with union $A \times B \in \mathcal{C}$, then

$$\lambda(A \times B) = \sum_{i=1}^{\infty} \mu(A_i)\nu(B_i). \tag{1}$$

Proof. Let $x \in A$. For each $y \in B$, there exists a unique i such that the pair $[x, y]$ belongs to $A_i \times B_i$ (because the rectangles are mutually disjoint). Thus, B decomposes as the disjoint union

$$B = \bigcup_{\{i; x \in A_i\}} B_i.$$

Therefore,

$$\nu(B) = \sum_{\{i; x \in A_i\}} \nu(B_i),$$

and so

$$\nu(B) I_A(x) = \sum_{i=1}^{\infty} \nu(B_i) I_{A_i}(x).$$

By Beppo Levi's theorem (1.16),

$$\begin{aligned} \lambda(A \times B) := \mu(A)\nu(B) &= \int_X \nu(B) I_A(x)\, d\mu \\ &= \sum_i \int \nu(B_i) I_{A_i}\, d\mu = \sum_i \lambda(A_i \times B_i). \end{aligned}$$

□

By the Caratheodory extension theorem, there exists a complete measure space, which we denote

$$(X \times Y, \mathcal{A} \times \mathcal{B}, \mu \times \nu),$$

and call *the product of the given measure spaces*, such that $\mathcal{C} \subset \mathcal{A} \times \mathcal{B}$ and

$$(\mu \times \nu)(A \times B) = \lambda(A \times B) := \mu(A)\nu(B)$$

for $A \times B \in \mathcal{C}$.

The central theorem of this section is the Fubini–Tonelli theorem, that relates the "double integral" (relative to $\mu \times \nu$) with the "iterated integrals" (relative to μ and ν in either order). We need first some technical lemmas.

Lemma 2.14. *For each $E \in \mathcal{C}_{\sigma\delta}$, the* sections $E_x := \{y \in Y; [x, y] \in E\}$ *($x \in X$) belong to $\mathcal{B}$.*

Proof. If $E = A \times B \in \mathcal{C}$, then E_x is either B (when $x \in A$) or $\emptyset$ (otherwise), so clearly it belongs to $\mathcal{B}$. If $E \in \mathcal{C}_\sigma$, then $E = \bigcup_i E_i$ with $E_i \in \mathcal{C}$; hence,

$$E_x = \bigcup_i (E_i)_x \in \mathcal{B}.$$

Similarly, if $E \in \mathcal{C}_{\sigma\delta}$, then $E = \bigcap_i E_i$ with $E_i \in \mathcal{C}_\sigma$, and therefore,

$$E_x = \bigcap_i (E_i)_x \in \mathcal{B}$$

for all $x \in X$. □

By the lemma, the function

$$g_E(x) := \nu(E_x) : X \to [0, \infty]$$

is well defined, for each $E \in \mathcal{C}_{\sigma\delta}$.

Lemma 2.15. *Suppose the measure space $(X, \mathcal{A}, \mu)$ is complete. For each $E \in \mathcal{C}_{\sigma\delta}$ with $(\mu \times \nu)(E) < \infty$, the function $g_E(x) := \nu(E_x)$ is $\mathcal{A}$-measurable, and*

$$\int_X g_E \, d\mu = (\mu \times \nu)(E). \tag{2}$$

Proof. For an arbitrary $E = A \times B \in \mathcal{C}$,

$$g_E = \nu(B) I_A$$

is clearly $\mathcal{A}$-measurable (since $A \in \mathcal{A}$), and (2) is trivially true.

If $E \in \mathcal{C}_\sigma$ (arbitrary), we may represent it as a *disjoint* union of $E_i \in \mathcal{C}$ ($i \in \mathbb{N}$), and therefore $g_E = \sum_i g_{E_i}$ is $\mathcal{A}$-measurable, and by the Beppo Levi theorem and the σ-additivity of $\mu \times \nu$,

$$\int_X g_E \, d\mu = \sum_i \int_X g_{E_i} \, d\mu = \sum_i (\mu \times \nu)(E_i) = (\mu \times \nu)(E).$$

Let now $E \in \mathcal{C}_{\sigma\delta}$ with $(\mu \times \nu)(E) < \infty$. Thus $E = \bigcap_i F_i$ with $F_i \in \mathcal{C}_\sigma$. By Lemma 2.10, there exists $G \in \mathcal{C}_\sigma$ such that $E \subset G$ and $(\mu \times \nu)(G) < (\mu \times \nu)(E) + 1 < \infty$. Then

$$E = E \cap G = \bigcap_i (F_i \cap G) = \bigcap_k E_k,$$

where (for $k = 1, 2, \ldots$)

$$E_k := \bigcap_{i=1}^{k} (F_i \cap G).$$

Since $\mathcal{C}_\sigma$ is an algebra, $E_k \in \mathcal{C}_\sigma$, $E_{k+1} \subset E_k$, and $E_1 \subset G$ has finite product measure. Therefore g_{E_k} is $\mathcal{A}$-measurable, and

$$\int_X g_{E_k} \, d\mu = (\mu \times \nu)(E_k) < \infty$$

for all k. In particular $g_{E_1} < \infty$ μ-a.e.

For x such that $g_{E_1}(x)(= \nu((E_1)_x)) < \infty$, we have by Lemma 1.11:

$$g_E(x) := \nu(E_x) = \nu\Big(\bigcap_k (E_k)_x\Big) = \lim_k \nu((E_k)_x) = \lim_k g_{E_k}(x).$$

Hence $g_{E_k} \to g_E$ μ-a.e. Since the measure space $(X, \mathcal{A}, \mu)$ is complete by hypothesis, it follows that g_E is $\mathcal{A}$-measurable. Also $0 \le g_{E_k} \le g_{E_1}$ for all k, and $\int_X g_{E_1}\, d\mu < \infty$. Therefore, by Lebesgue's dominated convergence theorem and Lemma 1.11,

$$\int_X g_E\, d\mu = \lim_k \int_X g_{E_k}\, d\mu = \lim_k (\mu \times \nu)(E_k) = (\mu \times \nu)(E).$$

□

We now extend this lemma to *all* $E \in \mathcal{A} \times \mathcal{B}$ with *finite* product measure.

Lemma 2.16. *Let $(X, \mathcal{A}, \mu)$ and $(Y, \mathcal{B}, \nu)$ be* complete *measure spaces. Let $E \in \mathcal{A} \times \mathcal{B}$ have* finite *product measure. Then the sections E_x are $\mathcal{B}$-measurable for μ-almost all x; the (μ-a.e. defined and finite) function $g_E(x) := \nu(E_x)$ is $\mathcal{A}$-measurable, and*

$$\int_X g_E\, d\mu = (\mu \times \nu)(E).$$

Proof. By Proposition 2.11, since E has finite product measure, there exists $F \in \mathcal{C}_{\sigma\delta}$ such that $E \subset F$ and $(\mu \times \nu)(F) = (\mu \times \nu)(E) < \infty$. Let $G := F - E$. Then $G \in \mathcal{A} \times \mathcal{B}$ has zero product measure (since E and F have equal *finite* product measure). Again by Proposition 2.11, there exists $H \in \mathcal{C}_{\sigma\delta}$ such that $G \subset H$ and $(\mu \times \nu)(H) = 0$. By Lemma 2.15, g_H is $\mathcal{A}$-measurable and $\int_X g_H\, d\mu = (\mu \times \nu)(H) = 0$. Therefore, $\nu(H_x) := g_H(x) = 0$ μ-a.e. Since $G_x \subset H_x$, it follows from the completeness of the measure space $(Y, \mathcal{B}, \nu)$ that, for μ-almost all x, G_x is $\mathcal{B}$-measurable and $\nu(G_x) = 0$. Since $E = F - G$, it follows that for μ-almost all x, E_x is $\mathcal{B}$-measurable and $\nu(E_x) = \nu(F_x)$, that is, $g_E = g_F$ (μ-a.e.) is $\mathcal{A}$-measurable (by Lemma 2.15), and

$$\int_X g_E\, d\mu = \int_X g_F\, d\mu = (\mu \times \nu)(F) = (\mu \times \nu)(E).$$

□

Note that for any $E \subset X \times Y$ and $x \in X$,

$$I_{E_x} = I_E(x, \cdot).$$

Therefore, if $E \in \mathcal{A} \times \mathcal{B}$ has finite product measure, then for μ-almost all x, the function $I_E(x, \cdot)$ is $\mathcal{B}$-measurable, with integral (over Y) equal to $\nu(E_x) := g_E(x) < \infty$, that is, for μ-almost all x,

$$I_E(x, \cdot) \in L^1(\nu), \tag{i}$$

and its integral ($= g_E$) is $\mathcal{A}$-measurable, with integral (over X) equal to $(\mu \times \nu)$ $(E) < \infty$, that is,

$$\int_Y I_E(x, \cdot)\, d\nu \in L^1(\mu), \tag{ii}$$

and

$$\int_X \left[\int_Y I_E(x, \cdot)\, d\nu \right] d\mu = \int_X g_E\, d\mu = (\mu \times \nu)(E)$$
$$= \int_{X \times Y} I_E\, d(\mu \times \nu). \tag{iii}$$

If f is a simple non-negative function in $L^1(\mu \times \nu)$, we may write $f = \sum c_k I_{E_k}$ (finite sum), with $c_k > 0$ and $(\mu \times \nu)(E_k) < \infty$. Then for μ-almost all x, $f(x, \cdot)$ is a linear combination of $L^1(\nu)$-functions (by (i)), and hence belongs to $L^1(\nu)$; its integral (over Y) is a linear combination of the $g_{E_k} \in L^1(\mu)$, and hence belongs to $L^1(\mu)$, and by (iii),

$$\int_X \left[\int_Y f(x, \cdot)\, d\nu \right] d\mu = \sum_k c_k \int_{X \times Y} I_{E_k}\, d(\mu \times \nu) = \int_{X \times Y} f\, d(\mu \times \nu).$$

If $f \in L^1(\mu \times \nu)$ is non-negative, by Theorem 1.8, we get simple measurable functions

$$0 \le s_1 \le s_2 \le \cdots \le f$$

such that $\lim s_n = f$. Necessarily, $s_n \in L^1(\mu \times \nu)$, so by the preceding conclusions, for μ-almost all x, $s_n(x, \cdot)$ are $\mathcal{B}$-measurable, and their integrals (over Y) are $\mathcal{A}$-measurable; therefore, for μ-almost all x, $f(x, \cdot)$ is $\mathcal{B}$-measurable, and by the monotone convergence theorem,

$$\int_Y f(x, \cdot)\, d\nu = \lim \int_Y s_n(x, \cdot)\, d\nu,$$

so that the integrals on the left are $\mathcal{A}$-measurable. Applying the monotone convergence theorem to the sequence on the right, we have by (iii) for s_n,

$$\int_X \left[\int_Y f(x, \cdot)\, d\nu \right] d\mu = \lim_n \int_X \left[\int_Y s_n(x, \cdot)\, d\nu \right] d\mu = \lim_n \int_{X \times Y} s_n\, d(\mu \times \nu)$$
$$= \int_{X \times Y} f\, d(\mu \times \nu) < \infty$$

(by another application of the monotone convergence theorem). In particular, $\int_Y f(x, \cdot)\, d\nu \in L^1(\mu)$ and therefore, $f(x, \cdot) \in L^1(\nu)$ for μ-almost all x.

For $f \in L^1(\mu \times \nu)$ complex, decompose $f = u^+ - u^- + iv^+ - iv^-$ to obtain the conclusions (i)–(iii) for f instead of I_E. Finally, we may interchange the roles of x and y. Collecting, we proved the following.

Theorem 2.17 (Fubini's theorem). *Let $(X, \mathcal{A}, \mu)$ and $(Y, \mathcal{B}, \nu)$ be complete (positive) measure spaces, and let $f \in L^1(\mu \times \nu)$. Then*

(i) for μ-almost all x, $f(x,\cdot) \in L^1(\nu)$ and for ν-almost all y, $f(\cdot\,,y) \in L^1(\mu)$;

(ii) $\int_Y f(x,\cdot)\,d\nu \in L^1(\mu)$ and $\int_X f(\cdot\,,y)\,d\mu \in L^1(\nu)$;

(iii) $\int_X [\int_Y f(x,\cdot)\,d\nu]\,d\mu = \int_{X\times Y} f\,d(\mu\times\nu) = \int_Y [\int_X f(\cdot\,,y)\,d\mu]\,d\nu$.

When we need to verify the hypothesis $f \in L^1(\mu\times\nu)$ (i.e., the *finiteness* of the integral $\int_{X\times Y} |f|\,d(\mu\times\nu)$), the following theorem on *non-negative* functions is useful.

Theorem 2.18 (Tonelli's theorem). *Let $(X,\mathcal{A},\mu)$ and $(Y,\mathcal{B},\nu)$ be complete σ-finite measure spaces, and let $f \geq 0$ be $\mathcal{A}\times\mathcal{B}$-measurable. Then (i) and (ii) in Fubini's theorem (2.17) are valid with the relation "$\in L^1(\cdots)$" replaced by the expression "is measurable", and (iii) is valid.*

Proof. The integrability of $f \geq 0$ was used in the preceding proof to guarantee that the measurable simple functions s_n be in $L^1(\mu\times\nu)$, that is, that they vanish outsides a measurable set of *finite* (product) measure, so that the preceding step, based on Lemma 2.16, could be applied. In our case, the product measure space $Z = X \times Y$ is σ-finite. Write $Z = \bigcup_n Z_n$ with $Z_n \in \mathcal{A}\times\mathcal{B}$ of finite product measure and $Z_n \subset Z_{n+1}$. With s_n as before, the "corrected" simple functions $s'_n := s_n I_{Z_n}$ meet the said requirements. □

Exercises

1. Calculate (with appropriate justification):

   (a) $\lim_{n\to\infty} \int_{\mathbb{R}} (\mathrm{e}^{-x^2/n})/(1+x^2)\,dx$.

   (b) $\lim_{t\to 0+} \int_0^{\pi/2} \sin[(\pi/2)\mathrm{e}^{-tx^2}]\cos x\,dx$.

   (c) $\int_0^1 \int_0^\infty [y\arctan(xy)]/[(1+x^2y^2)(1+y^2)]\,dy\,dx$.

2. Let $L^1(\mathbb{R})$ be the Lebesgue space with respect to the Lebesgue measure on $\mathbb{R}$. If $f \in L^1(\mathbb{R})$, define

$$F_u(t) = \int_{\mathbb{R}} \frac{\sin(t-s)u}{(t-s)u} f(s)\,ds \quad (u>0,\ t\in\mathbb{R}).$$

   Prove:

   (a) For each $u > 0$, the function $F_u : \mathbb{R} \to \mathbb{C}$ is well defined, continuous, and bounded by $\|f\|_1$.

   (b) $\lim_{u\to\infty} F_u = 0$ and $\lim_{u\to 0+} F_u = \int_{\mathbb{R}} f(s)\,ds$ pointwise.

3. Let $h : [0,\infty) \to [0,\infty)$ have a non-negative continuous derivative, $h(0) = 0$, and $h(\infty) = \infty$. Prove that

$$\int_0^\infty \int_{[h'\geq s]} \exp(-h(t)^2)\,dt\,ds = \sqrt{\pi}/2.$$

4. Let $(X, \mathcal{A}, \mu)$ and $(Y, \mathcal{B}, \nu)$ be complete σ-finite positive measure spaces, and $p \in [1, \infty)$ Consider the map

$$[f, g] \in L^p(\mu) \times L^p(\nu) \to F(x, y) := f(x)g(y).$$

Prove:

(a) $F \in L^p(\mu \times \nu)$ and $\|F\|_{L^p(\mu\times\nu)} = \|f\|_{L^p(\mu)}\|g\|_{L^p(\nu)}$.

(b) The map $[f, g] \to F$ is continuous from $L^p(\mu) \times L^p(\nu)$ to $L^p(\mu \times \nu)$.

5. Let $f : \mathbb{R}^2 \to \mathbb{C}$ be Lebesgue measurable, such that $|f(x, y)| \leq Me^{-x^2} I_{[-|x|,|x|]}(y)$ on $\mathbb{R}^2$, for some constant $M > 0$. Prove:

(a) $f \in L^p(\mathbb{R}^2)$ for all $p \in [1, \infty)$, and $\|f\|_{L^p(\mathbb{R}^2)} \leq M(2/p)^{1/p}$.

(b) Suppose $h : \mathbb{R} \to \mathbb{C}$ is continuous and vanishes outside the interval $[-1, 1]$. Define $f : \mathbb{R}^2 \to \mathbb{C}$ by $f(x, y) = e^{-x^2} h(y/x)$ for $x \neq 0$ and $f(0, y) = 0$. Then $\int_{\mathbb{R}^2} f\, dx\, dy = \int_{-1}^{1} h(t)\, dt$.

6. Let $f : \mathbb{R}^2 \to \mathbb{R}$. Prove:

(a) If $f(x, \cdot)$ is Borel for all real x and $f(\cdot\,, y)$ is continuous for all real y, then f is Borel on $\mathbb{R}^2$.

(b) If $f(x, \cdot)$ is Lebesgue measurable for all x in some dense set $E \subset \mathbb{R}$ and $f(\cdot\,, y)$ is continuous for almost all $y \in \mathbb{R}$, then f is Lebesgue measurable on $\mathbb{R}^2$.

Convolution and Fourier transform

7. If $E \subset \mathbb{R}$, denote

$$\tilde{E} := \{(x, y) \in \mathbb{R}^2; x - y \in E\}$$

and

$$\mathcal{S} := \{E \subset \mathbb{R};\ \tilde{E} \in \mathcal{B}(\mathbb{R}^2)\},$$

where $\mathcal{B}(\mathbb{R}^2)$ is the Borel σ-algebra on $\mathbb{R}^2$. Prove:

(a) $\mathcal{S}$ is a σ-algebra on $\mathbb{R}$ which contains the open sets (hence $\mathcal{B}(\mathbb{R}) \subset \mathcal{S}$).

(b) If f is a Borel function on $\mathbb{R}$, then $f(x - y)$ is a Borel function on $\mathbb{R}^2$.

(c) If f, g are integrable Borel functions on $\mathbb{R}$, then $f(x - y)g(y)$ is an integrable Borel function on $\mathbb{R}^2$ and its $L^1(\mathbb{R}^2)$-norm is equal to the product of the $L^1(\mathbb{R})$ norms of f and g.

(d) Let $L^1(\mathbb{R})$ and $L^1(\mathbb{R}^2)$ be the Lebesgue spaces for the Lebesgue measure spaces on $\mathbb{R}$ and $\mathbb{R}^2$ respectively. If $f, g \in L^1(\mathbb{R})$, then $f(x - y)g(y) \in L^1(\mathbb{R}^2)$,

$$\|f(x - y)g(y)\|_{L^1(\mathbb{R}^2)} = \|f\|_1\|g\|_1,$$

and

$$\int_{\mathbb{R}} |f(x-y)g(y)|\,dy < \infty \tag{1}$$

for almost all x.

(e) For x such that (1) holds, define

$$(f*g)(x) = \int_{\mathbb{R}} f(x-y)g(y)\,dy. \tag{2}$$

Show that the (almost everywhere defined and finite-valued) function $f*g$ (called the *convolution* of f and g) is in $L^1(\mathbb{R})$, and

$$\|f*g\|_1 \leq \|f\|_1\|g\|_1. \tag{3}$$

(f) For $f \in L^1(\mathbb{R})$, define its *Fourier transform* Ff by

$$(Ff)(t) = \int_{\mathbb{R}} f(x)\mathrm{e}^{-ixt}\,dx \quad (t \in \mathbb{R}). \tag{4}$$

Show that $Ff : \mathbb{R} \to \mathbb{C}$ is continuous, bounded by $\|f\|_1$, and $F(f*g) = (Ff)(Fg)$ for all $f, g \in L^1(\mathbb{R})$.

(g) If $f = I_{(a,b]}$ for $-\infty < a < b < \infty$, then

$$\lim_{|t|\to\infty} (Ff)(t) = 0. \tag{5}$$

(h) Show that the *step functions* (i.e., finite linear combinations of indicators of disjoint intervals $(a_k, b_k]$) are dense in $C_c(\mathbb{R})$ (the normed space of continuous complex functions on $\mathbb{R}$ with compact support, with pointwise operations and supremum norm), and hence also in $L^p(\mathbb{R})$ for any $1 \leq p < \infty$.

(i) Prove (5) for any $f \in L^1(\mathbb{R})$. (This is the *Riemann–Lebesgue lemma.*)

(j) Generalize the previous statements to functions on $\mathbb{R}^k$.

8. Let $p \in [1, \infty)$ and let q be its conjugate exponent. Let $K : \mathbb{R}^2 \to \mathbb{C}$ be Lebesgue measurable such that

$$\tilde{K}(y) := \int_{\mathbb{R}} |K(x,y)|\,dx \in L^q(\mathbb{R}).$$

Denote

$$(Tf)(x) = \int_{\mathbb{R}} K(x,y)f(y)\,dy$$

Prove:

(a) $\int_{\mathbb{R}}\int_{\mathbb{R}} |K(x,y)f(y)|\,dy\,dx \leq \|\tilde{K}\|_q\|f\|_p$ for all $f \in L^p(\mathbb{R})$. Conclude that $K(x,\cdot)f \in L^1(\mathbb{R})$ for almost all x, and therefore Tf is well defined a.e. (when $f \in L^p$).

(b) T is a continuous (linear) map of $L^p(\mathbb{R})$ into $L^1(\mathbb{R})$, and $\|Tf\|_1 \le \|\tilde{K}\|_q \|f\|_p$.

9. Apply Fubini's theorem to the function $e^{-xy} \sin x$ in order to prove the (Dirichlet) formula

$$\int_0^\infty \frac{\sin x}{x}\, dx = \pi/2.$$

3

Measure and topology

In this chapter, the space X will be a topological space, and we are interested in constructing a measure space $(X, \mathcal{M}, \mu)$ with a "natural" affinity to the given topology.

Denote by $C_c(X)$ the vector space of all complex-valued continuous functions with compact support on a locally compact Hausdorff space X. A linear functional ϕ on $C_c(X)$ is called *positive* if $\phi(f) \geq 0$ when $f \geq 0$. Measures provide a way to construct such functionals: indeed, if μ is a positive Borel measure on X that is finite on compact subsets of X, then the formula $\phi(f) := \int_X f \, d\mu$ defines a positive functional on $C_c(X)$. The central result of this chapter is the *Riesz–Markov theorem*, essentially saying that *all* positive functionals on $C_c(X)$ arise from positive measures this way. This result has a pivotal role in measure theory and functional analysis. In particular, it is the main ingredient in the proof of the Riesz representation theorem that we prove in Chapter 4. A simple application of the Riesz–Markov theorem is an alternative construction of the Lebesgue measure on $\mathbb{R}^k$.

We next prove a few results, including *Lusin's theorem*, concerning approximating various types of functions by elements of $C_c(X)$. A short section on the *support* of a measure follows.

The final section introduces *differentiability* of complex measures on $\mathbb{R}^k$, and proves that every such measure μ is differentiable m-a.e. where m is the Lebesgue measure on $\mathbb{R}^k$, and moreover, the resulting "derivative" function $D\mu$ is the Radon–Nikodym derivative of the absolutely continuous part of μ with respect to m. With some additional work this implies the (two parts of the) *Fundamental Theorem of Calculus*; the details are left to the reader in Exercise 4.

3.1 Partition of unity

We recall first some basic topological concepts.

Introduction to Modern Analysis. Second Edition. Shmuel Kantorovitz and Ami Viselter, Oxford University Press.
 DOI: 10.1093/oso/9780192849540.003.0003

A *Hausdorff space* (or T_2-space) is a topological space (X, τ) in which distinct points have disjoint open neighborhoods. A Hausdorff space X is *locally compact* if each point in X has a *compact neighborhood*. A Hausdorff space X can be imbedded (homeomorphically) as a dense subspace of a compact space Y (the *Alexandroff one-point compactification of* X), and Y is Hausdorff iff X is locally compact. In that case, Y is *normal* (as a compact Hausdorff space), and Urysohn's lemma is valid in Y, that is, given disjoint closed sets A, B in Y, there exists a continuous function $h : Y \to [0,1]$ such that $h(A) = \{0\}$ and $h(B) = \{1\}$. Theorem 3.1 "translates" this result to X. We need the following important concept: for any complex continuous function f on X, the *support* of f (denoted supp f) is defined as the *closure* of $[f^{-1}(\{0\})]^c$.

Theorem 3.1 (Urysohn's lemma for locally compact Hausdorff space). *Let X be a locally compact Hausdorff space, let $U \subset X$ be open, and let $K \subset U$ be compact.*

Then there exists a continuous function $f : X \to [0,1]$, with compact support such that supp $f \subset U$ *and* $f(K) = \{1\}$.

Proof. Let Y be the Alexandroff one-point compactification of X. The set U is open in X, hence in Y. The set K is compact in X, hence in Y, and is therefore closed in Y (since Y is Hausdorff). Since Y is normal, and the closed set K is contained in the open set U, there exists an open set V in Y such that $K \subset V$ and $\mathrm{cl}_Y(V) \subset U$ (where cl_Y denotes the closure operator in Y). Therefore, $K \subset V \cap X := W$, W is open in X, and $\mathrm{cl}_X(W) = \mathrm{cl}_Y(V) \cap X \subset U$.

In the sequel, all closures are *closures in* X.

Since X is locally compact, each $x \in K$ has an open neighborhood N_x with compact closure. Then $N_x \cap W$ is an open neighborhood of x, with closure contained in $\mathrm{cl}(N_x) \cap \mathrm{cl}(W)$, which is compact (since $\mathrm{cl}(N_x)$ is compact), and is contained in U. By compactness of K, we obtain finitely many points $x_i \in K$ such that

$$K \subset \bigcup_{i=1}^{p} (N_{x_i} \cap W) := N.$$

The open set N has closure equal to the union of the compact sets $\mathrm{cl}(N_{x_i} \cap W)$, which is compact and contained in U. We proved that whenever K is a compact subset of the open set U in X, there exists an open set W *with compact closure* such that

$$K \subset W \subset \mathrm{cl}(W) \subset U. \tag{1}$$

(We wrote W instead of N.)

The sets K and $Y - W$ are disjoint closed sets in Y. By Urysohn's lemma for normal spaces, there exists a continuous function $h : Y \to [0,1]$ such that $h = 0$ on $Y - W$ and $h(K) = \{1\}$. Let $f := h|_X$ (the restriction of h to X). Then $f : X \to [0,1]$ is continuous, $f(K) = \{1\}$, and since $[f \neq 0] \subset W$, we have by (1)

$$\mathrm{supp}\, f \subset \mathrm{cl}(W) \subset U.$$

In particular, f has compact support (since $\mathrm{supp}\, f$ is a closed subset of the compact set $\mathrm{cl}(W)$). □

Notation 3.2. We denote the space of all complex (real) continuous functions *with compact support* on the locally compact Hausdorff space X by $C_c(X)$ ($C_c^{\mathbb{R}}(X)$, respectively). By Theorem 3.1, this is a *non-trivial* normed vector space over $\mathbb{C}$ (over $\mathbb{R}$, respectively), with the *uniform norm*

$$\|f\|_u := \sup_X |f|.$$

The *positive cone*

$$C_c^+(X) := \{f \in C_c^{\mathbb{R}}(X); f \geq 0\}$$

will play a central role in this section.

Actually, Theorem 3.1 asserts that for any *open* set $U \neq \emptyset$, the set

$$\Omega(U) := \{f \in C_c^+(X); f \leq 1, \operatorname{supp} f \subset U\} \tag{2}$$

is $\neq \{0\}$.

The following theorem generalizes Theorem 3.1 to the case of a finite open cover of the compact set K. Any set of functions $\{h_1, \ldots, h_n\}$ with the properties described in Theorem 3.3 is called *a partition of unity in* $C_c(X)$ *subordinate to the open cover* $\{V_1, \ldots, V_n\}$ *of the compact set* K.

Theorem 3.3. *Let X be a locally compact Hausdorff space, let $K \subset X$ be compact, and let $V_1, \ldots, V_n$ be open subsets of X such that*

$$K \subset V_1 \cup \cdots \cup V_n.$$

Then there exist $h_i \in \Omega(V_i)$, $i = 1, \ldots, n$, such that $h_1 + \cdots + h_n \in \Omega(V_1 \cup \cdots \cup V_n)$ and

$$1 = h_1 + \cdots + h_n \quad \textit{on } K.$$

Proof. For each $x \in K$, there exists an index $i(x)$ (between 1 and n) such that $x \in V_{i(x)}$. By (1) (applied to the compact set $\{x\}$ contained in the open set $V_{i(x)}$), there exists an open set W_x *with compact closure* such that

$$x \in W_x \subset \operatorname{cl}(W_x) \subset V_{i(x)}. \tag{3}$$

By the compactness of K, there exist $x_1, \ldots, x_m \in K$ such that

$$K \subset W_{x_1} \cup \cdots \cup W_{x_m}.$$

Define for each $i = 1, \ldots, n$

$$H_i := \bigcup \{\operatorname{cl}(W_{x_j}); \operatorname{cl}(W_{x_j}) \subset V_i\}.$$

As a finite union of compact sets, H_i is compact, and contained in V_i. By Theorem 3.1, there exist $f_i \in \Omega(V_i)$, such that $f_i = 1$ on H_i. Take $h_1 = f_1$ and for $k = 2, \ldots, n$, consider the continuous functions

$$h_k = f_k \prod_{i=1}^{k-1} (1 - f_i).$$

An immediate induction on k (up to n) shows that

$$h_1 + \cdots + h_k = 1 - \prod_{i=1}^{k} (1 - f_i).$$

Since $f_i = 1$ on H_i, the product $\prod_{i=1}^n (1 - f_i)$ vanishes on $\bigcup_{i=1}^n H_i$, and this union contains the union of the $W_{x_j}, j = 1, \ldots, m$, hence contains K. Therefore, $h_1 + \cdots + h_n = 1$ on K. The support of h_k is contained in the support of f_k, hence in V_k, and $0 \leq h_k \leq 1$ trivially. The support of $h_1 + \cdots + h_n$ is thus contained in $V_1 \cup \cdots \cup V_n$. Moreover, since $h_1 + \cdots + h_n = 1 - \prod_{i=1}^n (1 - f_i)$ and $0 \leq f_i \leq 1$ for all i, we also have $0 \leq h_1 + \cdots + h_n \leq 1$. □

3.2 Positive linear functionals

Definition 3.4. A linear functional $\phi : C_c(X) \to \mathbb{C}$ is said to be *positive* if $\phi(f) \geq 0$ for all $f \in C_c^+(X)$.

This is clearly equivalent to the *monotonicity* condition: $\phi(f) \geq \phi(g)$ whenever $f \geq g$ $(f, g \in C_c^{\mathbb{R}}(X))$.

Let $V \in \tau$. The indicator I_V is continuous if and only if V is also closed. In that case, $I_V \in \Omega(V)$, and $f \leq I_V$ for all $f \in \Omega(V)$. By monotonicity of ϕ, $0 \leq \phi(f) \leq \phi(I_V)$ for all $f \in \Omega(V)$, and therefore $0 \leq \sup_{f \in \Omega(V)} \phi(f) \leq \phi(I_V)$. Since $I_V \in \Omega(V)$, we actually have an identity (in our special case). The set function $\phi(I_V)$ is a "natural" candidate for a measure of V (associated to the given functional). Since the supremum expression makes sense for arbitrary open sets, we take it as the definition of our "measure" (first defined on τ).

Definition 3.5. Let (X, τ) be a locally compact Hausdorff space, and let ϕ be a positive linear functional on $C_c(X)$. Set

$$\mu(V) := \sup_{f \in \Omega(V)} \phi(f) \quad (V \in \tau).$$

Note that by definition

$$\phi(f) \leq \mu(V)$$

whenever $f \in \Omega(V)$.

Lemma 3.6. *μ is non-negative, monotonic, and subadditive (on τ), and $\mu(\emptyset) = 0$.*

Proof. For each $f \in \Omega(V) \subset C_c^+(X)$, $\phi(f) \geq 0$, and therefore $\mu(V) \geq 0$ (for all $V \in \tau$). Since $\Omega(\emptyset) = \{0\}$ and $\phi(0) = 0$, we trivially have $\mu(\emptyset) = 0$. If $V \subset W$ (with $V, W \in \tau$), then $\Omega(V) \subset \Omega(W)$, so that $\mu(V) \leq \mu(W)$.

Next, let $V_i \in \tau, i = 1, \ldots, n$, with union V. Fix $f \in \Omega(V)$. Let then $h_1, \ldots, h_n$ be a partition of unity in $C_c(X)$ subordinate to the open covering $V_1, \ldots, V_n$ of the compact set supp f. Then

$$f = \sum_{i=1}^{n} h_i f \quad h_i f \in \Omega(V_i), \ i = 1, \ldots, n.$$

Therefore

$$\phi(f) = \sum_{i=1}^{n} \phi(h_i f) \leq \sum_{i=1}^{n} \mu(V_i).$$

Taking the supremum over all $f \in \Omega(V)$, we obtain

$$\mu(V) \leq \sum_{i=1}^{n} \mu(V_i).$$

□

We now extend the definition of μ to $\mathbb{P}(X)$.

Definition 3.7. For any $E \in \mathbb{P}(X)$, we set

$$\mu^*(E) = \inf_{E \subset V \in \tau} \mu(V).$$

If $E \in \tau$, then $E \subset E \in \tau$, so that $\mu^*(E) \leq \mu(E)$. On the other hand, whenever $E \subset V \in \tau$, $\mu(E) \leq \mu(V)$ (by Lemma 3.6), so that $\mu(E) \leq \mu^*(E)$. Thus $\mu^* = \mu$ on τ, and μ^* is indeed an extension of μ.

Lemma 3.8. *μ^* is an outer measure.*

Proof. First, since $\emptyset \in \tau$, $\mu^*(\emptyset) = \mu(\emptyset) = 0$.

If $E \subset F \subset X$, then $E \subset V \in \tau$ whenever $F \subset V \in \tau$, and therefore

$$\mu^*(E) := \inf_{E \subset V \in \tau} \mu(V) \leq \inf_{F \subset V \in \tau} \mu(V) := \mu^*(F),$$

proving the monotonicity of μ^*.

Let $\{E_i\}$ be any sequence of subsets of X. If $\mu^*(E_i) = \infty$ for *some* i, then $\sum_i \mu^*(E_i) = \infty \geq \mu^*(\bigcup_i E_i)$ trivially. Assume therefore that $\mu^*(E_i) < \infty$ for *all* i. Let $\epsilon > 0$. By Definition 3.7, there exist open sets V_i such that

$$E_i \subset V_i; \quad \mu(V_i) < \mu^*(E_i) + \epsilon/2^i, \qquad i = 1, 2, \ldots.$$

Let E and V be the unions of the sets E_i and V_i, respectively. If $f \in \Omega(V)$, then $\{V_i\}$ is an open cover of the compact set supp f, and there exists therefore $n \in \mathbb{N}$ such that supp $f \subset V_1 \cup \cdots \cup V_n$, that is, $f \in \Omega(V_1 \cup \cdots \cup V_n)$. Hence, by Lemma 3.6,

$$\phi(f) \leq \mu(V_1 \cup \cdots \cup V_n) \leq \mu(V_1) + \cdots + \mu(V_n) \leq \sum_{i=1}^{\infty} \mu^*(E_i) + \epsilon.$$

Taking the supremum over all $f \in \Omega(V)$, we get

$$\mu(V) \leq \sum_i \mu^*(E_i) + \epsilon.$$

Since $E \subset V \in \tau$, we get by definition

$$\mu^*(E) \leq \mu(V) \leq \sum_i \mu^*(E_i) + \epsilon,$$

and the σ-subadditivity of μ^* follows from the arbitrariness of ϵ. □

At this stage, we could appeal to Caratheodory's theory (cf. Chapter 2) to obtain the wanted measure space. We prefer, however, to give a construction independent of Chapter 2, and strongly linked to the topology of the space.

Denote by $\mathcal{K}$ the family of all *compact* subsets of X.

Lemma 3.9. *μ^* is finite and additive on $\mathcal{K}$.*

Proof. Let $K \in \mathcal{K}$. By Theorem 3.1, there exists $f \in C_c(X)$ such that $0 \le f \le 1$ and $f = 1$ on K. Let $V := [f > 1/2]$. Then $K \subset V \in \tau$, so that $\mu^*(K) \le \mu(V)$. On the other hand, for all $h \in \Omega(V)$, $h \le 1 < 2f$ on V, so that $\mu(V) := \sup_{h\in\Omega(V)} \phi(h) \le \phi(2f)$ (by monotonicity of ϕ), and therefore $\mu^*(K) \le \phi(2f) < \infty$.

Next, let $K_i \in \mathcal{K}(i = 1, 2)$ be disjoint, and let $\epsilon > 0$. Then $K_1 \subset K_2^c \in \tau$. By (1) in the proof of Theorem 3.1, there exists V_1 open with compact closure such that $K_1 \subset V_1$ and $\mathrm{cl}(V_1) \subset K_2^c$. Hence $K_2 \subset [\mathrm{cl}(V_1)]^c := V_2 \in \tau$, and $V_2 \subset V_1^c$. Thus, V_i are disjoint open sets containing K_i (respectively).

Since $K_1 \cup K_2$ is compact, $\mu^*(K_1 \cup K_2) < \infty$, and therefore, by definition of μ^*, there exists an open set W such that $K_1 \cup K_2 \subset W$ and

$$\mu(W) < \mu^*(K_1 \cup K_2) + \epsilon. \tag{1}$$

By definition of μ on open sets, there exist $f_i \in \Omega(W \cap V_i)$ such that

$$\mu(W \cap V_i) < \phi(f_i) + \epsilon, \quad i = 1, 2. \tag{2}$$

Since $K_i \subset W \cap V_i, i = 1, 2$, it follows from (2) that

$$\mu^*(K_1)+\mu^*(K_2) \le \mu(W\cap V_1)+\mu(W\cap V_2) < \phi(f_1)+\phi(f_2)+2\epsilon = \phi(f_1+f_2)+2\epsilon.$$

However, $f_1 + f_2 \in \Omega(W)$. Hence, by (1),

$$\mu^*(K_1) + \mu^*(K_2) < \mu(W) + 2\epsilon < \mu^*(K_1 \cup K_2) + 3\epsilon.$$

The arbitrariness of ϵ and the subadditivity of μ^* give $\mu^*(K_1 \cup K_2) = \mu^*(K_1) + \mu^*(K_2)$. □

Definition 3.10. The *inner measure* of $E \in \mathbb{P}(X)$ is defined by

$$\mu_*(E) := \sup_{\{K\in\mathcal{K};K\subset E\}} \mu^*(K).$$

By monotonicity of μ^*, we have $\mu_* \le \mu^*$, and equality (and finiteness) is valid on $\mathcal{K}$ (cf. Lemma 3.9). We consider then the family

$$\mathcal{M}_0 := \{E \in \mathbb{P}(X); \mu_*(E) = \mu^*(E) < \infty\}. \tag{3}$$

We just observed that

$$\mathcal{K} \subset \mathcal{M}_0. \tag{4}$$

Another important subfamily of $\mathcal{M}_0$ consists of the *open sets of finite measure* μ.

Lemma 3.11. $\tau_0 := \{V \in \tau; \mu(V) < \infty\} \subset \mathcal{M}_0$.

Proof. Let $V \in \tau_0$ and $\epsilon > 0$. Since $\mu(V) < \infty$, it follows from the definition of μ that there exists $f \in \Omega(V)$ such that $\mu(V) - \epsilon < \phi(f)$. Let $K := \text{supp } f$. Whenever $K \subset W \in \tau$, we have necessarily $f \in \Omega(W)$, and therefore $\phi(f) \leq \mu(W)$. Hence

$$\mu(V) - \epsilon < \phi(f) \leq \inf_{\{W \in \tau; K \subset W\}} \mu(W)$$

$$:= \mu^*(K) \leq \mu_*(V) \leq \mu^*(V) = \mu(V),$$

and so $\mu_*(V) = \mu(V)(= \mu^*(V))$ by the arbitrariness of ϵ. □

Lemma 3.12. *μ^* is σ-additive on $\mathcal{M}_0$, that is, for any sequence of mutually disjoint sets $E_i \in \mathcal{M}_0$ with union E,*

$$\mu^*(E) = \sum_i \mu^*(E_i). \tag{5}$$

Furthermore, $E \in \mathcal{M}_0$ if $\mu^*(E) < \infty$. In particular, $\mathcal{M}_0$ is closed under finite disjoint *unions.*

Proof. By Lemma 3.8, it suffices to prove the *inequality* $\geq$ in (5). Since this inequality is trivial when $\mu^*(E) = \infty$, we may assume that $\mu^*(E) < \infty$.

Let $\epsilon > 0$ be given. For all $i = 1, 2, \ldots$, since $E_i \in \mathcal{M}_0$, there exist $H_i \in \mathcal{K}$ such that $H_i \subset E_i$ and

$$\mu^*(E_i) < \mu^*(H_i) + \epsilon/2^i. \tag{6}$$

The sets H_i are necessarily mutually disjoint. Define

$$K_n = \bigcup_{i=1}^{n} H_i \quad n = 1, 2, \ldots.$$

Then $K_n \subset E$, and by Lemma 3.9 and (6),

$$\sum_{i=1}^{n} \mu^*(E_i) < \sum_{i=1}^{n} \mu^*(H_i) + \epsilon = \mu^*(K_n) + \epsilon \leq \mu^*(E) + \epsilon.$$

The arbitrariness of ϵ proves the wanted inequality $\geq$ in (5). Now $\mu^*(E)$ is the *finite* sum of the series $\sum \mu^*(E_i)$; hence, given $\epsilon > 0$, we may choose $n \in \mathbb{N}$ such that $\mu^*(E) < \sum_{i=1}^{n} \mu^*(E_i) + \epsilon$. For that n, if the compact set K_n is defined as before, we get $\mu^*(E) < \mu^*(K_n) + 2\epsilon$, and therefore $\mu^*(E) = \mu_*(E)$, that is, $E \in \mathcal{M}_0$. □

Lemma 3.13. *$\mathcal{M}_0$ is a* ring *of subsets of X, that is, it is closed under the operations $\cup, \cap, -$ between sets. Furthermore, if $E \in \mathcal{M}_0$, then for each $\epsilon > 0$, there exist $K \in \mathcal{K}$ and $V \in \tau_0$ such that*

$$K \subset E \subset V; \quad \mu(V - K) < \epsilon. \tag{7}$$

Proof. Note that $V-K$ is open, so that (7) makes sense. We prove it first. By definition of μ^* and $\mathcal{M}_0$, there exist $V \in \tau$ and $K \in \mathcal{K}$ such that $K \subset E \subset V$ and

$$\mu(V) - \epsilon/2 < \mu^*(E) < \mu^*(K) + \epsilon/2.$$

In particular, $\mu(V) < \infty$ and $\mu(V-K) \le \mu(V) < \infty$, so that $V, V-K \in \tau_0$. By Lemma 3.11, $V-K \in \mathcal{M}_0$. Since also $K \in \mathcal{K} \subset \mathcal{M}_0$, it follows from Lemma 3.12 that

$$\mu^*(K) + \mu^*(V-K) = \mu(V) < \mu^*(K) + \epsilon.$$

Since $\mu^*(K) < \infty$, we obtain $\mu^*(V-K) < \epsilon$.

Now, let $E_i \in \mathcal{M}_0, i = 1,2$. Given $\epsilon > 0$, pick K_i, V_i as in (7). Since

$$E_1 - E_2 \subset V_1 - K_2 \subset (V_1 - K_1) \cup (K_1 - V_2) \cup (V_2 - K_2),$$

and the sets on the right are disjoint sets in $\mathcal{M}_0$, we have by Lemma 3.12,

$$\mu^*(E_1 - E_2) < \mu^*(K_1 - V_2) + 2\epsilon.$$

Since $K_1 - V_2$ $(= K_1 \cap V_2^c)$ is a *compact* subset of $E_1 - E_2$, it follows that $\mu^*(E_1 - E_2) < \mu_*(E_1 - E_2) + 2\epsilon$ (and of course $\mu^*(E_1 - E_2) \le \mu^*(E_1) < \infty$), so that $E_1 - E_2 \in \mathcal{M}_0$.

Now $E_1 \cup E_2 = (E_1 - E_2) \cup E_2 \in \mathcal{M}_0$ as the disjoint union of sets in $\mathcal{M}_0$ (cf. Lemma 3.12), and $E_1 \cap E_2 = E_1 - (E_1 - E_2) \in \mathcal{M}_0$ since $\mathcal{M}_0$ is closed under difference. □

Definition 3.14.

$$\mathcal{M} = \{E \in \mathbb{P}(X); E \cap K \in \mathcal{M}_0 \text{ for all } K \in \mathcal{K}\}.$$

If E is a *closed* set, then $E \cap K \in \mathcal{K} \subset \mathcal{M}_0$ for all $K \in \mathcal{K}$, so that $\mathcal{M}$ contains all closed sets.

Lemma 3.15. *$\mathcal{M}$ is a σ-algebra containing the Borel algebra $\mathcal{B}$ (of X), and*

$$\mathcal{M}_0 = \{E \in \mathcal{M}; \mu^*(E) < \infty\}. \tag{8}$$

Furthermore, the restriction $\mu := \mu^*|_{\mathcal{M}}$ is a measure.

Proof. We first prove (8). If $E \in \mathcal{M}_0$, then since $\mathcal{K} \subset \mathcal{M}_0$, we surely have $E \cap K \in \mathcal{M}_0$ for all $K \in \mathcal{K}$ (by Lemma 3.13), so that $E \in \mathcal{M}$ (and of course $\mu^*(E) < \infty$ by definition).

On the other hand, suppose $E \in \mathcal{M}$ and $\mu^*(E) < \infty$. Let $\epsilon > 0$. By definition, there exists $V \in \tau$ such that $E \subset V$ and $\mu(V) < \mu^*(E) + 1 < \infty$. By Lemma 3.11, $V \in \mathcal{M}_0$. Applying Lemma 3.13 (7) to V, we obtain a set $K \in \mathcal{K}$ such that $K \subset V$ and $\mu^*(V-K) < \epsilon$. Since $E \cap K \in \mathcal{M}_0$ (by definition of $\mathcal{M}$), there exists $H \in \mathcal{K}$ such that $H \subset E \cap K$ and

$$\mu^*(E \cap K) < \mu^*(H) + \epsilon.$$

Now $E \subset (E \cap K) \cup (V - K)$, so that by Lemma 3.8,

$$\mu^*(E) \le \mu^*(E \cap K) + \mu^*(V - K) < \mu^*(H) + 2\epsilon \le \mu_*(E) + 2\epsilon.$$

The arbitrariness of ϵ implies that $\mu^*(E) \le \mu_*(E)$, so that $E \in \mathcal{M}_0$, and (8) is proved.

Since $\mathcal{M}$ contains all closed sets (see observation following Definition 3.14), we may conclude that $\mathcal{B} \subset \mathcal{M}$ once we know that $\mathcal{M}$ is a σ-algebra.

If $E \in \mathcal{M}$, then for all $K \in \mathcal{K}$,

$$E^c \cap K = K - (E \cap K) \in \mathcal{M}_0$$

by definition and Lemma 3.13. Hence $E^c \in \mathcal{M}$.

Let $E_i \in \mathcal{M}, i = 1, 2, \ldots$, with union E. Then for each $K \in \mathcal{K}$,

$$E \cap K = \bigcup_i E_i \cap K = \bigcup_i F_i,$$

where

$$F_i := (E_i \cap K) - \bigcup_{j<i} (E_j \cap K)$$

are mutually disjoint sets in $\mathcal{M}_0$ (by definition of $\mathcal{M}$ and Lemma 3.13). Since $\mu^*(E \cap K) \le \mu^*(K) < \infty$ (by Lemma 3.9), it follows from Lemma 3.12 that $E \cap K \in \mathcal{M}_0$, and we conclude that $E \in \mathcal{M}$.

Finally, let $E_i \in \mathcal{M}$ be mutually disjoint with union E. If $\mu^*(E_i) = \infty$ for some i, then also $\mu^*(E) = \infty$ by monotonicity, and $\mu^*(E) = \sum_i \mu^*(E_i)$ trivially. Suppose then that $\mu^*(E_i) < \infty$ for all i. By (8), it follows that $E_i \in \mathcal{M}_0$ for all i, and the wanted σ-additivity of $\mu = \mu^*|_{\mathcal{M}}$ follows from Lemma 3.12. □

We call $(X, \mathcal{M}, \mu)$ the *measure space associated with the positive linear functional* ϕ. Integration in the following discussion is performed over this measure space.

Lemma 3.16. *For all* $f \in C_c^+(X)$,

$$\phi(f) \le \int_X f \, d\mu.$$

Proof. Fix $f \in C_c^+(X)$, let K be its (compact) support, and let $0 \le a < b$ be such that $[a, b]$ contains the (compact) range of f. Given $\epsilon > 0$, choose points

$$0 \le y_0 \le a < y_1 < \cdots < y_n = b$$

such that $y_k - y_{k-1} < \epsilon$, and set

$$E_1 := [y_0 \le f \le y_1] \cap K; \quad E_k := [y_{k-1} < f \le y_k], \quad k = 2, \ldots, n.$$

Since f is continuous with support K, the sets E_k are disjoint Borel sets with union K. By definition of our measure space, there exist open sets V_k such that

$$E_k \subset V_k; \quad \mu(V_k) < \mu(E_k) + \epsilon/n$$

for $k = 1, \ldots, n$. Since $f \le y_k$ on E_k, it follows from the continuity of f that there exist open sets U_k such that

$$E_k \subset U_k; \quad f < y_k + \epsilon \text{ on } U_k.$$

Taking $W_k := V_k \cap U_k$, we have for all $k = 1, \ldots, n$

$$E_k \subset W_k; \quad \mu(W_k) < \mu(E_k) + \epsilon/n; \quad f < y_k + \epsilon \text{ on } W_k.$$

Let $\{h_k; k = 1, \ldots, n\}$ be a partition of unity in $C_c(X)$ subordinate to the open covering $\{W_k; k = 1, \ldots, n\}$ of K. Then

$$f = \sum_{k=1}^{n} h_k f,$$

and for all $k = 1, \ldots, n$,

$$h_k f \le h_k(y_k + \epsilon)$$

(since $h_k \in \Omega(W_k)$ and $f < y_k + \epsilon$ on W_k),

$$\phi(h_k) \le \mu(W_k)$$

(since $h_k \in \Omega(W_k)$), and

$$y_k = y_{k-1} + (y_k - y_{k-1}) < f + \epsilon \text{ on } E_k.$$

Therefore,

$$\begin{aligned}
\phi(f) &= \sum_{k=1}^{n} \phi(h_k f) \le \sum_k (y_k + \epsilon)\phi(h_k) \le \sum_k (y_k + \epsilon)\mu(W_k) \\
&\le \sum_k (y_k + \epsilon)[\mu(E_k) + \epsilon/n] \le \sum_k y_k \mu(E_k) + \epsilon\mu(K) + \sum_k (y_k + \epsilon)\epsilon/n \\
&\le \sum_k \int_{E_k} (f + \epsilon)\, d\mu + \epsilon\mu(K) + (b + \epsilon)\epsilon = \int_X f\, d\mu + \epsilon[2\mu(K) + b + \epsilon].
\end{aligned}$$

Since $\mu(K) < \infty$, the lemma follows from the arbitrariness of ϵ. □

Lemma 3.17. *For all* $f \in C_c(X)$,

$$\phi(f) = \int_X f\, d\mu.$$

Proof. By linearity, it suffices to prove the lemma for *real* $f \in C_c(X)$. Given such f, let K be its (compact) support, and let $M = \sup |f|$. For any $\epsilon > 0$, choose V open such that $K \subset V$ and $\mu(V) < \mu(K) + \epsilon$; then choose $h \in \Omega(V)$ such that $\mu(V) < \phi(h) + \epsilon$. By Urysohn's lemma, there is a function $k \in \Omega(V)$ such that $k = 1$ on K. Let $g = \max\{h, k\}$ $(= (1/2)(h + k + |h - k|) \in C_c^+(X))$.

Then $g \in \Omega(V)$, $g = 1$ on K, and $\mu(V) < \phi(g) + \epsilon$. Define $F = f + Mg$. Then $F \in C_c^+(X)$ and $F = f + M$ on K. By Lemma 3.16,

$$\phi(F) \leq \int_X F\, d\mu,$$

that is, since $g \in \Omega(V)$,

$$\begin{aligned} \phi(f) + M\phi(g) &\leq \int_X f\, d\mu + M \int_X g\, d\mu \leq \int_X f\, d\mu + M\mu(V) \\ &\leq \int_X f\, d\mu + M[\phi(g) + \epsilon]. \end{aligned}$$

Hence, by the arbitrariness of ϵ,

$$\phi(f) \leq \int_X f\, d\mu$$

for all *real* $f \in C_c(X)$. Replacing f by $-f$, we also have

$$-\phi(f) = \phi(-f) \leq \int_X (-f)\, d\mu = -\int_X f\, d\mu,$$

so that $\phi(f) = \int_X f\, d\mu$. □

3.3 The Riesz–Markov representation theorem

Theorem 3.18 (Riesz–Markov). *Let (X, τ) be a locally compact Hausdorff space, and let ϕ be a positive linear functional on $C_c(X)$. Let $(X, \mathcal{M}, \mu)$ be the measure space associated with ϕ. Then*

$$\phi(f) = \int_X f\, d\mu \quad f \in C_c(X). \tag{*}$$

In addition, the following properties are valid:

(1) $\mathcal{B}(X) \subset \mathcal{M}$.

(2) μ is finite on $\mathcal{K}$ (the compact subsets of X).

(3) $\mu(E) = \inf_{E \subset V \in \tau} \mu(V)$ for all $E \in \mathcal{M}$.

(4) $\mu(E) = \sup_{\{K \in \mathcal{K}; K \subset E\}} \mu(K)$ (i) for all $E \in \tau$, and (ii) for all $E \in \mathcal{M}$ with finite measure.

(5) the measure space $(X, \mathcal{M}, \mu)$ is complete.

Furthermore, the measure μ is uniquely determined on $\mathcal{M}$ by (*), (2), (3), and (4)-(i).

Proof. Properties (*), (1), (2), and (4)-(ii) are valid by Lemma 3.17, 3.15, 3.9, and 3.15 (together with Definition 3.10 and the following notation (3)), respectively. Property (3) follows from Definition 3.7, since $\mu := \mu^*|_{\mathcal{M}}$.

If $E \in \mathcal{M}$ has measure zero, and $F \subset E$, then $\mu^*(F) = 0$ and $\mu^*(K) = 0$ for all $K \in \mathcal{K}, K \subset F$ (by monotonicity), so that $\mu_*(F) = 0 = \mu^*(F) < \infty$, that is, $F \in \mathcal{M}_0 \subset \mathcal{M}$, and (5) is proved.

We prove (4)-(i). Let $V \in \tau$. If $\mu(V) < \infty$, then $V \in \mathcal{M}_0$ by Lemma 3.11, and (4)-(i) follows from the definition of $\mathcal{M}_0$. Assume then that $\mu(V) = \infty$. By Definition 3.5, for each $n \in \mathbb{N}$, there exists $f_n \in \Omega(V)$ such that $\phi(f_n) > n$. Let $K_n := \operatorname{supp}(f_n)$. Then for all n,

$$\mu_*(V) \geq \mu(K_n) \geq \int_{K_n} f_n \, d\mu = \phi(f_n) > n,$$

so that $\mu_*(V) = \infty = \mu(V)$, and (4)-(i) is valid for V.

Suppose ν is any positive measure on $\mathcal{M}$ satisfying Properties (*), (2), (3), and (4)-(i). Let $\epsilon > 0$ and $K \in \mathcal{K}$. By (2) and (3), there exists $V \in \tau$ such that $K \subset V$ and $\nu(V) < \nu(K) + \epsilon$. By Urysohn's lemma (3.1), there exists $f \in \Omega(V)$ such that $f = 1$ on K. Hence $I_K \leq f \leq I_V$, and therefore, by (*) for both μ and ν,

$$\mu(K) = \int_X I_K \, d\mu \leq \int_X f \, d\mu = \phi(f) = \int_X f \, d\nu \leq \int_X I_V \, d\nu = \nu(V) < \nu(K) + \epsilon.$$

Hence $\mu(K) \leq \nu(K)$, and so $\mu(K) = \nu(K)$ by symmetry. By (4)-(i), it follows that $\mu = \nu$ on τ, hence on $\mathcal{M}$, by (3). □

In case X is σ-compact, the following additional structural properties are valid for the measure space associated with ϕ.

Theorem 3.19. *Let X be a Hausdorff, locally compact, σ-compact space, and let $(X, \mathcal{M}, \mu)$ be the measure space associated with the positive linear functional ϕ on $C_c(X)$. Then:*

(1) For all $E \in \mathcal{M}$ and $\epsilon > 0$, there exist F closed and V open such that

$$F \subset E \subset V; \quad \mu(V - F) < \epsilon.$$

(2) Properties (3) and (4) in Theorem 3.18 are valid for all $E \in \mathcal{M}$ *(this fact is formulated by the expression: μ is* regular. *One says also that $\mu|_{\mathcal{B}(X)}$ is* a regular Borel measure*).*

(3) For all $E \in \mathcal{M}$, there exist an $\mathcal{F}_\sigma$ set A and a $\mathcal{G}_\delta$ set B such that

$$A \subset E \subset B; \quad \mu(B - A) = 0$$

(i.e., every set in $\mathcal{M}$ is the union of an $\mathcal{F}_\sigma$ set and a null set).

Proof. The σ-compactness hypothesis means that $X = \bigcup_i K_i$ with K_i compact. Let $\epsilon > 0$ and $E \in \mathcal{M}$. By 3.18(2), $\mu(K_i \cap E) \leq \mu(K_i) < \infty$, and therefore, by 3.18(3), there exist open sets V_i such that

$$K_i \cap E \subset V_i; \quad \mu(V_i - (K_i \cap E)) < \epsilon/2^{i+1}, \qquad i = 1, 2, \ldots.$$

Set $V = \bigcup_i V_i$. Then V is open, contains E, and

$$\mu(V - E) \le \mu\left(\bigcup_i (V_i - (K_i \cap E))\right) < \epsilon/2.$$

Replacing E by E^c, we obtain in the same fashion an open set W containing E^c such that $\mu(W - E^c) < \epsilon/2$. Setting $F := W^c$, we obtain a closed set contained in E such that $\mu(E - F) < \epsilon/2$, and (1) follows.

Next, for an arbitrary closed set F, we have $F = \bigcup_i (K_i \cap F)$. Let $H_n = \bigcup_{i=1}^n K_i \cap F$. Then H_n is compact for each n, $H_n \subset F$, and $\mu(H_n) \to \mu(F)$. Therefore, Property (4) in Theorem 3.18 is valid for *closed* sets. If $E \in \mathcal{M}$, the first part of the proof gives us a closed subset F of E such that $\mu(E - F) < 1$. If $\mu(E) = \infty$, also $\mu(F) = \infty$, and therefore

$$\sup_{\{K \in \mathcal{K}; K \subset E\}} \mu(K) \ge \sup_{\{K \in \mathcal{K}; K \subset F\}} \mu(K) = \mu(F) = \infty = \mu(E).$$

Together with (3) and (4)-(ii) in Theorem 3.18, this means that Properties (3) and (4) in 3.18 are valid for all $E \in \mathcal{M}$.

Finally, for any $E \in \mathcal{M}$, take $\epsilon = 1/n$ ($n = 1, 2, \ldots$) in (1); this gives us closed sets F_n and open sets V_n such that

$$F_n \subset E \subset V_n; \quad \mu(V_n - F_n) < 1/n, \qquad n = 1, 2, \ldots.$$

Set $A = \bigcup F_n$ and $B = \bigcap V_n$. Then $A \in \mathcal{F}_\sigma$, $B \in \mathcal{G}_\delta$, $A \subset E \subset B$, and since $B - A \subset V_n - F_n$, we have $\mu(B - A) < 1/n$ for all n, so that $\mu(B - A) = 0$. □

3.4 Lusin's theorem

For the measure space of Theorem 3.18, the relation between $\mathcal{M}$-measurable functions and continuous functions is described in the following.

Theorem 3.20 (Lusin). *Let X be a locally compact Hausdorff space, and let $(X, \mathcal{M}, \mu)$ be a measure space such that $\mathcal{B}(X) \subset \mathcal{M}$ and Properties (2), (3), and (4)-(ii) of Theorem 3.18 are satisfied. Let $A \in \mathcal{M}$, $\mu(A) < \infty$, and let $f : X \to \mathbb{C}$ be measurable and vanish on A^c. Then, for any $\epsilon > 0$, there exists $g \in C_c(X)$ such that $\mu([f \ne g]) < \epsilon$. In case f is bounded, one may choose g such that $\|g\|_u \le \|f\|_u$.*

Proof. Suppose the theorem proved for *bounded* functions f (satisfying the hypothesis of the theorem). For an arbitrary f (as in the theorem), the sets $E_n := [|f| \ge n]$, $n = 1, 2, \ldots$ form a decreasing sequence of measurable subsets of A. Since $\mu(A) < \infty$, it follows from Lemma 1.11 that $\lim \mu(E_n) = \mu(\bigcap E_n) = \mu(\emptyset) = 0$. Therefore, we may choose n such that $\mu(E_n) < \epsilon/2$. The function $f_n := f I_{E_n^c}$ satisfies the hypothesis of the theorem and is also bounded (by n). By our assumption, there exists $g \in C_c(X)$ such that $\mu([g \ne f_n]) < \epsilon/2$. Therefore,

$$\begin{aligned}\mu([g \ne f]) &= \mu([g \ne f] \cap E_n) + \mu([g \ne f_n] \cap E_n^c) \\ &\le \mu(E_n) + \mu([g \ne f_n]) < \epsilon.\end{aligned}$$

Next, we may restrict our attention to *non-negative* functions f as above. Indeed, in the general case, we may write $f = \sum_{k=0}^{3} i^k u_k$ with u_k non-negative, measurable, bounded, and vanishing on A^c. By the special case we assumed, there exist $g_k \in C_c(X)$ such that $\mu([g_k \neq u_k]) < \epsilon/4$. Let $E = \bigcup_{k=0}^{3}[g_k \neq u_k]$ and $g := \sum_{k=0}^{3} i^k g_k$. Then $g \in C_c(X)$, and since $[g \neq f] \subset E$, we have indeed $\mu([g \neq f]) < \epsilon$.

Let then $0 \leq f < M$ satisfy the hypothesis of the theorem. Replacing f by f/M, we may assume that $0 \leq f < 1$.

Since $\mu(A) < \infty$, Property (4)-(ii) gives us a *compact* set $K \subset A$ such that $\mu(A - K) < \epsilon/2$. Suppose the theorem is true for A compact. The function $f_K := f I_K$ is measurable with range in $[0, 1)$ and vanishes outside K. By the theorem for compact A, there exists $g \in C_c(X)$ such that $\mu([g \neq f_K]) < \epsilon/2$. Then

$$\begin{aligned}\mu([g \neq f]) &= \mu([g \neq f_K] \cap (K \cup A^c)) + \mu([g \neq f] \cap K^c \cap A) \\ &\leq \mu([g \neq f_K]) + \mu(A - K) < \epsilon.\end{aligned}$$

It remains to prove the theorem for f measurable with range in $[0, 1)$, that vanishes on the complement of a *compact* set A.

By Theorem 1.8, there exist measurable simple functions

$$0 \leq \phi_1 \leq \phi_2 \leq \cdots \leq f$$

such that $f = \lim \phi_n$. Therefore, $f = \sum_n \psi_n$, where $\psi_1 = \phi_1$, $\psi_n := \phi_n - \phi_{n-1} = 2^{-n} I_{E_n}$ (for $n > 1$), and E_n are measurable subsets of A (so that $\mu(E_n) < \infty$). Since A is a compact subset of the locally compact Hausdorff space X, there exists an open set V with compact closure such that $A \subset V$ (cf. (1) in the proof of Theorem 3.1). By Properties (3) and (4)-(ii) of the measure space (since $\mu(E_n) < \infty$), there exist K_n compact and V_n open such that

$$K_n \subset E_n \subset V_n \subset V,$$

and

$$\mu(V_n - K_n) < \epsilon/2^n, \quad n = 1, 2, \ldots.$$

By Urysohn's lemma (3.1), there exist $h_n \in C_c(X)$ such that $0 \leq h_n \leq 1$, $h_n = 1$ on K_n, and $h_n = 0$ on V_n^c. Set

$$g = \sum_n 2^{-n} h_n.$$

The series is majorized by the convergent series of constants $\sum 2^{-n}$, hence converges uniformly on X; therefore g is continuous. For all n, $V_n \subset \mathrm{cl}(V)$ and g vanishes on the set $\bigcap V_n^c = (\bigcup V_n)^c$, which contains $(\mathrm{cl}(V))^c$; thus the support of g is contained in the compact set $\mathrm{cl}(V)$, and so $g \in C_c(X)$. Since $2^{-n} h_n = \psi_n$ on $K_n \cup V_n^c$, we have

$$[g \neq f] \subset \bigcup_n [2^{-n} h_n \neq \psi_n] \subset \bigcup_n (V_n - K_n),$$

and therefore

$$\mu([g \neq f]) < \sum_n \epsilon/2^n = \epsilon.$$

We show finally how to "correct" g so that $\|g\|_u \leq \|f\|_u$ when f is a *bounded* function satisfying the hypothesis of the theorem. Suppose $g \in C_c(X)$ is such that $\mu([g \neq f]) < \epsilon$. Let $E = [|g| \leq \|f\|_u]$. Define

$$g_1 = gI_E + (g/|g|)\|f\|_u I_{E^c}.$$

Then g_1 is continuous (!), $\|g_1\|_u \leq \|f\|_u$, and since $g_1(x) = 0$ iff $g(x) = 0$, g_1 has compact support. Since $[g = f] \subset E$, we have $[g = f] \subset [g_1 = f]$, hence $\mu([g_1 \neq f]) \leq \mu([g \neq f]) < \epsilon$. □

Corollary 3.21. *Let $(X, \mathcal{M}, \mu)$ be a measure space as in Theorem 3.20. Then for each $p \in [1, \infty)$, $C_c(X)$ is dense in $L^p(\mu)$.*

In the terminology of Definition 1.28, Corollary 3.21 establishes that $L^p(\mu)$ is the *completion* of $C_c(X)$ in the $\|\cdot\|_p$-metric.

Proof. Since $\mathcal{B}(X) \subset \mathcal{M}$, Borel functions are $\mathcal{M}$-measurable; in particular, continuous functions are $\mathcal{M}$-measurable. If $f \in C_c(X)$ and $K := \operatorname{supp} f$, then $\int_X |f|^p \, d\mu \leq \|f\|_u^p \mu(K) < \infty$ by Property (2). Thus $C_c(X) \subset L^p(\mu)$ for all $p \in [1, \infty)$. By Theorem 1.27, it suffices to prove that for each *simple* measurable function ϕ vanishing outside a measurable set A of finite measure and for each $\epsilon > 0$, there exists $g \in C_c(X)$ such that $\|\phi - g\|_p < \epsilon$. By Theorem 3.20 applied to ϕ, there exists $g \in C_c(X)$ such that $\mu([\phi \neq g]) < (\epsilon/(2\|\phi\|_u))^p$ and $\|g\|_u \leq \|\phi\|_u$. Then

$$\|\phi - g\|_p^p = \int_{[\phi \neq g]} |\phi - g|^p \, d\mu \leq (2\|\phi\|_u)^p \mu([\phi \neq g]) < \epsilon^p,$$

as wanted. □

By Lemma 1.30, we obtain

Corollary 3.22. *Let $(X, \mathcal{M}, \mu)$ be a measure space as in Theorem 3.20. Let $f \in L^p(\mu)$ for some $p \in [1, \infty)$. Then there exists a sequence $\{g_n\} \subset C_c(X)$ that converges to f almost everywhere.*

In view of the observation following the statement of Corollary 3.21, it is interesting to find the completion of $C_c(X)$ with respect to the $\|\cdot\|_u$-metric. We start with a definition.

Definition 3.23. Let X be a locally compact Hausdorff space. Then $C_0(X)$ will denote the space of all complex continuous functions f on X with the following property:

(*) for each $\epsilon > 0$, there exists a compact subset $K \subset X$ such that $|f| < \epsilon$ on K^c.

A function with Property (*) is said to *vanish at infinity*.

Under pointwise operations, $C_0(X)$ is a complex vector space that contains $C_c(X)$. If $f \in C_0(X)$ and K is as in (*) with $\epsilon = 1$, then $\|f\|_u \le \sup_K |f|+1 < \infty$, and it follows that $C_0(X)$ is a normed space for the uniform norm.

Theorem 3.24. *$C_0(X)$ is the completion of $C_c(X)$.*

Proof. Let $\{f_n\} \subset C_0(X)$ be Cauchy. Then $f := \lim f_n$ exists pointwise uniformly on X, so that f is continuous on X and $\|f_n - f\|_u \to 0$. Given $\epsilon > 0$, let $n_0 \in \mathbb{N}$ be such that $\|f_n - f\|_u < \epsilon/2$ for all $n > n_0$. Fix $n > n_0$ and a compact set K such that $|f_n| < \epsilon/2$ on K^c (cf. (*)). Then $|f| \le |f - f_n| + |f_n| < \epsilon$ on K^c, so that $f \in C_0(X)$, and we conclude that $C_0(X)$ is complete.

Given $f \in C_0(X)$ and $\epsilon > 0$, let K be as in (*). By Urysohn's Lemma (3.1), there exists $h \in C_c(X)$ such that $0 \le h \le 1$ on X and $h = 1$ on K. Then $hf \in C_c(X)$, $|f - hf| = (1-h)|f| = 0$ on K, and $|f - hf| \le |f| < \epsilon$ on K^c, so that $\|f - hf\|_u < \epsilon$. This shows that $C_c(X)$ is dense in $C_0(X)$. □

Example 3.25. Consider the special case $X = \mathbb{R}^k$, the k-dimensional Euclidean space. If $f \in C_c(\mathbb{R}^k)$ and T is any closed cell containing supp f, let $\phi(f)$ be the Riemann integral of f on T. Then ϕ is a well-defined positive linear functional on $C_c(\mathbb{R}^k)$. Let $(\mathbb{R}^k, \mathcal{M}, m)$ be the associated measure space as in Theorem 3.18. Then, by Theorem 3.18, the integral $\int_{\mathbb{R}^k} f\, dm$ coincides with the Riemann integral of f for all $f \in C_c(\mathbb{R}^k)$.

For $n \in \mathbb{N}$ large enough and $a < b$ real, let $f_{n,a,b} : \mathbb{R} \to [0,1]$ denote the function equal to zero outside $[a + 1/n, b - 1/n]$, to 1 in $[a + 2/n, b - 2/n]$, and linear elsewhere. Then

$$\int_a^b f_{n,a,b}\, dx = b - a - 3/n.$$

If $T = \{x \in \mathbb{R}^k; a_i \le x_i \le b_i; i = 1, \dots, k\}$, consider the function $F_{n,T} = \prod_{i=1}^k f_{n,a_i,b_i} \in C_c(\mathbb{R}^k)$. Then

$$F_{n,T} \le I_T \le F_{n,T_n},$$

where $T_n = \{x \in \mathbb{R}^k; a_i - 2/n \le x_i \le b_i + 2/n\}$. Therefore, by Fubini's theorem for the Riemann integral on cells,

$$\prod_{i=1}^k (b_i - a_i - 3/n) = \int F_{n,T}\, dx_1 \dots dx_k \le m(T) \le \int F_{n,T_n} = \prod_{i=1}^k (b_i - a_i + 1/n).$$

Letting $n \to \infty$, we conclude that

$$m(T) = \prod_{i=1}^k (b_i - a_i) := \mathrm{vol}(T).$$

By Theorem 2.13 and the subsequent constructions of Lebesgue's measure on $\mathbb{R}$ and of the product measure, the measure m coincides with Lebesgue's measure on the Borel subsets of $\mathbb{R}^k$.

3.5 The support of a measure

Definition 3.26. Let $(X, \mathcal{M}, \mu)$ be as in Theorem 3.18. Let V be the union of all the *open* μ-null sets in X. The *support* of μ is the complement V^c of V, and is denoted by supp μ.

Since V is open, supp μ is a closed subset of X. Also, by Property (4)-(i) of μ (cf. Theorem 3.18),

$$\mu(V) = \sup_{K \in \mathcal{K}; K \subset V} \mu(K). \tag{1}$$

If K is a compact subset of V, the open μ-null sets are an open cover of K, and there exist therefore finitely many μ-null sets that cover K; hence $\mu(K) = 0$, and it follows from (1) that $\mu(V) = 0$. Thus $S = \operatorname{supp} \mu$ is *the smallest closed set with a μ-null complement.*

For any $f \in L^1(\mu)$, we have

$$\int_X f \, d\mu = \int_S f \, d\mu. \tag{2}$$

If $f \in C_c(X)$ is non-negative and $\int_X f \, d\mu = 0$, then $f = 0$ identically on the support S of μ. Indeed, suppose there exists $x_0 \in S$ such that $f(x_0) \neq 0$. Then there exists an open neighborhood U of x_0 such that $f \neq 0$ on U. Let K be any compact subset of U. Then $c := \min_K f > 0$, and

$$0 = \int_X f \, d\mu \geq \int_K f \, d\mu \geq c\mu(K).$$

Hence $\mu(K) = 0$, and therefore $\mu(U) = 0$ by Property (4)-(i) of μ (cf. Theorem 3.18). Thus, $U \subset S^c$, which implies the contradiction $x_0 \in S^c$.

Together with (2), this shows that $\int_X f \, d\mu = 0$ for a non-negative function $f \in C_c(X)$ iff f vanishes identically on $\operatorname{supp} \mu$.

3.6 Measures on $\mathbb{R}^k$; differentiability

Notation 3.27. If $E \subset \mathbb{R}^k$, we denote the diameter of E (i.e., $\sup_{x,y \in E} d(x, y)$) by $\delta(E)$. Let μ be a real or a positive Borel measure on $\mathbb{R}^k$, and let m denote the Lebesgue measure on $\mathbb{R}^k$. Fix $x \in \mathbb{R}^k$, and consider the quotients $\mu(E)/m(E)$ *for all open cubes E containing x*. The *upper derivative of μ at x* is defined by

$$(\bar{D}\mu)(x) = \limsup_{\delta(E) \to 0} \frac{\mu(E)}{m(E)} := \lim_{r \to 0} \sup_{\delta(E) < r} \frac{\mu(E)}{m(E)}.$$

The *lower derivative of μ at x*, denoted $(\underline{D}\mu)(x)$, is defined similarly by replacing lim sup and sup by lim inf and inf, respectively.

Since $\sup_{\delta(E) < r} \mu(E)/m(E)$ is an increasing function of r, $(\bar{D}\mu)(x)$ is well defined. The same is true of $(\underline{D}\mu)(x)$, and we have trivially $(\underline{D}\mu)(x) \leq (\bar{D}\mu)(x)$.

In case these quantities are *equal and finite*, one says that μ is *differentiable at* x; the common value is denoted $(D\mu)(x)$, and is called *the derivative of* μ *at* x.

If $f(x) := \sup_{x\in E;\delta(E)<r} \mu(E)/m(E) > c$ for some real c and some $r > 0$, there exists an *open* cube E_0 containing x with $\delta(E_0) < r$ such that $\mu(E_0)/m(E_0) > c$; this inequality is true for all $y \in E_0$, and therefore, for each $y \in E_0$, the shown supremum *over all open cubes* E *containing* y *with* $\delta(E) < r$ is $> c$. This shows that $[f > c]$ is open, and therefore f is a Borel function of x. Consequently, $\bar{D}\mu$ *is a Borel function.*

If μ_k $(k = 1,2)$ are real Borel measures with *finite* upper derivatives at x, then

$$\bar{D}(\mu_1 + \mu_2) \le \bar{D}\mu_1 + \bar{D}\mu_2$$

at every point x; for $\underline{D}$, the inequality is reversed. It follows in particular that if both μ_k are differentiable at x, the same is true of $\mu := \mu_1 + \mu_2$, and $(D\mu)(x) = (D\mu_1)(x) + (D\mu_2)(x)$.

The concepts of differentiability and derivative are extended to complex measures in the usual way.

The next theorem relates $D\mu$ to the Radon–Nikodym derivative $d\mu_a/dm$ of the absolutely continuous part μ_a of μ in its Lebesgue decomposition with respect to m (cf. Theorem 1.45).

Theorem 3.28. *Let* μ *be a complex Borel measure on* $\mathbb{R}^k$. *Then* μ *is differentiable* m*-a.e., and* $D\mu = d\mu_a/dm$ *(as elements of* $L^1(\mathbb{R}^k)$*).*

It follows in particular that $\mu \perp m$ iff $D\mu = 0$ m-a.e., and $\mu \ll m$ iff $\mu(E) = \int_E (D\mu)dm$ for all $E \in \mathcal{B} := \mathcal{B}(\mathbb{R}^k)$.

Proof. 1. Consider first a *positive* Borel measure μ *which is finite on compact sets.*

Fix $A \in \mathcal{B}$ and $c > 0$, and *assume* that the *Borel set*

$$A_c := A \cap [\bar{D}\mu > c] \tag{1}$$

(cf. Notation 3.27) has *positive* Lebesgue measure.

Since m is regular, there exists a compact set $K \subset A_c$ such that $m(K) > 0$. Fix $r > 0$. For each $x \in K$, there exists an open cube E with $\delta(E) < r$ such that $x \in E$ and $\mu(E)/m(E) > c$. By compactness of K, we may choose finitely many of these cubes, say $E_1, \ldots, E_n$, such that $K \subset \bigcup_i E_i$ and $\delta(E_i) \ge \delta(E_{i+1})$. We pick a disjoint subfamily of E_i as follows: $i_1 = 1$; i_2 is the first index $> i_1$ such that E_{i_2} does not meet E_{i_1}; i_3 is the first index $> i_2$ such that E_{i_3} does not meet E_{i_1} and E_{i_2}, etc. Let V_j be the closed ball centered at the center p_j of E_{i_j} with diameter $3\delta(E_{i_j})$. If γ_k denotes the ratio of the volumes of a ball and a cube in $\mathbb{R}^k$ with the same diameter, then $m(V_j) = \gamma_k 3^k m(E_{i_j})$.

For each $i = 1, \ldots, n$, there exists $i_j \le i$ such that E_i meets E_{i_j}, say, at some point q. Then for all $y \in E_i$,

$$d(y, p_j) \le d(y, q) + d(q, p_j) \le \delta(E_i) + \delta(E_{i_j})/2 \le 3\delta(E_{i_j})/2,$$

since $i_j \le i$ implies that $\delta(E_i) \le \delta(E_{i_j})$. Hence $E_i \subset V_j$, and

$$K \subset \bigcup_i E_i \subset \bigcup_j V_j.$$

Therefore,

$$m(K) \le \sum_j m(V_j) = \gamma_k 3^k \sum_j m(E_{i_j}) < \gamma_k 3^k c^{-1} \sum_j \mu(E_{i_j})$$
$$= \gamma_k 3^k c^{-1} \mu\Big(\bigcup_j E_{i_j}\Big).$$

Each E_{i_j} is an open cube of diameter $< r$ containing some point of K; therefore,

$$\bigcup_j E_{i_j} \subset \{y;\, d(y,K) < r\} := K_r.$$

The (open) set K_r has compact closure, and therefore $\mu(K_r) < \infty$ by hypothesis, and by the preceding calculation

$$m(K) \le \gamma_k 3^k c^{-1} \mu(K_r). \tag{2}$$

Take $r = 1/N$ ($N \in \mathbb{N}$); $\{K_{1/N}\}_{N\in\mathbb{N}}$ is a decreasing sequence of open sets of finite μ-measure with intersection K; therefore $\mu(K) = \lim_N \mu(K_{1/N})$, and it follows from (2) that $m(K) \le \gamma_k 3^k c^{-1} \mu(K)$. Hence

$$\mu(A_c) \ge \mu(K) \ge \gamma_k^{-1} 3^{-k} c\, m(K) > 0.$$

We proved therefore that $m(A_c) > 0$ implies $\mu(A_c) > 0$. Consequently, *if* $\mu(A) = 0$ (so that $\mu(A_c) = 0$ for all $c > 0$), then $m(A_c) = 0$ for all $c > 0$. Since $A \cap [\bar{D}\mu > 0] = \bigcup_{p=1}^{\infty} A_{1/p}$, it then follows that $m(A \cap [\bar{D}\mu > 0]) = 0$. But $\bar{D}\mu \ge 0$ since μ is a positive measure. Therefore $\bar{D}\mu = 0$ m-a.e. on A (for each $A \in \mathcal{B}$ with $\mu(A) = 0$). Hence $0 \le \underline{D}\mu \le \bar{D}\mu = 0$ m-a.e. on A, and we conclude that $D\mu$ exists and equals zero m-a.e. on A (if $\mu(A) = 0$).

If $\mu \perp m$, there exists $A \in \mathcal{B}$ such that $\mu(A) = 0$ and $m(A^c) = 0$.

Then $m([\bar{D}\mu > 0] \cap A) = 0$ and trivially $m([\bar{D}\mu > 0] \cap A^c) = 0$. Hence $m([\bar{D}\mu > 0]) = 0$, and therefore $D\mu = 0$ m-a.e.

If μ is a complex Borel measure, we use its canonical (Jordan) decomposition $\mu = \sum_{k=0}^{3} i^k \mu_k$, where μ_k are finite positive Borel measures. If $\mu \perp m$, also $\mu_k \perp m$ for all k, hence $D\mu_k = 0$ m-a.e. for $k = 0, \ldots, 3$, and consequently $D\mu = \sum_{k=0}^{3} i^k D\mu_k = 0$ m-a.e.

2. Let μ be a *real* Borel measure *absolutely continuous* with respect to m (restricted to $\mathcal{B}$), and let $h = d\mu/dm$ be the Radon–Nikodym derivative (h is real m-a.e., and since it is only determined m-a.e., we may assume that h is a *real* (Borel) function (in $L^1(\mathbb{R}^k)$). *We claim that*

$$m([h < \bar{D}\mu]) = 0. \tag{3}$$

Assuming the claim and replacing μ by $-\mu$ (so that h is replaced by $-h$), since $\bar{D}(-\mu) = -\underline{D}\mu$, we obtain $m([h > \underline{D}\mu]) = 0$. Consequently

$$h \le \underline{D}\mu \le \bar{D}\mu \le h \quad m\text{-a.e.},$$

that is, μ is differentiable and $D\mu = h$ m-a.e. The case of a complex Borel measure $\mu \ll m$ follows trivially from the real case. Finally, if μ is an arbitrary complex Borel measure, we use the Lebesgue decomposition $\mu = \mu_a + \mu_s$ as in Theorem 1.45. It follows that μ is differentiable and $D\mu = D\mu_a + D\mu_s = d\mu_a/dm$ m-a.e. (cf. Part 1 of the proof), as wanted.

To prove (3) it suffices to show that $E_r := [h < r < \bar{D}\mu] (= [h < r] \cap [\bar{D}\mu > r])$ is m-null for any rational number r, because $[h < \bar{D}\mu] = \bigcup_{r\in\mathbb{Q}} E_r$. Fix $r \in \mathbb{Q}$, and consider the *positive* Borel measure

$$\lambda(E) := \int_{E\cap[h\ge r]} (h - r)\, dm \quad (E \in \mathcal{B}). \tag{4}$$

Since $h \in L^1(m)$, λ is finite on compact sets, and $\lambda([h < r]) = 0$. By Part 1 of the proof, it follows that $D\lambda = 0$ m-a.e. on $[h < r]$. For any $E \in \mathcal{B}$,

$$\begin{aligned}\mu(E) &= \int_E h\, dm = \int_E [(h - r) + r]\, dm = \int_E (h - r) dm + rm(E)\\ &= \int_{E\cap[h>r]} (h - r) dm + \int_{E\cap[h\le r]} (h - r) dm + rm(E) \le \lambda(E) + rm(E).\end{aligned}$$

Given $x \in \mathbb{R}^k$, we have then for any open cube E containing x

$$\frac{\mu(E)}{m(E)} \le \frac{\lambda(E)}{m(E)} + r.$$

Taking the supremum over all such E with $\delta(E) < s$ and letting then $s \to 0$, we obtain

$$(\bar{D}\mu)(x) \le (\bar{D}\lambda)(x) + r = r$$

m-a.e. on $[h < r]$. Equivalently, $m([h < r] \cap [\bar{D}\mu > r]) = 0$. □

Corollary 3.29. *If $f \in L^1(\mathbb{R}^k)$, then*

$$\lim_{\delta(E)\to 0} m(E)^{-1} \int_E |f(y) - f(x)|\, dy = 0 \tag{5}$$

for almost all $x \in \mathbb{R}^k$. (The limit is over open cubes containing x.)

In particular, the averages of f over open cubes E containing x converge almost everywhere to $f(x)$ as $\delta(E) \to 0$.

Proof. For each $c \in \mathbb{Q} + i\mathbb{Q}$ and $N \in \mathbb{N}$, consider the finite positive Borel measure

$$\mu_N(E) := \int_{E\cap B(0,N)} |f - c|\, dy \quad (E \in \mathcal{B}),$$

where $B(0,N) = \{y \in \mathbb{R}^k; |y| < N\}$. By Theorem 3.28, $\mu_N(E)/m(E) \to |f(x) - c| I_{B(0,N)}(x)$ m-a.e. when the open cubes E containing x satisfy $\delta(E) \to 0$. Denote the "exceptional m-null set" by $G_{c,N}$, and let

$$G := \bigcup \{G_{c,N}; c \in \mathbb{Q} + i\mathbb{Q},\ N \in \mathbb{N}\}.$$

We have $m(G) = 0$, and the proof will be completed by showing that (5) is valid for each $x \notin G$.

Let $x \notin G$ and $\epsilon > 0$. By the density of $\mathbb{Q} + i\mathbb{Q}$ in $\mathbb{C}$, there exists $c \in \mathbb{Q} + i\mathbb{Q}$ such that $|f(x) - c| < \epsilon$. Choose $N > |x| + 1$. All open cubes containing x with diameter < 1 are contained in $B(0,N)$, and therefore $\mu_N(E)/m(E) \to |f(x) - c|$ when $\delta(E) \to 0$. Since

$$m(E)^{-1} \int_E |f(y) - f(x)|\, dy \le m(E)^{-1} \int_{E(=E \cap B(0,N))} |f(y) - c|\, dy$$

$$+ m(E)^{-1} \int_E |f(x) - c|\, dy \le \frac{\mu_N(E)}{m(E)} + \epsilon,$$

it follows that

$$\limsup_{\delta(E) \to 0} m(E)^{-1} \int_E |f(y) - f(x)|\, dy \le |f(x) - c| + \epsilon < 2\epsilon.$$

The arbitrariness of ϵ shows that the shown lim sup is 0 for all $x \notin G$, and therefore the limit of the averages exists and equals zero for all $x \notin G$. □

Exercises

Translations in L^p

1. Let L^p be the Lebesgue space on $\mathbb{R}^k$ with respect to Lebesgue measure. For each $t \in \mathbb{R}^k$, let

$$[T(t)f](x) = f(x + t) \quad (f \in L^p; x \in \mathbb{R}^k).$$

This so-called "translation operator" is a linear isometry of L^p onto itself. Prove that $T(t)f \to f$ in L^p-norm as $t \to 0$, for each $f \in L^p$ $(1 \le p < \infty)$. (Hint: use Corollary 3.21 and an "$\epsilon/3$ argument".)

Automatic regularity

2. Let X be a locally compact Hausdorff space in which every open set is σ-compact (e.g., an Euclidean space). Then every positive Borel measure λ which is finite on compact sets is regular. (Hint: consider the positive linear functional $\phi(f) := \int_X f\, d\lambda$. If $(X, \mathcal{M}, \mu)$ is the associated measure space as in Theorem 3.18, show that $\lambda = \mu$ on open sets and use Theorem 3.19.)

Hardy inequality

3. Let $1 < p < \infty$, and let $L^p(\mathbb{R}^+)$ denote the Lebesgue space for $\mathbb{R}^+ := (0, \infty)$ with respect to the Lebesgue measure. For $f \in L^p(\mathbb{R}^+)$, define

$$(Tf)(x) = (1/x) \int_0^x f(t)\,dt \quad (x \in \mathbb{R}^+).$$

Prove:

(a) Tf is well defined, and $|(Tf)(x)| \le x^{-1/p}\|f\|_p$.

(b) Denote by D, M, and I the differentiation, multiplication by x, and identity operators, respectively (on appropriate domains). Verify the identities

$$MDT = I - T \quad \text{on } C_c^+(\mathbb{R}^+), \tag{1}$$

where multiplication of operators is their composition.

$$\|Tf\|_p^p = q \int_0^\infty f(Tf)^{p-1}\,dx \tag{2}$$

for all $f \in C_c^+(\mathbb{R}^+)$, where q is the conjugate exponent of p. (Hint: integrate by parts.)

(c) $\|Tf\|_p \le q\,\|f\|_p \quad f \in C_c^+(\mathbb{R}^+)$.

(d) Extend the (Hardy) inequality (c) to all $f \in L^p(\mathbb{R}^+)$. (Hint: use Corollary 3.21.)

(e) Show that $\sup_{0 \ne f \in L^p} \|Tf\|_p / \|f\|_p = q$. (Hint: consider the functions $f_n(x) = x^{-1/p} I_{[1,n]}$.)

Absolutely continuous and singular functions

4. Recall that a function $f : \mathbb{R} \to \mathbb{C}$ has *bounded variation* if its *total variation function* v_f is *bounded*, where

$$v_f(x) := \sup_P \sum_k |f(x_k) - f(x_{k-1})| < \infty,$$

and $P = \{x_k; k = 0, \ldots, n\}$, $x_{k-1} < x_k$, $x_n = x$ (the supremum is taken over all such "partitions" P of $(-\infty, x]$).

The *total variation* of f is $V(f) := \sup_{\mathbb{R}} v_f$.

It follows from a theorem of Jordan that such a function has a "canonical" (Jordan) decomposition $f = \sum_{k=0}^3 i^k f_k$ where f_k are non-decreasing real function. Therefore, f has one-sided limits at every point. We say that f is *normalized* if it is left-continuous and $f(-\infty) = 0$.

(a) Let μ be a complex Borel measure on $\mathbb{R}$. Show that $f(x) := \mu((-\infty, x))$ is a normalized function of bounded variation (briefly, f is NBV).

(b) Conversely, if f is NBV and μ is the corresponding Lebesgue–Stieltjes measure (constructed through the Jordan decomposition of f as in Chapter 2, with left continuity replacing right continuity), then μ (restricted to $\mathcal{B} := \mathcal{B}(\mathbb{R})$) is a complex Borel measure such that $f(x) = \mu((-\infty, x))$ for all $x \in \mathbb{R}$. (Also $v_f(x) = |\mu|((-\infty, x))$ and $V(f) = \|\mu\|$.)

(c) $f : \mathbb{R} \to \mathbb{C}$ is *absolutely continuous* if for each $\epsilon > 0$ there exists $\delta > 0$ such that whenever $\{(a_k, b_k); k = 1, \ldots, n\}$ is a finite family of disjoint intervals of total length $< \delta$, we have $\sum_k |f(b_k) - f(a_k)| < \epsilon$. If f is NBV and μ is the Borel measure associated to f as in Part b., then $\mu \ll m$ iff f is absolutely continuous (cf. Theorem 3.28 and Exercise 8(f) in Chapter 1).

(d) Let $h \in L^1 := L^1(\mathbb{R})$, $f(x) = \int_{-\infty}^{x} h(t)\,dt$, and $\mu(E) = \int_E h(t)\,dt$ $(E \in \mathcal{B})$. Conclude from Parts (a) and (c) that f is absolutely continuous and $D\mu = h$ m-a.e. (cf. Theorem 3.28).

(e) Let μ and f be as in Part (a), and let $x \in \mathbb{R}$ be fixed. Show that $(D\mu)(x)$ exists iff $f'(x)$ exists and $f'(x) = (D\mu)(x)$. In particular, if $\mu \perp m$, then $f' = 0$ m-a.e. (such a function is called a *singular function*) (cf. Theorem 3.28).

(f) With h and f as in Part (d), conclude from Parts (d) and (e) (and Theorem 3.28) that $f' = h$ m-a.e.

(g) If f is NBV, show that f' exists m-a.e. and is in L^1, and $f(x) = f_s(x) + \int_{-\infty}^{x} f'(t)\,dt$ where f_s is a singular NBV function. (Apply Parts (b), (e), and (f), and the Lebesgue decomposition.)

Cantor functions

5. Let $\{r_n\}_{n=0}^{\infty}$ be a positive decreasing sequence with $r_0 = 1$. Denote $r = \lim_n r_n$. Let $C_0 = [0, 1]$, and for $n \in \mathbb{N}$, let C_n be the union of the 2^n disjoint closed intervals of length $r_n/2^n$ obtained by removing open intervals at the center of the 2^{n-1} intervals comprising C_{n-1} (note that the removed intervals have length $(r_{n-1} - r_n)/2^{n-1} > 0$ and $m(C_n) = r_n$). Let $C = \bigcap_n C_n$.

(a) C is a compact set of Lebesgue measure r.

(b) Let $g_n = r_n^{-1} I_{C_n}$ and $f_n(x) = \int_0^x g_n(t)\,dt$. Then f_n is continuous, non-decreasing, constant on each open interval comprising C_n^c, $f_n(0) = 0$, $f_n(1) = 1$, and f_n converge *uniformly* in $[0, 1]$ to some function f. The function f is continuous, non-decreasing, has range equal to $[0, 1]$, and $f' = 0$ on C^c. (In particular, if $r = 0$, $f' = 0$ m-a.e., but f is *not* constant. Such so-called *Cantor functions* are examples of continuous non-decreasing non-constant singular functions.)

Semi-continuity

6. Let X be a topological space. A function $f : X \to \overline{\mathbb{R}}$ is *lower semi-continuous* (l.s.c.) if $[f > c]$ is open for all real c; f is *upper semi-continuous* (u.s.c.) if $[f < c]$ is open for all real c. Prove:

   (a) f is continuous iff it is both l.s.c. and u.s.c.

   (b) If f is l.s.c. (u.s.c.) and α is a positive constant, then αf is l.s.c. (u.s.c., respectively). Also $-f$ is u.s.c. (l.s.c., respectively).

   (c) If f, g are l.s.c. (u.s.c.), then $f + g$ is l.s.c. (u.s.c., respectively).

   (d) The supremum (infimum) of any family of l.s.c. (u.s.c.) functions is l.s.c. (u.s.c., respectively).

   (e) If $\{f_n\}$ is a sequence of non-negative l.s.c. functions, then $f := \sum_n f_n$ is l.s.c.

   (f) The indicator I_A is l.s.c. (u.s.c.) if $A \subset X$ is open (closed, respectively).

   (g) The following conditions are equivalent:
   - f is l.s.c.;
   - for each net $\{x_\alpha\}_{\alpha \in A}$ in X converging to $x \in X$ such that $f(x_\alpha) \leq f(x)$ for each $\alpha \in A$, $\{f(x_\alpha)\}_{\alpha \in A}$ converges to $f(x)$;
   - for each net $\{x_\alpha\}_{\alpha \in A}$ in X converging to $x \in X$ we have $f(x) \leq \liminf_{\alpha \in A} f(x_\alpha)$;
   - for each net $\{x_\alpha\}_{\alpha \in A}$ in X converging to $x \in X$ such that $\{f(x_\alpha)\}_{\alpha \in A}$ converges in $\overline{\mathbb{R}}$ we have $f(x) \leq \lim_{\alpha \in A} f(x_\alpha)$.

   When X is first countable, nets can be replaced by sequences in these conditions.

7. Let $(X, \mathcal{M}, \mu)$ be a positive measure space as in the Riesz–Markov theorem.

   (a) Let $0 \leq f \in L^1(\mu)$ and $\epsilon > 0$. Represent $f = \sum_{j=1}^{\infty} c_j I_{E_j}$ as in Exercise 15 in Chapter 1, and choose K_j compact and V_j open such that $K_j \subset E_j \subset V_j$ and $\mu(V_j - K_j) < \epsilon / c_j 2^{j+1}$. Fix n such that $\sum_{j>n} c_j \mu(E_j) < \epsilon/2$ and define $u = \sum_{j=1}^{n} c_j I_{K_j}$ and $v = \sum_{j=1}^{\infty} c_j I_{V_j}$. Prove that u is u.s.c., v is l.s.c., $u \leq f \leq v$, and $\int_X (v - u)\, d\mu < \epsilon$.

   (b) Generalize the conclusion in (a) to any *real* function $f \in L^1(\mu)$. (This is the *Vitali–Caratheodory theorem.*) (Hint: Exercise 6)

Fundamental theorem of calculus

8. Let $f : [a, b] \to \mathbb{R}$ be differentiable at every point of $[a, b]$, and suppose $f' \in L^1 := L^1([a, b])$ (with respect to Lebesgue measure dt). Denote $\int_a^b f'(t)\, dt = c$ and fix $\epsilon > 0$. By Exercise 7, there exists v l.s.c. such that $f' \leq v$ and $\int_a^b v\, dt < c + \epsilon$. Fix a constant $r > 0$ such that $r(b - a) < c + \epsilon - \int_a^b v\, dt$,

and let $g = v + r$. Observe that g is l.s.c., $g > f'$, and $\int_a^b g\, dt < c + \epsilon$. By the l.s.c. property of g and the differentiability of f, we may associate to each $x \in [a, b)$ a number $\delta(x)$ such that $g(t) > f'(x)$ *and* $f(t) - f(x) < (t - x)[f'(x) + \epsilon]$ for all $t \in (x, x + \delta(x))$.

Define

$$F(x) = \int_a^x g(t)\, dt - f(x) + f(a) + \epsilon(x - a).$$

(F is clearly continuous and $F(a) = 0$.)

(a) Show that $F(t) > F(x)$ for all $t \in (x, x + \delta(x))$.

(b) Conclude that $F(b) \geq 0$, and consequently $f(b) - f(a) < c + \epsilon(1 + b - a)$. Hence $f(b) - f(a) \leq c$.

(c) Conclude that $\int_a^b f'(t)\, dt = f(b) - f(a)$. (Hint: replace f by $-f$ in the conclusion of Part b.)

Approximation almost everywhere by continuous functions

9. Let $(X, \mathcal{M}, \mu)$ be a positive measure space as in the Riesz–Markov theorem. Let $f : X \to \mathbb{C}$ be a bounded measurable function vanishing outside some measurable set of finite measure. Prove that there exists a sequence $\{g_n\} \subset C_c(X)$ such that $\|g_n\|_u \leq \|f\|_u$ and $g_n \to f$ almost everywhere. (Hint: Lusin and Exercise 16 of Chapter 1.)

4

Continuous linear functionals

The general form of continuous linear functionals on Hilbert spaces was described in Theorem 1.37. In this chapter, we shall obtain the general form of continuous linear functionals on some of the normed spaces we have encountered.

The first section introduces the starting point of operator theory: *bounded linear operators* and their norms, and we introduce them in the first section. The *conjugate space* X^* of a normed space X is the normed space of all bounded, equivalently continuous, linear functionals on X.

We prove two major representation theorems. The first describes the *conjugate of the Lebesgue spaces*. It says that there exists a natural isometric isomorphism between $L^p(\mu)^*$ and $L^q(\mu)$ for a positive measure μ, $1 < p < \infty$ and q its conjugate exponent (or $p = 1$ and $q = \infty$ under an additional assumption). The second, called the *Riesz representation theorem*, says that for a locally compact Hausdorff space X, every element of $C_c(X)^*$ (equivalently: $C_0(X)^*$) is induced by a *complex* Borel measure in the same way that a positive measure induces a positive (but not necessarily bounded!) functional on $C_c(X)$ as in the Riesz–Markov theorem (3.18).

The final section presents another celebrated application of the Riesz–Markov theorem: the construction of the Haar measure. Every locally compact topological group G is shown to admit a (unique up to scaling) non-zero left translation invariant positive linear functional on $C_c(G)$, thus a non-zero left invariant positive measure on G. The Haar measure is a far-reaching generalization of the Lebesgue measure on $\mathbb{R}$ and is one of the cornerstones of abstract harmonic analysis. Chapter 5 presents an alternative proof of the existence and uniqueness of the Haar measure for *compact* topological groups.

Introduction to Modern Analysis. Second Edition. Shmuel Kantorovitz and Ami Viselter, Oxford University Press.
 DOI: 10.1093/oso/9780192849540.003.0004

4.1 Linear maps

We consider first some basic facts about arbitrary linear maps between normed spaces.

Definition 4.1. Let X, Y be normed spaces (over $\mathbb{C}$, to fix the ideas), and let $T : X \to Y$ be a linear map (it is customary to write Tx instead of $T(x)$, and norms are denoted by $\|\cdot\|$ in any normed space, unless some distinction is absolutely necessary). One says that T is *bounded* if

$$\|T\| := \sup_{x \neq 0} \frac{\|Tx\|}{\|x\|} < \infty.$$

Equivalently, T is bounded iff there exists $M \geq 0$ such that

$$\|Tx\| \leq M \|x\| \quad (x \in X), \tag{1}$$

and $\|T\|$ is the smallest constant M for which (1) is valid. In particular

$$\|Tx\| \leq \|T\| \|x\| \quad (x \in X). \tag{2}$$

The homogeneity of T shows that the following conditions are equivalent:

(a) T is "bounded";

(b) the map T is bounded (in the usual sense) on the "unit ball" $B_X := \{x \in X; \|x\| < 1\}$;

(c) the map T is bounded on the "closed unit ball" $\bar{B}_X := \{x \in X; \|x\| \leq 1\}$;

(d) the map T is bounded on the "unit sphere" $S_X := \{x \in X; \|x\| = 1\}$.

In addition, one has (for T bounded):

$$\|T\| = \sup_{x \in B_X} \|Tx\| = \sup_{x \in \bar{B}_X} \|Tx\| = \sup_{x \in S_X} \|Tx\|. \tag{3}$$

By (3), the set $B(X, Y)$ of all bounded linear maps from X to Y is a complex vector space for the pointwise operations, and $\|\cdot\|$ is a norm on $B(X, Y)$, called the *operator norm* or the *uniform norm*.

Theorem 4.2. *Let X, Y be normed spaces, and $T : X \to Y$ be linear. Then the following properties are equivalent:*

(i) $T \in B(X, Y)$;

(ii) T is uniformly continuous on X;

(iii) T is continuous at some point $x_0 \in X$.

Proof. Assume (i). Then for all $x, y \in X$,

$$\|Tx - Ty\| = \|T(x - y)\| \leq \|T\| \|x - y\|,$$

which clearly implies (ii) (actually, this is the stronger property: T is *Lipschitz* with *Lipschitz constant* $\|T\|$).

Trivially, (ii) implies (iii). Finally, if (iii) holds, there exists $\delta > 0$ such that

$$\|Tx - Tx_0\| < 1$$

whenever $\|x - x_0\| < \delta$.

By linearity of T, this is equivalent to: $\|Tz\| < 1$ whenever $z \in X$ and $\|z\| < \delta$. Since $\|\delta x\| < \delta$ for all $x \in B_X$, it follows that $\delta\|Tx\| = \|T(\delta x)\| < 1$, that is, $\|Tx\| < 1/\delta$ on B_X, hence $\|T\| \leq 1/\delta$. □

Notation 4.3. Let X be a (complex) normed space. Then

$$B(X) := B(X, X);$$

$$X^* := B(X, \mathbb{C}).$$

Elements of $B(X)$ will be called bounded *operators* on X; elements of X^* will be called bounded *linear functionals* on X, and will be denoted usually by $x^*, y^*, \ldots$.

Since the norm on $\mathbb{C}$ is the absolute value, the norm of $x^* \in X^*$ as defined in Definition 4.1 takes the form

$$\|x^*\| = \sup_{x \neq 0} |x^* x|/\|x\| = \sup_{x \in B_X} |x^* x|;$$

also, (2) takes the form

$$|x^* x| \leq \|x^*\|\|x\| \quad (x \in X).$$

The normed space X^* is called the (normed) dual or the *conjugate space* of X.

Theorem 4.4. *Let X, Y be normed spaces. If Y is complete, then $B(X, Y)$ is complete.*

Proof. Suppose Y is complete, and let $\{T_n\} \subset B(X, Y)$ be a Cauchy sequence. For each $x \in X$,

$$\|T_n x - T_m x\| = \|(T_n - T_m)x\| \leq \|T_n - T_m\|\|x\| \to 0$$

when $n, m \to \infty$, that is, $\{T_n x\}$ is Cauchy in Y. Since Y is complete, the limit $\lim_n T_n x$ exists in Y. We denote it by Tx. By the basic properties of limits, the map $T : X \to Y$ is linear. Given $\epsilon > 0$, there exists $n_0 \in \mathbb{N}$ such that

$$\|T_n - T_m\| < \epsilon, \quad n, m > n_0.$$

Therefore

$$\|T_n x - T_m x\| < \epsilon\|x\|, \quad n, m > n_0, \ x \in X.$$

Letting $m \to \infty$, we get by continuity of the norm

$$\|T_n x - Tx\| \leq \epsilon\|x\|, \quad n > n_0, \ x \in X.$$

In particular $T_n - T \in B(X, Y)$, and thus $T = T_n - (T_n - T) \in B(X, Y)$, and $\|T_n - T\| \leq \epsilon$ for all $n > n_0$. This shows that $B(X, Y)$ is complete. □

Since $\mathbb{C}$ is complete, we have the following:

Corollary 4.5. *The conjugate space of any normed space is complete.*

4.2 The conjugates of Lebesgue spaces

Theorem 4.6.

(i) Let $(X, \mathcal{A}, \mu)$ be a positive measure space. Let $1 < p < \infty$, let q be the conjugate exponent, and let $\phi \in L^p(\mu)^*$. Then there exists a unique *element $g \in L^q(\mu)$ such that*

$$\phi(f) = \int_X fg\, d\mu \quad (f \in L^p(\mu)). \tag{1}$$

Moreover, the map $\phi \to g$ is an isometric isomorphism of $L^p(\mu)^*$ and $L^q(\mu)$.

(ii) In case $p = 1$, the result is valid if the measure space is σ-finite.

Proof. *Uniqueness.* If g, g' are as in the theorem, and $h := g - g'$, then $h \in L^q(\mu)$ and

$$\int_X fh\, d\mu = 0 \quad (f \in L^p(\mu)). \tag{2}$$

The function $\theta : \mathbb{C} \to \mathbb{C}$ defined by

$$\theta(z) = |z|/z \quad \text{for } z \neq 0; \qquad \theta(0) = 0$$

is Borel, so that, *in case* $1 < p < \infty$, the function $f := |h|^{q-1}\theta(h)$ is measurable, and

$$\int_X |f|^p\, d\mu = \int_X |h|^{(q-1)p}\, d\mu = \int_X |h|^q\, d\mu < \infty.$$

Hence by (2) for this function f,

$$0 = \int_X |h|^{q-1}\theta(h) h\, d\mu = \int_X |h|^q\, d\mu,$$

and consequently h is the zero element of $L^q(\mu)$.

In case $p = 1$, take in (2) $f = I_E$, where $E \in \mathcal{A}$ and $0 < \mu(E) < \infty$ (so that $f \in L^1(\mu)$). Then $0 = (1/\mu(E)) \int_E h\, d\mu$ for all such E, and therefore $h = 0$ a.e. by the Averages lemma (Lemma 1.38) (since we assume that X is σ-finite in case $p = 1$).

Existence. Let $(X, \mathcal{A}, \mu)$ be an *arbitrary* positive measure space, and $1 \le p < \infty$. If $g \in L^q(\mu)$ and we *define* $\psi(f)$ by the right-hand side of (1), then Holder's inequality (Theorem 1.26) implies that ψ is a well-defined linear functional on $L^p(\mu)$, and

$$|\psi(f)| \le \|g\|_q \|f\|_p \quad (f \in L^p(\mu)),$$

so that $\psi \in L^p(\mu)^*$ and

$$\|\psi\| \le \|g\|_q. \tag{3}$$

In order to prove the *existence* of g as in the theorem, it suffices to prove the following:

Claim. *There exists a complex measurable function g such that*

$$\|g\|_q \leq \|\phi\| \tag{4}$$

and

$$\phi(I_E) = \int_E g \, d\mu \quad (E \in \mathcal{A}_0), \tag{5}$$

where $\mathcal{A}_0 := \{E \in \mathcal{A}; \mu(E) < \infty\}$.

Indeed, Relation (5) means that (1) is valid for $f = I_E$, for all $E \in \mathcal{A}_0$; by linearity of ϕ and ψ, (1) is then valid for all simple functions in $L^p(\mu)$. Since these functions are dense in $L^p(\mu)$ (Theorem 1.27), the conclusion $\phi = \psi$ follows from the continuity of both functionals on $L^p(\mu)$, and the relation $\|g\|_q = \|\phi\|$ follows then from (3) and (4).

Proof of the claim. *Case of a finite measure space* $(X, \mathcal{A}, \mu)$. In that case, $I_E \in L^p(\mu)$ for any $E \in \mathcal{A}$, and $\|I_E\|_p = \mu(E)^{1/p}$. Consider the trivially additive set function

$$\lambda(E) := \phi(I_E) \quad (E \in \mathcal{A}).$$

If $\{E_k\} \subset \mathcal{A}$ is a sequence of mutually disjoint sets with union E, put $A_n = \bigcup_{k=1}^n E_k$. Then

$$\|I_{A_n} - I_E\|_p = \|I_{E-A_n}\|_p = \mu(E - A_n)^{1/p} \to 0$$

as $n \to \infty$, since $\{E - A_n\}$ is a decreasing sequence of measurable sets with empty intersection (cf. Lemma 1.11). Since ϕ is continuous on $L^p(\mu)$, it follows that

$$\begin{aligned}\lambda(E) &:= \phi(I_E) = \lim_n \phi(I_{A_n}) = \lim_n \lambda(A_n) \\ &= \lim_n \sum_{k=1}^n \lambda(E_k) = \sum_{k=1}^\infty \lambda(E_k),\end{aligned}$$

so that λ is a complex measure.

If $\mu(E) = 0$ for some $E \in \mathcal{A}$, then $\|I_E\|_p = 0$, and therefore $\lambda(E) := \phi(I_E) = 0$ by linearity of ϕ. This means that $\lambda \ll \mu$, and therefore, by the Radon–Nikodym theorem, there exists $g \in L^1(\mu)$ such that

$$\phi(I_E) = \int_E g \, d\mu = \int_X I_E g \, d\mu \quad (E \in \mathcal{A}).$$

Thus (5) is valid, with g *integrable.* We show that this *modified* version of (5) implies (4) (hence the claim).

By linearity of ϕ and the integral, it follows from (5) (modified version) that (1) is valid for all simple measurable functions f. If f is a *bounded measurable function,* there exists a sequence of simple measurable functions s_n such that $\|s_n - f\|_u \to 0$ (cf. Theorem 1.8). Then

$$\|s_n - f\|_p \leq \|s_n - f\|_u \mu(X)^{1/p} \to 0,$$

and therefore, by continuity of ϕ,

$$\phi(f) = \lim_n \phi(s_n) = \lim_n \int_X s_n g \, d\mu.$$

Also

$$\left| \int_X s_n g \, d\mu - \int_X f g \, d\mu \right| \leq \|s_n - f\|_u \|g\|_1 \to 0,$$

and we conclude that (1) is valid for all *bounded* measurable functions f.

Case $p = 1$. For any $E \in \mathcal{A}$ with $\mu(E) > 0$,

$$\left| \frac{1}{\mu(E)} \int_E g \, d\mu \right| = \frac{|\phi(I_E)|}{\mu(E)} \leq \|\phi\| \frac{\|I_E\|_1}{\mu(E)} = \|\phi\|.$$

Therefore, $|g| \leq \|\phi\|$ a.e. (by the averages Lemma), that is,

$$\|g\|_\infty \leq \|\phi\|,$$

as desired.

Case $1 < p < \infty$. Let $E_n := [|g| \leq n]$ $(n = 1, 2, \ldots)$. Define $f_n := I_{E_n} |g|^{q-1} \theta(g)$, with θ as in the beginning of the proof. Then f_n are *bounded* measurable functions, so that by (1) for such functions,

$$\int_{E_n} |g|^q \, d\mu = \int_X f_n g \, d\mu = \phi(f_n) = |\phi(f_n)| \leq \|\phi\| \|f_n\|_p.$$

However, since $|f_n|^p = I_{E_n} |g|^{(q-1)p} = I_{E_n} |g|^q$, it follows that

$$\|f_n\|_p^p = \int_{E_n} |g|^q \, d\mu.$$

Therefore

$$\left(\int_{E_n} |g|^q \, d\mu \right)^{1-1/p} \leq \|\phi\|,$$

that is,

$$\left(\int_X I_{E_n} |g|^q \, d\mu \right)^{1/q} \leq \|\phi\|.$$

Since $0 \leq I_{E_1} |g|^q \leq I_{E_2} |g|^q \leq \cdots$ and $\lim_n I_{E_n} |g|^q = |g|^q$, the monotone convergence theorem implies that $\|g\|_q \leq \|\phi\|$, as wanted.

Case of a σ-finite measure space; $1 \leq p < \infty$. We use the function w and the *equivalent finite measure* $d\nu = w \, d\mu$ (satisfying $\nu(X) = 1$), as defined in the proof of Theorem 1.40. Define

$$V_p : L^p(\nu) \to L^p(\mu)$$

by

$$V_p f = w^{1/p} f.$$

Then

$$\|V_p f\|^p_{L^p(\mu)} = \int_X |f|^p w\,d\mu = \int_X |f|^p\,d\nu = \|f\|^p_{L^p(\nu)},$$

so that V_p is a linear isometry of $L^p(\nu)$ onto $L^p(\mu)$. Consequently, $\phi \circ V_p \in L^p(\nu)^*$, and $\|\phi \circ V_p\| = \|\phi\|$ (where the norms are those of the respective dual spaces). Since ν is a *finite* measure, there exists (by the preceding case) a measurable function g_1 such that

$$\|g_1\|_{L^q(\nu)} \le \|\phi \circ V_p\| = \|\phi\|, \tag{6}$$

and

$$(\phi \circ V_p)(f) = \int_X f g_1\,d\nu \quad (f \in L^p(\nu)). \tag{7}$$

Thus, for all $E \in \mathcal{A}_0$,

$$\phi(I_E) = (\phi \circ V_p)(w^{-1/p} I_E) = \int_X w^{-1/p} I_E g_1\,d\nu = \int_E w^{1/q} g_1\,d\mu. \tag{8}$$

In case $p > 1$ (so that $1 < q < \infty$), set $g = w^{1/q} g_1 (= V_q g_1)$. Then (5) is valid, and by (6),

$$\|g\|_{L^q(\mu)} = \|g_1\|_{L^q(\nu)} \le \|\phi\|,$$

as desired.

In case $p = 1$ (so that $q = \infty$), we have by (8) $\phi(I_E) = \int_E g_1\,d\mu$. Thus (5) is valid with $g = g_1$, and since the measures μ and ν are equivalent, we have by (6)

$$\|g\|_{L^\infty(\mu)} = \|g_1\|_{L^\infty(\nu)} \le \|\phi\|,$$

as wanted.

Case of an arbitrary measure space; $1 < p < \infty$. For each $E \in \mathcal{A}_0$, consider the finite measure space $(E, \mathcal{A} \cap E, \mu)$, and let $L^p(E)$ be the corresponding L^p-space. We can identify $L^p(E)$ (isomorphically and isometrically) with the subspace of $L^p(\mu)$ of all elements vanishing on E^c, and therefore the restriction $\phi_E := \phi|_{L^p(E)}$ belongs to $L^p(E)^*$ and $\|\phi_E\| \le \|\phi\|$. By the finite measure case, there exists $g_E \in L^q(E)$ such that

$$\|g_E\|_{L^q(E)} = \|\phi_E\| (\le \|\phi\|)$$

and

$$\phi_E(f) = \int_E f g_E\,d\mu \quad \text{for all } f \in L^p(E).$$

If $E, F \in \mathcal{A}_0$, then for all measurable subsets G of $E \cap F$, $I_G \in L^p(E \cap F) \subset L^p(E)$, so that $\phi_E(I_G) = \phi_{E\cap F}(I_G)$, and therefore

$$\begin{aligned}\int_G (g_E - g_{E\cap F})\,d\mu &= \int_E I_G g_E\,d\mu - \int_{E\cap F} I_G g_{E\cap F}\,d\mu \\ &= \phi_E(I_G) - \phi_{E\cap F}(I_G) = 0.\end{aligned}$$

By Proposition 1.22 (applied to the *finite* measure space $(E\cap F, \mathcal{A}\cap(E\cap F), \mu)$) and by symmetry, $g_E = g_{E\cap F} = g_F$ a.e. on $E\cap F$. It follows that for any mutually disjoint sets $E, F \in \mathcal{A}_0$, $g_{E\cup F}$ coincides a.e. with g_E on E and with g_F on F, and therefore

$$\begin{aligned}\|g_{E\cup F}\|_q^q &= \int_{E\cup F} |g_{E\cup F}|^q \, d\mu \\ &= \int_E |g_E|^q \, d\mu + \int_F |g_F|^q \, d\mu = \|g_E\|_q^q + \|g_F\|_q^q,\end{aligned}$$

that is, $\|g_E\|_q^q$ is an additive function of E on $\mathcal{A}_0$. Let

$$K := \sup_{E\in\mathcal{A}_0} \|g_E\|_q \,(\leq \|\phi\|),$$

and let then $\{E_n\}$ be a non-decreasing sequence in $\mathcal{A}_0$ such that $\|\phi_{E_n}\| \to K$. Set $F := \bigcup_n E_n$.

If $E \in \mathcal{A}_0$ and $E\cap F = \emptyset$, then since E and E_n are disjoint for all n, it follows from the additivity of the set function $\|g_E\|_q^q$ that

$$\|g_E\|_q^q = \|g_{E\cup E_n}\|_q^q - \|g_{E_n}\|_q^q \leq K^q - \|\phi_{E_n}\|^q \to 0.$$

Hence, $\|g_E\|_q = 0$ for all $E \in \mathcal{A}_0$ disjoint from F, that is, $g_E = 0$ a.e. for such E. Consequently, for $E \in \mathcal{A}_0$ *arbitrary*, we have a.e. on $E - F$ $g_E = g_{E-F} = 0$, and therefore

$$\phi(I_E) = \phi_E(I_E) = \int_E g_E \, d\mu = \int_{E\cap F} g_E \, d\mu = \int_{E\cap F} g_{E\cap F} \, d\mu. \tag{9}$$

Since $g_{E_n} = g_{E_{n+1}}$ a.e. on E_n, the limit $g := \lim_n g_{E_n}$ exists a.e. and vanishes on F^c; it is measurable, and by the monotone convergence theorem,

$$\|g\|_{L^q(\mu)} = \lim_n \|g_{E_n}\|_{L^q(\mu)} = \lim_n \|\phi_{E_n}\| = K \leq \|\phi\|,$$

and (4) is verified. Fix n. For all $k \geq n$, $g_{E\cap F} = g_{E_k}$ a.e. on $(E\cap F)\cap E_k = E\cap E_k$, hence (a.e.) on $E \cap E_n$. Therefore, $g_{E\cap F} = g$ a.e. on $E \cap E_n$ for all n, hence (a.e.) on $E\cap F$, and consequently (5) follows from (9). This completes the proof of the claim. □

4.3 The conjugate of $C_c(X)$

Let X be a locally compact Hausdorff space, and consider the normed space $C_c(X)$ with the uniform norm

$$\|f\| = \|f\|_u := \sup_X |f| \quad (f \in C_c(X)).$$

If μ is a complex Borel measure on X, write $d\mu = h\, d|\mu|$, where $|\mu|$ is the total variation measure corresponding to μ and h is a uniquely determined Borel function with $|h| = 1$ (cf. Theorem 1.46). Set

$$\psi(f) := \int_X f \, d\mu := \int_X fh \, d|\mu| \quad (f \in C_c(X)). \tag{1}$$

Then

$$|\psi(f)| \leq \int_X |f| d|\mu| \leq |\mu|(X)\|f\|, \tag{2}$$

so that ψ is a well-defined, clearly linear, continuous functional on $C_c(X)$, with norm

$$\|\psi\| \leq \|\mu\| := |\mu|(X). \tag{3}$$

We shall prove that *every* continuous linear functional ϕ on $C_c(X)$ is of this form for a uniquely determined *regular* complex Borel measure μ, and $\|\phi\| = \|\mu\|$. This is done by using Riesz–Markov representation theorem 3.18 for *positive* linear functionals on $C_c(X)$. Our first step is to associate a positive linear functional $|\phi|$ to each given $\phi \in C_c(X)^*$.

Definition 4.7. Let $\phi \in C_c(X)^*$. The *total variation functional* $|\phi|$ is defined by

$$|\phi|(f) := \sup\{|\phi(h)|; h \in C_c(X), |h| \leq f\} \quad (0 \leq f \in C_c(X));$$

$$|\phi|(u + \mathrm{i}v) = |\phi|(u^+) - |\phi|(u^-) + \mathrm{i}|\phi|(v^+) - \mathrm{i}|\phi|(v^-) \quad (u, v \in C_c^R(X)).$$

Theorem 4.8. *The total variation functional $|\phi|$ of $\phi \in C_c(X)^*$ is a positive linear functional on $C_c(X)$, and it satisfies the inequality*

$$|\phi(f)| \leq |\phi|(|f|) \leq \|\phi\|\|f\| \quad (f \in C_c(X)).$$

Proof. Let $C_c^+(X) := \{f \in C_c(X); f \geq 0\}$. It is clear from Definition 4.7 that

$$0 \leq |\phi|(f) \leq \|\phi\|\|f\| < \infty, \tag{4}$$

$|\phi|$ is monotonic on $C_c^+(X)$ and $|\phi|(cf) = c|\phi|(f)$ (and in particular $|\phi|(0) = 0$) for all $c \geq 0$ and $f \in C_c^+(X)$. We show that $|\phi|$ is additive on $C_c^+(X)$.

Let $\epsilon > 0$ and $f_k \in C_c^+(X)$ be given ($k = 1, 2$). By definition, there exist $h_k \in C_c(X)$ such that $|h_k| \leq f_k$ and $|\phi|(f_k) \leq |\phi(h_k)| + \epsilon/2, k = 1, 2$. Therefore, writing the complex numbers $\phi(h_k)$ in polar form, we obtain

$$\begin{aligned}
0 &\leq |\phi|(f_1) + |\phi|(f_2) \leq |\phi(h_1)| + |\phi(h_2)| + \epsilon \\
&= \mathrm{e}^{-\mathrm{i}\theta_1}\phi(h_1) + \mathrm{e}^{-\mathrm{i}\theta_2}\phi(h_2) + \epsilon = \phi(\mathrm{e}^{-\mathrm{i}\theta_1}h_1 + \mathrm{e}^{-\mathrm{i}\theta_2}h_2) + \epsilon \\
&\leq |\phi|(f_1 + f_2) + \epsilon,
\end{aligned}$$

because

$$|\mathrm{e}^{-\mathrm{i}\theta_1}h_1 + \mathrm{e}^{-\mathrm{i}\theta_2}h_2| \leq |h_1| + |h_2| \leq f_1 + f_2.$$

Hence $|\phi|$ is "super-additive" on $C_c^+(X)$.

Next, let $h \in C_c(X)$ satisfy $|h| \leq f_1 + f_2 := f$. Let $V = [f > 0]$. Define for $k = 1, 2$

$$h_k = (f_k/f)h \quad \text{on } V; \qquad h_k = 0 \quad \text{on } V^c.$$

The functions h_k are continuous on V and V^c. If x is a boundary point of V, then $x \notin V$ (since V is open), so that $f(x) = 0$ and $h_k(x) = 0$. Let $\{x_\alpha\} \subset V$ be a net converging to x. Then by continuity of h, we have for $k = 1, 2$:

$$|h_k(x_\alpha)| \leq |h(x_\alpha)| \to |h(x)| \leq |f(x)| = 0,$$

so that $\lim_\alpha h_k(x_\alpha) = 0 = h_k(x)$. This shows that h_k are continuous on X. Trivially, $\operatorname{supp} h_k \subset \operatorname{supp} f_k$, so that $h_k \in C_c(X)$, and by definition, $|h_k| \le f_k$ and $h = h_1 + h_2$. Therefore

$$|\phi(h)| = |\phi(h_1) + \phi(h_2)| \le |\phi|(f_1) + |\phi|(f_2).$$

Taking the supremum over all $h \in C_c(X)$ such that $|h| \le f$, we obtain that $|\phi|$ is subadditive. Together with the super-additivity obtained before, this proves that $|\phi|$ is additive.

Next, consider $|\phi|$ over $C_c^R(X)$. The homogeneity over $\mathbb{R}$ is easily verified. Additivity is proved as in Theorem 1.19. Let $f = f^+ - f^-$ and $g = g^+ - g^-$ be functions in $C_c^R(X)$, and let $h = h^+ - h^- := f + g = f^+ - f^- + g^+ - g^-$. Then $h^+ + f^- + g^- = f^+ + g^+ + h^-$, so that by the additivity of $|\phi|$ on $C_c^+(X)$, we obtain

$$|\phi|(h^+) + |\phi|(f^-) + |\phi|(g^-) = |\phi|(f^+) + |\phi|(g^+) + |\phi|(h^-),$$

and since all summands are finite, it follows that

$$|\phi|(h) := |\phi|(h^+) - |\phi|(h^-) = |\phi|(f^+) - |\phi|(f^-) + |\phi|(g^+) - |\phi|(g^-) := |\phi|(f) + |\phi|(g).$$

The linearity of $|\phi|$ over $C_c(X)$ now follows easily from the definition. Thus $|\phi|$ is a positive linear functional on $C_c(X)$ (cf. (4)). By (4) for the function $|f|$, $|\phi|(|f|) \le \|\phi\| \|f\|$. Also, since $h = f$ belongs to the set of functions used in the definition of $|\phi|(|f|)$, we have $|\phi(f)| \le |\phi|(|f|)$. □

4.4 The Riesz representation theorem

Theorem 4.9. *Let X be a locally compact Hausdorff space, and let $\phi \in C_c(X)^*$. Then there exists a unique regular complex Borel measure μ on X such that*

$$\phi(f) = \int_X f \, d\mu \quad (f \in C_c(X)). \tag{1}$$

Furthermore,

$$\|\phi\| = \|\mu\|. \tag{2}$$

"Regularity" of the *complex* measure μ means *by definition* that its total variation measure $|\mu|$ is regular.

Proof. We apply Theorem 3.18 to the positive linear functional $|\phi|$. Denote by λ the positive Borel measure obtained by restricting the measure associated with $|\phi|$ (by Theorem 3.18) to the Borel algebra $\mathcal{B}(X) \subset \mathcal{M}$. Then

$$|\phi|(f) = \int_X f \, d\lambda \quad (f \in C_c(X)). \tag{3}$$

By Definition 3.5 and Theorem 4.8,

$$\lambda(X) = \sup\{|\phi|(f); 0 \le f \le 1\} \le \|\phi\|. \tag{4}$$

In particular, every Borel set in X has finite λ-measure, and therefore, by Theorem 3.18 (cf. (3) and (4)(ii)), λ is *regular*.

By Theorem 4.8 and (3), for all $f \in C_c(X)$,

$$|\phi(f)| \le |\phi|(|f|) = \int_X |f|\, d\lambda = \|f\|_{L^1(\lambda)}.$$

This shows that ϕ is a continuous linear functional on the subspace $C_c(X)$ of $L^1(\lambda)$, with norm ≤ 1. By Theorem 3.21, $C_c(X)$ is dense in $L^1(\lambda)$, and it follows that ϕ has a unique extension as an element of $L^1(\lambda)^*$ with norm ≤ 1. By Theorem 4.6, there exists a unique element $g \in L^\infty(\lambda)$ such that

$$\phi(f) = \int_X fg\, d\lambda \quad (f \in C_c(X)) \tag{5}$$

and $\|g\|_\infty \le 1$.

Define $d\mu = g\, d\lambda$. Then μ is a complex Borel measure satisfying (1). By Theorem 1.47 and (4),

$$|\mu|(X) = \int_X |g|\, d\lambda \le \lambda(X) \le \|\phi\|$$

By (3) of Section 4.3, the reversed inequality is a consequence of (1), so that (2) follows.

Gathering some of these inequalities, we have

$$\|\phi\| = |\mu|(X) = \int_X |g|\, d\lambda \le \lambda(X) \le \|\phi\|.$$

Thus, $\lambda(X) = \int_X |g|\, d\lambda$, that is, $\int_X (1 - |g|)\, d\lambda = 0$. Since $1 - |g| \ge 0$ λ-a.e., it follows that $|g| = 1$ a.e., and since g is only a.e.-determined, we may choose g such that $|g| = 1$ identically on X.

For all Borel sets E, $|\mu|(E) = \int_E |g|\, d\lambda = \lambda(E)$, which proves that $|\mu| = \lambda$. In particular, μ is regular.

In order to prove uniqueness, we observe that the sum ν of two finite positive regular Borel measures ν_k is regular. Indeed, given $\epsilon > 0$ and $E \in \mathcal{B}(X)$, there exist K_k compact and V_k open such that $K_k \subset E \subset V_k$ and

$$\nu_k(V_k) - \epsilon/2 \le \nu_k(E) \le \nu_k(K_k) + \epsilon/2.$$

Then $K := K_1 \cup K_2 \subset E \subset V := V_1 \cap V_2$, K is compact, V is open, and by monotonicity of positive measures,

$$\nu(V) - \epsilon \le \nu_1(V_1) + \nu_2(V_2) - \epsilon \le \nu(E) \le \nu_1(K_1) + \nu_2(K_2) + \epsilon \le \nu(K) + \epsilon.$$

Suppose now that the representation (1) is valid for the regular complex measures μ_1 and μ_2. Then $\int_X f\, d\mu = 0$ for all $f \in C_c(X)$, for $\mu = \mu_1 - \mu_2$. We must show that $\|\mu\| = 0$ (i.e., $\mu = 0$). Since $|\mu_k|$ are finite positive regular Borel measures, the positive Borel measure $\nu := |\mu_1| + |\mu_2|$ is regular. Write $d\mu = h\, d|\mu|$, where h

is a Borel function with $|h| = 1$ (cf. Theorem 1.46). Since ν is regular, it follows from Theorem 3.21 that there exists a sequence $\{f_n\} \subset C_c(X)$ that converges to $\bar{h}$ in the $L^1(\nu)$-metric. Since $\bar{h}h = 1$, $|\mu| = |\mu_1 - \mu_2| \leq |\mu_1| + |\mu_2| := \nu$, and $\int_X f_n h \, d|\mu| = \int_X f_n \, d\mu = 0$, we obtain

$$\|\mu\| := |\mu|(X) = \left| \int_X f_n h \, d|\mu| - \int_X \bar{h}h \, d|\mu| \right| = \left| \int_X (f_n - \bar{h}) h \, d|\mu| \right|$$
$$\leq \int_X |f_n - \bar{h}| d|\mu| \leq \int_X |f_n - \bar{h}| d\nu = \|f_n - \bar{h}\|_{L^1(\nu)} \to 0$$

as $n \to \infty$. Hence $\|\mu\| = 0$. □

Remark 4.10. If $S = \operatorname{supp}|\mu|$ (cf. Definition 3.26), we have

$$\|\phi\| = \|\mu\| := |\mu|(X) = |\mu|(S)$$

and

$$\int_X f \, d\mu = \int_S f \, d\mu \quad (f \in L^1(|\mu|)).$$

The second formula follows from Theorem 1.46 and Definition 3.26(2). Indeed, write $d\mu = h \, d|\mu|$ where h is a Borel measurable function with $|h| = 1$ on X (cf. Theorem 1.46). Then for all $f \in L^1(|\mu|)$, we have (cf. Definition 3.26(2))

$$\int_X f \, d\mu := \int_X fh \, d|\mu| = \int_S fh \, d|\mu| := \int_S f \, d\mu.$$

4.5 Haar measure

As an application of the Riesz–Markov representation theorem for positive linear functionals (Theorem 3.18), we shall construct a (left) translation-invariant positive measure on any locally compact topological group.

A *topological group* is a group G with a Hausdorff topology for which the group operations (multiplication and inverse) are continuous. It follows that for each fixed $a \in G$, the *left* (*right*) *translation* $x \to ax (x \to xa)$ is a homeomorphism of G onto itself. For any open neighborhood V of the identity e, the set aV (Va) is an open neighborhood of a.

Suppose G is locally compact, and $f, g \in C_c^+ := C_c^+(G) := \{f \in C_c(G); f \geq 0$, f not identically zero$\}$. Fix $0 < \alpha < \|g\| := \|g\|_u$. There exists $a \in G$ such that $g(a) > \alpha$, and therefore there exists an open neighborhood of e, V, such that $g(x) \geq \alpha$ for all $x \in aV$. By compactness of supp f, there exist $x_1, \ldots, x_n \in G$ such that supp $f \subset \bigcup_{k=1}^n x_k V$. Set $s_k := a x_k^{-1}$. Then for $x \in x_k V$, $s_k x \in aV$, and therefore $g(s_k x) \geq \alpha$. If $x \in$ supp f, there exists $k \in \{1, \ldots, n\}$ such that $x \in x_k V$, so that (for this k)

$$f(x) \leq \|f\| \leq \frac{\|f\|}{\alpha} g(s_k x) \leq \sum_{i=1}^n c_i g(s_i x), \tag{1}$$

where $c_i = \|f\|/\alpha$ for all $i = 1, \ldots, n$. Since (1) is trivial on $(\operatorname{supp} f)^c$, we see that there exist $n \in \mathbb{N}$ and $(c_1, \ldots, c_n, s_1, \ldots, s_n) \in (\mathbb{R}^+)^n \times G^n$ such that

$$f(x) \leq \sum_{i=1}^{n} c_i g(s_i x) \quad (x \in G). \tag{2}$$

Denote by $\Omega(f : g)$ the non-empty set of such rows (with n varying) and let

$$(f : g) = \inf \sum_{i=1}^{n} c_i, \tag{*}$$

where the infimum is taken over all $(c_1, \ldots, c_n)$ such that $(c_1, \ldots, c_n, s_1, \ldots, s_n) \in \Omega(f : g)$ for some n and s_i.

We verify some elementary properties of the functional $(f : g)$ for g fixed as above.

Let $f_s(x) := f(sx)$ for $s \in G$ fixed (f_s is the so-called *left s-translate of f*). If $(c_1, \ldots, c_n, s_1, \ldots, s_n) \in \Omega(f : g)$, then $f_s(x) = f(sx) \leq \sum_i c_i g(s_i s x)$ for all $x \in G$, hence $(c_1, \ldots, c_n, s_1 s, \ldots, s_n s) \in \Omega(f_s : g)$, and consequently $(f_s : g) \leq \sum_{i=1}^{n} c_i$. Taking the infimum over all such rows, we get $(f_s : g) \leq (f : g)$. But then $(f : g) = ((f_s)_{s^{-1}} : g) \leq (f_s : g)$, and we conclude that

$$(f_s : g) = (f : g) \tag{3}$$

for all $s \in G$ (i.e., the functional $(\cdot : g)$ is *left translation invariant*).

In the following arguments, ϵ denotes an arbitrary positive number.

If $c > 0$ and $(c_1, \ldots, c_n, s_1, \ldots, s_n) \in \Omega(f : g)$ is such that $\sum c_i < (f : g) + \epsilon$, then $(cf)(x) \leq \sum_i c c_i g(s_i x)$ for all $x \in G$, and therefore $(cf : g) \leq \sum_i c c_i < c(f : g) + c\epsilon$. The arbitrariness of ϵ implies that $(cf : g) \leq c(f : g)$. Applying this inequality to the function cf and the constant $1/c$ (instead of f and c, respectively), we obtain the reversed inequality. Hence

$$(cf : g) = c(f : g) \quad (c > 0). \tag{4}$$

Let $x_0 \in G$ be such that $f(x_0) = \max_G f$ (since f is continuous with compact support, such a point x_0 exists). Then for any $(c_1, \ldots, c_n, s_1, \ldots, s_n) \in \Omega(f : g)$,

$$\|f\| = f(x_0) \leq \sum_i c_i g(s_i x_0) \leq \|g\| \sum_i c_i.$$

Hence

$$\frac{\|f\|}{\|g\|} \leq (f : g). \tag{5}$$

Next, consider three functions $f_1, f_2, g \in C_c^+$. If $f_1 \leq f_2$, one has trivially $\Omega(f_2 : g) \subset \Omega(f_1 : g)$, and therefore

$$f_1 \leq f_2 \quad \text{implies} \quad (f_1 : g) \leq (f_2 : g). \tag{6}$$

There exist $(c_1, \ldots, c_n, s_1, \ldots, s_n) \in \Omega(f_1 : g)$ and $(d_1, \ldots, d_m, t_1, \ldots, t_m) \in \Omega(f_2 : g)$ such that $\sum c_i < (f_1 : g) + \epsilon/2$ and $\sum d_j < (f_2 : g) + \epsilon/2$. Then for all $x \in G$,

$$f_1(x) + f_2(x) \le \sum_i c_i g(s_i x) + \sum_j d_j g(t_j x) = \sum_{k=1}^{n+m} c'_k g(s'_k x),$$

where $c'_k = c_k, s'_k = s_k$ for $k = 1, \ldots, n$, and $c'_k = d_{k-n}, s'_k = t_{k-n}$ for $k = n+1, \ldots, n+m$. Thus $(c'_1, \ldots, c'_{n+m}, s'_1, \ldots, s'_{n+m}) \in \Omega(f_1 + f_2 : g)$, and therefore

$$(f_1 + f_2 : g) \le \sum_{k=1}^{n+m} c'_k = \sum_{k=1}^{n} c_k + \sum_{j=1}^{m} d_j < (f_1 : g) + (f_2 : g) + \epsilon.$$

This proves that

$$(f_1 + f_2 : g) \le (f_1 : g) + (f_2 : g). \tag{7}$$

Let $(c_1, \ldots, c_n, s_1, \ldots, s_n) \in \Omega(f : g)$ and $(d_1, \ldots, d_m, t_1, \ldots, t_m) \in \Omega(g : h)$, where $f, g, h \in C_c^+$. Then for all $x \in G$,

$$f(x) \le \sum_i c_i g(s_i x) \le \sum_i c_i \sum_j d_j h(t_j s_i x) = \sum_{i,j} c_i d_j h(t_j s_i x),$$

that is, $(c_i d_j, t_j s_i)_{i=1,\ldots,n; j=1,\ldots,m} \in \Omega(f : h)$, and consequently

$$(f : h) \le \sum_{i,j} c_i d_j = \left(\sum c_i\right)\left(\sum d_j\right).$$

Taking the infimum of the right-hand side over all the rows involved, we conclude that

$$(f : h) \le (f : g)(g : h). \tag{8}$$

With g fixed, denote

$$\Lambda_h f := \frac{(f : h)}{(g : h)}. \tag{9}$$

Since Λ_h is a constant multiple of $(f : h)$, the function $f \to \Lambda_h f$ satisfies (3), (4), (6), and (7).

By (8), $\Lambda_h f \le (f : g)$. Also

$$(g : f)\Lambda_h f = (g : f)\frac{(f : h)}{(g : h)} \ge \frac{(g : h)}{(g : h)} = 1.$$

Hence

$$\frac{1}{(g : f)} \le \Lambda_h f \le (f : g). \tag{10}$$

By (10), Λ_h is a point in the compact Hausdorff space

$$\Delta := \prod_{f \in C_c^+} \left[\frac{1}{(g : f)}, (f : g)\right]$$

(cf. Tychonoff's theorem). Consider the system $\mathcal{V}$ of all (open) neighborhoods of the identity. For each $V \in \mathcal{V}$, let Σ_V be the closure in Δ of the set $\{\Lambda_h; h \in C_V^+\}$, where $C_V^+ := C_V^+(G)$ consists of all $h \in C_c^+$ with support in V. Then Σ_V is a non-empty compact subset of Δ. If $V_1, \ldots, V_n \in \mathcal{V}$ and $V := \bigcap_{i=1}^n V_i$, then $C_V^+ \subset \bigcap_i C_{V_i}^+$, and therefore, $\Sigma_V \subset \bigcap_i \Sigma_{V_i}$. In particular, the family of compact sets $\{\Sigma_V; V \in \mathcal{V}\}$ has the finite intersection property, and consequently

$$\bigcap_{V \in \mathcal{V}} \Sigma_V \neq \emptyset.$$

Let Λ be any point in this intersection, and extend the functional Λ to $C_c := C_c(G)$ in the obvious way ($\Lambda 0 = 0$; $\Lambda f = \Lambda f^+ - \Lambda f^-$ for real $f \in C_c$, and $\Lambda(u + iv) = \Lambda u + i\Lambda v$ for real $u, v \in C_c$).

Theorem 4.11. *Λ is a non-zero left translation invariant positive linear functional on C_c.*

Proof. Since $\Lambda \in \Delta$, we have

$$\Lambda f \in \left[\frac{1}{(g:f)}, (f:g)\right]$$

for all $f \in C_c^+$, so that in particular $\Lambda f > 0$ for such f, and Λ is not identically zero.

For any $V \in \mathcal{V}$, we have $\Lambda \in \Sigma_V$; hence every basic neighborhood N of Λ in Δ meets the set $\{\Lambda_h; h \in C_V^+\}$. Recall that

$$N = N(\Lambda; f_1, \ldots, f_n; \epsilon) := \{\Phi \in \Delta; |\Phi f_i - \Lambda f_i| < \epsilon; i = 1, \ldots, n\},$$

where $f_i \in C_c^+$. Thus, for any $V \in \mathcal{V}$ and $f_1, \ldots, f_n \in C_c^+$, there exists $h \in C_V^+$ such that

$$|\Lambda_h f_i - \Lambda f_i| < \epsilon \quad (i = 1, \ldots, n). \tag{11}$$

Given $f \in C_c^+$ and $c > 0$, apply (11) with $f_1 = f$ and $f_2 = cf$. By Property (4) for Λ_h, we have

$$|\Lambda(cf) - c\Lambda f| \leq |\Lambda(cf) - \Lambda_h(cf)| + c|\Lambda_h f - \Lambda f| < (1 + c)\epsilon,$$

so that $\Lambda(cf) = c\Lambda f$ by the arbitrariness of ϵ.

A similar argument (using Relation (3) for Λ_h) shows that $\Lambda f_s = \Lambda f$ for all $f \in C_c^+$ and $s \in G$.

In order to prove the additivity of Λ on C_c^+, we use the following:

Lemma. *Let $f_1, f_2 \in C_c^+(G)$ and $\epsilon > 0$. Then there exists $V \in \mathcal{V}$ such that*

$$\Lambda_h f_1 + \Lambda_h f_2 \leq \Lambda_h(f_1 + f_2) + \epsilon$$

for all $h \in C_V^+(G)$.

Proof of lemma. Let $f = f_1 + f_2$, and fix $k \in C_c^+(G)$ such that $k = 1$ on $\{x \in G; f(x) > 0\}$. For g fixed as shown, let

$$\delta := \frac{\epsilon}{4(k:g)}; \qquad \eta := \min\left\{\frac{\epsilon}{4(f:g)}, 1/2\right\}.$$

Thus

$$2\eta(f:g) \leq \epsilon/2; \quad 2\eta \leq 1; \quad 2\delta(k:g) \leq \epsilon/2. \tag{12}$$

For $i = 1, 2$, let $h_i := f_i/F$, where $F := f + \delta k$ ($h_i = 0$ at points where $F = 0$). The functions h_i are well defined, and continuous with compact support; it follows that there exists $V \in \mathcal{V}$ such that

$$|h_i(x) - h_i(y)| < \eta \quad (i = 1, 2)$$

for all $x, y \in G$ such that $y^{-1}x \in V$ (uniform continuity of h_i!).

Let $h \in C_V^+(G)$. Let $(c_1, \ldots, c_n, s_1, \ldots, s_n) \in \Omega(F : h)$ and $x \in G$. If $j \in \{1, \ldots, n\}$ is such that $h(s_j x) \neq 0$, then $s_j x \in V$, and therefore $|h_i(x) - h_i(s_j^{-1})| < \eta$ for $i = 1, 2$. Hence

$$h_i(x) \leq \left|h_i(x) - h_i\left(s_j^{-1}\right)\right| + h_i\left(s_j^{-1}\right) < h_i\left(s_j^{-1}\right) + \eta.$$

Therefore, for $i = 1, 2$,

$$\begin{aligned} f_i(x) = F(x)h_i(x) &\leq \sum_{\{j; h(s_j x) \neq 0\}} c_j h(s_j x) h_i(x) \\ &\leq \sum_{\{j; h(s_j x) \neq 0\}} c_j \left[h_i\left(s_j^{-1}\right) + \eta\right] h(s_j x) \leq \sum_{j=1}^n c_j^i h(s_j x), \end{aligned}$$

where $c_j^i := c_j\left[h_i(s_j^{-1}) + \eta\right]$. Hence $(f_i : h) \leq \sum_j c_j^i$, and since $h_1 + h_2 = f/F \leq 1$, we obtain

$$(f_1 : h) + (f_2 : h) \leq \sum_j c_j(1 + 2\eta).$$

Taking the infimum of the right-hand side over all rows in $\Omega(F : h)$, we conclude that

$$\begin{aligned} (f_1 : h) + (f_2 : h) \leq (F : h)(1 + 2\eta) &\leq [(f : h) + \delta(k : h)](1 + 2\eta) \quad \text{by (7) and (4)} \\ &= (f : h) + 2\eta(f : h) + \delta(1 + 2\eta)(k : h). \end{aligned}$$

Dividing by $(g : h)$, we obtain

$$\Lambda_h f_1 + \Lambda_h f_2 \leq \Lambda_h f + 2\eta \Lambda_h f + \delta(1 + 2\eta)\Lambda_h k.$$

By (10) and (12), the second term on the right-hand side is $\leq 2\eta(f : g) \leq \epsilon/2$, and the third term is $\leq 2\delta(k : g) \leq \epsilon/2$, as desired.

We return to the proof of the theorem.

Given $\epsilon > 0$ and $f_1, f_2 \in C_c^+$, if $V \in \mathcal{V}$ is chosen as in the lemma, then for *any* $h \in C_V^+$, we have (by (7) for Λ_h)

$$|\Lambda_h(f_1 + f_2) - (\Lambda_h f_1 + \Lambda_h f_2)| \le \epsilon. \tag{13}$$

Apply (11) to the functions f_1, f_2, and $f_3 = f := f_1 + f_2$, with V as in the lemma. Then for h as in (11), it follows from (13) that

$$\begin{aligned}|\Lambda f - (\Lambda f_1 + \Lambda f_2)| &\le |\Lambda f - \Lambda_h f| + |\Lambda_h f - (\Lambda_h f_1 + \Lambda_h f_2)| \\ &\quad + |\Lambda_h f_1 - \Lambda f_1| + |\Lambda_h f_2 - \Lambda f_2| < 4\epsilon,\end{aligned}$$

and the additivity of Λ on C_c^+ follows from the arbitrariness of ϵ.

The desired properties of Λ on C_c follow as in the proof of Theorem 4.8. □

Theorem 4.12. *If Λ' is any left translation invariant positive linear functional on $C_c(G)$, then $\Lambda' = c\Lambda$ for some constant $c \ge 0$.*

Proof. If $\Lambda' = 0$, take $c = 0$. So we may assume $\Lambda' \neq 0$. Since both Λ and Λ' are uniquely determined by their values on C_c^+ (by linearity), it suffices to show that Λ'/Λ is constant on C_c^+. Thus, given $f, g \in C_c^+$ not in $\ker \Lambda$, we must show that $\Lambda' f/\Lambda f = \Lambda' g/\Lambda g$.

Let K be the (compact) support of f; since G is locally compact, there exists an open set W with compact closure such that $K \subset W$. For each $x \in K$, there exists $W_x \in \mathcal{V}$ such that the x-neighborhood xW_x is contained in W. By continuity of the group operation, there exists $V_x \in \mathcal{V}$ such that $V_x V_x \subset W_x$. By compactness of K, there exist $x_1, \dots, x_n \in K$ such that $K \subset \bigcup_{i=1}^n x_i V_{x_i}$. Let $V_1 = \bigcap_{i=1}^n V_{x_i}$. Then $V_1 \in \mathcal{V}$ and

$$KV_1 \subset \bigcup_{i=1}^n x_i V_{x_i} V_1 \subset \bigcup_{i=1}^n x_i V_{x_i} V_{x_i} \subset \bigcup_{i=1}^n x_i W_{x_i} \subset W.$$

Similarly, there exists $V_2 \in \mathcal{V}$ such that $V_2 K \subset W$.

Let $\epsilon > 0$. By uniform continuity of f, there exist $V_3, V_4 \in \mathcal{V}$ such that, for all $x \in G$,

$$|f(x) - f(sx)| < \epsilon/2 \quad \text{for all } s \in V_3$$

and

$$|f(x) - f(xt)| < \epsilon/2 \quad \text{for all } t \in V_4.$$

Let $U := \bigcap_{j=1}^4 V_j$ and $V := U \cap U^{-1}$ (where $U^{-1} := \{x^{-1}; x \in U\}$). Then $V \in \mathcal{V}$ has the following properties:

$$KV \subset W; \quad VK \subset W; \quad V^{-1} = V; \tag{14}$$

$$|f(sx) - f(xt)| < \epsilon \quad \text{for all } x \in G, \; s, t \in V. \tag{15}$$

We shall need to integrate (15) with respect to x over G; since the constant ϵ is not integrable (unless G is compact), we fix a function $k \in C_c^+$ such that $k = 1$ on W; necessarily

$$f(sx) = f(sx)k(x) \quad \text{and} \quad f(xs) = f(xs)k(x) \tag{16}$$

for all $x \in G$ and $s \in V$. (This is trivial for $x \in W$ since $k = 1$ on W. If $x \notin W$, then $x \notin KV$ and $x \notin VK$ by (14); if $sx \in K$ for some $s \in V$, then $s^{-1} \in V$, and consequently $x = s^{-1}(sx) \in VK$, a contradiction. Hence $sx \notin K$, and similarly $xs \notin K$, for all $s \in V$. Therefore, both relations in (16) reduce to $0 = 0$ when $x \notin W$ and $s \in V$.)

By (15) and (16)

$$|f(xs) - f(sx)| \leq \epsilon k(x) \tag{17}$$

for all $x \in G$ and $s \in V$.

Fix $h' \in C_V^+$ not in $\ker \Lambda$, and let $h(x) := h'(x) + h'(x^{-1})$. Clearly, $h \notin \ker \Lambda$, since $h \geq h'$ and Λ is a positive linear functional, hence monotone. Let μ, μ' be the unique positive measures associated with Λ and Λ', respectively (cf. Theorem 3.18). Since $h(x^{-1}y)f(y) \in C_c(G \times G) \subset L^1(\mu \times \mu')$, we have by Fubini's theorem and the relation $h(x^{-1}y) = h(y^{-1}x)$:

$$\iint h(y^{-1}x)f(y)\,d\mu'(x)\,d\mu(y) = \iint h(x^{-1}y)f(y)\,d\mu(y)\,d\mu'(x). \tag{18}$$

By left translation invariance of Λ', the left-hand side of (18) is equal to

$$\int \left(\int h(y^{-1}x)\,d\mu'(x) \right) f(y)\,d\mu(y) = \iint h(x)\,d\mu'(x) f(y)\,d\mu(y) = \Lambda' h \Lambda f. \tag{19}$$

By left translation invariance of Λ, the right-hand side of (18) is equal to

$$\int \left(\int h(y) f(xy)\,d\mu(y) \right) d\mu'(x),$$

and therefore $\Lambda' h \Lambda f$ equals this last integral. On the other hand, by left translation invariance of Λ',

$$\begin{aligned} \int \left(\int h(y) f(yx)\,d\mu'(x) \right) d\mu(y) &= \int h(y) \left(\int f(yx)\,d\mu'(x) \right) d\mu(y) \\ &= \int h(y) \int f(x)\,d\mu'(x)\,d\mu(y) = \Lambda h \Lambda' f. \end{aligned}$$

Since h has support in V, we conclude from these calculations and from (17) that

$$\begin{aligned} |\Lambda' h \Lambda f - \Lambda h \Lambda' f| &= \left| \int_{x \in G} \int_{y \in V} h(y)[f(xy) - f(yx)]\,d\mu(y)\,d\mu'(x) \right| \\ &\leq \epsilon \int_{x \in G} \int_{y \in V} h(y) k(x)\,d\mu(y)\,d\mu'(x) = \epsilon \Lambda h \Lambda' k. \end{aligned} \tag{20}$$

Similarly, for g instead of f, and k' associated to g as k was to f, we obtain

$$|\Lambda' h \Lambda g - \Lambda h \Lambda' g| \leq \epsilon \Lambda h \Lambda' k'. \tag{21}$$

By (20) and (21) divided, respectively, by the positive numbers $\Lambda h \Lambda f$ and $\Lambda h \Lambda g$, we have

$$\left|\frac{\Lambda' h}{\Lambda h} - \frac{\Lambda' f}{\Lambda f}\right| \leq \epsilon \frac{\Lambda' k}{\Lambda f}$$

and

$$\left|\frac{\Lambda' h}{\Lambda h} - \frac{\Lambda' g}{\Lambda g}\right| \leq \epsilon \frac{\Lambda' k'}{\Lambda g}.$$

Consequently

$$\left|\frac{\Lambda' f}{\Lambda f} - \frac{\Lambda' g}{\Lambda g}\right| \leq \epsilon \left(\frac{\Lambda' k}{\Lambda f} + \frac{\Lambda' k'}{\Lambda g}\right),$$

and the desired conclusion $\Lambda' f/\Lambda f = \Lambda' g/\Lambda g$ follows from the arbitrariness of ϵ.

□

Definition 4.13. The unique (up to a positive constant factor) non-zero left translation invariant positive linear functional Λ on $C_c(G)$ is called the (left) Haar functional for G. The measure μ corresponding to Λ through Theorem 3.18 is the (left) Haar measure for G.

Thus, μ is the unique (up to a positive constant factor) non-zero positive measure with the properties listed in Theorem 3.18 (for the locally compact Hausdorff space G) such that $\int_G f_t \, d\mu = \int_G f \, d\mu$ for all $f \in C_c(G)$ and $t \in G$. Equivalently, μ is the unique (up to a positive constant factor) non-zero positive measure with the properties listed in Theorem 3.18 such that for each measurable $E \subset G$ and $t \in G$, the set tE is also measurable and $\mu(tE) = \mu(E)$.

If G is compact, its (unique up to a positive constant factor) left Haar measure (which is finite by Theorem 3.18(2)) is *normalized* so that G has measure 1.

In an analogous way, there exists a unique (up to a positive constant factor) right translation invariant positive measure (as in Theorem 3.18) λ on G:

$$\int_G f^t \, d\lambda = \int_G f \, d\lambda \quad (f \in C_c(G); t \in G), \tag{22}$$

where $f^t(x) := f(xt)$.

Given the left Haar functional Λ on G and $t \in G$, define the functional Λ^t on C_c by $\Lambda^t f := \Lambda f^t$. Then Λ^t is a non-zero left translation invariant positive linear functional (because $\Lambda^t(f_s) = \Lambda(f_s)^t = \Lambda(f^t)_s = \Lambda(f^t) := \Lambda^t f$), and therefore, by Theorem 4.12,

$$\Lambda^t = \delta(t^{-1})\Lambda \tag{23}$$

for some positive number $\delta(t^{-1})$. The function $\delta(\cdot)$ is called the *modular function* of G. Since $(\Lambda^t)^s = (\Lambda)^{ts}$, we have

$$\delta((ts)^{-1})\Lambda = (\Lambda)^{ts} = \delta(t^{-1})\Lambda^s = \delta(t^{-1})\delta(s^{-1})\Lambda,$$

that is, $\delta(\cdot)$ is a homomorphism of G into the multiplicative group of positive reals. Furthermore, for each $f \in C_c$ the function $t \to \Lambda^t(f)$ is continuous by Lebesgue's dominated convergence theorem. Hence, by (23), δ is continuous.

We say that G is *unimodular* if $\delta(\cdot) = 1$. If G is compact, applying (23) to the function $1 \in C_c(G)$, we get $\delta(\cdot) = 1$. If G is abelian, we have $f^t = f_t$, hence $\Lambda^t f = \Lambda(f_t) = \Lambda f$ for all $f \in C_c$, and therefore $\delta(\cdot) = 1$. Thus, compact groups and (locally compact) abelian groups are unimodular. So are discrete groups; see Exercise 11.

For $f \in C_c$ let $\tilde{f}(x) := \delta(x^{-1})f(x^{-1})$ and define $\tilde{\Lambda}$ by $\tilde{\Lambda}f := \Lambda\tilde{f}$ $(f \in C_c)$. Then $\tilde{\Lambda}$ is a non-zero positive linear functional on C_c; it is left translation invariant because $\widetilde{(f_s)}(x) = \delta(x^{-1})f_s(x^{-1}) = \delta(x^{-1})f(sx^{-1}) = \delta(s^{-1})\delta((xs^{-1})^{-1})f((xs^{-1})^{-1}) = \delta(s^{-1})\tilde{f}(xs^{-1}) = \delta(s^{-1})(\tilde{f})^{s^{-1}}(x)$, and therefore by (23),

$$\tilde{\Lambda}(f_s) = \Lambda\widetilde{(f_s)} = \delta(s^{-1})\Lambda^{s^{-1}}\tilde{f} = \Lambda\tilde{f} = \tilde{\Lambda}f.$$

By Theorem 4.12, there exists a positive constant α such that $\tilde{\Lambda} = \alpha\Lambda$. Since $f = \tilde{\tilde{f}}$ for all f, we have $\tilde{\tilde{\Lambda}} = \Lambda$, hence $\alpha^2 = 1$, and therefore $\tilde{\Lambda} = \Lambda$. In terms of the Haar measure μ, this equality takes the form

$$\int_G \delta(x^{-1})f(x^{-1})\,d\mu(x) = \int_G f(x)\,d\mu(x) \quad (f \in C_c(G)). \tag{24}$$

As a result, if G is *unimodular*, then the left Haar functional Λ is also *inverse invariant*, that is, $\int_G f(x^{-1})\,d\mu(x) = \int_G f(x)\,d\mu(x)$ for all $f \in C_c(G)$.

Exercises

1. Let X, Y be Banach spaces, Z a dense subspace of X, and $T \in B(Z, Y)$. Then there exists a unique $\tilde{T} \in B(X, Y)$ such that $\tilde{T}|_Z = T$. Moreover, the map $T \to \tilde{T}$ is an isometric isomorphism of $B(Z, Y)$ onto $B(X, Y)$.

2. Let X be a locally compact Hausdorff space. Prove that $C_0(X)^*$ is isometrically isomorphic to $M_r(X)$, the space of all *regular* complex Borel measures on X. (Hint: Theorems 3.24 and 4.9, and Exercise 1.)

3. Let X_k $k = 1, \ldots, n$ be normed spaces, and consider $\prod_k X_k$ as a normed space with the norm $\|[x_1, \ldots, x_n]\| = \sum_k \|x_k\|$. Prove that there exists an isometric isomorphism of $(\prod_k X_k)^*$ and $\prod_k X_k^*$ with the norm $\|[x_1^*, \ldots, x_n^*]\| = \max_k \|x_n^*\|$. (Hint: given $\phi \in (\prod_k X_k)^*$, define $x_k^* x_k = \phi([0, \ldots, x_k, 0, \ldots, 0])$ for $x_k \in X_k$. Note that $\phi([x_1, \ldots, x_n]) = \sum_k x_k^* x_k$.)

4. Let X be a locally compact Hausdorff space. Let Y be a normed space, and $T \in B(C_c(X), Y)$. Prove that there exists a unique $P : \mathcal{B}(X) \to Y^{**} := (Y^*)^*$ such that $P(\cdot)y^* \in M_r(X)$ for each $y^* \in Y^*$ and

$$y^*Tf = \int_X f\,d(P(\cdot)y^*)$$

for all $f \in C_c(X)$ and $y^* \in Y^*$. Moreover $\|P(\cdot)y^*\| = \|y^* \circ T\|$ for the appropriate norms (for all $y^* \in Y^*$) and $\|P(\delta)\| \le \|T\|$ for all $\delta \in \mathcal{B}(X)$.

Convolution on L^p

5. Let L^p denote the Lebesgue spaces on $\mathbb{R}^k$ with respect to Lebesgue measure. Prove that if $f \in L^1$ and $g \in L^p$, then $f * g \in L^p$ and $\|f * g\|_p \leq \|f\|_1 \|g\|_p$. (Hint: use Theorems 4.6, 2.18, 1.33, and the translation invariance of Lebesgue measure; cf. Exercise 7, Chapter 2, in its $\mathbb{R}^k$ version.)

Approximate identities

6. Let m denote the normalized Lebesgue measure on $[-\pi, \pi]$. Let $K_n : [-\pi, \pi] \to [0, \infty)$ be Lebesgue measurable functions such that $\int_{-\pi}^{\pi} K_n \, dm = 1$ and
$$\sup_{\delta \leq |x| \leq \pi} K_n(x) \to 0 \tag{*}$$
as $n \to \infty$, for all $\delta > 0$. (Any sequence $\{K_n\}$ with these properties is called an *approximate identity*.) Extend K_n to $\mathbb{R}$ as 2π-periodic functions.

   Consider the convolutions
$$(K_n * f)(x) := \int_{-\pi}^{\pi} K_n(x-t) f(t) \, dm(t) = \int_{-\pi}^{\pi} f(x-t) K_n(t) \, dm(t)$$
with 2π-periodic functions f on $\mathbb{R}$. Prove:

   (a) If f is continuous, $K_n * f \to f$ uniformly on $[-\pi, \pi]$. (Hint: $\int_{-\pi}^{\pi} = \int_{|t|<\delta} + \int_{\delta \leq t \leq \pi}$.)

   (b) If $f \in L^p := L^p(-\pi, \pi)$ for some $p \in [1, \infty)$, then $K_n * f \to f$ in L^p. (Hint: use the density of $C([-\pi, \pi])$ in L^p, cf. Corollary 3.21, Part (a), and Exercise 5.)

   (c) If $f \in L^\infty$, then $K_n * f \to f$ in the *weak**-topology on L^∞ (cf. Theorem 4.6); this means that $\int (K_n * f) g \, dm \to \int fg \, dm$ for all $g \in L^1$.

Miscellaneous

7. Consider the measure space $(\mathbb{N}, \mathbb{P}(\mathbb{N}), \mu)$, where μ is the *counting measure* ($\mu(E)$ is the number of points in E if E is a finite subset of $\mathbb{N}$ and $= \infty$ otherwise). The space $l^p := L^p(\mathbb{N}, \mathbb{P}(\mathbb{N}), \mu)$ is the space of all complex sequences $x := \{x(n)\}$ such that $\|x\|_p := \left(\sum |x(n)|^p\right)^{1/p} < \infty$ (in case $p < \infty$) or $\|x\|_\infty := \sup |x(n)| < \infty$ (in case $p = \infty$). As a special case of Theorem 4.6, if $p \in [1, \infty)$ and q is its conjugate exponent, then $(l^p)^*$ is isometrically isomorphic to l^q through the map $x^* \in (l^p)^* \to y \in l^q$, where $y := \{y(n)\}$ is the unique element of l^q such that $x^* x = \sum x(n) y(n)$ for all $x \in l^p$. *Prove this directly!* (Hint: consider the unit vectors $e_m \in l^p$ with $e_m(n) = \delta_{n,m}$, the Kronecker delta.)

8. Consider $\mathbb{N}$ with the discrete topology, and let $c_0 := C_0(\mathbb{N})$ (this is the space of all complex sequences $x := \{x_n\} = \{x(n)\}$ with $\lim x_n = 0$). As a special case of Exercise 2, if $x^* \in c_0^*$, there exists a unique complex Borel measure μ on $\mathbb{N}$ such that $x^*x = \sum_n x(n)\mu(\{n\})$. Denote $y(n) = \mu(\{n\})$. Then $\|y\|_1 = \sum |\mu(\{n\})| \leq |\mu|(\mathbb{N}) = \|\mu\| = \|x^*\|$, that is, $y \in l^1$ and $\|y\|_1 \leq \|x^*\|$. The reversed inequality is trivial. This shows that c_0^* is isometrically isomorphic to l^1 through the map $x^* \to y$, where $x^*x = \sum_n x(n)y(n)$. *Prove this directly!*

9. Let c denote the space of all *convergent* complex sequences $x = \{x(n)\}$ with pointwise operations and the supremum norm. Show that c is a Banach space and c^* is isometrically isomorphic to l^1. (Hint: given $x^* \in c^*$, $x^*|_{c_0} \in c_0^*$; apply Exercise 8, and note that for each $x \in c$, $x - (\lim x)e \in c_0$, where $e(\cdot) = 1$.)

10. Let $(X, \mathcal{A}, \mu)$ be a positive measure space, $q \in (1, \infty]$, and $p = q/(q-1)$. Prove that for all $h \in L^q(\mu)$

$$\|h\|_q = \sup \left| \sum_k \alpha_k \int_{E_k} h \, d\mu \right|,$$

where the supremum is taken over all finite sums with $\alpha_k \in \mathbb{C}$ and $E_k \in \mathcal{A}$ with $0 < \mu(E_k) < \infty$, such that $\sum |\alpha_k|^p \mu(E_k) \leq 1$. (In case $q = \infty$, assume that the measure space is σ-finite.)

11. Describe the left Haar measure on a *discrete* group G (i.e., G with the discrete topology) and prove that G is unimodular.

5

Duality

We studied in preceding chapters the conjugate space X^* for various special normed spaces. Our purpose in Chapter 5 is to examine X^* and its relationship to X for an *arbitrary* normed space X. More generally, we study continuous linear functionals on *topological vector spaces*, which are (complex) vector spaces together with a Hausdorff topology making the vector space operations continuous.

The general results proved in this chapter are indispensable and invaluable to functional analysis and beyond, and in particular to operator theory.

First and foremost is the *Hahn–Banach lemma*, which is one of the "Three Basic Principles of Linear Functional Analysis" (see Chapter 6 for the other two). These principles are so fundamental and have so many consequences that they are often used tacitly. The Hahn–Banach lemma and its corollaries discuss extending linear functionals while preserving certain properties, and are often employed in separation arguments.

The Hahn–Banach *theorem* says that a bounded linear functional on a subspace of a normed space X can be extended to a linear functional on X *with the same norm*. One result of this is that X can be embedded isometrically as a subspace of its second dual $X^{**} := (X^*)^*$. We say that X is *reflexive* when this embedding is surjective. Examples and properties of reflexivity are discussed.

The next theme is *separation*. Given two disjoint, non-empty convex sets M, N in a vector space X, one is looking for a linear functional f on X that separates M and N, namely $\sup \Re f(M) \leq \inf \Re f(N)$ (sometimes strict inequality is desirable). We prove separation results under suitable assumptions in several contexts: the general one of vector spaces; the one of topological vector spaces, in which f is expected to be continuous; and the most restrictive one of *locally convex* topological vector spaces, which are particularly amenable to separation. Local convexity means that the topology of X has a base consisting of convex sets.

For a vector space X and a (separating) family of linear functionals Γ on X, the Γ-topology on X is the weakest topology making each functional in

Introduction to Modern Analysis. Second Edition. Shmuel Kantorovitz and Ami Viselter, Oxford University Press.
 DOI: 10.1093/oso/9780192849540.003.0005

Γ continuous. This topology turns X into a locally convex topological vector space. For a normed space X, we study in depth two distinguished cases: the X^*-topology on X, called the *weak topology*, and the X-topology on X^* (with X viewed as embedded in X^{**}), called the *weak*-topology*. In particular, we prove two results on the properties of the norm-closed unit ball vis-à-vis the *weak**-topology, namely Alaoglu's and Goldstine's theorems.

The *Krein–Milman theorem* asserts that a compact convex subset of a locally convex topological vector space can be reconstructed from its *extremal points*. This mighty result often yields astonishingly simple proofs of otherwise difficult theorems because extremal points tend to have favorable properties. As demonstration, we provide a short proof of the *Stone–Weierstrass theorem*, which gives sufficient conditions for a subalgebra of $C(X)$ to be dense in $C(X)$ where X is a compact Hausdorff space. In fact, we derive it from a more general result called *Bishop's antisymmetry theorem*. Finally, *Milman's theorem* about the "origin" of extremal points in compact convex sets is proved in Exercise 11.

We give another, more elementary (and classical) proof of the Stone–Weierstrass theorem. A generalization of the Stone–Weierstrass theorem to *locally* compact Hausdorff spaces is presented in Exercise 9.

Marcinkiewicz's interpolation theorem is then established with the aid of many results from the previous chapters.

The *fixed-point theorems* due to *Markov–Kakutani*, *Hahn*, and (consequently) *Kakutani* are proved. To give a small taste of the beauty and power of fixed-point theory we show how Kakutani's theorem yields a short proof of the existence of the Haar measure for *compact* topological groups. En route we prove the Arzelà–Ascoli theorem characterizing compactness in $C(X)$.

The last subject is the *bounded weak*-topology*. We prove theorems of *Dieudonné* and *Krein–Šmulian*. The latter says that, remarkably, if X is a Banach space, then for a linear functional on X^* to be *weak**-continuous it suffices that its restriction to the norm-closed unit ball of X^* be *weak**-continuous. In contrast to continuity of functionals with respect to the norm topology on X^*, this is not at all trivial.

5.1 The Hahn–Banach theorem

Let X be a vector space over $\mathbb{R}$. Suppose $p : X \to \mathbb{R}$ is subadditive and homogeneous for non-negative scalars. A linear functional f on a subspace Y of X is p-dominated if $f(y) \le p(y)$ for all $y \in Y$. The starting point of this section is the following:

Lemma 5.1 (The Hahn–Banach lemma). *Let f be a p-dominated linear functional on the subspace Y of X. Then there exists a p-dominated linear functional F on X such that $F|_Y = f$.*

Proof. A p-dominated extension of f is a p-dominated linear functional g on a subspace $D(g)$ of X containing Y, such that $g|_Y = f$. The family $\mathcal{F}$ of all p-dominated extensions of f is partially ordered by setting $g \le h$ (for $g, h \in \mathcal{F}$)

if h is an extension of g. Each totally ordered subfamily $\mathcal{F}_0$ of $\mathcal{F}$ has an upper bound in $\mathcal{F}$, namely, the functional w whose domain is the subspace $D(w) := \bigcup_{g\in\mathcal{F}_0} D(g)$, and for $x \in D(w)$ (so that $x \in D(g)$ for some $g \in \mathcal{F}_0$), $w(x) = g(x)$. Note that w is well defined, that is, its domain is indeed a subspace of X and the value $w(x)$ is independent of the particular g such that $x \in D(g)$, thanks to the *total* ordering of $\mathcal{F}_0$. By Zorn's lemma, $\mathcal{F}$ has a *maximal element* F. To complete the proof, we wish to show that $D(F) = X$. Suppose that $D(F)$ is a proper subspace of X, and let then $z_0 \in X - D(F)$. Let Z be the subspace spanned by $D(F)$ and z_0. The general element of Z has the form $z = u + \alpha z_0$ with $u \in D(F)$ and $\alpha \in \mathbb{R}$ *uniquely* determined (indeed, if $z = u' + \alpha' z_0$ is another such representation with $\alpha' \neq \alpha$, then $z_0 = (\alpha' - \alpha)^{-1}(u - u') \in D(F)$, a contradiction; thus $\alpha' = \alpha$, and therefore $u' = u$). For any choice of $\lambda \in \mathbb{R}$, the functional h with domain Z, defined by $h(z) = F(u) + \alpha\lambda$ is a well-defined linear functional such that $h|_{D(F)} = F$. If we show that λ *can be chosen* such that the corresponding h is p-dominated, then $h \in \mathcal{F}$ with domain Z *properly* containing $D(F)$, a contradiction to the maximality of the element F of $\mathcal{F}$.

Since F is p-dominated, we have for all $u', u'' \in D(F)$

$$\begin{aligned} F(u') + F(u'') &= F(u' + u'') \leq p(u' + u'') \\ &= p([u' + z_0] + [u'' - z_0]) \leq p(u' + z_0) + p(u'' - z_0), \end{aligned}$$

that is,

$$F(u'') - p(u'' - z_0) \leq p(u' + z_0) - F(u') \quad (u', u'' \in D(F)).$$

Pick *any* λ *between* the supremum of the numbers on the left-hand side and the infimum of the numbers on the right-hand side. Then for all $u', u'' \in D(F)$,

$$F(u') + \lambda \leq p(u' + z_0) \quad \text{and} \quad F(u'') - \lambda \leq p(u'' - z_0).$$

Taking $u' = u/\alpha$ if $\alpha > 0$ and $u'' = u/(-\alpha)$ if $\alpha < 0$ and multiplying the inequalities by α and $-\alpha$, respectively, it follows from the homogeneity of p for non-negative scalars that

$$F(u) + \alpha\lambda \leq p(u + \alpha z_0) \quad (u \in D(F))$$

for *all* real α, that is, $h(z) \leq p(z)$ for all $z \in Z$. □

Theorem 5.2 (The Hahn–Banach theorem). *Let Y be a subspace of the normed space X, and let $y^* \in Y^*$. Then there exists $x^* \in X^*$ such that $x^*|_Y = y^*$ and $\|x^*\| = \|y^*\|$.*

Proof. *Case of real scalar field*: Take

$$p(x) := \|y^*\| \|x\| \quad (x \in X).$$

This function is subadditive and homogeneous for non-negative scalars, and

$$y^* y \leq |y^* y| \leq \|y^*\| \|y\| := p(y) \quad (y \in Y).$$

By Lemma 5.1, there exists a p-dominated linear functional F on X such that $F|_Y = y^*$. Thus, for all $x \in X$,

$$F(x) \leq \|y^*\|\|x\|$$

and

$$-F(x) = F(-x) \leq p(-x) = \|y^*\|\|x\|,$$

that is,

$$|F(x)| \leq \|y^*\|\|x\|.$$

This shows that $F := x^* \in X^*$ and $\|x^*\| \leq \|y^*\|$. Since the reversed inequality is trivial for *any* linear extension of y^*, the theorem is proved in the case of real scalars.

Case of complex scalar field: Take $f := \Re y^*$ in Lemma 5.1. Then $f(\mathrm{i}y) = \Re[y^*(\mathrm{i}y)] = \Re[\mathrm{i}y^*y] = -\Im(y^*y)$, and therefore

$$y^*y = f(y) - \mathrm{i}f(\mathrm{i}y) \quad (y \in Y). \tag{1}$$

For p as before, the functional f is p-dominated and *linear* on the vector space Y *over the field* $\mathbb{R}$ (indeed, $f(y) \leq |y^*y| \leq p(y)$ for all $y \in Y$). By Lemma 5.1, there exists a p-dominated linear functional $F : X \to \mathbb{R}$ (over *real* scalars!) such that $F|_Y = f$. Define

$$x^*x := F(x) - \mathrm{i}F(\mathrm{i}x) \quad (x \in X).$$

By (1), $x^*|_Y = y^*$. Clearly, x^* is additive and homogeneous for *real* scalars. Also, for all $x \in X$,

$$x^*(\mathrm{i}x) = F(\mathrm{i}x) - \mathrm{i}F(-x) = \mathrm{i}[F(x) - \mathrm{i}F(\mathrm{i}x)] = \mathrm{i}x^*x,$$

and it follows that x^* is homogeneous over $\mathbb{C}$.

Given $x \in X$, write $x^*x = \rho\bar{\omega}$ with $\rho \geq 0$ and $\omega \in \mathbb{C}$ with modulus one. Then

$$\begin{aligned} |x^*x| &= \omega x^*x = x^*(\omega x) = \Re[x^*(\omega x)] \\ &= F(\omega x) \leq \|y^*\|\|\omega x\| = \|y^*\|\|x\|. \end{aligned}$$

Thus, $x^* \in X^*$ with norm $\leq \|y^*\|$ (hence $= \|y^*\|$, since x^* is an extension of y^*). □

Corollary 5.3. *Let Y be a subspace of the normed space X, and let $x \in X$ be such that*

$$d := d(x, Y) := \inf_{y \in Y} \|x - y\| > 0.$$

Then there exists $x^* \in X^*$ with $\|x^*\| = 1/d$, such that

$$x^*|_Y = 0, \qquad x^*x = 1.$$

Proof. Let Z be the linear span of Y and x. Since $d(x, Y) > 0, x \notin Y$, so that the general element $z \in Z$ has the *unique* representation $z = y + \alpha x$ with $y \in Y$

and $\alpha \in \mathbb{C}$. Define then $z^*z = \alpha$. This is a well-defined linear functional on Z, $z^*|_Y = 0$, and $z^*x = 1$. Also z^* is bounded, since

$$\|z^*\| := \sup_{0 \neq z \in Z} \frac{|\alpha|}{\|z\|}$$

$$= \sup_{\alpha \neq 0; y \in Y} \frac{1}{\|(y + \alpha x)/\alpha\|} = \frac{1}{\inf_{y \in Y} \|x - y\|} = 1/d.$$

By the Hahn–Banach theorem, there exists $x^* \in X^*$ with norm $= \|z^*\| = 1/d$ that extends z^*, whence $x^*|_Y = 0$ and $x^*x = 1$. □

Note that if Y is a *closed* subspace of X, the condition $d(x, Y) > 0$ is equivalent to $x \notin Y$. If $Y \neq X$, such an x exists, and therefore, by Corollary 5.3, there exists a *non-zero* $x^* \in X^*$ such that $x^*|_Y = 0$. Formally:

Corollary 5.4. *Let $Y \neq X$ be a closed subspace of the normed space X. Then there exists a non-zero $x^* \in X^*$ that vanishes on Y.*

For a not necessarily closed subspace Y, we apply the last corollary to its closure $\bar{Y}$ (which is a closed subspace). By continuity, vanishing of x^* on Y is equivalent to its vanishing on $\bar{Y}$, and we obtain, therefore, the following useful criterion for *non-density*.

Corollary 5.5. *Let Y be a subspace of the normed space X. Then Y is* not dense *in X iff there exists a non-zero $x^* \in X^*$ that vanishes on Y.*

For reference, we also state this criterion as a *density criterion*.

Corollary 5.6. *Let Y be a subspace of the normed space X. Then Y is dense in X iff the vanishing of an $x^* \in X^*$ on Y implies $x^* = 0$.*

Corollary 5.7. *Let X be a normed space, and let $0 \neq x \in X$. Then there exists $x^* \in X^*$ such that $\|x^*\| = 1$ and $x^*x = \|x\|$. In particular, X^** separates points, *that is, if x, y are distinct vectors in X, then there exists a functional $x^* \in X^*$ such that $x^*x \neq x^*y$.*

Proof. We take $Y = \{0\}$ in Corollary 5.3. Then $d(x, Y) = \|x\| > 0$, so that there exists $z^* \in X^*$ such that $\|z^*\| = 1/\|x\|$ and $z^*x = 1$. Let $x^* := \|x\|z^*$. Then $x^*x = \|x\|$ and $\|x^*\| = 1$ as wanted.

If x, y are distinct vectors, we apply the preceding result to the non-zero vector $x - y$; we then obtain $x^* \in X^*$ such that $\|x^*\| = 1$ and $x^*x - x^*y = x^*(x - y) = \|x - y\| \neq 0$. □

Corollary 5.8. *Let X be a normed space. Then for each $x \in X$,*

$$\|x\| = \sup_{x^* \in X^*; \|x^*\| = 1} |x^*x|.$$

Proof. The relation being trivial for $x = 0$, we assume $x \neq 0$, and apply Corollary 5.7 to obtain an x^* with unit norm such that $x^*x = \|x\|$. Therefore, the supremum shown is $\geq\|x\|$. Since the reverse inequality is a consequence of the definition of the norm of x^*, the result follows. □

Given $x \in X$, the functional

$$\kappa x : x^* \to x^* x$$

on X^* is linear, and Corollary 5.8 establishes that $\|\kappa x\| = \|x\|$. This means that $\hat{x} := \kappa x$ is a continuous linear functional on X^*, that is, an element of $(X^*)^* := X^{**}$.

The map $\kappa : X \to X^{**}$ is linear (since for all $x^* \in X^*, [\kappa(x + \alpha x')]x^* = x^*(x + \alpha x') = x^* x + \alpha x^* x' = [\kappa x + \alpha \kappa x']x^*$) and *isometric* (since $\|\kappa x - \kappa x'\| = \|\kappa(x - x')\| = \|x - x'\|$). The *isometric isomorphism* κ is called the *canonical* (or *natural*) *embedding* of X in the *second dual* X^{**}.

Note that X is complete iff its isometric image $\hat{X} := \kappa X$ is complete, and since conjugate spaces are always complete, κX is complete iff it is a *closed* subspace of X^{**}. Thus, a normed space is complete iff its canonical embedding κX is a *closed* subspace of X^{**}. In case $\kappa X = X^{**}$, we say that X is *reflexive.* Our observations show in particular that a reflexive space is *necessarily* complete.

5.2 Reflexivity

Theorem 5.9. *A closed subspace of a reflexive Banach space is reflexive.*

Proof. Let X be a reflexive Banach space and let Y be a closed subspace of X. The restriction map

$$\psi : x^* \to x^*|_Y \quad (x^* \in X^*)$$

is a norm-decreasing linear map of X^* into Y^*. For each $y^{**} \in Y^{**}$, the function $y^{**} \circ \psi$ belongs to X^{**}; we thus have the (continuous linear) map

$$\chi : Y^{**} \to X^{**} \quad \chi(y^{**}) = y^{**} \circ \psi.$$

Let κ denote the canonical imbedding of X *onto* X^{**} (recall that X is reflexive!), and consider the (continuous linear) map

$$\kappa^{-1} \circ \chi : Y^{**} \to X.$$

We claim that its range Z is *contained in* Y. Indeed, suppose $z \in Z$ but $z \notin Y$. Since Y is a closed subspace of X, there exists $x^* \in X^*$ such that $x^* Y = \{0\}$ and $x^* z = 1$. Then $\psi(x^*) = 0$, and since $z = (\kappa^{-1} \circ \chi)(y^{**})$ for some $y^{**} \in Y^{**}$, we have

$$1 = x^* z = (\kappa z)(x^*) = [\chi(y^{**})](x^*) = [y^{**} \circ \psi](x^*) = y^{**}(0) = 0,$$

a contradiction. Thus

$$\kappa^{-1} \circ \chi : Y^{**} \to Y.$$

Given $y^{**} \in Y^{**}$, consider then the element

$$y := [\kappa^{-1} \circ \chi](y^{**}) \in Y.$$

For any $y^* \in Y^*$, let $x^* \in X^*$ be an extension of y^* (cf. Hahn–Banach theorem). Then

$$y^{**}(y^*) = y^{**}(\psi(x^*)) = [\chi(y^{**})](x^*) = [\kappa(y)](x^*) = x^*(y) = y^*(y) = (\kappa_Y y)(y^*),$$

where κ_Y denotes the canonical imbedding of Y into Y^{**}. This shows that $y^{**} = \kappa_Y y$, so that κ_Y is *onto*, as wanted. □

Theorem 5.10. *If X and Y are isomorphic Banach spaces, then X is reflexive if and only if Y is reflexive.*

Proof. Let $T : X \to Y$ be an isomorphism (i.e., a linear homeomorphism). Assume Y reflexive; all we need to show is that X is reflexive.

Given $y^* \in Y^*$ and *any* $T \in B(X,Y)$, the composition $y^* \circ T$ is a continuous linear functional on X, which we denote T^*y^*. This defines a map $T^* \in B(Y^*, X^*)$, called the (Banach) *adjoint* of T. One verifies easily that if $T^{-1} \in B(Y, X)$, then $(T^*)^{-1}$ exists and equals $(T^{-1})^*$.

For simplicity of notation, we shall use the "hat notation" ($\hat{x}$ and $\hat{y}$) for elements of X and Y, without specifying the space in the hat symbol.

Let $x^{**} \in X^{**}$ be given. Then $x^{**} \circ T^* \in Y^{**}$, and since Y is reflexive, there exists a unique $y \in Y$ such that

$$x^{**} \circ T^* = \hat{y}.$$

Let $x = T^{-1}y$. Then for all $x^* \in X^*$,

$$\begin{aligned}\hat{x}x^* &= x^*x = x^*T^{-1}y = ((T^*)^{-1}x^*)y \\ &= \hat{y}((T^*)^{-1}x^*) = x^{**}[T^*((T^*)^{-1}x^*)] = x^{**}x^*,\end{aligned}$$

that is, $x^{**} = \hat{x}$. □

Theorem 5.11. *A Banach space is reflexive iff its conjugate is reflexive.*

Proof. Let X be a reflexive Banach space, and let κ be its canonical embedding *onto* X^{**}.

For any $\phi \in (X^*)^{**} = (X^{**})^*$ the map $\phi \circ \kappa$ is a continuous linear functional $x^* \in X^*$, and for any $x^{**} \in X^{**}$, letting $x := \kappa^{-1}x^{**}$, we have

$$\phi(x^{**}) = \phi(\kappa(x)) = x^*x = (\kappa x)(x^*) = x^{**}x^* = (\kappa_{X^*}x^*)(x^{**}).$$

This shows that κ_{X^*} is onto, that is, X^* is reflexive.

Conversely, if X^* is reflexive, then X^{**} is reflexive by the first part of the proof. Since κX is a *closed subspace* of X^{**}, it is reflexive by Theorem 5.9. Therefore, X is reflexive since it is isomorphic to κX, by Theorem 5.10. □

We show here that Hilbert space and L^p-spaces (for $1 < p < \infty$) are reflexive.

A map $T : X \to Y$ between complex vector spaces is said to be *conjugate-homogeneous* if

$$T(\lambda x) = \bar{\lambda} T x \quad (x \in X; \lambda \in \mathbb{C}).$$

An additive conjugate-homogeneous map is called a *conjugate-linear* map. In particular, we may talk of *conjugate-isomorphisms.*

Lemma 5.12. *If X is a Hilbert space (over $\mathbb{C}$), then there exists an isometric conjugate-isomorphism $V : X^* \to X$, such that*

$$x^* x = (x, V x^*) \quad (x \in X) \tag{1}$$

for all $x^* \in X^*$.

Proof. If $x^* \in X^*$, the "Little" Riesz representation theorem (Theorem 1.37) asserts that there exists a unique element $y \in X$ such that $x^* x = (x, y)$ for all $x \in X$. Denote $y = V x^*$, so that $V : X^* \to X$ is uniquely determined by the identity (1).

It follows from (1) that V is conjugate-linear. If $V x^* \neq 0$, we have by (1) and Schwarz's inequality

$$\begin{aligned} \|V x^*\| &= \frac{(V x^*, V x^*)}{\|V x^*\|} = \frac{|x^*(V x^*)|}{\|V x^*\|} \\ &\leq \|x^*\| = \sup_{x \neq 0} \frac{|x^* x|}{\|x\|} = \sup_{x \neq 0} \frac{|(x, V x^*)|}{\|x\|} \leq \|V x^*\|, \end{aligned}$$

so that $\|V x^*\| = \|x^*\|$ (this is trivially true also in case $V x^* = 0$, since then $x^* = 0$ by (1)).

Being *conjugate*-linear and norm-preserving, V is isometric, hence continuous and injective. It is also onto, because any $y \in X$ induces the functional x^* defined by $x^* x = (x, y)$ for all $x \in X$, and clearly $V x^* = y$ by the uniqueness of the Riesz representation. □

Theorem 5.13. *Hilbert space is reflexive.*

Proof. Denote by J the conjugation operator in $\mathbb{C}$.

Given $x^{**} \in X^{**}$ (for X a complex Hilbert space), the map

$$J \circ x^{**} \circ V^{-1} : X \to \mathbb{C}$$

is continuous and linear. Denote it by x_0^*. Let $x := V x_0^*$. Then for all $x^* \in X^*$,

$$\begin{aligned} \hat{x} x^* &= x^* x = (x, V x^*) = (V x_0^*, V x^*) = \overline{(V x^*, V x_0^*)} \\ &= (J \circ x_0^* \circ V) x^* = x^{**} x^*, \end{aligned}$$

that is, $x^{**} = \hat{x}$. □

In particular, finite dimensional spaces $\mathbb{C}^n$ are reflexive. Also $L^2(\mu)$ (for any positive measure space $(X, \mathcal{A}, \mu)$) is reflexive. Theorem 5.14 establishes the reflexivity of all L^p-spaces for $1 < p < \infty$.

Theorem 5.14. *Let $(X, \mathcal{A}, \mu)$ be a positive measure space. Then the space $L^p(\mu)$ is reflexive for $1 < p < \infty$.*

Proof. Let $q = p/(p-1)$. Write $L^p := L^p(\mu)$ and

$$\langle f, g \rangle := \int_X fg \, d\mu \quad (f \in L^p, g \in L^q).$$

By Theorem 4.6, there exists an isometric isomorphism

$$V_p : (L^p)^* \to L^q$$

such that

$$x^* f = \langle f, V_p x^* \rangle \quad (f \in L^p)$$

for all $x^* \in (L^p)^*$.

Given $x^{**} \in (L^p)^{**}$, the map $x^{**} \circ (V_p)^{-1}$ is a continuous linear functional on L^q; therefore

$$f := V_q \circ x^{**} \circ (V_p)^{-1} \in L^p.$$

Let $x^* \in (L^p)^*$, and write $g := V_p x^* \in L^q$; we have

$$\begin{aligned} \hat{f}(x^*) &= x^* f = \langle f, g \rangle = [(V_q)^{-1} f](g) \\ &= x^{**}(V_p^{-1} g) = x^{**} x^*. \end{aligned}$$

This shows that $x^{**} = \hat{f}$. $\square$

The theorem is false in general for $p = 1$ and $p = \infty$.

Also the space $C_0(X)$ (for a locally compact Hausdorff space X) is not reflexive in general. We shall not prove these facts here.

5.3 Separation

We now consider applications of the Hahn–Banach lemma to *separation* of convex sets in vector spaces.

Let X be a vector space over $\mathbb{C}$ or $\mathbb{R}$. A *convex combination* of vectors $x_k \in X$ $(k = 1, \ldots, n)$ is any vector of the form

$$\sum_{k=1}^{n} \alpha_k x_k \quad \left(\alpha_k \geq 0; \sum_k \alpha_k = 1\right).$$

A subset $K \subset X$ is *convex* if it contains the convex combinations of any two vectors in it.

Equivalently, a set $K \subset X$ is convex if it is invariant under the operation of taking convex combinations of its elements. Indeed, invariance under convex combinations of pairs of elements is precisely the definition of convexity. On the other hand, if K is convex, one can prove the said invariance by induction on the number n of vectors. Assuming invariance for $n \geq 2$ vectors, consider any convex combination $z = \sum_{k=1}^{n+1} \alpha_k x_k$ of vectors $x_k \in K$. If $\alpha := \sum_{k=1}^{n} \alpha_k = 0$, then $z = x_{n+1} \in K$ trivially. So assume $\alpha > 0$; since $\alpha_{n+1} = 1 - \alpha$, we have

$$z = \alpha \sum_{k=1}^{n} \frac{\alpha_k}{\alpha} x_k + (1-\alpha) x_{n+1} \in K,$$

by the induction hypothesis and the convexity of K.

The intersection of a family of convex sets is clearly convex. The *convex hull* of a set M (denoted $\mathrm{co}(M)$) is the intersection of all convex sets containing M. It is the smallest convex set containing M ("smallest" with respect to set inclusion), and consists of all the convex combinations of vectors in M.

If M, N are convex subsets of X and α, β are scalars, then $\alpha M + \beta N$ is convex. Also TM is convex for any linear map $T : X \to Y$ between vector spaces.

Let $M \subset X$; the point $x \in M$ is an *internal point of* M if for each $y \in X$, there exists $\epsilon = \epsilon(y)$ such that $x + \alpha y \in M$ for all $\alpha \in \mathbb{C}$ with $|\alpha| \leq \epsilon$. Clearly, an internal point of M is also an internal point of any N such that $M \subset N \subset X$.

Suppose 0 is an internal point of the *convex* set M. Then for each $y \in X$, there exists $\epsilon := \epsilon(y)$ such that $y/\rho \in M$ for all $\rho \geq 1/\epsilon$. If $y/\rho \in M$, then by convexity $y/(\rho/\alpha) = (1-\alpha)0 + \alpha(y/\rho) \in M$ for all $0 < \alpha \leq 1$. Since $\rho \leq \rho/\alpha < \infty$ for such α, this means that $y/\rho' \in M$ for all $\rho' \geq \rho$, that is, the set $\{\rho > 0; y/\rho \in M\}$ is a subray of $\mathbb{R}^+$ that contains $1/\epsilon$. Let $\kappa(y)$ be the left endpoint of that ray, that is,

$$\kappa(y) := \inf\{\rho > 0; y/\rho \in M\}.$$

Then $\kappa(y) \leq 1/\epsilon$, so that $0 \leq \kappa(y) < \infty$, and $\kappa(y) \leq 1$ for $y \in M$ (equivalently, if $\kappa(y) > 1$, then $y \in M^c$). If $\alpha > 0$,

$$\alpha\{\rho > 0; y/\rho \in M\} = \{\alpha\rho; (\alpha y)/(\alpha\rho) \in M\},$$

and it follows that $\kappa(\alpha y) = \alpha\kappa(y)$ (this is also trivially true for $\alpha = 0$, since $0 \in M$).

If $x, y \in X$ and $\rho > \kappa(x), \sigma > \kappa(y)$, then $x/\rho, y/\sigma \in M$, and since M is convex,

$$\frac{x+y}{\rho+\sigma} = \frac{\rho}{\rho+\sigma} x/\rho + \frac{\sigma}{\rho+\sigma} y/\sigma \in M.$$

Hence, $\kappa(x+y) \leq \rho + \sigma$, and therefore $\kappa(x+y) \leq \kappa(x) + \kappa(y)$.

We conclude that $\kappa : X \to [0, \infty)$ is a subadditive positive-homogeneous functional, referred to as the *Minkowski functional* of the convex set M.

Lemma 5.15. *Let $M \subset X$ be convex with 0 internal, and let κ be its Minkowski functional. Then $\kappa(x) < 1$ iff x is an internal point of M, and $\kappa(x) > 1$ iff x is an internal point of M^c.*

Proof. Let x be an internal point of M, and let $\epsilon(\cdot)$ be as in the definition. Then $x + \epsilon(x)x \in M$, and therefore

$$\kappa(x) \le \frac{1}{1+\epsilon(x)} < 1.$$

Conversely, suppose $\kappa(x) < 1$. Then $x = x/1 \in M$. Let $y \in X$. Since 0 is internal for M, there exists $\epsilon_0 > 0$ (depending on y) such that $\beta y \in M$ for $|\beta| \le \epsilon_0$. In particular $\kappa(\epsilon_0 \omega y) \le 1$ for all $\omega \in \mathbb{C}$ with $|\omega| = 1$. Now (with any $\alpha = |\alpha|\omega$),

$$\kappa(x + \alpha y) \le \kappa(x) + \kappa(\alpha y) = \kappa(x) + \frac{|\alpha|}{\epsilon_0}\kappa(\epsilon_0\omega y) \le \kappa(x) + |\alpha|/\epsilon_0 < 1$$

for $|\alpha| \le \epsilon$, with $\epsilon < [1 - \kappa(x)]\epsilon_0$. Hence, $x + \alpha y \in M$ for $|\alpha| \le \epsilon$, so that x is an internal point of M.

If x is an internal point of M^c, there exists $\epsilon > 0$ such that $x - \epsilon x \in M^c$. However, if $\kappa(x) \le 1$, then $1/(1-\epsilon) > 1 \ge \kappa(x)$, and therefore $x - \epsilon x \in M$, which is a contradiction. Thus $\kappa(x) > 1$.

Conversely, suppose $\kappa(x) > 1$. Let $y \in X$, and choose ϵ_0 as earlier. Then

$$\kappa(x + \alpha y) \ge \kappa(x) - \frac{|\alpha|}{\epsilon_0}\kappa(\epsilon_0\omega y) \ge \kappa(x) - |\alpha|/\epsilon_0 > 1$$

if $|\alpha| \le \epsilon$ with $\epsilon < [\kappa(x) - 1]\epsilon_0$. This shows that $x + \alpha y \in M^c$ for $|\alpha| \le \epsilon$, and so x is an internal point of M^c. □

By the lemma, $\kappa(x) = 1$ iff x is not internal for M and for M^c; such a point is called a *bounding point* for M (or for M^c).

We shall apply the Hahn–Banach lemma (Lemma 5.1) with $p = \kappa$ to obtain the following.

Theorem 5.16 (Separation theorem). *Let M, N be disjoint non-empty convex sets in the vector space X (over $\mathbb{C}$ or $\mathbb{R}$), and suppose M has an internal point. Then there exists a non-zero linear functional f on X such that*

$$\sup \Re f(M) \le \inf \Re f(N)$$

(one says that f separates *M and N).*

Proof. Suppose that the theorem is valid for vector spaces over $\mathbb{R}$. If X is a vector space over $\mathbb{C}$, we may consider it as a vector space over $\mathbb{R}$, and get an $\mathbb{R}$-linear non-zero functional $\phi : X \to \mathbb{R}$ such that $\sup \phi(M) \le \inf \phi(N)$. Setting $f(x) := \phi(x) - \mathrm{i}\phi(\mathrm{i}x)$ as in the proof of Theorem 5.2, we obtain a non-zero $\mathbb{C}$-linear functional on X such that $\sup \Re f(M) = \sup \phi(M) \le \inf \phi(N) = \inf \Re f(N)$, as wanted.

This shows that we need to prove the theorem for vector spaces over $\mathbb{R}$ only. We may also assume that 0 is an internal point of M. Indeed, suppose the theorem is valid in that case. By assumption, M has an internal point x. Thus for each $y \in X$, there exists $\epsilon > 0$ such that $x + \alpha y \in M$ for all $|\alpha| \leq \epsilon$. Equivalently, $0 + \alpha y \in M - x$ for all such α, that is, 0 is internal for $M - x$. The sets $N - x$ and $M - x$ are disjoint convex sets, and the theorem (for the special case 0 internal to $M - x$) implies the existence of a non-zero linear functional f such that $\sup f(M - x) \leq \inf f(N - x)$. Therefore, $\sup f(M) \leq \inf f(N)$ as desired.

Fix $z \in N$ and let $K := M - N + z$. Then K is convex, $M \subset K$, and therefore 0 is an internal point of K. Let κ be the Minkowski functional of K. Since M and N are disjoint, $z \notin K$, and therefore $\kappa(z) \geq 1$.

Define $f_0 : \mathbb{R}z \to \mathbb{R}$ by $f_0(\lambda z) = \lambda\kappa(z)$. Then f_0 is linear and κ-dominated (since for $\lambda \geq 0, f_0(\lambda z) := \lambda\kappa(z) = \kappa(\lambda z)$, and for $\lambda < 0, f_0(\lambda z) < 0 \leq \kappa(\lambda z)$). By the Hahn–Banach lemma (Lemma 5.1), there exists a κ-dominated linear extension $f : X \to \mathbb{R}$ of f_0. Then $f(z) = f_0(z) = \kappa(z) \geq 1$, and $f(x) \leq \kappa(x) \leq 1$ for all $x \in K$. This means that f is a non-zero linear functional on X such that

$$f(M) - f(N) + f(z) = f(M - N + z) = f(K) \leq 1 \leq f(z),$$

that is, $f(M) \leq f(N)$. □

5.4 Topological vector spaces

We consider next a vector space X with a Hausdorff topology such that the vector space operations are continuous (a *topological vector space*). The function

$$f : (x, y, \alpha) \in X \times X \times [0,1] \to \alpha x + (1-\alpha)y \in X$$

is continuous. The set $M \subset X$ is convex iff $f(M \times M \times [0,1]) \subset M$. Therefore, by continuity of f, if M is convex, we have

$$f(\bar{M} \times \bar{M} \times [0,1]) = f(\overline{M \times M \times [0,1]}) \subset \overline{f(M \times M \times [0,1])} \subset \bar{M},$$

which proves that the closure of a convex set is convex. A trivial modification of the proof shows that the closure of a subspace is a subspace.

Let M° denote the interior of M. We show that for $0 < \alpha < 1$ and $M \subset X$ convex,

$$\alpha M^{\circ} + (1-\alpha)\bar{M} \subset M^{\circ}. \tag{1}$$

In particular, it follows from (1) that M° is convex.

Let $x \in M^{\circ}$, and let V be a neighborhood of x contained in M. Since addition and multiplication by a non-zero scalar are homeomorphisms of X onto itself, $U := V - x$ is a neighborhood of 0 and $y + \beta U$ is a neighborhood of y for any $0 \neq \beta \in \mathbb{R}$. Therefore, if $y \in \bar{M}$, there exists $y_\beta \in M \cap (y + \beta U)$. Thus there exists $u \in U$ such that $y = y_\beta - \beta u$. Then, given $\alpha \in (0,1)$ and choosing $\beta = \alpha/(\alpha - 1)$, we have by convexity of M

$$\begin{aligned} \alpha x + (1-\alpha)y &= \alpha x + (1-\alpha)y_\beta + (\alpha - 1)\beta u \\ &= \alpha(x + u) + (1-\alpha)y_\beta \in \alpha V + (1-\alpha)y_\beta \\ &:= V_\beta \subset \alpha M + (1-\alpha)M \subset M, \end{aligned}$$

where V_β is clearly open. This proves that $\alpha x + (1-\alpha)y \in M^o$, as wanted.

If $M^o \neq \emptyset$ and we fix $x \in M^o$, the continuity of the vector space operations implies that

$$y = \lim_{\alpha \to 0+} \alpha x + (1-\alpha)y \quad (y \in \bar{M}),$$

and it follows from (1) that M^o is dense in M.

With notation as before, it follows from the continuity of multiplication by scalars that for any $y \in X$, there exists $\epsilon = \epsilon(y)$ such that $\alpha y \in U$ for all $\alpha \in \mathbb{C}$ with $|\alpha| \leq \epsilon$; thus, for these $\alpha, x + \alpha y \in x + U = V \subset M$. This shows that interior points of M are internal for M.

It follows that bounding points for M are boundary points of M.

Conversely, if $x \in M$ is internal and $M^o \neq \emptyset$, pick $m \in M^o$. Then there exists $\epsilon > 0$ such that

$$(1+\epsilon)x - \epsilon m = x + \epsilon(x - m) := m' \in M.$$

Therefore, by (1),

$$x = \frac{\epsilon}{1+\epsilon} m + \frac{1}{1+\epsilon} m' \in M^o.$$

Thus internal points for M are interior points of M (when the interior is not empty).

Still for M convex with non-empty interior, suppose y is a boundary point of M. Pick $x \in M^o$. For $0 < \alpha < 1, y + \alpha(x - y) = \alpha x + (1-\alpha)y \in M^o$ by (1), and therefore y is not internal for M^c. It is not internal for M as well (since internal points are interior!). Thus y is a bounding point for M. Collecting, we have:

Lemma 5.17. *Let M be a convex set with non-empty interior in a topological vector space. A point is internal (bounding) of M iff it is an interior (boundary) point of M.*

Lemma 5.18. *Let X be a topological vector space, let M, N be non-empty subsets of X with $M^o \neq \emptyset$. If f is a linear functional on X that separates M and N, then f is continuous.*

Proof. Since $(\Im f)(x) = -(\Re f)(ix)$ in the case of complex vector spaces, it suffices to show that $\Re f$ is continuous, and this reduces the complex case to the real case. Let then $f : X \to \mathbb{R}$ be linear such that $\sup f(M) := \delta \leq \inf f(N)$. Let $m \in M^o$ and $n \in N$. Let then U be a symmetric neighborhood of 0 (i.e., $-U = U$) such that $m + U \subset M$ (if V is any 0-neighborhood such that $m + V \subset M$, we may take $U = V \cap (-V)$). Then $0 \in -U = U \subset M - m$, and therefore, for any $u \in U$,

$$f(u) \leq \sup f(M) - f(m) = \delta - f(m) \leq f(n) - f(m),$$

and the same inequality holds for $-u$. In particular (taking $u = 0$), $f(n) - f(m) \geq 0$. Pick any $\rho > f(n) - f(m)$. Then $f(u) < \rho$ and also $-f(u) = f(-u) < \rho$, that is, $|f(u)| < \rho$ for all $u \in U$. Hence, given $\epsilon > 0$, the 0-neighborhood $(\epsilon/\rho)U$ is mapped by f into $(-\epsilon, \epsilon)$, which proves that f is continuous at 0. However, continuity at 0 is equivalent to continuity for linear maps between topological vector spaces, as is readily seen by translation. □

Combining Lemmas 5.17 and 5.18 with Theorem 5.16, we obtain the following *separation theorem* for topological vector spaces.

Theorem 5.19 (Separation theorem). *In a topological vector space, any two disjoint non-empty convex sets, one of which has non-empty interior, can be separated by a non-zero continuous linear functional.*

If we have *strict* inequality in Theorem 5.16, the functional f *strictly separates* the sets M and N (it is necessarily non-zero). A strict separation theorem is stated below for a *locally convex* topological vector space (t.v.s.), that is, a t.v.s. whose topology has a *base consisting of convex sets.*

Theorem 5.20 (Strict separation theorem). *Let X be a locally convex t.v.s. Let M, N be non-empty disjoint convex sets in X. Suppose M is compact and N is closed. Then there exists a continuous linear functional on X which strictly separates M and N.*

Proof. Observe first that $M - N$ is *closed.* Indeed, if a net $\{m_i - n_i\}$ ($m_i \in M; n_i \in N; i \in I$) converges to $x \in X$, then since M is compact, a subnet $\{m_{i'}\}$ converges to some $m \in M$. By continuity of vector space operations, the net $\{n_{i'}\} = \{m_{i'} - (m_{i'} - n_{i'})\}$ converges to $m - x$, and since N is closed, $m - x := n \in N$. Therefore, $x = m - n \in M - N$ and $M - N$ is closed. It is also convex.

Since M, N are disjoint, the point 0 is in the open set $(M - N)^{\mathrm{c}}$, and since X is locally convex, there exists a convex neighborhood of 0, U, disjoint from $M - N$. By Theorem 5.19 (applied to the sets $M - N$ and U), there exists a non-zero continuous linear functional f separating $M - N$ and U:

$$\sup \Re f(U) \leq \inf \Re f(M - N).$$

Since $f \neq 0$, there exists $y \in X$ such that $f(y) = 1$. By continuity of multiplication by scalars, there exists $\epsilon > 0$ such that $\epsilon y \in U$. Then

$$\epsilon = \Re f(\epsilon y) \leq \sup \Re f(U) \leq \inf \Re f(M - N),$$

that is, $\Re f(n) + \epsilon \leq \Re f(m)$ for all $m \in M$ and $n \in N$. Thus,

$$\sup \Re f(N) < \sup \Re f(N) + \epsilon \leq \inf \Re f(M).$$

□

Taking $M = \{p\}$, we get the following:

Corollary 5.21. *Let X be a locally convex t.v.s., let N be a (non-empty) closed convex set in X, and $p \notin N$. Then there exists a continuous linear functional f strictly separating p and N. In particular (with $N = \{q\}, q \neq p$), the continuous linear functionals on X separate the points of X (i.e., if p, q are any distinct points of X, then there exists a continuous linear functional f on X such that $f(p) \neq f(q)$).*

5.5 Weak topologies

Here we consider topologies induced on a given vector space X by families of linear functionals on it. Let Γ be a *separating* vector space of linear functionals on X. Equivalently, if $\Gamma x = \{0\}$, then $x = 0$. The Γ-*topology* of X is *the weakest* topology on X for which all $f \in \Gamma$ are continuous. A base for this topology consists of all sets of the form

$$N(x; \Delta, \epsilon) = \{y \in X; |f(y) - f(x)| < \epsilon \text{ for all } f \in \Delta\},$$

where $x \in X, \Delta \subset \Gamma$ is finite, and $\epsilon > 0$. The net $\{x_i; i \in I\}$ converges to x in the Γ-topology iff $f(x_i) \to f(x)$ for all $f \in \Gamma$. The vector space operations are Γ-continuous, and the sets in the basis are clearly convex, so that X with the Γ-topology (sometimes denoted X_Γ) is a locally convex t.v.s. Let X_Γ^* denote the space of all continuous linear functionals on X_Γ. By definition of the Γ-topology, $\Gamma \subset X_\Gamma^*$. We show here that we actually have equality between these sets.

Lemma 5.22. *Let $f_1, \ldots, f_n, g$ be linear functionals on the vector space X such that*

$$\bigcap_{i=1}^{n} \ker f_i \subset \ker g.$$

Then $g \in$ span $\{f_1, \ldots, f_n\}$ (:= the linear span of $f_1, \ldots, f_n$).

Proof. Consider the linear map

$$T : X \to \mathbb{C}^n, \quad Tx = (f_1(x), \ldots, f_n(x)) \in \mathbb{C}^n.$$

Define $\phi : TX \to \mathbb{C}$ by $\phi(Tx) = g(x)$ $(x \in X)$. If $Tx = Ty$, then $x - y \in \bigcap_i \ker f_i \subset \ker g$, hence $g(x) = g(y)$, which shows that ϕ is well defined. It is clearly linear, and has therefore an extension as a linear functional $\tilde{\phi}$ on $\mathbb{C}^n$. The form of $\tilde{\phi}$ is $\tilde{\phi}(\lambda_1, \ldots, \lambda_n) = \sum_i \alpha_i \lambda_i$ with $\alpha_i \in \mathbb{C}$. In particular, for all $x \in X$,

$$g(x) = \phi(Tx) = \tilde{\phi}(Tx) = \sum_i \alpha_i f_i(x).$$

□

Theorem 5.23. $X_\Gamma^* = \Gamma$.

Proof. It suffices to prove that if $0 \neq g$ is a Γ-continuous linear functional on X, then $g \in \Gamma$. Let U be the unit disc in $\mathbb{C}$. If g is Γ-continuous, there exists a basic neighborhood $N = N(0; f_1, \ldots, f_n; \epsilon)$ of zero (in the Γ-topology) such that $g(N) \subset U$. If $x \in \bigcap_i \ker f_i := Z$, then $x \in N$, and therefore $|g(x)| < 1$. But Z is a subspace of X, hence $kx \in Z$ for all $k \in \mathbb{N}$, and so $k|g(x)| < 1$ for all k. This shows that $Z \subset \ker g$, and therefore, by Lemma 5.22, $g \in$ span $\{f_1, \ldots, f_n\} \subset \Gamma$. □

The following special Γ-topologies are especially important:

(1) If X is a Banach space and X^* is its conjugate space, the X^*-topology for X is called *the weak topology for* X (the usual norm topology is also called *the strong topology*, and is clearly stronger than the weak topology).

(2) If X^* is the conjugate of the Banach space X, the $\hat{X}$-topology for X^* is called *the weak*-topology for* X^*. It is in general weaker than the weak topology (i.e., the X^{**}-topology) on X^*. The basis described earlier (in the case of the *weak**-topology) consists of the sets

$$N(x^*; x_1, \dots, x_n; \epsilon) = \{y^* \in X^*; |y^* x_k - x^* x_k| < \epsilon\}$$

with $x^* \in X^*, x_k \in X, \epsilon > 0, n \in \mathbb{N}$.

A net $\{x_i^*\}$ converges *weak** to x^* if and only if $x_i^* x \to x^* x$ for all $x \in X$ (this is pointwise convergence of the functions x_i^* to x^* on X!).

Theorem 5.24 (Alaoglu's theorem). *Let X be a Banach space. Then the (strongly) closed unit ball of X^**

$$S^* := \{x^* \in X^*; \|x^*\| \le 1\}$$

is compact in the weak* *topology.*

Proof. Let

$$\Delta(x) = \{\lambda \in \mathbb{C}; |\lambda| \le \|x\|\} \quad (x \in X),$$

and

$$\Delta = \prod_{x \in X} \Delta(x)$$

with the Cartesian product topology. By Tychonoff's theorem, Δ is compact.

If $f \in S^*, f(x) \in \Delta(x)$ for each $x \in X$, so that $f \in \Delta$, that is, $S^* \subset \Delta$. Convergence in the relative Δ-topology on S^* is pointwise convergence at all points $x \in X$, and this is precisely *weak**-convergence in S^*. The theorem will then follow if we show that S^* is closed in Δ. Suppose $\{f_i; i \in I\}$ is a net in X^* converging in Δ to some f. This means that $f_i(x) \to f(x)$ for all $x \in X$. Therefore, for each $x, y \in X$ and $\lambda \in \mathbb{C}$,

$$\begin{aligned} f(x + \lambda y) &= \lim_i f_i(x + \lambda y) = \lim_i [f_i(x) + \lambda f_i(y)] \\ &= \lim_i f_i(x) + \lambda \lim_i f_i(y) = f(x) + \lambda f(y), \end{aligned}$$

and since $|f(x)| \le \|x\|$, we conclude that $f \in S^*$. □

Theorem 5.25 (Goldstine's theorem). *Let S and S^{**} be the strongly closed unit balls in X and X^{**}, respectively, and let $\kappa : S \to S^{**}$ be the canonical embedding (cf. comments following Corollary 5.8). Then κS is weak*-dense in S^{**}.*

Proof. Let $\overline{\kappa S}$ denote the *weak**-closure of κS. Proceeding by contradiction, suppose $x^{**} \in S^{**}$ is not in $\overline{\kappa S}$. We apply Corollary 5.21 in the locally convex t.v.s. X^{**} with the *weak**-topology. There exists then a (*weak**-)continuous linear functional F on X^{**} and a real number λ such that

$$\Re F(x^{**}) > \lambda > \sup_{x \in S} \Re F(\hat{x}), \tag{1}$$

where $\hat{x} := \kappa x$.

The *weak**-topology on X^{**} is the Γ-topology on X^{**} where Γ consists of all the linear functionals on X^{**} of the form $x^{**} \to x^{**}x^*$, with $x^* \in X^*$. By Theorem 5.23, the functional F is of this form, for a suitable x^*. We can then rewrite (1) as follows:

$$\Re x^{**}x^* > \lambda > \sup_{x \in S} \Re x^* x. \tag{2}$$

For any $x \in S$, write $|x^*x| = \omega x^* x$ with $|\omega| = 1$. Then by (2), since $\omega x \in S$ whenever $x \in S$ and $\|x^{**}\| \leq 1$,

$$|x^*x| = x^*(\omega x) = \Re x^*(\omega x) < \lambda < \Re x^{**}x^* \leq |x^{**}x^*| \leq \|x^*\|.$$

Hence

$$\|x^*\| = \sup_{x \in S} |x^*x| \leq \lambda < \|x^*\|,$$

contradiction. □

In contrast to Theorem 5.24, we have:

Theorem 5.26. *The closed unit ball S of a Banach space X is weakly compact if and only if X is reflexive.*

Proof. Observe that $\kappa : S \to S^{**}$ is continuous when S and S^{**} are endowed with the (relative) *weak* and *weak** topologies, respectively. (The net $\{x_i\}_{i \in I}$ converges weakly to x in S if and only if $x^*x_i \to x^*x$ for all $x^* \in X^*$, i.e., $\hat{x}_i(x^*) \to \hat{x}(x^*)$ for all x^*, which is equivalent to $\kappa(x_i) \to \kappa(x)$ *weak**. This shows actually that κ is a homeomorphism of S with the relative *weak* topology and κS with the relative *weak** topology.) Therefore, if we assume that S is weakly compact, it follows that κS is *weak**-compact. Thus κS is *weak**-closed, and since it is *weak**-dense in S^{**} (by Theorem 5.25), it follows that $\kappa S = S^{**}$. By linearity of κ, we then conclude that $\kappa X = X^{**}$, so that X is reflexive.

Conversely, if X is reflexive, $\kappa S = S^{**}$ (since κ is norm-preserving). But S^{**} is *weak**-compact by Theorem 5.24 (applied to the conjugate space X^{**}). Since κ is a homeomorphism of S (with the weak topology) and κS (with the *weak** topology), as we observed earlier, it follows that S is weakly compact. □

It is natural to ask about compactness of S in the strong topology (i.e., the norm-topology). We have:

Theorem 5.27. *The strongly closed unit ball S of a Banach space X is strongly compact iff X is finite dimensional.*

Proof. If X is an n-dimensional Banach space, there exists a (linear) homeomorphism $\tau : X \to \mathbb{C}^n$. Then τS is closed and bounded in $\mathbb{C}^n$, hence compact, by the Heine–Borel theorem. Therefore S is compact in X.

Conversely, if S is (strongly) compact, its open covering $\{B(x,1/2); x \in S\}$ by balls has a finite subcovering $\{B(x_k,1/2); k=1,\dots,n\}$ ($x_k \in S$). Let Y be the linear span of the vectors x_k. Then Y is a closed subspace of X of dimension $\leq n$. Suppose $x^* \in X^*$ vanishes on Y. Given $x \in S$, there exists $k, 1 \leq k \leq n$, such that $x \in B(x_k, 1/2)$. Then

$$|x^*x| = |x^*(x-x_k)| \leq \|x^*\|\,\|x-x_k\| \leq \|x^*\|/2,$$

and therefore

$$\|x^*\| = \sup_{x \in S} |x^*x| \leq \|x^*\|/2,$$

this shows that $\|x^*\| = 0$, and so $X = Y$ by Corollary 5.4. □

It follows from Theorems 5.24 and 5.27 that the weak*-topology is strictly weaker than the strong topology on X^* when X (hence X^*) is infinite dimensional. Similarly, by Theorems 5.26 and 5.27, the weak topology on an infinite-dimensional reflexive Banach space is strictly weaker than the strong topology.

5.6 Extremal points

As an application of the strict separation theorem (cf. Corollary 5.21), we shall prove the Krein–Milman theorem on extremal points.

Let X be a vector space (over $\mathbb{C}$ or $\mathbb{R}$). If $x, y \in X$, denote

$$\overline{xy} := \{\alpha x + (1-\alpha)y;\ 0 < \alpha < 1\}.$$

Let $K \subset X$. A non-empty subset $A \subset K$ is *extremal* in K if

$$[x, y \in K;\ \overline{xy} \cap A \neq \emptyset] \quad \text{implies} \quad [x, y \in A].$$

If $A = \{a\}$ (a singleton) is extremal in K, we say that a is an *extremal point* of K: the criterion for this is

$$[x, y \in K; a \in \overline{xy}] \quad \text{implies} \quad [x = y = a].$$

Trivially, any non-empty K is extremal in itself. If B is extremal in A and A is extremal in K, then B is extremal in K. The non-empty intersection of a family of extremal sets in K is an extremal set in K.

From now on, let X be a *locally convex t.v.s.* and $K \subset X$. If A is a compact extremal set in K that contains no proper compact extremal subset, we call it a *minimal compact extremal set* in K.

Lemma 5.28. *A minimal compact extremal set A in K is a singleton.*

Proof. Suppose A contains two distinct points a, b. There exists $x^* \in X^*$ such that $f := \Re x^*$ assumes distinct values at these points (cf. Corollary 5.21). Let $\rho = \min_A f$; since A is compact, the minimum ρ is attained on a non-empty subset $B \subset A$, and $B \neq A$ since f is not constant on A ($f(a) \neq f(b)$!). The set B is a *closed* subset of the compact set A, and is therefore compact. We complete the proof by showing that B is extremal in A (hence in K, contradicting the minimality assumption on A). Let $x, y \in A$ be such that $\overline{xy} \cap B \neq \emptyset$. Then there exists $\alpha \in (0,1)$ such that

$$\rho = f(\alpha x + (1-\alpha)y) = \alpha f(x) + (1-\alpha)f(y). \tag{1}$$

We have $f(x) \geq \rho$ and $f(y) \geq \rho$; if either inequality is strict, we get the contradiction $\rho > \rho$ in (1). Hence, $f(x) = f(y) = \rho$, that is, $x, y \in B$. □

Lemma 5.29. *If $K \neq \emptyset$ is compact, then it has extremal points.*

Proof. Let $\mathcal{A}$ be the family of all compact extremal subsets of K. It is non-empty, since $K \in \mathcal{A}$, and partially ordered by set inclusion. If $\mathcal{B} \subset \mathcal{A}$ is totally ordered, then $\bigcap \mathcal{B}$ is a non-empty compact extremal set in K, that is, belongs to $\mathcal{A}$, and is a lower bound for $\mathcal{B}$. By Zorn's lemma, $\mathcal{A}$ has a minimal element, which is a singleton $\{a\}$ by Lemma 5.28. Thus, K has the extremal point a. □

If $E \subset X$, its *closed convex hull* $\overline{\text{co}}(E)$ is defined as the closure of its convex hull $\text{co}(E)$.

Theorem 5.30 (Krein–Milman's theorem). *Let X be a locally convex t.v.s., and let $K \subset X$ be compact. Let E be the set of extremal points of K. Then $K \subset \overline{\text{co}}(E)$.*

Proof. We may assume $K \neq \emptyset$, and therefore $E \neq \emptyset$ by Lemma 5.29. Hence $N := \overline{\text{co}}(E)$ is a non-empty closed convex set. Suppose there exists $x \in K$ such that $x \notin N$. By Corollary 5.21, there exists $x^* \in X^*$ such that $f(x) < \inf_N f$ (where $f = \Re x^*$). Let

$$B = \{k \in K; f(k) = \rho := \min_K f\}.$$

Then B is extremal in K (cf. proof of Lemma 5.28). Also B is a non-empty closed subset of the compact set K, hence is a non-empty compact set, and has therefore an extremal point b, by Lemma 5.29. Therefore, b is an extremal point of K, that is, $b \in E \subset N$. Hence

$$\rho = f(b) \geq \inf_N f > f(x) \geq \min_K f = \rho,$$

contradiction. □

Corollary 5.31. *(With assumptions and notation as in Theorem 5.30.)*

$$\overline{\text{co}}(K) = \overline{\text{co}}(E).$$

In particular, a compact convex set in a locally convex t.v.s. is the closed convex hull of its extremal points.

Consider for example the strongly closed unit ball S^* of the conjugate X^* of a normed space X. By Theorem 5.24, it is *weak**-compact and trivially convex. It is therefore the *weak**-closed convex hull of its extremal points.

This remark will be applied as follows. Let X be a compact Hausdorff space, let $C(X)$ be the space of all complex continuous functions on X, and let $M(X)$ be the space of all regular complex Borel measures on X (cf. Theorem 4.9). By Theorem 4.9 (for X compact!), the space $M(X)$ is isometrically isomorphic to the dual space $C(X)^*$. Its strongly closed unit ball $M(X)_1$ is then *weak**-compact (by Theorem 5.24).

If $\mathcal{A}$ is any subset of $C(X)$, let

$$Y := \left\{\mu \in M(X); \int_X f\,d\mu = 0 \quad (f \in \mathcal{A})\right\}. \tag{2}$$

Clearly Y is *weak**-closed, and therefore $K := Y \cap M(X)_1$ is *weak**-compact and trivially convex. It follows from Corollary 5.31 that K is the *weak**-closed convex hull of its extremal points.

If $\mathcal{A}$ is a closed subspace of $C(X)$ ($\mathcal{A} \neq C(X)$), it follows from Corollary 5.4 and Theorem 4.9 that $Y \neq \{0\}$ (and K is the strongly closed unit ball of the Banach space Y).

Lemma 5.32. *Let S be the strongly closed unit ball of a normed space $Y \neq \{0\}$, and let a be an extremal point of S. Then $\|a\| = 1$.*

Proof. Since $Y \neq \{0\}$, there exists $0 \neq y \in S$. Then $0 \neq -y \in S$ and $0 = (1/2)y + (1/2)(-y)$, so that 0 is not extremal for S. Therefore, $a \neq 0$. Define then $b = a/\|a\|(\in S)$. If $\|a\| < 1$, write $a = \|a\|b + (1 - \|a\|)0$, which is a proper convex combination of two elements of S distinct from a, contradicting the hypothesis that a is extremal. Hence, $\|a\| = 1$. □

With notations and hypothesis as in the paragraph preceding the lemma, let $\mu \in K$ be an extremal point of K (so that $\|\mu\| = 1$), and let $E = \operatorname{supp}|\mu|$ (cf. Definition 3.26). Then $E \neq \emptyset$ (since $\|\mu\| = 1$!), and by Remark 4.10,

$$\int_X f\,d\mu = \int_E f\,d\mu \quad (f \in C(X)). \tag{3}$$

Lemma 5.33. *Let $\mathcal{A} \neq C(X)$ be a closed subalgebra of $C(X)$ containing the identity 1. For K as mentioned, let μ be an extremal point of K, and let $E =$ $\operatorname{supp}|\mu|$. If $f \in \mathcal{A}$ is real on E, then f is constant on E.*

Proof. Assume first that $f \in \mathcal{A}$ has range in (0, 1) over E, and consider the measures $d\sigma = f\,d\mu$ and $d\tau = (1-f)\,d\mu$. Write $d\mu = h\,d|\mu|$ with h measurable and $|h| = 1$ (cf. Theorem 1.46). By Theorem 1.47,

$$d|\sigma| = |fh|\,d|\mu| = f\,d|\mu|; \quad d|\tau| = |(1-f)h|\,d|\mu| = (1-f)\,d|\mu|, \tag{4}$$

hence

$$\|\sigma\| = \int_X f\,d|\mu| = \int_E f\,d|\mu|, \tag{5}$$

and similarly

$$\|\tau\| = \int_E (1-f)\, d|\mu|. \tag{6}$$

Therefore

$$\|\sigma\| + \|\tau\| = \int_E d|\mu| = |\mu|(E) = \|\mu\| = 1. \tag{7}$$

Since f and $1-f$ do not vanish identically on the support E of $|\mu|$, it follows from (5) and (6) and the discussion in Section 3.26 that $\|\sigma\| > 0$ and $\|\tau\| > 0$. The measures $\sigma' = \sigma/\|\sigma\|$ and $\tau' = \tau/\|\tau\|$ are in $M(X)_1$, and for all $g \in \mathcal{A}$,

$$\int_X g\, d\sigma' = \frac{1}{\|\sigma\|}\int_X gf\; d\mu = 0,$$

and similarly

$$\int_X g\, d\tau' = \frac{1}{\|\tau\|}\int_X g(1-f)\, d\mu = 0,$$

since $\mathcal{A}$ is an algebra. This means that σ' and τ' are in K, and clearly

$$\mu = \|\sigma\|\sigma' + \|\tau\|\tau',$$

which is (by (7)) a proper convex combination. Since μ is an extremal point of K, it follows that $\mu = \sigma'$. Therefore

$$\int_X g(f - \|\sigma\|)\, d\mu = 0$$

for all bounded Borel functions g on X. Choose in particular $g = (f - \|\sigma\|)\bar{h}$. Then

$$\int_E (f - \|\sigma\|)^2\, d|\mu| = 0,$$

and consequently (cf. discussion following Definition 3.26) $f = \|\sigma\|$ identically on E.

If $f \in \mathcal{A}$ is real on E, there exist $\alpha \in \mathbb{R}$ and $\beta > 0$ such that $0 < \beta(f-\alpha) < 1$ on E. Since $1 \in \mathcal{A}$, the function $f_0 := \beta(f-\alpha)$ belongs to $\mathcal{A}$ and has range in $(0, 1)$. By the first part of the proof, f_0 is constant on E, and therefore f is constant on E. □

A non-empty subset $E \subset X$ with the property of the conclusion of Lemma 5.33 (i.e., any function of $\mathcal{A}$ that is real on E is necessarily constant on E) is called an *antisymmetric set (for $\mathcal{A}$)*. If $x, y \in X$ are contained in some antisymmetric set, they are said to be equivalent (with respect to $\mathcal{A}$). This is an equivalence relation. Let $\mathcal{E}$ be the family of all equivalence classes. If E is antisymmetric, and $\tilde{E}$ is the equivalence class of any $p \in E$, then $E \subset \tilde{E} \in \mathcal{E}$.

Theorem 5.34 (Bishop's antisymmetry theorem). *Let X be a compact Hausdorff space. Let $\mathcal{A}$ be a closed subalgebra of $C(X)$ containing the constant function 1. Define $\mathcal{E}$ as described. Suppose $g \in C(X)$ has the property:*

(B) *For each* $E \in \mathcal{E}$, *there exists* $f \in \mathcal{A}$ *such that* $g = f$ *on* E.

Then $g \in \mathcal{A}$.

Proof. We may assume that $\mathcal{A} \neq C(X)$. Suppose that $g \in C(X)$ satisfies Condition (B), but $g \notin \mathcal{A}$. By the Hahn–Banach theorem and the Riesz representation theorem, there exists $\nu \in M(X)_1$ such that $\int_X h\, d\nu = 0$ for all $h \in \mathcal{A}$ and $\int_X g\, d\nu \neq 0$. In particular, $K \neq \emptyset$ (since $\nu \in K$), and is the *weak**-closed convex hull of its extremal points. Let μ be an extremal point of K. By Lemma 5.33, the set $E := \mathrm{supp}|\mu|$ is antisymmetric for $\mathcal{A}$. Let $\tilde{E} \in \mathcal{E}$ be as defined previously. By Condition (B), there exists $f \in \mathcal{A}$ such that $g = f$ on $\tilde{E}$, hence on $\mathrm{supp}|\mu| := E \subset \tilde{E}$. Therefore,

$$\int_X g\, d\mu = \int_X f\, d\mu = 0,$$

since $\mu \in K \subset Y$ and $f \in \mathcal{A}$. Since K is the *weak**-closed convex hull of its extremal points and $\nu \in K$, it follows that $\int_X g\, d\nu = 0$, contradiction! □

We say that $\mathcal{A}$ is *selfadjoint* if $\bar{f} \in \mathcal{A}$ whenever $f \in \mathcal{A}$. This implies of course that $\Re f$ and $\Im f$ are in $\mathcal{A}$ whenever $f \in \mathcal{A}$ (and conversely); $\mathcal{A}$ *separates points* (*of* X) if whenever $x, y \in X$ are distinct points, there exists $f \in \mathcal{A}$ such that $f(x) \neq f(y)$.

Corollary 5.35 (Stone–Weierstrass theorem). *Let X be a compact Hausdorff space, and let $\mathcal{A}$ be a closed selfadjoint subalgebra of $C(X)$ containing 1 and separating points of X. Then $\mathcal{A} = C(X)$.*

Proof. If E is antisymmetric for $\mathcal{A}$ and $f \in \mathcal{A}$, the real functions $\Re f, \Im f \in \mathcal{A}$ are necessarily constant on E, and therefore f is constant on E. Since $\mathcal{A}$ separates points, it follows that E is a singleton. Hence, equivalent points must coincide, and so $\mathcal{E}$ consists of all singletons. But then Condition (B) is trivially satisfied by *any* $g \in C(X)$: given $\{p\} \in \mathcal{E}$, choose $f = g(p)1 \in \mathcal{A}$, then surely $g = f$ on $\{p\}$. □

5.7 The Stone–Weierstrass theorem

The Stone–Weierstrass theorem is one of the fundamental theorems of functional analysis, and it is worth giving it also an elementary proof, independent of the machinery developed previously in the chapter.

Let $C_R(X)$ denote the algebra (over $\mathbb{R}$) of all *real* continuous functions on X, and let $\mathcal{A}$ be a subalgebra (over $\mathbb{R}$). Since $h(u) := u^{1/2} \in C_R([0,1])$ and $h(0) = 0$, the classical Weierstrass approximation theorem establishes the existence of polynomials p_n without free coefficient, converging to h uniformly on $[0, 1]$. Given $f \in \mathcal{A}$, the function $u(x) := (f(x)^2/\|f\|^2) : X \to [0,1]$ belongs to $\mathcal{A}$, and therefore $p_n \circ u \in \mathcal{A}$ converge uniformly on X to $h(u(x)) = |f(x)|/\|f\|$. Hence $|f| = \|f\| \cdot (|f|/\|f\|) \in \bar{\mathcal{A}}$, where $\bar{\mathcal{A}}$ denotes the closure of $\mathcal{A}$ in $C_R(X)$ with respect to the uniform norm.

If $f, g \in \mathcal{A}$, since

$$\max(f,g) = \tfrac{1}{2}[f + g + |f - g|]; \qquad \min(f,g) = \tfrac{1}{2}[f + g - |f - g|],$$

it follows from the preceding conclusion that $\max(f,g)$ and $\min(f,g)$ belong to $\bar{\mathcal{A}}$ as well.

Formally:

Lemma 5.36. *If $\mathcal{A}$ is a subalgebra of $C_{\mathbb{R}}(X)$, then $|f|, \max(f,g)$ and $\min(f,g)$ belong to the uniform closure $\bar{\mathcal{A}}$, for any $f, g \in \mathcal{A}$.*

Lemma 5.37. *Let $\mathcal{A}$ be a separating subspace of $C_{\mathbb{R}}(X)$ containing 1, then for any distinct points $x_1, x_2 \in X$ and any $\alpha_1, \alpha_2 \in \mathbb{R}$, there exists $h \in \mathcal{A}$ such that $h(x_k) = \alpha_k, k = 1, 2$.*

Proof. By hypothesis, there exists $g \in \mathcal{A}$ such that $g(x_1) \neq g(x_2)$. Take

$$h(x) := \alpha_1 + \frac{\alpha_2 - \alpha_1}{g(x_2) - g(x_1)}[g(x) - g(x_1)].$$

□

We state now the Stone–Weierstrass theorem as an approximation theorem (real case first).

Theorem 5.38. *Let X be a compact Hausdorff space. Let $\mathcal{A}$ be a separating subalgebra of $C_{\mathbb{R}}(X)$ containing 1. Then $\mathcal{A}$ is dense in $C_{\mathbb{R}}(X)$.*

Proof. Let $f \in C_{\mathbb{R}}(X)$ and $\epsilon > 0$ be given. Fix $x_0 \in X$. For any $x' \in X$, there exists $f' \in \mathcal{A}$ such that

$$f'(x_0) = f(x_0); \qquad f'(x') \leq f(x') + \epsilon/2$$

(cf. Lemma 5.37).

By continuity of f and f', there exists an open neighborhood $V(x')$ of x' such that $f' \leq f + \epsilon$ on $V(x')$. By compactness of X, there exist $x_k \in X, k = 1, \ldots, n$, such that

$$X = \bigcup_{k=1}^{n} V(x_k).$$

Let $f_k \in \mathcal{A}$ be the function f' corresponding to the point $x' = x_k$ as shown, and let

$$g := \min(f_1, \ldots, f_n).$$

Then $g \in \bar{\mathcal{A}}$ by Lemma 5.36, and

$$g(x_0) = f(x_0); \quad g \leq f + \epsilon \quad \text{on } X.$$

By continuity of f and g, there exists an open neighborhood $W(x_0)$ of x_0 such that

$$g \geq f - \epsilon \quad \text{on } W(x_0).$$

We now vary x_0 (over X). The open cover $\{W(x_0); x_0 \in X\}$ of X has a finite subcover $\{W_1, \ldots, W_m\}$, corresponding to functions $g_1, \ldots, g_m \in \bar{\mathcal{A}}$ as shown. Thus

$$g_i \leq f + \epsilon \quad \text{on } X$$

and

$$g_i \geq f - \epsilon \quad \text{on } W_i.$$

Define

$$h = \max(g_1, \ldots, g_m).$$

Then $h \in \bar{\mathcal{A}}$ by Lemma 5.36, and

$$f - \epsilon \leq h \leq f + \epsilon \quad \text{on } X.$$

Therefore, $f \in \bar{\mathcal{A}}$. □

Theorem 5.39 (The Stone–Weierstrass Theorem). *Let X be a compact Hausdorff space, and let $\mathcal{A}$ be a separating selfadjoint subalgebra of $C(X)$ containing* 1. *Then $\mathcal{A}$ is dense in $C(X)$.*

Proof. Let $\mathcal{A}_R$ be the algebra (over $\mathbb{R}$) of all *real* functions in $\mathcal{A}$. It contains 1. Let x_1, x_2 be distinct points of X, and let then $h \in \mathcal{A}$ be such that $h(x_1) \neq h(x_2)$. Then either $\Re h(x_1) \neq \Re h(x_2)$ or $\Im h(x_1) \neq \Im h(x_2)$ (or both). Since $\Re h = (h + \bar{h})/2 \in \mathcal{A}_R$ (since $\mathcal{A}$ is selfadjoint), and similarly $\Im h \in \mathcal{A}_R$, it follows that $\mathcal{A}_R$ is separating.

Let $f \in C(X)$. Then $\Re f \in C_R(X)$, and therefore, by Theorem 5.38, there exists a sequence $g_n \in \mathcal{A}_R$ converging to $\Re f$ uniformly on X. Similarly, there exists a sequence $h_n \in \mathcal{A}_R$ converging to $\Im f$ uniformly on X. Then $g_n + ih_n \in \mathcal{A}$ converge to f uniformly on X. □

5.8 Operators between Lebesgue spaces: Marcinkiewicz's interpolation theorem

5.40. Weak and strong types. Let $(X, \mathcal{A}, \mu)$ and $(Y, \mathcal{B}, \nu)$ be measure spaces, and let $p, q \in [1, \infty]$ (presently, q is *not* the conjugate exponent of p!). We consider operators T defined on $L^p(\mu)$ such that Tf is a $\mathcal{B}$-measurable function on Y and

$$|T(f+g)| \leq |Tf| + |Tg| \quad (f, g \in L^p(\mu)). \tag{1}$$

We refer to such operators T as *sublinear operators.* (This includes linear operators with range as described.) Let n be the distribution function of $|Tf|$ for some *fixed non-zero element* f of $L^p(\mu)$, relative to the measure ν on $\mathcal{B}$, that is (cf. Definition 1.58)

$$n(y) := \nu([|Tf| > y]) \quad (y > 0).$$

We say that T *is of weak type* (p,q) if there exists a constant $0 < M < \infty$ independent of f and y such that

$$n(y)^{1/q} \leq M\,\|f\|_p/y \tag{2}$$

in case $q < \infty$, and

$$\|Tf\|_\infty \leq M\,\|f\|_p \tag{3}$$

in case $q = \infty$. We say that T *is of strong type* (p,q) if there exists M (as above) such that

$$\|Tf\|_q \leq M\,\|f\|_p. \tag{4}$$

The concepts of weak and strong types (p,∞) coincide by definition, while in general strong type (p,q) implies weak type (p,q), because by (2) in Definition 1.58 (for $q < \infty$)

$$n(y)^{1/q} \leq \|Tf\|_q/y \leq M\,\|f\|_p/y. \tag{5}$$

The infimum of all M for which (2) or (4) is satisfied is called the weak or strong (p,q)-norm of T, respectively. By (5), the weak (p,q)-norm of T is no larger than its strong (p,q) norm.

A *linear* operator T is of strong type (p,q) iff $T \in B(L^p(\mu), L^q(\nu))$, and in that case, the strong (p,q)-norm is the corresponding operator norm of T.

In the sequel, we consider only the case $q < \infty$. If T is of weak type (p,q), it follows from (2) that n is finite and $n(\infty) = 0$, and consequently, by Theorem 1.59,

$$\|Tf\|_q^q = q\int_0^\infty y^{q-1} n(y)\,dy, \tag{6}$$

where the two sides could be finite or infinite.

Let $u > 0$, and consider the decomposition $f = f_u + f'_u$ as in Technique 1.61. Since f_u and f'_u are both in $L^p(\mu)$ (for $f \in L^p(\mu)$), the given sublinear operator T is defined on them, and

$$|Tf| \leq |Tf_u| + |Tf'_u|. \tag{7}$$

Let n_u and n'_u denote the distribution functions of $|Tf_u|$ and $|Tf'_u|$, respectively. Since by (7)

$$[|Tf| > y] \subset [|Tf_u| > y/2] \cup [|Tf'_u| > y/2] \quad (y > 0),$$

we have

$$n(y) \leq n_u(y/2) + n'_u(y/2). \tag{8}$$

We now *assume* that T is of weak types (r,s) and (a,b), with $r \leq s$ and $a \leq b$, and $s \neq b$. *We consider the case* $r \neq a$ *(without loss of generality,* $r > a$*) and* $s > b$. Denote the respective weak norms of T by M and N, and

$$s/r := \sigma\ (\geq 1), \qquad b/a := \beta\ (\geq 1).$$

Let $a < p < r$ and $f \in L^p(\mu)$ such that $\|f\|_p > 0$. By Technique 1.61, $f_u \in L^r(\mu)$ and $f'_u \in L^a(\mu)$. Since T is of weak type (r, s) with weak (r, s)-norm M, we have

$$n_u(y) \leq M^s y^{-s} \|f_u\|_r^s. \tag{9}$$

Since T is of weak type (a, b) with weak (a, b)-norm N, we have

$$n'_u(y) \leq N^b y^{-b} \|f'_u\|_a^b. \tag{10}$$

By (8)–(10),

$$n(y) \leq (2M)^s y^{-s} \|f_u\|_r^s + (2N)^b y^{-b} \|f'_u\|_a^b. \tag{11}$$

Let $b < q < s$. By Theorem 1.59 and (11),

$$\begin{aligned}(1/q)\|Tf\|_q^q &= \int_0^\infty y^{q-1} n(y)\, dy \\ &\leq (2M)^s \int_0^\infty y^{q-s-1} \|f_u\|_r^s \, dy + (2N)^b \int_0^\infty y^{q-b-1} \|f'_u\|_a^b \, dy. \end{aligned}\tag{12}$$

Since $a < p < r$, we may apply Formulae (8) and (9) of Technique 1.61; we then conclude from (12) that

$$\begin{aligned}(1/q)\|Tf\|_q^q \leq{}& (2M)^s r^\sigma \int_0^\infty y^{q-s-1} \left(\int_0^u v^{r-1} m(v)\, dv \right)^\sigma dy \\ &+ (2N)^b a^\beta \int_0^\infty y^{q-b-1} \left(\int_u^\infty (v-u)^{a-1} m(v)\, dv \right)^\beta dy. \end{aligned}\tag{13}$$

In this formula, we may also take u dependent monotonically on y (integrability is then clear). Denote the two integrals in (13) by Φ and Ψ. Since $\sigma, \beta \geq 1$, it follows from Corollary 5.8 and Theorem 4.6 (applied to the spaces $L^\sigma(\mathbb{R}^+, \mathcal{M}, y^{q-s-1}\, dy)$ and $L^\beta(\mathbb{R}^+, \mathcal{M}, y^{q-b-1}\, dy)$ respectively, where $\mathcal{M}$ is the Lebesgue σ-algebra over $\mathbb{R}^+$) that

$$\Phi^{1/\sigma} = \sup \int_0^\infty y^{q-s-1} \left(\int_0^u v^{r-1} m(v)\, dv \right) g(y)\, dy, \tag{14}$$

where the supremum is taken over all measurable functions $g \geq 0$ on $\mathbb{R}^+$ such that $\int_0^\infty y^{q-s-1} g^{\sigma'}(y)\, dy \leq 1$ (σ' denotes here the conjugate exponent of σ).

Similarly

$$\Psi^{1/\beta} = \sup \int_0^\infty y^{q-b-1} \left(\int_u^\infty (v-u)^{a-1} m(v)\, dv \right) h(y)\, dy, \tag{15}$$

where the supremum is taken over all measurable function $h \geq 0$ on $\mathbb{R}^+$ such that $\int_0^\infty y^{q-b-1} h^{\beta'}(y)\, dy \leq 1$ (β' is the conjugate exponent of β).

We now choose $u = (y/c)^k$, where c, k are positive parameters to be determined later. By Tonelli's theorem, the integral in (14) is equal to

$$\int_0^\infty v^{r-1} m(v) \left(\int_{cv^{1/k}}^\infty y^{q-s-1} g(y)\, dy \right) dv.$$

By Holder's inequality on the measure space $(\mathbb{R}^+, \mathcal{M}, y^{q-s-1}dy)$, the inner integral is

$$\leq \left(\int_{cv^{1/k}}^{\infty} y^{q-s-1}\, dy\right)^{1/\sigma} \left(\int_0^{\infty} y^{q-s-1} g^{\sigma'}(y)\, dy\right)^{1/\sigma'}$$

$$\leq \left(\frac{y^{q-s}}{q-s}\bigg|_{cv^{1/k}}^{\infty}\right)^{1/\sigma} = (s-q)^{-1/\sigma} c^{(q-s)/\sigma} v^{(q-s)/k\sigma}.$$

Consequently

$$\Phi \leq (s-q)^{-1} c^{q-s} \left(\int_0^{\infty} v^{r-1+(q-s)/k\sigma} m(v)\, dv\right)^{\sigma}. \tag{16}$$

Similarly, the integral in (15) is equal to

$$\int_0^{\infty} \left(\int_0^{cv^{1/k}} y^{q-b-1} h(y)[v-(y/c)^k]^{a-1}\, dy\right) m(v)\, dv$$

$$\leq \int_0^{\infty} v^{a-1} m(v) \left(\int_0^{cv^{1/k}} y^{q-b-1} h(y)\, dy\right) dv$$

$$\leq \int_0^{\infty} v^{a-1} m(v) \left(\int_0^{cv^{1/k}} y^{q-b-1}\, dy\right)^{1/\beta} \left(\int_0^{cv^{1/k}} y^{q-b-1} h^{\beta'}(y) dy\right)^{1/\beta'} dv$$

$$\leq \int_0^{\infty} v^{a-1} m(v) \left(\frac{y^{q-b}}{q-b}\bigg|_0^{cv^{1/k}}\right)^{1/\beta} dv$$

$$= (q-b)^{-1/\beta} c^{(q-b)/\beta} \int_0^{\infty} v^{a-1+(q-b)/k\beta} m(v)\, dv.$$

Therefore

$$\Psi \leq (q-b)^{-1} c^{q-b} \left(\int_0^{\infty} v^{a-1+(q-b)/k\beta} m(v)\, dv\right)^{\beta}. \tag{17}$$

Since $b < q < s$, the integrals in (16) and (17) contain the terms $v^{\kappa-1}$ and $v^{\lambda-1}$ respectively, with $\kappa := r + (q-s)/k\sigma < r$ and $\lambda := a + (q-b)/k\beta > a$. Recall that we also have $a < p < r$. If we can choose the parameter k so that $\kappa = \lambda = p$, then by Corollary 1.60 both integrals will be equal to $(1/p)\|f\|_p^p$. Since $\beta = b/a$ and $\sigma = s/r$, the unique solutions for k of the equations $\kappa = p$ and $\lambda = p$ are

$$k = (a/b)\frac{q-b}{p-a} \quad \text{and} \quad k = (r/s)\frac{s-q}{r-p}, \tag{18}$$

respectively, so that the choice of k is possible iff the two expressions in (18) coincide. Multiplying both expressions by p/q, this *condition on* p, q can be rearranged as

$$\frac{(1/b)-(1/q)}{(1/a)-(1/p)} = \frac{(1/q)-(1/s)}{(1/p)-(1/r)},$$

or equivalently

$$\frac{(1/b)-(1/q)}{(1/q)-(1/s)}=\frac{(1/a)-(1/p)}{(1/p)-(1/r)}, \tag{19}$$

that is, $1/p$ and $1/q$ divide the segments $[1/r,1/a]$ and $[1/s,1/b]$, respectively, according to the same (positive, finite) ratio. Equivalently,

$$\frac{1}{p}=(1-t)\frac{1}{r}+t\frac{1}{a}; \qquad \frac{1}{q}=(1-t)\frac{1}{s}+t\frac{1}{b} \tag{20}$$

for some $t\in(0,1)$.

With the choice of k, it now follows from (13), (16), and (17) that

$$\begin{aligned}(1/q)\|Tf\|_q^q &\le (2M)^s(r/p)^\sigma(s-q)^{-1}c^{q-s}\|f\|_p^{p\sigma}\\ &\quad+(2N)^b(a/p)^\beta(q-b)^{-1}c^{q-b}\|f\|_p^{p\beta}. \end{aligned} \tag{21}$$

We now choose the parameter c in the form

$$c=(2M)^x(2N)^z\|f\|_p^w,$$

with x,z,w real parameters to be determined so that the two summands on the right-hand side of (21) contain the same powers of M,N, and $\|f\|_p$, respectively. This yields to the following equations for the unknown parameters x,z,w:

$$s+(q-s)x=(q-b)x; \quad b+(q-b)z=(q-s)z; \quad p\sigma+(q-s)w=p\beta+(q-b)w.$$

The unique solution is

$$x=\frac{s}{s-b}; \quad z=\frac{-b}{s-b}; \quad w=p\frac{\sigma-\beta}{s-b}.$$

With the choice of c, the right-hand side of (21) is equal to

$$[(r/p)^\sigma(s-q)^{-1}+(a/p)^\beta(q-b)^{-1}](2M)^{s(q-b)/(s-b)}(2N)^{b(s-q)/(s-b)}\|f\|_p^{p\gamma}, \tag{22}$$

where

$$\begin{aligned}\gamma=\beta+(q-b)\frac{\sigma-\beta}{s-b}&=\beta\frac{s-q}{s-b}+\sigma\frac{q-b}{s-b}\\ &=t(q/a)+(1-t)(q/r)=q/p,\end{aligned}$$

by the relations (20) and $\beta=b/a,\sigma=s/r$. By (20), the exponents of $2M$ and $2N$ in (22) are equal to $(1-t)q$ and tq, respectively. By (21) and (22), we conclude that

$$\|Tf\|_q\le KM^{1-t}N^t\|f\|_p,$$

where

$$K:=2q^{1/q}[(r/p)^{s/r}(s-q)^{-1}+(a/p)^{b/a}(q-b)^{-1}]^{1/q}$$

does not depend on f. Thus, T *is of strong type* (p,q), *with strong* (p,q)*-norm* $\le KM^{1-t}N^t$. Note that the constant K depends only on the parameters a,r,b,s and p,q; it tends to ∞ when q approachs either b or s.

A similar argument (which we omit) yields the same conclusion in case $s < b$. The result (which we proved for $r \neq a$ and *finite* b, s) is also valid when $r = a$ and one or both exponents b, s are infinite (again, we omit the details). We formalize our conclusion in the following:

Theorem 5.41 (Marcinkiewicz's interpolation theorem). *Let $1 \leq a \leq b \leq \infty$ and $1 \leq r \leq s \leq \infty$. For $0 < t < 1$, let p, q be such that*

$$\frac{1}{p} = (1-t)\frac{1}{r} + t\frac{1}{a} \quad \text{and} \quad \frac{1}{q} = (1-t)\frac{1}{s} + t\frac{1}{b}. \tag{23}$$

Suppose that the sublinear operator T is of weak types (r, s) and (a, b), with respective weak norms M and N. Then T is of strong type (p, q), with strong (p, q)-norm $\leq K M^{1-t} N^t$.

The constant K depends only on the parameters a, b, r, s, and t. For a, b, r, s fixed, $K = K(t)$ is *bounded* for t bounded away from the end points $0, 1$.

Corollary 5.42. *Let a, b, r, s, t be as in Theorem 5.41. For any $p, q \in [1, \infty]$, denote the strong (p, q)-norm of the sublinear operator T by $\|T\|_{p,q}$. If T is of strong types (r, s) and (a, b), then T is of strong type (p, q) whenever (23) is satisfied, and*

$$\|T\|_{p,q} \leq K \|T\|_{r,s}^{1-t} \|T\|_{a,b}^{t},$$

with K as in Theorem 5.41.

In particular,

$$\begin{aligned} &B(L^r(X, \mathcal{A}, \mu), L^s(Y, \mathcal{B}, \nu)) \cap B(L^a(X, \mathcal{A}, \mu), L^b(Y, \mathcal{B}, \nu)) \\ &\quad \subset B(L^p(X, \mathcal{A}, \mu), L^q(Y, \mathcal{B}, \nu)). \end{aligned}$$

5.9 Fixed points

Let X be a vector space and $K \subset X$ a convex subset. A map $T : K \to K$ is *affine* if

$$T(px + (1-p)y) = pTx + (1-p)Ty$$

for all $x, y \in K$ and all $p \in [0, 1]$. For example, for any fixed $a \in X$ and $S : X \to X$ linear, the map $T : X \to X$ given by

$$Tx = a + Sx \qquad (x \in X)$$

is affine, and if it maps K into itself, then $T|_K : K \to K$ is affine as well.

If $\mathcal{T}$ is any family of maps of K into itself, a *fixed point* for $\mathcal{T}$ is an element $x \in K$ such that $Tx = x$ for all $T \in \mathcal{T}$.

Theorem 5.43 (Markov–Kakutani's fixed point theorem). *Let X be a t.v.s. and let $K \subset X$ be a non-empty, compact, convex subset of X. Let $\mathcal{T}$ be a non-empty family of pairwise commuting, continuous, affine maps of K into itself. Then $\mathcal{T}$ has a fixed point.*

Proof. We combine two common tools: *averaging* and *compactness*.

For each $T \in \mathcal{T}$ and $n \in \mathbb{N}$, denote

$$T_n := \frac{1}{n} \sum_{j=0}^{n-1} T^j : K \to K$$

(the *average* of the powers $T^0, T^1, \ldots, T^{n-1}$), where $T^0 := I$, the identity map. This function indeed maps into K because K is convex. Observe:

(a) The sets $T_n K$ are compact for all $T \in \mathcal{T}$ and $n \in \mathbb{N}$, because K is compact and T_n is continuous.

(b) If $S, T, \ldots, V \in \mathcal{T}$ and $m, n, \ldots, r \in \mathbb{N}$, then $T_n \cdots V_r K \subset K$, and therefore

$$L := S_m T_n \cdots V_r K \subset S_m K.$$

By the commutativity assumption and the same fact applied to T_n,

$$L = T_n S_m \cdots V_r K \subset T_n K,$$

$$\vdots$$

$$L = V_r \cdots S_m T_n K \subset V_r K.$$

Hence

$$\emptyset \neq L \subset S_m K \cap T_n K \cap \ldots \cap V_r K.$$

We conclude from (a) and (b) that the family

$$\mathcal{K} := \{T_n K; T \in \mathcal{T}, n \in \mathbb{N}\}$$

is a non-empty family of compact subsets of K, which has the finite intersection property. Therefore, $\bigcap \mathcal{K} \neq \emptyset$. Let k be any element in this intersection. Then $k \in K$, and we can verify that $Tk = k$ for all $T \in \mathcal{T}$, that is, k is a fixed point of $\mathcal{T}$. Let U be a 0-neighborhood. Since K is compact and non-empty, $K - K$ is compact and contains 0. Therefore, there exists $n \in \mathbb{N}$ such that $K - K \subset nU$. Fix such an n, and let T be any map in $\mathcal{T}$. Since $k \in T_n K$, we can write $k = T_n k_1$ for some $k_1 \in K$. By the affinity property of T, we have

$$Tk - k = T(T_n k_1) - T_n k_1 = T(\frac{1}{n} \sum_{j=0}^{n-1} T^j k_1) - T_n k_1$$

$$= \frac{1}{n} \sum_{j=1}^{n} T^j k_1 - \frac{1}{n} \sum_{j=0}^{n-1} T^j k_1$$

$$= \frac{1}{n}(T^n k_1 - k_1) = \frac{1}{n}(k - k_1) \in \frac{1}{n}(K - K) \subset U.$$

Since U is an arbitrary 0-neighborhood, this proves that $Tk = k$ (for all $T \in \mathcal{T}$). □

In the next result, Kakutani's fixed point theorem, the family $\mathcal{T}$ is a *group* G (*not* necessarily commutative!) of affine maps. Another additional condition will be *equicontinuity.*

Definition 5.44. Let X be a topological space and Y be a t.v.s. Let $\mathcal{T}$ be a family of functions from X to Y and $x_0 \in X$. Say that $\mathcal{T}$ is *equicontinuous at* x_0 if for every 0-neighborhood $V \subset Y$ there is a x_0-neighborhood $U \subset X$ such that $U \subset f^{-1}(f(x_0) + V)$ for each $f \in \mathcal{T}$. Say that $\mathcal{T}$ is *equicontinuous* if it is equicontinuous at each point in X.

Equicontinuity of a family at a point evidently implies continuity of each of the individual functions at that point. The proof of the following assertions is left to the reader. Part (b) says that when X is compact, equicontinuity implies the stronger "uniform equicontinuity" condition.

Lemma 5.45. *Let X, Y, and $\mathcal{T}$ be as in Definition 5.44.*

(a) *Let $x_0 \in X$. Then $\mathcal{T}$ is equicontinuous at x_0 iff for every net $\{x_\alpha\}_{\alpha\in A}$ in X that converges to x_0 we have $\lim_{\alpha\in A} f(x_\alpha) = f(x_0)$ in Y uniformly in $f \in \mathcal{T}$, i.e., for every 0-neighborhood $V \subset Y$ there exists $\alpha_0 \in A$ such that for all $\alpha \geq \alpha_0$ we have $f(x_\alpha) - f(x_0) \in V$.*

(b) *Suppose that X is a compact and that $\mathcal{T}$ is equicontinuous:*

  (i) *If X is a metric space, then for every 0-neighborhood $V \subset Y$ there is $\delta > 0$ such that for every $x, y \in X$, if $d_X(x,y) < \delta$ then $f(x)-f(y) \in V$ for all $f \in \mathcal{T}$. Equivalently, for each nets $\{x_\alpha\}_{\alpha\in A}$ and $\{y_\alpha\}_{\alpha\in A}$ in X such that $\lim_{\alpha\in A} d_X(x_\alpha, y_\alpha) = 0$ we have $\lim_{\alpha\in A}(f(x_\alpha) - f(y_\alpha)) = 0$ uniformly in $f \in \mathcal{T}$.*

  (ii) *If X is a topological subspace of a t.v.s. X_1, then for every 0-neighborhood $V \subset Y$ there is a 0-neighborhood $U \subset X_1$ such that for every $x, y \in X$, if $x - y \in U$ then $f(x) - f(y) \in V$ for all $f \in \mathcal{T}$. Equivalently, for each nets $\{x_\alpha\}_{\alpha\in A}$ and $\{y_\alpha\}_{\alpha\in A}$ in X such that $\lim_{\alpha\in A}(x_\alpha - y_\alpha) = 0$ we have $\lim_{\alpha\in A}(f(x_\alpha) - f(y_\alpha)) = 0$ uniformly in $f \in \mathcal{T}$.*

Theorem 5.46 (Kakutani's fixed point theorem). *Let X be a locally convex t.v.s. and let $K \subset X$ be a non-empty, compact, convex set. Let $\mathcal{T}$ be an equicontinuous group of bijective affine maps of K onto itself. Then $\mathcal{T}$ has a fixed point.*

We derive the theorem from a more general one. A family of maps $\mathcal{T}$ from a topological space K into itself is called *distal* if whenever $x, y \in K$ and $\{T_\alpha\}_\alpha$ is a net in $\mathcal{T}$ such that $\lim_\alpha T_\alpha x$ and $\lim_\alpha T_\alpha y$ exist and are equal, we have $x = y$. For example, every family of linear isometries on a normed space is distal.

Theorem 5.47 (F. Hahn's fixed point theorem). *Let X be a locally convex t.v.s., and let $K \subset X$ be a non-empty, compact, convex subset of X. Let $\mathcal{T}$ be a non-empty distal semigroup of continuous affine maps of K into itself. Then $\mathcal{T}$ has a fixed point.*

Proof of F. Hahn's fixed point theorem. Let Ω be the family of all non-empty, compact, convex $\mathcal{T}$-invariant subsets of K, ordered by inclusion. Since $K \in \Omega$, Ω is non-empty. If $\Omega_0 \subset \Omega$ is totally ordered, every finite intersection of sets of the family Ω_0 is one of the sets of the family, hence non-empty. Thus, Ω_0 has the finite intersection property, and since the sets of the family are non-empty compact sets, the intersection $K_0 := \bigcap \Omega_0$ is a non-empty compact set. Clearly, $K_0 \subset K$ is convex and $\mathcal{T}$-invariant, that is $K_0 \in \Omega$. By definition, K_0 is a minimal element for Ω_0. By Zorn's lemma, Ω contains a minimal element H. It suffices to prove that H is a singleton.

Let $z \in H$. Then $H' := \overline{\mathrm{co}}(\overline{\mathcal{T}z}) = \overline{\mathrm{co}}(\mathcal{T}z)$ is a non-empty, compact, convex subset of H as H is compact, convex, and $\mathcal{T}$-invariant. It is also $\mathcal{T}$-invariant because $\mathcal{T}$ is a semigroup of affine continuous maps. Indeed, if $u \in \mathrm{co}(\mathcal{T}z)$, there exist $T_i \in \mathcal{T}$ and $t_i \in [0,1]$ $(i = 1, \dots, n)$ such that $\sum_{i=1}^n t_i = 1$ and $u = \sum_{i=1}^n t_i T_i z$. Hence, for all $T \in \mathcal{T}$, $Tu = \sum_{i=1}^n t_i T T_i z$ by affinity, and this vector belongs to $\mathrm{co}(\mathcal{T}z)$ because $TT_i \in \mathcal{T}$ for all i. This proves that $\mathrm{co}(\mathcal{T}z)$ is $\mathcal{T}$-invariant, and consequently so is $H' = \overline{\mathrm{co}}(\mathcal{T}z)$ by continuity. Thus, $H' \in \Omega$, and the minimality of H entails that

$$H' = H, \text{ that is, } \overline{\mathrm{co}}(\overline{\mathcal{T}z}) = H. \tag{1}$$

Assume that $x, y \in H$. Set $z := \frac{1}{2}(x+y)$. Then $z \in H$ by convexity, so we may apply the previous paragraph to this z. By the Krein–Milman theorem (5.30), H has an extremal point h. By (1) and Milman's theorem (Exercise 11), we must have $h \in \overline{\mathcal{T}z}$. Therefore, there is a net $\{T_\alpha\}_\alpha$ in $\mathcal{T}$ such that $h = \lim_\alpha T_\alpha z$. As H is compact we may assume, by passing to a subnet if necessary, that the limits $u := \lim_\alpha T_\alpha x$ and $v := \lim_\alpha T_\alpha y$ also exist (in H). Hence, the definition of z and the affinity of the maps T_α imply that $h = \frac{1}{2}(u + v)$. But h is an extremal point of H, so h equals both $u = \lim_\alpha T_\alpha x$ and $v = \lim_\alpha T_\alpha y$. Since $\mathcal{T}$ is distal, we infer that $x = y$. $\square$

Proof of Kakutani's fixed point theorem. To apply F. Hahn's theorem, all we need to do is to prove that the family of maps $\mathcal{T}$ is distal.

Suppose that $x, y \in K$ and $\{T_\alpha\}_\alpha$ is a net in $\mathcal{T}$ such that $\lim_\alpha T_\alpha x$ and $\lim_\alpha T_\alpha y$ exist and are equal. Let V be a 0-neighborhood of X. Since $\mathcal{T}$ is equicontinuous and K is compact, there exists a 0-neighborhood U of X such that for every $a, b \in K$ and $T \in \mathcal{T}$, if $a - b \in U$ then $Ta - Tb \in V$. Since $\lim_\alpha (T_\alpha x - T_\alpha y) = 0$, there is $\alpha_0 \in A$ such that for all $\alpha \geq \alpha_0$ we have $T_\alpha x - T_\alpha y \in U$, thus $TT_\alpha x - TT_\alpha y \in V$ when $T \in \mathcal{T}$. Taking $\alpha := \alpha_0$ and $T := T_{\alpha_0}^{-1}$ gives $x - y \in V$. Since V was an arbitrary 0-neighborhood, we conclude that $x = y$. $\square$

As an application of Kakutani's theorem, we shall give a proof of the existence of the Haar invariant measure for compact groups. While the proof is not "constructive" as the one given in Section 4.5, and applies only to compact groups, it is less technical and may be more elegant.

Preliminaries. Let G be a compact topological group, and let $C(G)$ be the Banach space of all complex-valued continuous functions on G. For each $s \in G$, the left (right) translation operator L_s (R_s) on $C(G)$ is defined by

$$(L_s f)(t) := f(st) \qquad ((R_s f)(t) := f(ts), \text{ resp.}) \qquad (t \in G).$$

The dual Banach space $C(G)^*$ may be identified with the space of all complex regular Borel measures on G with the total variation norm, through the Riesz representation theorem. A common notation is

$$\langle f, \mu \rangle := \mu(f) \qquad (f \in C(G),\, \mu \in C(G)^*).$$

For positive measures μ on G, $\|\mu\| = \mu(G)$; if $\mu(G) = 1$, μ is called a *probability measure*. Thus, the set $\mathcal{P}$ of regular Borel probability measures on G is a subset of the closed unit ball of $C(G)^*$. The latter is compact in the *weak** topology by Alaoglu's theorem (Theorem 5.24). Since $\mathcal{P}$ is closed in the *weak** topology, it follows that $\mathcal{P}$ is a compact subset of $C(G)^*$ (endowed with the *weak** topology). The (Dirac) point measure at the origin of G is clearly in $\mathcal{P}$, so that $\mathcal{P} \neq \emptyset$. The set $\mathcal{P}$ is clearly convex. The family $\{L_s; s \in G\}$ is a group (under composition) of linear operators on $C(G)$, since for all $s, t \in G$,

$$L_s L_t = L_{ts}; \qquad L_s L_{s^{-1}} = L_e = I,$$

where e is the identity in G and I is the identity operator on $C(G)$. Consequently, its family of Banach adjoints (see the proof of Theorem 5.10)

$$\{L_s^*; s \in G\},$$

consisting of the linear operators on $C(G)^*$ given by

$$\langle f, L_s^* \mu \rangle = \langle L_s f, \mu \rangle \qquad (f \in C(G),\, \mu \in C(G)^*,\, s \in G),$$

is a group (under composition), since for all $s, t \in G$,

$$L_s^* L_t^* = (L_t L_s)^* = L_{st}^*; \qquad L_s^* L_{s^{-1}}^* = L_e^* = I,$$

where now I is the identity operator on $C(G)^*$. Notice also that both the closed unit ball of $C(G)^*$ and $\mathcal{P}$ are $\{L_s^*; s \in G\}$-invariant.

We shall need the following well-known theorem. As preparation, suppose that Z is a topological space and $A \subset Z$. Recall that A is called *conditionally* (or *relatively*) *compact* if its closure $\overline{A}$ is compact. Also recall that if Z is a metric space, A is called *totally bounded* if there exist finitely many elements $a_1, \ldots, a_n$ in A such that the ϵ-balls centered at a_i, $i = 1, \ldots, n$ cover A:

$$A \subset \bigcup_{i=1}^{n} B(a_i, \epsilon).$$

Finally, if Z is a *complete* metric space, then A is conditionally compact iff it is totally bounded.

Theorem 5.48 (Arzelà–Ascoli). *Let X be a compact topological space, and let $\Phi \subset C(X)$. The following conditions are equivalent:*

(i) Φ *is pointwise bounded and equicontinuous;*

(ii) Φ *is bounded and equicontinuous;*

(iii) Φ *is conditionally compact (equivalently: totally bounded).*

The "pointwise boundedness" condition means that for each $x \in X$, the set $\{f(x); f \in \Phi\}$ is bounded.

Proof. We will show that (i) $\Longrightarrow$ (ii) $\Longrightarrow$ (iii), and leave the implication (iii) $\Longrightarrow$ (i) to the reader. For an alternative, "sequential", proof, see Exercise 3 in Chapter 9.

Assume that Φ is pointwise bounded and equicontinuous. Let $\epsilon > 0$. For each $x \in X$, the equicontinuity of Φ at x provides a 0-neighborhood V_x in X such that

$$\sup_{f\in\Phi} |f(y) - f(x)| < \frac{\epsilon}{3} \qquad (\forall y \in V_x). \tag{2}$$

The family $\{V_x; x \in X\}$ is an open cover of X. By compactness of X, there exist $x_i \in X,\ i = 1, \ldots, n$ such that

$$X = \bigcup_{i=1}^{n} V_{x_i}.$$

Since Φ is bounded pointwise, we have

$$M := \max_{i=1,\ldots,n} \sup_{f\in\Phi} |f(x_i)| + \frac{\epsilon}{3} < \infty. \tag{3}$$

Let $y \in X$. There exists i $(1 \le i \le n)$ such that $y \in V_{x_i}$. Therefore, for all $f \in \Phi$, we have by (2) and (3)

$$|f(y)| \le |f(x_i)| + |f(y) - f(x_i)| \le M.$$

Thus, Φ is a subset of the closed M-ball of $C(X)$, and in particular it is bounded.

The closed disc $D := \{\lambda \in \mathbb{C}; |\lambda| \le M\}$ is compact in $\mathbb{C}$; therefore, D^n is compact in $\mathbb{C}^n$. Define

$$\pi : \Phi \to D^n$$

by

$$\pi(f) := (f(x_1), \ldots, f(x_n)) \qquad (f \in \Phi).$$

As a subset of the compact set D^n, $\pi(\Phi)$ is totally bounded. Therefore, there exist $f_1, \ldots, f_m \in \Phi$ such that the $\epsilon/3$-balls in $\mathbb{C}^n$ centered at $\pi(f_k)$, $k = 1, \ldots, m$, cover $\pi(\Phi)$. Thus, for $f \in \Phi$, there exits k, $1 \le k \le m$, such that

$$|f(x_i) - f_k(x_i)| < \frac{\epsilon}{3} \qquad (\forall i = 1, \ldots, n). \tag{4}$$

Let $y \in X$. Let i $(1 \le i \le n)$ be such that $y \in V_{x_i}$. For this index i, we have

$$|f(y) - f_k(y)| \le |f(y) - f(x_i)| + |f(x_i) - f_k(x_i)| + |f_k(x_i) - f_k(y)| < \epsilon,$$

by (2) and (4). This shows that $\|f - f_k\| < \epsilon$ for all $f \in \Phi$, that is, the ϵ-balls centered at f_k, $k = 1, \dots, m$, cover Φ. □

Theorem 5.49 (Existence and uniqueness of the Haar measure). *Let G be a compact topological group. Then there exists a unique regular Borel probability measure μ on G that is left invariant, that is,*

$$\int_G (L_s f)\, d\mu = \int_G f\, d\mu \qquad (\forall f \in C(G),\, s \in G).$$

This measure μ is also right invariant, that is,

$$\int_G (R_s f)\, d\mu = \int_G f\, d\mu \qquad (\forall f \in C(G),\, s \in G),$$

and satisfies the identity

$$\int_G f(s^{-1})\, d\mu(s) = \int_G f(s)\, d\mu(s) \qquad (\forall f \in C(G)).$$

Proof. In applying Kakutani's theorem, we take X to be $C(G)^*$ with the *weak** topology. Consider the group $\{L_s^*; s \in G\}$ defined in the preliminaries here. We verify that its restriction to the closed unit ball $C(G)_1^*$ of the Banach space $C(G)^*$ is equicontinuous (with respect to the *weak** topology). Let V be the 0-neighborhood in $C(G)^*$ (with the *weak** topology) determined by $\epsilon > 0$ and $f_1, \dots, f_n \in C(G)$. We must find a 0-neighborhood U in $C(G)^*$ (with the *weak** topology), such that

$$L_s^*(q - p) \in V \qquad (\forall s \in G, \forall q, p \in C(G)_1^*,\ q - p \in U). \tag{5}$$

The *finite* set $\{f_1, \dots, f_n\}$ defining V is equicontinuous on the compact set G. Thus, given $\delta > 0$, there exists a neighborhood N of the identity in G such that

$$|f_j(u) - f_j(t)| < \delta \qquad (\forall j = 1, \dots, n)$$

whenever $t, u \in G$ are such that $u^{-1}t \in N$. For such t, u and all $s \in G$, $(su)^{-1}(st) = u^{-1}t \in N$, and therefore

$$|(L_s f_j)(u) - (L_s f_j)(t)| < \delta \qquad (\forall s \in G,\ j = 1, \dots, n).$$

The family $\Phi := \{L_s f_j;\ s \in G,\ j = 1, \dots, n\}$ is therefore equicontinuous on G, and it is evidently bounded. By the Arzelà–Ascoli theorem it is consequently totally bounded. Let then $g_1, \dots, g_m \in C(G)$ be such that the $\epsilon/4$-balls centered at these g_i cover Φ. Define U as the 0-neighborhood in $C(G)^*$ (with the *weak** topology) determined by $\epsilon/2$ and $g_1, \dots, g_m$. Let $q, p \in C(G)_1^*$. If $q - p \in U$, $s \in G$, and $1 \le j \le n$, let i $(1 \le i \le m)$ be such that the Φ-element $L_s f_j$ belong to the $\epsilon/4$-ball in $C(G)$ centered at g_i. Then (with norms understood by their context):

$$|\langle f_j, L_s^*(q - p)\rangle| = |\langle L_s f_j, q - p\rangle| \le |\langle L_s f_j - g_i, q - p\rangle| + |\langle g_i, q - p\rangle|$$

$$< \|L_s f_j - g_i\| \, \|q - p\| + \frac{\epsilon}{2} \le 2\frac{\epsilon}{4} + \frac{\epsilon}{2} = \epsilon.$$

In conclusion, $\{L_s^*|_{C(G)_1^*}; s \in G\}$ is equicontinuous, and therefore, so is $\mathcal{T} := \{L_s^*|_{\mathcal{P}}; s \in G\}$. We have verified that the hypothesis of Kakutani's theorem is satisfied by the group $\mathcal{T}$ (under composition) of affine maps, and the non-empty compact convex set $\mathcal{P}$ in $C(G)^*$ (with the *weak** topology). By the theorem, $\mathcal{T}$ has a fixed point $\mu \in \mathcal{P}$. This means that for all $s \in G$ and $f \in C(G)$,

$$\langle L_s f, \mu\rangle = \langle f, L_s^* \mu\rangle = \langle f, \mu\rangle,$$

that is,

$$\int_G (L_s f)\, d\mu = \int_G f\, d\mu \qquad (\forall f \in C(G),\, s \in G).$$

This proves the existence of a left-invariant regular Borel probability measure on G. A similar proof shows that there exists a "right-invariant" regular Borel probability measure μ_r on G, that is,

$$\int_G (R_t f)\, d\mu_r = \int_G f\, d\mu_r \qquad (\forall f \in C(G),\, t \in G).$$

Let μ_l and μ_r be any regular Borel probability measures on G, which are left and right invariant, respectively. By Fubini's theorem and the fact that μ_l, μ_r are probability measures, we have for all $f \in C(G)$,

$$\int_G f(t)\, d\mu_l(t) = \int_G \Big(\int_G f(st)\, d\mu_l(t)\Big)\, d\mu_r(s)$$

$$= \int_{G\times G} f(st)\, d(\mu_r \times \mu_l)(s,t) = \int_G \Big(\int_G f(st)\, d\mu_r(s)\Big)\, d\mu_l(t)$$

$$= \int_G f(s)\, d\mu_r(s).$$

The uniqueness of the Riesz representation implies that $\mu_l = \mu_r$. This proves the uniqueness of the left-invariant (probability) measure on G, and the fact that it is also right invariant. Denote it by μ.

Consider next the linear map T defined on $C(G)$ by

$$(Tf)(s) := f(s^{-1}) \qquad (f \in C(G),\, s \in G).$$

T is an isometric isomorphism of $C(G)$ onto itself. Let $\tilde{\mu} := T^*\mu = \mu \circ T$ ($\in C(G)^*$). If $f \in C(G)$ is non-negative, $\langle f, \tilde{\mu}\rangle = \langle Tf, \mu\rangle \ge 0$, because $Tf \ge 0$ and μ is a non-negative measure. Therefore $\tilde{\mu}$ is a positive measure. Also

$$\tilde{\mu}(G) = \langle 1, \tilde{\mu}\rangle = \langle T1, \mu\rangle = \langle 1, \mu\rangle = 1,$$

where we have denoted by 1 the constant function with value 1 on G. Thus $\tilde{\mu} \in \mathcal{P}$. Let $s \in G$. For all $f \in C(G)$ and $t \in G$,

$$(TL_s f)(t) = (L_s f)(t^{-1}) = f(st^{-1}) = f((ts^{-1})^{-1}) = (Tf)(ts^{-1}) = (R_{s^{-1}} Tf)(t),$$

proving that $TL_s = R_{s^{-1}}T$. Hence, by the right invariance of μ,

$$\tilde{\mu} \circ L_s = \mu \circ T \circ L_s = \mu \circ R_{s^{-1}} \circ T = \mu \circ T = \tilde{\mu}$$

for all $s \in G$, that is, $\tilde{\mu}$ is left invariant. By the uniqueness of the left-invariant (regular Borel) probability measure, we conclude that $\tilde{\mu} = \mu$, that is, for all $f \in C(G)$,

$$\int_G f(s^{-1})\, d\mu(s) = \langle Tf, \mu \rangle = \langle f, T^*\mu \rangle = \langle f, \tilde{\mu} \rangle = \langle f, \mu \rangle = \int_G f(s)\, d\mu(s).$$

This completes the proof. □

5.10 The bounded *weak**-topology

Let X be a normed space. A topology stronger than the *weak**-topology on X^* is the so-called "bounded *weak**-topology" (or *bweak**-topology for short). It is defined by letting $F \subset X^*$ be *bweak**-closed iff $F \cap aS^*$ is *weak**-closed for all $a > 0$ (where S^* denotes the closed unit ball of X^*). Note that since aS^* itself is *weak**-closed (it is even *weak**-compact, by Alaoglu's theorem!), every *weak**-closed set F is *bweak**-closed. Thus, the *bweak**-topology is indeed stronger than the *weak**-topology (both on X^*).

Given $x \in X$, set

$$x^{\sim} := \{x^* \in X^*; |x^*x| < 1\}.$$

If $A \subset X$, set $A^{\sim} := \bigcap_{x \in A} x^{\sim}$. Clearly, $x^{\sim}$ is a *weak**-open subset of X^*, so that $A^{\sim}$ is *weak**-open for *finite* subsets A of X, and we have:

Observation 1. The family

$$\{A^{\sim}; A \subset X \text{ finite}\}$$

is a base for the *weak** 0-neighborhoods in X^*.

A set $U \subset X^*$ is *bweak**-open iff U^c is *bweak**-closed, that is, iff $U^c \cap aS^*$ is *weak**-closed for all $a > 0$, that is, iff $(U^c \cap aS^*)^c = U \cup (aS^*)^c$ is *weak**-open for all $a > 0$. This implies that $(U \cup (aS^*)^c) \cap aS^*$ $(= U \cap aS^*)$ is relatively *weak**-open in aS^* for all $a > 0$. Conversely, if $U \cap aS^*$ relatively *weak**-open in aS^* for all $a > 0$, then for each $a > 0$, there exists a *weak**-open set V_a in X^* such that $U \cap aS^* = V_a \cap aS^*$. Then for all $a > 0$, $U \cup (aS^*)^c = (V_a \cap aS^*) \cup (aS^*)^c = V_a \cup (aS^*)^c$ is *weak**-open, because aS^* is *weak**-closed. Equivalently, U is *bweak**-open, as explained earlier. In conclusion:

Observation 2. A subset $U \subset X^*$ is *bweak**-open iff $U \cap aS^*$ is relatively *weak**-open in aS^* for all $a > 0$.

Given $A \subset X$ and $a > 0$, set

$$A_a := \left\{x \in A; \|x\| \geq \frac{1}{a}\right\}.$$

Trivially, $A_a \subset A$, so $A^\sim \subset A_a^\sim$. Hence

$$A^\sim \cap aS^* \subset A_a^\sim \cap aS^*.$$

On the other hand, if $0 \neq x^* \in A_a^\sim \cap aS^*$ and $x \in A \setminus A_a$ (i.e., $\|x\| < \frac{1}{a}$), then $|x^*x| \leq \|x^*\| \, \|x\| < \frac{\|x^*\|}{a} \leq 1$. This proves that $x^* \in A^\sim$. Hence:

Observation 3. For all $A \subset X$ and $a > 0$,

$$A^\sim \cap aS^* = A_a^\sim \cap aS^*.$$

In the following, a subset $A \subset X$ is called a norm-null sequence if it is the set of elements of a norm-null sequence in X.

Theorem 5.50 (Dieudonné's theorem). *Let X be a normed space. The family*

$$\{A^\sim; A \subset X \textit{ is a norm-null sequence}\}$$

is an open base of neighborhoods for the bweak-open 0-neighborhoods in X^*.*

Proof. Let A be a norm-null sequence in X. Then A_a is finite for each $a > 0$. Therefore, $A_a^\sim$ is *weak**-open, so that $A_a^\sim \cap aS^*$ is relatively *weak**-open in aS^*. By Observations 2 and 3, $A^\sim$ is *bweak**-open.

Conversely, let U be a *bweak**-open 0-neighborhoods in X^*. We proceed to define recursively a sequence of *finite* sets $A_n \subset X$ with the following properties:

(i) $A_1 \subset A_2 \subset A_3 \subset \ldots$;

(ii) if $x \in A_{n+1} \setminus A_n$, then $\|x\| \leq \frac{1}{n}$;

(iii) $A_n^\# \cap nS^* \subset U$, where

$$x^\# := \{x^* \in X^*; |x^*x| \leq 1\} \qquad (x \in X)$$

and $A^\# := \bigcap_{x \in A} x^\#$ for any $A \subset X$.

Note that $x^\#$ is *weak**-closed for each $x \in X$, and therefore $A^\#$ is *weak**-closed for every $A \subset X$.

By Observation 2, $U \cap S^*$ is relatively *weak**-open in S^*. Since, by Observation 1, $\{A^\sim \cap S^*; A \subset X \text{ finite}\}$ is a base for the relative *weak**-open 0-neighborhoods in S^*, there exists a finite set $B \subset X$ such that $B^\sim \cap S^* \subset U \cap S^*$. Let $A_1 := 2B$. If now $x^* \in A_1^\# \cap S^*$, then for all $x \in B$, we have $2x \in A_1$, hence, $|x^*x| = \frac{1}{2} |x^*(2x)| \leq \frac{1}{2} < 1$, that is, $x^* \in B^\sim \cap S^* \subset U$. Thus, $A_1^\# \cap S^* \subset U$, so A_1 has property (iii).

Suppose that $A_1, A_2, \ldots, A_n$ have been constructed with properties (i)–(iii) up to index n. Assume that for *all finite* sets $B \subset X$ with $\|B\| := \sup_{x \in B} \|x\| \leq \frac{1}{n}$ we have

$$(A_n \cup B)^\# \cap (n+1)S^* \cap U^c \neq \emptyset. \tag{1}$$

Since U^c is *bweak**-closed, $(n+1)S^* \cap U^c$ is *weak**-closed by the definition of the *bweak**-topology. Also, $(A_n \cup B)^\#$ is *weak**-closed (as observed previously), hence the sets in (1) are *weak**-closed subsets of the *weak**-compact set $(n+1)S^*$ (by Alaoglu's theorem). Consequently, the sets in (1) are *weak**-compact. They have the finite intersection property, because if $B_1, \ldots, B_r$ are finite subsets of X with $\|B_i\| \leq \frac{1}{n}$ for each $1 \leq i \leq r$, then $B := \bigcup_{i=1}^r B_i$ is a finite subset of X with $\|B\| \leq \frac{1}{n}$, and

$$\bigcap_{i=1}^r (A_n \cup B_i)^\# \cap (n+1)S^* \cap U^c = (A_n \cup B)^\# \cap (n+1)S^* \cap U^c \neq \emptyset$$

by the hypothesis (1). It follows that the intersection of all sets in (1) is non-empty. In particular

$$\bigcap_{x \in X, \|x\| \leq \frac{1}{n}} (A_n \cup \{x\})^\# \cap (n+1)S^* \cap U^c \neq \emptyset.$$

Since $(A_n \cup \{x\})^\# = A_n^\# \cap x^\#$, it follows that there exists $x^* \in A_n^\# \cap (n+1)S^* \cap U^c$ such that $|x^*x| \leq 1$ for all $x \in X$ with $\|x\| \leq \frac{1}{n}$. If $x \in X$ and $\|x\| \leq 1$, then $\left\|\frac{1}{n}x\right\| \leq \frac{1}{n}$, so that $\frac{1}{n}|x^*x| = \left|x^*(\frac{1}{n}x)\right| \leq 1$, proving that $\|x^*\| \leq n$. Therefore, $x^* \in A_n^\# \cap nS^* \cap U^c$, contradicting (iii) in the recursion hypothesis. This contradiction shows that our assumption that (1) holds for each finite $B \subset X$ with $\|B\| \leq \frac{1}{n}$ is false. That is, there exists a finite set $B_n \subset X$ such that $\|B_n\| \leq \frac{1}{n}$ and $(A_n \cup B_n)^\# \cap (n+1)S^* \subset U$. Define $A_{n+1} := A_n \cup B_n$. Evidently, properties (i)–(iii) are satisfied up to index $n+1$, and the recursive construction is complete.

Define now $A := \bigcup_{n=1}^\infty A_n$. Since each A_n is finite, (i) and (ii) show that A is a norm-null sequence.

Finally, if $x^* \in A^\sim$, then $x^* \in A_n^\sim$ for all $n \in \mathbb{N}$. Let $n \geq \|x^*\|$. Then

$$x^* \in A_n^\sim \cap nS^* \subset A_n^\# \cap nS^* \subset U$$

by (iii). Hence $A^\sim \subset U$, as desired. □

The meaning of Dieudonné's theorem is that a net $\{x_\alpha^*\}_{\alpha \in I}$ in X^* converges in the *bweak**-topology to $x^* \in X^*$ iff for every norm-null sequence $A \subset X$, $\{x_\alpha^*\}_{\alpha \in I}$ converges to x^* *uniformly* on A, that is, $\lim_{\alpha \in I} \left(\sup_{x \in A} |(x_\alpha^* - x^*)(x)|\right) = 0$.

It follows from Dieudonné's theorem that X^* with the *bweak**-topology is a topological vector space, which is locally convex as $A^\sim$ is convex for every $A \subset X$.

Theorem 5.51 (Krein–Šmulian's theorem). *Let X be a Banach space. A linear functional on X^* is bweak*-continuous iff it is weak*-continuous.*

Proof. Let $f : X^* \to \mathbb{C}$ be linear. Since the *bweak**-topology is stronger than the *weak**-topology, *weak**-continuity of f trivially implies *bweak**-continuity.

Conversely, suppose that f is *bweak**-continuous. Then

$$U := f^{-1}\left(\{\lambda \in \mathbb{C}; |\lambda| < 1\}\right)$$

is a *bweak**-open neighborhood of 0 in X^*, and contains therefore a basic neighborhood as described in Dieudonné's theorem. That is, there exists a null sequence $\{x_i; i \in \mathbb{N}\}$ in X such that $\{x_i; i \in \mathbb{N}\}^\sim \subset U$. Hence

$$|f(x^*)| < 1 \qquad (\forall x^* \in \{x_i; i \in \mathbb{N}\}^\sim). \tag{2}$$

Let $\{x_i; i \in \mathbb{N}\}^\perp := \{x^* \in X^*; x^* x_i = 0 \text{ for all } i \in \mathbb{N}\}$. Since $\{x_i; i \in \mathbb{N}\}^\perp \subset \{x_i; i \in \mathbb{N}\}^\sim$, we have by (2)

$$|f(x^*)| < 1 \qquad (\forall x^* \in \{x_i; i \in \mathbb{N}\}^\perp).$$

Let $x^* \in \{x_i; i \in \mathbb{N}\}^\perp$. For each $n \in \mathbb{N}$ we have $nx^* \in \{x_i; i \in \mathbb{N}\}^\perp$, and therefore $n\,|f(x^*)| = |f(nx^*)| < 1$. Hence $f(x^*) = 0$. This proves that

$$f|_{\{x_i; i \in \mathbb{N}\}^\perp} = 0. \tag{3}$$

Define

$$\tau : X^* \to c_0$$

by

$$\tau x^* := \{x^* x_i; i \in \mathbb{N}\} \qquad (x^* \in X^*).$$

If $x^*, y^* \in X^*$ are such that $x^* x_i = y^* x_i$ for all $i \in \mathbb{N}$, then $x^* - y^* \in \{x_i; i \in \mathbb{N}\}^\perp$, and therefore $f(x^* - y^*) = 0$ by (3). Thus, if $\tau x^* = \tau y^*$, also $f(x^*) = f(y^*)$. It follows that the map

$$\tilde{f} : \tau X^* \to \mathbb{C}$$

defined by

$$\tilde{f}(\tau x^*) = f(x^*)$$

is well defined. Its linearity is clear. If $\|\tau x^*\|_{c_0} < 1$, then $|x^* x_i| < 1$ for all $i \in \mathbb{N}$, hence $x^* \in \{x_i; i \in \mathbb{N}\}^\sim$, and therefore $|f(x^*)| < 1$ by (2), that is, $|\tilde{f}(\tau x^*)| < 1$. This shows that $\|\tilde{f}\| \leq 1$. By the Hahn–Banach theorem, there exists $g \in c_0^*$ such that $g|_{\tau X^*} = \tilde{f}$ and $\|g\| = \|\tilde{f}\| \leq 1$. By the "description" of c_0^* as l^1 (see Exercise 8 of Chapter 4), there exists a sequence $\{g_i; i \in \mathbb{N}\} \in l^1$ such that $g(\xi) = \sum_{i=1}^\infty g_i \xi_i$ for all $\xi \in c_0$ and $\sum_{i=1}^\infty |g_i| = \|\{g_i; i \in \mathbb{N}\}\|_{l^1} = \|g\|_{c_0^*} \leq 1$. Since $\{g_i; i \in \mathbb{N}\} \in l^1$ and $\{x_i; i \in \mathbb{N}\}$ is a norm-null sequence, the series $\sum_{i=1}^\infty g_i x_i$ converges (absolutely) in the Banach space X. Therefore, for all $x^* \in X^*$,

$$x^* \left(\sum_{i=1}^\infty g_i x_i \right) = \sum_{i=1}^\infty g_i x^* x_i = g(\tau x^*) = \tilde{f}(\tau x^*) = f(x^*).$$

Letting $x := \sum_{i=1}^\infty g_i x_i$, we infer that $f(x^*) = x^* x$ for all $x^* \in X^*$, that is, $f = \kappa x$, where κ is the canonical embedding of X in X^{**}. This proves that f is *weak**-continuous. □

We end the discussion with two equivalent restatements of the Krein–Šmulian theorem. The first follows from the definition of the *bweak**-topology. The second is obtained with the aid of Corollary 5.21 of the strict separation theorem.

Theorem 5.52. *Let X be a Banach space. A linear functional on X^* is weak*-continuous iff its restriction to S^* is continuous with respect to the relative weak*-topology on S^*.*

Theorem 5.53. *Let X be a Banach space. A convex set $C \subset X^*$ is weak*-closed iff it is bweak*-closed, that is, iff $C \cap aS^*$ is weak*-closed for all $a > 0$.*

Exercises

1. A Banach space X is *separable* if it contains a countable dense subset. Prove that if X^* is separable, then X is separable (but the converse is false). (Hint: let $\{x_n^*\}$ be a sequence of unit vectors dense in the unit sphere of X^*. Pick unit vectors x_n in X such that $|x_n^* x_n| > 1/2$. Use Corollary 5.5 to show that span $\{x_n\}$ is dense in X; the same is true when the scalars are complex numbers with *rational* real and imaginary parts.)

2. Consider the normed space
$$C_c^n(\mathbb{R}) := \{f \in C_c(\mathbb{R}); f^{(k)} \in C_c(\mathbb{R}), k = 1, \ldots, n\},$$
with the norm
$$\|f\| = \sum_{k=0}^{n} \|f^{(k)}\|_u.$$
Given $\phi \in C_c^n(\mathbb{R})$, prove that there exist complex Borel measures μ_k ($k = 0, \ldots, n$) such that
$$\phi(f) = \sum_k \int_{\mathbb{R}} f^{(k)}\, d\mu_k$$
for all $f \in C_c^n(\mathbb{R})$. (Hint: consider the subspace
$$Z = \{[f, f', \ldots, f^{(n)}]; f \in C_c^n(\mathbb{R})\}$$
of $C_c \times \cdots \times C_c$ ($n+1$ times). Define ψ on Z by $\psi([f, f', \ldots, f^{(n)}]) = \phi(f)$, cf. Exercise 3, Chapter 4.)

3. Let X be a Banach space, and let $\Gamma \subset X^*$ be (norm) bounded and *weak**-closed. Prove:
   (a) Γ is *weak**-compact.
   (b) If Γ is also convex, then it is the *weak**-closed convex hull of its extremal points.

4. Let X, Y be normed spaces and $T \in B(X, Y)$. Prove that T is continuous with respect to the weak topologies on X and Y, and $T^* : Y^* \to X^*$ is continuous with respect to the *weak**-topologies on Y^* and X^*. (Recall that the Banach adjoint T^* of $T \in B(X)$ is defined by means of the identity $(T^*y^*)x = y^*(Tx), x \in X, y^* \in Y^*$.)

5. Let $p, q \in [1, \infty]$ be conjugate exponents, and let $(X, \mathcal{A}, \mu)$ be a positive measure space. Let g be a complex measurable function on X such that $\|g\|_q \leq M$ for some constant M. Then $\|fg\|_1 \leq M\|f\|_p$ for all $f \in L^p(\mu)$ (by Theorems 1.26 and 1.33). *Prove the converse!*

Uniform convexity

6. Let X be a normed space, and let B and S denote its closed unit ball and its unit sphere, respectively. We say that X is *uniformly convex* (u.c.) if for each $\epsilon > 0$ there exists $\delta = \delta(\epsilon) > 0$ such that $\|x - y\| < \epsilon$ whenever $x, y \in B$ are such that $\|(x + y)/2\| > 1 - \delta$. Prove:

   (a) X is u.c. iff whenever $x_n, y_n \in S$ are such that $\|x_n + y_n\| \to 2$, it follows that $\|x_n - y_n\| \to 0$.

   (b) Every inner product space is u.c.

   (c) Let X be a u.c. normed space and $\{x_n\} \subset X$. Then $x_n \to x \in X$ strongly iff $x_n \to x$ weakly and $\|x_n\| \to \|x\|$. (Hint: suppose $x_n \to x$ weakly and $\|x_n\| \to \|x\|$. We may assume that $x_n, x \in S$. Pick $x_0^* \in X^*$ such that $x_0^* x = 1 = \|x_0^*\|$, cf. Corollary 5.7.)

   (d) The "distance theorem" (Theorem 1.35) is valid for a u.c. Banach space X.

   The following parts are steps in the proof of the result stated in Part (i) below.

   (e) Let X be a u.c. Banach space, and let ϵ, δ be as in the definition above. Denote by S^* and S^{**} the unit spheres of X^* and X^{**}, respectively. Given $x_0^{**} \in S^{**}$, there exists $x_0^* \in S^*$ such that $|x_0^{**} x_0^* - 1| < \delta$. Also there exists $x \in B$ such that $|x_0^* x - 1| < \delta$. Define

   $$E_\delta = \{x \in B; |x_0^* x - 1| < \delta\} (\neq \emptyset!).$$

   Show that $\|x - y\| < \epsilon$ for all $x, y \in E_\delta$.

   (f) In any normed space X, the set

   $$U := \{x^{**} \in X^{**}; |x^{**} x_0^* - 1| < \delta\}$$

   is a *weak**-neighborhood of x_0^{**}.

   (g) For any *weak**-neighborhood V of x_0^{**}, the *weak**-neighborhood $W := V \cap U$ of x_0^{**} meets κB. (κ denotes the canonical embedding of X in X^{**}.) Thus, $V \cap \kappa(E_\delta) \neq \emptyset$, and therefore x_0^{**} belongs to the *weak**-closure of $\kappa(E_\delta)$ (cf. Goldstine's theorem).

   (h) Fix $x \in E_\delta$. Then $x_0^{**} \in \kappa x + \epsilon B^{**}$, where B^{**} denotes the (norm) closed unit ball of X^{**}. (Hint: apply Parts (e) and (g), and the fact that B^{**} is *weak**-compact, hence *weak**-closed.)

   (i) Conclude from Part (h) that $d(x_0^{**}, \kappa B) = 0$, and therefore $x_0^{**} \in \kappa B$ since κB is norm-closed in X^{**} (cf. paragraph preceding Theorem 5.9).

This proves the following theorem: *uniformly convex Banach spaces are reflexive.*

Miscellaneous

7. Let $\{\beta_n\}_{n=0}^{\infty} \in l^{\infty}$ be such that there exists a positive constant K for which

$$\left|\sum_{n=0}^{N} \alpha_n \beta_n\right| \le K \max_{t\in[0,1]} \left|\sum_{n=0}^{N} \alpha_n t^n\right|$$

for all $\alpha_0, \ldots, \alpha_N \in \mathbb{C}$ and $N = 0, 1, 2, \ldots$. Prove that there exists a unique regular complex Borel measure μ on $[0,1]$ such that $\beta_n = \int_0^1 t^n \, d\mu$ for all $n = 0, 1, 2, \ldots$. Moreover $\|\mu\| \le K$. Formulate and prove the converse.

8. Prove the converse of Theorem 4.4.

9. Prove the following generalization of the Stone–Weierstrass theorem. Let X be a *locally* compact Hausdorff space, and let $\mathcal{A}$ be a separating selfadjoint subalgebra of $C_0(X)$ that *vanishes identically nowhere*, that is, for each $x \in X$ there is $f \in \mathcal{A}$ with $f(x) \neq 0$. Then $\mathcal{A}$ is dense in $C_0(X)$. Hint: if X is compact, prove that there are $0 < m \le M$ and $f \in \mathcal{A}$ with $m \le f \le M$, and use this to show that the constant function 1 belongs to the closure of $\mathcal{A}$; now the usual Stone–Weierstrass theorem for compact Hausdorff spaces applies. Next, assume that X is not compact and let $Y = X \cup \{\infty\}$ be the Alexandroff one-point compactification of X. Observe that $C_0(X)$ embeds in $C(Y)$ as the subspace of functions vanishing at the point ∞. Letting $\mathcal{B}$ be the image of $\mathcal{A}$ under this embedding, prove that the algebra $\mathcal{B} \oplus \mathbb{C}1 \subset C(Y)$ is dense in $C(Y)$ and deduce that $\mathcal{A}$ is dense in $C_0(X)$.

Milman's theorem

10. Let X be a t.v.s. and $A, B \subset X$ be compact and convex. Prove that $\mathrm{co}(A \cup B)$ (no closure required!) is compact.

11. Prove Milman's theorem, which complements (and does not rely on) the Krein–Milman theorem. Let X be a locally convex t.v.s., and let $K \subset X$ be closed such that $\overline{\mathrm{co}}(K)$ is compact. Then all extremal points of $\overline{\mathrm{co}}(K)$ belong to K. Hint: assume by contradiction that x_0 is an extremal point of $\overline{\mathrm{co}}(K)$ that does not belong to K. Prove that there exists a convex 0-neighborhood U such that $x_0 \notin \overline{K + U}$. Explain why there exist $x_1, \ldots, x_n \in K$ so that $K \subset \bigcup_{i=1}^{n} (x_i + U)$. For each $1 \le i \le n$, denote $K_i := \overline{\mathrm{co}}((x_i + U) \cap K)$. Then K_i is compact and contained in $\overline{x_i + U}$. Deduce that

$$\overline{\mathrm{co}}(K) = \overline{\mathrm{co}}\Big(\bigcup_{i=1}^{n} K_i\Big) = \mathrm{co}\Big(\bigcup_{i=1}^{n} K_i\Big).$$

Now, x_0 being an extremal point of $\overline{\text{co}}(K)$ implies that

$$x_0 \in \bigcup_{i=1}^{n} K_i \subset \bigcup_{i=1}^{n} \overline{x_i + U} \subset \overline{K + U},$$

which is a contradiction.

6

Bounded operators

This chapter is devoted to the basics of bounded operator theory. We recall that $B(X,Y)$ denotes the normed space of all bounded linear mappings from the normed space X to the normed space Y. The norm on $B(X,Y)$ is the *operator norm*

$$\|T\| := \sup_{\|x\|\le 1} \|Tx\| = \sup_{\|x\|<1} \|Tx\| = \sup_{\|x\|=1} \|Tx\| = \sup_{0\ne x\in X} \frac{\|Tx\|}{\|x\|}.$$

The elements of $B(X,Y)$ will be referred to as *operators*.

Two theorems about $B(X,Y)$, the *Uniform Boundedness Theorem* and the *Open Mapping Theorem*, are two of the "Three Basic Principles of Linear Functional Analysis" (the other one is the Hahn–Banach lemma, see Chapter 5). They are proved along with several of their consequences. The two theorems use so-called *category arguments* in their proofs, which are based on Baire's theorem about complete metric spaces, with which we open this chapter.

We mention here in particular one consequence of the open mapping theorem—the *Closed Graph Theorem*—which is an effectual tool for proving boundedness of linear maps between Banach spaces.

The quotient X/M of a normed space by a closed subspace M is defined, and proved to be a Banach space if X is. The converse, saying that if both M and X/M are Banach spaces then so is X, is given in Exercise 22.

The chapter ends with a section on two topologies that are weaker than the norm topology on $B(X,Y)$, namely, the *strong operator topology* and the *weak operator topology*. These are the topologies of pointwise convergence in norm, respectively weakly. We prove that a linear functional on $B(X,Y)$ is continuous in one of these topologies iff it is continuous in the other. These topologies play an important role in operator theory: for instance, see the exercises on Semigroups of Operators in Chapters 9 and 10 as well as Chapter 12 on von Neumann algebras.

Introduction to Modern Analysis. Second Edition. Shmuel Kantorovitz and Ami Viselter, Oxford University Press.
 DOI: 10.1093/oso/9780192849540.003.0006

6.1 Category

Theorem 6.1 (Baire's theorem). *Let X be a complete metric space (with metric d), and let $\{V_i\}$ be a sequence of open dense subsets of X. Then $V := \bigcap_{i=1}^{\infty} V_i$ is dense in X.*

Proof. Let U be a nonempty open subset of X. We must show that $U \cap V \neq \emptyset$.

Since V_1 is dense, it follows that the open set $U \cap V_1$ is nonempty, and we may then choose a closed ball $\bar{B}(x_1, r_1) := \{x \in X; d(x, x_1) \leq r_1\} \subset U \cap V_1$ with radius $r_1 < 1$. Let $B(x_1, r_1) := \{x \in X; d(x, x_1) < r_1\}$ be the corresponding open ball. Since V_2 is dense, it follows that the open set $B(x_1, r_1) \cap V_2$ is nonempty, and we may then choose a closed ball $\bar{B}(x_2, r_2) \subset B(x_1, r_1) \cap V_2$ with radius $r_2 < 1/2$. Continuing inductively, we obtain a sequence of balls $B(x_n, r_n)$ with $r_n < 1/n$, such that

$$\bar{B}(x_n, r_n) \subset B(x_{n-1}, r_{n-1}) \cap V_n \quad \text{for } n = 2, 3, \ldots .$$

If $i, j > n$, we have $x_i, x_j \in B(x_n, r_n)$, and therefore $d(x_i, x_j) < 2r_n < 2/n$. This means that $\{x_i\}$ is a Cauchy sequence. Since X is complete, the sequence converges to some $x \in X$. For $i \geq n$, $x_i \in \bar{B}(x_n, r_n)$ (a closed set!), and therefore $x \in \bar{B}(x_n, r_n) \subset U \cap V_n$ for all n, that is, $x \in U \cap V$. □

Definition 6.2. A subset E of a metric space X is *nowhere dense* if its closure $\bar{E}$ has empty interior. A countable union of nowhere dense sets in X is called a set of (*Baire's*) *first category in* X. A subset of X which is *not* of first category in X is said to be of (*Baire's*) *second category* in X.

The family of subsets of first category is closed under countable unions. Subsets of sets of first category are also of first category.

Using category terminology, Baire's theorem has the following variant form, which is the basis for the "category arguments" mentioned earlier.

Theorem 6.3 (Baire's category theorem). *A complete metric space is of Baire's second category in itself.*

Proof. Suppose the complete metric space X is of Baire's first category in itself. Then

$$X = \bigcup_i E_i$$

with E_i nowhere dense ($i = 1, 2, \ldots$). Hence $X = \bigcup_i \overline{E_i}$. Taking complements, we see that

$$\bigcap_i (\overline{E_i})^{\mathrm{c}} = \emptyset.$$

Since E_i are nowhere dense, the sets in the above intersection are open dense sets. By Baire's theorem, the intersection is dense, which is a contradiction. □

6.2 The uniform boundedness theorem

Theorem 6.4 (The uniform boundedness theorem, general version). *Let X, Y be normed spaces, and let $\mathcal{T} \subset B(X,Y)$. Suppose the subspace*

$$Z := \{x \in X; \sup_{T\in\mathcal{T}} \|Tx\| < \infty\}$$

is of Baire's second category in X. Then

$$\sup_{T\in\mathcal{T}} \|T\| < \infty.$$

Proof. Denote

$$r(x) := \sup_{T\in\mathcal{T}} \|Tx\| \quad (x \in Z).$$

If $S_Y := \overline{B_Y}(0,1)$ is the closed unit ball in Y, then for all $T \in \mathcal{T}, Tx \in r(x)S_Y \subset nS_Y$ if $x \in Z$ and n is an integer $\geq r(x)$. Thus, $T(x/n) \in S_Y$ for all $T \in \mathcal{T}$, that is,

$$x/n \in \bigcap_{T\in\mathcal{T}} T^{-1}S_Y := E$$

for $n \geq r(x)$. This shows that

$$Z \subset \bigcup_n nE.$$

Since Z is of Baire's second category in X and nE are closed sets (by continuity of each $T \in \mathcal{T}$), there exists n such that nE has nonempty interior. However, multiplication by scalars is a homeomorphism; therefore, E has nonempty interior E°. Thus, let $B_X(a,\delta) \subset E$. Then for all $T \in \mathcal{T}$,

$$\begin{aligned}\delta T B_X(0,1) &= TB_X(0,\delta) = T[B_X(a,\delta) - a] = TB_X(a,\delta) - Ta \\ &\subset TE - Ta \subset S_Y - S_Y \subset 2S_Y.\end{aligned}$$

Hence, $TB_X(0,1) \subset (2/\delta)S_Y$ for all $T \in \mathcal{T}$. This means that

$$\sup_{T\in\mathcal{T}} \|T\| \leq \frac{2}{\delta}.$$

□

If X is *complete*, it is of Baire's second category in itself by Theorem 6.3. This implies the nontrivial part of the following.

Corollary 6.5 (The uniform boundedness theorem). *Let X be a Banach space, Y a normed space, and $\mathcal{T} \subset B(X,Y)$. Then the following two statements are equivalent*

$$\sup_{T\in\mathcal{T}} \|Tx\| < \infty \quad \textit{for all } x \in X; \tag{i}$$

$$\sup_{T\in\mathcal{T}} \|T\| < \infty. \tag{ii}$$

Corollary 6.6. *Let X be a Banach space, Y a normed space, and let $\{T_n\}_{n\in\mathbb{N}} \subset B(X,Y)$ be such that*

$$\exists \lim_n T_n x := Tx \quad \textit{for all } x \in X.$$

Then $T \in B(X,Y)$ and $\|T\| \le \liminf_n \|T_n\| \le \sup_n \|T_n\| < \infty$.

Proof. The linearity of T is trivial. For each $x \in X, \sup_n \|T_n x\| < \infty$ (since $\lim_n T_n x$ exists). By Corollary 6.5, it follows that $\sup_n \|T_n\| := M < \infty$. For all unit vectors $x \in X$ and all $n \in \mathbb{N}$, $\|T_n x\| \le \|T_n\|$; therefore

$$\|Tx\| = \lim_n \|T_n x\| \le \liminf_n \|T_n\| \le M,$$

that is, $\|T\| \le \liminf_n \|T_n\| \le \sup_n \|T_n\| < \infty$. □

Corollary 6.7. *Let X be a normed space, and $E \subset X$. Then the following two statements are equivalent:*

$$\sup_{x\in E} |x^* x| < \infty \quad \textit{for all } x^* \in X^*. \tag{1}$$

$$\sup_{x\in E} \|x\| < \infty. \tag{2}$$

Proof. Let $\mathcal{T} := \kappa E \subset (X^*)^* = B(X^*, \mathbb{C})$, where κ denotes the canonical embedding of X into X^{**}. Then (1) is equivalent to

$$\sup_{\kappa x \in \mathcal{T}} |(\kappa x) x^*| < \infty \quad \text{for all } x^* \in X^*,$$

and since the *conjugate* space X^* is complete (cf. Corollary 4.5), Corollary 6.5 shows that this is equivalent to

$$\sup_{\kappa x \in \mathcal{T}} \|\kappa x\| < \infty.$$

Since κ is isometric, the last statement is equivalent to (2). □

Combining Corollaries 6.7 and 6.5, we obtain Corollary 6.8.

Corollary 6.8. *Let X be a Banach space, Y a normed space, and $\mathcal{T} \subset B(X,Y)$. Then the following two statements are equivalent*

$$\sup_{T\in\mathcal{T}} |x^* T x| < \infty \quad \textit{for all } x \in X, x^* \in X^*. \tag{3}$$

$$\sup_{T\in\mathcal{T}} \|T\| < \infty. \tag{4}$$

6.3 The open mapping theorem

Lemma 1. *Let X, Y be normed spaces and $T \in B(X,Y)$. Suppose the range TX of T is of Baire's second category in Y, and let $\epsilon > 0$ be given. Then there exists $\delta > 0$ such that*

$$B_Y(0,\delta) \subset \overline{TB_X(0,\epsilon)}. \tag{1}$$

Moreover, one has necessarily $\delta \le \|T\|\epsilon$.

The bar sign stands for the closure operation in Y.

Proof. We may write

$$X = \bigcup_{n=1}^{\infty} B_X(0, n\epsilon/2),$$

and therefore

$$TX = \bigcup_{n=1}^{\infty} TB_X(0, n\epsilon/2).$$

Since TX is of Baire's second category in Y, there exists n such that $\overline{TB_X(0,n\epsilon/2)}$ has nonempty interior. Therefore, $\overline{TB_X(0,\epsilon/2)}$ has nonempty interior (because T is homogeneous and multiplication by n is a homeomorphism of Y onto itself). Let then $B_Y(a,\delta)$ be a ball contained in it. If $y \in B_Y(0,\delta)$, then

$$a, a+y \in B_Y(a,\delta) \subset \overline{TB_X(0,\epsilon/2)},$$

and therefore there exist sequences

$$\{x'_k\}, \{x''_k\} \subset B_X(0,\epsilon/2)$$

such that

$$Tx'_k \to a+y; \quad Tx''_k \to a.$$

Let $x_k = x'_k - x''_k$. Then $\|x_k\| \le \|x'_k\| + \|x''_k\| < \epsilon$ and

$$Tx_k = Tx'_k - Tx''_k \to (a+y) - a = y,$$

that is, $\{x_k\} \subset B_X(0,\epsilon)$ and $y \in \overline{TB_X(0,\epsilon)}$. This proves (1).

We show finally that the relation $\delta \le \|T\|\epsilon$ follows necessarily from (1). Fix $y \in Y$ with $\|y\| = 1$ and $0 < t < 1$. Since $t\delta y \in B_Y(0,\delta)$, it follows from (1) that, for each $k \in \mathbb{N}$, there exists $x_k \in B_X(0,\epsilon)$ such that $\|t\delta y - Tx_k\| < 1/k$. Therefore,

$$\begin{aligned} t\delta = \|t\delta y\| &\le \|t\delta y - Tx_k\| + \|Tx_k\| \\ &< 1/k + \|T\|\,\epsilon. \end{aligned}$$

Letting $k \to \infty$ and then $t \to 1$, the conclusion follows. □

Lemma 2. *Let X be a Banach space, Y a normed space, and $T \in B(X,Y)$. Suppose TX is of Baire's second category in Y, and let $\epsilon > 0$ be given. Then there exists $\delta > 0$ such that*

$$B_Y(0,\delta) \subset TB_X(0,\epsilon). \tag{2}$$

Comparing the lemmas, we observe that the payoff for the added completeness hypothesis is the stronger conclusion (2) (instead of (1)).

Proof. We apply Lemma 1 with $\epsilon_n = \epsilon/2^{n+1}, n = 0, 1, 2, \ldots$. We then obtain $\delta_n > 0$ such that

$$B_Y(0,\delta_n) \subset \overline{TB_X(0,\epsilon_n)} \quad (n = 0, 1, 2, \ldots). \tag{3}$$

We shall show that (2) is satisfied with $\delta := \delta_0$.

Let $y \in B_Y(0,\delta)$.

By (3) with $n = 0$, there exists $x_0 \in B_X(0,\epsilon_0)$ such that

$$\|y - Tx_0\| < \delta_1,$$

that is, $y - Tx_0 \in B_Y(0,\delta_1)$.

By (3) with $n = 1$, there exists $x_1 \in B_X(0,\epsilon_1)$ such that

$$\|(y - Tx_0) - Tx_1\| < \delta_2.$$

Proceeding inductively, we obtain a sequence $\{x_n; n = 0, 1, 2, \ldots\}$ such that (for $n = 0, 1, 2, \ldots$)

$$x_n \in B_X(0,\epsilon_n) \tag{4}$$

and

$$\|y - T(x_0 + \cdots + x_n)\| < \delta_{n+1}. \tag{5}$$

Write

$$s_n = x_0 + \cdots + x_n.$$

Then for non-negative integers $n > m$

$$\begin{aligned} \|s_n - s_m\| &= \|x_{m+1} \cdots + x_n\| \le \|x_{m+1}\| + \cdots + \|x_n\| \\ &< \frac{\epsilon}{2^{m+2}} + \cdots + \frac{\epsilon}{2^{n+1}} < \frac{\epsilon}{2^{m+1}}, \end{aligned} \tag{6}$$

so that $\{s_n\}$ is a Cauchy sequence in X.

Since X is complete, $s := \lim s_n$ exists in X. By continuity of T and of the norm, the left-hand side of (5) converges to $\|y - Ts\|$ as $n \to \infty$. The right-hand side of (5) converges to 0 (cf. Lemma 1). Therefore, $y = Ts$. However, by (6) with $m = 0$, $\|s_n - s_0\| < \epsilon/2$ for all n; hence, $\|s - x_0\| \le \epsilon/2$, and by (4) $\|s\| \le \|x_0\| + \|s - x_0\| < \epsilon$. This shows that $y \in TB_X(0,\epsilon)$. □

Theorem 6.9 (The open mapping theorem). *Let X be a Banach space, and $T \in B(X,Y)$ for some normed space Y. Suppose TX is of Baire's second category in Y. Then T is an open mapping.*

Proof. Let V be a nonempty open subset of X, and let $y \in TV$. Let then $x \in V$ be such that $y = Tx$. Since V is open, there exists $\epsilon > 0$ such that $B_X(x, \epsilon) \subset V$. Let δ correspond to ϵ as in Lemma 2. Then

$$\begin{aligned} B_Y(y,\delta) &= y + B_Y(0,\delta) \subset Tx + TB_X(0,\epsilon) = T[x + B_X(0,\epsilon)] \\ &= TB_X(x,\epsilon) \subset TV. \end{aligned}$$

This shows that TV is an open set in Y. □

Corollary 6.10. *Let X, Y be Banach spaces, and let $T \in B(X, Y)$ be* onto. *Then T is an open map.*

Proof. Since T is onto, its range $TX = Y$ is a Banach space, and is therefore of Baire's second category in Y by Theorem 6.3. The result then follows from Theorem 6.9. □

Corollary 6.11 (Continuity of the inverse). *Let X, Y be Banach spaces, and let $T \in B(X, Y)$ be one-to-one and onto. Then $T^{-1} \in B(Y, X)$.*

Proof. By Corollary 6.10, T is a (linear) bijective continuous open map, that is, a (linear) homeomorphism. This means in particular that the inverse map is continuous. □

Corollary 6.12. *Suppose the vector space X is a Banach space under two norms $\|\cdot\|_k, k = 1, 2$. If there exists a constant $M > 0$ such that $\|x\|_2 \le M\,\|x\|_1$ for all $x \in X$, then there exists a constant $N > 0$ such that $\|x\|_1 \le N\,\|x\|_2$ for all $x \in X$.*

Norms satisfying inequalities of the form

$$\frac{1}{N}\|x\|_1 \le \|x\|_2 \le M\,\|x\|_1 \quad (x \in X)$$

for suitable constants $M, N > 0$ are said to be *equivalent.* They induce the same metric topology on X.

Proof. Let T be the identity map from the Banach space $(X, \|\cdot\|_1)$ to the Banach space $(X, \|\cdot\|_2)$. Then T is bounded (by the hypothesis on the norms), and clearly one-to-one and onto. The result then follows from Corollary 6.11. □

6.4 Graphs

For the next corollary, we consider the Cartesian product $X \times Y$ of two normed spaces, as a normed space with the usual operations and with the norm

$$\|[x, y]\| = \|x\| + \|y\| \quad ([x, y] \in X \times Y).$$

Clearly the sequence $\{[x_n, y_n]\}$ is Cauchy in $X \times Y$ iff both sequences $\{x_n\}$ and $\{y_n\}$ are Cauchy in X and Y, respectively, and it converges to $[x, y]$ in $X \times Y$ if and only if both $x_n \to x$ in X and $y_n \to y$ in Y. Therefore, $X \times Y$ is complete if and only if both X and Y are complete.

Let T be a linear map with domain $D(T) \subset X$ and range in Y. The domain is a subspace of X. The *graph* of T is the subspace of $X \times Y$ defined by

$$\Gamma(T) := \{[x, Tx]; x \in D(T)\}.$$

If $\Gamma(T)$ is a *closed* subspace of $X \times Y$, we say that T is a *closed* operator. Clearly T is closed iff whenever $\{x_n\} \subset D(T)$ is such that $x_n \to x$ and $Tx_n \to y$, then $x \in D(T)$ and $Tx = y$.

Corollary 6.13 (The closed graph theorem). *Let X, Y be Banach spaces and let T be a closed operator with $D(T) = X$ and range in Y. Then $T \in B(X, Y)$.*

Proof. Let P_X and P_Y be the projections of $X \times Y$ onto X and Y, respectively, *restricted to the closed subspace* $\Gamma(T)$ (which is a Banach space, as a closed subspace of the Banach space $X \times Y$). They are continuous, and P_X is one-to-one and onto. By Corollary 6.11, P_X^{-1} is continuous, and therefore the composition $P_Y \circ P_X^{-1}$ is continuous. However,

$$P_Y \circ P_X^{-1} x = P_Y[x, Tx] = Tx \quad (x \in X),$$

that is, T is continuous. □

Remark 6.14. The proof of Lemma 2 used the completeness of X to get the convergence of the sequence $\{s_n\}$, which is the sequence of partial sums of the series $\sum x_k$. The point of the argument is that, if X is complete, then the convergence of a series $\sum x_k$ in X follows from its *absolute convergence* (that is, the convergence of the series $\sum \|x_k\|$). This property actually characterizes completeness of normed spaces.

Theorem 6.15. *Let X be a normed space. Then X is complete iff absolute convergence implies convergence (of series in X).*

Proof. Suppose X is complete. If $\sum \|x_k\|$ converges and s_n denote the partial sums of $\sum x_k$, then for $n > m$

$$\|s_n - s_m\| = \left\| \sum_{k=m+1}^{n} x_k \right\| \le \sum_{k=m+1}^{n} \|x_k\| \to 0$$

as $m \to \infty$, and therefore $\{s_n\}$ converges in X.

Conversely, suppose absolute convergence implies convergence of series in X. Let $\{x_n\}$ be a Cauchy sequence in X. There exists a subsequence $\{x_{n_k}\}$ such that

$$\|x_{n_{k+1}} - x_{n_k}\| < \frac{1}{2^k}$$

(cf. proof of Lemma 1.30). The series

$$x_{n_1} + \sum_{k=1}^{\infty} (x_{n_{k+1}} - x_{n_k})$$

converges absolutely, and therefore converges in X. Its $(p-1)$-th partial sum is x_{n_p}, and so the *Cauchy sequence* $\{x_n\}$ has the convergent subsequence $\{x_{n_p}\}$; it follows that $\{x_n\}$ itself converges. □

6.5 Quotient space

If M is a closed subspace of the normed space X, the vector space X/M is a normed space for the *quotient norm*

$$\|[x]\| := \text{dist}\{0,[x]\} := \inf_{y\in[x]} \|y\|,$$

where $[x] := x+M$. The properties of the norm are easily verified; the assumption that M is *closed* is needed for the implication $\|[x]\| = 0$ *implies* $[x] = [0]$. For later use, we prove the following.

Theorem 6.16. *Let M be a closed subspace of the Banach space X. Then X/M is a Banach space.*

Proof. Let $\{[x_n]\}$ be a Cauchy sequence in X/M. It has a subsequence $\{[x'_n]\}$ such that

$$\|[x'_{n+1}] - [x'_n]\| < \frac{1}{2^{n+1}}$$

(cf. proof of Lemma 1.30). Pick $y_1 \in [x'_1]$ arbitrarily. Since

$$\inf_{y\in[x'_2]} \|y - y_1\| = \|[x'_2 - x'_1]\| < 1/4,$$

there exists $y_2 \in [x'_2]$ such that $\|y_2 - y_1\| < 1/2$. Assuming we found $y_k \in [x'_k]$, $k = 1,\ldots,n$, such that $\|y_{k+1} - y_k\| < 1/2^k$ for $k \le n-1$, since

$$\inf_{y\in[x'_{n+1}]} \|y - y_n\| = \|[x'_{n+1}] - [x'_n]\| < \frac{1}{2^{n+1}},$$

there exists $y_{n+1} \in [x'_{n+1}]$ such that

$$\|y_{n+1} - y_n\| < \frac{2}{2^{n+1}} = 1/2^n.$$

The series

$$y_1 + \sum_{n=1}^{\infty} (y_{n+1} - y_n)$$

converges absolutely, and therefore converges in X, since X is complete (cf. Theorem 6.15). Its partial sums y_n converge therefore to some $y \in X$. Then

$$\|[x'_n] - [y]\| = \|[y_n] - [y]\| = \|[y_n - y]\| \le \|y_n - y\| \to 0.$$

Since the *Cauchy sequence* $\{[x_n]\}$ has the *convergent subsequence* $\{[x'_n]\}$, it follows that it converges as well. □

Corollary 6.17. *The quotient map $\pi : x \to [x]$ of the normed space X onto the normed space X/M (for a given closed subspace M of X) maps the open unit ball of X onto that of X/M. Thus, π is an open mapping, and its norm is exactly* 1 *unless $M = X$.*

Proof. The map π is a norm-decreasing linear map. Thus, denoting by S_X^o and $S_{X/M}^o$ the open unit balls of X and X/M, respectively, we have $\pi(S_X^o) \subset S_{X/M}^o$. Conversely, if $\|[x]\| < 1$, i.e. $\inf_{y\in[x]} \|y\| < 1$, then there is $y \in [x]$ whose norm is less than 1; that is, $y \in S_X^o$ and $\pi(y) = [x]$. This proves that $S_{X/M}^o \subset \pi(S_X^o)$. The other assertions are easy consequences of the equality $\pi(S_X^o) = S_{X/M}^o$. □

6.6 Operator topologies

The norm topology on $B(X,Y)$ is also called the *uniform operator topology.* This terminology is motivated by the fact that a sequence $\{T_n\} \subset B(X,Y)$ converges in the norm topology of $B(X,Y)$ iff it converges (strongly) pointwise, uniformly on every bounded subset of X (that is, the sequence $\{T_n x\}$ converges strongly in Y, uniformly in x on any bounded subset of X). Indeed, if $\|T_n - T\| \to 0$, then for any bounded set $Q \subset X$,

$$\sup_{x\in Q} \|T_n x - Tx\| \le \|T_n - T\| \sup_{x\in Q} \|x\| \to 0,$$

so that $T_n x \to Tx$ strongly in Y, uniformly on Q. Conversely, if $T_n x$ converge strongly in Y uniformly for x in bounded subsets of X, this is true in particular for x in the unit ball $S = S_X$. Hence as $n, m \to \infty$,

$$\|T_n - T_m\| := \sup_{x\in S} \|(T_n - T_m)x\| = \sup_{x\in S} \|T_n x - T_m x\| \to 0.$$

If Y is complete, $B(X,Y)$ is complete by Theorem 4.4, and therefore T_n converge in the norm topology of $B(X,Y)$.

We consider two additional topologies on $B(X,Y)$, weaker than the uniform operator topology (u.o.t.). A *net* $\{T_j; j \in J\}$ converges to T in the *strong operator topology* (s.o.t.) of $B(X,Y)$ if $T_j x \to Tx$ strongly in Y, for each $x \in X$ (this is *strong pointwise convergence* of the functions T_j!). Since the uniformity requirement has been dropped, this convergence is clearly weaker than convergence in the u.o.t. If one requires that $T_j x$ converge *weakly* to Tx (rather than strongly!), for each $x \in X$, one gets a still-weaker convergence concept, called convergence *in the weak operator topology* (w.o.t.).

The s.o.t. and the w.o.t. may be defined by giving bases as follows.

Definition 6.18.

1. A base for the strong operator topology on $B(X,Y)$ consists of all the sets of the form

$$N(T, F, \epsilon) := \{S \in B(X,Y); \|(S - T)x\| < \epsilon, x \in F\},$$

where $T \in B(X,Y), F \subset X$ is *finite*, and $\epsilon > 0$.

2. A base for the weak operator topology on $B(X,Y)$ consists of all sets of the form

$$N(T,F,\Lambda,\epsilon) := \{S \in B(X,Y); |y^*(S-T)x| < \epsilon, x \in F, y^* \in \Lambda\},$$

where $T \in B(X,Y), F \subset X$ and $\Lambda \subset Y^*$ are finite sets, and $\epsilon > 0$.

The sets N are referred to as *basic neighborhoods of* T in the s.o.t. (w.o.t., respectively). It is clear that net convergence in these topologies is precisely as described.

Since the bases in Definition 6.18 consist of convex sets, it is clear that $B(X,Y)$ is a locally convex topological vector space (t.v.s.) for each of the above topologies. We denote by $B(X,Y)_{\text{s.o.}}$ and $B(X,Y)_{\text{w.o.}}$ the t.v.s. $B(X,Y)$ with the s.o.t. and the w.o.t., respectively.

Theorem 6.19. *Let X,Y be normed spaces. Then*

$$B(X,Y)^*_{s.o.} = B(X,Y)^*_{w.o.}.$$

Moreover, the general form of an element g of this (common) dual is

$$g(T) = \sum_k y_k^* T x_k \quad (T \in B(X,Y)),$$

where the sum is finite, $x_k \in X$, and $y_k^* \in Y^*$.

Proof. Let $g \in B(X,Y)^*_{\text{s.o.}}$. Since $g(0) = 0$, strong-operator continuity of g at zero implies the existence of $\epsilon > 0$ and of a finite set $F = \{x_1, \ldots, x_n\}$, such that $|g(T)| < 1$ for all $T \in N(0,F,\epsilon)$. Thus, the inequalities

$$\|Tx_k\| < \epsilon \quad (k = 1, \ldots, n) \tag{1}$$

imply $|g(T)| < 1$.

Consider the normed space Y^n with the norm $\|[y_1, \ldots, y_n]\| := \sum_k \|y_k\|$. One verifies easily that $(Y^n)^*$ is isomorphic to $(Y^*)^n$: given $\Gamma \in (Y^n)^*$, there exists a unique vector $[y_1^*, \ldots, y_n^*] \in (Y^*)^n$ such that

$$\Gamma([y_1, \ldots, y_n]) = \sum_k y_k^* y_k \tag{2}$$

for all $[y_1, \ldots, y_n] \in Y^n$.

With $x_1, \ldots, x_n$ as in (1), define the linear map

$$\Phi : B(X,Y) \to Y^n$$

by

$$\Phi(T) = [Tx_1, \ldots, Tx_n] \quad (T \in B(X,Y)).$$

On the range of Φ (a subspace of Y^n!), define Γ by

$$\Gamma(\Phi(T)) = g(T) \quad (T \in B(X,Y)).$$

If $T, S \in B(X,Y)$ are such that $\Phi(T) = \Phi(S)$, then $\Phi(m(T-S)) = 0$, so that $m(T-S)$ satisfies (1) for all $m \in \mathbb{N}$. Hence, $m|g(T-S)| = |g(m(T-S))| < 1$ for all m, and therefore $g(T) = g(S)$. This shows that Γ is well defined. It is clearly a linear functional on $range(\Phi)$. If $\|\Phi(T)\| < 1$, then $\|(\epsilon T)x_k\| < \epsilon$ for all k, hence, $|g(\epsilon T)| < 1$, that is, $|\Gamma(\Phi(T))| (= |g(T)|) < 1/\epsilon$. This shows that Γ is bounded, with norm $\leq 1/\epsilon$. By the Hahn–Banach theorem, Γ has an extension as an element $\tilde{\Gamma} \in (Y^n)^*$. As observed, it follows that there exist $y_1^*, \ldots, y_n^* \in Y^*$ such that $\tilde{\Gamma}([y_1, \ldots, y_n]) = \sum_k y_k^* y_k$. In particular,

$$g(T) = \Gamma([Tx_1, \ldots, Tx_n]) = \sum_k y_k^* T x_k \tag{3}$$

for all $T \in B(X,Y)$.

In particular, this representation shows that g is continuous with respect to the w.o.t. Since (linear) functionals continuous with respect to the w.o.t. are trivially continuous with respect to the s.o.t., the theorem follows. □

Corollary 6.20. *A* convex *subset of $B(X,Y)$ has the same closure in the w.o.t. and in the s.o.t.*

Proof. Let $K \subset B(X,Y)$ be convex (nonempty, without loss of generality), and denote by K_{s} and K_{w} its closures with respect to the s.o.t. and the w.o.t., respectively. Since the w.o.t. is weaker than the s.o.t., we clearly have $K_{\mathrm{s}} \subset K_{\mathrm{w}}$. Suppose there exists $T \in K_{\mathrm{w}}$ such that $T \notin K_{\mathrm{s}}$. By Corollary 5.21, there exists $f \in B(X,Y)^*_{\mathrm{s.o.}}$ such that

$$\Re f(T) < \inf_{S \in K_{\mathrm{s}}} \Re f(S). \tag{4}$$

By Theorem 6.19, $f \in B(X,Y)^*_{\mathrm{w.o.}}$. Since $T \in K_{\mathrm{w}}$ and $K \subset K_{\mathrm{s}}$, it follows that $\inf_{S \in K_{\mathrm{s}}} \Re f(S) \leq \inf_{S \in K} \Re f(S) \leq \Re f(T)$, which is a contradiction. □

Exercises

1. Let X be a Banach space, Y a normed space, and $T \in B(X,Y)$. Prove that if $TX \neq Y$, then TX is of Baire's first category in Y.

2. Let X, Y be normed spaces, and $T \in B(X,Y)$. Prove that

$$\|T\| = \sup\{|y^*Tx|; x \in X, y^* \in Y^*, \|x\| = \|y^*\| = 1\}.$$

3. Let $(S, \mathcal{A}, \mu)$ be a positive measure space, and let $p, q \in [1, \infty]$ be conjugate exponents. Let $T : L^p(\mu) \to L^p(\mu)$. Prove that

$$\|T\| = \sup\left\{\left|\int_S (Tf)g \, d\mu\right|; f \in L^p(\mu), g \in L^q(\mu), \|f\|_p = \|g\|_q = 1\right\}.$$

(In case $p = 1$ or $p = \infty$, assume that the measure space is σ-finite.)

4. Let X be a Banach space, $\{Y_\alpha; \alpha \in I\}$ a family of normed spaces, and $T_\alpha \in B(X, Y_\alpha)$, $(\alpha \in I)$. Define

$$Z = \left\{ x \in X; \sup_{\alpha \in I} \|T_\alpha x\| = \infty \right\}.$$

Prove that Z is either empty or a dense G_δ in X.

5. Let X be a Banach space, Y be a normed space, and $T : D(T) \subset X \to Y$ be a closed operator with range $R(T)$ of the second category in Y. Prove:

   (a) T is an open mapping and $R(T) = Y$. Hint: either use the fact that the graph of T is a Banach space or adapt the *proof* of Theorem 6.9.

   (b) There exists a constant $c > 0$ such that, for each $y \in Y$, there exists $x \in D(T)$ such that $y = Tx$ and $\|x\| \le c\|y\|$.

   (c) If T is one-to-one, then T^{-1} is bounded.

6. Let m denote Lebesgue measure on the interval $[0, 1]$, and let $1 \le p < r \le \infty$. Prove that the identity map of $L^r(m)$ into $L^p(m)$ is norm decreasing with range of Baire's first category in $L^p(m)$.

7. Let X be a Banach space, and let $T : D(T) \subset X \to X$ be a linear operator. Suppose there exists $\alpha \in \mathbb{C}$ such that $(\alpha I - T)^{-1} \in B(X)$. Let $p(\lambda) = \sum c_k \lambda^k$ be any polynomial (over $\mathbb{C}$) of degree $n \ge 1$. Prove that the operator $p(T) := \sum c_k T^k$ (with domain $D(T^n)$) is closed. (Hint: induction on n. Write $p(\lambda) = (\lambda - \alpha)q(\lambda) + r$, where the constant r may be assumed to be zero, without loss of generality, and q is a polynomial of degree $n - 1$.)

8. Let X, Y be Banach spaces. The operator $T \in B(X, Y)$ is *compact* if the set TB_X is conditionally compact in Y (where B_X denotes here the closed unit ball of X). Let $K(X, Y)$ be the set of all compact operators in $B(X, Y)$. Prove:

   (a) $K(X, Y)$ is a (norm-)closed subspace of $B(X, Y)$.

   (b) If Z is a Banach space, then

   $$K(X, Y)B(Z, X) \subset K(Z, Y) \quad \text{and} \quad B(Y, Z)K(X, Y) \subset K(X, Z).$$

   In particular, $K(X) := K(X, X)$ is a closed two-sided ideal in $B(X)$.

   (c) $T \in B(X, Y)$ is a *finite rank operator* if its range TX is finite dimensional. Prove that every finite rank operator is compact.

   (d) If Y is a Hilbert space, then the subspace of $B(X, Y)$ consisting of finite rank operators is dense in $K(X, Y)$.

Adjoints

9. Let X, Y be Banach spaces, and let $T : X \to Y$ be a linear operator with domain $D(T) \subset X$ and range $R(T)$. If T is one-to-one, the inverse map T^{-1} is a linear operator with domain $R(T)$ and range $D(T)$.

   If $D(T)$ is *dense* in X, the (Banach) adjoint T^* of T is defined as follows:

$$D(T^*) = \{y^* \in Y^*; y^* \circ T \text{ is continuous on } D(T)\}.$$

   Since $D(T)$ is dense in X, it follows that for each $y^* \in D(T^*)$ there exists a *unique* extension $x^* \in X^*$ of $y^* \circ T$ (cf. Exercise 1, Chapter 4); we set $x^* = T^*y^*$. Thus, T^* is uniquely defined on $D(T^*)$ by the relation

$$(T^*y^*)x = y^*(Tx) \quad (x \in D(T)).$$

   Prove:

   (a) T^* is closed. If T is closed, $D(T^*)$ is *weak**-dense in Y^*, and if Y is reflexive, $D(T^*)$ is strongly dense in Y^*.

   (b) If $T \in B(X,Y)$, then $T^* \in B(Y^*, X^*)$, and $\|T^*\| = \|T\|$. If $S, T \in B(X,Y)$, then $(\alpha S + \beta T)^* = \alpha S^* + \beta T^*$ for all $\alpha, \beta \in \mathbb{C}$. If $T \in B(X,Y)$ and $S \in B(Y,Z)$, then $(ST)^* = T^*S^*$.

   (c) If $T \in B(X,Y)$, then $T^{**} := (T^*)^* \in B(X^{**}, Y^{**})$, $T^{**}|_X = T$, and $\|T^{**}\| = \|T\|$. In particular, if X is reflexive, then $T^{**} = T$ (note that κX is identified with X).

   (d) If $T \in B(X,Y)$, then T^* is continuous with respect to the *weak**-topologies on Y^* and X^* (cf. Exercise 4, Chapter 5). Conversely, if $S \in B(Y^*, X^*)$ is continuous with respect to the *weak**-topologies on Y^* and X^*, then $S = T^*$ for some $T \in B(X,Y)$. Hint: given $x \in X$, consider the functional $\phi_x(y^*) = (Sy^*)x$ on Y^*.

   (e) $\overline{R(T)} = \bigcap\{\ker(y^*);\ y^* \in \ker(T^*)\}$. In particular, T^* is one-to-one iff $R(T)$ is dense in Y.

   (f) Let $x^* \in X^*$ and $M > 0$ be given. Then there exists $y^* \in D(T^*)$ with $\|y^*\| \le M$ such that $x^* = T^*y^*$ iff

$$|x^*x| \le M\|Tx\| \quad (x \in D(T)).$$

   In particular, $x^* \in R(T^*)$ iff

$$\sup_{x \in D(T),\, Tx \neq 0} \frac{|x^*x|}{\|Tx\|} < \infty.$$

   (Hint: Hahn–Banach.)

   (g) Let $T \in B(X,Y)$ and let S^* be the (norm-)closed unit ball of Y^*. Then T^*S^* is *weak**-compact.

(h) Let $T \in B(X,Y)$ have closed range TX. Suppose $x^* \in X^*$ vanishes on $\ker(T)$. Show that the map $\phi : TX \to \mathbb{C}$ defined by $\phi(Tx) = x^*x$ is a well-defined continuous linear functional, and therefore there exists $y^* \in Y^*$ such that $\phi = y^*|_{TX}$. (Hint: apply Corollary 6.10 to $T \in B(X,TX)$ to conclude that there exists $r > 0$ such that $\{y \in TX; \|y\| < r\} \subset TB_X(0,1)$, and deduce that $\|\phi\| \le (1/r)\|x^*\|$.)

(i) With T as in Part (h), prove that

$$T^*Y^* = \{x^* \in X^*; \ker(T) \subset \ker(x^*)\}.$$

In particular, T^* has (norm-)closed range in X^*.

10. Let X be a Banach space, and let T be a one-to-one linear operator with domain and range dense in X. Prove that $(T^*)^{-1} = (T^{-1})^*$, and T^{-1} is bounded (on its domain) iff $(T^*)^{-1} \in B(X^*)$.

11. Let $T : D(T) \subset X \to X$ have dense domain in the Banach space X. Prove:

(a) If the range $R(T^*)$ of T^* is *weak**-dense in X^*, then T is one-to-one.

(b) T^{-1} exists and is bounded (on its domain) iff $R(T^*) = X^*$.

12. Let X be a Banach space, and $T \in B(X)$. We say that T is *bounded below* if

$$\inf_{0 \neq x \in X} \frac{\|Tx\|}{\|x\|} > 0.$$

Prove:

(a) If T is bounded below, then it is one-to-one and has closed range.

(b) T is non-singular (that is, invertible in $B(X)$) iff it is bounded below and T^* is one-to-one.

Hilbert adjoint

13. Let X be a Hilbert space, and $T : D(T) \subset X \to X$ a linear operator with dense domain. The *Hilbert adjoint* T^* of T is defined in a way analogous to that of Exercise 9, through the Riesz representation:

$$D(T^*) := \{y \in X; x \to (Tx, y) \text{ is continuous on } D(T)\}.$$

Since $D(T)$ is dense, given $y \in D(T^*)$, there exists a unique vector in X, which we denote by T^*y, such that

$$(Tx, y) = (x, T^*y) \quad (x \in D(T)).$$

Prove:

(a) If $T \in B(X)$, then $T^* \in B(X)$, $\|T^*\| = \|T\|$, $T^{**} = T$, and $(\alpha T)^* = \bar{\alpha} T^*$ for all $\alpha \in \mathbb{C}$. Also $I^* = I$.

(b) If $S, T \in B(X)$, then $(S+T)^* = S^* + T^*$ and $(ST)^* = T^* S^*$.

(c) $T \in B(X)$ is called a *normal* operator if $T^*T = TT^*$. Prove that T is normal iff

$$(T^*x, T^*y) = (Tx, Ty) \quad (x, y \in X) \tag{1}$$

(d) If $T \in B(X)$ is normal, then $\|T^*x\| = \|Tx\|$ and $\|T^*Tx\| = \|T^2x\|$ for all $x \in X$. Conclude that $\|T^*T\| = \|T^2\|$ and $\|T^2\| = \|T\|^2$. (Hint: apply (1).)

14. Let X be a Hilbert space, and $T : D(T) \subset X \to X$ be a linear operator. T is *symmetric* if $(Tx, y) = (x, Ty)$ for all $x, y \in D(T)$. Prove that if T is symmetric and everywhere defined, then $T \in B(X)$ and $T = T^*$. (Hint: Corollary 6.13.)

Miscellaneous

15. Let X be a Hilbert space, and $B : X \times X \to \mathbb{C}$ be a sesquilinear form such that

$$|B(x,y)| \leq M\|x\| \, \|y\| \quad \text{and} \quad B(x,x) \geq m\|x\|^2$$

for all $x, y \in X$, for some constants $M < \infty$ and $m > 0$. Prove that there exists a unique nonsingular $T \in B(X)$ such that $B(x,y) = (x, Ty)$ for all $x, y \in X$. Moreover,

$$\|T\| \leq M \quad \text{and} \quad \|T^{-1}\| \leq 1/m.$$

(This is the *Lax–Milgram theorem.*) Hint: apply Theorem 1.37 to get T; show that $R(T)$ is closed and dense (cf. Theorem 1.36), and apply Corollary 6.11.

16. Let X, Y be normed spaces, and $T : X \to Y$ be linear. Prove that T is an open map iff $T\bar{B}_X(0,1)$ contains $\bar{B}_Y(0,r)$ for some $r > 0$. When this is the case, T is *onto.*

17. Let X be a Banach space, Y a normed space, and $T \in B(X, Y)$. Suppose the closure of $T\bar{B}_X(0,1)$ contains some ball $\bar{B}_Y(0,r)$. Prove that T is open. (Hint: adapt the *proof* of Lemma 2 in the proof of Theorem 6.9, and use Exercise 16.)

18. Let X be a Banach space, and let $P \in B(X)$ be such that $P^2 = P$. Such an operator is called a *projection.* Verify:

(a) $I - P$ is a projection (called *the complementary projection*).

(b) The ranges PX and $(I-P)X$ are *closed* subspaces such that $X = PX \oplus (I-P)X$. Moreover $PX = \ker(I-P) = \{x; Px = x\}$ and $(I-P)X = \ker P$.

(c) Conversely, if Y, Z are closed subspaces of X such that $X = Y \oplus Z$ ("complementary subspaces"), and $P : X \to Y$ is defined by $P(y+z) = y$ for all $y \in Y$, $z \in Z$, then P is a projection with $PX = Y$ and $\ker P = Z$. (Hint: Corollary 6.13.)

(d) If Y, Z are closed subspaces of X such that $Y \cap Z = \{0\}$, then $Y + Z$ is closed iff there exists a positive constant c such that $\|y\| \le c\|y+z\|$ for all $y \in Y$ and $z \in Z$.

19. Let X, Y be Banach spaces, and let $\{T_n\}_{n\in\mathbb{N}} \subset B(X,Y)$ be Cauchy in the s.o.t. (that is, $\{T_n x\}$ is Cauchy for each $x \in X$). Prove that $\{T_n\}$ is convergent in $B(X,Y)$ in the s.o.t.

20. Let X, Y be Banach spaces and $T \in B(X,Y)$. Prove that T is one-to-one with closed range iff there exists a positive constant c such that $\|Tx\| \ge c\|x\|$ for all $x \in X$. In that case, $T^{-1} \in B(TX, X)$.

21. Let X be a Banach space, and let $C \in B(X)$ be a *contraction*, that is, $\|C\| \le 1$. Prove:

(a) $e^{t(C-I)}$ (defined by means of the usual series) is a contraction for all $t \ge 0$.

(b) $\|C^m x - x\| \le m\|Cx - x\|$ for all $m \in \mathbb{N}$ and $x \in X$.

(c) Let $Q_n := e^{n(C-I)} - C^n$ $(n \in \mathbb{N})$. Then

$$\|Q_n x\| \le e^{-n} \sum_{k=0}^{\infty} (n^k/k!) \|C^{|k-n|}x - x\| \tag{2}$$

for all $n \in \mathbb{N}$ and $x \in X$. (Hint: note that $C^n x = e^{-n}\sum_k (n^k/k!) C^n x$; break the ensuing series for $Q_n x$ into series over $k \le n$ and over $k > n$).

(d) $\|Q_n x\| \le \sum_{k\ge 0} e^{-n}(n^k/k!)|k-n|\,\|Cx - x\|$.

(e) $\|Q_n x\| \le \sqrt{n}\|(C-I)x\|$ for all $n \in \mathbb{N}$ and $x \in X$. Hint: consider the *Poisson probability measure* μ (with "parameter" n) on $\mathbb{P}(\mathbb{N})$, defined by $\mu(\{k\}) = e^{-n}n^k/k!$; apply Schwarz's inequality in $L^2(\mu)$ and Part (d) to get the inequality

$$\|Q_n x\| \le \|k - n\|_{L^2(\mu)} \|Cx - x\| = \sqrt{n}\|Cx - x\|. \tag{3}$$

(f) Let $F : [0, \infty) \to B(X)$ be contraction-valued. For $t > 0$ fixed, set $A_n := (n/t)[F(t/n) - I]$, $n \in \mathbb{N}$. Suppose $\sup_n \|A_n x\| < \infty$ for all x in a dense subspace D of X. Then

$$\lim_{n\to\infty} \|e^{tA_n}x - F(t/n)^n x\| = 0 \tag{4}$$

for all $t > 0$ and $x \in X$. Hint: by Part (a), $\|e^{tA_n}\| \le 1$, and therefore $\|e^{tA_n} - F(t/n)^n\| \le 2$. By Part (e) with $C = F(t/n)$, the limit in (4) is 0 for all $x \in D$.

22. Let M be a closed subspace of a normed space X. Theorem 6.16 shows that if X is a Banach space, then so is the quotient space X/M. Conversely, show that if both M and the quotient space X/M are Banach spaces, then so is X.

23. Let X be a normed space and $M \subset X$. The *annihilator of M in X^** is

$$M^\perp := \{x^* \in X^*; x^*(M) = \{0\}\}.$$

Prove that $M^\perp$ is a *weak**-closed subspace of X^* and $M^\perp = (\overline{\operatorname{span}}\, M)^\perp$. Assuming that M is a closed subspace of X, prove:

(a) The map that takes $x^* + M^\perp$ to $x^*|_M$, $x^* \in X^*$, is an isometric isomorphism from $X^*/M^\perp$ onto M^*.

(b) The map that takes $x^* \in M^\perp$ to the functional on X/M given by $x + M \to x^*x$, $x \in X$, is an isometric isomorphism from $M^\perp$ onto $(X/M)^*$.

(c) $M = (M^\perp)_\perp$, where $(\cdot)_\perp$ is defined in the next exercise.

(d) The map that takes $x + M$ to $\kappa(x)|_{M^\perp}$, $x \in X$, is an isometric isomorphism from X/M onto its image $\{\kappa(x)|_{M^\perp}; x \in X\}$ inside $(M^\perp)^*$. Here κ is the canonical embedding of X into X^{**}.

24. Let X be a normed space and $N \subset X^*$. The *annihilator of N in X* is

$$N_\perp := \{x \in X; x^*x = 0 \text{ for all } x^* \in N\}.$$

Prove that $N_\perp$ is a closed subspace of X and $N_\perp = (\overline{\operatorname{span}}^{weak^*} N)_\perp$. Assuming that N is a *weak**-closed subspace of X^*, prove:

(a) $N = (N_\perp)^\perp$.

(b) There is a canonical isometric isomorphism from N onto $(X/N_\perp)^*$.

(c) There is a canonical isometric isomorphism from $X/N_\perp$ onto the subspace $\{\kappa(x)|_N; x \in X\}$ of N^*.

(d) The assertion of Part (c) actually holds for every subspace N of X^*, even when it is not *weak**-closed.

25. Let X, Y be normed spaces. Prove that the operator norm function from $B(X,Y)$ to $[0,\infty)$ is lower semi-continuous when $B(X,Y)$ is equipped with the w.o.t. (and thus also when equipped with the s.o.t.). For this, recall Exercise 6 in Chapter 3.

26. (Compare Exercise 8.) Let X, Y be Banach spaces. An operator $T \in B(X,Y)$ is called *weakly compact* if the set TB_X is conditionally weakly

compact in Y, that is, conditionally compact in the weak topology on Y. Prove:

(a) An operator $T \in B(X, Y)$ is weakly compact iff the image of $T^{**} : X^{**} \to Y^{**}$ is contained in Y (where we identify Y with κY). Hint: use Exercise 9 and some results from Section 5.5.

(b) If either X or Y is reflexive, then all elements of $B(X, Y)$ are weakly compact.

(c) The set of all weakly compact operators in $B(X, Y)$ is a closed subspace of $B(X, Y)$.

(d) If Z is another Banach space, $T \in B(X, Y)$ and $S \in B(Y, Z)$, and if either of T or S is weakly compact, then so is ST.

7

Banach algebras

Until this point our analysis focused mostly on single objects, for example, single functionals or, more generally, single operators. This chapter and Chapters 11–13 focus on algebras, more precisely: *Banach algebras*. These are Banach spaces with an additional product operation that turns them into algebras such that the norm is submultiplicative. Banach algebras extend and unify examples like $B(X)$ for a Banach space X, $C_0(X)$ for a topological space X, and $L^1(\mathbb{R})$ with convolution as multiplication.

The additional algebraic structure makes it possible to discuss *invertible elements*. This leads to the *spectrum* $\sigma(a)$ of an element a of a unital Banach algebra $\mathcal{A}$, which is the set of all scalars λ such that $\lambda e - a$ is *not* invertible in $\mathcal{A}$ (where e denotes the unit of $\mathcal{A}$). The notion of the spectrum is truly fundamental as demonstrated in every basic course in linear algebra. The first section of this chapter is chiefly concerned with the spectrum and related notions.

The next section restricts attention to *commutative* Banach algebras. To each such algebra $\mathcal{A}$ we associate a locally compact Hausdorff space Φ called the *Gelfand space* of $\mathcal{A}$ and a non-trivial homomorphism $\Gamma : \mathcal{A} \to C_0(\Phi)$ called the *Gelfand representation*, having crucial importance in Banach algebra theory. The Gelfand space Φ is the space of non-zero complex homomorphisms on $\mathcal{A}$, and $\Gamma(a)$ maps $\phi \in \Phi$ to $\phi(a)$. So this map views certain functions on $\mathcal{A}$ as "points" on which elements of $\mathcal{A}$ act, thus reversing the natural order. It turns out that there is a lot to gain from this perspective, as demonstrated in the so-called commutative Gelfand–Naimark theorem (Theorem 7.16). The discovery of this was a major breakthrough.

In the easiest case where $\mathcal{A} = C_0(X)$ for a locally compact Hausdorff space X, we have $\Phi = X$ and Γ is the identify map.

We then move on to a preliminary introduction of *C^*-algebras* spanning two sections. A C^*-algebra is a Banach algebra $\mathcal{A}$ having the additional algebraic structure of an *involution* and satisfying the so-called *C^*-identity*: $\|x^*x\| = \|x\|^2$ for all $x \in \mathcal{A}$. It is incredible how much comes out of this seemingly naive further assumption! Most of the C^*-algebra theory in this book appears in Chapters 11

Introduction to Modern Analysis. Second Edition. Shmuel Kantorovitz and Ami Viselter, Oxford University Press.
 DOI: 10.1093/oso/9780192849540.003.0007

and 13, while here our main goals are the celebrated *commutative Gelfand–Naimark theorem*, saying that the Gelfand representation of a *commutative C^*-algebra* is an isometric $*$-isomorphism, and the *continuous operational calculus of normal elements* and its features. We cover the non-commutative counterpart of the Gelfand–Naimark theorem in Chapter 11.

The chapter ends with a short discussion on the *Arens products*. For a Banach algebra $\mathcal{A}$ there are two natural products on its second dual $\mathcal{A}^{**}$ each of which makes $\mathcal{A}^{**}$ a Banach algebra in which $\mathcal{A}$ sits as a Banach subalgebra. We study these products and characterize the situation when they coincide, in which case $\mathcal{A}$ is called *Arens regular*.

7.1 Basics

This section introduces the theory of abstract Banach algebras, of which $B(X)$ and $C_0(X)$ are two important examples.

An *algebra* over a field $\mathbb{F}$ is a vector space $\mathcal{A}$ over $\mathbb{F}$ with a binary operation $\cdot : \mathcal{A} \times \mathcal{A} \to \mathcal{A}$ making $(\mathcal{A}, +, \cdot)$ a not-necessarily-unital ring, satisfying

$$\lambda(a \cdot b) = (\lambda a) \cdot b = a \cdot (\lambda b)$$

for all $a, b \in \mathcal{A}$ and $\lambda \in \mathbb{F}$. (Note that algebras are associative by definition, but are not necessarily commutative.) We will usually drop the product symbol and write ab for $a \cdot b$.

If X is a Banach space, the Banach space $B(X)$ (cf. Notation 4.3) is also an algebra under the composition of operators as multiplication. The operator norm (cf. Definition 4.1) clearly satisfies the relation

$$\|ST\| \leq \|S\|\|T\| \quad (S, T \in B(X)).$$

If the dimension of X is at least 2, it is immediate that $B(X)$ is not commutative.

On the other hand, if X is a topological space, the Banach space $C_0(X)$ of all complex continuous functions on X that vanish at infinity with pointwise operations and the supremum norm $\|f\| = \sup_X |f|$ is a *commutative* algebra, and again $\|fg\| \leq \|f\|\|g\|$ for all $f, g \in C_0(X)$. If X is compact, then the function with the constant value 1 on X is the *unit* of the algebra.

Definition 7.1. A (*complex*) *Banach algebra* is an algebra $\mathcal{A}$ over $\mathbb{C}$, which is a Banach space (as a vector space over $\mathbb{C}$) under a norm that is submultiplicative: $\|xy\| \leq \|x\|\|y\|$ for all $x, y \in \mathcal{A}$. We say that $\mathcal{A}$ is *unital* if it has an algebraic unit e.

If we omit the completeness requirement in Definition 7.1, $\mathcal{A}$ is called a *normed algebra*.

Note that the submultiplicativity of the norm implies the boundedness (i.e., the continuity) of the linear map of left multiplication by a, $L_a : x \to ax$ (for any given $a \in \mathcal{A}$), and clearly $\|L_a\| \leq \|a\|$ with equality when $\mathcal{A}$ is unital.

The same is true for the right multiplication map $R_a : x \to xa$. Actually, multiplication is continuous as a map from $\mathcal{A}^2$ to $\mathcal{A}$, since

$$\|xy - x'y'\| \le \|x\|\|y - y'\| + \|x - x'\|\|y'\| \to 0$$

as $[x', y'] \to [x, y]$ in $\mathcal{A}^2$.

Remark also that in the definition of unital Banach algebras some texts add the assumption that $\|e\| = 1$. We do not assume this, but the algebra can be renormed as in Exercise 2 with an equivalent norm so that this condition is satisfied.

Examples.

1. $B(X)$ is a *unital* Banach algebra for a (complex) Banach space X.
2. Let X be a topological space.
   i. $C_0(X)$ is a *commutative* Banach algebra for a topological space X. It is unital if X is compact. In particular, $c_0 := C_0(\mathbb{N})$ is a Banach algebra.
   ii. $C_b(X)$ is a *unital commutative* Banach algebra, where $C_b(X)$ is the Banach space of bounded continuous complex-valued functions on X with pointwise operations and the supremum norm.
3. $L^\infty(X, \mathbb{A}, \mu)$ is a *unital commutative* Banach algebra for a measure space $(X, \mathbb{A}, \mu)$ (with pointwise multiplication).
4. For $1 \le p \le \infty$, $l^p := L^p(\mathbb{N}, \mathbb{P}(\mathbb{N}), \text{counting measure})$ is a *commutative* Banach algebra with term-wise multiplication: if $x = \{x_n\}, y = \{y_n\} \in l^p$, then x is bounded and $\|x\|_\infty \le \|x\|_p$, so that

$$\|xy\|_p = \left(\sum_n |x_n y_n|^p\right)^{1/p} \le \left(\|x\|_\infty^p \sum_n |y_n|^p\right)^{1/p} = \|x\|_\infty \|y\|_p \le \|x\|_p \|y\|_p .$$

   l^p is *unital* iff $p = \infty$.
5. Let $\mathbb{D} := \{z \in \mathbb{C}; |z| < 1\}$ and $A(\mathbb{D}) := \{f \in C(\overline{\mathbb{D}}); f \text{ is holomorphic on } \mathbb{D}\}$. Then, with pointwise operations and the $C(\overline{\mathbb{D}})$-norm, $A(\mathbb{D})$ is a *unital commutative* Banach algebra, called the *disc algebra.*
6. $L^1(\mathbb{R})$ is a Banach algebra with respect to convolution; see Exercise 14. In fact, for an arbitrary locally compact topological group G, $L^1(G, \text{Haar measure})$ is a Banach algebra with respect to convolution. It is *unital* iff G is discrete, and *commutative* iff G is abelian.

Definition 7.2. Let $\mathcal{A}$ be a unital Banach algebra, and $a \in \mathcal{A}$. We say that a is *regular* (or *non-singular*) if it is invertible in $\mathcal{A}$, that is, if there exists $b \in \mathcal{A}$ such that $ab = ba = e$. If a is not regular, we say that it is *singular.*

If a is regular, the element b in Definition 7.2 is uniquely determined (if also b' satisfies the requirement, then $b' = b'e = b'(ab) = (b'a)b = eb = b$), and is called *the inverse* of a, denoted a^{-1}. Thus $aa^{-1} = a^{-1}a = e$. In particular, $a \neq 0$.

We denote by $G(\mathcal{A})$ the set of all regular elements of $\mathcal{A}$. It is a group under the multiplication of $\mathcal{A}$, and the map $x \to x^{-1}$ is an anti-automorphism of $G(\mathcal{A})$. Topologically, we have:

Theorem 7.3. *Let $\mathcal{A}$ be a unital Banach algebra, and let $G(\mathcal{A})$ be the group of regular elements of $\mathcal{A}$. Then $G(\mathcal{A})$ is open in $\mathcal{A}$, and the map $x \to x^{-1}$ is a homeomorphism of $G(\mathcal{A})$ onto itself.*

Proof. Let $y \in G := G(\mathcal{A})$ and $\delta := 1/\|y^{-1}\|$ (note that $\|y^{-1}\| \neq 0$, since $y^{-1} \in G$). We show that the ball $B(y, \delta)$ is contained in G (so that G is indeed open).

Let $x \in B(y, \delta)$, and set $a := y^{-1}x$. We have

$$\|e - a\| = \|y^{-1}(y - x)\| \leq \|y^{-1}\|\|y - x\| < \|y^{-1}\|\delta = 1. \tag{1}$$

Therefore, the geometric series $\sum_n \|e - a\|^n$ converges. By submultiplicativity of the norm, the series $\sum_n \|(e - a)^n\|$ converges as well, and since $\mathcal{A}$ is complete, it follows (cf. Theorem 6.15) that the series

$$\sum_{n=0}^{\infty} (e - a)^n$$

converges in $\mathcal{A}$ to some element $z \in \mathcal{A}$ ($v^0 = e$ by definition, for any $v \in \mathcal{A}$).

By continuity of L_a and R_a (with $a = y^{-1}x$),

$$az = \sum_n a(e - a)^n = \sum_n [e - (e - a)](e - a)^n = \sum_n [(e - a)^n - (e - a)^{n+1}] = e,$$

and similarly $za = e$. Hence $a \in G$ and $a^{-1} = z$. Since $x = ya$, also $x \in G$, as wanted.

Furthermore (for $x \in B(y, \delta)$!),

$$x^{-1} = a^{-1}y^{-1} = zy^{-1}, \tag{2}$$

and therefore by (1)

$$\begin{aligned}
\|x^{-1} - y^{-1}\| &= \|(z - e)y^{-1}\| = \left\| \sum_{n=1}^{\infty} (e - a)^n y^{-1} \right\| \\
&\leq \sum_{n=1}^{\infty} \|e - a\|^n \|y^{-1}\| = (1/\delta) \frac{\|e - a\|}{1 - \|e - a\|} \\
&\leq (1/\delta^2) \frac{\|x - y\|}{1 - (\|x - y\|/\delta)} \to 0
\end{aligned}$$

as $x \to y$. This proves the continuity of the map $x \to x^{-1}$ at $y \in G$. Since this map is its own inverse (on G), it is a homeomorphism. □

Remark 7.4. If we take in the preceding proof $y = e$ (so that $\delta = 1$ and $a = x$), we obtain in particular that $B(e, 1) \subset G$ and

$$x^{-1} = \sum_{n=0}^{\infty}(e - x)^n \quad (x \in B(e, 1)). \tag{3}$$

Since $B(e, 1) = e - B(0, 1)$, this is equivalent to

$$(e - u)^{-1} = \sum_{n=0}^{\infty} u^n \quad (u \in B(0, 1)). \tag{4}$$

Relation (4) is the abstract version of the elementary geometric series summation formula.

For $x \in \mathcal{A}$ arbitrary and λ complex with modulus $> \|x\|$, since $u := \frac{1}{\lambda}x \in B(0, 1)$, we then have $e - \frac{1}{\lambda}x \in G$, and

$$(e - \frac{1}{\lambda}x)^{-1} = \sum_{n=0}^{\infty} \frac{1}{\lambda^n} x^n.$$

Therefore $\lambda e - x = \lambda(e - \frac{1}{\lambda}x) \in G$ and

$$(\lambda e - x)^{-1} = \sum_{n=0}^{\infty} \frac{1}{\lambda^{n+1}} x^n \tag{5}$$

(for all complex λ with modulus $> \|x\|$).

Definition 7.5. The *resolvent set* of $x \in \mathcal{A}$ is the set

$$\rho(x) := \{\lambda \in \mathbb{C}; \lambda e - x \in G\} = f^{-1}(G),$$

where $f : \mathbb{C} \to \mathcal{A}$ is the continuous function $f(\lambda) := \lambda e - x$.

The complement of $\rho(x)$ in $\mathbb{C}$ is called the *spectrum* of x, denoted $\sigma(x)$. Thus $\lambda \in \sigma(x)$ iff $\lambda e - x$ is *singular*. The *spectral radius* of x, denoted $r(x)$, is defined by

$$r(x) = \sup\{|\lambda|; \lambda \in \sigma(x)\}.$$

By Theorem 7.3, $\rho(x)$ is *open*, as the inverse image of the open set G by the continuous function f. Therefore, $\sigma(x)$ is a *closed* subset of $\mathbb{C}$.

The *resolvent* of x, denoted $R(\cdot; x)$, is the function from $\rho(x)$ to G defined by

$$R(\lambda; x) = (\lambda e - x)^{-1} \quad (\lambda \in \rho(x)).$$

The series expansion (5) of the resolvent, valid for $|\lambda| > \|x\|$, is called the *Neumann expansion*.

Note the trivial but useful identity

$$xR(\lambda; x) = \lambda R(\lambda; x) - e \quad (\lambda \in \rho(x)).$$

If $x, y \in \mathcal{A}$ and $1 \in \rho(xy)$, then

$$(e - yx)[e + yR(1; xy)x] = (e - yx) + y(e - xy)R(1; xy)x = e,$$

and

$$[e + yR(1; xy)x](e - yx) = (e - yx) + yR(1; xy)(e - xy)x = e.$$

Therefore $1 \in \rho(yx)$ and

$$R(1; yx) = e + yR(1; xy)x.$$

Next, for any $\lambda \neq 0$, write $\lambda e - xy = \lambda[e - (\frac{1}{\lambda}x)y]$. If $\lambda \in \rho(xy)$, then $1 \in \rho((\frac{1}{\lambda}x)y)$; hence $1 \in \rho(y(\frac{1}{\lambda}x))$, and therefore $\lambda \in \rho(yx)$. By symmetry, this proves that

$$\sigma(xy) \cup \{0\} = \sigma(yx) \cup \{0\}.$$

Hence

$$r(yx) = r(xy).$$

With f as shown restricted to $\rho(x)$, $R(\lambda; x) = f(\lambda)^{-1}$; by Theorem 7.3, $R(\cdot; x)$ is therefore *continuous* on the open set $\rho(x)$.

By Remark 7.4,

$$\{\lambda \in \mathbb{C}; |\lambda| > \|x\|\} \subset \rho(x) \tag{6}$$

and

$$R(\lambda; x) = \sum_{n=0}^{\infty} \frac{1}{\lambda^{n+1}} x^n \quad (|\lambda| > \|x\|). \tag{7}$$

Thus

$$\sigma(x) \subset \Delta(0, \|x\|) := \{\lambda \in \mathbb{C}; |\lambda| \leq \|x\|\} \tag{8}$$

and so

$$r(x) \leq \|x\|. \tag{9}$$

The spectrum of x is closed and bounded (since it is contained in $\Delta(0, \|x\|)$). Thus $\sigma(x)$ is a *compact* subset of the plane.

Theorem 7.6. *Let $\mathcal{A}$ be a unital Banach algebra and $x \in \mathcal{A}$. Then $\sigma(x)$ is a non-empty compact set, and $R(\cdot; x)$ is an analytic function on $\rho(x)$ that vanishes at ∞.*

Proof. We observed already that $\sigma(x)$ is compact. By continuity of the map $y \to y^{-1}$, $(e - \lambda^{-1}x)^{-1} \to e^{-1} = e$ as $\lambda \to \infty$, and therefore

$$\lim_{\lambda \to \infty} R(\lambda; x) = \lim_{\lambda \to \infty} \lambda^{-1}(e - \lambda^{-1}x)^{-1} = 0.$$

For $\lambda \in \rho(x)$, since $\lambda e - x$ and x commute, also the inverse $R(\lambda; x)$ commutes with x. If also $\mu \in \rho(x)$, writing $R(\cdot) := R(\cdot; x)$, we have

$$R(\mu) = (\lambda e - x)R(\lambda)R(\mu) = \lambda R(\lambda)R(\mu) - xR(\lambda)R(\mu)$$

and

$$R(\lambda) = R(\lambda)(\mu e - x)R(\mu) = \mu R(\lambda)R(\mu) - xR(\lambda)R(\mu).$$

Subtracting, we obtain the so-called *resolvent identity*

$$R(\mu) - R(\lambda) = (\lambda - \mu)R(\lambda)R(\mu). \tag{10}$$

For $\mu \neq \lambda$ in $\rho(x)$, rewrite (10) as

$$\frac{1}{\mu - \lambda}(R(\mu) - R(\lambda)) = -R(\lambda)R(\mu).$$

Since $R(\cdot)$ is continuous on $\rho(x)$, we have

$$\exists \lim_{\mu \to \lambda} \frac{1}{\mu - \lambda}(R(\mu) - R(\lambda)) = -R(\lambda)^2.$$

This shows that $R(\cdot)$ is analytic on $\rho(x)$, and $R'(\cdot) = -R(\cdot)^2$.

For any $x^* \in \mathcal{A}^*$, it follows that $x^*R(\cdot)$ is a complex analytic function in $\rho(x)$. If $\sigma(x)$ is empty, $x^*R(\cdot)$ is entire and vanishes at ∞. By Liouville's theorem, $x^*R(\cdot)$ is identically 0, for all $x^* \in \mathcal{A}^*$. Therefore, $R(\cdot) = 0$, which is absurd since $R(\cdot)$ has values in $G(\mathcal{A})$. This shows that $\sigma(x) \neq \emptyset$. □

Corollary 7.7 (The Gelfand–Mazur theorem). *A unital Banach algebra that is a division algebra equals* $\mathbb{C}e$.

Proof. Suppose the unital Banach algebra $\mathcal{A}$ is a division algebra. If $x \in \mathcal{A}, \sigma(x) \neq \emptyset$ (by Theorem 7.6); pick then $\lambda \in \sigma(x)$. Since $\lambda e - x$ is singular, and $\mathcal{A}$ is a division algebra, we must have $\lambda e - x = 0$. Hence $x = \lambda e$ and therefore $\mathcal{A} = \mathbb{C}e$. □

Let $\mathcal{A}$ be a unital Banach algebra. If $p(\lambda) = \sum \alpha_k \lambda^k$ is a polynomial with complex coefficients, and $x \in \mathcal{A}$, we denote as usual $p(x) := \sum \alpha_k x^k$ (where $x^0 := e$). The map $p \to p(x)$ is an *algebra homomorphism* τ of the algebra of polynomials (over $\mathbb{C}$) into $\mathcal{A}$, that sends 1 to e and λ to x.

Theorem 7.8 (The spectral mapping theorem for polynomials). *Let $\mathcal{A}$ be a unital Banach algebra. For any polynomial p (over $\mathbb{C}$) and any $x \in \mathcal{A}$*

$$\sigma(p(x)) = p(\sigma(x)).$$

Proof. Let $\mu = p(\lambda_0)$. Then λ_0 is a root of the polynomial $\mu - p$, and therefore

$$\mu - p(\lambda) = (\lambda - \lambda_0)q(\lambda)$$

for some polynomial q over $\mathbb{C}$. Applying the homomorphism τ, we get

$$\mu e - p(x) = (x - \lambda_0 e)q(x) = q(x)(x - \lambda_0 e).$$

If $\mu \in \rho(p(x))$, it follows that

$$(x - \lambda_0 e)(q(x)R(\mu; p(x))) = (R(\mu; p(x))q(x))(x - \lambda_0 e) = e,$$

so that $\lambda_0 \in \rho(x)$. Therefore, if $\lambda_0 \in \sigma(x)$, it follows that $\mu := p(\lambda_0) \in \sigma(p(x))$. This shows that

$$p(\sigma(x)) \subset \sigma(p(x)).$$

On the other hand, factor the polynomial $\mu - p$ into linear factors

$$\mu - p(\lambda) = \alpha \prod_{k=1}^{n} (\lambda - \lambda_k).$$

Note that $\mu = p(\lambda_k)$ for all $k = 1, \ldots, n$. Applying the homomorphism τ, we get

$$\mu e - p(x) = \alpha \prod_{k=1}^{n} (x - \lambda_k e).$$

If $\lambda_k \in \rho(x)$ for all k, then the product here is in $G(\mathcal{A})$, and therefore $\mu \in \rho(p(x))$. Consequently, if $\mu \in \sigma(p(x))$, there exists $k \in \{1, \ldots, n\}$ such that $\lambda_k \in \sigma(x)$, and therefore $\mu = p(\lambda_k) \in p(\sigma(x))$. This shows that $\sigma(p(x)) \subset p(\sigma(x))$. □

Theorem 7.9 (The Beurling–Gelfand spectral radius formula). *For any element x of a unital Banach algebra $\mathcal{A}$,*

$$\exists \lim_n \|x^n\|^{1/n} = r(x) = \inf_{n \in \mathbb{N}} \|x^n\|^{1/n}.$$

Proof. By Theorem 7.8 with the polynomial $p(\lambda) = \lambda^n$ $(n \in \mathbb{N})$,

$$\sigma(x^n) = \sigma(x)^n := \{\lambda^n; \lambda \in \sigma(x)\}.$$

Hence, by (8) applied to x^n, $|\lambda^n| \leq \|x^n\|$ for all n and $\lambda \in \sigma(x)$. Thus, $|\lambda| \leq \|x^n\|^{1/n}$ for all n, and therefore

$$|\lambda| \leq \inf_{n \in \mathbb{N}} \|x^n\|^{1/n} \leq \liminf_n \|x^n\|^{1/n}$$

for all $\lambda \in \sigma(x)$. Taking the supremum over all such λ, we obtain

$$r(x) \leq \inf_{n \in \mathbb{N}} \|x^n\|^{1/n} \leq \liminf_n \|x^n\|^{1/n}. \tag{11}$$

For each $x^* \in \mathcal{A}^*$, the complex function $x^* R(\cdot)$ is analytic in $\rho(x)$, and since $\sigma(x)$ is contained in the closed disc around 0 with radius $r(x)$, $\rho(x)$ contains the open "annulus" $r(x) < |\lambda| < \infty$. By Laurent's theorem, $x^* R(\cdot)$ has a unique Laurent series expansion in this annulus. In the possibly smaller annulus $\|x\| < |\lambda| < \infty$ (cf. (9)), this function has the expansion (cf. (7))

$$x^* R(\lambda) = \sum_{n=0}^{\infty} \frac{x^*(x^n)}{\lambda^{n+1}}.$$

This is a Laurent expansion; by uniqueness, this is *the* Laurent expansion of $x^* R(\cdot)$ in the full annulus $r(x) < |\lambda| < \infty$. The convergence of the series implies in particular that

$$\sup_n |x^*(\frac{1}{\lambda^{n+1}} x^n)| < \infty \quad (|\lambda| > r(x))$$

for all $x^* \in \mathcal{A}^*$. By Corollary 6.7, it follows that (whenever $|\lambda| > r(x)$)

$$\sup_n \left\| \frac{1}{\lambda^{n+1}} x^n \right\| := M_\lambda < \infty.$$

Hence, for all $n \in \mathbb{N}$ and $|\lambda| > r(x)$,

$$\|x^n\| \leq M_\lambda |\lambda|^{n+1},$$

so that

$$\limsup_n \|x^n\|^{1/n} \leq |\lambda|,$$

and therefore

$$\limsup_n \|x^n\|^{1/n} \leq r(x). \tag{12}$$

The conclusion of the theorem follows from (11) and (12). □

Definition 7.10. An element x of a unital Banach algebra is said to be *quasi-nilpotent* if $\lim \|x^n\|^{1/n} = 0$.

By Theorem 7.9, the element x is quasi-nilpotent iff $r(x) = 0$, that is, iff $\sigma(x) = \{0\}$.

In particular, *nilpotent* elements ($x^n = 0$ for some n) are quasi-nilpotent.

We consider now the *boundary* points of the open set $G(\mathcal{A})$.

Theorem 7.11. *Let $\mathcal{A}$ be a unital Banach algebra and x be a boundary point of $G(\mathcal{A})$. Then x is a (two-sided)* topological divisor of zero*, that is, there exist sequences of unit vectors $\{x_n\}$ and $\{x'_n\}$ such that $x_n x \to 0$ and $x x'_n \to 0$.*

Proof. Let $x \in \partial G$ (:= the boundary of $G := G(\mathcal{A})$). Since G is open, there exists a sequence $\{y_n\} \subset G$ such that $y_n \to x$ and $x \notin G$.

If $\{\|y_n^{-1}\|\}$ is bounded (say by $0 < M < \infty$), and n is so large that $\|x - y_n\| < 1/M$, then

$$\|y_n^{-1}x - e\| = \|y_n^{-1}(x - y_n)\| \leq \|y_n^{-1}\|\|x - y_n\| < 1,$$

and therefore $z := y_n^{-1}x \in G$ by Remark 7.4. Hence, $x = y_n z \in G$, contradiction. Thus, $\{\|y_n^{-1}\|\}$ is unbounded, and has therefore a subsequence $\{\|y_{n_k}^{-1}\|\}$ diverging to infinity. Define

$$x_k := \frac{1}{\|y_{n_k}^{-1}\|} y_{n_k}^{-1} \quad (k \in \mathbb{N}).$$

Then $\|x_k\| = 1$ and

$$\|x_k x\| = \|x_k y_{n_k} + x_k(x - y_{n_k})\| \leq \frac{1}{\|y_{n_k}^{-1}\|} + \|x_k\|\|x - y_{n_k}\| \to 0,$$

and similarly $x x_k \to 0$. □

Theorem 7.12. *Let $\mathcal{B}$ be a Banach subalgebra of the unital Banach algebra $\mathcal{A}$ such that $e_{\mathcal{A}} \in \mathcal{B}$. If $x \in \mathcal{B}$, denote the spectrum of x as an element of $\mathcal{B}$ by $\sigma_{\mathcal{B}}(x)$. Then*

(1) $\sigma(x) \subset \sigma_{\mathcal{B}}(x)$ and

(2) $\partial\sigma_{\mathcal{B}}(x) \subset \partial\sigma(x)$.

Proof. The first inclusion is trivial, since $G(\mathcal{B}) \subset G(\mathcal{A})$, so that $\rho_{\mathcal{B}}(x) \subset \rho(x)$.

Let $\lambda \in \partial\sigma_{\mathcal{B}}(x)$. Then $\lambda e - x \in \partial G(\mathcal{B})$, and therefore, by Theorem 7.11, $\lambda e - x$ is a topological divisor of zero in $\mathcal{B}$, hence in $\mathcal{A}$. In particular, $\lambda e - x \notin G(\mathcal{A})$, that is, $\lambda \in \sigma(x)$. This shows that $\partial\sigma_{\mathcal{B}}(x) \subset \sigma(x)$. Since $\rho_{\mathcal{B}}(x) \subset \rho(x)$, we obtain (using part (1)):

$$\begin{aligned}\partial\sigma_{\mathcal{B}}(x) &= \overline{\rho_{\mathcal{B}}(x)} \cap \sigma_{\mathcal{B}}(x) \subset [\overline{\rho(x)} \cap \sigma_{\mathcal{B}}(x)] \cap \sigma(x) \\ &= \overline{\rho(x)} \cap \sigma(x) = \partial\sigma(x).\end{aligned}$$ □

Corollary 7.13. *Let $\mathcal{B}$ be a Banach subalgebra of the unital Banach algebra $\mathcal{A}$ such that $e_{\mathcal{A}} \in \mathcal{B}$, and $x \in \mathcal{B}$. Then $\sigma_{\mathcal{B}}(x) = \sigma(x)$ if* either *$\sigma_{\mathcal{B}}(x)$ is* nowhere dense or *$\rho(x)$ is* connected.

Proof. If $\sigma_{\mathcal{B}}(x)$ is nowhere dense, Theorem 7.12 implies that

$$\sigma_{\mathcal{B}}(x) = \partial\sigma_{\mathcal{B}}(x) \subset \partial\sigma(x) \subset \sigma(x) \subset \sigma_{\mathcal{B}}(x),$$

and the conclusion follows.

If $\rho(x)$ is connected and $\sigma(x)$ is a *proper* subset of $\sigma_{\mathcal{B}}(x)$, there exists $\lambda \in \sigma_{\mathcal{B}}(x) \cap \rho(x)$, and it can be connected with the point at ∞ by a continuous curve lying in $\rho(x)$. Since $\sigma_{\mathcal{B}}(x)$ is compact, the curve meets $\partial\sigma_{\mathcal{B}}(x)$ at some point, and therefore $\partial\sigma_{\mathcal{B}}(x) \cap \rho(x) \neq \emptyset$, contradicting Statement 2. of Theorem 7.12. □

The definition of the spectrum naturally requires the algebra to be unital. This can be "fixed" as follows. Let $\mathcal{A}$ be a non-unital Banach algebra. The Cartesian product Banach space $\mathcal{A}^{\#} := \mathcal{A} \times \mathbb{C}$ with the norm $\|[x, \lambda]\| = \|x\| + |\lambda|$ and the multiplication

$$[x, \lambda]\,[y, \mu] = [xy + \lambda y + \mu x, \lambda\mu]$$

is a unital Banach algebra with the identity $e := [0, 1]$ in which $\mathcal{A}$ is (isometrically) naturally embedded. It is called the *unitization* of $\mathcal{A}$ (see Exercise 1). For $a \in \mathcal{A}$, define $\sigma(a), \rho(a), R(\cdot; a)$ to be $\sigma_{\mathcal{A}^{\#}}(a), \rho_{\mathcal{A}^{\#}}(a), R_{\mathcal{A}^{\#}}(\cdot; a)$, respectively (the last one has values in $\mathcal{A}^{\#}$). Then it is immediate that Theorems 7.6, 7.8, and 7.9 hold for general Banach algebras.

Recall that an *ideal* in an algebra $\mathcal{A}$ is a linear subspace $M \subset \mathcal{A}$ that is $\mathcal{A}$-invariant, that is, $\mathcal{A}M, M\mathcal{A} \subset M$. In this case, the quotient space $\mathcal{A}/M$ *is an algebra.* Say that M is *proper* if it is not equal to $\mathcal{A}$. If $\mathcal{A}$ is a Banach algebra and M is *closed*, the quotient norm on the Banach space $\mathcal{A}/M$ (cf. Theorem 6.16) is submultiplicative, that is, $\mathcal{A}/M$ *is a Banach algebra.* If, moreover, $\mathcal{A}$ is unital and M is proper, then $\mathcal{A}/M$ is also unital, with $e + M$ as its algebraic unit. Section 7.2 discusses the case of *commutative* Banach algebras in more detail.

7.2 Commutative Banach algebras

Let $\mathcal{A}$ be a unital *commutative* Banach algebra.

1. Let $x \in \mathcal{A}$. Then $x \in G(\mathcal{A})$ iff $x\mathcal{A} = \mathcal{A}$. Equivalently, x is singular iff $x\mathcal{A} \neq \mathcal{A}$, that is, iff x is contained in a proper ideal $(x\mathcal{A})$. Proper ideals are contained therefore in the *closed* set $G(\mathcal{A})^c$, and it follows that *the closure of a proper ideal is a proper ideal.*

2. A *maximal* ideal M (in $\mathcal{A}$) is a proper ideal in $\mathcal{A}$ with the property that if N is a proper ideal in $\mathcal{A}$ containing M, then $N = M$. Since the closure of M is a proper ideal containing M, it follows that *maximal ideals are closed.* In particular $\mathcal{A}/M$ is a Banach algebra, and *is also a field* (by a well-known elementary algebraic characterization of maximal ideals). By Theorem 7.7, $\mathcal{A}/M$ is isomorphic (and isometric) to $\mathbb{C}$. Composing the natural homomorphism $\mathcal{A} \to \mathcal{A}/M$ with this isomorphism, we obtain a (norm-decreasing) homomorphism ϕ_M of $\mathcal{A}$ onto $\mathbb{C}$, *whose kernel is* M. Thus, for any $x \in \mathcal{A}$, $\phi_M(x)$ is the unique scalar λ such that $x + M = \lambda e + M$. Equivalently, $\phi_M(x)$ is uniquely determined by the relation $\phi_M(x)e - x \in M$.

Let $\Phi = \Phi(\mathcal{A})$ denote the set of all homomorphisms of $\mathcal{A}$ onto $\mathbb{C}$. Note that $\phi \in \Phi$ iff ϕ is a homomorphism of $\mathcal{A}$ *into* $\mathbb{C}$ such that $\phi(e) = 1$ (equivalently, iff ϕ is a *non-zero* homomorphism of $\mathcal{A}$ into $\mathbb{C}$). The elements of Φ are called the *characters* of $\mathcal{A}$.

The mapping $M \to \phi_M$ described here is a mapping of the set $\mathcal{M}$ of all maximal ideals into Φ.

On the other hand, if $\phi \in \Phi$, and $M := \ker\phi$, then (by Noether's "first homomorphism theorem") $\mathcal{A}/M$ is isomorphic to $\mathbb{C}$, and is therefore a field. By the algebraic characterization of maximal ideals mentioned, it follows that M is a maximal ideal. We have $\ker\phi_M = M = \ker\phi$. For any $x \in \mathcal{A}, x - \phi(x)e \in \ker\phi = \ker\phi_M$, hence $0 = \phi_M(x - \phi(x)e) = \phi_M(x) - \phi(x)$. This shows that $\phi = \phi_M$, that is, the mapping $M \to \phi_M$ is *onto*. It is clearly one-to-one, because if $M, N \in \mathcal{M}$ are such that $\phi_M = \phi_N$, then $M = \ker\phi_M = \ker\phi_N = N$. We conclude that the mapping $M \to \phi_M$ is a bijection of $\mathcal{M}$ onto Φ, with the inverse mapping $\phi \to \ker\phi$.

3. If J is a proper ideal, then $e \notin J$. The set $\mathcal{U}$ of all proper ideals containing J is partially ordered by inclusion, and every totally ordered subset $\mathcal{U}_0$ has the upper bound $\bigcup\mathcal{U}_0$ (which is *proper* because the identity does *not* belong to it) in $\mathcal{U}$. By Zorn's lemma, $\mathcal{U}$ has a maximal element, which is clearly a maximal ideal containing J. Thus, *every proper ideal is contained in a maximal ideal.* Together with 1., this shows that *an element x is singular iff it is contained in a maximal ideal M*. By the bijection established between $\mathcal{M}$ and Φ, this means that x is singular iff $\phi(x) = 0$ for some $\phi \in \Phi$. Therefore, for any $x \in \mathcal{A}, \lambda \in \sigma(x)$ iff $\phi(\lambda e - x) = 0$ for some $\phi \in \Phi$, that is, iff $\lambda = \phi(x)$ for some ϕ. Thus

$$\sigma(x) = \{\phi(x); \phi \in \Phi\}. \tag{1}$$

Therefore

$$\sup_{\phi\in\Phi} |\phi(x)| = \sup_{\lambda\in\sigma(x)} |\lambda| := r(x). \tag{2}$$

By (9) in Section 7.1, it follows in particular that $|\phi(x)| \leq \|x\|$, so that the homomorphism ϕ is necessarily continuous, with norm ≤ 1. If we assume that $\|e\| = 1$, then since $\|\phi\| \geq |\phi(e)| = 1$, we have $\|\phi\| = 1$ (for all $\phi \in \Phi$).

We have then $\Phi \subset S^*$, where S^* is the strongly closed unit ball of $\mathcal{A}^*$. By Theorem 5.24, S^* is compact in the *weak** topology on $\mathcal{A}^*$. If ϕ_α is a net in Φ converging *weak** to $h \in S^*$, then

$$h(xy) = \lim_\alpha \phi_\alpha(xy) = \lim_\alpha \phi_\alpha(x)\phi_\alpha(y) = h(x)h(y)$$

for all $x, y \in \mathcal{A}$ and $h(e) = \lim_\alpha \phi_\alpha(e) = 1$, so that $h \in \Phi$. This shows that Φ is a *closed* subset of the compact space S^* (with the *weak** topology), hence Φ is compact (in this topology). Since the *weak** topology on $\mathcal{A}^*$ is Hausdorff, we conclude that Φ endowed with the relative *weak** topology (called the *Gelfand topology on* Φ) is a *compact Hausdorff space*. It has several names: the *Gelfand space*, the *character space*, the *structure space*, and the *spectrum* of $\mathcal{A}$.

4. For any $x \in \mathcal{A}$, let $\hat{x} := (\kappa x)\big|_\Phi$, the restriction of $\kappa x : \mathcal{A}^* \to \mathbb{C}$ to Φ (where κ is the canonical embedding of $\mathcal{A}$ in its second dual). By definition of the *weak** topology, κx is continuous on $\mathcal{A}^*$ (with the *weak** topology); therefore its restriction $\hat{x}$ to Φ (with the Gelfand topology) is continuous. The function $\hat{x} \in C(\Phi)$ is called *the Gelfand transform* of x. By definition

$$\hat{x}(\phi) = \phi(x) \quad (\phi \in \Phi), \tag{3}$$

and therefore, by (1),

$$\hat{x}(\Phi) = \sigma(x) \tag{4}$$

and

$$\|\hat{x}\|_{C(\Phi)} = r(x). \tag{5}$$

Note that the subalgebra $\hat{\mathcal{A}} := \{\hat{x}; x \in \mathcal{A}\}$ of $C(\Phi)$ *contains* $1 = \hat{e}$ *and separates the points of* Φ (if $\phi \neq \psi$ are elements of Φ, there exists $x \in \mathcal{A}$ such that $\phi(x) \neq \psi(x)$, that is, $\hat{x}(\phi) \neq \hat{x}(\psi)$, by (3)).

5. It is also customary to consider $\mathcal{M}$ with the Gelfand topology of Φ transferred to it through the bijection $M \to \phi_M$. In this case the compact Hausdorff space $\mathcal{M}$ (with this Gelfand topology) is called *the maximal ideal space* of $\mathcal{A}$, and $\hat{x}$ is considered as defined on $\mathcal{M}$ through the described bijection, that is, we write $\hat{x}(M)$ instead of $\hat{x}(\phi_M)$, so that

$$\hat{x}(M) = \phi_M(x) \quad (M \in \mathcal{M}). \tag{6}$$

The basic neighborhoods for the Gelfand topology on $\mathcal{M}$ are of the form

$$N(M_0; x_1, \ldots, x_n; \epsilon) := \{M \in \mathcal{M}; |\hat{x}_k(M) - \hat{x}_k(M_0)| < \epsilon, k = 1, \ldots, n\}.$$

6. The mapping $\Gamma : x \to \hat{x}$ is clearly a *representation* of the algebra $\mathcal{A}$ into the algebra $C(\Phi)$ (or $C(\mathcal{M})$), that is, an algebra homomorphism sending e to 1:

$$[\Gamma(x+y)](\phi) = \phi(x+y) = \phi(x) + \phi(y) = (\Gamma x + \Gamma y)(\phi)$$

for all $\phi \in \Phi$, etc. It is called the *Gelfand representation* of $\mathcal{A}$. By (5), the map Γ is also norm-decreasing (hence continuous), and (cf. also (3))

$$\begin{aligned}\ker\Gamma &= \{x \in \mathcal{A}; r(x) = 0\} = \{x \in \mathcal{A}; \sigma(x) = \{0\}\} \\ &= \{x \in \mathcal{A}; x \text{ is quasi-nilpotent}\} = \bigcap \mathcal{M}. \qquad (7)\end{aligned}$$

We now describe succinctly the changes that need to be made to the foregoing discussion to construct the Gelfand representation when the commutative Banach algebra $\mathcal{A}$ is *not necessarily unital.*

A. An ideal M in $\mathcal{A}$ is called *modular* if the quotient algebra $\mathcal{A}/M$ is unital, that is, if there exists $u \in \mathcal{A}$ such that $au - a, ua - a \in M$ for all $a \in \mathcal{A}$ (such u is called a modular, or relative, unit for M). Evidently, if $\mathcal{A}$ itself is unital, then every ideal is modular. Also, an ideal containing a modular ideal is also modular (with the same modular unit). Hence, Zorn's lemma implies that every proper modular ideal is contained in a *maximal modular ideal*, and every such maximal modular ideal is a maximal ideal. One shows, mimicking the proof of Theorem 7.3, that the closure of a proper modular ideal is, again, a proper modular ideal. In particular, a maximal modular ideal is closed. Once again, a *character* of $\mathcal{A}$ is a homomorphism of $\mathcal{A}$ onto $\mathbb{C}$, or equivalently, a non-zero homomorphism of $\mathcal{A}$ into $\mathbb{C}$. The kernel of a character is a maximal modular ideal in $\mathcal{A}$. The map $M \to \phi_M$ defined previously is a bijection between the set $\mathcal{M}$ of maximal modular ideals and the set Φ of all characters of $\mathcal{A}$ (note that $\mathcal{A}/M$ is unital because M is modular, so the Gelfand–Mazur theorem indeed applies).

B. If $\mathcal{A}$ is not unital, then every $\phi \in \Phi(\mathcal{A})$ has a unique extension to an element $\phi^{\#} \in \Phi(\mathcal{A}^{\#})$, which is given by $\phi^{\#}(x + \lambda e) := \phi(x) + \lambda$ for all $x \in \mathcal{A}$ and $\lambda \in \mathbb{C}$. We have $\Phi(\mathcal{A}^{\#}) = \Phi(\mathcal{A})^{\#} \cup \{\phi_\infty\}$, where $\phi_\infty(x + \lambda e) = \lambda$ for all $x \in \mathcal{A}$ and $\lambda \in \mathbb{C}$. As a result, for each $x \in \mathcal{A}$,

$$\sigma(x) = \sigma_{\mathcal{A}^{\#}}(x) = \{\phi(x); \phi \in \Phi(\mathcal{A}^{\#})\} = \{\phi(x); \phi \in \Phi(\mathcal{A})\} \cup \{0\}.$$

In particular, $\Phi \subset S^*$. The set $\Phi \cup \{0\}$ is closed in the *weak** topology of $\mathcal{A}^*$. Hence, giving Φ the relative *weak** topology we obtain the *Gelfand space*, which is a *locally* compact Hausdorff space.

C. The Gelfand transform $\hat{x} : \Phi \to \mathbb{C}$ of an element $x \in \mathcal{A}$ is defined precisely as for unital $\mathcal{A}$. It is plainly continuous, and it also vanishes at infinity, so $\hat{x} \in C_0(\Phi)$. As before, $\|\hat{x}\|_{C_0(\Phi)} = r(x)$. The *Gelfand representation* $\Gamma : \mathcal{A} \to C_0(\Phi)$ given by $x \to \hat{x}$ is a norm-decreasing algebra homomorphism. Its image $\hat{\mathcal{A}}$ separates the points of Φ, and it also *vanishes identically at no point of* Φ in the sense that it is not possible for all $\hat{x}$, $x \in \mathcal{A}$, to vanish at a particular $\phi \in \Phi$. Finally, (7) is valid too.

Example. Let X be a locally compact Hausdorff space and consider $\mathcal{A} := C_0(X)$. For each $t \in X$, consider the "evaluation at t" character ϕ_t given by $\phi_t(f) := f(t)$ for all $f \in C_0(X)$. One can show that $\Phi(C_0(X)) = \{\phi_t; t \in X\}$, and that the map $t \to \phi_t$ is a homeomorphism of X onto $\Phi(C_0(X))$ (see Exercise 17). Upon identifying these two topological spaces, the Gelfand representation of $C_0(X)$ is just the identity map!

Consequently, $\sigma(f) = \overline{f(X)}$ for each $f \in C_0(X)$. This can also be proved by a straightforward argument.

Let $\mathcal{A}$ be an arbitrary commutative Banach algebea. Since Γ is a homomorphism, it follows from (5) that for all $x, y \in \mathcal{A}$,

$$r(x+y) \leq r(x) + r(y); \quad r(xy) \leq r(x)r(y). \tag{8}$$

It follows that $\mathcal{A}$ is a normed algebra for the so-called *spectral norm* $r(\cdot)$ iff $r(x) = 0$ implies $x = 0$, that is (in view of (7)!), iff the so called *radical of* $\mathcal{A}$, $\operatorname{rad}\mathcal{A} := \ker\Gamma$, is trivial. In that case we say that $\mathcal{A}$ is *semi-simple*. By (7) and (3), equivalent characterizations of semi-simplicity are:

(i) The Gelfand representation Γ of $\mathcal{A}$ is injective.

(ii) $\mathcal{A}$ contains no non-zero quasi-nilpotent elements.

(iii) $\mathcal{A}$ is a normed algebra for the spectral norm.

(iv) The maximal modular ideals of $\mathcal{A}$ have trivial intersection.

(v) Φ separates the points of $\mathcal{A}$.

Example. It is clear that for a locally compact Hausdorff space X, $C_0(X)$ is semi-simple.

Theorem 7.14. *Let $\mathcal{A}$ be a commutative Banach algebra. Then $\mathcal{A}$ is semi-simple and $\hat{\mathcal{A}}$ is closed in $C_0(\Phi)$ iff there exists $K > 0$ such that*

$$\|x\|^2 \leq K\|x^2\| \quad (x \in \mathcal{A}).$$

In that case, the spectral norm is equivalent to the given norm on $\mathcal{A}$ and Γ is a homeomorphism of $\mathcal{A}$ onto $\hat{\mathcal{A}}$. Γ is isometric iff $K = 1$ (i.e., $\|x\|^2 = \|x^2\|$ for all $x \in \mathcal{A}$).

Proof. If $\mathcal{A}$ is semi-simple and $\hat{\mathcal{A}}$ is closed, Γ is a one-to-one continuous linear map of the Banach space $\mathcal{A}$ onto the *Banach* space $\hat{\mathcal{A}}$. By Corollary 6.11, Γ is a homeomorphism. The continuity of Γ^{-1} means that there exists a constant $K > 0$ such that $\|x\| \leq \sqrt{K}\|\hat{x}\|_{C_0(\Phi)}$ for all $x \in \mathcal{A}$. Therefore

$$\begin{aligned} \|x\|^2 \leq K \left(\sup_{\phi \in \Phi} |\hat{x}(\phi)| \right)^2 &= K \sup_{\phi \in \Phi} |\widehat{x^2}(\phi)| \\ &= K\|\widehat{x^2}\|_{C_0(\Phi)} \leq K\|x^2\|. \end{aligned} \tag{9}$$

Conversely, if there exists $K > 0$ such that $\|x\|^2 \leq K\|x^2\|$ for all $x \in \mathcal{A}$, it follows by induction that

$$\|x\| \leq K^{(1/2)+\dots+(1/2^n)}\|x^{2^n}\|^{1/2^n}$$

for all $n \in \mathbb{N}$ and $x \in \mathcal{A}$. Letting $n \to \infty$, it follows that

$$\|x\| \leq K r(x) = K\|\hat{x}\|_{C_0(\Phi)} \quad (x \in \mathcal{A}). \tag{10}$$

Hence, $\ker\Gamma = \{0\}$, that is, $\mathcal{A}$ is semi-simple, and Γ is a homeomorphism of $\mathcal{A}$ onto $\hat{\mathcal{A}}$. Since $\mathcal{A}$ is complete, so is $\hat{\mathcal{A}}$, that is, $\hat{\mathcal{A}}$ is closed in $C_0(\Phi)$.

If $K = 1$ (i.e., if $\|x\|^2 = \|x^2\|$ for all $x \in \mathcal{A}$), it follows from (10) that $\|x\| = r(x) = \|\hat{x}\|_{C_0(\Phi)}$ and Γ is isometric. Conversely, if Γ is isometric, it follows from (9) (with $K = 1$ and equality throughout) that $\|x\|^2 = \|x^2\|$ for all x. □

7.3 Involutions and C^*-algebras

Let $\mathcal{A}$ be a *semi-simple* commutative Banach algebra. Since Φ is a locally compact Hausdorff space (with the Gelfand topology), and $\hat{\mathcal{A}}$ is a *separating* subalgebra of $C_0(\Phi)$ that *vanishes identically nowhere*, it follows from Theorem 5.39 or its generalization in Exercise 9 of Chapter 5 that $\hat{\mathcal{A}}$ is *dense* in $C_0(\Phi)$ if it is *selfadjoint*. In that case, if $J : f \to \overline{f}$ is the conjugation conjugate automorphism of $C_0(\Phi)$, define $\mathcal{J} : \mathcal{A} \to \mathcal{A}$ by

$$\mathcal{J} := \Gamma^{-1} J \Gamma. \tag{1}$$

Since $J\hat{\mathcal{A}} \subset \hat{\mathcal{A}}$ and Γ maps $\mathcal{A}$ bijectively onto $\hat{\mathcal{A}}$ (when $\mathcal{A}$ is semi-simple), $\mathcal{J}$ is well defined. As a composition of two isomorphisms and the conjugate isomorphism J, $\mathcal{J}$ is a conjugate isomorphism of $\mathcal{A}$ onto itself such that $\mathcal{J}^2 = I$ (because $J^2 = I$, where I denotes the identity operator in the relevant space). Such a map $\mathcal{J}$ is called an *involution*. In the non-commutative case, multiplicativity of the involution is replaced by *anti-multiplicativity*:

$$\mathcal{J}(xy) = \mathcal{J}(y)\mathcal{J}(x).$$

So, to be precise, an involution on an algebra $\mathcal{A}$ is a conjugate-linear anti-multiplicative map $\mathcal{J} : \mathcal{A} \to \mathcal{A}$ satisfying $\mathcal{J}^2 = I$. It is customary to denote $x^* := \mathcal{J}x$ whenever $\mathcal{J}$ is an involution on $\mathcal{A}$ (not to be confused with elements of the conjugate space!). An algebra with an involution is then called a *$*$-algebra*. A subalgebra of a $*$-algebra that is closed under the involution operation is said to be *selfadjoint* or a *$*$-subalgebra*.

If $\mathcal{A}$ and $\mathcal{B}$ are $*$-algebras, a $*$-homomorphism (or isomorphism) $f : \mathcal{A} \to \mathcal{B}$ is a homomorphism (or isomorphism) such that $f(x^*) = f(x)^*$ for all $x \in \mathcal{A}$.

An element x in a $*$-algebra is *normal* if it commutes with its *adjoint* x^*. Special normal elements are the *selfadjoint* ($x^* = x$) and the *unitary* ($x^* = x^{-1}$) elements (the latter is relevant when $\mathcal{A}$ is unital). The identity is necessarily selfadjoint and unitary, because

$$e^* = ee^* = e^{**}e^* = (ee^*)^* = e^{**} = e.$$

Every element x can be uniquely written as $x = a + ib$ with $a, b \in \mathcal{A}$ selfadjoint: we have $a = \Re x := (x + x^*)/2$, $b = \Im x := (x - x^*)/2\mathrm{i}$, and $x^* = a - ib$. Clearly, x is normal iff a, b commute.

The "canonical involution" $\mathcal{J}$ defined by (1) on a *semi-simple commutative* Banach algebra is *uniquely determined* by the natural relation $\Gamma\mathcal{J} = J\Gamma$, that is, by the relation

$$\widehat{x^*} = \overline{\hat{x}} \quad (x \in \mathcal{A}), \tag{2}$$

which is equivalent to the property that $\hat{a}$ *is real whenever* a *is selfadjoint.*

In case the Gelfand representation Γ is *isometric*, Relation (2) implies the norm-identity:

$$\|x^*x\| = \|x\|^2 \quad (x \in \mathcal{A}). \tag{3}$$

Indeed

$$\begin{aligned}\|x^*x\| &= \|\Gamma(x^*x)\|_{C_0(\Phi)} = \|\overline{\hat{x}}\hat{x}\|_{C_0(\Phi)} = \||\hat{x}|^2\|_{C_0(\Phi)} \\ &= \|\hat{x}\|^2_{C_0(\Phi)} = \|x\|^2.\end{aligned}$$

A Banach algebra with an involution satisfying the norm-identity (3) (called the *C^*-identity*) is called a *C^*-algebra.* If X is any locally compact Hausdorff space, $C_0(X)$ is a (commutative) C^*-algebra for the involution J. Theorem 7.16 establishes that this is a universal model, up to C^*-algebra isomorphism, for *commutative* C^*-algebras.

An example of a generally non-commutative C^*-algebra is the Banach algebra $B(X)$ of all bounded linear operators on a *Hilbert* space X. The involution is the *Hilbert adjoint* operation $T \to T^*$. Given $y \in X$, the map $x \in X \to (Tx, y)$ is a continuous linear functional on X. By the "Little" Riesz representation theorem (Theorem 1.37), there exists a unique vector (depending on T and y, hence denoted T^*y) such that $(Tx, y) = (x, T^*y)$ for all $x, y \in X$. The uniqueness implies that $T^* : X \to X$ is linear, and (with the following suprema taken over all *unit* vectors x, y)

$$\|T^*\| = \sup |(x, T^*y)| = \sup |(Tx, y)| = \|T\| < \infty.$$

Thus, $T^* \in B(X)$, and an easy calculation shows that the map $T \to T^*$ is an (isometric) involution on $B(X)$. Moreover

$$\begin{aligned}\|T\|^2 = \|T^*\|\|T\| \geq \|T^*T\| &= \sup_{x,y} |(T^*Tx, y)| \\ &= \sup_{x,y} |(Tx, Ty)| \geq \sup_{x} \|Tx\|^2 = \|T\|^2.\end{aligned}$$

Therefore, $\|T^*T\| = \|T\|^2$, and $B(X)$ is indeed a C^*-algebra. Every closed selfadjoint subalgebra of $B(X)$ (that is, any C^*-subalgebra of $B(X)$) is likewise an example of a generally non-commutative C^*-algebra. The second Gelfand–Naimark theorem (see Chapter 11) establishes that this example is, up to C^*-algebra isomorphism, the most general example of a C^*-algebra.

Note that in *any C^*-algebra*, the C^*-identity implies that the involution is *isometric*:

> Since $\|x\|^2 = \|x^*x\| \leq \|x^*\|\,\|x\|$, we have $\|x\| \leq \|x^*\|$, hence $\|x^*\| \leq \|x^{**}\| = \|x\|$, and therefore $\|x^*\| = \|x\|$.

It follows in particular that $\|\Re x\| = \|(x + x^*)/2\| \leq \|x\|$, and similarly $\|\Im x\| \leq \|x\|$. Furthermore, if the C^*-algebra is unital, then $\|e\|^2 = \|e^*e\| = \|e\|$, yielding that $\|e\| = 1$.

Assume that $\mathcal{A}$ is a non-unital C^*-algebra, and consider the unital algebra $\mathcal{A}^\#$ defined in Section 7.2. We turned it into a unital Banach algebra by giving it the norm $\|x + \lambda e\| = \|x\| + |\lambda|$ $(x \in \mathcal{A}, \lambda \in \mathbb{C})$. However, this norm does not make $\mathcal{A}^\#$ a C^*-algebra. To this end we give $\mathcal{A}^\#$ the different norm

$$\|x + \lambda e\| := \sup_{y \in \mathcal{A}, \|y\| \leq 1} \|xy + \lambda y\| \quad (x \in \mathcal{A}, \lambda \in \mathbb{C}).$$

Together with the involution $(x + \lambda e)^* := x^* + \overline{\lambda} e$, $\mathcal{A}^\#$ indeed becomes a unital C^*-algebra in which $\mathcal{A}$ is (isometrically) naturally embedded; see Exercise 21. It is called the *unitization* of $\mathcal{A}$ (as a C^*-algebra). Henceforth, $\mathcal{A}^\#$ denotes this C^*-algebra. Note that the new norm on $\mathcal{A}^\#$ does not change anything in the construction or properties of the Gelfand space and representation in the non-unital case (points A. to C. in the previous section). For convenience, if $\mathcal{A}$ is a unital C^*-algebra, we set $\mathcal{A}^\# := \mathcal{A}$.

Lemma 7.15. *Let $\mathcal{A}$ be a C^*-algebra.*

(i) Every normal element $x \in \mathcal{A}$ satisfies $\|x\|^2 = \|x^2\|$ and $\|x\| = r(x)$.

(ii) If $\mathcal{A}$ is commutative, then it is semi-simple, and its involution coincides with the canonical involution.

Proof. (i) Applying the C^*-identity successively to x, x^*x, and x^2 and using the normality of x, we have

$$\|x\|^4 = \|x^*x\|^2 = \|(x^*x)^*(x^*x)\| = \|(x^2)^*x^2\| = \|x^2\|^2.$$

Thus $\|x\|^2 = \|x^2\|$. By induction we get $\|x\|^{2^n} = \|x^{2^n}\|$ for all $n \in \mathbb{N}$, so from Theorem 7.9 we infer that $r(x) = \lim_n \|x^n\|^{1/n} = \lim_n \|x^{2^n}\|^{1/2^n} = \|x\|$.

(ii) Since $\mathcal{A}$ is commutative, each of its elements is normal. By part (i), Theorem 7.14 implies that Γ is isometric. In particular $\mathcal{A}$ is semi-simple, so that the canonical involution $\mathcal{J}$ is well-defined (and uniquely determined by the relation $\Gamma\mathcal{J} = J\Gamma$). The conclusion of the lemma will follow if we prove that the given involution satisfies (2), or equivalently, if we show that $\hat{a}$ is *real* whenever $a \in \mathcal{A}$ is selfadjoint (with respect to the given involution).

Suppose then that $a \in \mathcal{A}$ is selfadjoint, but $\beta := \Im\hat{a}(\phi) \neq 0$ for some $\phi \in \Phi$. We may assume that $\mathcal{A}$ is unital, for otherwise replace it by $\mathcal{A}^\#$ and replace ϕ by $\phi^\# \in \Phi(\mathcal{A}^\#)$. Let $\alpha := \Re\hat{a}(\phi)$ and $b := (1/\beta)(a - \alpha e)$. Then b is selfadjoint, and $\hat{b}(\phi) = \mathrm{i}$. For any *real* λ, since ϕ is contractive,

$$\begin{aligned}(1+\lambda)^2 &= |(1+\lambda)\mathrm{i}|^2 = |\phi(b + \mathrm{i}\lambda e)|^2 \\ &\leq \|b + \mathrm{i}\lambda e\|^2 = \|(b + \mathrm{i}\lambda e)^*(b + \mathrm{i}\lambda e)\| \\ &= \|(b - \mathrm{i}\lambda e)(b + \mathrm{i}\lambda e)\| = \|b^2 + \lambda^2 e\| \leq \|b^2\| + \lambda^2.\end{aligned}$$

Therefore, $2\lambda < \|b^2\|$, which is absurd since λ is arbitrary. □

The last lemma has the following neat consequence: a $*$-algebra admits at most one norm that makes it into a C^*-algebra. Indeed, for every element x, since

x^*x is selfadjoint and hence normal, the C^*-identity implies $\|x\|^2 = \|x^*x\| = r(x^*x)$, and the spectral radius depends only on the algebraic structure of $\mathcal{A}$.

Putting together all the ingredients accumulated, we obtain the following important result.

Theorem 7.16 (The commutative Gelfand–Naimark theorem). *Let $\mathcal{A}$ be a commutative C^*-algebra. Then the Gelfand representation Γ is an isometric $*$-isomorphism of $\mathcal{A}$ onto $C_0(\Phi)$.*

Proof. It was observed in the proof of Lemma 7.15 that Γ is a $*$-preserving isometry of $\mathcal{A}$ onto $\hat{\mathcal{A}}$. It follows in particular that $\hat{\mathcal{A}}$ is a closed selfadjoint subalgebra of $C_0(\Phi)$. As observed in the beginning of this section, commutativity and semi-simplicity of $\mathcal{A}$ and selfadjointness of $\hat{\mathcal{A}}$ imply that $\hat{\mathcal{A}}$ is dense in $C_0(\Phi)$ by virtue of the Stone–Weierstrass theorem. Thus, these two algebras coincide. □

Let $\mathcal{A}$ be any (not necessarily commutative!) C^*-algebra, and let $x \in \mathcal{A}$ be *selfadjoint*. Denote by $[x]$ the closure in $\mathcal{A}^\#$ of the set $\{p(x); p \text{ complex polynomial}\}$. Then $[x]$ is clearly the unital C^*-subalgebra of $\mathcal{A}^\#$ generated by x and $e_{\mathcal{A}^\#}$, and it is commutative. Let $\Gamma : y \to \hat{y}$ be the Gelfand representation of $[x]$. Since it is a $*$-isomorphism and x is selfadjoint, $\hat{x}$ is *real*. As we know from Section 7.2, the spectrum $\sigma_{[x]}(x)$ is either equal to the image of $\hat{x}$ or to this image with 0 added to it. Either way, we obtain that $\sigma_{[x]}(x) \subset \mathbb{R}$. Since $\sigma(x) \subset \sigma_{[x]}(x)$ by Theorem 7.12, we get $\sigma(x) \subset \mathbb{R}$. This proves the following:

Theorem 7.17. *The spectrum of a selfadjoint element of a C^*-algebra is real.*

In fact, the spectrum of an arbitrary element of a C^*-algebra does not change when passing to a subalgebra (provided there is no change in the unit). Precisely:

Theorem 7.18. *Let $\mathcal{B}$ be a C^*-subalgebra of the unital C^*-algebra $\mathcal{A}$ with $e_{\mathcal{A}} \in \mathcal{B}$. Then $G(\mathcal{B}) = G(\mathcal{A}) \cap \mathcal{B}$ and $\sigma_{\mathcal{B}}(x) = \sigma(x)$ for all $x \in \mathcal{B}$.*

Proof. If $b \in \mathcal{B}$ is selfadjoint then $\sigma_{\mathcal{B}}(b) = \sigma(b)$ by Corollary 7.13: either apply Theorem 7.17 to obtain that $\sigma_{\mathcal{B}}(b)$ is real, thus nowhere dense in $\mathbb{C}$, or to obtain that $\sigma(b)$ is real, thus $\rho(b)$ is connected.

Since $G(\mathcal{B}) \subset G(\mathcal{A}) \cap \mathcal{B}$ trivially, we must show that if $x \in \mathcal{B}$ has an inverse $x^{-1} \in \mathcal{A}$, then $x^{-1} \in \mathcal{B}$. The element $x^*x \in \mathcal{B}$ is selfadjoint, and has clearly the inverse $x^{-1}(x^{-1})^*$ in $\mathcal{A}$:

$$[x^{-1}(x^{-1})^*][x^*x] = x^{-1}(xx^{-1})^*x = x^{-1}x = e_{\mathcal{A}},$$

and similarly for multiplication in reversed order. Thus $0 \notin \sigma(x^*x) = \sigma_{\mathcal{B}}(x^*x)$ by the foregoing. Hence, $x^*x \in G(\mathcal{B})$, and therefore the inverse $x^{-1}(x^{-1})^*$ belongs to $G(\mathcal{B})$. Consequently $x^{-1} = [x^{-1}(x^{-1})^*]x^* \in \mathcal{B}$, as wanted.

It now follows that $\rho_{\mathcal{B}}(x) = \rho(x)$, hence $\sigma_{\mathcal{B}}(x) = \sigma(x)$, for all $x \in \mathcal{B}$. □

Finally, let us derive a few obvious consequences of the previous results. By Lemma 7.15, a normal quasi-nilpotent element of a C^*-algebra is necessarily zero, and the spectrum of a normal element x contains a complex number with modulus equal to $\|x\|$. In particular, if x is selfadjoint, $\sigma(x)$ is contained in the closed interval $[-\|x\|, \|x\|]$ (cf. (8) following Definition 7.5 and Theorem 7.17), so either $\|x\|$ or $-\|x\|$ (or both) belong to $\sigma(x)$.

7.4 Normal elements

Terminology 7.19. If x is a *normal* element of the arbitrary C^*-algebra $\mathcal{A}$, we still denote by $[x]$ the unital C^*-subalgebra of $\mathcal{A}^\#$ generated by x and $e_{\mathcal{A}^\#}$, that is, the closure in $\mathcal{A}^\#$ of all complex polynomials in x *and* x^*, $\sum \alpha_{kj} x^k (x^*)^j$ (finite sums, with $\alpha_{kj} \in \mathbb{C}$). Since x is normal, it is clear that $[x]$ is a *commutative* C^*-algebra. By the commutative Gelfand–Naimark theorem, the Gelfand representation Γ of $[x]$ is an isometric $*$-isomorphism of $[x]$ onto $C(\Phi)$, where Φ denotes the space of all characters (= non-zero complex homomorphisms) of $[x]$ with the Gelfand topology.

If $\phi, \psi \in \Phi$ are such that $\hat{x}(\phi) = \hat{x}(\psi)$, then $\widehat{x^*}(\phi) = \overline{\hat{x}(\phi)} = \overline{\hat{x}(\psi)} = \widehat{x^*}(\psi)$, and therefore $\hat{y}(\phi) = \hat{y}(\psi)$, that is, $\phi(y) = \psi(y)$, for all $y \in [x]$, namely $\phi = \psi$. It follows that $\hat{x} : \Phi \to \sigma(x)$ (cf. (4) in Section 7.2) is a continuous bijective map. Since both Φ and $\sigma(x)$ are compact Hausdorff spaces, the map $\hat{x}$ is a homeomorphism. It induces the isometric $*$-isomorphism

$$\Xi : f \in C(\sigma(x)) \to f \circ \hat{x} \in C(\Phi).$$

The composition $\tau := \Gamma^{-1} \circ \Xi$ is a unital isometric $*$-isomorphism of $C(\sigma(x))$ onto $[x]$, which carries the identity function $f_1(\lambda) = \lambda$ onto $\Gamma^{-1}(\hat{x}) = x$. This isometric $*$-isomorphism is called the $C(\sigma(x))$- (or *continuous*) *operational calculus* for the normal element x of $\mathcal{A}$. It is customary to write $f(x)$ instead of $\tau(f)$. Note that $f(x)$ is a normal element of $\mathcal{A}^\#$, for each $f \in C(\sigma(x))$. Its adjoint is $\overline{f}(x)$. The isometricity of τ means that

$$\|f(x)\| = \|f\|_{C(\sigma(x))} \qquad (\forall f \in C(\sigma(x))),$$

and the fact that it is onto means that $[x] = \{f(x); f \in C(\sigma(x))\}$.

The $C(\sigma(x))$-operational calculus τ sends any polynomial $p(\lambda) = \sum \alpha_{kj} \lambda^k (\overline{\lambda})^j$ to $p(x) := \sum \alpha_{kj} x^k (x^*)^j \in \mathcal{A}^\#$ (the constant function 1 is mapped to $e_{\mathcal{A}^\#}$). Denote by $\mathcal{P}$ the algebra of these polynomials, that is, in a complex variable and its conjugate. It is dense in $C(\sigma(x))$ by the Stone–Weierstrass theorem (cf. Theorem 5.39). As a result, the $C(\sigma(x))$-operational calculus for x is uniquely determined by the following weaker property: $\tau : C(\sigma(x)) \to \mathcal{A}^\#$ is a continuous unital $*$-homomorphism that sends f_1 to x. Indeed, such τ sends any $p \in \mathcal{P}$ to $p(x)$, and by continuity, τ is then uniquely determined on $C(\sigma(x))$.

For every $f \in C(\sigma(x))$, it follows from the foregoing that $f(x)$ is the limit in $\mathcal{A}^\#$ of a sequence of polynomials in x and x^*. If $\mathcal{A}$ is not unital and $f(0) = 0$, then one can approximate f by elements of $\mathcal{P}$ that vanish at 0, that is, without a free term (why?), and since $p(x) \in \mathcal{A}$ for such p, we have $f(x) \in \mathcal{A}$.

Remark that in order to prove the existence of the continuous operational calculus, it suffices to extend the spectral mapping theorem to elements of $\mathcal{P}$, that is, to show that for $p \in \mathcal{P}$ we have $\sigma(p(x)) = p(\sigma(x))$ (recall that x is normal!). Indeed, by Lemma 7.15 this implies that $\|p(x)\| = r(p(x)) = \|p\|_{C(\sigma(x))}$. As a result, the map that sends $p \in \mathcal{P}$ to $p(x) \in \mathcal{A}^{\#}$ is an *isometric* unital $*$-homomorphism. The density of $\mathcal{P}$ in $C(\sigma(x))$ implies that this map extends to all of $C(\sigma(x))$, and the extension has the desired properties.

The continuous operational calculus has numerous applications. For instance, suppose that $\mathcal{A}$ is unital and $u \in \mathcal{A}$ is unitary. Then u is normal, and by the properties of the continuous operational calculus, uu^* being equal to e is equivalent to $f_1\overline{f_1} = |f_1|^2$ being equal to 1 identically on $\sigma(u)$, that is, to $\sigma(u)$ being contained in the complex unit circle $\{z \in \mathbb{C}; |z| = 1\}$.

Theorem 7.20 (The spectral mapping and composition theorems). *Let x be a normal element of the C^*-algebra $\mathcal{A}$, and let $f \to f(x)$ be its $C(\sigma(x))$-operational calculus. Then for all $f \in C(\sigma(x))$, $\sigma(f(x)) = f(\sigma(x))$, and furthermore, for all $g \in C(\sigma(f(x)))$ (so that necessarily $g \circ f \in C(\sigma(x))$), the identity $g(f(x)) = (g \circ f)(x)$ is valid.*

If $x, y \in \mathcal{A}$ are normal, and $f \in C(\sigma(x) \cup \sigma(y))$ is injective, then $f(x) = f(y)$ implies $x = y$.

Proof. Let $\mu \in \mathbb{C}$. Since τ is a unital isomorphism of $C(\sigma(x))$ and $[x]$, it follows that $\mu e_{\mathcal{A}^{\#}} - f(x) = \tau(\mu - f)$ is singular in $[x]$ iff $\mu - f$ is singular in $C(\sigma(x))$, that is, iff there exists $\lambda \in \sigma(x)$ such that $\mu = f(\lambda)$. Since $\sigma(f(x)) = \sigma_{\mathcal{A}^{\#}}(f(x))$ equals $\sigma_{[x]}(f(x))$ by Theorem 7.18, we conclude that $\mu \in \sigma(f(x))$ iff $\mu \in f(\sigma(x))$.

The maps $g \to g \circ f$ and $h \to h(x)$ are unital isometric $*$-homomorphisms of $C(\sigma(f(x))) = C(f(\sigma(x)))$ into $C(\sigma(x))$ and of $C(\sigma(x))$ into $\mathcal{A}^{\#}$, respectively. Their composition $g \to (g \circ f)(x)$ is a unital isometric $*$-homomorphism of $C(\sigma(f(x)))$ into $\mathcal{A}^{\#}$ that carries f_1 to $f(x)$. By the uniqueness of the continuous operational calculus for the normal element $f(x)$, we have $(g \circ f)(x) = g(f(x))$ for all $g \in C(\sigma(f(x)))$.

The last statement of the theorem follows by taking the continuous function $g := f^{-1}$ (whose domain is the image of f) in the last formula applied to both x and y:

$$x = f_1(x) = (g \circ f)(x) = g(f(x)) = g(f(y)) = (g \circ f)(y) = f_1(y) = y. \qquad \square$$

7.5 The Arens products

Let $\mathcal{A}$ be a Banach algebra. This section shows how to construct two products on the second dual $\mathcal{A}^{**}$ of $\mathcal{A}$, each of which turns $\mathcal{A}^{**}$ into a Banach algebra.

It will be convenient to adopt the following convention. If V is a vector space (such as $\mathcal{A}$ or $\mathcal{A}^*$), $x \in V$ and f is a linear functional on V, we write $\langle x, f\rangle$ or $\langle f, x\rangle$ for $f(x)$.

The two Arens products on $\mathcal{A}^{**}$ are defined in three steps:

(A) Let $\omega \in \mathcal{A}^*$ and $a \in \mathcal{A}$. Define elements $a\omega, \omega a \in \mathcal{A}^*$ by

$$\langle \omega a, b \rangle := \langle \omega, ab \rangle \text{ and } \langle b, a\omega \rangle := \langle ba, \omega \rangle$$

("moving a to the other side"), that is, $(\omega a)(b) := \omega(ab)$ and $(a\omega)(b) := \omega(ba)$, for each $b \in \mathcal{A}$. It is clear that $a\omega, \omega a$ are indeed bounded linear functionals on $\mathcal{A}$, namely elements of $\mathcal{A}^*$, of norm at most $\|a\| \, \|\omega\|$.

(B) For $\omega \in \mathcal{A}^*$ and $X \in \mathcal{A}^{**}$, define elements $X\omega, \omega X \in \mathcal{A}^*$ by

$$\langle a, \omega X \rangle := \langle a\omega, X \rangle \text{ and } \langle X\omega, a \rangle := \langle X, \omega a \rangle$$

("moving ω to the other side"), that is, $(\omega X)(a) := X(a\omega)$ and $(X\omega)(a) := X(\omega a)$, for each $a \in \mathcal{A}$. It is clear that $\omega X, X\omega$ are indeed bounded linear functionals on $\mathcal{A}$, namely, elements of $\mathcal{A}^*$, of norm at most $\|X\| \, \|\omega\|$.

(C) Finally, for $X, Y \in \mathcal{A}^{**}$, define elements $X \,\Box\, Y, X \Diamond Y \in \mathcal{A}^{**}$ by

$$\langle X \,\Box\, Y, \omega \rangle := \langle X, Y\omega \rangle \text{ and } \langle \omega, X \Diamond Y \rangle := \langle \omega X, Y \rangle,$$

that is, $(X \,\Box\, Y)(\omega) := X(Y\omega)$ and $(X \Diamond Y)(\omega) := Y(\omega X)$, for each $\omega \in \mathcal{A}^*$. It is clear that $X \,\Box\, Y, X \Diamond Y$ are indeed bounded linear functionals on $\mathcal{A}^*$, namely, elements of $\mathcal{A}^{**}$, of norm at most $\|X\| \, \|Y\|$.

The notation we used in (A) and (B) is suggestive, because:

(i) The left and right operations defined in (A) turn $\mathcal{A}^*$ into an $\mathcal{A}$-bimodule.

(ii) The operations of $\mathcal{A}^{**}$ on $\mathcal{A}^*$ defined in (B) "extend" the operations of $\mathcal{A}$ on $\mathcal{A}^*$ defined in (A): for $\omega \in \mathcal{A}^*$ and $a \in \mathcal{A}$ we have $\omega\hat{a} = \omega a$ and $\hat{a}\omega = a\omega$, where κ is the canonical embedding of $\mathcal{A}$ in $\mathcal{A}^{**}$ and $\hat{a} = \kappa(a)$.

Each of the binary operations $\Box$ and $\Diamond$ turns $\mathcal{A}^{**}$ into an algebra. They are called the first and the second *Arens products* on $\mathcal{A}^{**}$, respectively. As the plural form suggests, *these products do not always agree.* When they do agree, we say that $\mathcal{A}$ is *Arens regular.* See the exercises for examples and non-examples of Arens regularity. We prove in an exercise in Chapter 12 that all C^*-algebras are Arens regular. However, generally speaking, Arens regularity is a rare property.

Each of the Arens products is submultiplicative, and thus turns $\mathcal{A}^{**}$ into a Banach algebra. Furthermore:

(iii) Each of the Arens products "extends" the product in $\mathcal{A}$: $\hat{a} \,\Box\, \hat{b} = \widehat{ab} = \hat{a} \Diamond \hat{b}$ for all $a, b \in \mathcal{A}$.

(iv) The Arens products agree when one of the multiplicands comes from $\mathcal{A}$: $X \,\Box\, \hat{a} = X \Diamond \hat{a}$ and $\hat{a} \,\Box\, X = \hat{a} \Diamond X$ for all $X \in \mathcal{A}^{**}$ and $a \in \mathcal{A}$.

(v) The operations of $\mathcal{A}^{**}$ on $\mathcal{A}^*$ defined in (B) turn $\mathcal{A}^*$ into a left $(\mathcal{A}^{**}, \Box)$-module and a right $(\mathcal{A}^{**}, \Diamond)$-module.

From (iii) we see that each of the Banach algebras $(\mathcal{A}^{**}, \Box)$ and $(\mathcal{A}^{**}, \Diamond)$ contains $\mathcal{A}$ as a Banach subalgebra.

Separate continuity in the *weak**-topology on $\mathcal{A}^{**}$ (that is, the $\mathcal{A}^*$-topology) is subtle:

(vi) The following maps on $\mathcal{A}^{**}$ are continuous with respect to the *weak**-topology:

- for each fixed $Y \in \mathcal{A}^{**}$, the map $X \to X \,\Box\, Y$;
- for each fixed $X \in \mathcal{A}^{**}$, the map $Y \to X \diamondsuit Y$.

Facts (i)–(vi) follow easily from the definitions.

Let $X, Y \in \mathcal{A}^{**}$, and let $\{a_i\}$,$\{b_j\}$ be nets in $\mathcal{A}$ such that $\{\widehat{a_i}\}$, $\{\widehat{b_j}\}$ converge to X, Y, respectively, in the *weak**-topology on $\mathcal{A}^{**}$. By (iii), (iv), and (vi) we have

$$X \,\Box\, Y = \lim_i \widehat{a_i} \,\Box\, Y = \lim_i \widehat{a_i} \diamondsuit Y = \lim_i \left(\lim_j \widehat{a_i} \diamondsuit \widehat{b_j}\right) = \lim_i \left(\lim_j \widehat{a_i b_j}\right)$$

and

$$X \diamondsuit Y = \lim_j X \diamondsuit \widehat{b_j} = \lim_j X \,\Box\, \widehat{b_j} = \lim_j \left(\lim_i \widehat{a_i} \,\Box\, \widehat{b_j}\right) = \lim_j \left(\lim_i \widehat{a_i b_j}\right),$$

where all limits are in the *weak**-topology. These formulae express the difference between the two Arens products lucidly.

We proceed to establish a criterion for Arens regularity. As preparation, fix $\omega \in \mathcal{A}^*$ and consider the module operation maps $\lambda, \rho : \mathcal{A} \to \mathcal{A}^*$ defined by

$$\lambda(a) := a\omega, \quad \rho(a) := \omega a \qquad (a \in \mathcal{A}).$$

These are bounded linear maps whose Banach adjoints $\lambda^*, \rho^* : \mathcal{A}^{**} \to \mathcal{A}^*$ satisfy

$$\lambda^*(X) := \omega X, \quad \rho^*(X) := X\omega \qquad (\forall X \in \mathcal{A}^{**}).$$

Indeed, $\lambda^*(X) = X \circ \lambda$ maps $a \in \mathcal{A}$ to $\langle a\omega, X\rangle = \langle a, \omega X\rangle$ and thus equals ωX, and similarly for ρ^*. Taking the adjoint again, we get the maps $\lambda^{**}, \rho^{**} : \mathcal{A}^{**} \to \mathcal{A}^{***}$, which satisfy

$$(\lambda^{**}(Y))(X) = \langle X \diamondsuit Y, \omega\rangle, \quad (\rho^{**}(X))(Y) = \langle X \,\Box\, Y, \omega\rangle \qquad (\forall X, Y \in \mathcal{A}^{**}).$$

Indeed, $\lambda^{**}(Y) = Y \circ \lambda^*$ maps X to $\langle \omega X, Y\rangle = \langle \omega, X \diamondsuit Y\rangle$, and similarly for ρ^{**}.

Theorem 7.21. *Let $\mathcal{A}$ be a Banach algebra and $\omega \in \mathcal{A}^*$. The following conditions are equivalent:*

1. *For each $X, Y \in \mathcal{A}^{**}$ we have $\langle X \,\Box\, Y, \omega\rangle = \langle X \diamondsuit Y, \omega\rangle$.*
2. *For each fixed $X \in \mathcal{A}^{**}$, the linear functional $Y \to \langle X \,\Box\, Y, \omega\rangle$ on $\mathcal{A}^{**}$ is weak*-continuous (that is, continuous in the $\mathcal{A}^*$-topology on $\mathcal{A}^{**}$).*
3. *For each fixed $Y \in \mathcal{A}^{**}$, the linear functional $X \to \langle X \diamondsuit Y, \omega\rangle$ on $\mathcal{A}^{**}$ is weak*-continuous.*

4. *The set $\{\omega a; a \in \mathcal{A}, \|a\| \leq 1\}$ has compact closure in the weak topology of $\mathcal{A}^*$ (that is, the $\mathcal{A}^{**}$-topology).*
5. *The set $\{a\omega; a \in \mathcal{A}, \|a\| \leq 1\}$ has compact closure in the weak topology of $\mathcal{A}^*$.*

Proof. The equivalences 1 $\iff$ 2 and 1 $\iff$ 3 follow from (iv) and (vi). Indeed, 1 implies both 2 and 3 by (vi). Conversely, if, for instance, 2 holds and $X, Y \in \mathcal{A}^{**}$, then taking nets $\{a_i\}$, $\{b_j\}$ in $\mathcal{A}$ such that $\{\widehat{a_i}\}, \{\widehat{b_j}\}$ converge to X, Y, respectively in the *weak**-topology, we get

$$\langle X \,\square\, Y, \omega\rangle = \lim_j \left\langle X \,\square\, \widehat{b_j}, \omega \right\rangle = \lim_j \left\langle X \diamond \widehat{b_j}, \omega \right\rangle = \langle X \diamond Y, \omega\rangle$$

by (iv) and (vi), proving 1.

In the rest of the proof we use the maps $\lambda, \rho : \mathcal{A} \to \mathcal{A}^*$ associated with ω as in the paragraph preceding the theorem. We require the notion of weak compactness of operators (see Exercise 26 of Chapter 6). Notice that 4 and 5 mean weak compactness of ρ and ω, respectively.

Condition 2 precisely means that for all $X \in \mathcal{A}^{**}$, the functional $\rho^{**}(X) \in \mathcal{A}^{***}$ is continuous in the $\mathcal{A}^*$-topology on $\mathcal{A}^{**}$, which, by Theorem 5.23, means that $\rho^{**}(X)$ belongs to the canonical image of $\mathcal{A}^*$ in $\mathcal{A}^{***}$. By Part (a) of Exercise 26 of Chapter 6, this is equivalent to weak compactness of ρ, namely, to 4. Similarly, 3 is equivalent to 5. $\square$

A linear functional $\omega \in \mathcal{A}^*$ satisfying the equivalent conditions of Theorem 7.21 is called *weakly almost periodic.* Plainly, $\mathcal{A}$ is Arens regular iff all elements of $\mathcal{A}^*$ are weakly almost periodic.

Exercises

1. Let $\mathcal{A}$ be a non-unital Banach algebra. Consider then the Cartesian product Banach space $\mathcal{A}^\# := \mathcal{A} \times \mathbb{C}$ with the norm $\|[x, \lambda]\| = \|x\| + |\lambda|$ and the multiplication

$$[x, \lambda]\,[y, \mu] = [xy + \lambda y + \mu x, \lambda\mu].$$

Prove that $\mathcal{A}^\#$ is a unital Banach algebra with the identity $e := [0, 1]$, commutative if $\mathcal{A}$ is commutative, and the map $x \in \mathcal{A} \to [x, 0] \in \mathcal{A}^\#$ is an isometric isomorphism of $\mathcal{A}$ onto a maximal ideal (identified with $\mathcal{A}$) in $\mathcal{A}^\#$. (With this identification, we have $\mathcal{A}^\# = \mathcal{A} + \mathbb{C}e$.)

If ϕ is a homomorphism of the commutative Banach algebra $\mathcal{A}$ into $\mathbb{C}$, it extends uniquely to a homomorphism (also denoted by ϕ) of $\mathcal{A}^\#$ into $\mathbb{C}$ by the identity $\phi([x, \lambda]) = \phi(x) + \lambda$. Conclude that $\|\phi\| = 1$.

2. The requirement $\|xy\| \leq \|x\|\|y\|$ in the definition of a Banach algebra implies the joint continuity of multiplication. Prove:

(a) If $\mathcal{A}$ is a Banach space and also an algebra for which multiplication is *separately* continuous, then multiplication is jointly continuous. (Hint: consider the bounded operators $L_x : y \to xy$ and $R_y : x \to xy$ on $\mathcal{A}$ and use the uniform boundedness theorem.)

(b) We use the notation as in Part (a). The norm $|x| := \|L_x\|$ (where the norm on the right is the $B(\mathcal{A})$-norm) is equivalent to the given norm on $\mathcal{A}$, and satisfies the submultiplicativity requirement $|xy| \leq |x||y|$. If $\mathcal{A}$ has an algebraic identity e, then $|e| = 1$.

3. Let $\mathcal{A}$ be a unital complex Banach algebra. If $F \subset \mathcal{A}$ consists of commuting elements, denote by $\mathcal{C}_F$ the maximal commutative Banach subalgebra of $\mathcal{A}$ containing F. Prove:

   (a) $\sigma_{\mathcal{C}_F}(a) = \sigma(a)$ for all $a \in \mathcal{C}_F$.

   (b) If $a, b \in \mathcal{A}$ commute, then

$$\sigma(a+b) \subset \sigma(a) + \sigma(b) \quad \text{and} \quad \sigma(ab) \subset \sigma(a)\sigma(b).$$

   Conclude that

$$r(a+b) \leq r(a) + r(b) \quad \text{and} \quad r(ab) \leq r(a)r(b).$$

   (c) For all $a \in \mathcal{A}$ and $\lambda \in \rho(a)$,

$$r(R(\lambda; a)) = \frac{1}{d(\lambda, \sigma(a))}.$$

   (Hint: use the Gelfand representation of $\mathcal{C}_{\{a,b\}}$.)

4. Let $\mathcal{A}$ be a commutative unital Banach algebra (over $\mathbb{C}$). A set $E \subset \mathcal{A}$ *generates* $\mathcal{A}$ if the minimal closed subalgebra of $\mathcal{A}$ containing E and the identity e coincides with $\mathcal{A}$. In that case, prove that the maximal ideal space of $\mathcal{A}$ is homeomorphic to a closed subset of the Cartesian product $\prod_{a \in E} \sigma(a)$.

5. Let $\mathcal{A}$ be a unital Banach algebra, and let G be the group of regular elements of $\mathcal{A}$. Suppose $\{a_n\} \subset G$ has the following properties:

   (i) $a_n \to a$ and $a_n a = a a_n$ for all n;

   (ii) the sequence $\{r(a_n^{-1})\}$ is bounded.

   Prove that $a \in G$. (Hint: observe that $r(e - a_n^{-1}a) \leq r(a_n^{-1})r(a_n - a) \to 0$, hence, $1 - \sigma(a_n^{-1}a) = \sigma(e - a_n^{-1}a) \subset B(0, 1/2)$ for n large enough, and therefore $0 \notin \sigma(a_n^{-1}a)$.)

6. Let $\mathcal{A}$ be a unital Banach algebra, $a \in \mathcal{A}$, and $\lambda \in \rho(a)$. Prove that $\lambda \in \rho(b)$ for all $b \in \mathcal{A}$ for which the series $s(\lambda) := \sum_n [(b-a)R(\lambda; a)]^n$ converges in $\mathcal{A}$, and for such b, $R(\lambda; b) = R(\lambda; a)s(\lambda)$.

7. Let $\mathcal{A}$ and a be as in Exercise 6, and let V be an open set in $\mathbb{C}$ such that $\sigma(a) \subset V$. Prove that there exists $\delta > 0$ such that $\sigma(b) \subset V$ for all b in the ball $B(a, \delta)$. (Hint: if M is a bound for $R(\cdot; a)$ on the complement of V in the Riemann sphere, take $\delta = 1/M$ and apply Exercise 6.)

8. Let ϕ be a non-zero linear functional on the unital Banach algebra $\mathcal{A}$. Trivially, if ϕ is *multiplicative*, then $\phi(e) = 1$ and $\phi \neq 0$ on $G(\mathcal{A})$. The following steps provide a proof of the *converse*. Suppose $\phi(e) = 1$ and $\phi \neq 0$ on $G(\mathcal{A})$. Denote $N = \ker \phi$ (note that $N \cap G(\mathcal{A}) = \emptyset$). Prove:

   (a) $d(e, N) = 1$. (Hint: if $\|e - x\| < 1$, then $x \in G(\mathcal{A})$, hence $x \notin N$.)

   (b) $\phi \in \mathcal{A}^*$ and has norm 1. (Hint: if $a \notin N$, $a_1 := e - \phi(a)^{-1}a \in N$, hence $d(e, a_1) \geq 1$ by Part (a).)

   (c) Fix $a \in N$ with norm 1, and let $f(\lambda) := \phi(\exp(\lambda a))$ (where the exponential is defined by means of the usual power series, converging absolutely in $\mathcal{A}$ for all $\lambda \in \mathbb{C}$). Then f is an entire function with no zeros such that $f(0) = 1$, $f'(0) = 0$, and $|f(\lambda)| \leq e^{|\lambda|}$.

   (d) (*This is a result about entire functions.*) If f has the properties listed in Part (c), then $f = 1$ identically. *Sketch of proof*: since f has no zeros, it can be represented as $f = e^g$ with g entire; necessarily $g(0) = g'(0) = 0$, so that $g(\lambda) = \lambda^2 h(\lambda)$ with h entire, and $\Re g(\lambda) \leq |\lambda|$. For any $r > 0$, verify that $|2r - g| \geq |g|$ in the disc $|\lambda| \leq r$ and $|2r - g| > 0$ in the disc $|\lambda| < 2r$. Therefore, $F(\lambda) := [r^2 h(\lambda)]/[2r - g(\lambda)]$ is analytic in $|\lambda| < 2r$, and $|F| \leq 1$ on the circle $|\lambda| = r$, hence in the disc $|\lambda| \leq r$ by the maximum modulus principle. Thus

$$\frac{|h|}{|2 - g/r|} \leq 1/r \quad (|\lambda| < r).$$

   Given λ, let $r \to \infty$ to conclude that $h = 0$.

   (e) If $a \in N$, then $a^2 \in N$. (Hint: apply Parts (c) and (d) and look at the coefficient of λ^2 in the series for f.)

   (f) $\phi(x^2) = \phi(x)^2$ for all $x \in \mathcal{A}$. (Represent $x = x_1 + \phi(x)e$ with $x_1 \in N$ and apply Part (e).) In particular, $x \in N$ iff $x^2 \in N$.

   (g) If either x or y belong to N, then (i) $xy + yx \in N$; (ii) $(xy)^2 + (yx)^2 \in N$; and (iii) $xy - yx \in N$. (For (i), apply Part (f) to $x + y$; for (ii), apply (i) to yxy instead of y, when $x \in N$; for (iii), write $(xy - yx)^2 = 2[(xy)^2 + (yx)^2] - (xy + yx)^2$ and use Part (f).) Conclude that N is a two-sided ideal in $\mathcal{A}$ and ϕ is multiplicative (use the representation $x = x_1 + \phi(x)e$ with $x_1 \in N$).

9. Let $\mathcal{A}$ be a unital Banach algebra. For $a, b \in \mathcal{A}$, denote $C(a,b) = L_a - R_b$ (cf. Section 7.1), and consider the series

$$b_{\mathrm{L}}(\lambda) = \sum_{j=0}^{\infty} (-1)^j R(\lambda; a)^{j+1} [C(a,b)^j e];$$

$$b_{\mathrm{R}}(\lambda) = \sum_{j=0}^{\infty} [C(b,a)^j e] R(\lambda; a)^{j+1}$$

for $\lambda \in \rho(a)$. Prove that if $b_{\mathrm{L}}(\lambda)$ ($b_{\mathrm{R}}(\lambda)$) converges in $\mathcal{A}$ for some $\lambda \in \rho(a)$, then its sum is a left inverse (right inverse, respectively) for $\lambda e - b$. In particular, if $\lambda \in \rho(a)$ is such that both series converge in $\mathcal{A}$, then $\lambda \in \rho(b)$ and $R(\lambda; b) = b_{\mathrm{L}}(\lambda) = b_{\mathrm{R}}(\lambda)$.

10. Use notation as in Exercise 9. Set

$$r(a,b) = \limsup_n \|C(a,b)^n e\|^{1/n},$$

and consider the compact subsets of $\mathbb{C}$

$$\begin{aligned} \sigma_{\mathrm{L}}(a,b) &= \{\lambda \in \mathbb{C};\ d(\lambda, \sigma(a)) \le r(a,b)\}; \\ \sigma_{\mathrm{R}}(a,b) &= \{\lambda \in \mathbb{C};\ d(\lambda, \sigma(a)) \le r(b,a)\}; \\ \sigma(a,b) &= \sigma_{\mathrm{L}}(a,b) \cup \sigma_{\mathrm{R}}(a,b). \end{aligned}$$

Prove that the series $b_{\mathrm{L}}(\lambda)$ ($b_{\mathrm{R}}(\lambda)$) converge absolutely and uniformly on compact subsets of $\sigma_{\mathrm{L}}(a,b)^{\mathrm{c}}$ ($\sigma_{\mathrm{R}}(a,b)^{\mathrm{c}}$, respectively). In particular, $\sigma(b) \subset \sigma(a,b)$, and $R(\cdot; b) = b_{\mathrm{L}} = b_{\mathrm{R}}$ on $\sigma(a,b)^{\mathrm{c}}$.

11. Use notation as in Exercise 10. Set

$$d(a,b) = \max\{r(a,b),\ r(b,a)\},$$

so that trivially

$$\sigma(a,b) = \{\lambda;\ d(\lambda, \sigma(a)) \le d(a,b)\}$$

and $\sigma(a,b) = \sigma(a)$ iff $d(a,b) = 0$. In this case, it follows from Exercise 10 (and symmetry) that $\sigma(b) = \sigma(a)$ (for this reason, elements a, b such that $d(a,b) = 0$ are said to be *spectrally equivalent*).

12. Let D be a *derivation* on a unital Banach algebra $\mathcal{A}$, that is, a linear map $D : \mathcal{A} \to \mathcal{A}$ such that $D(ab) = (Da)b + a(Db)$ for all $a, b \in \mathcal{A}$. (Example: given $s \in \mathcal{A}$, the map $D_s := L_s - R_s$ is a derivation; it is called an *inner derivation*.) Prove:

    (a) If D is a derivation on $\mathcal{A}$ and Dv commutes with v for some v, then $Df(v) = f'(v)Dv$ for all polynomials f.

(b) Let $s \in \mathcal{A}$. The element $v \in \mathcal{A}$ is *s-Volterra* if $D_s v = v^2$. (Example: in $\mathcal{A} = B(L^p([0,1]))$, take $S : f(t) \to tf(t)$ and $V : f(t) \to \int_0^t f(u)du$, the so-called *classical Volterra operator.*) Prove: (i) $D_s v^n = nv^{n+1}$; (ii) $v^{n+1} = D_s^n v/n!$; and (iii) $C(s+\alpha v, s+\beta v)^n e = (-1)^n n! \binom{\beta-\alpha}{n} v^n$, for all $n \in \mathbb{N}$ and $\alpha, \beta \in \mathbb{C}$.

(c) If $v \in \mathcal{A}$ is s-Volterra, then (i) $\|v^n\|^{1/n} = O(1/n)$. In particular, v is quasi-nilpotent. (ii) $r(s+\alpha v, s+\beta v) = 0$ if $\beta - \alpha \in \mathbb{N} \cup \{0\}$, and $= \limsup(n!\|v^n\|)^{1/n}$ otherwise. (iii) $d(s+\alpha v, s+\beta v) = \limsup(n!\|v^n\|)^{1/n})$ if $\alpha \neq \beta$. (iv) For $\alpha \neq \beta$, $s+\alpha v$ and $s+\beta v$ are spectrally equivalent iff $\|v^n\|^{1/n} = o(1/n)$. (v) $d(s+\alpha v, s+\beta v) \leq \operatorname{diam} \sigma(s)$. (vi) $d(S+\alpha V, S+\beta V) = 1$ when $\alpha \neq \beta$ (cf. Part (b) for notation). In particular, $S+\alpha V$ and $S+\beta V$ are spectrally equivalent iff $\alpha = \beta$ (however, they all have the same spectrum, but do not try to prove this here!). Note that if $\beta - \alpha \in \mathbb{N}$, then $r(S+\alpha V, S+\beta V) = 0$ while $r(S+\beta V, S+\alpha V) = 1$.

(d) If v is s-Volterra, then

$$R(\lambda; v) = \lambda^{-1}e + \lambda^{-2}\exp(s/\lambda)v\exp(-s/\lambda) \quad (\lambda \neq 0).$$

(e) If v is s-Volterra, then for all $\alpha, \lambda \in \mathbb{C}$

$$\exp[\lambda(s+\alpha v)] = \exp(\lambda s)(e+\lambda v)^\alpha = (e-\lambda v)^{-\alpha}\exp(\lambda v),$$

where the binomials are defined by means of the usual series (note that v is quasi-nilpotent, so that the binomial series converge for all complex λ).

(f) If v is s-Volterra and $\rho(s)$ is connected, then $\sigma(s+kv) \subset \sigma(s)$ for all $k \in \mathbb{Z}$. For all $\lambda \in \rho(s)$,

$$\begin{aligned} R(\lambda; s+kv) &= \sum_{j=0}^{k} \binom{k}{j} j! R(\lambda; s)^{j+1} v^j \quad (k \geq 0); \\ &= \sum_{j=0}^{|k|} (-1)^j \binom{|k|}{j} j! v^j R(\lambda; s)^{j+1} \quad (k < 0). \end{aligned}$$

(Apply Exercise 9.) If $\rho(s)$ and $\rho(s+kv)$ are both connected for some integer k, then $\sigma(s+kv) = \sigma(s)$. In particular, if $\sigma(s) \subset \mathbb{R}$, then $\sigma(s+kv) = \sigma(s)$ for all $k \in \mathbb{Z}$.

13. Let $\mathcal{A}$ be a unital Banach algebra, and let $a, b, c \in \mathcal{A}$ be such that $C(a,b)c = 0$ (i.e., $ac = cb$). Prove:

(a) $C(\mathrm{e}^a, \mathrm{e}^b)c = 0$ (i.e., $\mathrm{e}^a c = c\mathrm{e}^b$, where the exponential function e^a is defined by the usual absolutely convergent series; the base of the exponential should not be confused with the identity of $\mathcal{A}$!).

(b) If $\mathcal{A}$ is a C^*-algebra, then e^{x-x^*} is unitary, for any $x \in \mathcal{A}$. (In particular, $\|\mathrm{e}^{x-x^*}\| = 1$.)

(c) If $\mathcal{A}$ is a C^*-algebra and a, b are normal elements (such that $ac = cb$, as before!), then

$$\mathrm{e}^{a^*} c\, \mathrm{e}^{-b^*} = \mathrm{e}^{a^*-a} c\, \mathrm{e}^{b-b^*};$$

$$\|\mathrm{e}^{a^*} c\, \mathrm{e}^{-b^*}\| \leq \|c\|.$$

(d) For a, b, c as in Part (c), define

$$f(\lambda) = \mathrm{e}^{\lambda a^*} c\, \mathrm{e}^{-\lambda b^*} \quad (\lambda \in \mathbb{C}).$$

Prove that $\|f(\lambda)\| \leq \|c\|$ for all $\lambda \in \mathbb{C}$, and conclude that $f(\lambda) = c$ for all λ (i.e., $\mathrm{e}^{\lambda a^*} c = c\, \mathrm{e}^{\lambda b^*}$ for all $\lambda \in \mathbb{C}$).

(e) If $\mathcal{A}$ is a C^*-algebra, and a, b are normal elements of $\mathcal{A}$ such that $ac = cb$ for some $c \in \mathcal{A}$, then $a^* c = cb^*$. (Consider the coefficient of λ in the last identity in Part (d).)

In particular, if c commutes with a normal element a, it commutes also with its adjoint; this is *Fuglede's theorem.*

14. Consider $L^1(\mathbb{R})$ (with respect to Lebesgue measure) *with convolution as multiplication.* Prove that $L^1(\mathbb{R})$ is a non-unital commutative Banach algebra, and the Fourier transform F is a contractive (i.e., norm-decreasing) homomorphism of $L^1(\mathbb{R})$ into $C_0(\mathbb{R})$ (cf. Exercise 7, Chapter 2).

15. Let ϕ be a non-zero homomorphism of the Banach algebra $L^1 = L^1(\mathbb{R})$ into $\mathbb{C}$, that is, a character of L^1 (cf. Exercise 14). Prove:

(a) There exists a unique $h \in L^\infty = L^\infty(\mathbb{R})$ such that $\phi(f) = \int fh\, dx$ for all $f \in L^1$, and $\|h\|_\infty = 1$. Moreover,

$$\phi(f_y)\phi(g) = \phi(f)\phi(g_y) \quad (f, g \in L^1;\ y \in \mathbb{R}),$$

where $f_y(x) = f(x-y)$.

(b) For any $f \in L^1$ such that $\phi(f) \neq 0$,

(i) $h(y) = \phi(f_y)/\phi(f)$ a.e. (in particular, h may be chosen to be continuous).

(ii) $\phi(f_y) \neq 0$ for all $y \in \mathbb{R}$.

(iii) $|h(y)| = 1$ for all $y \in \mathbb{R}$.

(iv) $h(x+y) = h(x)h(y)$ for all $x, y \in \mathbb{R}$ and $h(0) = 1$.

Conclude that $h(y) = \mathrm{e}^{-ity}$ for some $t \in \mathbb{R}$ (for all y) and that $\phi(f) = (Ff)(t)$, where F is the Fourier transform.

Conversely, each $t \in \mathbb{R}$ determines the homomorphism $\phi_t(f) = (Ff)(t)$. Conclude that the map $t \to \phi_t$ is a homeomorphism of $\mathbb{R}$ onto the Gelfand space Φ of L^1 (recall that it is the space of characters of L^1 with the Gelfand topology). (Hint: the Gelfand topology is Hausdorff and is weaker than the metric topology on $\mathbb{R}$.)

16. Let $\mathcal{A}, \mathcal{B}$ be commutative Banach algebras, $\mathcal{B}$ semi-simple. Let $\tau : \mathcal{A} \to \mathcal{B}$ be an algebra homomorphism. Prove that τ is continuous. (Hint: for each $\phi \in \Phi(\mathcal{B})$ (the Gelfand space of $\mathcal{B}$), $\phi \circ \tau \in \Phi(\mathcal{A})$. Use the closed graph theorem.)

17. (a) Let X be a compact Hausdorff space, and let $\mathcal{A} = C(X)$. Prove that the Gelfand space Φ of $\mathcal{A}$ is homeomorphic to X. (Hint: consider the map $t \in X \to \phi_t \in \Phi$, where $\phi_t(f) = f(t)$, $(f \in C(X))$ (this is the "evaluation at t" homomorphism). If $\exists \phi \in \Phi$ such that $\phi \neq \phi_t$ for all $t \in X$ and $M = \ker \phi$, then for each $t \in X$ there exists $f_t \in M$ such that $f_t(t) \neq 0$. Use continuity of the functions and compactness of X to get a finite set $\{f_{t_j}\} \subset M$ such that $h := \sum |f_{t_j}|^2 > 0$ on X, hence $h \in G(\mathcal{A})$; however, $h \in M$, contradiction. Thus, $t \to \phi_t$ is onto Φ, and one-to-one (by Urysohn's lemma). Identifying Φ with X through this map, observe that the Gelfand topology is weaker than the given topology on X and is Hausdorff.)

    (b) Extend part (a) to *locally* compact Hausdorff spaces X and $\mathcal{A} := C_0(X)$, proving again that the map $t \in X \to \phi_t \in \Phi$ is a homeomorphism. (Hint: assuming by contradiction that $\phi \in \Phi$ is not an evaluation character and writing $M := \ker \phi$, for every compact $K \subset X$ there is a $h \in M$ with $h \geq 1$ on K. Use this and Urysohn's lemma to deduce that $C_c(X)$ is contained in M, so density of $C_c(X)$ in $C_0(X)$ and closedness of M imply that $M = C_0(X)$, a contradiction. Finally, Urysohn's lemma can be useful again when comparing the two topologies on X.)

18. Let U be the open unit disc in $\mathbb{C}$. For $n \in \mathbb{N}$, let $\mathcal{A} = \mathcal{A}(U^n)$ denote the Banach algebra of all complex functions analytic in $U^n := U \times \cdots \times U$ and continuous on the closure $\overline{U^n}$ of U^n in $\mathbb{C}^n$, with pointwise operations and supremum norm $\|f\|_u := \sup\{|f(z)|; z \in \overline{U^n}\}$. Let Φ be the Gelfand space of $\mathcal{A}$. Given $f \in \mathcal{A}$ and $0 < r < 1$, denote $f_r(z) = f(rz)$ and $Z(f) = \{z \in \overline{U^n};\ f(z) = 0\}$. Prove:

    (a) f_r is the sum of an absolutely and uniformly convergent power series in $\overline{U^n}$. Conclude that the polynomials (in n variables) are dense in $\mathcal{A}$.

    (b) Each $\phi \in \Phi$ is an "evaluation homomorphism" ϕ_w for some $w \in \overline{U^n}$, where $\phi_w(f) = f(w)$. (Hint: consider the polynomials $p_j(z) = z_j$ (where $z = (z_1, \ldots, z_n)$). Then $w := (\phi(p_1), \ldots, \phi(p_n)) \in \overline{U^n}$ and $\phi(p_j) = p_j(w)$. Hence $\phi(p) = p(w)$ for all polynomials p. Apply

Part (a) to conclude that $\phi = \phi_w$. The map $w \to \phi_w$ is the wanted homeomorphism of $\overline{U^n}$ onto Φ.)

(c) Given $f_1, \ldots, f_m \in \mathcal{A}$ such that $\bigcap_{k=1}^m Z(f_k) = \emptyset$, there exist $g_1, \ldots, g_m \in \mathcal{A}$ such that $\sum_k f_k g_k = 1$ (on $\overline{U^n}$). (Hint: otherwise, the ideal J generated by $f_1, \ldots, f_m$ is proper, and therefore there exists $\phi \in \Phi$ vanishing on J. Apply Part (b) to reach a contradiction.)

19. Let $\mathcal{A}$ be a unital Banach algebra such that

$$K := \sup_{0 \neq a \in \mathcal{A}} \frac{\|a\|^2}{\|a^2\|} < \infty.$$

Prove:

(a) $\|a\| \leq K\, r(a)$ for all $a \in \mathcal{A}$.

(b) $\|p(a)\| \leq K\, \|p\|_{C(\sigma(a))}$ for all $p \in \mathcal{P}$, where $\mathcal{P}$ denotes the algebra of all polynomials of one complex variable over $\mathbb{C}$.

(c) If $a \in \mathcal{A}$ has the property that $\mathcal{P}$ is dense in $C(\sigma(a))$, then there exists a continuous algebra homomorphism (with norm $\leq K$) $\tau : C(\sigma(a)) \to \mathcal{A}$ such that $\tau(p) = p(a)$ for all $p \in \mathcal{P}$.

20. Let $K \subset \mathbb{C}$ be compact $\neq \emptyset$, and let $C(K)$ be the corresponding Banach algebra of continuous functions with the supremum norm $\|f\|_K := \sup_K |f|$. Denote $\mathcal{P}_1 := \{p \in \mathcal{P};\ \|p\|_K \leq 1\}$ (cf. Exercise 19 b).

Let X be a Banach space, and $T \in B(X)$. For $x \in X$, denote

$$\|x\|_T := \sup_{p \in \mathcal{P}_1} \|p(T)x\|;$$

$$Z_T := \{x \in X;\ \|x\|_T < \infty\}.$$

Prove:

(a) Z_T is a Banach space for the norm $\|\cdot\|_T$ (which is greater than the given norm on X).

(b) $\|p(T)\|_{B(Z_T)} \leq 1$.

(c) If the compact set K is such that $\mathcal{P}$ is dense in $C(K)$, there exists a contractive algebra homomorphism $\tau : C(K) \to B(Z_T)$ such that $\tau(p) = p(T)$ for all $p \in \mathcal{P}$.

21. Let $\mathcal{A}$ be a non-unital C^*-algebra. Prove that the unitization $\mathcal{A}^\#$ defined in Section 7.3 is a (unital) C^*-algebra in which $\mathcal{A}$ is isometrically embedded. (Hint: the proof should show that the norm on $\mathcal{A}^\#$ is indeed a norm, that it extends that of $\mathcal{A}$, that it is submultiplicative and that it satisfies the C^*-identity, and that $\mathcal{A}^\#$ is complete. For the last part, note that the quotient $\mathcal{A}^\#/\mathcal{A}$ is one dimensional and use Exercise 22 of Chapter 6.)

22. Let X be a non-compact, locally compact Hausdorff space and let Y be its Alexandroff one-point compactification. Find natural isometric $*$-isomorphisms $C_0(X)^\# \cong \{f + \lambda 1; f \in C_0(X)\} \cong C(Y)$, where $C_0(X)^\#$ is the unitization of the C^*-algebra $C_0(X)$ defined in Section 7.3 and the middle algebra is viewed as a C^*-subalgebra of $C_b(X)$.

23. Let $\mathcal{A}$ be a Banach algebra with an involution such that $\|a\|^2 \leq \|a^*a\|$ for all $a \in \mathcal{A}$. Prove that $\mathcal{A}$ is a C^*-algebra.

24. Let u be a unitary element in a unital C^*-algebra. Without using the operational calculus, explain why $\sigma(u), \sigma(u^{-1}) \subset \{z \in \mathbb{C}; |z| \leq 1\}$, and use the relation $\sigma(u^{-1}) = \sigma(u)^{-1} := \{z^{-1}; z \in \sigma(u)\}$ to conclude that $\sigma(u) \subset \{z \in \mathbb{C}; |z| = 1\}$. (We proved this in a less elementary fashion.)

25. Let $\mathcal{A}$ be a C^*-algebra and $0 \neq p \in \mathcal{A}$ be a selfadjoint idempotent: $p^2 = p = p^*$. Prove that $p\mathcal{A}p := \{pap; a \in \mathcal{A}\}$ is a C^*-subalgebra of $\mathcal{A}$ with p as its unit (such a subalgebra is called a *corner* of $\mathcal{A}$). Assuming that p is not a unit of $\mathcal{A}$, prove that $\sigma_{\mathcal{A}}(x) = \sigma_{p\mathcal{A}p}(x) \cup \{0\}$ for all $x \in p\mathcal{A}p$ (this part of the exercise is purely algebraic).

26. This exercise supplements Theorem 7.18. Let $\mathcal{B}$ be a C^*-subalgebra of a C^*-algebra $\mathcal{A}$. Prove that if $\mathcal{B}$ has a unit $e_{\mathcal{B}}$ which is not a unit for $\mathcal{A}$, then $\sigma_{\mathcal{A}}(x) = \sigma_{\mathcal{B}}(x) \cup \{0\}$ for all $x \in \mathcal{B}$; while in any other case, $\sigma_{\mathcal{B}}(x) = \sigma_{\mathcal{A}}(x)$ for all $x \in \mathcal{B}$. Hint: if $\mathcal{A}$ and $\mathcal{B}$ have the same unit, we are on the premises of Theorem 7.18. If $\mathcal{B}$ is not unital, note that $\mathcal{B}^\#$ is algebraically isomorphic to the C^*-algebra $\operatorname{span}(\mathcal{B} \cup \{e_{\mathcal{A}^\#}\})$. Finally, assume that $\mathcal{B}$ has a unit $e_{\mathcal{B}}$ which is not a unit for $\mathcal{A}$, and consider the unital C^*-algebra $e_{\mathcal{B}}\mathcal{A}e_{\mathcal{B}}$, which contains $\mathcal{B}$.

27. Let $\mathcal{A}$ be a C^*-algebra and $x \in \mathcal{A}$ be normal. Prove:

    (a) If $\sigma(x) \subset \mathbb{R}$ then x is selfadjoint.

    (b) If $\mathcal{A}$ is unital and $\sigma(x) \subset \{z \in \mathbb{C}; |z| = 1\}$, then x is unitary.

28. Let $\mathcal{A}$ be a unital C^*-algebra and $u \in \mathcal{A}$ be unitary. Suppose that $\sigma(u) \neq \{z \in \mathbb{C}; |z| = 1\}$. Prove that there is a selfadjoint $a \in \mathcal{A}$ such that $u = \mathrm{e}^{\mathrm{i}a}$. Compare Part (a) of Exercise 5 in Chapter 9.

29. Let $\mathcal{A}$ be a unital C^*-algebra. Suppose that $x \in \mathcal{A}$ is normal and μ is an isolated point of $\sigma(x)$. Define $e_\mu \in C(\sigma(x))$ to be 1 at μ and 0 elsewhere. Prove that $e_\mu(x)$ is a non-zero selfadjoint idempotent and that $xe_\mu(x) = \mu e_\mu(x)$. Compare Theorem 9.8 and Section 9.11.

30. Let $\mathcal{A}$ be a non-unital C^*-algebra and $x \in \mathcal{A}$ be normal.

    (a) Prove that the (generally non-unital) C^*-subalgebra $\mathcal{B}$ of $\mathcal{A}$ generated by x is equal to $\{f(x); f \in C(\sigma(x)), f(0) = 0)\}$. (Hint: we already explained one inclusion.)

(b) Deduce that the map $f \to f(x)$ is an isometric $*$-isomorphism from the C^*-subalgebra $\{f \in C(\sigma(x)); f(0) = 0\}$ of $C(\sigma(x))$ onto $\mathcal{B}$.

31. Let $\mathcal{A}$ be a Banach algebra.

(a) Prove that the set of all weakly almost periodic elements in $\mathcal{A}^*$ is a closed linear subspace of $\mathcal{A}^*$.

(b) Prove that for $\omega \in \mathcal{A}^*$, the following conditions are equivalent:

(i) ω is weakly almost periodic;

(ii) whenever $\{a_i\}$ and $\{b_j\}$ are bounded nets in $\mathcal{A}$, the equality

$$\lim_i \left(\lim_j \omega(a_i b_j) \right) = \lim_j \left(\lim_i \omega(a_i b_j) \right)$$

holds provided that the limits in both sides exist; and

(iii) same as (ii), but with sequences in lieu of nets.

(Hint for (iii) $\implies$ (i): assume that (i) is false and construct by recursion two sequences that make (iii) false.)

32. In this exercise we identify c_0^* with l^1 as in Exercise 8 of Chapter 4 and $(l^1)^*$ with l^∞ as in Theorem 4.6, and thus also c_0^{**} with l^∞ (all identifications are by isometric isomorphisms).

(a) Prove that for every $a \in c_0$ the canonical image $\kappa(a) \in l^\infty$ equals a, viewed as an element of l^∞. That is, κ is the inclusion map.

(b) Let $a \in c_0$, $\omega \in l^1$ and $X \in l^\infty$. Using the notation of Section 7.5, what are $a\omega, \omega a \in l^1$, $X\omega, \omega X \in l^1$ and $X \,\Box\, Y, X \lozenge Y \in l^\infty$? Show that c_0 is Arens regular.

33. The Banach space $l^1(\mathbb{Z})$ becomes a Banach algebra with the product being convolution $*$: the convolution of $a, b \in l^1(\mathbb{Z})$ is $a * b \in l^1(\mathbb{Z})$ given by $(a * b)_k := \sum_{l \in \mathbb{Z}} a_{k-l} b_l$ $(k \in \mathbb{Z})$.

(a) Define sequences $\{p_n\}_{n=1}^\infty$ and $\{q_m\}_{m=1}^\infty$ in $\mathbb{Z}$ by $p_n := 2^{2n}$, $q_m := 2^{2m+1}$ $(n, m \in \mathbb{N})$. Observe that

$$\{p_n + q_m; n < m\} \cap \{p_n + q_m; n > m\} = \emptyset. \tag{1}$$

(b) Let $\{p_n\}_{n=1}^\infty$ and $\{q_m\}_{m=1}^\infty$ be sequences in $\mathbb{Z}$ satisfying (1). For $n, m \in \mathbb{N}$, let $a_n, b_m \in l^1(\mathbb{Z})$ be the indicator functions of the singletons $\{p_n\}$ and $\{q_m\}$, respectively. Let $\omega \in l^\infty(\mathbb{Z})$ be the indicator function of the set $\{p_n + q_m; n < m\}$. Identifying $l^\infty(\mathbb{Z})$ with $(l^1(\mathbb{Z}))^*$, prove that ω does not satisfy Condition (iii) of Part (b) of Exercise 31, so that it is not weakly almost periodic. This proves that $l^1(\mathbb{Z})$ is not Arens regular.

34. Use the idea of the Exercise 33 to show that the Banach algebra $L^1(\mathbb{R})$ (with convolution as product, see Exercise 14) is not Arens regular.

8

Hilbert spaces

Recall that a Banach space is called a Hilbert space when its norm is induced by an inner product. In contrast to general Banach spaces, whose geometry can be very complicated, Hilbert spaces have a clear and simple geometry. This is demonstrated, first and foremost, by the results proved in our first look at Hilbert spaces in Section 1.7, namely, the distance theorem, the orthogonal decomposition theorem, and the "Little" Riesz representation theorem.

This chapter continues the study of Hilbert spaces, the first central notions being *orthonormal sets and bases*. We give several characterizations of orthonormal bases, and prove that they always exist and that all orthonormal bases of a specific Hilbert space X have the same cardinality, called the Hilbert dimension of X. Along the way we introduce projections and particularly orthogonal projections and see how they are related to orthonormal sets.

We define and characterize *Hilbert space isomorphisms* and find a *canonical model for Hilbert spaces*: every Hilbert space is isomorphic to some L^2-space.

Two basic ways to construct Hilbert spaces are presented, namely *direct sums* and *tensor products*. The latter construction is essential in operator algebras and we rely on it in Chapter 13.

8.1 Orthonormal sets

We recall that the vectors x, y in an inner product space X are (mutually) *orthogonal* if $(x, y) = 0$ (cf. Theorem 1.35). The set $A \subset X$ is orthogonal if any two distinct vectors of A are orthogonal; it is *orthonormal* if it is orthogonal and all vectors in A are unit vectors, that is, if

$$(a, b) = \delta_{a,b} \quad (a, b \in A)$$

where $\delta_{a,b}$ is *Kronecker's delta*, which equals zero for $a \neq b$ and equals one for $a = b$.

A classical example is the set $A = \{e^{int}; n \in \mathbb{N}\}$ in the Hilbert space $X = L^2([0, 2\pi])$ with the normalized Lebesgue measure $dt/2\pi$.

Introduction to Modern Analysis. Second Edition. Shmuel Kantorovitz and Ami Viselter, Oxford University Press.
 DOI: 10.1093/oso/9780192849540.003.0008

Lemma 8.1 (Pythagoras's theorem). *Let $\{x_k; k = 1, \dots, n\}$ be an orthogonal subset of the inner product space X. Then*

$$\|\sum_{k=1}^{n} x_k\|^2 = \sum_{k=1}^{n} \|x_k\|^2.$$

Proof. By "sesqui-linearity" of the inner product, the left-hand side equals

$$\left(\sum_k x_k, \sum_j x_j\right) = \sum_{k,j} (x_k, x_j).$$

Since $(x_k, x_j) = 0$ for $k \neq j$, the last sum equals $\sum_k (x_k, x_k) = \sum_k \|x_k\|^2$. □

If $A := \{a_1, \dots, a_n\} \subset X$ is *orthonormal* and $\{\lambda_1, \dots, \lambda_n\} \subset \mathbb{C}$, then (taking $x_k = \lambda_k a_k$ in Lemma 8.1),

$$\|\sum_{k=1}^{n} \lambda_k a_k\|^2 = \sum_{k=1}^{n} |\lambda_k|^2. \tag{1}$$

Theorem 8.2. *Let $\{a_k; k = 1, 2, \dots\}$ be an orthonormal sequence in the Hilbert space X, and let $\Lambda := \{\lambda_k; k = 1, 2, \dots\}$ be a complex sequence. Then*

(a) The series $\sum_k \lambda_k a_k$ converges in X iff $\|\Lambda\|_2^2 := \sum_k |\lambda_k|^2 < \infty$.

(b) In this case, the shown series converges unconditionally *in X, and $\|\sum_k \lambda_k a_k\| = \|\Lambda\|_2$.*

Proof. By (1) applied to the orthonormal set $\{a_{m+1}, \dots, a_n\}$ and the set of scalars $\{\lambda_{m+1}, \dots, \lambda_n\}$ with $n > m \geq 0$,

$$\left\|\sum_{k=m+1}^{n} \lambda_k a_k\right\|^2 = \sum_{k=m+1}^{n} |\lambda_k|^2. \tag{2}$$

This means that the series $\sum \lambda_k a_k$ satisfies Cauchy's condition iff the series $\sum |\lambda_k|^2$ satisfies Cauchy's condition. Since X is complete, this is equivalent to Statement (a) of the theorem.

Suppose now that $\|\Lambda\|_2 < \infty$, and let then $s \in X$ denote the sum of the series $\sum_k \lambda_k a_k$. Taking $m = 0$ and letting $n \to \infty$ in (2), we obtain $\|s\| = \|\Lambda\|_2$.

Since $(\cdot\,, x)$ is a continuous linear functional (for any fixed $x \in X$), we have

$$(s, x) = \sum_{k=1}^{\infty} \lambda_k (a_k, x). \tag{3}$$

If $\pi : \mathbb{N} \to \mathbb{N}$ is any permutation of $\mathbb{N}$, the series $\sum_k |\lambda_{\pi(k)}|^2$ converges to $\|\Lambda\|_2^2$ by a well-known property of positive series. Therefore, by what we already proved, the series $\sum_k \lambda_{\pi(k)} a_{\pi(k)}$ converges in X; denoting its sum by t, we also have $\|t\| = \|\Lambda\|_2$, and by (3), for any $x \in X$,

$$(t, x) = \sum_k \lambda_{\pi(k)} (a_{\pi(k)}, x).$$

Choose $x = a_j$ for $j \in \mathbb{N}$ fixed. By orthonormality, we get $(t, a_j) = \lambda_j$, and therefore, by (3) with $x = t$,

$$(t, s) = \overline{(s, t)} = \overline{\sum_k \lambda_k (a_k, t)} = \sum_k \overline{\lambda_k} (t, a_k) = \sum_k |\lambda_k|^2 = \|\Lambda\|_2^2.$$

Hence $\|t - s\|^2 = \|t\|^2 - 2\Re(t, s) + \|s\|^2 = 0$, and $t = s$. □

Lemma 8.3 (Bessel's inequality). *Let $\{a_1, \ldots, a_n\}$ be an orthonormal set in the inner product space X. Then for all $x \in X$,*

$$\sum_k |(x, a_k)|^2 \le \|x\|^2.$$

Proof. Given $x \in X$, denote $y := \sum_k (x, a_k) a_k$. Then by (1)

$$(y, x - y) = (y, x) - (y, y) = \sum_k (x, a_k)(a_k, x) - \|y\|^2 = \sum_k |(x, a_k)|^2 - \|y\|^2 = 0.$$

Therefore, by Lemma 8.1,

$$\|x\|^2 = \|(x - y) + y\|^2 = \|x - y\|^2 + \|y\|^2 \ge \|y\|^2 = \sum_k |(x, a_k)|^2.$$

□

Corollary 8.4. *Let $\{a_k; k = 1, 2, \ldots\}$ be an orthonormal sequence in the Hilbert space X. Then for any $x \in X$, the series $\sum_k (x, a_k) a_k$ converges unconditionally in X to an element Px, and $P \in B(X)$ has the following properties:*

(a) $\|P\| = 1$;

(b) $P^2 = P$;

(c) *the ranges PX and $(I - P)X$ are orthogonal;*

(d) $P^* = P$.

Proof. Let $x \in X$. By Bessel's inequality (Lemma 8.3), the partial sums of the positive series $\sum_k |(x, a_k)|^2$ are bounded by $\|x\|^2$; the series therefore converges, and consequently $\sum_k (x, a_k) a_k$ converges unconditionally to an element $Px \in X$, by Theorem 8.2. The linearity of P is trivial, and by Theorem 8.2, $\|Px\|^2 = \sum_k |(x, a_k)|^2 \le \|x\|^2$, so that $P \in B(X)$ and $\|P\| \le 1$. By (3)

$$(Px, a_j) = \sum_{k=1}^{\infty} (x, a_k)(a_k, a_j) = (x, a_j). \tag{4}$$

Therefore

$$P^2x = P(Px) = \sum_j (Px, a_j)a_j = \sum_j (x, a_j)a_j = Px.$$

This proves Property (b)

By Property (b), $\|P\| = \|P^2\| \le \|P\|^2$. Since $P \neq 0$ (e.g., $Pa_j = a_j \neq 0$ for all $j \in \mathbb{N}$), it follows that $\|P\| \ge 1$, and therefore, $\|P\| = 1$, by our previous inequality. This proves Property (a). By (3) with the proper choices of scalars and vectors, we have for all $x, y \in X$

$$(Py, x) = \sum_k (y, a_k)(a_k, x) = \overline{\sum_k (x, a_k)(a_k, y)} = \overline{(Px, y)} = (y, Px).$$

This proves Property (d), and therefore, by Property (b),

$$(Px, (I-P)y) = (x, P(I-P)y) = (x, (P-P^2)y) = 0,$$

which verifies Property (c). □

8.2 Projections

This section discusses projections and, more specifically, orthogonal projections. The existence of general projections in Banach spaces was studied in Exercise 18 of Chapter 6.

Terminology 8.5.

(a) Any $P \in B(X)$ satisfying Property (b) in Corollary 8.4 is called a *projection* (X could be any Banach space!). If P is a projection, so is $I - P$ (because $(I-P)^2 = I - 2P + P = I - P$); $I - P$ is called the *complementary* projection of P. Note that the complementary projections P and $I - P$ commute and have product equal to zero.

(b) For any projection P on a Banach space X,

$$PX = \{x \in X; Px = x\} = \ker(I - P) \tag{1}$$

(if $x = Py$ for some $y \in X$, then $Px = P^2y = Py = x$). Since $I - P$ is continuous, it follows from (1) that the range PX is closed. The closed subspaces PX and $(I-P)X$ have trivial intersection (if x is in the intersection, then $x = Px$, and $x = (I-P)x = x - Px = 0$), and their sum is X (every x can be written as $Px + (I-P)x$). This means that X is the *direct sum of the closed subspaces* PX *and* $(I-P)X$.

A closed subspace $M \subset X$ is T-*invariant* (for a given $T \in B(X)$) if $TM \subset M$. By (1), the closed subspace PX is T-invariant iff $P(TPx) = TPx$ for all $x \in X$, that is, iff $TP = PTP$. Applying this to the complementary projection $I-P$, we conclude that $(I-P)X$ is T-invariant iff $T(I-P) = (I-P)T(I-P)$, that is (expand and cancel!), iff $PT = PTP$. Therefore *both* complementary subspaces PX and $(I-P)X$ are T-invariant iff P *commutes* with T. One says in this case that PX is a *reducing subspace for* T (or that P reduces T).

(c) When X is a Hilbert space and a projection P in X satisfies Property (c) in Corollary 8.4, it is called an *orthogonal projection.* In that case the direct sum decomposition $X = PX \oplus (I-P)X$ is an *orthogonal decomposition.* Conversely, if Y is any closed subspace of X, we may use the orthogonal decomposition $X = Y \oplus Y^{\perp}$ (Theorem 1.36) to define an orthogonal projection P in X with range equal to Y: given any $x \in X$, it has the unique orthogonal decomposition $x = y + z$ with $y \in Y$ and $z \in Y^{\perp}$; define $Px = y$. It is easy to verify that P is the wanted (selfadjoint!) projection; it is called *the orthogonal projection onto* Y. Given $T \in B(X)$, Y is a reducing subspace for T iff P commutes with T (by Point (b)); since P is selfadjoint, the relations $PT = TP$ and $T^*P = PT^*$ are equivalent (since one follows from the other by taking adjoints). Thus Y reduces T iff it reduces T^*. If Y is invariant for both T and T^*, then (cf. Point (b)) $TP = PTP$ and $T^*P = PT^*P$; taking adjoints in the second relation, we get $PT = PTP$, hence $TP = PT$. As observed before, this last relation implies in particular that Y is invariant for both T and T^*. Thus, if Y is a closed subspace of the Hilbert space X and P is the corresponding orthogonal projection, then for any $T \in B(X)$, the following propositions are equivalent:

(i) Y reduces T;

(ii) Y is invariant for both T and T^*;

(iii) P commutes with T.

(d) In the proof of Corollary 8.4, we deduced Property (c) from Property (d) Conversely, if P is an orthogonal projection, then it is selfadjoint. Indeed, for any $x, y \in X$, $(Px, (I-P)y) = 0$, that is, $(Px, y) = (Px, Py)$. Interchanging the roles of x and y and taking complex adjoints, we get $(x, Py) = (Px, Py)$, and therefore $(Px, y) = (x, Py)$. Thus, *a projection in Hilbert space is orthogonal iff it is selfadjoint.*

We consider now an *arbitrary* orthonormal set A in the Hilbert space X.

Lemma 8.6. *Let X be an inner product space and $A \subset X$ be orthonormal. Let $\delta > 0$ and $x \in X$ be given. Then the set*

$$A_\delta(x) := \{a \in A; |(x, a)| > \delta\}$$

is finite (it has at most $\left[\|x\|^2/\delta^2\right]$ elements).

Proof. We may assume that $A_\delta(x) \neq \emptyset$ (otherwise there is nothing to prove). Let then $a_1, \ldots, a_n$ be $n \geq 1$ distinct elements in $A_\delta(x)$. By Bessel's inequality,

$$n\delta^2 \leq \sum_{k=1}^{n} |(x, a_k)|^2 \leq \|x\|^2,$$

so that $n \le \|x\|^2/\delta^2$, and the conclusion follows. □

Theorem 8.7. *Let X be an inner product space, and $A \subset X$ be orthonormal. Then for any given $x \in X$, the set*

$$A(x) := \{a \in A; (x,a) \neq 0\}$$

is at most countable.

Proof. Since

$$A(x) = \bigcup_{m=1}^{\infty} A_{1/m}(x),$$

this is an immediate consequence of Lemma 8.6. □

Notation 8.8. Let A be any orthonormal subset in the *Hilbert space* X. Given $x \in X$, write $A(x)$ as a (finite or infinite) sequence $A(x) = \{a_k\}$. Let Px be defined as before with respect to the orthonormal sequence $\{a_k\}$. Since the convergence is unconditional, the definition is *independent* of the particular representation of $A(x)$ as a sequence, and one may use the following notation that ignores the sequential representation:

$$Px = \sum_{a \in A(x)} (x,a)a.$$

Since $(x,a) = 0$ for all $a \in A$ not in $A(x)$, one may add to the earlier sum the zero terms $(x,a)a$ for all such vectors a, that is,

$$Px = \sum_{a \in A} (x,a)a.$$

By Corollary 8.4, P is an orthogonal projection.

Lemma 8.9. *The ranges of the orthogonal projections P and $I - P$ are $\overline{\text{span}(A)}$ (the closed span of A, i.e., the closure of the linear span of A) and $A^{\perp}$, respectively.*

Proof. Since $Pb = b$ for any $b \in A$, we have $A \subset PX$, and since PX is a closed subspace, it follows that $\overline{\text{span}(A)} \subset PX$. On the other hand, given $x \in X$, represent $A(x) = \{a_k\}$; then $Px = \sum_k (x,a_k)a_k = \lim_n \sum_{k=1}^n (x,a_k)a_k \in \overline{\text{span}(A)}$, and the first statement of the lemma follows.

By uniqueness of the orthogonal decomposition, we have

$$(I-P)X = \left(\overline{\text{span}(A)}\right)^{\perp}. \tag{2}$$

Clearly, $x \in A^{\perp}$ iff $A \subset \ker(\cdot\,,x)$. Since $(\cdot\,,x)$ is a continuous linear functional, its kernel is a closed subspace, and therefore, the last inclusion is equivalent to $\overline{\text{span}(A)} \subset \ker(\cdot\,,x)$, that is, to $x \in \left(\overline{\text{span}(A)}\right)^{\perp}$. This shows that the set on the right of (2) is equal to $A^{\perp}$. □

8.3 Orthonormal bases

Theorem 8.10. *Let A be an orthonormal set in the Hilbert space X, and let P be the associated projection. Then the following statements are equivalent:*

(1) $A^{\perp} = \{0\}$.

(2) If $A \subset B \subset X$ and B is orthonormal, then $A = B$.

(3) $X = \overline{\operatorname{span}(A)}$.

(4) $P = I$.

(5) Every $x \in X$ has the representation

$$x = \sum_{a \in A} (x, a)a.$$

(6) For every $x, y \in X$, one has

$$(x, y) = \sum_{a \in A} (x, a)\overline{(y, a)},$$

where the series, which has at most countably many non-zero terms, converges absolutely.

(7) For every $x \in X$, one has

$$\|x\|^2 = \sum_{a \in A} |(x, a)|^2.$$

Proof.

(1) *implies* (2). Suppose $A \subset B$ with B orthonormal. If $B \neq A$, pick $b \in B$, $b \notin A$. For any $a \in A$, the vectors a, b are distinct (unit) vectors in the orthonormal set B, and therefore $(b, a) = 0$. Hence, $b \in A^{\perp}$, and therefore $b = 0$ by (1), contradicting the fact that b is a unit vector. Hence $B = A$.

(2) *implies* (3). If (3) does not hold, then by the orthogonal decomposition theorem and Lemma 8.9, the subspace $A^{\perp}$ is non-trivial, and contains therefore a unit vector b. Then $b \notin A$, so that the *orthonormal* set $B := A \cup \{b\}$ contains A *properly*, contradicting (2).

(3) *implies* (4). By (3) and Lemma 8.9, the range of P is X, hence, $Px = x$ for all $x \in X$ (cf. Point (b) in Terminology 8.5), that is, $P = I$.

(4) *implies* (5). By (4), for all $x \in X$, $x = Ix = Px = \sum_{a \in A} (x, a)a$.

(5) *implies* (6). Given $x, y \in A$ and representing $A(x) = \{a_k\}$, we have by the Cauchy–Schwarz inequality in the Hilbert space $\mathbb{C}^n$ (for any $n \in \mathbb{N}$):

$$\sum_{k=1}^{n} |(x, a_k)\overline{(y, a_k)}| \leq \left(\sum_{k=1}^{n} |(x, a_k)|^2 \right)^{1/2} \left(\sum_{k=1}^{n} |(y, a_k)|^2 \right)^{1/2} \leq \|x\|\|y\|,$$

where we used Bessel's inequality for the last step.

Since the partial sums of the positive series $\sum_k |(x, a_k)\overline{(y, a_k)}|$ are bounded, the series $\sum_k (x, a_k)\overline{(y, a_k)}$ converges absolutely, hence unconditionally, and may be written without specifying the particular ordering of $A(x)$; after adding zero terms, we finally write it in the form $\sum_{a \in A} (x, a)\overline{(y, a)}$. Now since $(\cdot\,, y)$ is a continuous linear functional, this sum is equal to

$$\sum_k (x, a_k)(a_k, y) = \left(\sum_k (x, a_k)a_k, y \right) = (x, y),$$

where we used (5) for the last step.

(6) *implies* (7). Take $x = y$ in (6).

(7) *implies* (1). If $x \in A^{\perp}$, $(x, a) = 0$ for all $a \in A$, and therefore $\|x\| = 0$ by (7).

□

Terminology 8.11. Let A be an orthonormal set in X. If A has Property (1), it is called a *complete orthonormal set.* If it has Property (2), it is called *a maximal orthonormal set.* If it has Property (3), one says that A *spans* X. We express Property (5) by saying that *every* $x \in X$ *has a generalized Fourier expansion with respect to* A. Properties (6) and (7) are the *Parseval* and *Bessel* identities for A, respectively.

If A has any (*and therefore all*) of these seven properties, it is called an *orthonormal basis* or a *Hilbert basis* for X.

Example. (1) Let Γ denote the unit circle in $\mathbb{C}$, and let $\mathcal{A}$ be the subalgebra of $C(\Gamma)$ consisting of the restrictions to Γ of all the functions in $\text{span}\{z^k; k \in \mathbb{Z}\}$. Since $\bar{z} = z^{-1}$ on Γ, $\mathcal{A}$ is selfadjoint. The function $z \in \mathcal{A}$ assumes distinct values at distinct points of Γ, so that $\mathcal{A}$ is separating, and contains $1 = z^0$. By Theorem 5.39, $\mathcal{A}$ is dense in $C(\Gamma)$.

Given $f \in L^2(\Gamma)$ (with the arc-length measure) and $\epsilon > 0$, let $g \in C(\Gamma)$ be such that $\|f - g\|_2 < \epsilon/2$ (by density of $C(\Gamma)$ in $L^2(\Gamma)$). Next let $p \in \mathcal{A}$ be such that $\|g - p\|_{C(\Gamma)} < \epsilon/2\sqrt{2\pi}$ (by density of $\mathcal{A}$ in $C(\Gamma)$). Then $\|g - p\|_2 < \epsilon/2$, and therefore $\|f - p\|_2 < \epsilon$. This shows that $\mathcal{A}$ is dense in $L^2(\Gamma)$. Equivalently, writing $z = e^{ix}$ with $x \in [-\pi, \pi]$, we proved that the span of the (obviously) orthonormal sequence

$$\left\{ \frac{e^{ikx}}{\sqrt{2\pi}}; k \in \mathbb{Z} \right\} \tag{*}$$

in the Hilbert space $L^2(-\pi, \pi)$ is dense, that is, the sequence (*) *is an orthonormal basis for* $L^2(-\pi, \pi)$. In particular, every $f \in L^2(-\pi, \pi)$ has the unique so-called $L^2(-\pi, \pi)$-convergent *Fourier expansion*

$$f = \sum_{k \in \mathbb{Z}} c_k e^{ikx},$$

with

$$c_k = (1/2\pi) \int_{-\pi}^{\pi} f(x) e^{-ikx}\, dx.$$

By Bessel's identity,

$$\sum_{k \in \mathbb{Z}} |c_k|^2 = (1/2\pi) \|f\|_2^2. \tag{**}$$

Take, for example, $f = I_{(0,\pi)}$. A simple calculation shows that $c_0 = 1/2$ and $c_k = (1 - e^{-ik\pi})/2\pi i k$ for $k \neq 0$. Therefore, for $k \neq 0$, $|c_k|^2 = (1/4\pi^2 k^2)(2 - 2\cos k\pi)$ vanishes for k even and equals $1/\pi^2 k^2$ for k odd. Substituting in (**), we get $1/4 + (1/\pi^2) \sum_{k\,\text{odd}} (1/k^2) = 1/2$, hence

$$\sum_{k\,\text{odd}} (1/k^2) = \pi^2/4.$$

If $a := \sum_{k=1}^{\infty} (1/k^2)$, then

$$\sum_{k\,\text{even} \geq 2} (1/k^2) = \sum_{j=1}^{\infty} (1/4j^2) = a/4,$$

and therefore

$$a = \sum_{k\,\text{odd} \geq 1} (1/k^2) + a/4 = \pi^2/8 + a/4.$$

Solving for a, we get $a = \pi^2/6$.

(2) Let

$$f_k(x) = \frac{e^{ikx}}{\sqrt{2\pi}} I_{(-\pi,\pi)}(x) \quad (k \in \mathbb{Z}, x \in \mathbb{R}).$$

The sequence $\{f_k\}$ is clearly orthonormal in $L^2(\mathbb{R})$. If f belongs to the closure of span $\{f_k\}$ and $\epsilon > 0$ is given, there exists h in the span such that $\|f - h\|_2^2 < \epsilon$. Since $h = 0$ on $(-\pi, \pi)^c$, we have

$$\int_{(-\pi,\pi)^c} |f|^2\, dx = \int_{(-\pi,\pi)^c} |f - h|^2\, dx < \epsilon,$$

and the arbitrariness of epsilon implies that the integral on the left-hand side vanishes. Hence $f = 0$ a.e. outside $(-\pi, \pi)$ (since f represents an equivalence class of functions, we may say that $f = 0$ identically outside $(-\pi, \pi)$). On the other hand, the density of the span of $\{e^{ikx}\}$ in $L^2(-\pi, \pi)$ means that *every* $f \in L^2(\mathbb{R})$ vanishing outside $(-\pi, \pi)$ is in the closure of span $\{f_k\}$ in $L^2(\mathbb{R})$. Thus $\{f_k\}$ is an orthonormal sequence in $L^2(\mathbb{R})$ which is *not* an orthonormal basis for the space; the closure of its span consists precisely of all $f \in L^2(\mathbb{R})$ *vanishing outside* $(-\pi, \pi)$.

Theorem 8.12 (Existence of orthonormal bases). *Every (non-trivial) Hilbert space has an orthonormal base.*

More specifically, given any orthonormal set A_0 in the Hilbert space X, it can be completed to an orthonormal base A.

Proof. Since a non-trivial Hilbert space contains a unit vector a_0, the first statement of the theorem follows from the second with the orthonormal set $A_0 = \{a_0\}$.

To prove the second statement, consider the family $\mathcal{A}$ of all orthonormal sets in X containing the given set A_0. It is non-empty (since $A_0 \in \mathcal{A}$) and partially ordered by inclusion. If $\mathcal{A}' \subset \mathcal{A}$ is totally ordered, it is clear that $\bigcup \mathcal{A}'$ is an *orthonormal set* (because of the total order!) containing A_0, that is, it is an element of $\mathcal{A}$, and is an upper bound for all the sets in $\mathcal{A}'$. Therefore, by Zorn's lemma, the family $\mathcal{A}$ contains a maximal element A; A is a maximal orthonormal set (hence an orthonormal base, cf. Section 8.11) containing A_0. □

8.4 Hilbert dimension

Theorem 8.13 (Equi-cardinality of orthonormal bases). *All orthonormal bases of a given Hilbert space have the same cardinality.*

Proof. Note first that an orthonormal set A is necessarily linearly independent. Indeed, if the finite linear combination $x := \sum_{k=1}^{n} \lambda_k a_k$ of vectors $a_k \in A$ vanishes, then $\lambda_j = (x, a_j) = (0, a_j) = 0$ for all $j = 1, \ldots, n$.

We consider two orthonormal bases A, B of the Hilbert space X.

Case 1. At least one of the bases is finite.

We may assume that A is finite, say $A = \{a_1, \ldots, a_n\}$. By Property (5) of the orthonormal base A, the vectors $a_1, \ldots, a_n$ span X (in the algebraic sense!), and are linearly independent by the previous remark. This means that $\{a_1, \ldots, a_n\}$ is a base (in the algebraic sense) for the vector space X, and therefore the algebraic dimension of X is n. Since $B \subset X$ is linearly independent (by the previous remark), its cardinality $|B|$ is at most n, that is, $|B| \leq |A|$. In particular, B is finite, and the preceding conclusion applies with B and A interchanged, that is, $|A| \leq |B|$, and the equality $|A| = |B|$ follows.

Case 2. Both bases are infinite.

With notation as in Theorem 8.7, we claim that

$$A = \bigcup_{b \in B} A(b).$$

Indeed, suppose some $a \in A$ is *not* in the shown union. Then $a \notin A(b)$ for all $b \in B$, that is, $(a, b) = 0$ for all $b \in B$. Hence, $a \in B^{\perp} = \{0\}$ (by Property (1) of the orthonormal base B), but this is absurd since a is a unit vector.

By Theorem 8.7, each set $A(b)$ is at most countable; therefore the union above has cardinality $\leq \aleph_0 \times |B|$. Also $\aleph_0 \leq |B|$ (since B is infinite). Hence $|A| \leq |B|^2 = |B|$ (the last equation is a well-known property of *infinite* cardinalities). By symmetry of the present case with respect to A and B, we also have $|B| \leq |A|$, and the equality of the cardinalities follows. □

Definition 8.14. The cardinality of any (hence of all) orthonormal bases of the Hilbert space X is called the *Hilbert dimension of* X, denoted $\dim_H X$.

8.5 Isomorphism of Hilbert spaces

Two Hilbert spaces X and Y are *isomorphic* if there exists an algebraic isomorphism $V : X \to Y$ that preserves the inner product: $(Vp, Vq) = (p, q)$ for all $p, q \in X$ (the same notation is used for the inner product in both spaces).

The map V is necessarily isometric (since it is linear and norm-preserving). Conversely, by the polarization identity (cf. (11) following Definition 1.34), any bijective isometric linear map between Hilbert spaces is an isomorphism (of Hilbert spaces). Such an isomorphism is also called a *unitary equivalence*; accordingly, isomorphic Hilbert spaces are said to be unitarily equivalent.

The isomorphism relation between Hilbert spaces is clearly an equivalence relation. Each equivalence class is completely determined by the Hilbert dimension:

Theorem 8.15. *Two Hilbert spaces are isomorphic iff they have the same Hilbert dimension.*

Proof. If the Hilbert spaces X, Y have the same Hilbert dimension, and A, B are orthonormal bases for X and Y, respectively, then since $|A| = |B|$, we can choose an index set J to index the elements of *both* A and B:

$$A = \{a_j; j \in J\}; \qquad B = \{b_j; j \in J\}.$$

By Property (5) of orthonormal bases, there is a unique continuous linear map $V : X \to Y$ such that $Va_j = b_j$ for all $j \in J$ (namely $Vx = \sum_{j\in J}(x, a_j)b_j$ for all $x \in X$). There is also a unique continuous linear map $W : Y \to X$ such that $Wb_j = a_j$ for all $j \in J$. Clearly, $VW = WV = I$, where I denotes the identity map in both spaces. Therefore, V is bijective, and by Parseval's identity

$$(Vx, Vy) = \left(\sum_{j\in J}(x, a_j)b_j, \sum_{j\in J}(y, a_j)b_j\right) = \sum_{j\in J}(x, a_j)\overline{(y, a_j)} = (x, y)$$

for all $x, y \in X$. Thus V is a Hilbert space isomorphism of X onto Y.

Conversely, suppose the Hilbert spaces X and Y are isomorphic, and let then $V : X \to Y$ be an isomorphism (of Hilbert spaces). If A is an orthonormal base for X, then VA is an orthonormal base for Y. Indeed, VA is orthonormal, because

$$(Vs, Vt) = (s, t) = \delta_{s,t} = \delta_{Vs,Vt}$$

for all $s, t \in A$ (the last equality follows from the injectiveness of V). In order to show completeness, let $y \in (VA)^\perp$ (in Y), and write $y = Vx$ for some $x \in X$ (since V is onto). Then for all $a \in A$

$$(x, a) = (Vx, Va) = (y, Va) = 0,$$

that is, $x \in A^{\perp} = \{0\}$ (by Property (1) of the orthonormal base A). Hence, $y = 0$, and we conclude that VA is indeed an orthonormal base for Y (cf. Section 8.11).

Now by Theorem 8.13 and the fact that $V : A \to VA$ is bijective, $\dim_H Y = |VA| = |A| = \dim_H X$. □

8.6 Direct sums

Let $(X_s, (\cdot,\cdot)_s)_{s\in\mathcal{S}}$ be a family of Hilbert spaces. Let X be the vector space (under pointwise operations) of all $f \in \prod_{s\in\mathcal{S}} X_s$ such that $\mathcal{S}(f) := \{s \in \mathcal{S}; f(s) \neq 0\}$ is at most countable, and

$$\|f\|^2 := \sum_{s\in\mathcal{S}} \|f(s)\|_s^2 < \infty.$$

By the Cauchy–Schwarz inequality in $\mathbb{C}^n$, if $f, g \in X$, the series

$$(f,g) := \sum_{s\in\mathcal{S}} (f(s), g(s))_s$$

converges absolutely (hence unconditionally), and defines an inner product on X with induced norm $\|\cdot\|$. Let $\{f_n\}$ be a Cauchy sequence with respect to this norm. For all $s \in \mathcal{S}$, the inequality $\|f\| \geq \|f(s)\|_s$ $(f \in X)$ implies that $\{f_n(s)\}$ is a Cauchy sequence in the Hilbert space X_s. Let $f(s) := \lim_n f_n(s) \in X_s$. Then $f \in \prod_{s\in\mathcal{S}} X_s$ and $\mathcal{S}(f) \subset \bigcup_n \mathcal{S}(f_n)$ is at most countable. Given $\epsilon > 0$, let $n_0 \in \mathbb{N}$ be such that $\|f_n - f_m\| < \epsilon$ for all $n, m > n_0$. By Fatou's lemma for the counting measure on $\mathcal{S}$, for $n, m > n_0$,

$$\begin{aligned}\|f_n - f\|^2 &= \sum_{s\in\mathcal{S}} \|f_n(s) - f(s)\|_s^2 = \sum_{s\in\mathcal{S}} \liminf_m \|f_n(s) - f_m(s)\|_s^2 \\ &\leq \liminf_m \sum_{s\in\mathcal{S}} \|(f_n - f_m)(s)\|_s^2 = \liminf_m \|f_n - f_m\|^2 \leq \epsilon^2.\end{aligned}$$

This shows that $f = f_n - (f_n - f) \in X$ and $f_n \to f$ in the X-norm, so that X is a Hilbert space. It is usually called *the direct sum of the Hilbert spaces* $(X_s)_{s\in\mathcal{S}}$ and is denoted

$$X = \sum_{s\in\mathcal{S}} \oplus X_s.$$

The elements of X are usually denoted by $\sum_{s\in\mathcal{S}} \oplus x_s$ (rather than the functional notation f). In this section we maintain the preceding notation for simplicity of symbols.

Suppose that for every $s \in \mathcal{S}$, T_s is an element of $B(X_s)$, and that the family $\{T_s\}_{s\in\mathcal{S}}$ is uniformly bounded, that is, $\sup_{s\in\mathcal{S}} \|T_s\|_{B(X_s)} < \infty$. We define a new map $T : X \to X$ by

$$(Tf)(s) := T_s(f(s)) \quad (f \in X, s \in \mathcal{S}).$$

Clearly $Tf \in \prod_{s\in\mathcal{S}} X_s$ and $\mathcal{S}(Tf) \subset \mathcal{S}(f)$ is at most countable. Also,

$$\|Tf\|^2 = \sum_{s\in\mathcal{S}} \|T_s(f(s))\|_s^2 \le \sum_{s\in\mathcal{S}} \|T_s\|_{B(X_s)}^2 \|f(s)\|_s^2 \le \left(\sup_{s\in\mathcal{S}} \|T_s\|\right)^2 \|f\|^2 < \infty.$$

Therefore $Tf \in X$, and T is a bounded linear operator on X with operator norm $\le \sup_{s\in\mathcal{S}} \|T_s\|$. Actually, we have $\|T\| = \sup_{s\in\mathcal{S}} \|T_s\|$. Indeed, for each $s \in \mathcal{S}$ and $v \in X_s$, consider the function $f_{s,v} \in X$ defined by

$$f_{s,v}(s) := v; \qquad f_{s,v}(t) := 0 \quad (t \in \mathcal{S}, t \neq s).$$

Then $\|f_{s,v}\| = \|v\|_s$ and $\|Tf_{s,v}\| = \|T_s v\|_s$. Hence, $\|T\| \ge \|T_s\|$ for all $s \in \mathcal{S}$, and therefore $\|T\| \ge \sup_{s\in\mathcal{S}} \|T_s\|$. Together with the reverse inequality, we obtain the desired equality $\|T\| = \sup_{s\in\mathcal{S}} \|T_s\|$.

The usual notation for the operator T is $\sum_{s\in\mathcal{S}} \oplus T_s$. It is called *the direct sum of the operators* $(T_s)_{s\in\mathcal{S}}$.

8.7 Canonical model

Given any cardinality γ, choose a set J with this cardinality. Consider the vector space $l^2(J)$ of all functions $f : J \to \mathbb{C}$ with $J(f) := \{j \in J; f(j) \neq 0\}$ at most countable and

$$\|f\|_2 := \left(\sum_{j\in J} |f(j)|^2\right)^{1/2} < \infty.$$

The inner product associated with the norm $\|\cdot\|_2$ is defined by the absolutely convergent series

$$(f, g) := \sum_{j\in J} f(j)\overline{g(j)} \quad (f, g \in l^2(J)).$$

The space $l^2(J)$ is the Hilbert direct sum of copies $\mathbb{C}_j$ $(j \in J)$ of the Hilbert space $\mathbb{C}$ (cf. previous section). In particular, $l^2(J)$ is a Hilbert space (it is actually the L^2 space of the measure space $(J, \mathbb{P}(J), \mu)$, with μ the *counting measure*, that is, equal to one on singletons). For each $j \in J$, let $a_j \in l^2(J)$ be defined by $a_j(i) = \delta_{i,j} (i, j \in J)$. Then $A := \{a_j; j \in J\}$ is clearly orthonormal, and $(f, a_j) = f(j)$ for all $f \in l^2(J)$ and $j \in J$. In particular $A^\perp = \{0\}$, so that A is an orthonormal base for $l^2(J)$. Since $|A| = |J| = \gamma$, the space $l^2(J)$ has Hilbert dimension γ. By Theorem 8.15, every Hilbert space with Hilbert dimension γ is isomorphic to the "canonical model" $l^2(J)$.

8.8 Tensor products

8.8.1 An interlude: tensor products of vector spaces

Fix vector spaces U, V over the same field $\mathbb{F}$. Recall that their direct sum is a vector space $U \oplus V$ over $\mathbb{F}$ containing "copies" of U, V (as subspaces) whose sum equals $U \oplus V$ and is direct: $U \cap V = \{0\}$. Similarly, the tensor product vector space $U \otimes V$ will be spanned by formal products of the form $u \otimes v$ in such a way that the map $U \times V \ni (u,v) \to u \otimes v \in U \otimes V$ be bilinear, without any further relation.

An equivalent point of view is that of bases. Recall that a subset $A \subset U$ is called a *Hamel base* of U if it is linearly independent, namely, every finite subset of A is linearly independent, and spanning, that is, that each element of U can be expressed as the linear combination of finitely many elements of A; equivalently, each element of U can be expressed uniquely as the linear combination of finitely many elements of A. If A, B are Hamel bases of U, V, respectively, recall that the disjoint union $A \cup B$ is a Hamel basis of $U \oplus V$; it will follow from our definition that $A \otimes B := \{a \otimes b; a \in A, b \in B\}$ is a Hamel basis of $U \otimes V$.

Consider the set $T(U,V)$ of all formal linear combinations of elements of the cartesian product $U \times V$:

$$\sum_{u \in U, v \in V} \alpha_{u,v}\,[u,v]$$

with $\alpha_{u,v} \in \mathbb{F}$ and the support $\{[u,v] \in U \times V : \alpha_{u,v} \neq 0\}$ being finite. It becomes a vector space over $\mathbb{F}$ when endowed with the following operations:

$$\beta \cdot \sum \alpha_{u,v}\,[u,v] := \sum \beta\alpha_{u,v}\,[u,v]$$

$$\sum \alpha^1_{u,v}\,[u,v] + \sum \alpha^2_{u,v}\,[u,v] := \sum \left(\alpha^1_{u,v} + \alpha^2_{u,v}\right)[u,v]\,.$$

Note that the zero element is $\sum 0\,[u,v]$. For convenience, given $u \in U$ and $v \in V$, write $[u,v]$ for $\sum_{u,v} \delta_{u,u'}\delta_{v,v'}\,[u',v'] \in T(U,V)$, where δ stands for Kronecker's delta.

Define $T_0(U,V)$ to be the linear span in $T(U,V)$ of all elements of the form

$$\left[\sum_{i=1}^n \alpha_i u_i, \sum_{j=1}^m \beta_j v_j\right] - \sum_{i=1}^n \sum_{j=1}^m \alpha_i\beta_j\,[u_i, v_j]$$

with $n, m \in \mathbb{N}$, $u_1, \dots, u_n \in U$, $v_1, \dots, v_m \in V$, and $\alpha_1, \dots, \alpha_n, \beta_1, \dots, \beta_m \in \mathbb{F}$, which equals the linear span of all elements of one of the following forms:

$$\alpha\,[u,v] - [\alpha u, v]\,, \qquad \alpha\,[u,v] - [u, \alpha v]\,,$$
$$[u+u', v] - [u,v] - [u',v]\,, \qquad [u, v+v'] - [u,v] - [u,v']\,,$$

for $u, u' \in U$, $v, v' \in V$, and $\alpha, \beta \in \mathbb{F}$. The quotient vector space $T(U,V)/T_0(U,V)$ is called the *tensor product* of U and V and is denoted by $U \otimes V$. We write $u \otimes v$ for the coset of $[u,v]$ in $U \otimes V$, and call such elements *simple tensors.*

Thus, $U \otimes V$ is spanned by the set of simple tensors and we have the relations

$$\left(\sum_{i=1}^n \alpha_i u_i\right) \otimes \left(\sum_{j=1}^m \beta_j v_j\right) = \sum_{i=1}^n \sum_{j=1}^m \alpha_i\beta_j (u_i \otimes v_j).$$

We list the basic properties of tensor products of vector spaces. Recall that if $\{U_\alpha\}_{\alpha\in A}$ is a family of vector spaces over the same field $\mathbb{F}$, then their direct sum is the set

$$\sum_{\alpha\in A} \oplus U_\alpha := \Big\{ \{u_\alpha\}_{\alpha\in A} \in \prod_{\alpha\in A} U_\alpha; u_\alpha = 0 \text{ for all but finitely many } \alpha \in A \Big\}$$

with the pointwise operations. We will denote $\{u_\alpha\}_{\alpha\in A} \in \sum_{\alpha\in A} \oplus U_\alpha$ by $\sum_{\alpha\in A} \oplus u_\alpha$.

Proposition 8.16. *Let U, V, W, Z be vector spaces over the same field $\mathbb{F}$.*

(i) Commutativity: the map $u\otimes v \to v\otimes u$ ($u \in U$, $v \in V$) extends to a canonical isomorphism between the vector spaces $U\otimes V$ and $V\otimes U$.

(ii) Associativity: the map $(u\otimes v)\otimes w \to u\otimes(v\otimes w)$ ($u \in U$, $v \in V$, $w \in W$) extends to a canonical isomorphism between the vector spaces $(U\otimes V)\otimes W$ and $U\otimes(V\otimes W)$.

(iii) Distributivity: the map $(u+v)\otimes w \to u\otimes w + v\otimes w$ ($u \in U$, $v \in V$, $w \in W$) extends to a canonical isomorphism between the vector spaces $(U\oplus V)\otimes W$ and $(U\otimes W)\oplus(V\otimes W)$.

(iv) For each pair of linear maps $T : U \to W$ and $S : V \to Z$ there exists a unique linear map $T\otimes S : U\otimes V \to W\otimes Z$ that maps $u\otimes v$ to $Tu\otimes Sv$ for each $u \in U$ and $v \in V$.

(v) For every linear functional ϕ on U there exists a unique linear map $\phi\otimes I : U\otimes V \to V$ that maps $u\otimes v$ to $\phi(u)v$ for each $u \in U$ and $v \in V$.

(vi) Let $u_1, \ldots, u_n \in U$ be linearly independent and $v_1, \ldots, v_n \in V$. Then $\sum_{i=1}^n u_i\otimes v_i = 0$ iff $v_1 = v_2 = \ldots = v_n = 0$.

(vii) Let $\{e_\alpha\}_{\alpha\in A}$ be a Hamel base of U. The map $\sum_{\alpha\in A}\oplus V \to U\otimes V$ given by $\sum_{\alpha\in A}\oplus v_\alpha \to \sum_{\alpha\in A} e_\alpha\otimes v_\alpha$ is an isomorphism of vector spaces.

(viii) If $\{e_\alpha\}_{\alpha\in A}$ and $\{f_\beta\}_{\beta\in B}$ are Hamel bases of U and V, respectively, then $\{e_\alpha\otimes f_\beta\}_{(\alpha,\beta)\in A\times B}$ is a Hamel base of $U\otimes V$. In particular, $\dim(U\otimes V) = \dim(U)\cdot\dim(V)$.

(ix) Let $u_1, \ldots, u_n \in U$ and $v_1, \ldots, v_n \in V$. Then $\sum_{i=1}^n u_i\otimes v_i = 0$ iff there exists a matrix $(\alpha_{ij})_{1\le i,j\le n} \in M_n(\mathbb{F})$ such that

$$\sum_{j=1}^n \alpha_{ij}u_j = u_i \quad (\forall 1 \le i \le n), \tag{1}$$

$$\sum_{i=1}^n \alpha_{ij}v_i = 0 \quad (\forall 1 \le j \le n). \tag{2}$$

Proof. (i), (ii), and (iii) are left to the reader as simple exercises.

(iv) Consider the linear map from $T(U,V)$ to $T(W,Z)$ sending $\sum \alpha_{u,v}\,[u,v]$ to $\sum \alpha_{u,v}\,[Tu, Sv]$ (where as usual, the coefficient family $\{\alpha_{u,v}\}_{u\in U, v\in V}$ has finite support). It maps $T_0(U,V)$ into $T_0(W,Z)$. Consequently, it induces a linear map from $U \otimes V = T(U,V)/T_0(U,V)$ to $W \otimes Z = T(W,Z)/T_0(W,Z)$, which has the desired property. Uniqueness is a result of the simple tensors spanning the tensor product.

(v) Follows readily from Part (iv) and the natural identification of $\mathbb{F} \otimes V$ with V.

(vi) Sufficiency is clear. Assume that $\sum_{i=1}^n u_i \otimes v_i = 0$. By linear independence of $u_1, \ldots, u_n$ there exist linear functionals $\phi_1, \ldots, \phi_n$ on U such that $\phi_j(u_i) = \delta_{ij}$ for all $1 \leq i, j \leq n$. Applying $\phi_j \otimes I$ of Part (v) to both sides of the equation $\sum_{i=1}^n u_i \otimes v_i = 0$ gives $v_j = 0$.

(vii) The map is clearly linear and onto. It is injective by Part (vi).

(viii) Straightforward either from Part (vi) or from Part (vii).

(ix) Sufficiency is clear. To prove necessity, assume that $\sum_{i=1}^n u_i \otimes v_i = 0$. Let $\{e_j\}_{j=1}^m$ be a Hamel base of span$\{u_1, \ldots, u_n\}$ in U (so $m \leq n$). Let $B = (\beta_{ik})_{\substack{1 \leq i \leq n \\ 1 \leq k \leq m}} \in M_{n \times m}(\mathbb{F})$ and $C = (\gamma_{ki})_{\substack{1 \leq k \leq m \\ 1 \leq i \leq n}} \in M_{m \times n}(\mathbb{F})$ be such that $u_i = \sum_{k=1}^m \beta_{ik} e_k$ for all $1 \leq i \leq n$ and $e_k = \sum_{i=1}^n \gamma_{ki} u_i$ for all $1 \leq k \leq m$. Set $A = (\alpha_{ij})_{1 \leq i,j \leq n} := BC \in M_n(\mathbb{F})$. Then evidently (1) holds. Furthermore, $\sum_{i=1}^n u_i \otimes v_i = 0$ entails that

$$0 = \sum_{i=1}^n \left(\sum_{k=1}^m \beta_{ik} e_k \right) \otimes v_i = \sum_{k=1}^m e_k \otimes \left(\sum_{i=1}^n \beta_{ik} v_i \right)$$

by the definition of the tensor product. Hence, using Part (vi) we obtain $\sum_{i=1}^n \beta_{ik} v_i = 0$ for every $1 \leq k \leq m$. This implies (2), as for $1 \leq j \leq n$,

$$\sum_{i=1}^n \alpha_{ij} v_i = \sum_{i=1}^n \sum_{k=1}^m \beta_{ik} \gamma_{kj} v_i = \sum_{k=1}^m \gamma_{kj} \left(\sum_{i=1}^n \beta_{ik} v_i \right) = 0. \qquad \square$$

8.8.2 Tensor products of Hilbert spaces

Here we present the Hilbert space counterpart of the tensor product of vector spaces.

Henceforth, the tensor product of two inner product spaces X, Y as complex vector spaces, as defined in Section 8.8.1, will be called their *algebraic* tensor product and will be denoted by $X \otimes_{\text{alg}} Y$.

Proposition 8.17. *Let X, Y be inner product spaces. There exists on $X \otimes_{\text{alg}} Y$ a unique inner product $(\cdot,\cdot)$ satisfying $(x \otimes y, x' \otimes y') := (x, x')_X \cdot (y, y')_Y$ for all $x, x' \in X$ and $y, y' \in Y$.*

Proof. Uniqueness is obvious as the algebraic tensor product is spanned by simple tensors. The existence of a sesquilinear map satisfying the required condition is easy: one defines it first at the level of $T(X,Y)$, and makes sure that the result is zero when either variable belongs to $T_0(X,Y)$.

To prove positive definiteness, note that every element of $X \otimes_{\text{alg}} Y$ can be written as $w = \sum_{i=1}^n \sum_{j=1}^m \alpha_{ij} e_i \otimes f_j$, where $\{e_i\}_{i=1}^n$ and $\{f_j\}_{j=1}^m$ are orthonormal in X and Y, respectively, and the α_{ij} are scalars. We then have $(w,w) = \sum_{i=1}^{n}\sum_{j=1}^{m}|\alpha_{ij}|^2$. This number is evidently non-negative, and equals zero only when $w = 0$. □

Let X, Y be Hilbert spaces. The *tensor product* Hilbert space $X \otimes Y$ is the completion of the inner product space $X \otimes_{\text{alg}} Y$.

Recall the Hilbert space direct sum construction of Section 8.6.

Proposition 8.18. *Let X, Y, Z, W be Hilbert spaces.*

(i) Commutativity: the map $x \otimes y \to y \otimes x$ ($x \in X$, $y \in Y$) extends to a canonical isomorphism between the Hilbert spaces $X \otimes Y$ and $Y \otimes X$.

(ii) Associativity: the map $(x \otimes y) \otimes z \to x \otimes (y \otimes z)$ ($x \in X$, $y \in Y$, $z \in Z$) extends to a canonical isomorphism between the Hilbert spaces $(X \otimes Y) \otimes Z$ and $X \otimes (Y \otimes Z)$.

(iii) Distributivity: the map $(x + y) \otimes z \to x \otimes z + y \otimes z$ ($x \in X$, $y \in Y$, $z \in Z$) extends to a canonical isomorphism between the Hilbert spaces $(X \oplus Y) \otimes Z$ and $(X \otimes Z) \oplus (Y \otimes Z)$.

(iv) Let $\{e_\alpha\}_{\alpha \in A}$ be an orthonormal base of X. The map $\sum_{\alpha \in A} \oplus Y \to X \otimes Y$ given by $\sum_{\alpha \in A} \oplus y_\alpha \to \sum_{\alpha \in A} e_\alpha \otimes y_\alpha$ is an isomorphism of Hilbert spaces.

(v) If $\{e_\alpha\}_{\alpha \in A}$ and $\{f_\beta\}_{\beta \in B}$ are orthonormal bases of X and Y, respectively, then $\{e_\alpha \otimes y_\beta\}_{(\alpha,\beta) \in A \times B}$ is an orthonormal base of $X \otimes Y$. In particular, $\dim_H(X \otimes Y) = \dim_H(X) \cdot \dim_H(Y)$.

(vi) For all $T \in B(X, W)$ and $S \in B(Y, Z)$ there exists a unique map $T \otimes S \in B(X \otimes Y, W \otimes Z)$ that maps $x \otimes y$ to $Tx \otimes Sy$ for each $x \in X$ and $y \in Y$. We have $\|T \otimes S\| = \|T\|\,\|S\|$.

Proof. We leave (i), (ii), and (iii) as exercises.

(iv) Let $\sum_{\alpha \in A} \oplus y_\alpha \in \sum_{\alpha \in A} \oplus Y$. Recall that (it is countably supported and) $\sum_{\alpha \in A} \|y_\alpha\|^2 < \infty$. The family $\{e_\alpha \otimes y_\alpha\}_{\alpha \in A}$ in $X \otimes Y$ is orthogonal and $\sum_{\alpha \in A} \|e_\alpha \otimes y_\alpha\|^2 = \sum_{\alpha \in A} \|y_\alpha\|^2$. This implies that the map $\Phi : \sum_{\alpha \in A} \oplus Y \to X \otimes Y$ given by $\sum_{\alpha \in A} \oplus y_\alpha \to \sum_{\alpha \in A} e_\alpha \otimes y_\alpha$ is well defined and isometric. On the other hand, every simple tensor $x \otimes y$ can be written as $\sum_{\alpha \in A} (x, e_\alpha)\, e_\alpha \otimes y = \sum_{\alpha \in A} e_\alpha \otimes (x, e_\alpha)\, y$, so the image of Φ contains the dense subspace $X \otimes_{\text{alg}} Y$ of $X \otimes Y$. But this image is closed because Φ is isometric and $\sum_{\alpha \in A} \oplus Y$ is a Hilbert space. Hence, Φ is onto.

(v) Follows from Part (iv).

(vi) The linear map from $\sum_{\alpha \in A} \oplus Y$ to $\sum_{\alpha \in A} \oplus Z$ given by $\sum_{\alpha \in A} \oplus y_\alpha \to \sum_{\alpha \in A} \oplus S y_\alpha$ is well defined and bounded by $\|S\|$, because $\sum_{\alpha \in A} \|S y_\alpha\|^2 \leq \|S\|^2 \sum_{\alpha \in A} \|y_\alpha\|^2 = \|S\|^2 \left\|\sum_{\alpha \in A} \oplus y_\alpha\right\|^2$ (in fact, its norm is precisely $\|S\|$). Using the Hilbert space isomorphism of Part (iv) we obtain a map $I \otimes S$ with the desired property. By symmetry, we also get $T \otimes I$. Composing them yields the map $T \otimes S$ and we have $\|T \otimes S\| \leq \|T\|\,\|S\|$. Since $\|x \otimes y\| = \|x\|\,\|y\|$ for all $x \in X$ and $y \in Y$, the reverse inequality $\|T \otimes S\| \geq \|T\|\,\|S\|$ is obvious. □

Exercises

The trigonometric Hilbert basis of L^2

1. Let m denote the normalized Lebesgue measure on $[-\pi, \pi]$. (The integral of $f \in L^1(m)$ over this interval will be denoted by $\int f\,dm$.) Let $e_k(x) = e^{ikx}$ ($k \in \mathbb{Z}$), and denote

$$s_n = \sum_{|k| \le n} e_k \quad (n = 0, 1, 2, \ldots)$$

$$\sigma_n = (1/n) \sum_{j=0}^{n-1} s_j \quad (n \in \mathbb{N}).$$

Note that $\sigma_1 = s_0 = e_0 = 1$, $\{e_k; k \in \mathbb{Z}\}$ is orthonormal in the Hilbert space $L^2(m)$, and

$$\int s_n\,dm = (s_n, e_0) = \sum_{|k| \le n} (e_k, e_0) = 1$$

$$\int \sigma_n\,dm = (1/n) \sum_{j=0}^{n-1} \int s_j\,dm = 1.$$

Prove

(a)

$$s_n(x) = \frac{\cos(nx) - \cos(n+1)x}{1 - \cos x} = \frac{\sin(n+1/2)x}{\sin(x/2)}.$$

(b)

$$\sigma_n(x) = (1/n) \frac{1 - \cos(nx)}{1 - \cos x} = (1/n) \left(\frac{\sin(nx/2)}{\sin(x/2)} \right)^2.$$

(c) $\{\sigma_n\}$ is an "approximate identity" in the sense of Exercise 6, Chapter 4. Consequently, $\sigma_n * f \to f$ uniformly on $[-\pi, \pi]$ if $f \in C([-\pi, \pi])$ and $f(\pi) = f(-\pi)$, and in L^p-metric if $f \in L^p := L^p(m)$ (for each $p \in [1, \infty)$) (cf. Exercise 6, Chapter 4).

(d) Consider the orthogonal projections P and P_n associated with the orthonormal sets $\{e_k; k \in \mathbb{Z}\}$ and $\{e_k; |k| \le n\}$ respectively, and denote $Q_n = (1/n) \sum_{j=0}^{n-1} P_j$ for $n \in \mathbb{N}$.

Terminology: $Pf := \sum_{k \in \mathbb{Z}} (f, e_k) e_k$ is called the *(formal) Fourier series* of f for *any* integrable f (it converges in L^2 if $f \in L^2$); (f, e_k) is the kth *Fourier coefficient* of f; $P_n f$ is the nth *partial sum* of the Fourier series for f; $Q_n f$ is the nth *Cesaro mean* of the Fourier series of f.

Observe that $P_n f = s_n * f$ and $Q_n f = \sigma_n * f$ for any integrable function f. Consequently, $Q_n f \to f$ uniformly in $[-\pi, \pi]$ if $f \in C_T :=$

$\{f \in C([-\pi,\pi]); f(\pi) = f(-\pi)\}$, and in L^p-norm if $f \in L^p$ (for each $p \in [1,\infty)$). If $f \in L^\infty := L^\infty(-\pi,\pi)$, $Q_n f \to f$ in the *weak**-topology on L^∞ (cf. Exercise 6, Chapter 4).

(e) $\{e_k; k \in \mathbb{Z}\}$ is a Hilbert basis for $L^2(m)$. (Note that $Q_n f \in \operatorname{span}\{e_k\}$ and use Part (d).)

Fourier coefficients

2. Use notation as in Exercise 1. Given $k \in \mathbb{Z} \to c_k \in \mathbb{C}$, denote $g_n = \sum_{|k|\le n} c_k e_k$ $(n = 0,1,2,\ldots)$ and $G_n = (1/n)\sum_{j=0}^{n-1} g_j$ $(n \in \mathbb{N})$. Note that $(g_n, e_m) = c_m$ for $n \ge |m|$ and $= 0$ for $n < |m|$, and consequently $(G_n, e_k) = (1 - |k|/n)c_k$ for $n > |k|$ and $= 0$ for $n \le |k|$.

(a) Let $p \in (1,\infty]$. Prove that if

$$M := \sup_n \|G_n\|_p < \infty,$$

then there exists $f \in L^p$ such that $c_k = (f, e_k)$ for all $k \in \mathbb{Z}$, and conversely. (Hint: the ball $\bar{B}(0,M)$ in L^p is *weak**-compact, cf. Theorems 5.24 and 4.6. For the converse, see Exercise 1.)

(b) If $\{G_n\}$ converges in L^1-norm, then there exists $f \in L^1$ such that $c_k = (f, e_k)$ for all $k \in \mathbb{Z}$, and conversely.

(c) If $\{G_n\}$ converges uniformly in $[-\pi,\pi]$, then there exists $f \in C_T$ such that $c_k = (f, e_k)$ for all $k \in \mathbb{Z}$, and conversely.

(d) If $\sup_n \|G_n\|_1 < \infty$, there exists a complex Borel measure μ on $[-\pi,\pi]$ with $\mu(\{\pi\}) = \mu(\{-\pi\})$ (briefly, $\mu \in M_T$) such that $c_k = \int e_k \, d\mu$ for all k, and conversely. (Hint: consider the measures $d\mu_n = G_n \, dm$, and apply Theorems 4.9 and 5.24.)

(e) If $G_n \ge 0$ for all $n \in \mathbb{N}$, there exists a finite *positive* Borel measure μ as in Part (d), and conversely.

Poisson integrals

3. Use notation as in Exercise 1. Let D be the open unit disc in $\mathbb{C}$.

(a) Verify that $(e_1 + z)/(e_1 - z) = 1 + 2\sum_{k\in\mathbb{N}} e_{-k} z^k$ for all $z \in D$, where the series converges absolutely and uniformly in z in any compact subset of D. Conclude that for any complex Borel measure μ on $[-\pi,\pi]$,

$$g(z) := \int \frac{e_1 + z}{e_1 - z} \, d\mu = \mu([-\pi,\pi]) + 2\sum_{k\in\mathbb{N}} c_k z^k, \tag{3}$$

where $c_k = \int e_{-k} \, d\mu$ and integration is over $[-\pi,\pi]$. In particular, g is analytic in D, and if μ is *real*, $\Re g(z) = \int \Re((e_1 + z)/(e_1 - z)) \, d\mu$ is (real) harmonic in D. Verify that the "kernel" in the last integral has the form $P_r(\theta - t)$, where $z = re^{i\theta}$ and

$$P_r(\theta) := \frac{1 - r^2}{1 - 2r\cos\theta + r^2}$$

is the classical *Poisson kernel* (for the disc). Thus

$$(\Re g)(r\,e^{i\theta}) = (P_r * \mu)(\theta) := \int P_r(\theta - t)\,d\mu(t). \tag{4}$$

(b) Let μ be a complex Borel measure on $[-\pi, \pi]$. Then $P_r * \mu$ is a complex harmonic function in D (as a function of $z = r\,e^{i\theta}$). (This is true in particular for $P_r * f$, for any $f \in L^1(m)$.)

(c) Verify that $\{P_r; 0 < r < 1\}$ is an "approximate identity" for L^1 in the sense of Exercise 6, Chapter 4 (with the continuous parameter r instead of the discrete n). Consequently, as $r \to 1$,

(i) if $f \in L^p$ for some $p < \infty$, then $P_r * f \to f$ in L^p-norm;

(ii) if $f \in C_T$, then $P_r * f \to f$ uniformly in $[-\pi, \pi]$;

(iii) if $f \in L^\infty$, then $P_r * f \to f$ in the *weak**-topology on L^∞;

(iv) if $\mu \in M_T$, then $(P_r * \mu)dm \to d\mu$ in the *weak**-topology.

(d) For $\mu \in M_T$, denote $F(t) = \mu([-\pi, t))$ and verify the identity

$$(P_r * \mu)(\theta) = r \int K_r(t) \frac{F(\theta + t) - F(\theta - t)}{2\sin t}\,dt, \tag{5}$$

where

$$K_r(t) = \frac{(1 - r^2)\sin^2 t}{(1 - 2r\cos t + r^2)^2}. \tag{6}$$

Verify that $\{K_r; 0 < r < 1\}$ is an approximate identity for $L^1(m)$ in the sense of Exercise 6, Chapter 4. (Hint: integration by parts.)

(e) Let $G_\theta(t)$ denote the function integrated against $K_r(t)$ in (3). If F is differentiable at the point θ, $G_\theta(\cdot)$ is continuous at 0 and $G_\theta(0) = F'(\theta)$. Conclude from Part (d) that $P_r * \mu \to 2\pi F' (= 2\pi D\mu = d\mu/dm)$ as $r \to 1$ at all points θ where F is differentiable, that is, m-almost everywhere in $[-\pi, \pi]$. (Cf. Theorem 3.28 with $k = 1$ and Exercise 4e, Chapter 3; note that here m is *normalized* Lebesgue measure on $[-\pi, \pi]$.) This is the "radial limit" version of *Fatou's theorem* on "Poisson integrals". (The same conclusion is true with "non-tangential convergence" of $re^{i\theta}$ to points of the unit circle.)

(f) State and prove the analogue of Exercise 2 for the representation of harmonic functions in D as Poisson integrals.

4. *Poisson integrals in the right half-plane.* Let $\mathbb{C}^+$ denote the right half-plane, and

$$P_x(y) := \pi^{-1} \frac{x}{x^2 + y^2} \quad (x > 0; y \in \mathbb{R}).$$

(This is the so-called *Poisson kernel of the right half-plane.*) Prove:

(a) $\{P_x; x > 0\}$ is an approximate identity for $L^1(\mathbb{R})$ (as $x \to 0+$) (cf. Exercise 6, Chapter 4). Consequently, as $x \to 0+$, $P_x * f \to f$ uniformly on $\mathbb{R}$ if $f \in C_c(\mathbb{R})$, and in L^p-norm if $f \in L^p(\mathbb{R})$ $(1 \le p < \infty)$.

(b) $(P_x * f)(y)$ is a harmonic function of (x, y) in $\mathbb{C}^+$.

(c) If $f \in L^p(\mathbb{R})$, then for each $\delta > 0$, $(P_x * f)(y) \to 0$ uniformly for $x \ge \delta$ as $x^2 + y^2 \to \infty$. (Hint: use Holder's inequality with the probability measure $(1/\pi)P_x(y-t)dt$ for x, y fixed.)

(d) If $f \in L^1(dt/(1+t^2))$, then $P_x * f \to f$ as $x \to 0+$ pointwise a.e. on $\mathbb{R}$. (Hint: revisit the argument in Parts (d)–(e) of the preceding exercise, or transform the disc onto the half-plane and use Fatou's theorem for the disc.)

5. Let μ be a complex Borel measure on $[-\pi, \pi]$. Show that

$$\lim_{n\to\infty} \int e_{-n}\, d\mu = 0 \tag{7}$$

iff

$$\lim_{n\to\infty} \int e_{-n} d|\mu| = 0. \tag{8}$$

(Hint: (5) for the measure μ implies (5) for the measure $d\nu = h\, d\mu$ for any trigonometric polynomial h; use a density argument and the relation $d|\mu| = h\, d\mu$ for an appropriate h.)

Divergence of Fourier series

6. Use notation as in Exercise 1. Consider the partial sums $P_n f$ of the Fourier series of $f \in C_T$. Let

$$\phi_n(f) := (P_n f)(0) = (s_n * f)(0) \quad (f \in C_T). \tag{9}$$

Prove:

(a) For each $n \in \mathbb{N}$, ϕ_n is a bounded linear functional on C_T with norm $\|s_n\|_1$ (the $L^1(m)$-norm of s_n). (Hint: $\|\phi_n\| \le \|s_n\|_1$ trivially. Consider real functions $f_j \in C_T$ such that $\|f_j\|_u \le 1$ and $f_j \to \operatorname{sgn} s_n$ a.e., cf. Exercise 9, Chapter 3. Then $\phi_n(f_j) \to \|s_n\|_1$.)

(b) $\lim_n \|s_n\|_1 = \infty$. (Use the fact that the *Dirichlet integral* $\int_0^\infty ((\sin t)/t)dt$ does *not* converge *absolutely*.)

(c) The subspace

$$Z := \{f \in C_T; \sup_n |\phi_n(f)| < \infty\}$$

is of Baire's first category in the Banach space C_T. Conclude that the subspace of C_T consisting of all $f \in C_T$ with convergent Fourier series at 0 is of Baire's first category in C_T. (Hint: assume Z is of Baire's second category in C_T and apply Theorem 6.4 and Parts (a) and (b).)

Fourier coefficients of L^1 functions

7. Use notation as in Exercise 1.

(a) If $f \in L^1 := L^1(m)$, prove that

$$\lim_{|k|\to\infty} (f, e_k) = 0. \tag{10}$$

Hint: if $f = e_n$ for some $n \in \mathbb{Z}$, $(f, e_k) = 0$ for all k such that $|k| > |n|$, and (8) is trivial. Hence, (8) is true for $f \in \text{span}\{e_n; n \in \mathbb{Z}\}$, and the general case follows by density of this span in L^1, cf. Exercise 1, Part (d).

(b) Consider $\mathbb{Z}$ with the discrete topology (this is a locally compact Hausdorff space!), and let $c_0 := C_0(\mathbb{Z})$; cf. Definition 3.23. Consider the map

$$F : f \in L^1 \to \{(f, e_k)\} \in c_0$$

(cf. Part (a)). Then $F \in B(L^1, c_0)$ is one-to-one. (Hint: if $(f, e_k) = 0$ for all $k \in \mathbb{Z}$, then $(f, g) = 0$ for all $g \in \text{span}\{e_k\}$, hence for all $g \in C_T$ by Exercise 1(d), hence for $g = I_E$ for any measurable subset E of $[-\pi, \pi]$; cf. Exercise 9, Chapter 3, and Proposition 1.22.)

(c) Prove that the range of F is of Baire's first category in the Banach space c_0 (in particular, F is *not onto*). (Hint: if the range of F is of Baire's second category in c_0, F^{-1} with domain range F is continuous, by Theorem 6.9. Therefore there exists $c > 0$ such that $\|Ff\|_u \geq c\|f\|_1$ for all $f \in L^1$. Get a contradiction by choosing $f = s_n$; cf. Exercise 6(b).)

Miscellaneous

8. Let $\{a_n\}$ and $\{b_n\}$ be two orthonormal sequences in the Hilbert space X such that

$$\sum_n \|b_n - a_n\|^2 < 1.$$

Prove that $\{a_n\}$ is a Hilbert basis for X iff this is true for $\{b_n\}$.

9. Let $\{a_k\}$ be a Hilbert basis for the Hilbert space X. Define $T \in B(X)$ by $Ta_k = a_{k+1}$, $k \in \mathbb{N}$. Prove:

(a) T is isometric.

(b) $T^n \to 0$ in the weak operator topology.

10. If $\{a_n\}$ is an orthonormal (infinite) sequence in an inner product space, then $a_n \to 0$ weakly (however, $\{a_n\}$ has no strongly convergent subsequence, because $\|a_n - a_m\|^2 = 2$; this shows in particular that the closed unit ball of an infinite dimensional Hilbert space is not stongly compact).

11. Let $\{x_\alpha\}$ be a net in the inner product space X. Then $x_\alpha \to x \in X$ strongly iff $x_\alpha \to x$ weakly and $\|x_\alpha\| \to \|x\|$.

12. Let A be an orthonormal basis for the Hilbert space X. Prove:

(a) If $f : A \to X$ is any map such that $(f(a), a) = 0$ for all $a \in A$, then it does not necessarily follow that $f = 0$ (the zero map on A).

(b) If $T \in B(X)$ is such that $(Tx, x) = 0$ for all $x \in X$, then $T = 0$ (the zero operator).

(c) If $S, T \in B(X)$ are such that $(Tx, x) = (Sx, x)$ for all $x \in X$, then $S = T$.

13. Let X be a Hilbert space, and let $\mathcal{N}$ be the set of normal operators in $B(X)$. Prove that the adjoint operation $T \to T^*$ is continuous on $\mathcal{N}$ in the s.o.t.

14. Let X be a Hilbert space, and $T \in B(X)$. Denote by $P(T)$ and $Q(T)$ the orthogonal projections onto the closed subspaces $\ker T$ and $\overline{TX}$, respectively. Prove:

(a) The complementary orthogonal projections of $P(T)$ and $Q(T)$ are $Q(T^*)$ and $P(T^*)$, respectively.

(b) $P(T^*T) = P(T)$ and $Q(T^*T) = Q(T^*)$.

15. For any (non-empty) set A, denote by $\mathbb{B}(A)$ the C^*-algebra of all bounded complex functions on A with pointwise operations, the involution $f \to \bar{f}$ (complex conjugation), and the supremum norm $\|f\|_u = \sup_A |f|$.

Let A be an orthonormal basis of the Hilbert space X. For each $f \in \mathbb{B}(A)$ and $x \in X$, let

$$T_f x = \sum_{a \in A} f(a)(x, a)a.$$

Prove:

(a) The map $f \to T_f$ is an isometric $*$-isomorphism of the C^*-algebra $\mathbb{B}(A)$ into $B(X)$. (In particular, T_f is a normal operator.)

(b) T_f is selfadjoint (positive, unitary) iff f is real-valued ($f \geq 0, |f| = 1$, respectively).

16. Let $(S, \mathcal{A}, \mu)$ be a σ-finite positive measure space. Consider $L^\infty(\mu)$ as a C^*-algebra with pointwise multiplication and complex conjugation as involution. Let $p \in [1, \infty)$. For each $f \in L^\infty(\mu)$ define

$$T_f g = fg \quad (g \in L^p(\mu)).$$

Prove:

(a) The map $f \to T_f$ is an isometric isomorphism of $L^\infty(\mu)$ into $B(L^p(\mu))$; in case $p = 2$, the map is an isometric $*$-isomorphism of $L^\infty(\mu)$ onto a commutative C^*-algebra of (normal) operators on $L^2(\mu)$.

(b) (Case $p = 2$.) T_f is selfadjoint (positive, unitary) iff f is real-valued ($f \geq 0, |f| = 1$, respectively) almost everywhere.

17. Let X be a Hilbert space. Show that multiplication in $B(X)$ is *not* (jointly) continuous in the w.o.t., even on the norm-closed unit ball $B(X)_1$ of $B(X)$ in the relative w.o.t., unless X is finite dimensional; however, it is continuous on $B(X)_1$ in the relative s.o.t.

18. Use notation as in Exercise 17. Prove that $B(X)_1$ is compact in the w.o.t., but *not* in the s.o.t. (unless X is finite dimensional).

19. Let X, Y be Hilbert spaces, and $T \in B(X, Y)$. Imitate the definition of the Hilbert adjoint of an operator in $B(X)$ to define the (Hilbert) adjoint $T^* \in B(Y, X)$. Observe that T^*T is a positive operator in $B(X)$. Let $\{a_n\}_{n \in \mathbb{N}}$ be an orthonormal basis for X and let $Vx = \{(x, a_n)\}$. We know that V is a Hilbert space isomorphism of X onto l^2. What are V^*, V^{-1}, and V^*V?

20. Let $\{a_n\}_{n \in \mathbb{N}}$ be an orthonormal basis for the Hilbert space X, and let $Q \in B(X)$ be invertible in $B(X)$. Let $b_n = Qa_n$, $n \in \mathbb{N}$. Prove that there exist positive constants A, B such that

$$A \sum |\lambda_k|^2 \leq \|\sum \lambda_k b_k\|^2 \leq B \sum |\lambda_k|^2$$

for all finite sets of scalars λ_k.

21. Let X be a Hilbert space. A sequence $\{a_n\} \subset X$ is *upper (lower) Bessel* if there exists a positive constant B(resp. A) such that

$$\sum |(x, a_n)|^2 \leq B\|x\|^2 \quad (\geq A\|x\|^2, \text{resp.})$$

for all $x \in X$. The sequence is *two-sided Bessel* if it is both upper and lower Bessel (e.g., an orthonormal sequence is upper Bessel with $B = 1$, and is two-sided Bessel with $A = B = 1$ iff it is an orthonormal basis for X).

(a) Let $\{a_n\}$ be an upper Bessel sequence in X, and define V as in Exercise 19. Then $V \in B(X, l^2)$ and $\|V\| \leq B^{1/2}$. On the other hand, for any $\{\lambda_n\} \in l^2$, the series $\sum \lambda_n a_n$ converges in X and its sum equals $V^*\{\lambda_n\}$. The operator $S := V^*V \in B(X)$ is a positive operator with norm $\leq B$.

(b) If $\{a_n\}$ is two-sided Bessel, $S - AI$ is positive, and therefore $\sigma(S) \subset [A, B]$. In particular, S is *onto*. Conclude that every $x \in X$ can be represented as $x = \sum(x, S^{-1}a_n)a_n$ (convergent in X).

Tensor products of vector spaces

Here we use the notation from Section 8.8.1. In particular, the symbol $\otimes$ denotes the tensor product of vector spaces. Also, the letters U, V, W, Z denote vector spaces over the same field $\mathbb{F}$. In addition, we write $L(U, V)$ for the vector space of all linear maps from U to V.

22. A *bilinear map* from $U \times V$ to W is a function $B : U \times V \to W$ satisfying

$$B\left(\sum_{i=1}^{n} \alpha_i u_i, \sum_{j=1}^{m} \beta_j v_j\right) = \sum_{i=1}^{n}\sum_{j=1}^{m} \alpha_i \beta_j B(u_i, v_j)$$

for all $n, m \in \mathbb{N}$, $u_1, \ldots, u_n \in U$, $v_1, \ldots, v_m \in V$, and $\alpha_1, \ldots, \alpha_n, \beta_1, \ldots, \beta_m \in \mathbb{F}$. The set of all such bilinear maps, denoted by $L(U \times V, W)$, becomes a vector space when endowed with the pointwise operations.

Prove that there is a linear isomorphism from the space $L(U \times V, W)$ onto the space $L(U \otimes V, W)$ mapping B in the former to T in the latter given by $T(u \otimes v) = B(u, v)$ for all $u \in U$ and $v \in V$ (T is called the *linearization* of B).

23. Prove that if $T \in L(U, W)$ and $S \in L(V, Z)$ are injective, then the map $T \otimes S \in L(U \otimes V, W \otimes Z)$ of Part (iv) of Proposition 8.16 is also injective.

24. (a) Prove that there exists a unique linear map from $L(U, W) \otimes L(V, Z)$ into $L(U \otimes V, W \otimes Z)$ sending $T \otimes S \in L(U, W) \otimes L(V, Z)$ to the element of $L(U \otimes V, W \otimes Z)$ with the same notation defined by Part (iv) of Proposition 8.16.

(b) Prove that this map is injective.

Tensor products of Hilbert spaces

Here we use the notation from Section 8.8.2. In particular, the symbols $\otimes_{\text{alg}}, \otimes$ denote the tensor product of vector spaces and the tensor product of Hilbert spaces, respectively. Unless otherwise specified, X, Y, Z, W are Hilbert spaces.

25. (a) Let $T \in B(X,Y)$. Prove that the value in $[0,\infty]$ of the positive sum $\sum_{\alpha\in A}\|Te_\alpha\|^2$, with $\{e_\alpha\}_{\alpha\in A}$ being an orthonormal base of X, is independent of the choice of $\{e_\alpha\}_{\alpha\in A}$. If this sum is finite we say that T is a *Hilbert–Schmidt operator* from X to Y. Denote the set of all these operators by $\mathrm{HS}(X,Y)$.

   (b) Prove that $\mathrm{HS}(X,Y)$ is a linear subspace of $B(X,Y)$.

   (c) For $T,S \in \mathrm{HS}(X,Y)$, prove that the sum $\sum_{\alpha\in A}(Te_\alpha, Se_\alpha)$, with $\{e_\alpha\}_{\alpha\in A}$ being an orthonormal base of X, converges, and its sum is independent of the choice of $\{e_\alpha\}_{\alpha\in A}$. Denote this number by $(T,S)_{\mathrm{HS}}$.

   (d) Prove that $(\mathrm{HS}(X,Y),(\cdot,\cdot)_{\mathrm{HS}})$ is a Hilbert space. Explain why the norm $\|\cdot\|_{\mathrm{HS}}$ induced by $(\cdot,\cdot)_{\mathrm{HS}}$ dominates the operator norm: $\|T\|_{B(X,Y)} \leq \|T\|_{\mathrm{HS}}$ for all $T \in \mathrm{HS}(X,Y)$.

   (e) For $x \in X$ and $y \in Y$, let $\theta_{y,x} : X \to Y$ be the linear operator $(\cdot,x)\,y$, that is, $X \ni w \to (w,x)\,y$. Prove that $\mathrm{span}\,\{\theta_{y,x}; x \in X, y \in Y\}$, which equals the set of finite rank operators in $B(X,Y)$, is dense in the Hilbert space $\mathrm{HS}(X,Y)$. Compare Part (d) of Exercise 8 in Chapter 6.

   (f) The *conjugate Hilbert space* $\overline{X}$ of X is defined as follows. Its underlying set is isomorphic to that of X and its elements are denoted by $\overline{x}$, $x \in X$. It becomes a complex vector space by letting $\overline{x}+\overline{y} := \overline{x+y}$ and $\alpha\cdot\overline{x} := \overline{\overline{\alpha}\cdot x}$ for $x,y \in X$ and $\alpha \in \mathbb{C}$. Endowed with the inner product given by $(\overline{x},\overline{y}) := (y,x)$ for $x,y \in X$ it becomes a Hilbert space. Note that the map $\overline{X} \to X^*$ given by $\overline{x} \to (\cdot,x)$ is an isometric isomorphism of Banach spaces.

   Prove that the map that for $x \in X$ and $y \in Y$ takes $\overline{x}\otimes y$ to $\theta_{y,x}$ extends (uniquely) to an isomorphism of the Hilbert spaces $\overline{X}\otimes Y$ and $\mathrm{HS}(X,Y)$.

   (g) Prove that every Hilbert–Schmidt operator from X to Y is compact. (This part is not related to tensor products.)

26. Prove that if $T \in B(X,W)$ and $S \in B(Y,Z)$ are injective, then the map $T\otimes S \in B(X\otimes Y, W\otimes Z)$ of Part (vi) of Proposition 8.18 is also injective.

27. (a) Prove that there is a unique linear map from $B(X,W)\otimes_{\mathrm{alg}} B(Y,Z)$ into $B(X\otimes Y, W\otimes Z)$ sending $T\otimes S \in B(X,W)\otimes_{\mathrm{alg}} B(Y,Z)$ to the element of $B(X\otimes Y, W\otimes Z)$ with the same notation defined by Part (vi) of Proposition 8.18.

   (b) Prove that this map is injective.

28. Let $(X,\mathcal{A},\mu)$ and $(Y,\mathcal{B},\nu)$ be complete σ-finite positive measure spaces. In this exercise we prove that $L^2(\mu)\otimes L^2(\nu)$ is "equal" to $L^2(\mu\times\nu)$.

   For functions $f : X \to \mathbb{C}$ and $g : Y \to \mathbb{C}$, define $f\times g : X\times Y \to \mathbb{C}$ by $(f\times g)(x,y) \to f(x)g(y)$ $(x \in X,\ y \in Y)$. Recall from Exercise 4 of Chapter 2 that if $f \in L^2(X,\mathcal{A},\mu)$ and $g \in L^2(Y,\mathcal{B},\nu)$, then $f\times g \in L^2(X\times Y, \mathcal{A}\times\mathcal{B}, \mu\times\nu)$.

Consider the Hilbert spaces $L^2(\mu) := L^2(X, \mathcal{A}, \mu)$, $L^2(\nu) := L^2(Y, \mathcal{B}, \nu)$ and $L^2(\mu \times \nu) := L^2(X \times Y, \mathcal{A} \times \mathcal{B}, \mu \times \nu)$. Prove that there exists a unique map in $B(L^2(\mu) \otimes L^2(\nu), L^2(\mu \times \nu))$ that sends $f \otimes g$ to $f \times g$ for all $f \in L^2(\mu)$ and $g \in L^2(\nu)$, and that this map is an isomorphism of Hilbert spaces.

(Hint: first prove that the map exists and preserves the inner product, and then use the averages lemma to prove that it is surjective.)

29. Let $(\Omega, \mathcal{A}, \mu)$ be a positive measure space and X be a Hilbert space. In this exercise we introduce the Hilbert space $L^2((\Omega, \mathcal{A}, \mu), X)$ and prove that it is "equal" to $L^2(\Omega, \mathcal{A}, \mu) \otimes X$.

A function $f : \Omega \to X$ is called:

- *weakly measurable* if for every $x^* \in X^*$, the function $x^* \circ f : \Omega \to \mathbb{C}$ is measurable;
- *almost separably-valued* if there exists a null set $E \in \mathcal{A}$ such that $f(\Omega \backslash E)$ is separable in X;
- *strongly measurable* if it is weakly measurable and almost separably-valued.

(a) Prove that if $f : \Omega \to X$ is strongly measurable, then the scalar-valued function $\|f(\cdot)\| : \Omega \to [0, \infty)$ (given by $\omega \to \|f(\omega)\|$) is measurable.

(b) Let $L^2((\Omega, \mathcal{A}, \mu), X)$ be the set of all equivalence classes (under equality a.e.) of strongly measurable functions $f : \Omega \to X$ satisfying $\int_\Omega \|f(\omega)\|^2 \, d\mu(\omega) < \infty$. Prove that $L^2((\Omega, \mathcal{A}, \mu), X)$ is a Hilbert space with respect to the pointwise operations and the inner product

$$(f, g) := \int_\Omega (f(\omega), g(\omega))_X \, d\mu(\omega) \qquad (f, g \in L^2((\Omega, \mathcal{A}, \mu), X))$$

(the fact that this integral exists also requires proof!).
(Hint: mimic the proof of Theorem 1.29.)

(c) For $f : \Omega \to \mathbb{C}$ and $x \in X$, let $fx : \Omega \to X$ be given by $\omega \to f(\omega)x$. Prove that if $f \in L^2(\Omega, \mathcal{A}, \mu)$ then $fx \in L^2((\Omega, \mathcal{A}, \mu), X)$.

(d) Prove that there exists a unique map in $B(L^2(\Omega, \mathcal{A}, \mu) \otimes X, L^2((\Omega, \mathcal{A}, \mu), X))$ that sends $f \otimes x \in L^2(\Omega, \mathcal{A}, \mu) \otimes X$ to $fx \in L^2((\Omega, \mathcal{A}, \mu), X)$ for every $f \in L^2(\Omega, \mathcal{A}, \mu)$ and $x \in X$, and that this map is an isomorphism of Hilbert spaces.

Remark that for two positive measure spaces as in Exercise 28, Exercises 28 and 29 provide two different "realizations" of the tensor product of their L^2-spaces.

9

Integral representation

A well-known elementary fact about any selfadjoint matrix T with complex entries is that it is unitarily equivalent to a diagonal matrix, with the eigenvalues of T as its diagonal entries. Such a diagonal matrix is clearly the linear combination $\sum \lambda_i E_i$, where λ_i are the eigenvalues and E_i are diagonal matrices with 1 at one spot of the diagonal and 0 elsewhere. If Q is the unitary matrix of the equivalence, then the given selfadjoint matrix T is the linear combination $\sum \lambda_i F_i$, where $F_i = QE_iQ^*$ are selfadjoint projections, just like E_i. Shifting to the corresponding operators, we get a version that suggests a straightforward generalization to infinite-dimensional complex Hilbert spaces, with the sum "representation" replaced by an integral representation over the spectrum of T. This chapter is mostly concerned with such representations from a rather general point of view. After a short discussion of spectral measures on a "Banach subspace" Z of a Banach space X and the corresponding spectral integral, we prove the classical *spectral theorem for a normal operator* T on a Hilbert space X, which establishes that every continuous function $f(T)$ of T is the "spectral" integral of f with respect to a unique spectral measure on X supported by the spectrum of T. The spectral measure is then shown to allow a "representation" of $f(T)$ as a "multiplication by f" operator on a suitable direct sum of L^2 spaces. A *renorming method* is used in the general Banach space setting to construct a maximal Banach subspace Z, called the *semi-simplicity space* of the given operator T, on which T admits an operational calculus $f \to f(T)$ with continuous functions f, thus generating an integral representation of $f(T)$ with respect to a spectral measure on Z when X is reflexive.

The more restrictive *analytic operational calculus* for elements of an arbitrary Banach algebra is then defined and studied, and is used to develop the *Riesz–Schauder theory* of compact operators in arbitrary Banach spaces.

Introduction to Modern Analysis. Second Edition. Shmuel Kantorovitz and Ami Viselter, Oxford University Press.
 DOI: 10.1093/oso/9780192849540.003.0009

9.1 Spectral measure on a Banach subspace

Let X be a Banach space. A *Banach subspace* Z of X is a subspace of X in the algebraic sense, which is a Banach space for a norm $\|\cdot\|_Z$ *larger than or equal to the given norm* $\|\cdot\|$ *of* X. Clearly, if Z and X are Banach subspaces of each other, then they coincide as Banach spaces (with equality of norms).

Let K be a compact subset of the complex plane $\mathbb{C}$, and let $C(K)$ be the Banach algebra of all complex continuous functions on K with the supremum norm $\|f\| := \sup_K |f|$.

Let $T \in B(X)$, and suppose Z is a T-invariant Banach subspace of X, such that $T|_Z$, the restriction of T to Z, belongs to $B(Z)$.

A *(contractive) C(K)-operational calculus for T on Z* is a *(contractive) C(K)-operational calculus for* $T|_Z$ in $B(Z)$, that is, a (norm-decreasing) continuous algebra-homomorphism $\tau : C(K) \to B(Z)$ such that $\tau(f_0) = I|_Z$ and $\tau(f_1) = T|_Z$, where $f_k(\lambda) = \lambda^k, k = 0, 1$. When such τ exists, we say that T is *of (contractive) class C(K) on Z* (or that $T|_Z$ is of (contractive) class $C(K)$).

If the complex number β is not in K, the function $g_\beta(\lambda) = (\beta - \lambda)^{-1}$ belongs to $C(K)$ and $(\beta - \lambda)g_\beta(\lambda) = 1$ on K. Since τ is an algebra homomorphism of $C(K)$ into $B(Z)$, it follows that $(\beta I - T)|_Z$ has the inverse $\tau(g_\beta)$ in $B(Z)$. In particular, $\sigma_{B(Z)}(T|_Z) \subset K$.

Let f be any rational function with poles off K. Write

$$f(\lambda) = \alpha \prod_{k,j} (\alpha_k - \lambda)(\beta_j - \lambda)^{-1},$$

where $\alpha \in \mathbb{C}, \alpha_k$ are the zeroes of f and β_j are its poles (reduced decomposition). Since τ is an algebra homomorphism, $\tau(f)$ is uniquely determined as

$$\tau(f) = \alpha \prod_{k,j} (\alpha_k I - T)|_Z (\beta_j I - T)|_Z^{-1}. \tag{0}$$

If K has planar Lebesgue measure zero, the rational functions with poles off K are dense in $C(K)$, by the Hartogs–Rosenthal theorem (cf. Exercise 2). The continuity of τ implies then that the $C(K)$-operational calculus for T on Z is *unique* (when it exists). The operator $\tau(f) \in B(Z)$ is usually denoted by $f(T|_Z)$, for $f \in C(K)$.

A *spectral measure on* Z is a map E of the Borel algebra $\mathcal{B}(\mathbb{C})$ of $\mathbb{C}$ into $B(Z)$, such that:

(1) For each $x \in Z$ and $x^* \in X^*$, $x^*E(\cdot)x$ is a regular complex Borel measure; and

(2) $E(\mathbb{C}) = I|_Z$, and $E(\delta \cap \epsilon) = E(\delta)E(\epsilon)$ for all $\delta, \epsilon \in \mathcal{B}(\mathbb{C})$.

The spectral measure E is *contractive* if $\|E(\delta)\|_{B(Z)} \leq 1$ for all $\delta \in \mathcal{B}(\mathbb{C})$.

By Property (2), $E(\delta)$ is a projection in Z; therefore a contractive spectral measure satisfies $\|E(\delta)\|_{B(Z)} = 1$ whenever $E(\delta) \neq 0$. The closed subspaces $E(\delta)Z$ of Z satisfy the relation

$$E(\delta)Z \cap E(\epsilon)Z = E(\delta \cap \epsilon)Z$$

for all $\delta, \epsilon \in \mathcal{B}(\mathbb{C})$. In particular, $E(\delta)Z \cap E(\epsilon)Z = \{0\}$ if $\delta \cap \epsilon = \emptyset$. Therefore, for any partition $\{\delta_k; k \in \mathbb{N}\}$ of $\mathbb{C}, Z$ has the direct sum decomposition $Z = \sum_k \oplus E(\delta_k)Z$ (cf. Property (2) of E). This is equivalent to the decomposition $I|_Z = \sum_k E(\delta_k)$, with projections $E(\delta_k) \in B(Z)$ such that $E(\delta_k)E(\delta_j) = 0$ for $k \neq j$. For that reason, E is also called a *resolution of the identity on* Z.

Property (1) of E together with Theorem 1.43 and Corollary 6.7 imply that $E(\cdot)x$ is a Z-valued additive set function and

$$M_x := \sup_{\delta \in \mathcal{B}(\mathbb{C})} \|E(\delta)x\| < \infty \tag{1}$$

for each $x \in Z$.

9.2 Integration

If μ is a *real* Borel measure on $\mathbb{C}$ (to fix the ideas), and $\{\delta_k\}$ is a partition of $\mathbb{C}$, let $J = \{k; \mu(\delta_k) > 0\}$. Then

$$\sum_k |\mu(\delta_k)| = \sum_{k \in J} \mu(\delta_k) - \sum_{k \in \mathbb{N}-J} \mu(\delta_k)$$

$$= \mu\Big(\bigcup_{k \in J} \delta_k\Big) - \mu\Big(\bigcup_{k \in \mathbb{N}-J} \delta_k\Big) \leq 2M,$$

where $M := \sup_{\delta \in \mathcal{B}(\mathbb{C})} |\mu(\delta)| (\leq \|\mu\|$, the total variation norm of μ).

If μ is a complex Borel measure, apply the shown inequality to the real Borel measures $\Re\mu$ and $\Im\mu$ to conclude that $\sum_k |\mu(\delta_k)| \leq 4M$ for all partitions, hence $\|\mu\| \leq 4M$.

By (1) of Section 9.1, for each $x^* \in X^*$ and $x \in Z$,

$$\sup_{\delta \in \mathcal{B}(\mathbb{C})} |x^* E(\delta)x| \leq M_x \|x^*\|.$$

Therefore

$$\|x^* E(\cdot)x\| \leq 4M_x \|x^*\|. \tag{1}$$

The Banach algebra of all bounded complex Borel functions on $\mathbb{C}$ or $\mathbb{R}$ is denoted by $\mathbb{B}(\mathbb{C})$ or $\mathbb{B}(\mathbb{R})$, respectively (briefly, $\mathbb{B}$); $\mathbb{B}_0(\mathbb{C})$ and $\mathbb{B}_0(\mathbb{R})$ (briefly $\mathbb{B}_0$) are the respective dense subalgebras of simple Borel functions. The norm on $\mathbb{B}$ is the supremum norm (denoted $\|\cdot\|$).

Integration with respect to the vector measure $E(\cdot)x$ is defined on simple Borel functions as in Definition 1.12. It follows from (1) that for $f \in \mathbb{B}_0$, $x \in Z$, and $x^* \in X^*$,

$$\left| x^* \int_{\mathbb{C}} f \, dE(\cdot)x \right| = \Big| \int_{\mathbb{C}} f \, dx^* E(\cdot)x \Big| \leq \|f\| \|x^* E(\cdot)x\| \leq 4M_x \|f\| \|x^*\|.$$

Therefore

$$\Big\| \int_{\mathbb{C}} f \, dE(\cdot)x \Big\| \leq 4M_x \|f\|,$$

and we conclude that the map $f \to \int_{\mathbb{C}} f dE(\cdot)x$ is a continuous linear map from $\mathbb{B}_0$ to X with norm $\leq 4M_x$. It extends uniquely (by density of $\mathbb{B}_0$ in $\mathbb{B}$) to a continuous linear map (same notation!) of $\mathbb{B}$ into X, with the same norm ($\leq 4M_x$).

It follows clearly from the definition that the vector $\int_{\mathbb{C}} f\, dE(\cdot)x$ (belonging to X, and not necessarily in Z!) satisfies the relation

$$x^* \int_{\mathbb{C}} f\, dE(\cdot)x = \int_{\mathbb{C}} f\, dx^* E(\cdot)x \tag{2}$$

for all $f \in \mathbb{B}(\mathbb{C}), x \in Z$, and $x^* \in X^*$.

As for scalar measures, the *support of* E, $\operatorname{supp} E$, is the complement in $\mathbb{C}$ of the union of all open set δ such that $E(\delta) = 0$. The support of each complex measure $x^*E(\cdot)x$ is then contained in the support of E. One has

$$\int_{\mathbb{C}} f\, dE(\cdot)x = \int_{\operatorname{supp} E} f\, dE(\cdot)x$$

for all $f \in \mathbb{B}$ and $x \in Z$ (where as usual the right-hand side is defined as the integral over $\mathbb{C}$ of $f\chi_{\operatorname{supp} E}$, and χ_V denotes *in this chapter* the indicator of $V \subset \mathbb{C}$) (cf. (2) of Section 3.5). The right-hand side of the last equation can be used to extend the definition of the integral to complex Borel function *that are only bounded on* $\operatorname{supp} E$.

In case $Z = X$, it follows from (1) of Section 9.1 and the uniform boundedness theorem that $M := \sup_{\delta \in \mathcal{B}(\mathbb{C})} \|E(\delta)\| < \infty, M_x \leq M\|x\|$, and (1) takes the form $\|x^*E(\cdot)x\| \leq 4M\|x\|\|x^*\|$ for all $x \in X$ and $x^* \in X^*$. It then follows that the *spectral integral* $\int_{\mathbb{C}} f\, dE$, *defined by*

$$\left(\int_{\mathbb{C}} f\, dE\right) x := \int_{\mathbb{C}} f\, dE(\cdot)x \quad (f \in \mathbb{B}) \tag{3}$$

belongs to $B(X)$ and has operator norm $\leq 4M\|f\|$.

If X is a *Hilbert space* and $Z = X$, one may use the Riesz representation for X^* to express Property (1) of E and Relation (2) in the following form:

For each $x, y \in X$, $(E(\cdot)x, y)$ is a regular complex measure.

For each $x, y \in X$ and $f \in \mathbb{B}(\mathbb{C})$,

$$\left(\int_{\mathbb{C}} f\, dE(\cdot)x, y\right) = \int_{\mathbb{C}} f\, d(E(\cdot)x, y). \tag{4}$$

In the Hilbert space context, it is particularly significant to consider *selfadjoint* spectral measures on X, $E(\cdot)$, that is, $E(\delta)^* = E(\delta)$ for all $\delta \in \mathcal{B}(\mathbb{C})$. The operators $E(\delta)$ are then *orthogonal projections*, and any partition $\{\delta_k; k \in \mathbb{N}\}$ gives an *orthogonal decomposition* $X = \sum_k \oplus E(\delta_k)X$ into mutually orthogonal closed subspaces (by Property (2) of $E(\cdot)$). Equivalently, the identity operator is the sum of the *mutually orthogonal projections* $E(\delta_k)$ (the adjective "orthogonal" is transferred from the subspace to the corresponding projection). For this reason, E is also called a (selfadjoint) *resolution of the identity*.

9.3 Case $Z = X$

The relationship between $C(K)$-operational calculi and spectral measures is especially simple in case $Z = X$.

Theorem 9.1.

(1) *Let E be a spectral measure on the Banach space X, supported by the compact subset K of $\mathbb{C}$, and let $\tau(f)$ be the associated spectral integral, for $f \in \mathbb{B} := \mathbb{B}(\mathbb{C})$. Then $\tau : \mathbb{B} \to B(X)$ is a continuous representation of $\mathbb{B}$ on X with norm $\le 4M$. The restriction of τ to (Borel) functions continuous on K defines a $C(K)$-operational calculus for $T := \tau(f_1) = \int_{\mathbb{C}} \lambda\, dE(\lambda)$.*

If X is a Hilbert space and the spectral measure E is selfadjoint, then τ is a norm-decreasing $*$-representation of $\mathbb{B}$, and $\tau|_{C(K)}$ is a contractive $C(K)$-operational calculus on X for T (sending adjoints to adjoints).

(2) *Conversely, Let τ be a $C(K)$-operational calculus for a given operator T on the reflexive Banach space X. Then there exists a unique spectral measure E commuting with T with support in K, such that $\tau(f) = \int_{\mathbb{C}} f\, dE$ for all $f \in C(K)$ (and the spectral integral on the right-hand side extends τ to a continuous representation of $\mathbb{B}$ on X).*

If X is a Hilbert space and τ sends adjoints to adjoints, then E is a contractive selfadjoint spectral measure on X.

Proof.

(1) A calculation shows that Property (2) of E implies the multiplicativity of τ on $\mathbb{B}_0$. Since $\tau : \mathbb{B} \to B(X)$ was shown to be linear and continuous (with norm at most $4M$), it follows from the density of $\mathbb{B}_0$ in $\mathbb{B}$ that τ is a continuous algebra homomorphism of $\mathbb{B}$ into $B(X)$ and $\tau(1) = E(\mathbb{C}) = I$.

If X is a Hilbert space and E is selfadjoint, then τ restricted to $\mathbb{B}_0$ sends adjoints to adjoints, because if $f = \sum \alpha_k \chi_{\delta_k}$, then $\tau(\bar{f}) = \sum \overline{\alpha_k} E(\delta_k) = [\sum \alpha_k E(\delta_k)]^* = \tau(f)^*$. By continuity of τ and of the involutions, and by density of $\mathbb{B}_0$ in $\mathbb{B}$, $\tau(\bar{f}) = \tau(f)^*$ for all $f \in \mathbb{B}$.

Let $\{\delta_k\}$ be a partition of $\mathbb{C}$. For any $x \in X$, the sequence $\{E(\delta_k)x\}$ is orthogonal with sum equal to x. Therefore

$$\sum_k \|E(\delta_k)x\|^2 = \|x\|^2. \tag{1}$$

By Schwarz's inequality for X and for l^2, we have for all $x, y \in X$ (since $E(\delta_k)$ are selfadjoint projections):

$$\sum_k |(E(\delta_k)x, y)| = \sum_k |(E(\delta_k)x, E(\delta_k)y)| \le \sum_k \|E(\delta_k)x\| \|E(\delta_k)y\|$$

$$\le \left(\sum_k \|E(\delta_k)x\|^2 \right)^{1/2} \left(\sum_k \|E(\delta_k)y\|^2 \right)^{1/2} = \|x\| \|y\|.$$

Hence

$$\|(E(\cdot)x,y)\| \le \|x\|\|y\| \quad (x,y \in X). \tag{2}$$

Therefore,

$$|(\tau(f)x,y)| = \left| \int_{\mathbb{C}} f\, d(E(\cdot)x,y) \right| \le \|f\|\|x\|\|y\|,$$

that is, $\|\tau(f)\| \le \|f\|$ for all $f \in \mathbb{B}$.

(2) Let τ be a $C(K)$-operational calculus on X for $T \in B(X)$. For each $x \in X$ and $x^* \in X^*$, $x^*\tau(\cdot)x$ is a continuous linear functional on $C(K)$ with norm $\le \|\tau\|\|x\|\|y\|$ (where $\|\tau\|$ denotes the norm of the bounded linear map $\tau : C(K) \to B(X)$). By the Riesz representation theorem, there exists a unique regular complex Borel measure $\mu = \mu(\cdot; x, x^*)$ on $\mathcal{B}(\mathbb{C})$, with support in K, such that

$$x^*\tau(f)x = \int_K f\, d\mu(\cdot; x, x^*) \tag{3}$$

and

$$\|\mu(\cdot; x, x^*)\| \le \|\tau\|\|x\|\|x^*\| \tag{4}$$

for all $f \in C(K), x \in X$, and $x^* \in X^*$.

For each fixed $\delta \in \mathcal{B}(\mathbb{C})$ and $x \in X$, it follows from the uniqueness of the Riesz representation, the linearity of the left-hand side of (3) with respect to x^*, and (4), that the map $\mu(\delta; x, \cdot)$ is a continuous linear functional on X^*, with norm $\le \|\tau\|\|x\|$. Since X is assumed reflexive, there exists a unique vector in X, which we denote $E(\delta)x$, such that

$$\mu(\delta; x, x^*) = x^*E(\delta)x.$$

A routine calculation using the uniqueness properties mentioned and the linearity of the left-hand side of (3) with respect to x, shows that $E(\delta) : X \to X$ is linear. By (4), $E(\delta)$ is bounded with operator norm $\le \|\tau\|$. By definition, E satisfies Property (1) of spectral measures on X, and has support in K. Therefore, the integral $\int_{\mathbb{C}} f\, dE$ makes sense (see mentioned construction) for any Borel function $f : \mathbb{C} \to \mathbb{C}$ bounded on K, and (by (2) of Section 9.2 and (3))

$$\tau(f) = \int_{\mathbb{C}} f\, dE \quad (f \in C(K)). \tag{5}$$

For all $f, g \in C(K)$ and $x \in X$,

$$\int_K f\, dE(\cdot)\tau(g)x = \tau(f)\tau(g)x = \tau(g)\tau(f)x = \int_K f\, d\tau(g)E(\cdot)x$$
$$= \tau(fg)x = \int_K fg\, dE(\cdot)x.$$

The uniqueness of the Riesz representation implies that $E(\cdot)\tau(g) = \tau(g)E(\cdot)$ (in particular, E commutes with $\tau(f_1) = T$) and $dE(\cdot)\tau(g)x = g\, dE(\cdot)x$. Therefore, for all $\delta \in \mathcal{B}(\mathbb{C})$ and $g \in C(K)$,

$$\int_K g\,dE(\cdot)E(\delta)x = \tau(g)E(\delta)x = E(\delta)\tau(g)x$$

$$= \int_K \chi_\delta\,dE(\cdot)\tau(g)x = \int_K \chi_\delta g\,dE(\cdot)x. \tag{6}$$

By uniqueness of the Riesz representation, we get $dE(\cdot)E(\delta)x = \chi_\delta\,dE(\cdot)x$. Thus, for all $\epsilon, \delta \in \mathcal{B}(\mathbb{C})$ and $x \in X$,

$$E(\epsilon)E(\delta)x = \int_K \chi_\epsilon\,dE(\cdot)E(\delta)x = \int_K \chi_\epsilon\chi_\delta\,dE(\cdot)x$$

$$= \int_K \chi_{\epsilon\cap\delta}\,dE(\cdot)x = E(\epsilon\cap\delta)x.$$

Taking $f = f_0(= 1)$ in (5), we get $E(\mathbb{C}) = \tau(f_0) = I$, so that E satisfies Property (2) of spectral measures. Relation (5) provides the wanted relation between the operational calculus and the spectral measure E. The uniqueness of E follows from Property (1) of spectral measures and the uniqueness of the Riesz representation. Finally, if X is a Hilbert space and τ sends adjoints to adjoints, then by (4) of Section 9.2

$$\int_K f\,d(E(\cdot)x, y) = (\tau(f)x, y) = (x, \tau(f)^*y)$$

$$= (x, \tau(\bar{f})y) = \overline{(\tau(\bar{f})y, x)} = \overline{\int_K \bar{f}\,d(E(\cdot)y, x)}$$

$$= \int_K f\,d\overline{(E(\cdot)y, x)} = \int_K f\,d(x, E(\cdot)y)$$

for all $f \in C(K)$ and $x, y \in X$. Therefore (by uniqueness of the Riesz representation), $(E(\delta)x, y) = (x, E(\delta)y)$ for all δ, that is, E is a *selfadjoint* spectral measure. By (2), it is necessarily contractive. □

Terminology 9.2. Given a spectral measure E on X with compact support, the bounded operator $T := \int_{\mathbb{C}} \lambda\,dE(\lambda)$ is the associated *scalar operator*. By Theorem 9.1, T is of class $C(K)$ on X for any compact set K containing the support of E, and $\tau : f \to \int_{\mathbb{C}} f\,dE$ ($f \in C(K)$) is a $C(K)$-operational calculus for T. Conversely, if X is reflexive, then any operator of class $C(K)$, for a given compact set K, is a scalar operator (the scalar operator associated with the spectral measure E in Theorem 9.1, Part 2).

If E and F are spectral measures on X with support in the compact set K, and their associated scalar operators coincide, it follows from Theorem 9.1 (Part 1) that their associated spectral integrals coincide for all rational functions with poles off K. In case K has planar Lebesgue measure zero, these rational functions are dense in $C(K)$ (by the Hartogs–Rosenthal theorem), and the continuity of the spectral integrals on $C(K)$ (cf. Theorem 9.1) implies that they coincide on $C(K)$. It then follows that $E = F$, by the uniqueness of the Riesz representation. The uniqueness of the spectral measure associated with a scalar operator can be proved without the "planar measure zero" condition on K, but this will not be done here. The unique spectral measure with compact support associated with the scalar operator T is called the *resolution of the identity for T.*

For each $\delta \in \mathcal{B}(\mathbb{C})$, the projection $E(\delta)$ commutes with T (cf. Theorem 9.1, Part 2), and therefore the closed subspace $E(\delta)X$ reduces T. If μ is a complex number not in the closure $\bar{\delta}$ of δ, the function $h(\lambda) := \chi_\delta(\lambda)/(\mu - \lambda)$ belongs to $\mathbb{B}$ ($\|h\| \leq 1/\text{dist}(\mu, \delta)$) and $(\mu - \lambda)h(\lambda) = \chi_\delta(\lambda)$. Applying the $\mathbb{B}$-operational calculus, we get $(\mu I - T)\tau(h) = E(\delta)$. Restricting to $E(\delta)X$, this means that $\mu \in \rho(T|_{E(\delta)X})$. Hence

$$\sigma(T|_{E(\delta)X}) \subset \bar{\delta} \quad (\delta \in \mathcal{B}(\mathbb{C})). \tag{7}$$

Remark 9.3. A bounded operator T for which there exists a spectral measure E on X, commuting with T and with support in $\sigma(T)$, such that (7) is satisfied, is called a *spectral operator*. It turns out that E is uniquely determined; it is called the resolution of the identity for T, as before. If S is the scalar operator associated with E, it can be proved that T has the (unique) *Jordan decomposition* $T = S + N$ *with N quasi-nilpotent commuting with S*. Conversely, any operator with such a Jordan decomposition is spectral. The Jordan canonical form for complex matrices establishes the fact that *every* linear operator on $\mathbb{C}^n$ is spectral (for any finite n).

Combining 7.19 and Theorem 9.1, we obtain the following.

9.4 The spectral theorem for normal operators

Theorem 9.4. *Let T be a normal operator on the Hilbert space X, and let $\tau : f \to f(T)$ be its $C(\sigma(T))$-operational calculus (cf. Terminology 7.19). Then there exists a unique selfadjoint spectral measure E on X, commuting with T and with support in $\sigma(T)$, such that*

$$f(T) = \int_{\sigma(T)} f \, dE$$

for all $f \in C(\sigma(T))$. The spectral integral above extends the $C(\sigma(T))$-operational calculus to a norm-decreasing $*$-representation $\tau : f \to f(T)$ of $\mathbb{B}(\mathbb{C})$ on X.

Remark 9.5. We can use Terminology 9.2 to restate Theorem 9.4 in the form: normal operators are scalar, and their resolutions of the identity are selfadjoint, with support in the spectrum. The converse is also true, since the map τ associated with a selfadjoint spectral measure with compact support is a $*$-representation. If we consider spectral measures that are not necessarily selfadjoint, it can be shown that a bounded operator T in Hilbert space is scalar *if and only if it is similar to a normal operator*, that is, iff there exists a non-singular $Q \in B(X)$ such that QTQ^{-1} is normal.

Theorem 9.6. *Let T be a normal operator on the Hilbert space X, and let E be its resolution of the identity. Then*

(1) *For each $x \in X$, the measure $(E(\cdot)x, x) = \|E(\cdot)x\|^2$ is a positive regular Borel measure bounded by $\|x\|^2$, and*

$$\|f(T)x\|^2 = \int_{\sigma(T)} |f|^2 \, d(E(\cdot)x, x)$$

for all Borel functions f bounded on $\sigma(T)$.

(2) *If $\{\delta_k\}$ is a sequence of mutually disjoint Borel subsets of $\mathbb{C}$ with union δ, then for all $x \in X$,*

$$E(\delta)x = \sum_k E(\delta_k)x,$$

where the series converges strongly in X (this is "strong σ-additivity" of E).

(3) *If $\{f_n\}$ is a sequence of Borel functions, uniformly bounded on $\sigma(T)$, such that $f_n \to f$ pointwise on $\sigma(T)$, then $f_n(T) \to f(T)$ in the s.o.t.*

(4) *$\operatorname{supp} E = \sigma(T)$.*

Proof. (1) Since $E(\delta)$ is a self-adjoint projection for each $\delta \in \mathcal{B}(\mathbb{C})$, we have

$$(E(\delta)x, x) = (E(\delta)^2 x, x) = (E(\delta)x, E(\delta)x) = \|E(\delta)x\|^2 \le \|x\|^2,$$

and the stated properties of the measure follow from Property (1) of spectral measures in Hilbert space.

If f is a Borel function bounded on $\sigma(T)$, we have for all $x \in X$

$$\begin{aligned}\|f(T)x\|^2 &= (f(T)x, f(T)x) = (f(T)^* f(T)x, x) = ((\bar{f}f)(T)x, x) \\ &= \int_{\sigma(T)} |f|^2 \, d\|E(\cdot)x\|^2.\end{aligned}$$

(2) The sequence $\{E(\delta_k)x\}$ is orthogonal. Therefore, for each $n \in \mathbb{N}$,

$$\sum_{k=1}^{n} \|E(\delta_k)x\|^2 = \|\sum_{k=1}^{n} E(\delta_k)x\|^2 = \|E(\delta)x\|^2 \le \|x\|^2.$$

This shows that the series $\sum_k \|E(\delta_k)x\|^2$ converges. Therefore

$$\|\sum_{k=m}^{n} E(\delta_k)x\|^2 = \sum_{k=m}^{n} \|E(\delta_k)x\|^2 \to 0$$

when $m, n \to \infty$. By completeness of X, it follows that $\sum_k E(\delta_k)x$ converges strongly in X. Hence, for all $y \in X$, Property (1) of spectral measures gives

$$(E(\delta)x, y) = \sum_k (E(\delta_k)x, y) = \left(\sum_k E(\delta_k)x, y\right),$$

and Part (2) follows.

(3) For all $x \in X$, we have by Part (1)

$$\|f_n(T)x - f(T)x\|^2 = \|(f_n - f)(T)x\|^2 = \int_{\sigma(T)} |f_n - f|^2 \, d\|E(\cdot)x\|^2,$$

and Part (3) then follows from Lebesgue's dominated convergence theorem for the *finite* positive measure $\|E(\cdot)x\|^2$.

(4) We already know that $\operatorname{supp} E \subset \sigma(T)$. So we need to prove that $(\operatorname{supp} E)^c \subset \rho(T)$. Let $\mu \in (\operatorname{supp} E)^c$. By definition of the support, there exists an open neighborhood δ of μ such that $E(\delta) = 0$. Then $r := \operatorname{dist}(\mu, \delta^c) > 0$. Let $g(\lambda) := \chi_{\delta^c}(\lambda)/(\mu - \lambda)$. Then $g \in \mathbb{B}$ (in fact, $\|g\| = 1/r < \infty$), and since $(\mu - \lambda)g(\lambda) = \chi_{\delta^c}(\lambda)$, the operational calculus for T implies that $(\mu I - T)g(T) = g(T)(\mu I - T) = E(\mathbb{C}) - E(\delta) = I$. Hence $\mu \in \rho(T)$. □

Note that the proof of Part (4) of Theorem 9.6 is valid for any scalar operator in Banach space.

9.5 Parts of the spectrum

Definition 9.7. Let X be a Banach space, and $T \in B(X)$.

(1) The set of all $\lambda \in \mathbb{C}$ such that $\lambda I - T$ is not injective is called the *point spectrum* of T, and is denoted by $\sigma_p(T)$; any $\lambda \in \sigma_p(T)$ is called an *eigenvalue* of T, and the non-zero subspace $\ker(\lambda I - T)$ is the corresponding *eigenspace*. The non-zero vectors x in the eigenspace are the *eigenvectors* of T corresponding to the eigenvalue λ (briefly, the λ-eigenvectors of T): they are the non-trivial solutions of the equation $Tx = \lambda x$.

(2) The set of all $\lambda \in \mathbb{C}$ such that $\lambda I - T$ is injective, and $\lambda I - T$ *has range dense but not equal to* X, is called the *continuous spectrum* of T, and is denoted by $\sigma_c(T)$.

(3) The set of all $\lambda \in \mathbb{C}$ such that $\lambda I - T$ is injective and $\lambda I - T$ *has range not dense in* X is called the *residual spectrum* of T, and is denoted by $\sigma_r(T)$.

Clearly, the three sets defined here are mutually disjoint subsets of $\sigma(T)$. If λ is *not* in their union, then $\lambda I - T$ is bijective, hence invertible in $B(X)$ (cf. Corollary 6.11). This shows that

$$\sigma(T) = \sigma_p(T) \cup \sigma_c(T) \cup \sigma_r(T).$$

Theorem 9.8. *Let X be a Hilbert space and $T \in B(X)$. Then:*

(1) If $\lambda \in \sigma_r(T)$, then $\bar{\lambda} \in \sigma_p(T^*)$.

(2) $\lambda \in \sigma_p(T) \cup \sigma_r(T)$ iff $\bar{\lambda} \in \sigma_p(T^*) \cup \sigma_r(T^*)$.

(3) If T is normal, then $\lambda \in \sigma_p(T)$ iff $\bar{\lambda} \in \sigma_p(T^*)$, and in that case the λ-eigenspace of T coincides with the $\bar{\lambda}$-eigenspace of T^*.

(4) If T is normal, then $\sigma_r(T) = \emptyset$.

(5) *If T is normal and E is its resolution of the identity, then*

$$\sigma_{\mathrm{p}}(T) = \{\mu \in \mathbb{C}; E(\{\mu\}) \neq 0\}$$

and the range of $E(\{\mu\})$ coincides with the μ-eigenspace of T, for each $\mu \in \sigma_p(T)$.

Proof. (1) Let $\lambda \in \sigma_{\mathrm{r}}(T)$. Since the range of $\lambda I - T$ is not dense in X, the orthogonal decomposition theorem (Theorem 1.36) implies the existence of $y \neq 0$ orthogonal to this range. Hence, for all $x \in X$

$$(x, [\bar{\lambda} I - T^*]y) = ((\lambda I - T)x, y) = 0,$$

and therefore $y \in \ker(\bar{\lambda} I - T^*)$ and $\bar{\lambda} \in \sigma_p(T^*)$.

(2) Let $\lambda \in \sigma_{\mathrm{p}}(T)$ and let then x be an eigenvector of T corresponding to λ. Then for all $y \in X, (x, (\bar{\lambda} I - T^*)y) = ((\lambda I - T)x, y) = 0$, which implies that the range of $\bar{\lambda} I - T^*$ is not dense in X (because $x \neq 0$ is orthogonal to it). Therefore $\bar{\lambda}$ belongs to $\sigma_{\mathrm{r}}(T^*)$ or to $\sigma_{\mathrm{p}}(T^*)$ if $\bar{\lambda} I - T^*$ is injective or not, respectively. Together with Part (1), this shows that if $\lambda \in \sigma_{\mathrm{p}}(T) \cup \sigma_{\mathrm{r}}(T)$, then $\bar{\lambda} \in \sigma_{\mathrm{p}}(T^*) \cup \sigma_{\mathrm{r}}(T^*)$. Applying this to $\bar{\lambda}$ and T^*, we get the reverse implication (because $T^{**} = T$).

(3) For any normal operator S, $\|S^*x\| = \|Sx\|$ (for all x) because

$$\|S^*x\|^2 = (S^*x, S^*x) = (SS^*x, x) = (S^*Sx, x) = (Sx, Sx) = \|Sx\|^2.$$

If T is normal, so is $S := \lambda I - T$, and therefore $\|(\bar{\lambda} I - T^*)x\| = \|(\lambda I - T)x\|$ (for all $x \in X$ and $\lambda \in \mathbb{C}$). This implies Part (3).

(4) If $\lambda \in \sigma_{\mathrm{r}}(T)$, then Part (1) implies that $\lambda \in \sigma_{\mathrm{p}}(T^*)$, and therefore, by Part (3), $\lambda \in \sigma_{\mathrm{p}}(T)$, a contradiction. Hence $\sigma_{\mathrm{r}}(T) = \emptyset$.

(5) If $E(\{\mu\}) \neq 0$, let $x \neq 0$ be in the range of this projection. Then $E(\cdot)x = E(\cdot)E(\{\mu\})x = E(\cdot \cap \{\mu\})x$ is the point mass measure at μ (with total mass 1) multiplied by x. Hence,

$$Tx = \int_{\sigma(T)} \lambda \, dE(\lambda)x = \mu x,$$

so that $\mu \in \sigma_{\mathrm{p}}(T)$ and each $x \neq 0$ in $E(\{\mu\})X$ is a μ-eigenvector for T.

On the other hand, if $\mu \in \sigma_{\mathrm{p}}(T)$ and x is a μ-eigenvector for T, let $\delta_n = \{\lambda \in \mathbb{C}; |\lambda - \mu| > 1/n\} (n = 1, 2, \ldots)$ and $f_n(\lambda) := \chi_{\delta_n}(\lambda)/(\mu - \lambda)$. Then $f_n \in \mathbb{B}$ (it is clearly Borel, and bounded by n), and therefore $f_n(T) \in B(X)$. Since $(\mu - \lambda) f_n(\lambda) = \chi_{\delta_n}(\lambda)$, we have

$$E(\delta_n)x = f_n(T)(\mu I - T)x = f_n(T)0 = 0.$$

Hence, by σ-subadditivity of the positive measure $\|E(\cdot)x\|^2$,

$$\|E(\{\lambda; \lambda \neq \mu\})x\|^2 = \|E\Big(\bigcup_n \delta_n\Big)x\|^2 \leq \sum_n \|E(\delta_n)x\|^2 = 0.$$

Therefore, $E(\{\lambda; \lambda \neq \mu\})x = 0$ and so $E(\{\mu\})x = Ix = x$, that is, the non-zero vector x belongs to the range of the projection $E(\{\mu\})$. $\square$

9.6 Spectral representation

Construction 9.9. The spectral theorem for normal operators has an interesting interpretation through isomorphisms of Hilbert spaces. Let T be a (bounded) normal operator on the Hilbert space X, and let E be its resolution of the identity. Given $x \in X$, define the *cycle of* x (relative to T) as the closed linear span of the vectors $T^n(T^*)^m x, n, m = 0, 1, 2, \ldots$. This is clearly a reducing subspace for T. It follows from the Stone–Weierstrass theorem that the cycle of x, $[x]$, coincides with the closure (in X) of the subspace

$$[x]_0 := \{g(T)x; g \in C(\sigma(T))\}.$$

Define $V_0 : [x]_0 \to C(\sigma(T))$ by

$$V_0 g(T)x = g.$$

By Theorem 9.6, Part (1), the map V_0 is a *well-defined* linear isometry of $[x]_0$ onto the *subspace* $C(\sigma(T))$ of $L^2(\mu)$, where $\mu = \mu_x := \|E(\cdot)x\|^2$ is a regular finite positive Borel measure with support in $\sigma(T)$. By Corollary 3.21, $C(\sigma(T))$ is dense in $L^2(\mu)$, and therefore V_0 extends uniquely as a linear isometry V of the cycle $[x]$ onto $L^2(\mu)$. If g is a Borel function bounded on $\sigma(T)$, we may apply Theorem 3.20 to get a sequence $g_n \in C(\sigma(T))$, uniformly bounded on the spectrum by $\sup_{\sigma(T)} |g|$, such that $g_n \to g$ pointwise μ-almost everywhere on $\sigma(T)$. It follows from the proof of Part (3) in Theorem 9.6 that $g(T)x = \lim_n g_n(T)x$, hence $g(T)x \in [x]$. Also $g_n \to g$ in $L^2(\mu)$ by dominated convergence, and therefore

$$Vg(T)x = V \lim_n g_n(T)x = \lim_n V_0 g_n(T)x = \lim_n g_n = g,$$

where the last two limits are in the space $L^2(\mu)$ and equalities are between $L^2(\mu)$-elements.

For each $y = g(T)x \in [x]_0$ and each Borel function f bounded on $\sigma(T)$, the function fg (restricted to $\sigma(T)$) is a bounded Borel function on the spectrum, hence $(fg)(T)x$ makes sense and equals $f(T)g(T)x = f(T)y$. Applying V, we get

$$Vf(T)y = V(fg)(T)x = fg = fVg(T)x = fVy$$

(in $L^2(\mu)$). Let M_f denote the *multiplication by* f operator in $L^2(\mu)$, defined by $M_f g = fg$. Since $Vf(T) = M_f V$ on the dense subspace $[x]_0$ of $[x]$, and both operators are continuous, we conclude that the last relation is valid on $[x]$. We express this by saying that *the isomorphism* V *intertwines* $f(T)|_{[x]}$ *and* M_f (for all Borel functions f bounded on $\sigma(T)$). Equivalently, $f(T)|_{[x]} = V^{-1} M_f V$, that is, restricted to the cycle $[x]$, the operators $f(T)$ are unitarily equivalent (through V!) to the multiplication operators M_f on $L^2(\mu)$ with $\mu := \|E(\cdot)x\|^2$. This is particularly interesting when T possess a *cyclic vector*, that is, a vector x such that $[x] = X$. Using such a cyclic vector in the shown construction, we conclude that the abstract operator $f(T)$ is unitarily equivalent to the concrete operator M_f acting in the concrete Hilbert space $L^2(\mu)$, through the Hilbert isomorphism $V : X \to L^2(\mu)$, for each Borel function f bounded on the spectrum; in particular, taking $f(\lambda) = f_1(\lambda) := \lambda$, we have $T = V^{-1}MV$, where $M := M_{f_1}$.

The construction is generalized to an arbitrary normal operator T by considering a maximal family of non-zero mutually orthogonal cycles $\{[x_j], j \in J\}$ for T (J denotes some index set). Such a family exists by Zorn's lemma.

Let $\mu_j := \|E(\cdot)x_j\|^2$ and let $V_j : [x_j] \to L^2(\mu_j)$ be the Hilbert isomorphism constructed earlier, for each $j \in J$. If $\sum_j \oplus[x_j] \neq X$, pick $x \neq 0$ orthogonal to the orthogonal sum; then the non-zero cycle $[x]$ is orthogonal to all $[x_j]$ (since the cycles are reducing subspaces for T), contradicting the maximality of the family shown. Hence $X = \sum_j \oplus[x_j]$. Consider the operator

$$V := \sum_{j \in J} \oplus V_j$$

operating on $X = \sum_j \oplus[x_j]$. Recall that if $y = \sum_j \oplus y_j$ is any vector in X, then

$$Vy := \sum_j \oplus V_j y_j.$$

Then V is a linear isometry of X onto $\sum_j \oplus L^2(\mu_j)$:

$$\|Vy\|^2 = \sum_j \|V_j y_j\|^2 = \sum_j \|y_j\|^2 = \|y\|^2.$$

Since each cycle reduces $f(T)$ for all Borel function f bounded on $\sigma(T)$, we have

$$Vf(T)y = \sum_j \oplus V_j f(T) y_j = \sum_j \oplus f V_j y_j = M_f V y,$$

where M_f is now defined as the orthogonal sum $M_f := \sum_j \oplus M_f^j$, with M_f^j equal to the multiplication by f operator on the space $L^2(\mu_j)$. The Hilbert isomorphism V is usually referred to as a *spectral representation* of X (relative to T). Formally:

Theorem 9.10. *Let T be a bounded normal operator on the Hilbert space X, and let E be its resolution of the identity. Then there exists an isomorphism V of X onto $\sum_{j \in J} \oplus L^2(\mu_j)$ with $\mu_j := \|E(\cdot)x_j\|^2$ for suitable non-zero mutually orthogonal vectors x_j, such that $f(T) = V^{-1}M_f V$ for all Borel functions f bounded on $\sigma(T)$. The operator M_f acts on $\sum \oplus L^2(\mu_j)$ by $M_f \sum_j \oplus g_j := \sum \oplus f g_j$. In case T has a cyclic vector x, then there exists an isomorphism V of X onto $L^2(\mu)$ with $\mu := \|E(\cdot)x\|^2$, such that $f(T) = V^{-1}M_f V$ for all f as described; here M_f is the ordinary multiplication by f operator on $L^2(\mu)$.*

9.7 Renorming method

In the following, X denotes a given Banach space, and $B(X)$ is the Banach algebra of all bounded linear operators on X. The identity operator is denoted by I. If $\mathcal{A} \subset B(X)$, the *commutant* $\mathcal{A}'$ of $\mathcal{A}$ consists of all $S \in B(X)$ that commute with every $T \in \mathcal{A}$.

Given an operator T, we shall construct a *maximal Banach subspace* Z of X such that T has a contractive $C(K)$-operational calculus on Z, where K is an adequate compact subset of $\mathbb{C}$ containing the spectrum of T. In case X is reflexive, this construction will associate with T a contractive spectral measure on Z such that $f(T|_Z)$ is the corresponding spectral integral for each $f \in C(K)$. This "maximal" spectral integral representation is a generalization of the spectral theorem for normal operators in Hilbert space. The construction is based on the following.

Theorem 9.11 (Renorming theorem). *Let $\mathcal{A} \subset B(X)$ be such that its strong closure contains I. Let*

$$\|x\|_{\mathcal{A}} := \sup_{T \in \mathcal{A}} \|Tx\| \quad (x \in X),$$

and

$$Z = Z(\mathcal{A}) := \{x \in X; \|x\|_{\mathcal{A}} < \infty\}.$$

Then:

(i) Z with the norm $\|\cdot\|_{\mathcal{A}}$ is a Banach subspace of X.

(ii) For any $S \in \mathcal{A}'$, $SZ \subset Z$ and $S|_Z \in B(Z)$ with $\|S|_Z\|_{B(Z)} \leq \|S\|$.

(iii) If $\mathcal{A}$ is a multiplicative semigroup (for operator multiplication), then Z is $\mathcal{A}$-invariant, and $\mathcal{A}|_Z := \{T|_Z; T \in \mathcal{A}\}$ is contained in the closed unit ball $B_1(Z)$ of $B(Z)$. Moreover, Z is maximal with this property, that is, if W is an $\mathcal{A}$-invariant Banach subspace of X such that $\mathcal{A}|_W \subset B_1(W)$, then W is a Banach subspace of Z.

Proof. Subadditivity and homogeneity of $\|\cdot\|_{\mathcal{A}}$ are clear.

Let $\epsilon > 0$. The neighborhood of I

$$N := N(I, \epsilon, x) := \{T \in B(X); \|(T - I)x\| < \epsilon\}$$

(in the strong operator topology) meets $\mathcal{A}$ for each $x \in X$. Fix x, and pick then $T \in \mathcal{A} \cap N$. Then

$$\|x\|_{\mathcal{A}} \geq \|Tx\| = \|x + (T - I)x\| \geq \|x\| - \epsilon.$$

Therefore, $\|x\|_{\mathcal{A}} \geq \|x\|$ for all $x \in X$, and it follows in particular that $\|\cdot\|_{\mathcal{A}}$ is a norm on $\mathcal{A}$.

Let $\{x_n\} \subset Z$ be $\|\cdot\|_{\mathcal{A}}$-Cauchy. In particular, it is $\|\cdot\|_{\mathcal{A}}$-bounded; let then $K = \sup_n \|x_n\|_{\mathcal{A}}$. Since $\|\cdot\| \leq \|\cdot\|_{\mathcal{A}}$, the sequence is $\|\cdot\|$-Cauchy; let x be its X-limit. Given $\epsilon > 0$, let $n_0 \in \mathbb{N}$ be such that

$$\|x_n - x_m\|_{\mathcal{A}} < \epsilon \quad (n, m > n_0).$$

Then

$$\|Tx_n - Tx_m\| < \epsilon \quad (n, m > n_0; T \in \mathcal{A}).$$

Letting $m \to \infty$, we see that

$$\|Tx_n - Tx\| \leq \epsilon \quad (n > n_0; T \in \mathcal{A}).$$

Hence

$$\|x_n - x\|_{\mathcal{A}} \leq \epsilon \quad (n > n_0). \tag{1}$$

Fixing $n > n_0$, we see that

$$\|x\|_{\mathcal{A}} \leq \|x_n\|_{\mathcal{A}} + \|x - x_n\|_{\mathcal{A}} \leq K + \epsilon < \infty,$$

that is, $x \in Z$, and $x_n \to x$ in $(Z, \|\cdot\|_{\mathcal{A}})$ by (1). This proves Part (i).

If $S \in \mathcal{A}'$, then for all $x \in Z$ and $T \in \mathcal{A}$,

$$\|T(Sx)\| = \|S(Tx)\| \leq \|S\|\|Tx\| \leq \|S\|\|x\|_{\mathcal{A}}.$$

Therefore, $\|Sx\|_{\mathcal{A}} \leq \|S\|\|x\|_{\mathcal{A}} < \infty$, that is, $SZ \subset Z$ and $S|_Z \in B(Z)$ with $\|S|_Z\|_{B(Z)} \leq \|S\|$.

In case $\mathcal{A}$ is a multiplicative sub-semigroup of $B(X)$, we have for all $x \in Z$ and $T, U \in \mathcal{A}$

$$\|U(Tx)\| = \|(UT)x\| \leq \sup_{V \in \mathcal{A}} \|Vx\| = \|x\|_{\mathcal{A}}.$$

Hence, $\|Tx\|_{\mathcal{A}} \leq \|x\|_{\mathcal{A}}$, so that Z is $\mathcal{A}$-invariant and $\mathcal{A}|_Z \subset B_1(Z)$.

Finally, if W is as stated in the theorem, and $x \in W$, then for all $T \in \mathcal{A}$,

$$\|Tx\| \leq \|Tx\|_W \leq \|T\|_{B(W)}\|x\|_W \leq \|x\|_W,$$

and therefore $\|x\|_{\mathcal{A}} \leq \|x\|_W$. □

9.8 Semi-simplicity space

Let $\Delta \subset \mathbb{R}$ be compact, and let $\mathcal{P}_1(\Delta)$ denote the set of all complex polynomials p with

$$\|p\|_\Delta := \sup_\Delta |p| \leq 1.$$

Given an arbitrary bounded operator T, let $\mathcal{A}$ be the (multiplicative) semigroup

$$\mathcal{A} := \{p(T); p \in \mathcal{P}_1(\Delta)\},$$

and let $Z = Z(\mathcal{A})$.

By Theorem 9.11, Part (iii), $p(T)Z \subset Z$ and

$$\|p(T)|_Z\|_{B(Z)} \leq 1$$

for all $p \in \mathcal{P}_1(\Delta)$. This means that the polynomial operational calculus $\tau : p \to p(T)|_Z (= p(T|_Z))$ is *norm-decreasing* as a homomorphism of $\mathcal{P}(\Delta)$, the subalgebra of $C(\Delta)$ of all (complex) polynomials restricted to Δ, into $B(Z)$. Since $\mathcal{P}(\Delta)$ is dense in $C(\Delta)$, τ extends uniquely as a norm decreasing homomorphism of the algebra $C(\Delta)$ into $B(Z)$, that is, T is *of contractive class* $C(\Delta)$ *on* Z. The Banach subspace Z in this application of the renorming theorem is called the *semi-simplicity space for* T.

On the other hand, suppose W is a T-invariant Banach subspace of X, such that T is of contractive class $C(\Delta)$ on W. Then for each $p \in \mathcal{P}_1(\Delta)$ and $w \in W$,

$$\|p(T)w\| \leq \|p(T)|_W\|_{B(W)}\|w\|_W \leq \|w\|_W,$$

and therefore $w \in Z$ and $\|w\|_{\mathcal{A}} \leq \|w\|_W$. This shows that W is a Banach subspace of Z, and concludes the proof of the following.

Theorem 9.12. *Let* $\Delta \subset \mathbb{R}$ *be compact,* $T \in B(X)$*, and let* Z *be the* semi-simplicity space for T, *that is,* $Z = Z(\mathcal{A})$, *where* $\mathcal{A}$ *is the (multiplicative) semigroup*

$$\{p(T); p \in \mathcal{P}_1(\Delta)\}.$$

Then T *is of contractive class* $C(\Delta)$ *on* Z.

Moreover, Z *is* maximal *in the following sense: if* W *is a* T*-invariant Banach subspace of* X *such that* T *is of contractive class* $C(\Delta)$ *on* W*, then* W *is a Banach subspace of* Z.

Remark 9.13. (1) If X is a Hilbert space and T is a bounded selfadjoint operator on X, it follows from Remark (2) in Terminology 7.19 that $\|p(T)\| = \|p\|_{\sigma(T)} \leq 1$ for all $p \in \mathcal{P}_1(\Delta)$ for any given compact subset Δ of $\mathbb{R}$ containing $\sigma(T)$. Therefore $\|x\|_{\mathcal{A}} \leq \|x\|$ for all $x \in X$. Hence $Z = X$ (with equality of norms) in this case.

Another way to see this is by observing that the selfadjoint operator T has a contractive $C(\Delta)$-operational calculus on X (see Terminology 7.19, Remark (2)). Therefore, X is a Banach subspace of Z by the maximality statement in Theorem 9.12. Hence $Z = X$.

(2) If T is a normal operator (with spectrum in a compact set $\Delta \subset \mathbb{C}$), we may take

$$\mathcal{A} = \{p(T, T^*); p \in \mathcal{P}_1(\Delta)\},$$

where $\mathcal{P}_1(\Delta)$ is now the set of all complex polynomials p in λ and $\bar{\lambda}$ with $\|p\|_\Delta := \sup_{\lambda \in \Delta} |p(\lambda, \bar{\lambda})| \leq 1$.

Then $\mathcal{A}$ is a commutative semigroup (for operator multiplication), and since polynomials in λ and $\bar{\lambda}$ are dense in $C(\Delta)$ (by Theorem 5.39), we conclude as before that $Z := Z(\mathcal{A})$ coincides with X (with equality of norms).

(3) In the general Banach space setting, we took $\Delta \subset \mathbb{R}$ in the construction leading to Theorem 9.12 to ensure the *density* of $\mathcal{P}(\Delta)$ in $C(\Delta)$. If Δ is *any compact* subset of $\mathbb{C}$, Lavrentiev's theorem (cf. Theorem 8.7 in Gamelin, 1969) states that the wanted density occurs if and only if Δ is *nowhere dense and has connected complement.* Theorem 9.12 is then valid for such Δ, with the same statement and proof.

(4) When the complement of Δ is *not* connected, the choice of $\mathcal{A}$ may be adapted in some cases so that Theorem 9.12 remains valid.

An important case of this kind is the unit circle:

$$\Delta = \Gamma := \{\lambda \in \mathbb{C}; |\lambda| = 1\}.$$

Suppose $T \in B(X)$ has spectrum in Γ. Consider the algebra $\mathcal{R}(\Gamma)$ of restrictions to Γ of all complex polynomials in λ and $\lambda^{-1}(= \bar{\lambda}$ on $\Gamma)$. Following our previous notation, let

$$\mathcal{R}_1(\Gamma) = \left\{ p \in \mathcal{R}(\Gamma); \|p\|_\Gamma := \sup_\Gamma |p| \leq 1 \right\}.$$

Since $\sigma(T) \subset \Gamma, T$ is invertible in $B(X)$. Let $p \in \mathcal{R}(\Gamma)$. Writing p as the finite sum

$$p(\lambda) = \sum_k \alpha_k \lambda^k \quad (\alpha_k \in \mathbb{C}, k \in \mathbb{Z}),$$

it makes sense to define

$$p(T) = \sum_k \alpha_k T^k.$$

The map $\tau : p \to p(T)$ is the unique algebra homomorphism of $\mathcal{R}(\Gamma)$ into $B(X)$ such that $\tau(p_0) = I$ and $\tau(p_1) = T$, where $p_k(\lambda) = \lambda^k$. As before, we choose $\mathcal{A}$ to be the (multiplicative) semigroup

$$\mathcal{A} = \{p(T); p \in \mathcal{R}_1(\Gamma)\},$$

and we call $Z = Z(\mathcal{A})$ the *semi-simplicity space for* T.

By Theorem 9.11, Part (iii), Z is a T-invariant Banach subspace of X, and

$$\|p(T)|_Z\|_{B(Z)} \leq 1 \quad (p \in \mathcal{R}_1(\Gamma)). \tag{1}$$

By Theorem 5.39, $\mathcal{R}(\Gamma)$ is dense in $C(\Gamma)$. Consequently, it follows from (1) that

$$\tau_Z : p \to p(T)|_Z (= p(T|_Z))$$

extends uniquely to a norm-decreasing algebra homomorphism of $C(\Gamma)$ into $B(Z)$, that is, T is of contractive class $C(\Gamma)$ on Z.

The maximality of Z is proved word for word as in Theorem 9.12.

We restate Theorem 9.12 in the present case for future reference.

Theorem 9.14. *Let $T \in B(X)$ have spectrum on the unit circle Γ. Let Z be its* semi-simplicity space, *as defined in Remark 9.13 (4). Then Z is a T-invariant Banach subspace of X such that T is of contractive class $C(\Gamma)$ on Z.*

Moreover, Z is maximal *in the following sense: if W is a T-invariant Banach subspace of X such that T is of contractive class $C(\Gamma)$ on W, then W is a Banach subspace of Z.*

Remark 9.15. With a minor change in the choice of $\mathcal{A}$, Theorem 9.14 generalizes (with identical statement and proof) to the case when Γ is an *arbitrary* compact subset of the plane with *planar Lebesgue measure zero*. In this case, let

$\mathcal{R}(\Gamma)$ denote the algebra of all restrictions to Γ of rational functions with poles off Γ. Each $f \in \mathcal{R}(\Gamma)$ can be written as a finite product

$$f(\lambda) = \gamma \prod_{j,k} (\lambda - \alpha_j)(\lambda - \beta_k)^{-1},$$

with $\gamma, \alpha_j \in \mathbb{C}$ and poles $\beta_k \notin \Gamma$. Since $\beta_k \in \rho(T)$, we may define

$$f(T) := \gamma \prod_{j,k} (T - \alpha_j I)(T - \beta_k I)^{-1}.$$

As before, we choose $\mathcal{A}$ to be the (multiplicative) semigroup

$$\mathcal{A} = \{f(T); f \in \mathcal{R}_1(\Gamma)\},$$

where $\mathcal{R}_1(\Gamma)$ is the set of all $f \in \mathcal{R}(\Gamma)$ with $\|f\|_\Gamma \leq 1$.

The corresponding space $Z = Z(\mathcal{A})$ (the *semi-simplicity space for* T) is a T-invariant Banach subspace of X such that $\|f(T)|_Z\|_{B(Z)} \leq 1$ for all $f \in \mathcal{R}_1(T)$. Since $\mathcal{R}(\Gamma)$ is dense in $C(\Gamma)$ by the Hartogs–Rosenthal theorem (cf. Exercise 2), the map $f \to f(T)|_Z = f(T|_Z)$ has a unique extension as a norm decreasing algebra homomorphism of $C(\Gamma)$ into $B(Z)$, that is, T is of contractive class $C(\Gamma)$ on Z.

9.9 Resolution of the identity on Z

Theorem 9.16. *Let Γ be any compact subset of $\mathbb{C}$ for which the semi-simplicity space Z was defined earlier, for a given bounded operator T on the Banach space X, with spectrum in Γ (recall that, in the general case, Γ could be any compact subset of $\mathbb{C}$ with planar Lebesgue measure zero). Then T is of contractive class $C(\Gamma)$ on Z, and Z is* maximal *with this property.*

Moreover, if X is reflexive, *there exists a unique* contractive spectral measure on Z, E, *with support in Γ and values in T'', such that*

$$f(T|_Z)x = \int_\Gamma f \, dE(\cdot)x \tag{1}$$

for all $x \in Z$ and $f \in C(\Gamma)$, where $f \to f(T|_Z)$ denotes the (unique) $C(\Gamma)$-operational calculus for $T|_Z$.

The integral (1) extends the $C(\Gamma)$-operational calculus for $T|_Z$ in $B(Z)$ to a contractive $\mathbb{B}(\Gamma)$-operational calculus $\tau : f \in \mathbb{B}(\Gamma) \to f(T|_Z) \in B(Z)$, where $\mathbb{B}(\Gamma)$ stands for the Banach algebra of all bounded complex Borel functions on Γ with the supremum norm, and $\tau(f)$ commutes with every $U \in B(X)$ that commutes with T.

Proof. The first statement of the theorem was verified in the preceding sections.

Suppose then that X is a *reflexive* Banach space, and let $f \to f(T|_Z)$ denote the (unique) $C(\Gamma)$-operational calculus for $T|_Z$ (in $B(Z)$). For each $x \in Z$ and $x^* \in X^*$, the map

$$f \in C(\Gamma) \to x^* f(T|_Z)x \in \mathbb{C}$$

is a continuous linear functional on $C(\Gamma)$ with norm $\leq \|x\|_Z \|x^*\|$ (where we denote the Z-norm by $\|\cdot\|_Z$ rather than $\|\cdot\|_{\mathcal{A}}$). By the Riesz representation theorem, there exists a unique regular complex Borel measure $\mu = \mu(\cdot; x, x^*)$ on $\mathcal{B}(\mathbb{C})$, supported by Γ, such that

$$x^* f(T|_Z)x = \int_\Gamma f \, d\mu(\cdot; x, x^*) \tag{2}$$

and

$$\|\mu(\cdot; x, x^*)\| \leq \|x\|_Z \|x^*\| \tag{3}$$

for all $f \in C(\Gamma), x \in Z$, and $x^* \in X^*$.

For each $\delta \in \mathcal{B}(\mathbb{C})$ and $x \in Z$, the map $x^* \in X^* \to \mu(\delta; x, x^*)$ is a continuous linear functional on X^* with norm $\leq \|x\|_Z$ (by (3)). Since X is reflexive, there exists a unique vector in X, which we denote $E(\delta)x$, such that $\mu(\delta; x, x^*) = x^* E(\delta)x$. The map $E(\delta) : Z \to X$ is clearly linear and norm decreasing.

If $U \in T'$, then U commutes with $f(T)$ for all $f \in \mathcal{R}(\Gamma)$ (notation as in Section 9.8), hence $U \in \mathcal{A}'$. Let $f \in C(\Gamma)$, and let $f_n \in \mathcal{R}(\Gamma)$ converge to f uniformly on Γ. Since $UZ \subset Z$ and $\|U\|_{B(Z)} \leq \|U\|_{B(X)}$ (cf. Theorem 9.11, Part (ii)), we have

$$\begin{aligned}
\|Uf(T|_Z) - f(T|_Z)U\|_{B(Z)} &\leq \|U[f(T|_Z) - f_n(T)]\|_{B(Z)} \\
&\quad + \|[f_n(T) - f(T|_Z)]U\|_{B(Z)} \leq 2\|U\|_{B(X)} \|f(T|_Z) - f_n(T)\|_{B(Z)} \\
&\leq 2\|U\|_{B(X)} \|f - f_n\|_{C(\Gamma)} \to 0.
\end{aligned}$$

Thus, U commutes with $f(T|_Z)$ for all $f \in C(\Gamma)$. Therefore, for all $x \in Z$, $x^* \in X^*$, and $f \in C(\Gamma)$,

$$\begin{aligned}
\int_\Gamma f \, dx^* UE(\cdot)x &= \int_\Gamma f \, d(U^* x^*)E(\cdot)x = (U^* x^*) f(T|_Z)x \\
&= x^* U f(T|_Z)x = x^* f(T|_Z)Ux = \int_\Gamma f \, dx^* E(\cdot)Ux.
\end{aligned}$$

By uniqueness of the Riesz representation, it follows that $UE(\delta) = E(\delta)U$ for all $\delta \in \mathcal{B}(\mathbb{C})$ (i.e., E has values in T''). The last relation is true in particular for all $U \in \mathcal{A}$. Since $\mathcal{A} \subset B_1(Z)$ (by Theorem 9.11, Part (iii)) and $E(\delta) : Z \to X$ is norm decreasing, we have for each $x \in Z$ and $\delta \in \mathcal{B}(\mathbb{C})$,

$$\begin{aligned}
\|E(\delta)x\|_Z &:= \sup_{U \in \mathcal{A}} \|UE(\delta)x\| = \sup_{U \in \mathcal{A}} \|E(\delta)Ux\| \\
&\leq \sup_{U \in \mathcal{A}} \|Ux\|_Z \leq \|x\|_Z.
\end{aligned}$$

Thus, $E(\delta) \in B_1(Z)$ for all $\delta \in \mathcal{B}(\mathbb{C})$.

Since $x^*[E(\cdot)x] = \mu(\cdot; x, x^*)$ is a regular countably additive complex measure on $\mathcal{B}(\mathbb{C})$ for each $x^* \in X^*, E$ satisfies Condition (1) of a spectral measure on Z and has support in Γ. We may then rewrite (2) in the form

$$f(T|_Z)x = \int_\Gamma f \, dE(\cdot)x \quad f \in C(\Gamma), \; x \in Z, \tag{4}$$

where the integral is defined as in Section 9.2.

Taking $f = f_0(=1)$ in (4), we see that $E(\mathbb{C}) = E(\Gamma) = I|_Z$.

Since $f \to f(T|_Z)$ is an algebra homomorphism of $C(\Gamma)$ into $B(Z)$, we have for all $f, g \in C(\Gamma)$ and $x \in Z$ (whence $g(T|_Z)x \in Z$):

$$\int_\Gamma f\, dE(\cdot)[g(T|_Z)x] = f(T|_Z)[g(T|_Z)x]$$
$$= (fg)(T|_Z)x = \int_\Gamma fg\, dE(\cdot)x.$$

By uniqueness of the Riesz representation, it follows that

$$dE(\cdot)[g(T|_Z)x] = g\, dE(\cdot)x.$$

This means that for all $\delta \in \mathcal{B}(\mathbb{C}), g \in C(\Gamma)$, and $x \in Z$,

$$E(\delta)[g(T|_Z)x] = \int_\Gamma \chi_\delta g\, dE(\cdot)x, \tag{5}$$

where χ_δ denotes the characteristic function of δ.

We observed that $E(\delta)$ commutes with every $U \in B(X)$ that commutes with T. In particular, $E(\delta)$ commutes with $g(T)$ for all $g \in \mathcal{R}(\Gamma)$. If $g \in C(\Gamma)$ and $g_n \in \mathcal{R}(\Gamma) \to g$ uniformly on Γ, then since $E(\delta) \in B_1(Z)$, we have

$$\begin{aligned}
&\|E(\delta)g(T|_Z) - g(T|_Z)E(\delta)\|_{B(Z)} \\
&\quad \le \|E(\delta)[g(T|_Z) - g_n(T|_Z)]\|_{B(Z)} + \|[g_n(T|_Z) - g(T|_Z)]E(\delta)\|_{B(Z)} \\
&\quad \le 2\|g - g_n\|_{C(\Gamma)} \to 0.
\end{aligned}$$

Thus $E(\delta)$ commutes with $g(T|_Z)$ for all $g \in C(\Gamma)$ and $\delta \in \mathcal{B}(\mathbb{C})$.

We can then rewrite (5) in the form

$$\int_\Gamma g\chi_\delta\, dE(\cdot)x = \int_\Gamma g\, dE(\cdot)E(\delta)x$$

for all $x \in Z, \delta \in \mathcal{B}(\mathbb{C})$, and $g \in C(\Gamma)$.

Again, by uniqueness of the Riesz representation, it follows that

$$\chi_\delta dE(\cdot)x = dE(\cdot)E(\delta)x.$$

Therefore, for each $\epsilon \in \mathcal{B}(\mathbb{C})$ and $x \in Z$,

$$E(\epsilon)E(\delta)x = \int_\Gamma \chi_\epsilon\, dE(\cdot)E(\delta)x = \int_\Gamma \chi_\epsilon \chi_\delta dE(\cdot)x$$
$$= \int_\Gamma \chi_{\epsilon\cap\delta}\, dE(\cdot)x = E(\epsilon \cap \delta)x.$$

We consider now the map $f \in \mathbb{B}(\Gamma) \to \int_\Gamma f\,dE(\cdot)x \in X$. Denote the integral by $\tau(f)x$. We have for all $x \in Z$ and $f \in \mathbb{B}(\Gamma)$

$$\|\tau(f)x\| = \sup_{x^* \in X^*; \|x^*\|=1} \left|\int_\Gamma f\,dx^*E(\cdot)x\right| \leq \|x\|_Z \sup_\Gamma |f|. \tag{6}$$

If $U \in T'$, we saw that $UE(\delta) = E(\delta)U$ for all $\delta \in \mathcal{B}(\mathbb{C})$. It follows from the definition of the integral with respect to $E(\cdot)x$ (cf. Section 9.2) that for each $x \in Z$

$$U\tau(f)x = \tau(f)Ux.$$

In particular, this is true for all $U \in \mathcal{A}$. Hence by (6) and Theorem 9.11, Part (iii),

$$\|\tau(f)x\|_Z = \sup_{U\in\mathcal{A}} \|U\tau(f)x\| = \sup_{U\in\mathcal{A}} \|\tau(f)Ux\| \leq \sup_\Gamma |f| \sup_{U\in\mathcal{A}} \|Ux\|_Z \leq \|f\|_{\mathbb{B}(\Gamma)}\|x\|_Z.$$

Thus $\tau(f) \in B(Z)$ for all $f \in \mathbb{B}(\Gamma)$. Furthermore, the map $\tau : \mathbb{B}(\Gamma) \to B(Z)$ is linear and norm-decreasing, and clearly multiplicative on the simple Borel functions on Γ. A routine density argument proves the multiplicativity of τ on $\mathbb{B}(\Gamma)$.

The uniqueness statement about E is an immediate consequence of the uniqueness of the Riesz representation. □

Remark 9.17. Let Δ be a closed interval. The semi-simplicity space for $T \in B(X)$ is the smallest Banach subspace W_0 in the increasing scale $\{W_m; m = 0, 1, 2, \ldots\}$ of Banach subspaces defined next.

Let $C^m(\Delta)$ denote the Banach algebra of all complex functions with continuous derivatives in Δ up to the mth order, with pointwise operations and norm

$$\|f\|_m := \sum_{j=0}^m \sup_\Delta |f^{(j)}|/j!.$$

We apply Theorem 9.11, Part (iii), to the multiplicative semigroup of operators

$$\mathcal{A}_m := \{p(T); p \in \mathcal{P}, \|p\|_m \leq 1\},$$

where $\mathcal{P}$ denotes the polynomial algebra $\mathbb{C}[t]$ and $m = 0, 1, 2, \ldots$.

Fix $m \in \mathbb{N} \cup \{0\}$. By Theorem 9.11, the Banach subspace $W_m := Z(\mathcal{A}_m)$ is $\mathcal{A}_m$-invariant and $\mathcal{A}_m|W_m \subset B_1(W_m)$, that is,

$$\|p(T)x\|_{W_m} \leq \|x\|_{W_m}\|p\|_m$$

for all $p \in \mathcal{P}$ and $x \in W_m$. By density of $\mathcal{P}$ in $C^m(\Delta)$, it follows that *$T|_{W_m}$ is of contractive class $C^m(\Delta)$*, that is, there exists a norm-decreasing algebra homomorphism of $C^m(\Delta)$ into $B(W_m)$ that extends the usual polynomial operational calculus. Moreover, W_m is maximal with this property.

9.10 Analytic operational calculus

Let $K \subset \mathbb{C}$ be a non-empty compact set, and denote by $H(K)$ the (complex) algebra of all complex functions analytic in some neighborhood of K (depending on the function), with pointwise operations. A net $f_\alpha \subset H(K)$ converges to f if all f_α are analytic in some fixed neighborhood Ω of K, and $f_\alpha \to f$ pointwise uniformly on every compact subset of Ω. When this is the case, f is analytic in Ω (hence $f \in H(K)$), and one verifies easily that the operations in $H(K)$ are continuous relative to the described convergence concept (or, equivalently, relative to the topology associated with the described convergence concept). Thus, $H(K)$ is a so-called *topological algebra*.

Throughout this section, $\mathcal{A}$ denotes a fixed (complex) Banach algebra with unit e. Let $\mathcal{F}$ be a topological algebra of complex functions defined on some subset of $\mathbb{C}$, with pointwise operations, such that $f_k(\lambda) := \lambda^k \in \mathcal{F}$ for $k = 0, 1$. Given $a \in \mathcal{A}$, an *$\mathcal{F}$-operational calculus for a* is a continuous representation $\tau : f \to \tau(f)$ of $\mathcal{F}$ into $\mathcal{A}$ such that $\tau(f_1) = a$ (in the present context, a *representation* is an algebra homomorphism sending the identity f_0 of $\mathcal{F}$ to the identity e of $\mathcal{A}$).

Notation 9.18. Let $K \subset \Omega \subset \mathbb{C}, K$ compact and Ω open. There exists an open set Δ with boundary Γ consisting of finitely many positively oriented rectifiable Jordan curves, such that $K \subset \Delta$ and $\Delta \cup \Gamma \subset \Omega$. Let $\Gamma(K, \Omega)$ denote the family of all Γs with these properties.

If F is an $\mathcal{A}$-valued function analytic in $\Omega \cap K^c$ and $\Gamma, \Gamma' \in \Gamma(K, \Omega)$, it follows from (the vector-valued version of) Cauchy's theorem that

$$\int_\Gamma F(\lambda)\, d\lambda = \int_{\Gamma'} F(\lambda)\, d\lambda.$$

We apply this observation to the function $F = f(\cdot)R(\cdot; a)$ when $\sigma(a) \subset K$ and $f \in H(K)$, to conclude that the so-called *Riesz–Dunford integral*

$$\tau(f) := \frac{1}{2\pi i} \int_\Gamma f(\lambda) R(\lambda; a)\, d\lambda \tag{1}$$

(with $\Gamma \in \Gamma(K, \Omega)$ and f analytic in the open neighborhood Ω of K) is a well-defined element of $\mathcal{A}$ (independent on the choice of Γ!).

Theorem 9.19. *The element $a \in \mathcal{A}$ has an $H(K)$-operational calculus iff $\sigma(a) \subset K$. In that case, the $H(K)$-operational calculus for a is unique, and is given by (1).*

Proof. Suppose $\sigma(a) \subset K$. For $f \in H(K)$, define $\tau(f)$ by (1). As observed, $\tau : H(K) \to \mathcal{A}$ is well defined and clearly linear. Let $f, g \in H(K)$; suppose both are analytic in the open neighborhood Ω of K. Let $\Gamma' \in \Gamma(K, \Omega)$, and let Δ' be the open neighborhood of K with boundary Γ'. Choose $\Gamma \in \Gamma(K, \Delta')$ (hence $\Gamma \in \Gamma(K, \Omega)$). By Cauchy's integral formula,

$$\frac{1}{2\pi i} \int_{\Gamma'} \frac{g(\mu)}{\mu - \lambda}\, d\mu = g(\lambda) \quad (\lambda \in \Gamma), \tag{2}$$

and

$$\int_\Gamma \frac{f(\lambda)}{\lambda - \mu}\, d\lambda = 0 \quad (\mu \in \Gamma'). \tag{3}$$

Therefore, by the resolvent identity for $R(\cdot) := R(\cdot; a)$, Fubini's theorem, and Relations (2) and (3), we have

$$\begin{aligned}
(2\pi i)^2 \tau(f)\tau(g) &= \int_\Gamma f(\lambda)R(\lambda)\, d\lambda \int_{\Gamma'} g(\mu)R(\mu)\, d\mu \\
&= \int_\Gamma \int_{\Gamma'} f(\lambda)g(\mu) \frac{R(\lambda) - R(\mu)}{\mu - \lambda}\, d\mu\, d\lambda \\
&= \int_\Gamma f(\lambda)R(\lambda) \int_{\Gamma'} \frac{g(\mu)}{\mu - \lambda}\, d\mu\, d\lambda + \int_{\Gamma'} g(\mu)R(\mu) \int_\Gamma \frac{f(\lambda)}{\lambda - \mu}\, d\lambda\, d\mu \\
&= 2\pi i \int_\Gamma f(\lambda)g(\lambda)R(\lambda)\, d\lambda = (2\pi i)^2 \tau(fg).
\end{aligned}$$

This proves the multiplicativity of τ.

Let $\{f_\alpha\}$ be a net in $H(K)$ converging to f, let Ω be an open neighborhood of K in which all f_α (and f) are analytic, and let $\Gamma \in \Gamma(K, \Omega)$. Since $R(\cdot)$ is continuous on Γ, we have $M := \sup_\Gamma \|R(\cdot)\| < \infty$, and therefore, denoting the (finite) length of Γ by $|\Gamma|$, we have

$$\|\tau(f_\alpha) - \tau(f)\| = \|\tau(f_\alpha - f)\| \le \frac{M|\Gamma|}{2\pi} \sup_\Gamma |f_\alpha - f| \to 0,$$

since $f_\alpha \to f$ uniformly on Γ. This proves the continuity of $\tau : H(K) \to \mathcal{A}$.

If f is analytic in the disc $\Delta_r := \{\lambda; |\lambda| < r\}$ and $K \subset \Delta_r$, the (positively oriented) circle $C_\rho = \{\lambda; |\lambda| = \rho\}$ is in $\Gamma(K, \Delta_r)$ for suitable $\rho < r$, and for any $a \in \mathcal{A}$ with spectrum in K, the Neumann series expansion of $R(\lambda)$ converges uniformly on C_ρ. Integrating term by term, we get

$$\tau(f) = \sum_{n=0}^\infty a^n \frac{1}{2\pi i} \int_{C_\rho} \frac{f(\lambda)}{\lambda^{n+1}}\, d\lambda = \sum_{n=0}^\infty \frac{f^{(n)}(0)}{n!} a^n. \tag{4}$$

In particular, $\tau(\lambda^k) = a^k$ for all $k = 0, 1, 2, \ldots$, and we conclude that τ is an $H(K)$-operational calculus for a.

If τ' is any $H(K)$-operational calculus for a, then necessarily $\tau'(\lambda^n) = a^k$ for all $k = 0, 1, 2, \ldots$, that is, τ' coincides with τ on all polynomials. If f is a rational function with poles in K^c, then $f \in H(K)$. Writing $f = p/q$ with polynomials p, q such that $q \neq 0$ on K (so that $1/q \in H(K)$!), we have $e = \tau'(1) = \tau'(q \cdot (1/q)) = \tau'(q)\tau'(1/q)$, hence, $\tau'(q)$ is non-singular, with inverse $\tau'(1/q)$. Therefore $\tau'(f) = \tau'(p)\tau'(1/q) = \tau'(p)\tau'(q)^{-1} = \tau(p)\tau(q)^{-1} = \tau(f)$. By Runge's theorem (cf. Exercise 1), the rational functions with poles in K^c are dense in $H(K)$, and the conclusion $\tau'(f) = \tau(f)$ for all $f \in H(K)$ follows from the continuity of both τ' and τ. This proves the uniqueness of the $H(K)$-operational calculus for a.

Finally, suppose $a \in \mathcal{A}$ has an $H(K)$-operational calculus τ, and let $\mu \in K^c$. The polynomial $q(\lambda) := \mu - \lambda$ does not vanish on K, so that (as was proved earlier) $\tau(q)(= \mu e - a)$ is non-singular, and $R(\mu; a) = \tau(1/(\mu - \lambda))$. In particular, $\sigma(a) \subset K$. □

Let τ be the $H(\sigma(a))$-operational calculus for a. Since $\tau(f) = f(a)$ when f is a polynomial, it is customary to use the notation $f(a)$ instead of $\tau(f)$ for all $f \in H(\sigma(a))$.

Theorem 9.20 (Spectral mapping theorem). *Let $a \in \mathcal{A}$ and $f \in H(\sigma(a))$. Then $\sigma(f(a)) = f(\sigma(a))$.*

Proof. Let $\mu = f(\lambda)$ with $\lambda \in \sigma(a)$. Since λ is a zero of the analytic function $\mu - f$, there exists $h \in H(\sigma(a))$ such that

$$\mu - f(\zeta) = (\lambda - \zeta)h(\zeta)$$

in a neighborhood of $\sigma(a)$. Applying the $H(\sigma(a))$-operational calculus, we get

$$\mu e - f(a) = (\lambda e - a)h(a) = h(a)(\lambda e - a).$$

If $\mu \in \rho(f(a))$, and $v = R(\mu; f(a))$, then

$$(\lambda e - a)[h(a)v] = e \text{ and } [vh(a)](\lambda e - a) = e,$$

that is, $\lambda \in \rho(a)$, contradiction. Hence, $\mu \in \sigma(f(a))$, and we proved the inclusion $f(\sigma(a)) \subset \sigma(f(a))$.

If $\mu \notin f(\sigma(a))$, then $\mu - f \neq 0$ on $\sigma(a)$, and therefore $g := 1/(\mu - f) \in H(\sigma(a))$. Since $(\mu - f)g = 1$ in a neighborhood of $\sigma(a)$, we have $(\mu e - f(a))g(a) = e$, that is, $\mu \in \rho(f(a))$. This proves the inclusion $\sigma(f(a)) \subset f(\sigma(a))$. □

Theorem 9.21 (Composite function theorem). *Let $a \in \mathcal{A}, f \in H(\sigma(a))$, and $g \in H(f(\sigma(a)))$. Then $g \circ f \in H(\sigma(a))$ and $(g \circ f)(a) = g(f(a))$.*

Proof. By Theorem 9.20, $g(f(a))$ is well defined.

Let Ω be an open neighborhood of $K := f(\sigma(a)) = \sigma(f(a))$ in which g is analytic, and let $\Gamma \in \Gamma(K, \Omega)$. Then

$$g(f(a)) = \frac{1}{2\pi i} \int_\Gamma g(\mu) R(\mu; f(a))\, d\mu. \tag{5}$$

Since $\Gamma \subset K^c$, for each fixed $\mu \in \Gamma$, the function $\mu - f$ does not vanish on $\sigma(a)$, and consequently $k_\mu := 1/(\mu - f) \in H(\sigma(a))$. The relation $(\mu - f)k_\mu = 1$ (valid in a neighborhood of $\sigma(a)$) implies through the operational calculus for a that $k_\mu(a) = R(\mu; f(a))$, that is,

$$R(\mu; f(a)) = \frac{1}{2\pi i} \int_{\Gamma'} k_\mu(\lambda) R(\lambda; a)\, d\lambda \tag{6}$$

for a suitable Γ'. We now substitute (6) in (5), interchange the order of integration, and use Cauchy's integral formula:

$$(2\pi i)^2 g(f(a)) = \int_\Gamma g(\mu) \int_{\Gamma'} k_\mu(\lambda) R(\lambda; a)\, d\lambda\, d\mu = \int_{\Gamma'} \int_\Gamma \frac{g(\mu)}{\mu - f(\lambda)}\, d\mu R(\lambda; a)\, d\lambda$$

$$= 2\pi i \int_{\Gamma'} g(f(\lambda)) R(\lambda; a)\, d\lambda = (2\pi i)^2 (g \circ f)(a).$$

□

9.11 Isolated points of the spectrum

Construction 9.22. Let μ be an isolated point of $\sigma(a)$. There exists then a function $e_\mu \in H(\sigma(a))$ that equals 1 in a neighborhood of μ and 0 in a neighborhood of $\sigma_\mu := \sigma(a) \cap \{\mu\}^c$. Set $E_\mu = e_\mu(a)$. The element E_μ is independent of the choice of the function e_μ, and since $e_\mu^2 = e_\mu$ in a neighborhood of $\sigma(a)$, it is an idempotent commuting with a.

Let $\delta = \operatorname{dist}(\mu, \sigma_\mu)$. By Laurent's theorem (whose classical proof applies word for word to vector-valued functions), we have for $0 < |\lambda - \mu| < \delta$:

$$R(\lambda; a) = \sum_{k=-\infty}^{\infty} a_k (\mu - \lambda)^k, \tag{1}$$

where

$$a_k = -\frac{1}{2\pi i} \int_\Gamma (\mu - \lambda)^{-k-1} R(\lambda; a)\, d\lambda, \tag{2}$$

and Γ is a positively oriented circle centered at μ with radius $r < \delta$. Choosing a function e_μ as shown that equals 1 in a neighborhood of the corresponding closed disc, we can add the factor e_μ to the integrand in (2). For $k \in -\mathbb{N}$, the new integrand is analytic in a neighborhood Ω of $\sigma(a)$, and therefore, by Cauchy's theorem, the circle Γ may be replaced by any $\Gamma' \in \Gamma(\sigma(a), \Omega)$. By the multiplicativity of the analytic operational calculus, it follows that

$$a_{-k} = -(\mu e - a)^{k-1} E_\mu \quad (k \in \mathbb{N}). \tag{3}$$

In particular, it follows from (3) that $a_{-k} = 0$ for all $k \geq k_0$ iff $a_{-k_0} = 0$. Consequently the point μ is a pole of order m of $R(\cdot; a)$ iff

$$(\mu e - a)^m E_\mu = 0 \quad \text{and} \quad (\mu e - a)^{m-1} E_\mu \neq 0. \tag{4}$$

Similarly, $R(\cdot; a)$ has a removable singularity at μ iff $E_\mu = 0$. In this case, the relation $(\lambda e - a) R(\lambda; a) = R(\lambda; a)(\lambda e - a) = e$ extends by continuity to the point $\lambda = \mu$, so that $\mu \in \rho(a)$, contradicting our hypothesis. Consequently $E_\mu \neq 0$.

These observations have a particular significance when $\mathcal{A} = B(X)$ for a Banach space X. If μ is an isolated point of the spectrum of $T \in B(X)$ and E_μ is the corresponding idempotent, the non-zero projection E_μ (called the *Riesz projection at μ for T*) commutes with T, so that its range $X_\mu \neq \{0\}$ is a reducing subspace for T (cf. Terminology 8.5 (2)).

Let $T_\mu := T|_{X_\mu}$. If $\zeta \neq \mu$, the function $h(\lambda) := e_\mu(\lambda)/(\zeta - \lambda)$ belongs to $H(\sigma(T))$ for a proper choice of e_μ, and $(\zeta - \lambda)h(\lambda) = e_\mu$. Applying the analytic operational calculus, we get

$$(\zeta I - T)h(T) = h(T)(\zeta I - T) = E_\mu,$$

and therefore (since $h(T)X_\mu \subset X_\mu$),

$$(\zeta I - T_\mu)h(T)x = h(T)(\zeta I - T_\mu)x = x \quad (x \in X_\mu).$$

Hence, $\zeta \in \rho(T_\mu)$, and consequently $\sigma(T_\mu) \subset \{\mu\}$. Since $X_\mu \neq \{0\}$, the spectrum of T_μ in non-empty (cf. Theorem 7.6), and therefore

$$\sigma(T_\mu) = \{\mu\}. \tag{5}$$

Consider the complementary projection $E'_\mu := I - E_\mu$, and let $X'_\mu := E'_\mu X$ and $T'_\mu := T|_{X'_\mu}$. The argument (with $h(\lambda) := (1-e_\mu)/(\zeta-\lambda)$ for a fixed $\zeta \notin \sigma_\mu$) shows that $\sigma(T'_\mu) \subset \sigma_\mu$. If the inclusion is strict, pick $\zeta \in \sigma_\mu \cap \rho(T'_\mu)$. Then $\zeta \neq \mu$, so that $\exists R(\zeta; T_\mu)$ (by (5)), and of course $\exists R(\zeta; T'_\mu)$. Let

$$V := R(\zeta; T_\mu)E_\mu + R(\zeta; T'_\mu)E'_\mu. \tag{6}$$

Clearly, $V \in B(X)$, and a simple calculation shows that $(\zeta I - T)V = V(\zeta I - T) = I$. Hence, $\zeta \in \rho(T)$, contradicting the fact that $\zeta \in \sigma_\mu (\subset \sigma(T))$. Consequently

$$\sigma(T'_\mu) = \sigma_\mu. \tag{7}$$

We also read from (6) (and the shown observation) that $V = R(\zeta; T)$ for all $\zeta \notin \sigma(T)$, that is,

$$R(\zeta; T_\mu) = R(\zeta; T)|_{X_\mu} \quad (\zeta \neq \mu); \tag{8}$$

$$R(\zeta; T'_\mu) = R(\zeta; T)|_{X'_\mu} \quad (\zeta \notin \sigma_\mu). \tag{9}$$

(Rather than discussing an isolated point, we could consider any closed subset σ of the spectrum, whose complement in the spectrum is also closed; such a set is called a *spectral set*. The previous arguments and conclusions go through with very minor changes.)

By (4), the isolated point μ of $\sigma(T)$ is a pole of order m of $R(\cdot; T)$ iff

$$(\mu I - T)^m X_\mu = \{0\} \text{ and } (\mu I - T)^{m-1} X_\mu \neq \{0\}. \tag{10}$$

In this case, any non-zero vector in the latter space is an eigenvector of T for the eigenvalue μ, that is, $\mu \in \sigma_p(T)$.

By (10), $X_\mu \subset \ker(\mu I - T)^m$. Let $x \in \ker(\mu I - T)^m$. Since E_μ commutes with T,

$$(\mu I - T'_\mu)^m (I - E_\mu)x = (I - E_\mu)(\mu I - T)^m x = 0.$$

By (7), $\mu \in \rho(T'_\mu)$, so that $\mu I - T'_\mu$ is one-to-one; hence, $(I - E_\mu)x = 0$, and therefore, $x \in X_\mu$, and we conclude that

$$X_\mu = \ker(\mu I - T)^m. \tag{11}$$

9.12 Compact operators

Definition 9.23. Let X be a Banach space, and denote by S its closed unit ball. An operator $T \in B(X)$ is compact if the set TS is conditionally compact.

Equivalently, T is compact iff it maps bounded sets onto conditionally compact sets.

In terms of sequences, the compactness of T is characterized by the property: if $\{x_n\}$ is a bounded sequence, then $\{Tx_n\}$ has a convergent subsequence.

Denote by $K(X)$ the set of all compact operators on X.

Proposition 9.24.

(i) $K(X)$ *is a closed two-sided ideal in* $B(X)$.

(ii) $K(X) = B(X)$ *iff* X *has finite dimension.*

(iii) The restriction of a compact operator to a closed invariant subspace is compact.

Proof. (i) $K(X)$ is trivially stable under linear combinations. Let $T \in K(X), A \in B(X)$, and let $\{x_n\}$ be a bounded sequence. Let then $\{Tx_{n_k}\}$ be a convergent subsequence of $\{Tx_n\}$. Then $\{ATx_{n_k}\}$ converges (by continuity of A), that is, $AT \in K(X)$. Also $\{Ax_n\}$ is bounded (by boundedness of A), and therefore $\{TAx_{n'_k}\}$ converges for some subsequence, that is, $TA \in K(X)$, and we conclude that $K(X)$ is a two-sided ideal in $B(X)$.

Suppose $\{T_m\} \subset K(X)$ converges in $B(X)$ to T. Let $\{x_n\} \subset X$ be bounded, say, $\|x_n\| < M$ for all n. By a Cantor diagonal process, we can select a subsequence $\{x_{n_k}\}$ of $\{x_n\}$ such that $\{T_m x_{n_k}\}_k$ converges *for all* m. Given $\epsilon > 0$, let $m_0 \in \mathbb{N}$ be such that $\|T - T_m\| < \epsilon/(4M)$ for all $m > m_0$. Fix $m > m_0$, and then $k_0 = k_0(m)$ such that $\|T_m x_{n_k} - T_m x_{n_j}\| < \epsilon/2$ for all $k, j > k_0$. Then for all $k, j > k_0$,

$$\begin{aligned}\|Tx_{n_k} - Tx_{n_j}\| &\leq \|(T - T_m)(x_{n_k} - x_{n_j})\| + \|T_m x_{n_k} - T_m x_{n_j}\| \\ &< [\epsilon/(4M)]2M + \epsilon/2 = \epsilon,\end{aligned}$$

and we conclude that $\{Tx_{n_k}\}$ converges to some element y.
Hence,

$$\limsup_k \|Tx_{n_k} - y\| \leq \epsilon,$$

and therefore, $Tx_{n_k} \to y$ by the arbitrariness of ϵ.

(ii) If X has finite dimension, any linear operator T on X maps bounded sets onto bounded sets, and a bounded set in X is conditionally compact.

Conversely, if $K(X) = B(X)$, then, equivalently, the identity operator I is compact, and therefore the closed unit ball $S = IS$ is compact. Hence, X has finite dimension, by Theorem 5.27.

The proof of (iii) is trivial. □

Theorem 9.25 (Schauder). *$T \in K(X)$ iff $T^* \in K(X^*)$.*

Proof. (1) Let $T \in K(X)$, and let $\{x_n^*\}$ be a bounded sequence in X^*, say, $\|x_n^*\| \le M$ for all n. Then, for all n, $|x_n^* x| \le M\|T\|$ for all $x \in \overline{TS}$ and $|x_n^* x - x_n^* y| \le M\|x-y\|$ for all $x, y \in X$, that is, the sequence of functions $\{x_n^*\}$ is uniformly bounded and equicontinuous on the compact metric space $\overline{TS}$. By Theorem 5.48 (also cf. Exercise 3), there exists a subsequence $\{x_{n_k}^*\}$ of $\{x_n^*\}$ converging uniformly on $\overline{TS}$. Hence,

$$\sup_{x \in S} |x_{n_k}^*(Tx) - x_{n_j}^*(Tx)| \to 0 \quad (k, j \to \infty),$$

that is, $\|T^* x_{n_k}^* - T^* x_{n_j}^*\| \to 0$ as $k, j \to \infty$, and consequently $\{T^* x_{n_k}^*\}$ converges (strongly) in X^*. This proves that T^* is compact.

(2) Let T^* be compact. By Part (1) of the proof, T^{**} is compact. Let $\{x_n\} \subset S$. Then $\|\hat{x}_n\| = \|x_n\| \le 1$, and therefore, $T^{**}\hat{x}_{n_k}$ converges in X^{**} for some $1 \le n_1 < n_2 < \cdots$, that is,

$$\sup_{\|x^*\|=1} |(T^{**}\hat{x}_{n_k})x^* - (T^{**}\hat{x}_{n_j})x^*| \to 0 \quad (k, j \to \infty).$$

Equivalently,

$$\sup_{\|x^*\|=1} |x^* T x_{n_k} - x^* T x_{n_j}| \to 0,$$

that is, $\|Tx_{n_k} - Tx_{n_j}\| \to 0$ as $k, j \to \infty$, and consequently T is compact. □

Lemma 9.26. *Let Y be a proper closed subspace of the Banach space X. Then $\sup_{x \in X_1} d(x, Y) = 1$. ($X_1$ denotes the unit sphere of X.)*

Proof. Let $1 > \epsilon > 0$. If $d(x, Y) = 0$ for all $x \in X_1$, then since Y is closed, $X_1 \subset Y$, and therefore, $X \subset Y$ (because Y is a subspace), contrary to the assumption that Y is a *proper* subspace of X. Thus there exists $x_1 \in X_1$ such that $\delta := d(x_1, Y) > 0$. By definition of $d(x_1, Y)$, there exists $y_1 \in Y$ such that $(\delta \le) d(x_1, y_1) < (1+\epsilon)\delta$. Let $u = x_1 - y_1$ and $x = u/\|u\|$. Then $x \in X_1$, $\|u\| < (1+\epsilon)\delta$, and for all $y \in Y$

$$(1+\epsilon)\delta\|x-y\| \ge \|u\|\|x-y\| = \big\|u - \|u\|y\big\| = \big\|x_1 - (y_1 + \|u\|y)\big\| \ge \delta.$$

Hence, $\|x - y\| \ge 1/(1+\epsilon) > 1 - \epsilon$ for all $y \in Y$, and therefore $d(x, Y) \ge 1 - \epsilon$. Since we have trivially $d(x, Y) \le 1$, the conclusion of the lemma follows. □

Theorem 9.27 (Riesz–Schauder). *Let T be a compact operator on the Banach space X. Then*

(i) $\sigma(T)$ is at most countable. If $\{\mu_n\}$ is a sequence of distinct non-zero points of the spectrum, then $\mu_n \to 0$.

(ii) Each non-zero point $\mu \in \sigma(T)$ is an isolated point of the spectrum, and is an eigenvalue of T and a pole of the resolvent of T. If m is the order of the pole μ, and E_μ is the Riesz projection for T at μ, then its range

$E_\mu X$ equals $\ker(\mu I - T)^m$ *and is finite dimensional. In particular, the μ-eigenspace of T is finite dimensional.*

Proof. (1) Let μ be a non-zero complex number, and let $\{x_n\} \subset X$ be such that $(\mu I - T)x_n$ converge to some y. If $\{x_n\}$ is unbounded, say $0 < \|x_n\| \to \infty$ without loss of generality (w.l.o.g.), consider the unit vectors $z_n := x_n/\|x_n\|$. Since T is compact, there exist $1 \le n_1 < n_2, \cdots$ such that $Tz_{n_k} \to v \in X$. Then

$$\mu z_{n_k} = \frac{1}{\|x_{n_k}\|}(\mu I - T)x_{n_k} + Tz_{n_k} \to 0y + v = v. \tag{*}$$

Hence

$$\mu v = \lim_k T(\mu z_{n_k}) = Tv.$$

If $\mu \notin \sigma_p(T)$, we must have $v = 0$. Then by (*) $|\mu| = \|\mu z_{n_k}\| \to 0$, a contradiction. Therefore (if $\mu \notin \sigma_p(T)$!), the sequence $\{x_n\}$ is bounded, and has therefore a subsequence $\{x_{n_k}\}$ such that $\exists \lim_k Tx_{n_k} := u$. Then as shown

$$x_{n_k} = \mu^{-1}[(\mu I - T)x_{n_k} + Tx_{n_k}] \to \mu^{-1}(y + u) := x,$$

and therefore $y = \lim_k (\mu I - T)x_{n_k} = (\mu I - T)x \in (\mu I - T)X$. Thus $(\mu I - T)X$ is closed. This proves that a non-zero μ *is either in $\sigma_p(T)$ or else the range of $\mu I - T$ is closed.* In the later case, if this range is *dense* in X, $\mu I - T$ is onto (and one-to-one!), and therefore $\mu \in \rho(T)$. If the range is not dense in X, it is a *proper closed subspace* of X; by Corollary 5.4. there exists $x^* \neq 0$ such that $x^*(\mu I - T)x = 0$ for all $x \in X$, that is, $(\mu I - T^*)x^* = 0$. Thus $\mu \in \sigma_p(T^*)$. In conclusion, *if $\mu \in \sigma(T)$ is not zero, then $\mu \in \sigma_p(T) \cup \sigma_p(T^*)$.*

(2) Suppose $\mu_n, n = 1, 2, \ldots,$ are distinct eigenvalues of T that do not converge to zero. By passing if necessary to a subsequence, we may assume that $|\mu_n| \ge \epsilon$ for all n, for some positive ϵ. Let x_n be an eigenvector of T corresponding to the eigenvalue μ_n. Then $\{x_n\}$ is necessarily linearly independent (an elementary linear algebra fact!). Setting $Y_n := \operatorname{span}\{x_1, \ldots, x_n\}, Y_{n-1}$ is therefore, a *proper* closed T-invariant subspace of the Banach space Y_n , and clearly $(\mu_n I - T)Y_n \subset Y_{n-1}$ (for all $n > 1$). By Lemma 9.26, there exists $y_n \in Y_n$ such that $\|y_n\| = 1$ and $d(y_n, Y_{n-1}) > 1/2$, for each $n > 1$. Set $z_n = y_n/\mu_n$. Since $\|z_n\| \le 1/\epsilon$, there exist $1 < n_1 < n_2 < \cdots$ such that Tz_{n_k} converges. However, for $j > k$,

$$\|Tz_{n_j} - Tz_{n_k}\| = \|y_{n_j} - [(\mu_{n_j} I - T)z_{n_j} + Tz_{n_k}]\| > 1/2,$$

since the vector is square brackets belongs to $Y_{n_j - 1}$, contradiction. This proves that *if $\{\mu_n\}$ is a sequence of distinct eigenvalues of T, then $\mu_n \to 0$.*

(3) Suppose $\mu \in \sigma(T), \mu \neq 0$, is not an isolated point of the spectrum, and let then $\mu_n, (n \in \mathbb{N})$ be distinct non-zero points of the spectrum converging to μ. By the conclusion of Part (1) of the proof, $\{\mu_n\} \subset \sigma_p(T) \cup \sigma_p(T^*)$. Since the set $\{\mu_n\}$ is infinite, at least one of its intersections with $\sigma_p(T)$ and $\sigma_p(T^*)$ is infinite. This infinite intersection converges to zero, by Part (2) of the proof (since both T and T^* are compact, by Theorem 9.25). Hence $\mu = 0$, a contradiction! This

shows that the non-zero points of $\sigma(T)$ are isolated points of the spectrum. Since $\sigma(T)$ is compact, it then follows that it is at most countable.

(4) Let $\mu \neq 0, \mu \in \sigma(T)$, and let E_μ be the Riesz projection for T at (the isolated point) μ. As before, let $X_\mu = E_\mu X$ and $T_\mu = T|_{X_\mu}$. Let S_μ denote the closed unit ball of X_μ. Since $\sigma(T_\mu) = \{\mu\}$ (cf. (5) of Section 9.11), we have $0 \in \rho(T_\mu)$, that is, $\exists T_\mu^{-1} \in B(X_\mu)$, and consequently $T_\mu^{-1} S_\mu$ is bounded. The latter's image by the compact operator T_μ (cf. Proposition 9.24 (iii)) is then conditionally compact; this image is the closed set S_μ, hence S_μ is compact, and therefore X_μ is finite dimensional (by Theorem 5.27). Since $\sigma(\mu I - T_\mu) = \mu - \sigma(T_\mu) = \{0\}$ by (5) of Section 9.11, the operator $\mu I - T_\mu$ on the *finite dimensional space* X_μ is *nilpotent*, that is, there exists $m \in \mathbb{N}$ such that $(\mu I - T_\mu)^m = 0$ but $(\mu I - T_\mu)^{m-1} \neq 0$. Equivalently,

$$(\mu I - T)^m E_\mu = 0 \quad \text{and} \quad (\mu I - T)^{m-1} E_\mu \neq 0.$$

By (4) of Section 9.11, μ is a pole of order m of $R(\cdot;T)$, hence an eigenvalue of T (cf. observation following (10) of Section 9.11), and $\ker(\mu I - T)^m = X_\mu$ by (11) of Section 9.11. □

Exercises

[The first two exercises provide the proofs of theorems used in this chapter.]

Runge's theorem

1. Let $S^2 = \bar{\mathbb{C}}$ denote the Riemann sphere, and let $K \subset \mathbb{C}$ be compact. Fix a point a_j in each component V_j of $S^2 - K$, and let $\mathcal{R}(\{a_j\})$ denote the set of all rational functions with poles in the set $\{a_j\}$.

   If μ is a complex Borel measure on K, we define its *Cauchy transform* $\tilde{\mu}$ by

$$\tilde{\mu}(z) = \int_K \frac{d\mu(w)}{w - z} \quad (z \in S^2 - K). \tag{1}$$

   Prove

   (a) $\tilde{\mu}$ is analytic in $S^2 - K$.

   (b) For $a_j \neq \infty$, let $d_j = d(a_j, K)$ and fix $z \in B(a_j, r) \subset V_j$ (necessarily $r < d_j$). Observe that

$$\frac{1}{w - z} = \sum_{n=0}^{\infty} \frac{(z - a_j)^n}{(w - a_j)^{n+1}}, \tag{2}$$

   and the series converges uniformly for $w \in K$.

   For $a_j = \infty$, we have

$$\frac{1}{w - z} = -\sum_{n=0}^{\infty} \frac{w^n}{z^{n+1}} \quad (|z| > r), \tag{3}$$

   and the series converges uniformly for $w \in K$.

(c) If $\int_K h\,d\mu = 0$ for all $h \in \mathcal{R}(\{a_j\})$, then $\tilde{\mu}(z) = 0$ for all $z \in B(a_j, r)$, hence, for all $z \in V_j$, for all j, and therefore $\tilde{\mu} = 0$ on $S^2 - K$.

(d) Let $\Omega \subset \mathbb{C}$ be open such that $K \subset \Omega$. If f is analytic in Ω and μ is as in Part (c), then $\int_K f\,d\mu = 0$. (Hint: represent $f(z) = (1/2\pi i)\int_\Gamma f(w)/(w-z)dw$ for all $z \in K$, where $\Gamma \in \Gamma(K,\Omega)$, cf. Notation 9.18, and use Fubini's theorem.)

(e) Prove that $\mathcal{R}(\{a_j\})$ is $C(K)$-dense in $H(\Omega)$ (the subspace of $C(K)$ consisting of the analytic functions in Ω restricted to K). (Hint: Theorem 4.9, Corollary 5.3, and Part (d).) The result in Part (e) is *Runge's theorem*. In particular, the rational functions with poles off K are $C(K)$-dense in $H(\Omega)$.

(f) If $S^2 - K$ is *connected*, the polynomials are $C(K)$-dense in $H(\Omega)$. (Hint: apply Part (e) with $a = \infty$ in the single component of $S^2 - K$.)

Hartogs–Rosenthal's theorem

2. Use notation as in Exercise 1. Let m denote the $\mathbb{R}^2$-Lebesgue measure.

(a) The integral defining the Cauchy transform $\tilde{\mu}$ converges absolutely m-a.e. (Hint: show that

$$\int_{\mathbb{R}^2}\int_K \frac{d|\mu|(w)}{|w-z|}dx\,dy < \infty$$

by using Tonelli's theorem and polar coordinates.)

(b) Let $\mathcal{R}(K)$ denote the space of rational functions with poles off K. Then $\int_K h\,d\mu = 0$ for all $h \in \mathcal{R}(K)$ iff $\tilde{\mu} = 0$ off K. (Hint: use Cauchy's formula and Fubini's theorem for the non-trivial implication.)

(c) It can be shown that if $\tilde{\mu} = 0$ m-a.e., then $\mu = 0$. Conclude that if $m(K) = 0$ and μ is a complex Borel measure on K such that $\int_K h\,d\mu = 0$ for all $h \in \mathcal{R}(K)$, then $\mu = 0$. Consequently, if $m(K) = 0$, then $\mathcal{R}(K)$ is dense in $C(K)$ (cf. Theorem 4.9 and Corollary 5.6). This is the Hartogs–Rosenthal theorem.

Arzelà–Ascoli's theorem

We give an alternative, "sequential" proof of Theorem 5.48, which is closer to the sequential nature of the proof of Theorem 9.25.

3. Let X be a compact metric space. Recall from Section 5.9 that a set $\mathcal{F} \subset C(X)$ is *equicontinuous* if for each $\epsilon > 0$, there exists $\delta > 0$ such that $|f(x) - f(y)| < \epsilon$ for all $f \in \mathcal{F}$ and $x, y \in X$ such that $d(x,y) < \delta$, and it is *pointwise bounded* if $\sup_{f\in\mathcal{F}} |f(x)| < \infty$ for each

$x \in X$. Prove that if $\mathcal{F}$ is pointwise bounded and equicontinuous, then it is relatively compact in $C(X)$. Sketch: X is necessarily separable. Let $\{a_k\}$ be a countable dense set in X. Let $\{f_n\} \subset \mathcal{F}$. $\{f_n(a_1)\}$ is a bounded complex sequence; therefore there is a subsequence $\{f_{n,1}\}$ of $\{f_n\}$ converging at a_1; $\{f_{n,1}(a_2)\}$ is a bounded complex sequence, and therefore there is a subsequence $\{f_{n,2}\}$ of $\{f_{n,1}\}$ converging at a_2 (and a_1). Continuing inductively, we get subsequences $\{f_{n,r}\}$ such that the $(r+1)$-th subsequence is a subsequence of the rth subsequence, and the rth subsequence converges at the points $a_1, \ldots, a_r$. The diagonal subsequence $\{f_{n,n}\}$ converges at *all* the points a_k. Use the compactness of X and an $\epsilon/5$ argument to show that $\{f_{n,n}\}$ is Cauchy in $C(X)$.

Compact normal operators

4. Let X be a Hilbert space, and $T \in K(X)$ be normal. Prove that there exist a sequence $\{\lambda_n\} \in c_0$ and a sequence $\{E_n\}$ of pairwise orthogonal finite rank projections such that $\sum_{n=1}^{N} \lambda_n E_n \to T$ in $B(X)$ as $N \to \infty$. This is the spectral theorem for compact normal operators.

Logarithms of Banach algebra elements

5. Let $\mathcal{A}$ be a unital Banach algebra, and let $x \in \mathcal{A}$. Prove:

   (a) If 0 belongs to the unbounded component V of $\rho(x)$, then $x \in \exp \mathcal{A} (:= \{e^a; a \in \mathcal{A}\})$ (that is, x has a logarithm in $\mathcal{A}$). Hint: $\Omega := V^c$ is a simply connected open subset of $\mathbb{C}$ containing $\sigma(x)$, and the analytic function $f_1(\lambda) = \lambda$ does not vanish on Ω. Therefore, there exists g analytic in Ω such that $e^g = f_1$ (cf. Exercise 28 of Chapter 7).

   (b) The group generated by $\exp \mathcal{A}$ is an open subset of $\mathcal{A}$.

6. Let $\mathcal{A}$ be a unital Banach algebra, and let G_e denote the component of $G := G(\mathcal{A})$ containing the identity e. Prove:

   (a) G_e is open.

   (b) G_e is a normal subgroup of G.

   (c) $\exp \mathcal{A} \subset G_e$.

   (d) $\bigcup \exp \mathcal{A} \cdots \exp \mathcal{A}$ (the union of all finite products) is an open subset of G_e (cf. Exercise 5(b)).

   (e) Let H be the group generated by $\exp \mathcal{A}$. Then H is an open and closed subset of G_e. Conclude that $H = G_e$.

   (f) If $\mathcal{A}$ is commutative, then $G_e = \exp \mathcal{A}$.

Non-commutative Taylor theorem

7. Use notation as in Exercise 10, Chapter 7. Let $\mathcal{A}$ be a unital Banach algebra, and let $a, b \in \mathcal{A}$. Prove the following *non-commutative Taylor theorem* for each $f \in H(\sigma(a,b))$:

$$\begin{aligned} f(b) &= \sum_{j=0}^{\infty} (-1)^j \frac{f^{(j)}(a)}{j!}[C(a,b)^j e] \\ &= \sum_{j=0}^{\infty} [C(a,b)^j e] \frac{f^{(j)}(a)}{j!}. \end{aligned}$$

In particular, if a, b commute,

$$f(b) = \sum \frac{f^{(j)}(a)}{j!}(b-a)^j \tag{*}$$

for all $f \in H(\sigma(a,b))$, where (in this special case)

$$\sigma(a,b) = \{\lambda \in \mathbb{C}; d(\lambda, \sigma(a)) \le r(b-a)\}.$$

If $b - a$ is quasi-nilpotent, (*) is valid for all $f \in H(\sigma(a))$.

Positive operators

8. Let X be a Hilbert space. Recall that $T \in B(X)$ is *positive* (in symbols, $T \ge 0$) iff $(Tx, x) \ge 0$ for all $x \in X$. Prove:

   (a) The positive operator T is non-singular (i.e., invertible in $B(X)$) iff $T - \epsilon I \ge 0$ for some $\epsilon > 0$ (one can also write $T \ge \epsilon I$ to express the last relation).

   (b) The (arbitrary) operator T is non-singular iff *both* $TT^* \ge \epsilon I$ and $T^*T \ge \epsilon I$ for some $\epsilon > 0$.

9. Let X be a Hilbert space, $T \in B(X)$. Prove:

   (a) If T is positive, then

   $$|(Tx, y)|^2 \le (Tx, x)(Ty, y) \text{ for all } x, y \in X.$$

   (b) Let $\{T_k\} \subset B(X)$ be a sequence of positive operators. Then $T_k \to 0$ in the s.o.t. iff it does so in the w.o.t.

   (c) If $0 \le T_k \le T_{k+1} \le K\,I$ for all k (for some positive constant K), then $\{T_k\}$ converges in $B(X)$ in the s.o.t.

Analytic functions operate on $\hat{\mathcal{A}}$

10. Let $\mathcal{A}$ be a unital commutative Banach algebra, and $a \in \mathcal{A}$. Let $f \in H(\sigma(a))$. Prove that there exists $b \in \mathcal{A}$ such that $\hat{b} = f \circ \hat{a}$ ($\hat{a}$ denotes the Gelfand transform of a.) In particular, if $\hat{a}$ vanishes nowhere, there exists $b \in \mathcal{A}$ such that $\hat{b} = 1/\hat{a}$. (This is *Wiener's theorem*.) (Hint: use the analytic operational calculus.)

Polar decomposition

11. Let X be a Hilbert space, and let $T \in B(X)$ be non-singular. Prove that there exist a unique pair of operators S, U such that S is non-singular and positive, U is unitary, and $T = US$. If T is normal, the operators S, U commute with each other and with T. (Hint: assuming the result, find out how to define S and $U|_{SX}$; verify that U is isometric on SX, etc. See Chapter 12 for an extension of this exercise.)

Cayley transform

12. Let X be a Hilbert space, and let $T \in B(X)$ be selfadjoint. Prove:

    (a) The operator $V := (T + \mathrm{i}I)(T - \mathrm{i}I)^{-1}$ (called the *Cayley transform* of T) is unitary and $1 \notin \sigma(V)$.

    (b) Conversely, every unitary operator V such that $1 \notin \sigma(V)$ is the Cayley transform of some selfadjoint operator $T \in B(X)$.

Riemann integrals of operator functions

13. Let X be a Banach space, and let $T(\cdot) : [a, b] \to B(X)$ be *strongly continuous* (that is, continuous with respect to the s.o.t. on $B(X)$). Prove:

    (a) $\|T(\cdot)\|$ is bounded and lower semi-continuous (l.s.c.) (cf. Exercise 6, Chapter 3).

    (b) For each $x \in X$, the Riemann integral $\int_a^b T(t)x\,dt$ is a well-defined element of X with norm $\leq \int_a^b \|T(t)\|\,dt\|x\|$. Therefore the operator $\int_a^b T(t)\,dt$ *defined by* $(\int_a^b T(t)\,dt)x = \int_a^b T(t)x\,dt$ has norm $\leq \int_a^b \|T(t)\|\,dt$. For each $S \in B(X), ST(\cdot)$ and $T(\cdot)S$ are strongly continuous on $[a, b]$, and $S\int_a^b T(t)\,dt = \int_a^b ST(t)\,dt$; $(\int_a^b T(t)\,dt)S = \int_a^b T(t)S\,dt$.

    (c) $(\int_a^t T(s)\,ds)'(c) = T(c)$ (derivative in the s.o.t.).

(d) If $T(\cdot) = V'(\cdot)$ (derivative in the s.o.t.) for some operator function V, then $\int_a^b T(t)\,dt = V(b) - V(a)$.

(e) If $T(\cdot) : [a, \infty) \to B(X)$ is strongly continuous and $\int_a^\infty \|T(t)\|\,dt < \infty$, then $\lim_{b\to\infty} \int_a^b T(t)\,dt := \int_a^\infty T(t)\,dt$ exists in the norm topology of $B(X)$, and $\|\int_a^\infty T(t)\,dt\| \le \int_a^\infty \|T(t)\|\,dt$. (Note that $\|T(\cdot)\|$ is l.s.c. by Part (a), and the integral on the right makes sense as the integral of a non-negative Borel function.)

Semigroups of operators

14. Let X be a Banach space, and let $T(\cdot) : [0, \infty) \to B(X)$ be such that $T(t + s) = T(t)T(s)$ for all $t, s \ge 0$ and $T(0) = I$. (Such a function is called a *semigroup of operators*.) Assume $T(\cdot)$ is (right) continuous at 0 in the s.o.t. (briefly, $T(\cdot)$ is a C_0*-semigroup*). Prove:

(a) $T(\cdot)$ is right continuous on $[0, \infty)$, in the s.o.t.

(b) Let $c_n := \sup\{\|T(t)\|; 0 \le t \le 1/n\}$. Then there exists n such that $c_n < \infty$. (Fix such an n and let $c := c_n (\ge 1)$.) (Hint: the uniform boundedness theorem.)

(c) With n and c as in Part (b), $\|T(t)\| \le M e^{at}$ on $[0, \infty)$, where $M := c^n (\ge 1)$ and $a := \log M (\ge 0)$.

(d) $T(\cdot)$ is strongly continuous on $[0, \infty)$.

(e) Let $V(t) := \int_0^t T(s)ds$. Then

$$T(h)V(t) = V(t + h) - V(h) \quad (h, t > 0).$$

Conclude that $(1/h)(T(h) - I)V(t) \to T(t) - I$ in the s.o.t., as $h \to 0+$ (i.e., the strong right derivative of $T(\cdot)V(t)$ at 0 exists and equals $T(t) - I$, for each $t > 0$). (Hint: Exercise 13, Part (c).)

(f) Let $\omega := \inf_{t>0} t^{-1} \log \|T(t)\|$. Then $\omega = \lim_{t\to\infty} t^{-1} \log \|T(t)\| (< \infty)$ (cf. Part (c)). (Hint: fix $s > 0$ and $r > s^{-1} \log \|T(s)\|$. Given $t > 0$, let $n = [t/s]$. Then $t^{-1} \log \|T(t)\| < rns/t + t^{-1} \sup_{[0,s]} \log \|T(\cdot)\|$.) ($\omega$ is called the *type* of the semigroup $T(\cdot)$.)

(g) Let ω be the type of $T(\cdot)$. Then the spectral radius of $T(t)$ is $e^{\omega t}$, for each $t \ge 0$.

More on the spectral theorem

Let T be a normal operator on a Hilbert space X and E be its resolution of the identity.

15. Prove that every isolated point in $\sigma(T)$ is an eigenvalue of T. (Remark that the converse is false.)

16. (a) Prove that for each $r > 0$, the range of $E(\{\lambda \in \mathbb{C}; |\lambda| \geq r\})$ is contained in the range $T(X)$ of T.

    (b) Prove that the range of $E(\mathbb{C}\setminus\{0\})$ is equal to $\overline{T(X)}$.

17. Assume that the set of eigenvectors of T spans a dense subspace of X. Prove that

$$T = \sum_{\lambda \in \sigma_{\mathrm{p}}(T)} \lambda E(\{\lambda\})$$

in the strong operator topology (s.o.t.); that is, denoting by $\mathcal{F}$ the set of all finite subsets of $\sigma_{\mathrm{p}}(T)$ ordered by inclusion, the net $\left(\sum_{\lambda \in F} \lambda E(\{\lambda\})\right)_{F \in \mathcal{F}}$ converges to T in the s.o.t. (Hint: to prove that the series converges use (the net version of) Part (c) of Exercise 9. To prove that the sum equals T use Theorem 9.8.)

18. Let $S \in B(X)$. Prove that the following conditions are equivalent:

    (i) T, S commute: $TS = ST$.

    (ii) for all Borel sets $\delta \subset \mathbb{C}$ the operators $E(\delta), S$ commute.

    (iii) for all $f \in \mathbb{B}(\mathbb{C})$ the operators $f(T), S$ commute.

    (Hint for (i) $\Longrightarrow$ (ii): use Fuglede's theorem (Exercise 13 in Chapter 7) to obtain that T^*, S commute. Thus, every polynomial in T and in T^* commutes with S. Deduce with the aid of Section 7.4 that $f(T), S$ commute for all $f \in C(\sigma(T))$. This implies that $E(\delta), S$ commute at least for certain Borel sets δ.)

19. Continuing Exercise 18, assume that S too is normal and has resolution of the identity F. Prove that each of the following conditions is equivalent to T, S commuting:

    (iv) for all Borel sets $\delta, \epsilon \subset \mathbb{C}$ the projections $E(\delta), F(\epsilon)$ commute.

    (v) for all $f, g \in \mathbb{B}(\mathbb{C})$ the operators $f(T), g(S)$ commute.

10

Unbounded operators

This chapter develops the basic structure theory of unbounded operators, mostly on Hilbert spaces. This beautiful and deep theory is the work of many hands, including Carleman, Friedrichs, Hilbert, von Neumann, Riesz, Stone, and Sz.-Nagy. Naturally, unbounded operators are less well-behaved than bounded operators. Nevertheless, they "arise in nature"—both mathematical and physical, for example, quantum mechanics—and they have enough structure to allow for intriguing results and applications.

The first two sections provide some preliminaries, including the *spectrum* of unbounded operators and the *Hilbert adjoint* of densely defined unbounded operators on Hilbert spaces.

The *spectral theorem* and the *operational calculus* for unbounded selfadjoint operators are established next. They are the perfect extensions to unbounded operators of the results on bounded operators in Chapter 9 carrying the same names. In contrast to the bounded case and for simplicity, we stick here to selfadjoint operators rather than normal ones. We also work out the theory of the *semi-simplicity space* of unbounded operators with real spectrum on Banach spaces. The method of proof of these results involves reducing to bounded operators.

The chapter then discusses the theory of symmetric, and in particular selfadjoint, *extensions of symmetric operators* on Hilbert spaces. It has applications in differential equations and quantum mechanics, as well as other areas of research.

The chapter ends with the subject of *quadratic forms*. We prove *the representation theorem*, linking between closed densely defined quadratic forms and positive selfadjoint operators. We also characterize closedness and closability of quadratic forms in terms of lower semi-continuity. These results are very helpful in operator theory since often it is more convenient to talk about the quadratic form associated to an operator than about the operator itself.

In the exercises we continue the study of semigroups of operators begun in the exercises of Chapter 9, starting with the (generally unbounded) *generator*. We

Introduction to Modern Analysis. Second Edition. Shmuel Kantorovitz and Ami Viselter, Oxford University Press.
 DOI: 10.1093/oso/9780192849540.003.0010

also use quadratic forms to prove that every densely defined positive operator on a Hilbert space possesses a positive selfadjoint extension; the particular extension we construct is called the *Friedrichs extension.*

10.1 Basics

This chapter deals with (linear) operators T with domain $D(T)$ and range $R(T)$ in a Banach space X; $D(T)$ and $R(T)$ are (linear) subspaces of X. The operators S, T are *equal* if $D(S) = D(T)$ and $Sx = Tx$ for all x in the (common) domain of S and T. If S, T are operators such that $D(S) \subset D(T)$ and $T|_{D(S)} = S$, we say that T is an *extension* of S (notation: $S \subset T$).

The algebraic operations between unbounded operators are defined with the obvious restrictions on domains. Both sum and product are associative, but the distributive laws take the form

$$AB + AC \subset A(B + C); \quad (A + B)C = AC + BC.$$

The *graph* of T is the subspace of $X \times X$ given by

$$\Gamma(T) := \{[x, Tx]; x \in D(T)\}.$$

The operator T is *closed* if $\Gamma(T)$ is a *closed* subspace of $X \times X$.

A convenient elementary criterion for T being closed is the following condition:

If $\{x_n\} \subset D(T)$ is such that $x_n \to x$ and $Tx_n \to y$, then $x \in D(T)$ and $Tx = y$.

Clearly, if $D(T)$ is closed and T is continuous on $D(T)$, then T is a closed operator. In particular, every $T \in B(X)$ is closed. Conversely, if T is a closed operator with closed domain (hence a Banach space!), then T is continuous on $D(T)$, by the *closed graph theorem.* Also if T is closed and continuous (on its domain), then it has a closed domain.

If $B \in B(X)$ and T is closed, then $T + B$ and TB (with their "maximal domains" $D(T)$ and $\{x \in X; Bx \in D(T)\}$, respectively) are closed operators. In particular, the operators $\lambda I - T$ and λT are closed, for any $\lambda \in \mathbb{C}$.

If $B \in B(X)$ is non-singular and T is closed, then BT (with domain $D(T)$) is closed.

Usually, the norm taken on $X \times X$ is $\|[x, y]\| = \|[x, y]\|_1 := \|x\| + \|y\|$, or in case X is a Hilbert space, $\|[x, y]\| = \|[x, y]\|_2 := \sqrt{\|x\|^2 + \|y\|^2}$. These norms are equivalent, since

$$\|[x, y]\|_2 \le \|[x, y]\|_1 \le \sqrt{2}\|[x, y]\|_2.$$

If X is a Hilbert space, the space $X \times X$ (also denoted $X \oplus X$) is a Hilbert space with the inner product

$$([x, y], [u, v]) := (x, u) + (y, v),$$

and the norm induced by this inner product is indeed $\|[x,y]\| := \|[x,y]\|_2$.

The graph norm on $D(T)$ is defined by

$$\|x\|_T := \|[x,Tx]\| \quad (x \in D(T)).$$

We shall denote by $[D(T)]$ the space $D(T)$ with the graph norm. The space $[D(T)]$ is *complete* iff T is a closed operator.

If the operator S has a closed extension T, it clearly satisfies the property

$$\text{If } \{x_n\} \subset D(S) \text{ is such that } x_n \to 0 \text{ and } Sx_n \to y, \quad \text{then} \quad y = 0.$$

An operator S with this property is said to be *closable*. Conversely, if S is closable, then the $X \times X$-closure of its graph, $\overline{\Gamma(S)}$, is the graph of a (necessarily closed) operator $\bar{S}$, called the *closure* of S. Indeed, if $[x,y],[x,y'] \in \overline{\Gamma(S)}$, there exist sequences $\{[x_n, Sx_n]\}$ and $\{[x'_n, Sx'_n]\}$ in $\Gamma(S)$ converging respectively to $[x,y]$ and $[x,y']$ in $X \times X$. Then $x_n - x'_n \in D(S) \to 0$ and $S(x_n - x'_n) \to y - y'$. Therefore, $y - y' = 0$ since S is closable. Consequently, the map $\bar{S} : x \to y$ is well defined, clearly linear, and by definition,

$$\Gamma(\bar{S}) = \overline{\Gamma(S)}.$$

Hence the closable operator S has the (minimal) closed extension $\bar{S}$.

By definition, $D(\bar{S}) = \{x \in X; \exists \{x_n\} \subset D(S)$ such that $x_n \to x$ and $\exists \lim Sx_n\}$ and $\bar{S}x$ is equal to the shown limit for $x \in D(\bar{S})$.

A *core* for a closed operator T is a linear subspace D of $D(T)$ such that $\overline{T|_D} = T$; that is, $\Gamma(T|_D)$ is dense in $\Gamma(T)$; equivalently, D is dense in $[D(T)]$. This means that to each $x \in D(T)$ there is a sequence $\{x_n\}$ in D converging to x such that $\{Tx_n\}$ converges to Tx.

If T is one-to-one, the inverse operator T^{-1} with domain $R(T)$ and range $D(T)$ has the graph

$$\Gamma(T^{-1}) = J\Gamma(T),$$

where J is the isometric automorphism of $X \times Y$ given by $J[x,y] = [y,x]$. Therefore T is closed iff T^{-1} is closed. In particular, if $T^{-1} \in B(X)$, then T is closed.

The *resolvent set* $\rho(T)$ of T is the set of all $\lambda \in \mathbb{C}$ such that $\lambda I - T$ *has an inverse in* $B(X)$; the inverse operator is called the *resolvent* of T, and is denoted by $R(\lambda;T)$ (or $R(\lambda)$, when T is understood). If $\rho(T) \neq \emptyset$, and λ is any point in $\rho(T)$, then $R(\lambda)^{-1}$ (with domain $D(T)$) is closed, and therefore $T = \lambda I - R(\lambda;T)^{-1}$ is closed. On the other hand, if T is closed and λ is such that $\lambda I - T$ is bijective, then $\lambda \in \rho(T)$ (because $(\lambda I - T)^{-1}$ is closed and everywhere defined, hence belongs to $B(X)$, by the closed graph theorem).

By definition, $TR(\lambda) = \lambda R(\lambda) - I \in B(X)$, while $R(\lambda)T = (\lambda R(\lambda) - I)|_{D(T)}$.

The complement of $\rho(T)$ in $\mathbb{C}$ is the *spectrum* of T, $\sigma(T)$. By the preceding remark, the spectrum of the closed operator T is the disjoint union of the following sets:

- the *point spectrum* of T, $\sigma_p(T)$, which consists of all scalars λ for which $\lambda I - T$ is not one-to-one.

- the *continuous spectrum* of T, $\sigma_c(T)$, which consists of all λ for which $\lambda I - T$ is one-to-one but *not* onto, and its range is dense in X.
- the *residual spectrum* of T, $\sigma_r(T)$, which consists of all λ for which $\lambda I - T$ is one-to-one, and its range is *not dense in* X.

Theorem 10.1. *Let T be any (unbounded) operator. Then $\rho(T)$ is open, and $R(\cdot)$ is analytic on $\rho(T)$ and satisfies the "resolvent identity"*

$$R(\lambda) - R(\mu) = (\mu - \lambda)R(\lambda)R(\mu) \quad (\lambda, \mu \in \rho(T)).$$

In particular, $R(\lambda)$ commutes with $R(\mu)$.

Moreover, $\|R(\lambda)\| \geq 1/d(\lambda, \sigma(T))$.

Proof. We assume without loss of generality that the resolvent set is non-empty. Let then $\lambda \in \rho(T)$, and denote $r = \|R(\lambda)\|^{-1}$. We wish to prove that the disc $B(\lambda, r)$ is contained in $\rho(T)$. This will imply that $\rho(T)$ is open and $d(\lambda, \sigma(T)) \geq r$ (i.e., $\|R(\lambda)\| \geq 1/d(\lambda, \sigma(T))$).

For $\mu \in B(\lambda, r)$, the series

$$S(\mu) = \sum_{k=0}^{\infty} [(\lambda - \mu)R(\lambda)]^k$$

converges in $B(X)$, commutes with $R(\lambda)$, and satisfies the identity

$$(\lambda - \mu)R(\lambda)S(\mu) = S(\mu) - I.$$

For $x \in D(T)$,

$$S(\mu)R(\lambda)(\mu I - T)x = S(\mu)R(\lambda)[(\lambda I - T) - (\lambda - \mu)I]x = S(\mu)x - [S(\mu) - I]x = x$$

by the above identity, and similarly, for all $x \in X$,

$$(\mu I - T)R(\lambda)S(\mu)x = [(\lambda I - T) - (\lambda - \mu)I]R(\lambda)S(\mu)x = S(\mu)x - [S(\mu) - I]x = x.$$

This shows that $\mu \in \rho(T)$ and $R(\mu) = R(\lambda)S(\mu)$ for all $\mu \in B(\lambda, r)$.

In particular, $R(\cdot)$ is analytic in $\rho(T)$ (since it is locally the sum of a $B(X)$-convergent power series).

Finally, for $\lambda, \mu \in \rho(T)$, we have on X:

$$\begin{aligned}
&(\lambda I - T)[R(\lambda) - R(\mu) - (\mu - \lambda)R(\lambda)R(\mu)] \\
&= I - [(\lambda - \mu)I + (\mu I - T)]R(\mu) - (\mu - \lambda)R(\mu) \\
&= I - (\lambda - \mu)R(\mu) - I + (\lambda - \mu)R(\mu) = 0.
\end{aligned}$$

Since $\lambda I - T$ is one-to-one, the resolvent identity follows. □

Theorem 10.2. *Let T be an unbounded operator in the Banach space X, with $\rho(T) \neq \emptyset$. Fix $\alpha \in \rho(T)$ and let $h(\lambda) = 1/(\alpha - \lambda)$. Then h maps $\sigma(T) \cup \{\infty\}$ onto $\sigma(R(\alpha))$.*

Proof. (In order to reduce the number of brackets in the following formulae, we shall write R_λ instead of $R(\lambda)$.)

Taking complements in $\mathbb{C} \cup \{\infty\}$, we must show that h maps $\rho(T)$ onto $\rho(R_\alpha) \cup \{\infty\}$. Since $h(\alpha) = \infty$, we consider $\lambda \neq \alpha$ in $\rho(T)$, and define

$$V := (\alpha - \lambda)[I + (\alpha - \lambda)R_\lambda].$$

Then V commutes with $h(\lambda)I - R_\alpha$ and by the resolvent identity (cf. Theorem 10.1)

$$[h(\lambda)I - R_\alpha]V = I + (\alpha - \lambda)[R_\lambda - R_\alpha - (\alpha - \lambda)R_\alpha R_\lambda] = I.$$

This shows that $h(\lambda) \in \rho(R_\alpha)$ and $R(h(\lambda); R_\alpha) = V$. Hence h maps $\rho(T)$ into $\rho(R_\alpha) \cup \{\infty\}$.

Next, let $\mu \in \rho(R_\alpha)$. If $\mu = 0$, $T = \alpha I - (\alpha I - T) = \alpha I - R_\alpha^{-1} \in B(X)$, contrary to our hypothesis. Hence, $\mu \neq 0$, and let then $\lambda = \alpha - 1/\mu$ (so that $h(\lambda) = \mu$). Let

$$W := \mu R_\alpha R(\mu; R_\alpha).$$

Then W commutes with $\lambda I - T$ and

$$\begin{aligned}(\lambda I - T)W &= [(\lambda - \alpha)I + (\alpha I - T)]W = \mu[(\lambda - \alpha)R_\alpha + I]R(\mu; R_\alpha)\\ &= (\mu I - R_\alpha)R(\mu; R_\alpha) = I.\end{aligned}$$

Thus $\lambda \in \rho(T)$, and we conclude that h maps $\rho(T)$ onto $\rho(R_\alpha) \cup \{\infty\}$. □

10.2 The Hilbert adjoint

Terminology 10.3. Let T be an operator with *dense domain* $D(T)$ in the *Hilbert space* X. For $y \in X$ fixed, consider the function

$$\phi(x) = (Tx, y) \quad (x \in D(T)). \tag{1}$$

If ϕ is continuous, it has a unique extension as a continuous linear functional on X (since $D(T)$ is dense in X), and there exists therefore a unique $z \in X$ such that

$$\phi(x) = (x, z) \quad (x \in D(T)). \tag{2}$$

(Conversely, if there exists $z \in X$ such that (2) holds, then ϕ is continuous on $D(T)$.)

Let $D(T^*)$ denote the subspace of all y for which ϕ is continuous on $D(T)$ (equivalently, for which $\phi = (\cdot, z)$ for some $z \in X$). Define $T^* : D(T^*) \to X$ by $T^*y = z$ (the map T^* is well defined, by the uniqueness of z for given y). The defining identity for T^* is then

$$(Tx, y) = (x, T^*y) \quad (x \in D(T),\ y \in D(T^*)). \tag{3}$$

It follows clearly from (3) that T^* (with domain $D(T^*)$) is a linear operator. It is called the *adjoint operator* of T.

If S is another operator on X and $T \subset S$, then (S has dense domain and) evidently $S^* \subset T^*$.

By (2), $[y, z] \in \Gamma(T^*)$ iff $(Tx, y) = (x, z)$ for all $x \in D(T)$, that is, iff

$$([Tx, -x], [y, z]) = 0 \quad \text{for all } x \in D(T).$$

Consider the isometric automorphism of $X \times X$ defined by

$$Q[x, y] = [y, -x].$$

The preceding statement means that $[y, z] \in \Gamma(T^*)$ iff $[y, z]$ is orthogonal to $Q\Gamma(T)$ in $X \times X$. Hence

$$\Gamma(T^*) = (Q\Gamma(T))^{\perp}. \tag{4}$$

In particular, it follows from (4) that T^* is *closed.*

One verifies easily that if $B \in B(X)$, then

$$(T + B)^* = T^* + B^* \text{ and } (BT)^* = T^* B^*.$$

It follows in particular (or directly) that $(\lambda T)^* = [(\lambda I)T]^* = \bar{\lambda} T^*$.

If $T = T^*$, the operator T is called a *selfadjoint* operator. Since T^* is closed, a selfadjoint operator is necessarily *closed and densely defined.* An everywhere defined selfadjoint operator is necessarily bounded by the closed graph theorem.

The operator T is *symmetric* if

$$(Tx, y) = (x, Ty) \quad (x, y \in D(T)). \tag{5}$$

If T is densely defined (so that T^* exists), Condition (5) is equivalent to $T \subset T^*$. If T is everywhere defined, it is symmetric iff it is selfadjoint. Therefore, a symmetric everywhere defined operator is a bounded selfadjoint operator.

Selfadjoint operators are *maximal symmetric*: if T, S are operators on X with T selfadjoint, S symmetric, and $T \subset S$, then $S \subset S^* \subset T^* = T$, thus $T = S$.

If T is one-to-one with *domain and range both dense in* X, the adjoint operators T^* and $(T^{-1})^*$ both exist. If $T^* y = 0$ for some $y \in D(T^*)$, then for all $x \in D(T)$

$$(Tx, y) = (x, T^* y) = (x, 0) = 0, \tag{6}$$

and therefore $y = 0$ since $R(T)$ is dense. Thus, T^* is one-to-one, and $(T^*)^{-1}$ exists. By (4)

$$\begin{aligned}\Gamma((T^{-1})^*) &= [Q\Gamma(T^{-1})]^{\perp} = [QJ\Gamma(T)]^{\perp} \\ &= [-JQ\Gamma(T)]^{\perp} = J[Q\Gamma(T)]^{\perp} = J\Gamma(T^*) = \Gamma((T^*)^{-1}),\end{aligned}$$

since $(JA)^{\perp} = JA^{\perp}$ for any $A \subset X \times X$. Therefore

$$(T^{-1})^* = (T^*)^{-1}. \tag{7}$$

It follows that if T is densely defined then

$$R(\lambda; T)^* = R(\bar{\lambda}; T^*) \quad (\lambda \in \rho(T)). \tag{8}$$

In particular, if T is *selfadjoint*,

$$R(\lambda, T)^* = R(\bar{\lambda}; T), \tag{9}$$

and therefore $R(\lambda; T)$ is a *bounded normal* operator for each $\lambda \in \rho(T)$ (cf. Theorem 10.1).

Note that (6) also shows that for any T with dense domain, $\ker(T^*) \subset R(T)^\perp$. On the other hand, if $y \in R(T)^\perp$, then $(Tx, y) = 0$ for all $x \in D(T)$. In particular, the function $x \to (Tx, y)$ is continuous on $D(T)$, so that $y \in D(T^*)$, and $(x, T^*y) = (Tx, y) = 0$ for all $x \in D(T)$. Since $D(T)$ is dense, it follows that $T^*y = 0$, and we conclude that

$$\ker(T^*) = R(T)^\perp. \tag{10}$$

Theorem 10.4. *Let T be a symmetric operator. Then for any non-real $\lambda \in \mathbb{C}$, $\lambda I - T$ is one-to-one and*

$$\|(\lambda I - T)^{-1}y\| \le |\Im\lambda|^{-1}\|y\| \quad (y \in R(\lambda I - T)). \tag{11}$$

If T is closed, the range $R(\lambda I - T)$ is closed, and coincides with X if T is selfadjoint. In the latter case, every non-real λ is in $\rho(T)$, $R(\lambda; T)$ is a bounded normal operator, and

$$\|R(\lambda; T)\| \le 1/|\Im\lambda|. \tag{12}$$

Proof. If T is symmetric, (Tx, x) is real for all $x \in D(T)$ (since $\overline{(Tx, x)} = (x, Tx) = (Tx, x)$). Therefore $(Tx, \mathrm{i}\beta x)$ is pure imaginary for $\beta \in \mathbb{R}$. Since $\alpha I - T$ is symmetric for any $\alpha \in \mathbb{R}$, $((\alpha I - T)x, \mathrm{i}\beta x)$ is pure imaginary for $\alpha, \beta \in \mathbb{R}$. Hence, for all $x \in D(T)$ and $\lambda = \alpha + \mathrm{i}\beta$,

$$\begin{aligned}\|(\lambda I - T)x\|^2 &= \|(\alpha I - T)x + \mathrm{i}\beta x\|^2 \\ &= \|(\alpha I - T)x\|^2 + 2\Re((\alpha I - T)x, \mathrm{i}\beta x) + \beta^2\|x\|^2 \\ &= \|(\alpha I - T)x\|^2 + \beta^2\|x\|^2 \ge \beta^2\|x\|^2.\end{aligned}$$

Hence

$$\|(\lambda I - T)x\| \ge |\Im\lambda|\,\|x\|. \tag{13}$$

If λ is non-real, it follows from (13) that $\lambda I - T$ is one-to-one, and (11) holds.

If T is also closed, $(\lambda I - T)^{-1}$ is closed and continuous on its domain $R(\lambda I - T)$ (by (11)), and therefore this domain is closed (for non-real λ).

If T is selfadjoint,

$$R(\lambda I - T)^\perp = \ker((\lambda I - T)^*) = \ker(\bar{\lambda} I - T) = \{0\}$$

since $\bar{\lambda}$ is non-real. Therefore, $(\lambda I - T)^{-1}$ is everywhere defined, with operator norm $\le 1/|\Im\lambda|$, by (11). This shows that every non-real λ is in $\rho(T)$, that is, $\sigma(T) \subset \mathbb{R}$. □

We end this section with a few more properties of the adjoint.

Suppose that, in addition to being densely defined, T is closable. Then $\bar{T}^* = T^*$, because since Q is an isometry we have $Q\Gamma(\bar{T}) = Q\overline{\Gamma(T)} = \overline{Q\Gamma(T)}$, so (4) implies that $\Gamma(\bar{T}^*) = (Q\Gamma(\bar{T}))^\perp = (Q\Gamma(T))^\perp = \Gamma(T^*)$. Furthermore, T^* is now also densely defined, because if $w \in D(T^*)^\perp$ then $[w, 0] \in \Gamma(T^*)^\perp = (Q\Gamma(T))^{\perp\perp} = \overline{Q\Gamma(T)} = Q\Gamma(\bar{T})$ by the orthogonal decomposition theorem, which by the definition of Q means that $[0, w] \in \Gamma(\bar{T})$; since $\Gamma(\bar{T})$ is the graph of an operator, this means that $w = 0$.

So if the densely defined operator T is closable, its adjoint T^* also has an adjoint, which we denote by T^{**}. Using (4) again we deduce that

$$\Gamma(T^{**}) = (Q\Gamma(T^*))^\perp = \Big(Q\big((Q\Gamma(T))^\perp\big)\Big)^\perp.$$

Since Q is a unitary equivalence of $X \times X$ with itself, we have $Q(A^\perp) = (QA)^\perp$ for all $A \subset X \times X$. This and the fact that $Q^2 = -I_{X\times X}$ yields

$$\Gamma(T^{**}) = \overline{Q^2\Gamma(T)} = \overline{\Gamma(T)} = \Gamma(\bar{T}),$$

proving that $T^{**} = \bar{T}$. In particular, if T is closed then $T^{**} = T$.

Conversely, if a densely defined operator T is such that T^* is densely defined, then T must be closable, because if $\{x_n\} \subset D(T)$ converges to 0 and $\{Tx_n\}$ converges to $u \in X$, then for $y \in D(T^*)$ we have $(Tx_n, y) = (x_n, T^*y)$ for each n. The left-hand side converges to (u, y) while the right-hand side converges to 0, so the density of $D(T^*)$ implies that $u = 0$ indeed. (Alternatively, we could use the reasoning of the previous paragraph to show that $\overline{\Gamma(T)}$ is the graph of an operator, namely, T^{**}, thus T is closable.)

10.3 The spectral theorem for unbounded selfadjoint operators

Theorem 10.5. *Let T be a selfadjoint operator on the Hilbert space X. Then there exists a unique regular selfadjoint spectral measure E on $\mathcal{B} := \mathcal{B}(\mathbb{C})$, supported by $\sigma(T) \subset \mathbb{R}$, such that*

$$D(T) = \left\{x \in X;\ \int_{\sigma(T)} \lambda^2\, d\|E(\lambda)x\|^2 < \infty\right\}$$
$$= \left\{x \in X;\ \lim_{n\to\infty} \int_{-n}^{n} \lambda\, dE(\lambda)x \ \textit{exists}\right\} \tag{1}$$

and

$$Tx = \lim_{n\to\infty} \int_{-n}^{n} \lambda\, dE(\lambda)x \quad (x \in D(T)). \tag{2}$$

(The limits shown here are strong limits in X.)

Proof. By Theorem 10.4, every non-real α (to be fixed from now on) is in $\rho(T)$, and $R_\alpha := R(\alpha; T)$ is a bounded normal operator. Let F be its resolution of the identity, and define

$$E(\delta) = F(h(\delta)) \quad (\delta \in \mathcal{B}), \tag{3}$$

where h is as in Theorem 10.2.

By Theorem 9.8 (Part 5), $F(\{0\})X = \ker R_\alpha = \{0\}$, and therefore

$$E(\mathbb{C}) = F(\{0\}^c) = I - F(\{0\}) = I. \tag{4}$$

We conclude that E is a selfadjoint regular spectral measure from the corresponding properties of F.

By Theorem 10.2,

$$E(\sigma(T)) := F(h(\sigma(T))) = F(\sigma(R_\alpha)) - F(\{0\}) = I,$$

hence E is supported by $\sigma(T)$ (by (4)).

Denote the sets in (1) by D_0 and D_1.

If $\delta \in \mathcal{B}$ is *bounded*, then for all $x \in X$,

$$\int_{\sigma(T)} \lambda^2 \, d\|E(\lambda)E(\delta)x\|^2 = \int_{\delta\cap\sigma(T)} \lambda^2 \, d\|E(\lambda)x\|^2 < \infty, \tag{5}$$

since λ^2 is bounded on $\delta\cap\sigma(T)$. Hence $E(\delta)X \subset D_0$. Moreover, by Theorem 9.6, the last integral in (5) equals $\| \int_{\delta\cap\sigma(T)} \lambda dE(\lambda)x\|^2$. For positive integers $n > m$, take $\delta = [-n, -m] \cup [m, n]$. Then

$$\begin{aligned}\left\| \int_{-n}^{n} \lambda \, dE(\lambda)x - \int_{-m}^{m} \lambda \, dE(\lambda)x \right\|^2 \\ = \int_{-n}^{n} \lambda^2 \, d\|E(\lambda)x\|^2 - \int_{-m}^{m} \lambda^2 \, d\|E(\lambda)x\|^2.\end{aligned}$$

It follows that $D_0 = D_1$.

Let $x \in D(T)$. We may then write $x = R_\alpha y$ for a unique $y \in X$, and therefore

$$\begin{aligned}\int_{-n}^{n} \lambda \, dE(\lambda)x &= \int_{-n}^{n} \lambda \, dE(\lambda) \int_{\mathbb{C}} \mu \, dF(\mu)y \\ &= \int_{-n}^{n} \lambda \, dE(\lambda) \int_{\mathbb{R}} h(\lambda) \, dE(\lambda)y \\ &= \int_{-n}^{n} \lambda h(\lambda) \, dE(\lambda)y \to \int_{\mathbb{R}} \lambda h(\lambda) \, dE(\lambda)y.\end{aligned}$$

(The limit exists in X because $\lambda h(\lambda)$ is *bounded*.) Thus, $x \in D_0$, and we proved that $D(T) \subset D_0$.

Next, let $x \in D_0(= D_1)$, and denote $z = \lim_n \int_{-n}^{n} \lambda \, dE(\lambda)x$. Consider the sequence $x_n := E([-n, n])x$. Then $x_n \to x$ in X,

$$x_n = R_\alpha \int_{-n}^{n} (\alpha - \lambda) \, dE(\lambda)x \in R_\alpha X = D(T), \tag{6}$$

and by (6)

$$(\alpha I - T)x_n = \int_{-n}^{n} (\alpha - \lambda) \, dE(\lambda)x \to \alpha x - z.$$

Since $\alpha I - T$ (with domain $D(T)$) is closed, it follows that $x \in D(T)$ and $(\alpha I - T)x = \alpha x - z$. Hence $D_0 \subset D(T)$ (and so $D(T) = D_0$), and (2) is valid.

For each bounded $\delta \in \mathcal{B}$, the restriction of T to the reducing subspace $E(\delta)X$ is the *bounded* selfadjoint operator $\int_\delta \lambda \, dE(\lambda)$. By the uniqueness of the resolution of the identity for bounded selfadjoint operators, E is uniquely determined on the bounded Borel sets, and therefore on all Borel sets, by Theorem 9.6 (Part 2). □

10.4 The operational calculus for unbounded selfadjoint operators

The unique spectral measure E of Theorem 10.5 is called the *resolution of the identity for T*.

The map $f \to f(T) := \int_{\mathbb{R}} f \, dE$ of $\mathbb{B} := \mathbb{B}(\mathbb{R})$ into $B(X)$ is a norm-decreasing $*$-representation of $\mathbb{B}$ on X (cf. Theorem 9.1). The map is extended to arbitrary complex Borel functions f on $\mathbb{R}$ as follows. Let χ_n be the indicator of the set $[|f| \le n]$, and consider the "truncations" $f_n := f\chi_n \in \mathbb{B}$, $n \in \mathbb{N}$. The operator $f(T)$ has domain $D(f(T))$ equal to the set of all $x \in X$ for which the strong limit $\lim_n f_n(T)x$ exists, and $f(T)x$ is defined as this limit for $x \in D(f(T))$.

Note that if f is bounded, then $f_n = f$ for all $n \ge \|f\|_u$, and therefore the new definition of $f(T)$ coincides with the previous one for $f \in \mathbb{B}$. In particular, $f(T) \in B(X)$. For general Borel functions, we have the following.

Theorem 10.6. *Let T be an unbounded selfadjoint operator on the Hilbert space X, and let E be its resolution of the identity. For $f : \mathbb{R} \to \mathbb{C}$ Borel, let $f(T)$ be defined as shown. Then:*

(a) $D(f(T)) = \{x \in X; \int_{\mathbb{R}} |f|^2 \, d\|E(\cdot)x\|^2 < \infty\}$;

(b) $f(T)$ is a closed densely defined operator; and

(c) $f(T)^* = \bar{f}(T)$.

Proof. (a) Let D denote the set on the right-hand side of (a).

Since $f_n(x) = f(x)$ for all $n \ge |f(x)|$, $f_n \to f$ pointwise. If $x \in D$,

$$|f_n - f_m|^2 \le 4|f|^2 \in L^1(\|E(\cdot)x\|^2)$$

and $|f_n - f_m|^2 \to 0$ pointwise when $n, m \to \infty$. Therefore, by Theorem 9.6 and Lebesgue's dominated convergence theorem,

$$\|f_n(T)x - f_m(T)x\|^2 = \|(f_n - f_m)(T)x\|^2 = \int_{\mathbb{R}} |f_n - f_m|^2 \, d\|E(\cdot)x\|^2 \to 0$$

as $n, m \to \infty$. Hence $x \in D(f(T))$. On the other hand, if $x \in D(f(T))$, we have by Fatou's lemma

$$\int_{\mathbb{R}} |f|^2 \, d\|E(\cdot)x\|^2 \le \liminf_n \int_{\mathbb{R}} |f_n|^2 \, d\|E(\cdot)x\|^2$$
$$= \liminf_n \|f_n(T)x\|^2 = \|f(T)x\|^2 < \infty,$$

that is, $x \in D$, and (a) has been verified.

(b) Let $x \in X$, and $\delta_n = [|f| \le n]$. Clearly $\delta_{n+1}^c \subset \delta_n^c$ and $\bigcap \delta_n^c = \emptyset$. Since $\|E(\cdot)x\|^2$ is a finite positive measure,

$$\lim_n \|E(\delta_n^c)x\|^2 = \left\| E\left(\bigcap \delta_n^c\right)x \right\|^2 = 0,$$

that is,

$$\lim \|x - E(\delta_n)x\| = 0 \quad (x \in X). \tag{1}$$

Now

$$\int_{\mathbb{R}} |f|^2 \, d\|E(\cdot)E(\delta_n)x\|^2 = \int_{\delta_n} |f|^2 \, d\|E(\cdot)x\|^2 \le n^2 \|x\|^2 < \infty,$$

that is, $E(\delta_n)x \in D(f(T))$, by Part (a). This proves that $D(f(T))$ is dense in X.

Fix $x \in D(f(T))$ and $m \in \mathbb{N}$. Since $E(\delta_m)$ is a bounded operator, we have by the operational calculus for *bounded* Borel functions and the relation $\chi_{\delta_m} f_n = f_m$ for all $n \ge m$,

$$E(\delta_m)f(T)x = \lim_n E(\delta_m)f_n(T)x = \lim_n f_m(T)x = f_m(T)x. \tag{2}$$

Similarly

$$f(T)E(\delta_m)x = f_m(T)x \quad (x \in X). \tag{3}$$

In order to show that $f(T)$ is closed, let $\{x_n\}$ be any sequence in $D(f(T))$ such that $x_n \to x$ and $f(T)x_n \to y$. By (2) applied to $x_n \in D(f(T))$,

$$E(\delta_m)y = \lim_n E(\delta_m)f(T)x_n = \lim_n f_m(T)x_n = f_m(T)x,$$

since $f_m(T) \in B(X)$. Letting $m \to \infty$, we see that $f_m(T)x \to y$ (by (1)). Hence $x \in D(f(T))$, and $f(T)x := \lim_m f_m(T)x = y$. This proves (b).

(c) By the operational calculus for *bounded* Borel functions, $(f_n(T)x, y) = (x, \bar{f}_n(T)y)$ for all $x, y \in X$. When $x, y \in D(f(T)) = D(\bar{f}(T))$ (cf. (a)), letting $n \to \infty$ implies the relation $(f(T)x, y) = (x, \bar{f}(T)y)$. Hence $\bar{f}(T) \subset f(T)^*$. On the other hand, if $y \in D(f(T)^*)$, we have by (3) (for all $x \in X$)

$$\begin{aligned}(x, \bar{f}_m(T)y) &= (f_m(T)x, y) = (f(T)E(\delta_m)x, y) \\ &= (E(\delta_m)x, f(T)^*y) = (x, E(\delta_m)f(T)^*y),\end{aligned}$$

that is,

$$\bar{f}_m(T)y = E(\delta_m)f(T)^*y.$$

The right-hand side converges to $f(T)^*y$ when $m \to \infty$. Hence, $y \in D(\bar{f}(T))$, and (c) follows. $\square$

10.5 The semi-simplicity space for unbounded operators in Banach space

Let T be an *unbounded* operator with *real spectrum* on the Banach space X. Its *Cayley transform*

$$V := (\mathrm{i}I - T)(\mathrm{i}I + T)^{-1} = -2\mathrm{i}R(-\mathrm{i}; T) - I$$

belongs to $B(X)$.

By Theorem 10.2 with $\alpha = -\mathrm{i}$ and the corresponding h,

$$\sigma(R(-\mathrm{i}; T)) = h(\sigma(T) \cup \{\infty\}),$$

where ∞ denotes the point at infinity of the Riemann sphere. Therefore,

$$\begin{aligned}\sigma(V) &= -2\mathrm{i}h(\sigma(T) \cup \{\infty\}) - 1 \subset -2\mathrm{i}h(\mathbb{R} \cup \{\infty\}) - 1\\ &= \left\{\frac{\mathrm{i} - \lambda}{\mathrm{i} + \lambda}; \lambda \in \mathbb{R}\right\} \cup \{-1\} \subset \Gamma,\end{aligned}$$

where Γ denotes the unit circle.

Definition 10.7. Let T be an unbounded operator with real spectrum, and let V be its Cayley transform. The semi-simplicity space for the unbounded operator T is defined as the semi-simplicity space Z for the bounded operator V with spectrum in Γ (cf. Remark 9.13, (4)).

The function

$$\phi(s) := \frac{\mathrm{i} - s}{\mathrm{i} + s}$$

is a homeomorphism of $\bar{\mathbb{R}} := \mathbb{R} \cup \{\infty\}$ onto Γ, with the inverse $\phi^{-1}(\lambda) = \mathrm{i}(1 - \lambda)/(1 + \lambda)$.

For any $g \in C(\bar{\mathbb{R}})$, we have $g \circ \phi^{-1} \in C(\Gamma)$, and therefore, by Theorem 9.14, the operator $(g \circ \phi^{-1})(V|_Z)$ belongs to $B(Z)$, with $B(Z)$-norm $\leq \|g \circ \phi^{-1}\|_{C(\Gamma)} = \|g\|_{C(\bar{\mathbb{R}})}$.

The restriction $V|_Z$ is the Cayley transform of T_Z, which is the restriction of T to the domain

$$D(T_Z) := \{x \in D(T) \cap Z; Tx \in Z\}.$$

The operator T_Z is called *the part of* T *in* Z.

It is therefore natural to *define*

$$g(T_Z) := (g \circ \phi^{-1})(V|_Z) \quad (g \in C(\bar{\mathbb{R}})). \tag{1}$$

The map $\tau : g \to g(T_Z)$ is a norm-decreasing algebra homomorphism of $C(\bar{\mathbb{R}})$ into $B(Z)$ such that $f_0(T_Z) = I|_Z$ and $\phi(T_Z) = V|_Z$. We call a map τ with the shown properties a *contractive* $C(\bar{\mathbb{R}})$*-operational calculus for* T *on* Z; *when such* τ *exists, we say that* T *is of contractive class* $C(\bar{\mathbb{R}})$ *on* Z.

If W is a Banach subspace of X such that T_W is of contractive class $C(\bar{\mathbb{R}})$ on W, then the map

$$f \in C(\Gamma) \to f(V|_W) := (f \circ \phi)(T_W) \in B(W)$$

is a contractive $C(\Gamma)$-operational calculus for $V|_W$ in $B(W)$ (note that $(f_1 \circ \phi)(T_W) = \phi(T_W) = V|_W$). Therefore, W is a Banach subspace of Z, by Theorem 9.14. We formalize these observations as:

Theorem 10.8. *Let T be an unbounded operator with real spectrum, and let Z be its semi-simplicity space. Then T is of contractive class $C(\bar{\mathbb{R}})$ on Z, and Z is maximal with this property (in the sense detailed in Theorem 9.14).*

For X reflexive, we obtain a spectral integral representation for T_Z.

Theorem 10.9. *Let T be an unbounded operator with real spectrum on the reflexive Banach space X, and let Z be its semi-simplicity space. Then there exists a contractive spectral measure on Z*

$$F : \mathcal{B}(\mathbb{R}) \to B(Z),$$

such that

(1) F commutes with every $U \in B(X)$ which commutes with T;

(2) $D(T_Z)$ is the set Z_1 of all $x \in Z$ such that the integral

$$\int_{\mathbb{R}} s\, dF(s)x := \lim_{a \to -\infty, b \to \infty} \int_a^b s\, dF(s)x$$

exists in X and belongs to Z;

(3) $Tx = \int_{\mathbb{R}} s\, dF(s)x$ for all $x \in D(T_Z)$; and

(4) For all non-real $\lambda \in \mathbb{C}$ and $x \in Z$,

$$R(\lambda; T)x = \int_{\mathbb{R}} \frac{1}{\lambda - s}\, dF(s)x.$$

Proof. We apply Theorem 9.16 to the Cayley transform V. Let then E be the unique contractive spectral measure on Z, with support on the unit circle Γ, such that

$$f(V|_Z)x = \int_{\Gamma} f\, dE(\cdot)x \tag{2}$$

for all $x \in Z$ and $f \in C(\Gamma)$.

If $E(\{-1\}) \neq 0$, each $x \neq 0$ in $E(\{-1\})Z$ is an eigenvector for V, corresponding to the eigenvalue -1 (the argument is the same as in the proof of Theorem 9.8, Part 5, first paragraph). However, since $V = -2\mathrm{i}R(-\mathrm{i}; T) - I$, we have the relation

$$R(-\mathrm{i}; T) = (\mathrm{i}/2)(I + V), \tag{3}$$

from which it is evident that -1 is *not* an eigenvalue of V (since $R(-\mathrm{i};T)$ is one-to-one). Thus

$$E(\{-1\}) = 0. \tag{4}$$

Define

$$F(\delta) = E(\phi(\delta)) \quad (\delta \in \mathcal{B}(\mathbb{R})).$$

Then F is a contractive spectral measure on Z defined on $\mathcal{B}(\mathbb{R})$ (note that the requirement $F(\mathbb{R}) = I|_Z$ follows from (4):

$$F(\mathbb{R}) = E(\Gamma - \{-1\}) = E(\Gamma) = I|_Z.)$$

If $U \in B(X)$ commutes with T, it follows that U commutes with $V = -2\mathrm{i}R(-\mathrm{i};T) - I$, and therefore U commutes with E, hence with F. By (2)

$$f(V|_Z)x = \int_{\mathbb{R}} f \circ \phi \, dF(\cdot)x \tag{5}$$

for all $x \in Z$ and $f \in C(\Gamma)$. By definition, the left-hand side of (5) is $(f\circ\phi)(T_Z)x$ for $f \in C(\Gamma)$. We may then rewrite (5) in the form

$$g(T_Z)x = \int_{\mathbb{R}} g \, dF(\cdot)x \quad (x \in Z) \tag{6}$$

for all $g \in C(\bar{\mathbb{R}})$. Taking in particular $g = \phi$, we get (since $\phi(T_Z) = V|_Z$)

$$Vx = \int_{\mathbb{R}} \phi \, dF(\cdot)x \quad (x \in Z). \tag{7}$$

By (3) and (7), we have for all $x \in Z$

$$R(-\mathrm{i};T)x = (\mathrm{i}/2)\int_{\mathbb{R}} (1+\phi)\, dF(\cdot)x = \int_{\mathbb{R}} \frac{1}{-\mathrm{i}-s}\, dF(s)x. \tag{8}$$

Observe that

$$D(T_Z) = R(-\mathrm{i};T)Z. \tag{9}$$

Indeed, if $x \in D(T_Z)$, then $x \in D(T) \cap Z$ and $Tx \in Z$, by definition. Therefore, $z := (-\mathrm{i}I - T)x \in Z$, and $x = R(-\mathrm{i};T)z \in R(-\mathrm{i};T)Z$. On the other hand, if $x = R(-\mathrm{i};T)z$ for some $z \in Z$, then $x \in D(T) \cap Z$ (because Z is invariant for $R(-\mathrm{i};T)$), and $Tx = -\mathrm{i}x - z \in Z$, so that $x \in D(T_Z)$.

Now let $x \in D(T_Z)$, and write $x = R(-\mathrm{i};T)z$ for a suitable $z \in Z$ (by (9)). The spectral integral on the right-hand side of (6) defines a norm-decreasing algebra homomorphism τ of $\mathbb{B}(\mathbb{R})$ into $B(Z)$, which extends the $C(\bar{\mathbb{R}})$-operational calculus for T on Z (cf. Theorem 9.16). For real $a < b$, take $g(s) = s\chi_{[a,b]}(s) \in \mathbb{B}(\mathbb{R})$. By (8)

$$\int_a^b s\, dF(s)x = \tau(g)\tau(1/(-\mathrm{i}-s))z = \tau\left(\frac{g(s)}{-\mathrm{i}-s}\right)z$$

$$= \int_a^b \frac{s}{-\mathrm{i}-s}\, dF(s)z \to \int_{\mathbb{R}} \frac{s}{-\mathrm{i}-s}\, dF(s)z$$

as $a \to -\infty$ and $b \to \infty$ (convergence in X of the last integral follows from the boundedness of the integrand on $\mathbb{R}$). Thus, the integral $\int_{\mathbb{R}} s\,dF(s)x$ exists in X (in the sense stated in the theorem). Writing $s/(-\mathrm{i}-s) = [-\mathrm{i}/(-\mathrm{i}-s)] - 1$, the last relation and (8) show that

$$\int_{\mathbb{R}} s\,dF(s)x = -\mathrm{i}R(-\mathrm{i};T)z - z = TR(-\mathrm{i};T)z = Tx \in Z. \tag{10}$$

This proves that $D(T_Z) \subset Z_1$ and Statement 3 of the theorem is valid.

On the other hand, if $x \in Z_1$, consider the well-defined element of Z given by $z := \int_{\mathbb{R}} s\,dF(s)x$. Since $R(-\mathrm{i};T) \in B(X)$ commutes with T (hence with F) and $x \in Z$, we have by (8) and the multiplicativity of τ on $\mathbb{B}(\mathbb{R})$

$$\begin{aligned} R(-\mathrm{i};T)z &= \lim_{a\to-\infty, b\to\infty} \int_a^b sR(-\mathrm{i};T)\,dF(s)x = \lim_{a,b} \int_a^b s\,dF(s)R(-\mathrm{i};T)x \\ &= \lim_{a,b} \int_a^b \frac{s}{-\mathrm{i}-s}\,dF(s)x = \int_{\mathbb{R}} \frac{s}{-\mathrm{i}-s}\,dF(s)x \\ &= \int_{\mathbb{R}} \left(\frac{-\mathrm{i}}{-\mathrm{i}-s} - 1\right) dF(s)x = -\mathrm{i}R(-\mathrm{i};T)x - x. \end{aligned}$$

Hence, $x = -R(-\mathrm{i};T)(\mathrm{i}x + z) \in R(-\mathrm{i};T)Z = D(T_Z)$, and we proved that $D(T_Z) = Z_1$.

For any non-real $\lambda \in \mathbb{C}$, the function $g_\lambda(s) := (\lambda - s)^{-1}$ belongs to $C(\bar{\mathbb{R}})$, so that $g_\lambda(T_Z)$ is a well-defined operator in $B(Z)$ and by (6)

$$g_\lambda(T_Z)x = \int_{\mathbb{R}} \frac{1}{\lambda - s}\,dF(s)x \quad (x \in Z). \tag{11}$$

Fix $x \in Z$, and let $y := g_\lambda(T_Z)x$ ($\in Z$). By the multiplicativity of $\tau : \mathbb{B}(\mathbb{R}) \to B(Z)$ and (10),

$$\begin{aligned} \int_a^b s\,dF(s)y &= \int_a^b \frac{s}{\lambda - s}\,dF(s)x \to \int_{\mathbb{R}} \frac{s}{\lambda - s}\,dF(s)x \\ &= \int_{\mathbb{R}} \left(\frac{\lambda}{\lambda - s} - 1\right) dF(s)x = \lambda y - x \in Z. \end{aligned}$$

(The limit is the X-limit as $a \to -\infty$ and $b \to \infty$, and it exists because $s/(\lambda - s)$ is a *bounded* continuous function on $\mathbb{R}$.) Thus, $y \in D(T_Z)$ and $Ty = \lambda y - x$ (by Statements 2 and 3 of the theorem). Hence, $(\lambda I - T)y = x$, and since $\lambda \in \rho(T)$, it follows that $y = R(\lambda;T)x$, and Statement 4 is verified. □

10.6 Symmetric operators in Hilbert space

In this section, T will be an unbounded *densely defined* operator on a given Hilbert space X. The adjoint operator T^* is then a well-defined *closed* operator, to which we associate the *Hilbert space* $[D(T^*)]$ with the T^*-graph norm $\|\cdot\|^*$ and the inner product

$$(x,y)^* := (x,y) + (T^*x, T^*y) \quad (x,y \in D(T^*)).$$

We also consider the continuous sesquilinear form on $[D(T^*)]$

$$\phi(x,y) := i[(x,T^*y) - (T^*x,y)] \quad (x,y \in D(T^*)).$$

Recall that T is symmetric iff $T \subset T^*$. In particular, a symmetric operator T is *closable* (since it has the closed extension T^*). If S is a symmetric extension of T, then $T \subset S \subset S^* \subset T^*$, so that $S = T^*|_D$, where $D = D(S)$, and $D(T) \subset D \subset D(T^*)$. Clearly $\phi(x,y) = 0$ for all $x,y \in D$. (Call such a subspace D of $[D(T^*)]$ a *symmetric* subspace.) By the polarization formula for the sesquilinear form ϕ, D is symmetric iff $\phi(x,x)\ (= 2\Im(T^*x,x)) = 0$ on D, that is, iff (T^*x,x) *is real on* D. Since $T^* \in B([D(T^*)],X)$, the $[D(T^*)]$-closure $\bar{D}$ of a symmetric subspace D is symmetric.

If D is a symmetric subspace such that $D(T) \subset D \subset D(T^*)$, then D is the domain of the *symmetric* extension $S := T^*|_D$ of T. Together with the previous remarks, this shows that the symmetric extensions S of T are precisely the restrictions of T^* to symmetric subspaces of $[D(T^*)]$.

We verify easily that S is *closed* iff D is a *closed* (symmetric) subspace of $[D(T^*)]$ (Suppose S is closed and $x_n \in D \to x$ in $[D(T^*)]$, i.e., $x_n \to x$ and $Sx_n (= T^*x_n) \to T^*x$ in X. Since S is closed, it follows that $x \in D$, and so D is closed in $[D(T^*)]$. Conversely, if D is closed in $[D(T^*)]$, $x_n \to x$ and $Sx_n \to y$ in X, then $T^*x_n \to y$, and since T^* is closed, $y = T^*x$, i.e., $x_n \to x$ in $[D(T^*)]$. Hence $x \in D$, and $Sx = T^*x = y$, i.e., S is closed.)

Let S be a symmetric extension of the symmetric operator T. Since $D(S)$ is then a symmetric subspace of $[D(T^*)]$, so is its $[D(T^*)]$-closure $\overline{D(S)}$; therefore, the restriction of T^* to $\overline{D(S)}$ is a closed symmetric extension of S, which is precisely the *closure* $\bar{S}$ of the closable operator S. (If $x \in D(\bar{S})$, there exist $x_n \in D(S) \subset D(T^*)$ such that $x_n \to x$ and $Sx_n \to \bar{S}x$. Since $\bar{S} \subset T^*$, we have $x_n \to x$ in $[D(T^*)]$, hence $x \in \overline{D(S)}$. Conversely, if $x \in \overline{D(S)}$, there exist $x_n \in D(S)$ such that $x_n \to x$ in $[D(T^*)]$, that is, $x_n \to x$ and $Sx_n\ (= T^*x_n) \to T^*x$, hence $x \in D(\bar{S})$. This shows that $D(\bar{S}) = \overline{D(S)}$, and $\bar{S}$ is the restriction of T^* to this domain.)

Clearly, $\bar{S}$ is the *minimal* closed symmetric extension of S, and S is closed iff $S = \bar{S}$.

Note that T and $\overline{T}$ have equal adjoints, since

$$\Gamma(\bar{T}^*) = (Q\Gamma(\bar{T}))^{\perp} = (Q\overline{\Gamma(T)})^{\perp} = (\overline{Q\Gamma(T)})^{\perp} = (Q\Gamma(T))^{\perp} = \Gamma(T^*).$$

(The $\perp$ signs and the closure signs in the third and fourth expressions refer to the Hilbert space $X \times X$.)

Therefore, T and $\bar{T}$ have the *same family* of closed symmetric extensions (namely, the restrictions of T^* to closed symmetric subspaces of $[D(T^*)]$).

We are interested in *the family of selfadjoint extensions of* T, which is contained in the family of closed symmetric extensions of T. We may then assume without loss of generality that T is a *closed* symmetric operator.

By the orthogonal decomposition theorem for the Hilbert space $[D(T^*)]$,

$$[D(T^*)] = D(T) \oplus D(T)^{\perp}. \tag{1}$$

Definition 10.10. Let T be a closed densely defined symmetric operator. The kernels

$$D^+ := \ker(I + iT^*); \quad D^- := \ker(I - iT^*)$$

are called the positive and negative deficiency spaces of T (respectively). Their (Hilbert) dimensions n^+ and n^- (in the Hilbert space $[D(T^*)]$) are called the deficiency indices of T.

Note that

$$D^+ = \{y \in D(T^*); T^*y = iy\}; \quad D^- = \{y \in D(T^*); T^*y = -iy\}. \tag{2}$$

In particular, $(x,y)^* = 2(x,y)$ on D^+ and on D^-, so that the Hilbert dimensions n^+ and n^- may be taken with respect to X (the deficiency spaces are also closed *in* X, as can be seen from their definition and the fact that T^* is a closed operator).

We have $D^+ \perp D^-$, because if $x \in D^+$ and $y \in D^-$, then

$$(x,y)^* = (x,y) + (T^*x, T^*y) = (x,y) + (ix, -iy) = 0.$$

If $y \in D^+$, then for all $x \in D(T)$,

$$\begin{aligned}(x,y)^* &= (x,y) + (T^*x, T^*y) = (x,y) + (Tx, iy) = (x,y) + (x, iT^*y) \\ &= (x,y) - (x,y) = 0,\end{aligned}$$

and similarly for $y \in D^-$. Hence

$$D^+ \oplus D^- \subset D(T)^{\perp}. \tag{3}$$

On the other hand, if $y \in D(T)^{\perp}$, we have

$$0 = (x,y)^* = (x,y) + (Tx, T^*y) \quad (x \in D(T)),$$

hence, $(Tx, T^*y) = -(x,y)$ is a continuous function of x on $D(T)$, that is, $T^*y \in D(T^*)$ and $T^*(T^*y) = -y$. It follows that

$$(I - iT^*)(I + iT^*)y = (I + iT^*)(I - iT^*)y = 0. \tag{4}$$

Therefore

$$y - iT^*y \in \ker(I + iT^*) := D^+; \quad y + iT^*y \in \ker(I - iT^*) := D^-.$$

Consequently

$$y = (1/2)(y - iT^*y) + (1/2)(y + iT^*y) \in D^+ \oplus D^-.$$

This shows that $D(T)^\perp \subset D^+ \oplus D^-$, and we conclude from (3) and (1) that

$$D(T)^\perp = D^+ \oplus D^-, \tag{5}$$

and

$$[D(T^*)] = D(T) \oplus D^+ \oplus D^-. \tag{6}$$

It follows trivially from (6) that T is selfadjoint iff $n^+ = n^- = 0$.

Let D be a closed symmetric subspace of $[D(T^*)]$ containing $D(T)$. By the orthogonal decomposition theorem for the Hilbert space D with respect to its closed subspace $D(T)$, $D = D(T) \oplus W$, where $W = D \ominus D(T) := D \cap D(T)^\perp$ is a closed symmetric subspace of $D(T)^\perp$. Conversely, given such a subspace W, the subspace $D := D(T) \oplus W$ is a closed symmetric subspace of $D(T^*)$. By (5), the problem of finding all the closed symmetric extensions S of T is now reduced to the problem of finding all the closed symmetric subspaces W of $D^+ \oplus D^-$. Let $x_k, k = 1, 2$ be the components of $x \in W$ in D^+ and D^- ($x = x_1 + x_2$ corresponds as usual to the element $[x_1, x_2] \in D^+ \times D^-$). The symmetry of D means that (T^*x, x) is real on W. However

$$\begin{aligned}(T^*x, x) = (T^*x_1 + T^*x_2, x_1 + x_2) &= \mathrm{i}(x_1 - x_2, x_1 + x_2)\\ &= \mathrm{i}(\|x_1\|^2 - \|x_2\|^2) - 2\Im(x_1, x_2)\end{aligned}$$

is real iff $\|x_1\| = \|x_2\|$. Thus, (T^*x, x) is real on W iff the map $U : x_1 \to x_2$ is a (linear) isometry of a (closed) subspace $D(U)$ of D^+ onto a (closed) subspace $R(U)$ of D^-. Thus, W is a closed symmetric subspace of $D(T)^\perp$ iff

$$W = \{[x_1, Ux_1]; x_1 \in D(U)\}$$

is the graph of a linear isometry U as shown. (Note that since $\|x\|^* = \sqrt{2}\|x\|$ on D^+ and D^-, U is an isometry in both Hilbert spaces X and $[D(T^*)]$.)

Suppose $D(U)$ is a proper (closed) subspace of D^+. Let then $0 \neq y \in D^+ \cap D(U)^\perp$. Necessarily, $y \in D(S)^\perp$, so that for all $x \in D(S)$

$$0 = (x, y)^* = (x, y) + (Sx, T^*y) = (x, y) - \mathrm{i}(Sx, y).$$

Hence, $(Sx, y) = -\mathrm{i}(x, y)$ is a continuous function of x on $D(S)$, that is, $y \in D(S^*)$. Since $0 \neq y \in D(S)^\perp$, this shows that $S \neq S^*$. The same conclusion is obtained if $R(U)$ is a proper subspace of D^- (same argument!). In other words, a *necessary condition* for S to be selfadjoint is that U be an *isometry of* D^+ *onto* D^-. Thus, if T has a selfadjoint extension, there exists a (linear) isometry of D^+ onto D^- (equivalently, $n^+ = n^-$).

On the other hand, if there exists a (linear) isometry U of D^+ onto D^-, define S as the restriction of T^* to $D(S) := D(T) \oplus \Gamma(U)$. Since this domain $D(S)$ is a closed symmetric subspace of $D(T^*)$ (containing $D(T)$), S is a closed symmetric extension of T. In particular, $S \subset S^*$, and we have the decomposition (6) for S

$$D(S^*) = D(S) \oplus D^+(S) \oplus D^-(S). \tag{7}$$

Since $S^* \subset T^*$, the graph inner products for S^* and T^* coincide on $D(S^*)$, $D^+(S) \subset D^+$, and $D^-(S) \subset D^-$.

If $S \neq S^*$, it follows from (7) that there exists $0 \neq x \in D^+(S)$ (or $\in D^-(S)$). Hence $x + Ux \in D(S)$ (or $U^{-1}x + x \in D(S)$, respectively). Therefore, by (7), since $Ux \in D^-$ and $x \in D^+$ ($U^{-1}x \in D^+$ and $x \in D^-$, respectively),

$$0 = (x + Ux, x)^* = (x, x)^* = 2\|x\|^2 > 0$$

($0 = (U^{-1}x + x, x)^* = (x, x)^* = 2\|x\|^2 > 0$, respectively), contradiction. Hence $S = S^*$.

We proved the following theorem of von Neumann.

Theorem 10.11. *Let T be a closed densely defined symmetric operator on the Hilbert space X. Then the closed symmetric extensions of T are the restrictions of T^* to the closed subspaces of $[D(T^*)]$ of the form $D(T) \oplus \Gamma(U)$, where U is a linear isometry of a closed subspace of D^+ onto a closed subspace of D^-, and $\Gamma(U)$ is its graph. Such a restriction is selfadjoint if and only if U is an isometry of D^+ onto D^-. In particular, T has a selfadjoint extension iff $n^+ = n^-$ and has no proper closed symmetric extensions iff at least one of its deficiency indices vanishes.*

10.7 Quadratic forms

We start with the following well-known result from linear algebra, given here without proof.

Lemma 10.12. *Let X be a complex vector space.*

1. *There exists a 1–1 correspondence between the set of functions $q(\cdot,\cdot) : X \times X \to \mathbb{C}$ that are linear in the left variable and conjugate linear in the right variable, and the set of functions $q(\cdot) : X \to \mathbb{C}$ satisfying*

$$\begin{aligned} q(\alpha x) &= |\alpha|^2 q(x) \\ q(x+y) + q(x-y) &= 2\left(q(x) + q(y)\right) \end{aligned} \qquad (\forall x, y \in X, \alpha \in \mathbb{C}). \tag{1}$$

It given by

$$q(x) = q(x, x) \qquad (\forall x \in X)$$

and

$$q(x, y) = \frac{1}{4}\sum_{k=0}^{3} i^k q(x + i^k y) \qquad (\forall x, y \in X).$$

2. *Let $q(\cdot)$ and $q(\cdot,\cdot)$ be related as above. Then $q(\cdot)$ is real valued if and only if $q(\cdot,\cdot)$ is Hermitian, namely, $q(y, x) = \overline{q(x, y)}$ for all $x, y \in X$; and $q(\cdot)$ is non-negative valued iff $q(\cdot,\cdot)$ is positive semi-definite, namely, $q(x, x) \geq 0$ for all $a \in X$, that is: $q(\cdot,\cdot)$ is a semi-inner product on X.*

Definition 10.13. A *quadratic form* on a complex vector space X is a function $q : X \to \mathbb{C}$ satisfying (1). The associated function $X \times X \to \mathbb{C}$ will be denoted by $q(\cdot, \cdot)$ as shown.

We will only treat *non-negative* quadratic forms, and thus omit this adjective.

Definition 10.14. Let X be a Hilbert space. A quadratic form q on X is a quadratic form on a linear subspace $D(q)$ of X. We say that q is densely defined if $D(q)$ is dense in X.

In the sequel, X is a Hilbert space.

Example 1. A (generally unbounded) operator S on X induces a quadratic form on X given by $\|S\cdot\|^2$, that is, its domain is $D(S)$ and it maps $x \in D(S)$ to $\|Sx\|^2$.

Example 2. A (generally unbounded) operator S on X is called *positive* if $(Sx, x) \geq 0$ for all $x \in D(S)$. Such an operator induces a quadratic form on X with domain $D(S)$ mapping $x \in D(S)$ to (Sx, x).

A quadratic form q on X induces an inner product $(\cdot, \cdot)_q$ on $D(q)$ given by

$$(x, y)_q := (x, y) + q\,(x, y) \qquad (x, y \in D(q)).$$

Equivalently, $(\cdot, \cdot)_q$ is the inner product associated with the quadratic form $\|\cdot\|^2 + q(\cdot)$ on $D(q)$. The associated norm $\|\cdot\|_q$ clearly dominates $\|\cdot\|$.

Definition 10.15. Let q be a quadratic form on X.

- Say that q is *closed* if the inner product space $(D(q), (\cdot, \cdot)_q)$ is complete. Equivalently, if whenever $\{x_n\}_{n=1}^{\infty}$ is a sequence in $D(q)$ converging in X to some $x \in X$ and satisfying $q(x_n - x_m) \xrightarrow[n,m\to\infty]{} 0$, we have $x \in D(q)$ and $q(x_n - x) \xrightarrow[n\to\infty]{} 0$.
- Say that q is *closable* if whenever $\{x_n\}_{n=1}^{\infty}$ is a sequence in $D(q)$ converging in X to 0 and satisfying $q(x_n - x_m) \xrightarrow[n,m\to\infty]{} 0$, we have $q(x_n) \xrightarrow[n\to\infty]{} 0$.
- An *extension* of q is a quadratic form q' on X that is an extension of q as a function, that is, $D(q) \subset D(q')$ and $q'|_{D(q)} = q$. This means that $(D(q), (\cdot, \cdot)_q)$ is an inner product subspace of $(D(q'), (\cdot, \cdot)_{q'})$.

Theorem 10.16. *A quadratic form q on X is closable iff it admits a closed extension. In that case, it has a smallest closed extension $\overline{q}$, called the* closure *of q. Its domain $D(\overline{q})$ consists of all $x \in X$ for which there is a sequence $\{x_n\}_{n=1}^{\infty}$ in $D(q)$ converging to x and satisfying $q(x_n - x_m) \xrightarrow[n,m\to\infty]{} 0$, in which case $\lim_{n\to\infty} q(x_n)$ exists and equals $\overline{q}(x)$.*

Proof. It is clear that if q admits a closed extension then it is closable.

Conversely, assume that q is closable. We will express the completion of $(D(q), (\cdot, \cdot)_q)$ as the Hilbert space coming from a quadratic form.

Let $x \in D(\overline{q})$ (the latter is defined in the theorem's statement). Pick a sequence $\{x_n\}_{n=1}^{\infty}$ in $D(q)$ converging to x and satisfying $q(x_n - x_m) \xrightarrow[n,m\to\infty]{} 0$. Since $q(\cdot)^{1/2}$ is a semi-norm on $D(q)$, the triangle inequality shows that $\left|q(x_n)^{1/2} - q(x_m)^{1/2}\right| \leq q(x_n - x_m)^{1/2}$ for all $n, m \in \mathbb{N}$. Consequently, $\{q(x_n)\}_{n=1}^{\infty}$ is a Cauchy sequence, so it converges. If $\{x'_n\}_{n=1}^{\infty}$ is another such sequence, then $\{x_n - x'_n\}_{n=1}^{\infty}$ converges to 0 and $q\left((x_n - x'_n) - (x_m - x'_m)\right)^{1/2} \leq q\left(x_n - x_m\right)^{1/2} + q\left(x'_n - x'_m\right)^{1/2} \xrightarrow[n,m\to\infty]{} 0$ by the triangle inequality, thus $q(x_n - x'_n) \xrightarrow[n\to\infty]{} 0$ as q is closable, from which it follows that $\lim_{n\to\infty} q(x_n) = \lim_{n\to\infty} q(x'_n)$ by the triangle inequality again.

To conclude, the extension $\overline{q}$ of q introduced in the theorem's statement is well defined, and it is readily seen to be a quadratic form (in particular, $D(\overline{q})$ is a linear subspace of X). Additionally, if $x \in D(\overline{q})$ and $\{x_n\}_{n=1}^{\infty}$ in $D(q)$ is as above, then for every $n \in \mathbb{N}$, $\overline{q}(x_n - x) = \lim_{m\to\infty} q(x_n - x_m)$ by the definition of $\overline{q}$, and since the right-hand side converges to 0 as $n \to \infty$, so does the left-hand side, proving that $x_n \xrightarrow[n\to\infty]{} x$ in $(D(\overline{q}), (\cdot,\cdot)_{\overline{q}})$. Hence, $D(q)$ is dense in the inner product space $(D(\overline{q}), (\cdot,\cdot)_{\overline{q}})$ and every Cauchy sequence in $(D(\overline{q}), (\cdot,\cdot)_{\overline{q}})$ all of whose elements are in $D(q)$ converges in $(D(\overline{q}), (\cdot,\cdot)_{\overline{q}})$. This entails that $(D(\overline{q}), (\cdot,\cdot)_{\overline{q}})$ is complete, namely, $\overline{q}$ is closed. Finally, $\overline{q}$ is plainly the smallest closed extension of q. □

Definition 10.17. For a closed quadratic form q on X, a linear subspace D of $D(q)$ is a *core* for q if $\overline{q|_D} = q$, equivalently: if D is dense in $(D(q), (\cdot,\cdot)_q)$.

Example 3. If S is a (generally unbounded) operator on X, then S is closed (resp., closable) iff the quadratic form $\|S\cdot\|^2$ of Example 1 is closed (resp., closable). If S is closed, then a linear subspace D of $D(S)$ is a core for S (that is, $\overline{S|_D} = S$) iff it is a core for the quadratic form $\|S\cdot\|^2$.

A (generally unbounded) selfadjoint operator T on X is positive iff $\sigma(T) \subset [0, \infty)$. In this case, its square root $T^{1/2}$ makes sense by means of the operational calculus, and it is also a positive selfadjoint operator on X. See Exercises 12 and 15.

Example 4. A positive selfadjoint operator T on X induces the closed densely-defined quadratic form $q_T := \left\|T^{1/2}\cdot\right\|^2$ on X (take $S := T^{1/2}$ in Example 1).

We will now see that the construction of the previous example is exhaustive.

Theorem 10.18 (The representation theorem). *For every closed, densely defined quadratic form q on X there exists a unique positive selfadjoint operator T on X such that $q = q_T$. In addition, $D(T)$ is a core for q.*

Proof. Closedness of q means that the inner product space $(D(q), (\cdot,\cdot)_q)$ is a Hilbert space. For $x \in X$, the linear functional $(\cdot, x)|_{D(q)}$ belongs to $(D(q), (\cdot,\cdot)_q)^*$ and has norm at most $\|x\|$, because for each $y \in D(q)$, the

Cauchy–Schwarz inequality yields $|(y,x)| \leq \|y\| \, \|x\| \leq \|y\|_q \, \|x\|$. The "Little" Riesz representation theorem thus gives a unique vector $Bx \in D(q)$ such that

$$(x,y) = (Bx,y)_q \qquad (\forall y \in D(q)), \tag{2}$$

and we have $\|Bx\|_q \leq \|x\|$.

This construction produces a contractive (i.e., of norm at most 1) operator $B : X \to (D(q), (\cdot,\cdot)_q)$ satisfying (2) for all $x \in X$. The positive definiteness of $(\cdot,\cdot)$ and the density of $D(q)$ in X imply that B is injective and has dense range in $(D(q), (\cdot,\cdot)_q)$. Treating B as an operator $C : X \to X$, it is also contractive. For all $x \in X$ we have $(x, Cx) = (Bx, Bx)_q \geq 0$ by (2). Therefore, C is positive.

Define $T := C^{-1} - I$ (with domain $CX \subset D(q)$). It is a positive selfadjoint operator as C is positive and bounded. Using (2), for $x \in D(T)$ and $y \in D(q)$ we have

$$q(x,y) = (x,y)_q - (x,y) = \left(CC^{-1}x, y\right)_q - (x,y) = \left(C^{-1}x, y\right) - (x,y) = (Tx, y)\,.$$

In particular, for x in $D(T)$ (which is contained in $D(T^{1/2})$) we have $q(x) = (Tx, x) = \left(T^{1/2}x, T^{1/2}x\right) = \left\|T^{1/2}x\right\|^2 = q_T(x)$, proving that $q|_{D(T)} = q_T|_{D(T)}$.

But $D(T)$ is a core for both closed quadratic forms q_T and q. Indeed, on one hand, $D(T)$ is a core for $T^{1/2}$, equivalently: for q_T (see Exercise 15 and Example 3). On the other hand, the range of B, which equals $D(T)$, is dense in $(D(q), (\cdot,\cdot)_q)$, that is: $D(T)$ is a core for q. As a result, $q_T = q$.

The proof of uniqueness is left to the reader. □

Remark 10.19. Recall that if T, S are selfadjoint operators on X and $T \subset S$, then $T = S$. In sharp contrast, it is possible for one closed, densely defined quadratic form on X to strictly extend another one; see Exercise 25. Extension of closed, densely defined quadratic forms does *not* entail extension of the positive selfadjoint operators associated to them by the representation theorem.

A quadratic form $q : D(q) \to [0,\infty)$ on a Hilbert space X can be extended to a function $\widetilde{q} : X \to [0,\infty]$ assigning the value ∞ to all elements of $X \backslash D(q)$. Note that this extension satisfies (1) (with $\widetilde{q}$ in place of q) for all $x, y \in X$ and $\alpha \in \mathbb{C}$ because $D(q)$ is a linear subspace of X. Conversely, if $\widetilde{q} : X \to [0,\infty]$ satisfies (1) for all $x, y \in X$ and $\alpha \in \mathbb{C}$, then $D(q) := \widetilde{q}^{-1}([0,\infty))$ is a linear subspace of X and $\widetilde{q}|_{D(q)}$ is a quadratic form on X.

Recall from Exercise 6 in Chapter 3 that for a topological space X, a function $f : X \to [-\infty,\infty]$ is called lower semi-continuous if it satisfies one of the following equivalent conditions: for each $c \in \mathbb{R}$, $\{x \in X; f(x) > c\}$ is open; for each net $\{x_\alpha\}_{\alpha\in A}$ in X converging to $x \in X$ we have $f(x) \leq \liminf_{\alpha\in A} f(x_\alpha)$; and for each net $\{x_\alpha\}_{\alpha\in A}$ in X converging to $x \in X$ such that $\{f(x_\alpha)\}_{\alpha\in A}$ converges in $[-\infty,\infty]$ we have $f(x) \leq \lim_{\alpha\in A} f(x_\alpha)$. If X is first countable, nets can be replaced by sequences in the foregoing.

Theorem 10.20. *Let q be a quadratic form on X. Then q is closed (resp., closable) iff $\widetilde{q}$ (resp., q) is lower semi-continuous.*

Proof. Assume that $\widetilde{q}$ (resp., q) is lower semi-continuous. Let $\{x_n\}_{n=1}^{\infty}$ be a sequence in $D(q)$ converging in X to some $x \in X$ (resp., to 0) and satisfying $q(x_n - x_m) \xrightarrow[n,m\to\infty]{} 0$. Since, for each n, the sequence $\{x_n - x_m\}_{m=1}^{\infty}$ converges to $x_n - x$ (resp., to x_n), lower semi-continuity of $\widetilde{q}$ (resp., q) implies that

$$\widetilde{q}(x_n - x) \text{ (resp., } q(x_n)) \text{ is } \leq \liminf_{m\to\infty} q(x_n - x_m) \xrightarrow[n\to\infty]{} 0.$$

In particular, x belongs to $D(q)$ as the latter is a linear subspace of X eventually containing $x_n - x$. This proves that $q(x_n - x) \xrightarrow[n\to\infty]{} 0$ (resp., $q(x_n) \xrightarrow[n\to\infty]{} 0$), that is, q is closed (resp., closable).

Assume that q is closable. Let q' be a closed extension of q. Then $\widetilde{q'}$ is lower semi-continuous by the next paragraph. So $q = \widetilde{q'}|_{D(q)}$ is lower semi-continuous.

Assume that q is closed. We can further assume that q is densely defined without loss of generality. By the representation theorem, $q = q_T = \left\|T^{1/2}\cdot\right\|^2$ for some positive selfadjoint operator T on X. Write E for its resolution of the identity. Recall that $D(T^{1/2})$ consists of all $x \in X$ such that $\int_{[0,\infty)} \lambda \, d\left\|E(\lambda)x\right\|^2 < \infty$, in which case $\left\|T^{1/2}x\right\|^2 = \int_{[0,\infty)} \lambda \, d\left\|E(\lambda)x\right\|^2$. Hence, for each $x \in X$, $\widetilde{q_T}(x)$ equals $\int_{[0,\infty)} \lambda \, d\left\|E(\lambda)x\right\|^2$, which is the supremum of the ascending sequence of non-negative numbers $\left\{ \int_{[0,n]} \lambda \, d\left\|E(\lambda)x\right\|^2 = \left\|T^{1/2}E([0,n])x\right\|^2 \right\}_{n=1}^{\infty}$. Of course, $T^{1/2}E([0,n]) \in B(X)$ for all $n \in \mathbb{N}$. For every bounded operator $B \in B(X)$, the function $\left\|B\cdot\right\|^2$ is continuous, thus lower semi-continuous. From all this and Part (d) of Exercise 6 in Chapter 3 it follows that $\widetilde{q_T}$ is lower semi-continuous. □

Exercises

The generator of a semigroup

1. Use notation as in Exercise 14, Chapter 9. The *generator* A of the C_0-semigroup $T(\cdot)$ is its strong right derivative at 0 with *maximal domain* $D(A)$: denoting the (right) differential ratio at 0 by A_h, that is, $A_h := h^{-1}[T(h) - I]$ $(h > 0)$, we have

$$Ax = \lim_{h\to 0+} A_h x \quad x \in D(A) = \{x \in X; \lim_h A_h x \text{ exists}\}.$$

Prove:

(a) $\bigcup_{t>0} V(t)X \subset D(A)$, and for each $t > 0$ and $x \in X$, $AV(t)x = T(t)x - x$. (Hint: Exercise 14(e), Chapter 9.)

(b) $D(A)$ is dense in X. (Hint: by Part (a), $V(t)x \in D(A)$ for any $t > 0$ and $x \in X$ and $AV(t)x = T(t)x - x$. Apply Exercises 14(d) and 13(c) in Chapter 9.)

(c) For $x \in D(A)$ and $t > 0$, $T(t)x \in D(A)$ and

$$AT(t)x = T(t)Ax = (d/dt)T(t)x,$$

where the right-hand side denotes the strong derivative at t of $u := T(\cdot)x$. Therefore $u : [0,\infty) \to D(A)$ is a solution of class C^1 of the *abstract Cauchy problem* (ACP)

$$\text{(ACP)} \qquad u' = Au \quad u(0) = x.$$

Also

$$\int_0^t T(s)Ax\,ds = T(t)x - x \quad (x \in D(A)). \tag{*}$$

(Hint: for left derivation, use Exercise 14(c), Chapter 9.)

(d) A is a closed operator. (Hint: use the identity

$$V(t)Ax = AV(t)x = T(t)x - x \quad (x \in D(A); t > 0)$$

(cf. Part (a) and Exercise 13(c)), Chapter 9.)

(e) If $v : [0,\infty) \to D(A)$ is a solution of class C^1 of ACP, then $v = T(\cdot)x$. (This is the *uniqueness* of the solution of ACP when A is the generator of a C_0-semigroup.) In particular, the generator A determines the semigroup $T(\cdot)$ uniquely. (Hint: apply Exercise 13(d), Chapter 9, to $V := T(\cdot)v(s-\cdot)$ on the interval $[0,s]$.)

Semigroups continuous in the u.o.t.

2. Use notation as in Exercise 1. Suppose $T(h) \to I$ in the u.o.t. (i.e., $\|T(h) - I\| \to 0$ as $h \to 0+$). Prove:

(a) $V(h)$ is non-singular for h small enough (which we fix from now on). Define $A := [T(h) - I]V(h)^{-1}$ ($\in B(X)$).

(b) $T(t) - I = V(t)A$ for all $t \geq 0$ (with A as above). Conclude that A is the generator of $T(\cdot)$ (in particular, the generator is a bounded operator).

(c) Conversely, if the generator A of $T(\cdot)$ is a bounded operator, then $T(t) = \mathrm{e}^{tA}$ (defined by the usual absolutely convergent series in $B(X)$) and $T(h) \to I$ in the u.o.t. (Hint: the exponential is a continuous semigroup (in the u.o.t.) with generator A; use the uniqueness statement in Exercise 1(e).)

The resolvent of a semigroup generator

3. Let $T(\cdot)$ be a C_0-semigroup on the Banach space X. Let A be its generator, and ω its type (cf. Exercise 14(f), Chapter 9). Fix $a > \omega$. Prove:

(a) The *Laplace transform*

$$L(\lambda) := \int_0^\infty \mathrm{e}^{-\lambda t} T(t)\,dt$$

converges absolutely (in $B(X)$) and $\|L(\lambda)\| = O(1/(\Re\lambda - a))$ for $\Re\lambda > a$ (cf. Exercises 13(e) and 14(c), Chapter 9).

(b) $L(\lambda)(\lambda I - A)x = x$ for all $x \in D(A)$ and $\Re\lambda > a$.

(c) $L(\lambda)X \subset D(A)$, and $(\lambda I - A)L(\lambda) = I$ for $\Re\lambda > a$.

(d) Conclude that $\sigma(A) \subset \{\lambda \in \mathbb{C}; \Re\lambda \le \omega\}$ and $R(\lambda; A) = L(\lambda)$ for $\Re\lambda > \omega$.

(e) For any $\lambda_k > a$ $(k = 1, \ldots, m)$,

$$\left\| \prod_k (\lambda_k - a) R(\lambda_k; A) \right\| \le M, \tag{1}$$

where M is a positive constant depending only on a and $T(\cdot)$. In particular

$$\|R(\lambda)^m\| \le \frac{M}{(\lambda - a)^m} \quad (\lambda > a; m \in \mathbb{N}). \tag{2}$$

(Hint: apply Part (d), and the multiple integral version of Exercise 13(e), Chapter 9.)

(f) Let A be any closed densely defined operator on X whose resolvent set contains a ray (a, ∞) and whose resolvent $R(\cdot)$ satisfies $\|R(\lambda)\| \le M/(\lambda - a)$ for $\lambda > \lambda_0$ (for some $\lambda_0 \ge a$). (Such an A is sometimes called an *abstract potential*.) Consider the function $A(\cdot) : (a, \infty) \to B(X)$:

$$A(\lambda) := \lambda A R(\lambda) = \lambda^2 R(\lambda) - \lambda I.$$

Then, as $\lambda \to \infty$,

$$\lim A(\lambda)x = Ax \quad (x \in D(A));$$

$$\lim \lambda R(\lambda) = I \quad \text{and} \quad \lim AR(\lambda) = 0 \quad \text{in the s.o.t.}$$

Note that these conclusions are valid if A is the generator of a C_0-semigroup, with $a > \omega$ fixed (cf. Exercise 3, Parts (d) and (e)).

4. Let A be a closed densely defined operator on the Banach space X such that $(a, \infty) \subset \rho(A)$ and (2) in Exercise 3(e) is satisfied. Define $A(\cdot)$ as in Exercise 3(f) and denote $T_\lambda(t) := e^{tA(\lambda)}$ (the usual power series). Prove:

(a) $\|T_\lambda(t)\| \le M \exp(t(a\lambda/(\lambda - a))$ for all $\lambda > a$. Conclude that

$$\|T_\lambda(t)\| \le M\, e^{2at} \quad (\lambda > 2a) \tag{3}$$

and

$$\limsup_{\lambda \to \infty} \|T_\lambda(t)\| \le M\, e^{at}. \tag{4}$$

(b) If $x \in D(A)$, then uniformly for t in bounded intervals,

$$\lim_{2a < \lambda, \mu \to \infty} \|T_\lambda(t)x - T_\mu(t)x\| = 0. \tag{5}$$

(Hint: apply Exercise 13(d), Chapter 9, to the function $V(s) := T_\lambda(t-s)T_\mu(s)$ on the interval $[0,t]$; Exercise 1(c) to the semigroups $T_\lambda(\cdot)$ and $T_\mu(\cdot)$; Part (a), and Exercise 3(f)).

(c) For each $x \in X$, $\{T_\lambda(t)x; \lambda \to \infty\}$ is Cauchy (uniformly for t in bounded intervals). (Use Part (b), the density of $D(A)$, and (3) in Part (a)). *Define* then

$$T(t) = \lim_{\lambda\to\infty} T_\lambda(t)$$

in the s.o.t. Then $T(\cdot)$ is a strongly continuous semigroup such that $\|T(t)\| \le M\,\mathrm{e}^{at}$ and

$$T(t)x - x = \int_0^t T(s)Ax\,ds \quad (x \in D(A)). \tag{6}$$

(Hint: use (*) in Exercise 1(c) for the semigroup $\mathrm{e}^{tA(\lambda)}$, and apply Exercise 3(f)).

(d) If A' is the generator of the semigroup $T(\cdot)$ defined in Part (c), then $A \subset A'$. Since $\lambda I - A$ and $\lambda I - A'$ are both one-to-one and onto for $\lambda > a$ and coincide on $D(A)$, conclude that $A' = A$.

(e) An operator A with domain $D(A) \subset X$ is the generator of a C_0-semigroup satisfying $\|T(t)\| \le M\,\mathrm{e}^{at}$ for some real a iff it is closed, densely defined, $(a,\infty) \subset \rho(A)$ and (2) is satisfied. (Collect information from earlier!) This is the *Hille-Yosida theorem.* In particular (case $M = 1$ and $a = 0$), A is the generator of a contraction semigroup iff it is closed, densely defined, and $\lambda R(\lambda)$ exist and are contractions for all $\lambda > 0$. (Terminology: the bounded operators $A(\lambda)$ are called the *Hille–Yosida approximations* of the generator A.)

Core for the generator

5. Let $T(\cdot)$ be a C_0-semigroup on the Banach space X, and let A be its generator. Prove:

(a) $T(\cdot)$ is a C_0-semigroup on the Banach space $[D(A)]$. (Recall that the norm on $[D(A)]$ is the graph norm $\|x\|_A := \|x\| + \|Ax\|$.)

(b) Let D be a $T(\cdot)$-invariant subspace of $D(A)$, dense in X. For each $x \in D$, consider $V(t)x := \int_0^t T(s)x\,ds$ (defined in the Banach space $\bar{D}$, the closure of D in $[D(A)]$). Given $x \in D(A)$, let $x_n \in D$ be such that $x_n \to x$ (in X, by density of D in X). Then $V(t)x_n \to V(t)x$ in the graph-norm. Conclude that $V(t)x \in \bar{D}$ for each $t > 0$, and therefore $x \in \bar{D}$, that is, *D is dense in $[D(A)]$*. (Recall that a dense subspace of $[D(A)]$ is called a *core* for A.) Thus *a $T(\cdot)$-invariant subspace of $D(A)$ that is dense in X is a core for A.* (On the other hand, a core

D for A is trivially dense in X, since $D(A)$ is dense in X and D is $\|\cdot\|_A$-dense in $D(A)$.)

(c) A C^∞-vector for A is a vector $x \in X$ such that $T(\cdot)x$ is of class C^∞ ("strongly") on $[0,\infty)$. Let D^∞ denote the space of all C^∞-vectors for A. Then

$$D^\infty = \bigcap_{n=1}^{\infty} D(A^n). \tag{7}$$

(d) Let $\phi_n \in C_c^\infty(\mathbb{R})$ be non-negative, with support in $(0,1/n)$ and integral equal to 1. Given $x \in X$, let $x_n = \int \phi_n(t)T(t)x\,dt$. Then:

(i) $x_n \to x$ in X;

(ii) $x_n \in D(A)$ and $Ax_n = -\int \phi_n'(t)T(t)x\,dt$; and

(iii) $x_n \in D(A^k)$ and $A^k x_n = (-1)^k \int \phi_n^{(k)}(t)T(t)x\,dt$ for all $k \in \mathbb{N}$. In particular, $x_n \in D^\infty$.

Conclude that D^∞ is dense in X and is a core for A (cf. Part (b).)

The Hille–Yosida space of an arbitrary operator

6. Let A be an unbounded operator on the Banach space X with $(a,\infty) \subset \rho(A)$, for some real a. Denote its resolvent by $R(\cdot)$. Let $\mathcal{A}$ be the multiplicative semigroup generated by the set $\{(\lambda - a)R(\lambda); \lambda > a\}$. Let $Z := Z(\mathcal{A})$ (cf. Theorem 9.11), and consider A_Z, the part of A in Z. The *Hille–Yosida space* for A, denoted W, is the closure of $D(A_Z)$ in the Banach subspace Z. Prove:

(a) W is $R(\lambda)$-invariant for each $\lambda > a$ and $R(\lambda; A_W) = R(\lambda)|_W$. In particular, A_W is closed as an operator in the Banach space W.

(b) $\|R(\lambda; A_W)^m\|_{B(W)} \le 1/(\lambda - a)^m$ for all $\lambda > a$ and $m \in \mathbb{N}$.

(c) $\lim_{\lambda\to\infty} \lambda R(\lambda; A_W)w = w$ in the Z-norm. Conclude that $D(A_W)$ is dense in W.

(d) A_W *generates a C_0-semigroup $T(\cdot)$ on the Banach space W*, such that $\|T(t)\|_{B(W)} \le e^{at}$.

(e) If Y is a Banach subspace of X such that A_Y generates a C_0-semigroup on Y with the growth condition $\|T(t)\|_{B(Y)} \le e^{at}$, then Y is a Banach subspace of W. (This is the *maximality* of the Hille–Yosida space.)

Convergence of semigroups

7. Let $\{T_s(\cdot); 0 \le s < c\}$ be a family of C_0-semigroups on the Banach space X, such that
$$\|T_s(t)\| \le M\,\mathrm{e}^{at} \quad (t \ge 0; 0 \le s < c) \tag{8}$$
for some $M \ge 1$ and $a \ge 0$. Let A_s be the generator of $T_s(\cdot)$, and denote $T(\cdot) = T_0(\cdot)$ and $A = A_0$. Note that (8) implies that
$$\|R(\lambda; A_s)\| \le M/(\lambda - a) \quad (\lambda > a; s \in [0, c)). \tag{9}$$
Fix a core D for A. We say that A_s *graph-converge on* D to A (as $s \to 0$) if for each $x \in D$, there exists a vector function $s \in (0, c) \to x_s \in X$ such that $x_s \in D(A_s)$ for each s and $[x_s, A_s x_s] \to [x, Ax]$ in $X \times X$. Prove:

   (a) A_s graph-converge to A on D iff, for each $\lambda > a$ and $y \in (\lambda I - A)D$, there exists a vector function $s \to y_s$ such that $[y_s, R(\lambda; A_s)y] \to [y, R(\lambda; A)y]$ in $X \times X$ (as $s \to 0$). (Hint: $y_s = (\lambda I - A_s)x_s$ and (9).)

   (b) If A_s graph-converge to A, then as $s \to 0$, $R(\lambda; A_s) \to R(\lambda; A)$ in the s.o.t. for all $\lambda > a$ (the later property is called *resolvents strong convergence*). (Hint: show that $(\lambda I - A)D$ is dense in X, and use Part (a) and (9).)

   (c) Conversely, resolvents strong convergence implies graph-convergence on D. (Given $y \in (\lambda I - A)D$, choose $y_s = y$ constant!)

   (d) If $T'(\cdot)$ is also a C_0-semigroup satisfying (8), and A' is its generator, then
$$R(\lambda; A')[T'(t) - T(t)]R(\lambda; A) = \int_0^t T'(t-u)[R(\lambda; A') - R(\lambda; A)]T(u)\,du \tag{10}$$
for $\Re\lambda > a$ and $t \ge 0$. (Hint: verify that the integrand in (10) is the derivative with respect to u of the function $-T'(t - u)R(\lambda; A')T(u)R(\lambda; A)$.)

   (e) Resolvents strong convergence implies *semigroups strong convergence*, that is, for each $0 < \tau < \infty$,
$$\sup_{t \le \tau} \|T_s(t)x - T(t)x\| \to 0 \tag{11}$$
as $s \to 0$. (Hint: by (8), it suffices to consider $x \in D(A) = R(\lambda; A)X$. Write $[T_s(t) - T(t)]R(\lambda; A)y = R(\lambda; A_s)[T_s(t) - T(t)]y + T_s(t)[R(\lambda; A) - R(\lambda; A_s)]y + [R(\lambda; A_s) - R(\lambda; A)]T(t)y$. Estimate the norm of the first summand for $y \in D(A)$ (hence $y = R(\lambda; A)x$) using (10), and use the density of $D(A)$ and (8)–(9). The second summand $\to 0$ strongly, uniformly for $t \le \tau$, by (8)–(9). For the third summand, consider again $y \in D(A)$, for which one can use the relation $T(t)y = y + \int_0^t T(u)Ay\,du$; cf. Exercise 13(b), Chapter 9, and the dominated convergence theorem.)

(f) Conversely, semigroups strong convergence implies resolvents strong convergence. (Hint: use the Laplace integral representation of the resolvents.)

Collecting, we conclude that *generators graph-convergence, resolvents strong convergence, and semigroups strong convergence are equivalent* (when Condition (8) is satisfied).

Exponential formulas

8. Let A be the generator of a C_0-semigroup $T(\cdot)$ of contractions on the Banach space X.

   Let $F : [0, \infty) \to B(X)$ be contraction-valued, such that $F(0) = I$ and the (strong) right derivative of $F(\cdot)x$ at 0 coincides with Ax, for all x in a core D for A. Prove:

   (a) Fix $t > 0$ and define A_n as in Exercise 21(f), Chapter 6. Then $\mathrm{e}^{tA_n} - F(t/n)^n \to 0$ in the s.o.t. as $n \to \infty$.

   (b) $s \to \mathrm{e}^{sA_n}$ is a (uniformly continuous) contraction semigroup, for each $n \in \mathbb{N}$ (cf. Exercise 21(a), Chapter 6).

   (c) Suppose $T(\cdot)$ is a *contraction* C_0-semigroup. As $n \to \infty$, the semigroups e^{sA_n} converge strongly to the semigroup $T(s)$, uniformly on compact intervals (cf. conclusion of previous exercise (7); note that $A_n x \to Ax$ for all $x \in D$). Conclude that $F(t/n)^n \to T(t)$ in the s.o.t., for each $t \geq 0$.

   (d) Let $T(\cdot)$ be a C_0-semigroup such that $\|T(t)\| \leq \mathrm{e}^{at}$, and consider the *contraction* semigroup $S(t) := \mathrm{e}^{-at}T(t)$ (with generator $A - aI$; $a \geq 0$). Choose F as follows: $F(0) = I$ and for $0 < s < 1/a$,

   $$F(s) := (s^{-1} - a)R(s^{-1}; A) = (s^{-1} - a)R(s^{-1} - a; A - aI).$$

   Verify that F satisfies the hypothesis stated at the beginning of the exercise, and conclude that

   $$T(t) = \lim_{n\to\infty} \left[\frac{n}{t} R\left(\frac{n}{t}; A\right)\right]^n \tag{12}$$

   in the s.o.t., for each $t > 0$.

   (e) Let $T(\cdot)$ be *any* C_0-semigroup. By Exercise 14(c), Chapter 9, $\|T(t)\| \leq M\,\mathrm{e}^{at}$ for some $M \geq 1$ and $a \geq 0$. Consider the equivalent norm

   $$|x| := \sup_{t\geq 0} \mathrm{e}^{-at}\|T(t)x\| \quad (x \in X).$$

   Then $|T(t)x| \leq \mathrm{e}^{at}|x|$, and therefore (12) is valid over $(X, |\cdot|)$, hence over X (since the two norms are equivalent). Relation (12) (true for any C_0-semigroup!) is called *the exponential formula* for semigroups.

(f) Let A, B, C generate contraction C_0-semigroups $S(\cdot), T(\cdot), U(\cdot)$, respectively, and suppose $C = A + B$ on a core D for C. Then

$$U(t) = \lim_{n\to\infty} [S(t/n)T(t/n)]^n \quad (t \geq 0) \tag{13}$$

in the s.o.t. (Hint: choose $F(t) = S(t)T(t)$ in Part (c).)

Groups of operators

9. A group of operators on the Banach space X is a map $T(\cdot) : \mathbb{R} \to B(X)$ such that

$$T(s+t) = T(s)T(t) \quad (s, t \in \mathbb{R}).$$

We assume that it is of class C_0, that is, *the semigroup* $T(\cdot)|_{[0,\infty)}$ is of class C_0. Let A be the generator of this semigroup. Prove:

(a) The semigroup $S(t) := T(-t), t \geq 0$, is of class C_0, and has the generator $-A$.

(b) $\sigma(A)$ is contained in the strip

$$\Omega : -\omega' \leq \Re\lambda \leq \omega,$$

where ω, ω' are the types of the semigroups $T(\cdot)$ and $S(\cdot)$, respectively. Fix $a > \omega$ and $a' > \omega'$, and let

$$\Omega' = \{\lambda \in \mathbb{C}; -a' \leq \Re\lambda \leq a\}.$$

For $\lambda \notin \Omega'$,

$$\|R(\lambda; A)^n\| \leq \frac{M}{d(\lambda, \Omega')}. \tag{14}$$

If A generates a *bounded* C_0-group, then $\sigma(A) \subset \mathrm{i}\mathbb{R}$ and

$$\|R(\lambda; A)^n\| \leq \frac{M}{|\Re\lambda|^n}$$

where M is a bound for $\|T(\cdot)\|$.

(c) An operator A generates a C_0-group of operators iff it is closed, densely defined, has spectrum in a strip Ω as in Part (b), and (14) is satisfied for all *real* $\lambda \notin [-a', a]$. (Hint: apply the Hille–Yosida theorem (cf. Exercise 4(e)) separately in the half-planes $\Re\lambda > a$ and $\Re\lambda > a'$.)

(d) Let $T(\cdot)$ be a C_0-group of *unitary* operators on a Hilbert space X. Let $H = -\mathrm{i}A$, where A is the generator of $T(\cdot)$. Then H is a (closed, densely defined) *symmetric* operator with *real* spectrum. In particular, $\mathrm{i}I - H$ and $-\mathrm{i}I - H$ are both *onto*, so that the deficiency indices of H are both zero. Therefore H is selfadjoint (cf. (6) following Definition 10.10).

(e) Define e^{itH} by means of the operational calculus for the selfadjoint operator H. This is a C_0-group with generator $iH = A$, and therefore $T(t) = e^{itH}$ (cf. Exercise 1(e): the generator determines the semigroup uniquely). This representation of unitary groups is *Stone's theorem.*

Unbounded operators on Hilbert spaces

All operators are unbounded unless stated otherwise.
Let X be a Hilbert space. Unless stated otherwise, T is a selfadjoint operator on X with resolution of the identity E.

10. Let $f, g : \mathbb{R} \to \mathbb{C}$ be Borel functions. Prove:

    (a) $\|f(T)x\|^2 = \int_{\mathbb{R}} |f|^2 \, d\|E(\cdot)x\|^2$ for all $x \in D(f(T))$.

    (b) $(f(T)x, y) = \int_{\mathbb{R}} f \, d(E(\cdot)x, y)$ for all $x \in D(f(T))$ and $y \in X$. (Hint: the equality is not difficult after proving that the integral exists. One way to do that is to show that the total variation of the complex measure $(E(\cdot)x, y)$ is bounded by the function $\|E(\cdot)x\| \, \|E(\cdot)y\|$.)

    (c) $f(T) + g(T) \subset (f + g)(T)$.

    (d) $f(T)g(T) \subset (fg)(T)$ and $D(f(T)g(T)) = D((fg)(T)) \cap D(g(T))$.

    (e) For every $n \in \mathbb{N}$, T^n equals $f(T)$ where $f(\lambda) := \lambda^n$ for each $\lambda \in \mathbb{R}$.

11. Let $f : \mathbb{R} \to \mathbb{R}$ be Borel. Show that the resolution of the identity of the selfadjoint operator $f(T)$ maps a Borel set $\sigma \subset \mathbb{R}$ to $E(f^{-1}(\sigma))$.

12. Recall from Section 10.7 that an operator S on X is called *positive* if $(Sx, x) \geq 0$ for all $x \in D(S)$. Prove that the *selfadjoint* operator T is positive iff $\sigma(T) \subset [0, \infty)$ iff E is supported in $[0, \infty)$.

13. Prove that T is injective iff the range of T is dense in X iff $E(\{0\}) = 0$.

14. Assuming that T is positive selfadjoint, prove that $(Tx, x) > 0$ for all $0 \neq x \in D(T)$ iff T is injective. In this case T is called *strictly positive.*

15. Assume that T is a positive selfadjoint operator on X. The *square root* of T, denoted by $T^{1/2}$, is $f(T)$ for a Borel function $f : \mathbb{R} \to \mathbb{R}$ with $f(\lambda) = \lambda^{1/2}$ when $\lambda \geq 0$ (the values of $f(\lambda)$ for $\lambda < 0$ do not matter). Prove:

    (a) $T^{1/2}$ is the unique positive selfadjoint operator on X whose square is T, namely $(T^{1/2})^2 = T$. (Hint: Exercises 10 and 11.)

    (b) $D(T) \subset D(T^{1/2})$ and $D(T)$ is a core for $T^{1/2}$.

16. Let S be a closed densely defined operator on X. We will prove that S^*S is a positive selfadjoint operator (in particular, it is densely defined!) and that $D(S^*S)$ is a core for S. See Exercise 26 for a different proof.

(a) Prove: S^*S is positive, $((I + S^*S)x, x) \geq (x, x)$ for all $x \in D(S^*S)$, and $I + S^*S$ is one-to-one.

(b) Recall from (4) of Section 10.2 that $\Gamma(S^*) = (Q\Gamma(S))^\perp$. Prove that the Hilbert space $X \times X$ equals $Q\Gamma(S) \oplus \Gamma(S^*)$.

(c) Prove that for each $w, z \in X$ there exist unique elements $x \in D(S)$ and $y \in D(S^*)$ solving the equations $w = Sx + y$ and $z = -x + S^*y$.

(d) Prove that for each $z \in X$ there exist unique elements $Bz \in D(S)$ and $Cz \in D(S^*)$ such that $SBz = Cz$ and $z = Bz + S^*Cz$.

(e) Part (d) constructs functions $B, C : X \to X$. Prove: B, C are linear operators, $\|B\|, \|C\| \leq 1$, B is positive, and $I = (I + S^*S)B$. The last equality together with Part (a) imply that $B = (I + S^*S)^{-1}$.

(f) Use the selfadjointness of B to conclude the selfadjointness of S^*S.

(g) Prove that $D(S^*S)$ is a core for S. (Hint: assume by contradiction that this is false, i.e., that $\Gamma(S|_{D(S^*S)})$ is not dense in $\Gamma(S)$. Let $[x, Sx] \in \Gamma(S)$ be orthogonal to $\Gamma(S|_{D(S^*S)})$. Show that x is orthogonal to $(I + S^*S)X = X$.)

17. Let S be a closed densely defined operator on X. By Exercise 16, S^*S is a positive selfadjoint operator and $D(S^*S)$ is a core for S. The *absolute value of S* is the positive selfadjoint operator $|S| := (S^*S)^{1/2}$, where the right-hand side is given by Exercise 15. Prove that $D(|S|) = D(S)$ and $\||S|\,x\| = \|Sx\|$ for all x in the common domain. (Hint: prove this first for $x \in D(S^*S)$ and then employ $D(S^*S)$ being a core for both $|S|$ and S.)

18. Let $B \in B(X)$. Prove that the following conditions are equivalent; when they hold we say that T, B *commute*:

(i) $BT \subset TB$;

(ii) for all Borel sets $\delta \subset \mathbb{R}$ the bounded operators $E(\delta), B$ commute;

(iii) for all $f \in \mathbb{B}(\mathbb{R})$ the bounded operators $f(T), B$ commute;

(iv) for some (resp., all) $\lambda \in \rho(T)$ the bounded operators $R(\lambda; T), B$ commute; and

(v) for all $t \in \mathbb{R}$ the unitary operator e^{itT} commutes with B.

Show also that in this case BT is closable and $\overline{BT} = TB$.
(Hints: use Exercise 18 of Chapter 9 to prove that (iv) $\Longrightarrow$ (ii). Use the following fact to prove that (v) $\Longrightarrow$ (ii): the function that maps a regular complex Borel measure μ on $\mathbb{R}$ to its Fourier–Stieltjes transform $\hat{\mu} : \mathbb{R} \to \mathbb{C}$ given by $\hat{\mu}(y) := \int_{\mathbb{R}} \mathrm{e}^{-ixy}\, d\mu(x)$ is injective (see Section II.3.3).)

19. Commutativity of *two unbounded* operators is not trivial to define. Let T, S be selfadjoint operators on X with resolutions of the identity E, F, respectively.

(a) Prove that the following conditions are equivalent; when they hold we say that T, S *commute (strongly)*:

(i) for all Borel sets $\delta, \epsilon \subset \mathbb{R}$ the projections $E(\delta), F(\epsilon)$ commute;

(ii) for all $f, g \in \mathbb{B}(\mathbb{R})$ the bounded operators $f(T), g(S)$ commute;

(iii) for some (resp., all) $\lambda \in \rho(T)$ and $\mu \in \rho(S)$ the bounded operators $R(\lambda; T)$ and $R(\mu; S)$ commute; and

(iv) for all $t, s \in \mathbb{R}$ the unitary operators $\mathrm{e}^{itT}, \mathrm{e}^{isS}$ commute.

(b) Prove that if T, S commute, then so do the (selfadjoint, generally unbounded) operators $f(T), g(S)$ for all Borel functions $f, g : \mathbb{R} \to \mathbb{R}$.

20. Assume that the selfadjoint operators T, S on X commute. Prove that:

(a) T, S admit a common core: there exists a dense subspace D of X that is contained in $D(T) \cap D(S)$ and such that $\overline{T|_D} = T$ and $\overline{S|_D} = S$.

(b) If T, S agree on a dense subspace of X (contained in $D(T) \cap D(S)$), then $T = S$.

(c) $T + S, TS, ST$ are closable, their closures are selfadjoint and commute with one another and with T and S, and $\overline{TS} = \overline{ST}$.

21. Let T be a positive densely defined operator on X. Assume that $TD(T) \subset D(T)$ and that $I + T$ is bijective as a map from $D(T)$ to itself. Prove that $\bar{T}$ is selfadjoint. (Hint: show that $(I + T)^{-1}$ is bounded and positive.)

22. A *conjugation* on X is a conjugate-linear map $J : X \to X$ such that $(Jx, Jy) = (y, x)$ for all $x, y \in X$ and $J^2 = I$.
Prove that a closed densely defined symmetric operator T on X that commutes with some conjugation J in the simplest sense that $TJ = JT$ has equal deficiency indices, and thus has a selfadjoint extension.

23. Let T be a closed densely defined operator on X. Let $\{x_\alpha\}_{\alpha \in A}$ be a net in $D(T)$ that converges weakly in X to some x such that $\{\|Tx_\alpha\|\}_{\alpha \in A}$ is bounded by some $C \geq 0$. Prove that $x \in D(T)$ and $\|Tx\| \leq C$.

Quadratic forms

Let X denote a Hilbert space.

24. Let S be a densely defined positive operator on X. Consider the densely defined form q associated to S as in Example 2 of Section 10.7, that is, $D(q) = D(S)$ and $q(x) = (Sx, x)$ for all $x \in D(S)$. Show that q is closable, hence there exists a positive selfadjoint operator T on X such that $\overline{q} = q_T$. Prove that T is an extension of S. It is called the *Friedrichs extension* of S. This exercise proves that every densely defined positive operator possesses a positive selfadjoint extension.

25. (Continuing Exercise 24.) Let S be a densely defined positive operator on X, let T be the Friedrichs extension of S, and let T' be another positive selfadjoint extension of S. Prove that $q_{T'}$ extends q_T. (Remark: generally T' does *not* extend T, for that is equivalent to $T = T'$; cf. Remark 10.19.)

26. (A different approach to Exercise 16.) Let S be a closed densely defined operator on X. Consider the closed densely defined form $\|S\cdot\|^2$ associated to S as in Example 1 of Section 10.7, and let T be the positive selfadjoint operator on X such that $\|S\cdot\|^2 = q_T$ (see the representation theorem 10.18). Prove that $T = S^*S$ and that $D(S^*S)$ is a core for S.

27. (a) Let q_1, q_2 be closed quadratic forms on X. Prove that $q_1 + q_2$, with maximal domain $D(q_1) \cap D(q_2)$, is a closed quadratic form on X.

    (b) Let A, B be positive selfadjoint operators on X such that $D(A^{1/2}) \cap D(B^{1/2})$ is dense in X. Prove that there exists a unique positive selfadjoint operator C on X such that $D(C^{1/2}) = D(A^{1/2}) \cap D(B^{1/2})$ and for every x in this subspace we have $\|C^{1/2}x\|^2 = \|A^{1/2}x\|^2 + \|B^{1/2}x\|^2$. The operator C is called the *quadratic form sum* of A and B.

28. For quadratic forms q_1, q_2 on X, write $q_1 \le q_2$ if $D(q_2) \subset D(q_1)$ and $q_1(x) \le q_2(x)$ for every $x \in D(q_2)$. This is equivalent to $\widetilde{q_1} \le \widetilde{q_2}$. For positive selfadjoint operators A_1, A_2 on X, write $A_1 \le A_2$ if $q_{A_1} \le q_{A_2}$, that is, $D(A_2^{1/2}) \subset D(A_1^{1/2})$ and $\|A_1^{1/2}x\| \le \|A_2^{1/2}x\|$ for every $x \in D(A_2^{1/2})$.

    (a) Let $\{q_n\}_{n=1}^{\infty}$ be an ascending sequence of quadratic forms. Prove that

    $$D(q) := \left\{ x \in \bigcap_{n=1}^{\infty} D(q_n);\ \lim_{n\to\infty} q_n(x) \text{ exists in } [0, \infty) \right\}$$

    is a linear subspace of X, and that the pointwise limit $q : D(q) \to [0, \infty)$, $q(\cdot) := \lim_{n\to\infty} q_n(\cdot)$, is a quadratic form. Note that $\widetilde{q}(\cdot) = \lim_{n\to\infty} \widetilde{q_n}(\cdot)$.

    (b) Suppose that $\{q_n\}_{n=1}^{\infty}$ as in 28(a) consists of closed forms. Prove that q is also closed.

    (c) Let $\{A_n\}_{n=1}^{\infty}$ be an ascending sequence of positive selfadjoint operators on X. Suppose that

    $$D := \left\{ x \in \bigcap_{n=1}^{\infty} D(A_n^{1/2});\ \lim_{n\to\infty} \|A_n^{1/2}x\| \text{ exists in } [0, \infty) \right\}$$

    is dense in X. Prove that there exists a unique positive selfadjoint operator A on X such that $D(A^{1/2}) = D$ and for every x in this subspace we have $\|A^{1/2}x\| = \lim_{n\to\infty} \|A_n^{1/2}x\|$.

29. Take $X := L^2(\mathbb{R})$ with respect to the Lebesgue measure. Define $q : C_c(\mathbb{R}) \to [0, \infty)$ by $q(f) := |f(0)|^2$, $f \in C_c(\mathbb{R})$. Prove that q is a non-closable, densely defined quadratic form on X.

11

C^*-algebras

This chapter continues the discussion on Banach algebras started in Chapter 7 and is devoted exclusively to C^*-algebras, whose theory is vast and interacts with numerous areas of mathematics.

One of the highlights of Chapter 7 was the commutative Gelfand–Naimark theorem (7.16), roughly saying that every commutative C^*-algebra "is" $C_0(\Omega)$ for some locally compact Hausdorff space Ω. This is the first of many reasons why the theory of general C^*-algebras is often referred to as "*non-commutative topology*". The reader should bear this in mind throughout the chapter. One of the highlights here is the *non-commutative Gelfand–Naimark theorem* (11.26), which roughly says that every ("abstract") C^*-algebra "is" a C^*-subalgebra of $B(X)$ for some Hilbert space X (this is a "concrete" C^*-algebra).

The chapter is structured as follows. After fixing the notation and conventions, we proceed to exploit further the continuous operational calculus of normal elements introduced in Section 7.4. One remarkable and useful result is that every $*$-homomorphism between C^*-algebras is *automatically continuous* (even contractive), and that it is isometric if it is injective. Such results, in which algebraic properties imply analytic properties, are particularly satisfying.

Positivity, of elements and of functionals, is among the foundations of C^*-algebra theory. A selfadjoint element of a C^*-algebra with non-negative spectrum is called positive. These elements have various interesting properties. For instance, they form a cone and admit (positive) square roots. Moreover, every element of the form x^*x is positive—a key fact in C^*-algebras. We show that every C^*-algebra possesses an increasing contractive *approximate identity* consisting of positive elements.

Approximate identities, in turn, are employed to establish two facts about (closed) *ideals* in C^*-algebras: first, that they are automatically selfadjoint, thus C^*-algebras by themselves; and second, that the quotients are also C^*-algebras.

Recall that by the Riesz representation theorem, the space $C_0(\Omega)^*$ of all bounded linear functionals on $C_0(\Omega)$, where Ω is a locally compact Hausdorff space, is isometrically isomorphic to the space of regular complex Borel measures

Introduction to Modern Analysis. Second Edition. Shmuel Kantorovitz and Ami Viselter, Oxford University Press.
 DOI: 10.1093/oso/9780192849540.003.0011

on Ω. Hence, the bounded linear functionals on an arbitrary C^*-algebra should be viewed as a non-commutative version of complex regular Borel measures.

A linear functional on a C^*-algebra is called *positive* if it maps positive elements of the algebra to non-negative numbers. Such functionals are the non-commutative version of regular finite positive Borel measures. It turns out that every positive functional ω is automatically continuous and that its norm satisfies $\|\omega\| = \lim_\alpha \omega(e_\alpha)$ for every approximate identity $\{e_\alpha\}$; and conversely, a bounded linear functional satisfying this equality for some approximate identity is positive.

A positive linear functional of norm 1 is called a *state*. This terminology, courtesy of I. E. Segal, is taken from quantum mechanics. States are plentiful: there are enough of them to compute the norm of every element of the algebra.

Representations of C^*-algebras on Hilbert spaces are of the utmost importance. As in other areas of mathematics, their role is to match concrete objects (here, bounded operators on some Hilbert space) to abstract objects (here, elements of the C^*-algebra). The *Gelfand–Naimark–Segal (GNS) construction* assigns to every state of a C^*-algebra $\mathcal{A}$ some cyclic representation of $\mathcal{A}$. This construction and the richness of states in C^*-algebras imply the *non-commutative Gelfand–Naimark theorem*: every C^*-algebra has a *faithful* representation on some Hilbert space! This cardinal theorem of Gelfand and Naimark was the starting point of the whole theory of C^*-algebras.

In the final section we study convexity in positive linear functionals on C^*-algebras. The set of states of a C^*-algebra is convex and its extremal points are called *pure states*. The representation assigned to a state by the GNS theorem is irreducible iff the state is pure. It also presents two decomposition theorems, which, loosely speaking, are non-commutative versions of two results/constructions on measures: the Jordan decomposition of a real measure and the total variation measure of a complex measure (see Section 1.9).

Last, a word about terminology is in order. What we presently call a C^*-algebra was initially called a *B^*-algebra* (after Banach). Back then, a C^*-algebra was a B^*-algebra satisfying the extra assumption that x^*x is positive for every x. Gelfand and Naimark conjectured that this assumption was redundant. Kaplansky proved their conjecture based on results of Fukamiya and of Kelley and Vaught. The first edition of this book used the old term "B^*-algebras".

11.1 Notation and examples

Recall that a *C^*-algebra* is a Banach algebra $\mathcal{A}$ with involution $x \to x^*$ satisfying the so-called C^*-identity:

$$\|x^*x\| = \|x\|^2 \qquad (\forall x \in \mathcal{A}).$$

A C^*-algebra need not be unital; if it is, we denote its unit by $\mathbb{1}$ (in contrast to Chapters 7 and 9, where the unit was denoted by e), and it is then selfadjoint ($\mathbb{1}^* = \mathbb{1}$) and of norm 1. The unitization of a C^*-algebra $\mathcal{A}$ (see Section 7.3) is again denoted by $\mathcal{A}^\#$, and it equals $\mathcal{A}$ if the latter is unital.

Let us start with some examples and non-examples.

Examples. We continue the examples from the beginning of Section 7.1.

1. For a locally compact Hausdorff space Ω, $C_0(\Omega)$ and $C_b(\Omega)$ are commutative C^*-algebras with respect to the pointwise operations, including pointwise complex conjugation as involution, and the uniform norm. So is $L^\infty(\Omega, \mathbb{A}, \mu)$ for a measure space $(\Omega, \mathbb{A}, \mu)$ with respect to the pointwise operations and the L^∞-norm.
2. For a Hilbert space X, the Banach algebra $B(X)$ of all bounded linear operators on X is a C^*-algebra with the Hilbert adjoint as involution. The set $K(X)$ of compact operators on X is a closed subalgebra (indeed, an ideal) of $B(X)$ and it is also closed under the adjoint operation. Hence, it is a C^*-algebra of $B(X)$.
3. The Banach algebra $A(\mathbb{D})$ admits an involution given by $f^*(z) := \overline{f(\overline{z})}$ ($f \in A(\mathbb{D}), z \in \overline{\mathbb{D}}$), but it is *not* a C^*-algebra. Indeed, the function $f \in A(\mathbb{D})$ given by $f(z) := z + i$, $z \in \overline{\mathbb{D}}$, satisfies $\|f^*f\| = 2 = \|f\|$, so $\|f^*f\| \neq \|f\|^2$.

Henceforth we write $\mathcal{A}_{\mathrm{sa}}$ for the subset of a C^*-algebra $\mathcal{A}$ consisting of its selfadjoint elements; it is a real subspace of $\mathcal{A}$. We also write $\mathbb{1}$ for $\mathbb{1}_{\mathcal{A}^\#}$ (even when $\mathcal{A}$ is not unital) for brevity. This should not lead to any confusion.

In this chapter and Chapter 12, an element p of a C^*-algebra $\mathcal{A}$ is called a *projection* if it is a *selfadjoint idempotent*: $p^2 = p = p^*$. In particular, when $\mathcal{A} \subset B(X)$ for a Hilbert space X, a projection P in $\mathcal{A}$ is what we called in Section 8.2 an "orthogonal projection" (characterized by the ranges PX and $(I-P)X$ being orthogonal). For a Hilbert space X and a closed subspace $X_0 \subset X$, the unique projection in $B(X)$ whose range is X_0 is called the projection of X onto X_0.

11.2 The continuous operational calculus continued

This section presents some further results pertaining to the continuous operational calculus of normal elements in C^*-algebras. Let $\mathcal{A}$ be a C^*-algebra. Recall that the operational calculus for a normal element $x \in \mathcal{A}$ is the (unique) unital $*$-isomorphism from $C(\sigma(x))$ onto $[x]$ (:= the unital C^*-subalgebra of $\mathcal{A}^\#$ generated by x and $\mathbb{1}$) mapping the identity function f_1 to x.

First, if $\mathcal{A}$ is unital, then every element of $\mathcal{A}$ is a linear combination of at most four unitary elements of $\mathcal{A}$. Indeed, let first $x \in \mathcal{A}_{\mathrm{sa}}$. We may assume that $\|x\| \leq 1$, so that $\sigma(x) \subset [-1,1]$. Define $f \in C(\sigma(x))$ by $f(\lambda) := \lambda + \mathrm{i}\sqrt{1-\lambda^2}$, $\lambda \in \sigma(x)$. Then $f\overline{f} = 1 = \overline{f}f$ identically and $\frac{1}{2}(f+\overline{f}) = f_1$. Since the operational calculus is a unital $*$-homomorphism it follows that $u := f(x)$ satisfies $uu^* = \mathbb{1} = u^*u$, that is, u is unitary, and $\frac{1}{2}(u+u^*) = x$. This completes the proof because every element of $\mathcal{A}$ can be written as $x + \mathrm{i}y$ with $x, y \in \mathcal{A}_{\mathrm{sa}}$.

Next, normal elements of unital C^*-algebras possess two operational calculi: continuous and analytic (Section 9.10). They coincide:

Theorem 11.1. *Suppose that $\mathcal{A}$ is a unital C^*-algebra and $a \in \mathcal{A}$ is normal. For every $f \in H(\sigma(a))$, the two meanings assigned to the symbol $f(a)$, namely, the analytic and the continuous operational calculi, coincide.*

Proof. Denote by τ the continuous operational calculus of a, which is a $*$-isomorphism from $C(\sigma(a))$ onto $[a]$. Observe that for each $\lambda \in \rho(a)$, $R(\lambda; a)$ equals the image under τ of the function $\mu \to (\lambda - \mu)^{-1}$, thus $R(\lambda; a) \in [a]$

Let Ω be an open neighborhood of $\sigma(a)$ in which f is analytic, let $\Gamma \in \Gamma(\sigma(a), \Omega)$, and write

$$b := \frac{1}{2\pi i} \int_\Gamma f(\lambda) R(\lambda; a)\, d\lambda.$$

We should prove that $b = \tau(f)$. From the first paragraph of the proof it follows that $b \in [a]$. From the continuity of τ^{-1} we get

$$\tau^{-1}(b) = \frac{1}{2\pi i} \int_\Gamma f(\lambda) \tau^{-1}(R(\lambda; a))\, d\lambda$$

(the integral converges in $C(\sigma(a))$), so for every $\mu \in \sigma(a)$,

$$\begin{aligned} (\tau^{-1}(b))(\mu) &= \frac{1}{2\pi i} \int_\Gamma f(\lambda) [\tau^{-1}(R(\lambda; a))](\mu)\, d\lambda \\ &= \frac{1}{2\pi i} \int_\Gamma \frac{f(\lambda)}{\lambda - \mu}\, d\lambda = f(\mu) \end{aligned}$$

by Cauchy's theorem. Thus $b = \tau(f)$, as desired. □

We end this section with the following surprising and fundamental result, which says that the *algebraic* properties of being a $*$-homomorphism between two C^*-algebras and, respectively, being an injective one, imply the *analytic* properties of being contractive and, respectively, isometric. Such results are particularly satisfying. We later improve this result further in Theorem 11.14.

Theorem 11.2. *Let $\mathcal{A}, \mathcal{B}$ be C^*-algebras and $\varphi : \mathcal{A} \to \mathcal{B}$ be a $*$-homomorphism.*

(i) φ is contractive ($\|\varphi\| \leq 1$); in particular, φ is continuous.

(ii) If φ is injective, then it is isometric. Consequently, $\varphi(\mathcal{A})$ is a C^*-subalgebra of $\mathcal{B}$.

Proof. We let the reader explain why we may assume that $\mathcal{A}, \mathcal{B}$ are unital and so is φ. This implies that for every $a \in \mathcal{A}$, $\sigma_\mathcal{B}(\varphi(a)) \subset \sigma_\mathcal{A}(a)$, and if φ is an isomorphism from $\mathcal{A}$ onto $\mathcal{B}$, then $\sigma_\mathcal{B}(\varphi(a)) = \sigma_\mathcal{A}(a)$.

(i) Let $a \in \mathcal{A}$. By normality of a^*a and $\varphi(a^*a)$, Lemma 7.15 implies that $\|a\|^2 = \|a^*a\| = r_\mathcal{A}(a^*a)$ and $\|\varphi(a)\|^2 = \|\varphi(a)^*\varphi(a)\| = \|\varphi(a^*a)\| = r_\mathcal{B}(\varphi(a^*a))$. Since $\sigma_\mathcal{B}(\varphi(a^*a)) \subset \sigma_\mathcal{A}(a^*a)$, we obtain $r_\mathcal{B}(\varphi(a^*a)) \leq r_\mathcal{A}(a^*a)$, so $\|\varphi(a)\| \leq \|a\|$.

Before proceeding, notice that if $c \in \mathcal{A}$ is normal and $f \in C(\sigma(c))$, then $\varphi(f(c)) = f(\varphi(c))$. Indeed, this holds for f being a polynomial

$f(\lambda) = \sum \alpha_{kj}\lambda^k(\overline{\lambda})^j$ because φ is a $*$-homomorphism, and then for every $f \in C(\sigma(c))$ by approximation and the continuity of φ.

(ii) Assume that φ is injective. Let $c \in \mathcal{A}$ be normal. We already know that $\sigma_{\mathcal{B}}(\varphi(c)) \subset \sigma_{\mathcal{A}}(c)$. If strict inclusion occurs, then by Urysohn's lemma (Theorem 3.1) there is a non-zero $f \in C(\sigma_{\mathcal{A}}(c))$ vanishing on $\sigma_{\mathcal{B}}(\varphi(c))$. Thus,

$$\begin{aligned} &f(c) \neq 0 \text{ (by injectivity of the operational calculus)} \\ \text{but} \quad &\varphi(f(c)) = f(\varphi(c)) = 0, \end{aligned}$$

in contrast to φ being assumed injective. Thus $\sigma_{\mathcal{B}}(\varphi(c)) = \sigma_{\mathcal{A}}(c)$ whenever $c \in \mathcal{A}$ in normal.

Finally, for $a \in \mathcal{A}$ we have $\sigma_{\mathcal{B}}(\varphi(a^*a)) = \sigma_{\mathcal{A}}(a^*a)$, so $r_{\mathcal{B}}(\varphi(a^*a)) = r_{\mathcal{A}}(a^*a)$ and $\|\varphi(a)\| = \|a\|$ as in the proof of (i). □

11.3 Positive elements

Positivity plays an essential role in C^*-algebra theory. We introduce two forms of positivity: of elements of C^*-algebras in this section and of functionals on C^*-algebras in Section 11.6. Let $\mathcal{A}$ be a C^*-algebra.

An element $x \in \mathcal{A}$ is called *positive* if it is selfadjoint and $\sigma(x) \subset \mathbb{R}^+ := [0, \infty)$. Denote by $\mathcal{A}_+$ the set of all positive elements of $\mathcal{A}$.

If $\mathcal{B}$ is a C^*-subalgebra of $\mathcal{A}^\#$ containing x and $\mathbb{1}$, then x is positive in $\mathcal{A}$ iff it is positive in $\mathcal{B}$ by Theorem 7.18. We later strengthen this assertion.

Since the operational calculus for a normal element x is a $*$-isomorphism of $C(\sigma(x))$ and $[x]$ (= the unital C^*-subalgebra of $\mathcal{A}^\#$ generated by x and $\mathbb{1}$), the element $f(x)$ is selfadjoint iff f is *real* on $\sigma(x)$, and by the spectral mapping theorem (Theorem 7.20), it is positive iff $f(\sigma(x)) \subset \mathbb{R}^+$, that is, iff $f \geq 0$ on $\sigma(x)$. In particular, for x selfadjoint, decompose the real function $f_1(\lambda) = \lambda$ $(\lambda \in \sigma(x) \subset \mathbb{R})$ as $f_1 = f_1^+ - f_1^-$, so that $x = x^+ - x^-$, where $x^+ := f_1^+(x)$ and $x^- := f_1^-(x)$ are both positive elements of $\mathcal{A}$ (since $f_1^+, f_1^- \geq 0$ on $\sigma(x)$) and $x^+x^- = x^-x^+ = 0$ (since $f_1^+f_1^- = 0$ on $\sigma(x)$). We call x^+ and x^- the *positive part* and the *negative part* of x respectively.

Since every element of $\mathcal{A}$ can be written as $x + iy$ with $x, y \in \mathcal{A}_{\text{sa}}$, we deduce that every element of $\mathcal{A}$ can be written as the linear combination of at most 4 elements of $\mathcal{A}_+$.

If $x \in \mathcal{A}_+, \sigma(x) \subset [0, \|x\|]$ and $\|x\| \in \sigma(x)$ (cf. the last paragraph of Section 7.3).

For a real scalar $\alpha \geq \|x\|, \alpha\mathbb{1} - x$ is selfadjoint, and by Theorem 7.20,

$$\sigma(\alpha\mathbb{1} - x) = \alpha - \sigma(x) \subset \alpha - [0, \|x\|] = [\alpha - \|x\|, \alpha] \subset [0, \alpha].$$

Therefore, by Lemma 7.15,

$$\|\alpha\mathbb{1} - x\| = r(\alpha\mathbb{1} - x) \leq \alpha \tag{1}$$

(the norm is in $\mathcal{A}^{\#}$). Conversely, if x is selfadjoint and (1) is satisfied, then

$$\alpha - \sigma(x) = \sigma(\alpha\mathbb{1} - x) \subset [-\alpha, \alpha],$$

hence,

$$\sigma(x) \subset \alpha + [-\alpha, \alpha] = [0, 2\alpha],$$

and therefore $x \in \mathcal{A}_+$. This proves the following

Lemma 11.3. *Let $x \in \mathcal{A}$ be selfadjoint and fix $\alpha \geq \|x\|$. Then x is positive iff $\|\alpha\mathbb{1} - x\| \leq \alpha$.*

Theorem 11.4. *Let $\mathcal{A}$ be a C^*-algebra. $\mathcal{A}_+$ is a closed positive cone in $\mathcal{A}$ (i.e., a closed subset of $\mathcal{A}$, closed under addition and multiplication by non-negative scalars, such that $\mathcal{A}_+ \cap (-\mathcal{A}_+) = \{0\}$).*

Proof. We may assume without loss of generality that $\mathcal{A}$ is unital. Let $\{x_n\}_{n=1}^{\infty}$ in $\mathcal{A}_+$ converge to $x \in \mathcal{A}$. Then x is selfadjoint (because x_n are selfadjoint and the involution is continuous). By Lemma 11.3 with $\alpha_n := \|x_n\|$ $(\to \|x\| =: \alpha)$,

$$\|\alpha\mathbb{1} - x\| = \lim_n \|\alpha_n\mathbb{1} - x_n\| \leq \lim_n \alpha_n = \alpha,$$

hence $x \in \mathcal{A}_+$ (by the same lemma).

Let $x_n \in \mathcal{A}_+$, $\alpha_n := \|x_n\|$, $n = 1, 2$, $x := x_1 + x_2$, and $\alpha := \alpha_1 + \alpha_2$ $(\geq \|x\|)$. Again by Lemma 11.3,

$$\|\alpha\mathbb{1} - x\| = \|(\alpha_1\mathbb{1} - x_1) + (\alpha_2\mathbb{1} - x_2)\| \leq \|\alpha_1\mathbb{1} - x_1\| + \|\alpha_2\mathbb{1} - x_2\| \leq \alpha_1 + \alpha_2 = \alpha,$$

hence $x \in \mathcal{A}_+$.

If $x \in \mathcal{A}_+$ and $\alpha \geq 0$, then αx is selfadjoint and $\sigma(\alpha x) = \alpha\sigma(x) \subset \mathbb{R}^+$, so that $\alpha x \in \mathcal{A}_+$.

Finally, if $x \in \mathcal{A}_+ \cap (-\mathcal{A}_+)$, then $\sigma(x) \subset \mathbb{R}^+ \cap (-\mathbb{R}^+) = \{0\}$, so that x is both selfadjoint and quasi-nilpotent, hence $x = 0$ (see the last paragraph of Section 7.3). □

Let $f : \mathbb{R}^+ \to \mathbb{R}^+$ be the positive square root function. Since it belongs to $C(\sigma(x))$ for any $x \in \mathcal{A}_+$, it "operates" on each element $x \in \mathcal{A}_+$ through the $C(\sigma(x))$-operational calculus. The element $f(x) \in \mathcal{A}^{\#}$ is positive (since $f \geq 0$ on $\sigma(x)$), and $f(x)^2 = x$ (since the operational calculus is a homomorphism). It is called *the positive square root of* x, denoted $x^{1/2}$. Note that $x^{1/2} \in [x]$, which means that it is the limit of polynomials in x, $p_n(x)$, where $p_n \to f$ uniformly on $\sigma(x)$. In fact we have $x^{1/2} \in \mathcal{A}$, because if $\mathcal{A}$ is not unital, then since $f(0) = 0$ we may choose the polynomials to satisfy $p_n(0) = 0$ for every n. Suppose also $y \in \mathcal{A}_+$ satisfies $y^2 = x$. The polynomials $q_n(\lambda) = p_n(\lambda^2)$ converge uniformly to $f(\lambda^2) = \lambda$ on $\sigma(y)$ (since $\lambda^2 \in \sigma(y)^2 = \sigma(x)$ when $\lambda \in \sigma(y)$). Therefore (by continuity of the operational calculus) $q_n(y) \to y$. But $q_n(y) = p_n(y^2) = p_n(x) \to x^{1/2}$. Hence, $y = x^{1/2}$, which means that the positive square root is *unique*. Note that conversely, if $x \in \mathcal{A}$ equals y^2 for a selfadjoint $y \in \mathcal{A}$, then x is positive. We show two applications of this in the next proposition.

Proposition 11.5. *Let $\mathcal{A}$ be a C^*-algebra.*

(i) The decomposition $x = x^+ - x^-$ of a selfadjoint element $x \in \mathcal{A}$, with $x^+, x^- \in \mathcal{A}_+$ and $x^+x^- = 0$, is unique.

(ii) If $\mathcal{B}$ is a C^*-subalgebra of $\mathcal{A}$, then $\mathcal{B}_+ = \mathcal{B} \cap \mathcal{A}_+$. That is, positivity is not affected by the C^*-algebra is question.

Proof. (i) If $u, v \in \mathcal{A}_+$, $u - v = x$ and $uv = 0 = vu$, then $(u+v)^2 = u^2 + v^2 = (u-v)^2 = x^2$. Since $x^2, u+v \in \mathcal{A}_+$ (see Theorem 11.4), we have $u+v = (x^2)^{1/2}$. Since $u - v = x$, we get $u = \frac{1}{2}(x + (x^2)^{1/2})$ and $v = \frac{1}{2}((x^2)^{1/2} - x)$.

(ii) If $x \in \mathcal{B}_+$, it equals y^2 for a selfadjoint $y \in \mathcal{B} \subset \mathcal{A}$, so $x \in \mathcal{A}_+$. Conversely, if $x \in \mathcal{A}_+$, then $x^{1/2}$ belongs to the Banach algebra generated by x (in $\mathcal{A}$, without the unit $\mathbb{1}_{\mathcal{A}^\#}$ of $\mathcal{A}^\#$), which is contained in $\mathcal{B}$. Thus $x = (x^{1/2})^2$ is in $\mathcal{B}_+$.

(An alternative, straightforward proof of (ii) is via Exercise 26 of Chapter 7.) □

The representation $x = y^2$ (with $y \in \mathcal{A}_+$) of a positive element x shows in particular that $x = y^*y$ (since y is selfadjoint). This last property characterizes positive elements by Part (i) of Theorem 11.6, proved by Kaplansky based on results of Fukamiya and of Kelley and Vaught:

Theorem 11.6. *Let $\mathcal{A}$ be a C^*-algebra.*

(i) The element $x \in \mathcal{A}$ is positive iff $x = y^*y$ for some $y \in \mathcal{A}$.

(ii) If x is positive, then z^*xz is positive for all $z \in \mathcal{A}$.

(iii) If $\mathcal{A}$ is a C^*-subalgebra of $B(X)$ for some Hilbert space X, then $T \in \mathcal{A}$ is positive iff $(Tx, x) \geq 0$ for all $x \in X$.

Before we give the proof, recall that in every Banach algebra $\mathcal{A}$,

$$\sigma(xy) \cup \{0\} = \sigma(yx) \cup \{0\} \qquad (\forall x, y \in \mathcal{A}) \tag{2}$$

(see the remarks following Definition 7.5).

Proof of Theorem 11.6.

(i) The preceding remarks show that we only need to prove that $x := y^*y$ is positive for any $y \in \mathcal{A}$. Since it is trivially selfadjoint, we decompose it as $x = x^+ - x^-$, and we only need to show that $x^- = 0$. Let $z = yx^-$. Then since $x^+x^- = 0$,

$$z^*z = x^-y^*yx^- = x^-xx^- = x^-(x^+ - x^-)x^- = -(x^-)^3. \tag{3}$$

But $(x^-)^3$ is positive; therefore

$$-z^*z \in \mathcal{A}_+. \tag{4}$$

Write $z = a + ib$ with a, b selfadjoint elements of $\mathcal{A}$. Then $a^2, b^2 \in \mathcal{A}_+$, and therefore, by Theorem 11.4,

$$z^*z + zz^* = 2a^2 + 2b^2 \in \mathcal{A}_+. \tag{5}$$

By (2) and (4),

$$\sigma(-zz^*) \subset \sigma(-z^*z) \cup \{0\} \subset \mathbb{R}^+.$$

Thus, $-zz^* \in \mathcal{A}_+$, and so by (5) (cf. Theorem 11.4)

$$z^*z = (z^*z + zz^*) + (-zz^*) \in \mathcal{A}_+.$$

Together with (4), this shows that $z^*z \in \mathcal{A}_+ \cap (-\mathcal{A}_+)$, hence $z^*z = 0$ by Theorem 11.4, and therefore $z = 0$ because $\|z\|^2 = \|z^*z\| = 0$. By (3), we conclude that $x^- = 0$ because x^- is both selfadjoint and nilpotent, as wanted.

(ii) Write the positive element x in the form $x = y^*y$ with $y \in \mathcal{A}$ (by (i)). Then $z^*xz = z^*(y^*y)z = (yz)^*(yz) \in \mathcal{A}_+$ again by (i).

(iii) If T is positive, write $T = S^*S$ for some $S \in \mathcal{A}$ (by (i)). Then $(Tx, x) = (Sx, Sx) \geq 0$ for all $x \in X$.

Conversely, if $(Tx, x) \in \mathbb{R}$ for all $x \in X$, then $(T^*x, x) = (x, Tx) = \overline{(Tx, x)} = (Tx, x)$ for all x, and by polarization (cf. identity (11) following Definition 1.34) $(T^*x, y) = (Tx, y)$ for all $x, y \in X$, hence $T^* = T$.

For any $\delta > 0$, we have

$$\|(-\delta I - T)x\|^2 = \|\delta x + Tx\|^2 = \delta^2\|x\|^2 + 2\Re[\delta(x, Tx)] + \|Tx\|^2 \geq \delta^2\|x\|^2,$$

because $(x, Tx) \geq 0$. Therefore

$$\|(-\delta I - T)x\| \geq \delta\|x\| \qquad (\forall x \in X). \tag{6}$$

This implies that $T_\delta := -\delta I - T$ is injective (trivially) and has closed range $:= Y$ (indeed, if $T_\delta x_n \to y$, then

$$\|x_n - x_m\| \leq \delta^{-1}\|T_\delta(x_n - x_m)\| \to 0;$$

hence $\exists \lim x_n =: x$, and $y = \lim_n T_\delta x_n = T_\delta x \in Y$).

If $z \in Y^\perp$, then for all $x \in X$, since T_δ is selfadjoint,

$$(x, T_\delta z) = (T_\delta x, z) = 0.$$

Hence, $T_\delta z = 0$, and therefore $z = 0$ since T_δ is injective. Consequently, $Y = X$, by Theorem 1.36. This shows that T_δ is bijective. We have $\|T_\delta^{-1}\| \leq \delta^{-1}$ by (6), and therefore $T_\delta^{-1} \in B(X)$. Thus, $-\delta \in \rho_{B(X)}(T)$. Also, since T is selfadjoint, $\sigma_{B(X)}(T) \subset \mathbb{R}$ by Theorem 7.17. Therefore $\sigma_{B(X)}(T) \subset \mathbb{R}^+$ and hence $T \in B(X)_+$. From Proposition 11.5 we conclude that $T \in \mathcal{A}_+$. □

For $x \in \mathcal{A}_+$ and $\alpha > 0$, define $x^\alpha := f_\alpha(x)$ for $f_\alpha \in C(\sigma(x))$ given by $f_\alpha(\lambda) := \lambda^\alpha$, $\lambda \in \sigma(a)$. Then x^α is also positive, and it belongs to $\mathcal{A}$ (and not "just" to $\mathcal{A}^\#$) because $0^\alpha = 0$. When $\mathcal{A}$ is unital and $x \in \mathcal{A}_+$ is invertible, we similarly define x^α for each $\alpha \in \mathbb{R}$. Now x^α is both positive and invertible.

Give $\mathcal{A}_{sa}$ the partial order induced by the positive cone $\mathcal{A}_+$ (Theorem 11.4): for $a, b \in \mathcal{A}_{sa}$, $a \geq b$ iff $a - b \in \mathcal{A}_+$. If $a, b \in \mathcal{A}_{sa}$ and $x \in \mathcal{A}$, then $x^*ax, x^*bx \in \mathcal{A}_{sa}$, and $a \geq b$ implies $x^*ax \geq x^*bx$ by Theorem 11.6.

We will use repeatedly the following observations. If $a \in \mathcal{A}_{sa}$ and $m \in \mathbb{R}$, then $a \leq m\mathbb{1}$ (resp., , $m\mathbb{1} \leq a$) in $\mathcal{A}^\#$ iff $\max \sigma(a) \leq m$ (resp., $m \leq \min \sigma(a)$). Indeed, $m\mathbb{1} - a = (m - f_1)(a)$, so the former is positive iff the latter is positive on $\sigma(a)$, which means that $\max \sigma(a) \leq m$. Therefore, for $m \geq 0$, we have $-m\mathbb{1} \leq a \leq m\mathbb{1}$ iff $\|a\| \leq m$, and if $a \in \mathcal{A}_+$, then $a \leq m\mathbb{1}$ iff $\|a\| \leq m$.

Theorem 11.7. *Let $\mathcal{A}$ be a C^*-algebra and $a, b \in \mathcal{A}_{sa}$.*

(i) If $-b \leq a \leq b$, then $\|a\| \leq \|b\|$.

(ii) If $0 \leq a \leq b$, then $0 \leq a^{1/2} \leq b^{1/2}$.

(iii) If $0 \leq a \leq b$, $\mathcal{A}$ is unital and a is invertible, then b is also invertible and $0 \leq b^{-1} \leq a^{-1}$.

Proof. We may assume that $\mathcal{A}$ is unital.

(i) From the observations preceding the theorem's statement we have

$$-\|b\|\mathbb{1} \leq -b \leq a \leq b \leq \|b\|\mathbb{1} \implies \sigma(a) \subset [-\|b\|, \|b\|] \implies \|a\| \leq \|b\|.$$

(ii) and (iii): Suppose that $0 \leq a \leq b$ and a is invertible. For $\lambda_0 := \min \sigma(a) > 0$ we have $\lambda_0 \mathbb{1} \leq a \leq b$, so $\lambda_0 \leq \min \sigma(b)$, thus b is also invertible. Additionally,

$$0 \leq b^{-1/2}ab^{-1/2} \leq b^{-1/2}bb^{-1/2} = \mathbb{1}.$$

Hence, by the C^*-identity,

$$\begin{aligned} 1 \geq \left\|b^{-1/2}ab^{-1/2}\right\| &= \left\|(a^{1/2}b^{-1/2})^*(a^{1/2}b^{-1/2})\right\| = \left\|a^{1/2}b^{-1/2}\right\|^2 \\ &= \left\|(a^{1/2}b^{-1/2})^*\right\|^2 = \left\|(a^{1/2}b^{-1/2})(a^{1/2}b^{-1/2})^*\right\| = \left\|a^{1/2}b^{-1}a^{1/2}\right\|. \end{aligned}$$

Since $a^{1/2}b^{-1}a^{1/2} \in \mathcal{A}_{sa}$, we get $a^{1/2}b^{-1}a^{1/2} \leq \mathbb{1}$. Therefore

$$b^{-1} = a^{-1/2}a^{1/2}b^{-1}a^{1/2}a^{-1/2} \leq a^{-1/2}\mathbb{1}a^{-1/2} = a^{-1},$$

proving (iii). Furthermore, since $b^{-1/4}a^{1/2}b^{-1/4} \in \mathcal{A}_{sa}$, from (2) we obtain that

$$\begin{aligned} \left\|b^{-1/4}a^{1/2}b^{-1/4}\right\| &= r(b^{-1/4}a^{1/2}b^{-1/4}) = r(a^{1/2}b^{-1/4}b^{-1/4}) \\ &= r(a^{1/2}b^{-1/2}) \leq \left\|a^{1/2}b^{-1/2}\right\| \leq 1, \end{aligned}$$

thus $b^{-1/4}a^{1/2}b^{-1/4} \leq \mathbb{1}$, so

$$a^{1/2} = b^{1/4}b^{-1/4}a^{1/2}b^{-1/4}b^{1/4} \leq b^{1/4}\mathbb{1}b^{1/4} = b^{1/2},$$

proving (ii).

In the general case when a is not necessarily invertible, for every $r > 0$ we have $0 \leq a + r\mathbb{1} \leq b + r\mathbb{1}$ and $a + r\mathbb{1}$ is invertible, thus $(a+r\mathbb{1})^{1/2} \leq (b+r\mathbb{1})^{1/2}$. Letting $r \to 0^+$ we conclude that $a^{1/2} \leq b^{1/2}$ by the continuity of the continuous operational calculus. □

We end this section with the observation that a $*$-homomorphism from a C^*-algebra $\mathcal{A}$ to a C^*-algebra $\mathcal{B}$ preserves positivity: it maps $\mathcal{A}_+$ into $\mathcal{B}_+$ (see Exercise 2).

11.4 Approximate identities

A bounded *approximate identity* for a Banach algebra $\mathcal{A}$ is a bounded net $\{e_\alpha\}$ of elements of $\mathcal{A}$ such that for every $a \in \mathcal{A}$, $\lim_\alpha e_\alpha a = a$ and $\lim_\alpha a e_\alpha = a$. One-sided approximate identities are defined analogously. If $\mathcal{A}$ is a C^*-algebra, then we require, in addition, that e_α be positive and of norm at most 1 for every α, and that $\{e_\alpha\}$ be increasing. (In this case, $\lim_\alpha e_\alpha a = a$ for every $a \in \mathcal{A}$ iff $\lim_\alpha a e_\alpha = a$ for every $a \in \mathcal{A}$, as $(a^* e_\alpha)^* = e_\alpha a$.)

In this section we prove that every C^*-algebra has an approximate identity.

Theorem 11.8. *Let $\mathcal{A}$ be a C^*-algebra. Set $\Lambda := \{a \in \mathcal{A}_+; \|a\| < 1\}$ and give Λ the order inherited from $\mathcal{A}_{\mathrm{sa}}$. Let $e_\alpha := \alpha$ for $\alpha \in \Lambda$. Then $\{e_\alpha\}_{\alpha \in \Lambda}$ is an approximate identity for $\mathcal{A}$.*

Lemma 11.9. *The map $T : \mathcal{A}_+ \ni x \to x(\mathbb{1} + x)^{-1}$ is an order-preserving isomorphism of $\mathcal{A}_+$ onto Λ. That is: T maps into Λ, it is 1–1 and onto, and if $x, y \in \mathcal{A}_+$, then $x \leq y \iff x(\mathbb{1} + x)^{-1} \leq y(\mathbb{1} + y)^{-1}$.*

Proof. Let $x \in \mathcal{A}_+$. Note that $f \in C(\sigma(x))$ given by $\lambda \to \frac{\lambda}{1+\lambda}$ has $f(0) = 0$ if $0 \in \sigma(x)$. Thus, although the computations are in $\mathcal{A}^\#$, we have $x(\mathbb{1} + x)^{-1} = f(x) \in \mathcal{A}$. Also $f(x) \in \Lambda$ because $f(\sigma(x))$ is a compact subset of $[0, 1)$ and the continuous operational calculus is isometric. The map $\Lambda \ni a \to a(\mathbb{1} - a)^{-1}$ is the inverse of T (check!), so T is 1–1 and onto.

Let $x, y \in \mathcal{A}_+$. By Theorem 11.7, $x \leq y \iff \mathbb{1} + x \leq \mathbb{1} + y$ (and both are invertible!) $\iff (\mathbb{1} + y)^{-1} \leq (\mathbb{1} + x)^{-1} \iff x(\mathbb{1} + x)^{-1} = \mathbb{1} - (\mathbb{1} + x)^{-1} \leq \mathbb{1} - (\mathbb{1} + y)^{-1} = y(\mathbb{1} + y)^{-1}$. This proves that T is order preserving. □

Corollary 11.10. *Λ is upwards directed: for every $a, b \in \Lambda$ there exists $c \in \Lambda$ such that $a, b \leq c$.*

This follows since the partial orders Λ and $\mathcal{A}_+$ are isomorphic by the lemma, and the latter is upwards directed: if $x, y \in \mathcal{A}_+$, then $x+y \in \mathcal{A}_+$ and $x, y \leq x+y$.

Proof of Theorem 11.8. To prove that the net $\{e_\alpha\}_{\alpha \in \Lambda}$ is an approximate identity for $\mathcal{A}$ we need only show that $\lim_{\alpha \in \Lambda} e_\alpha a = a$ for every $a \in \mathcal{A}$.

If $\mathcal{A}$ is commutative, this is simple. Indeed, by the commutative Gelfand–Naimark theorem, we can assume that $\mathcal{A} = C_0(\Omega)$ for a locally compact Hausdorff space Ω. Fix $f \in C_0(\Omega)$ and $\epsilon > 0$. The set $K := \{x \in \Omega; |f(x)| \geq \epsilon\}$

is compact in Ω, so by Urysohn's lemma there is $g : \Omega \to [0,1]$ in $C_0(\Omega)$ with $g|_K \equiv 1$. Then for $0 < r < 1$ large enough we have $\|f - rgf\| \le \epsilon$ and $rg \in \Lambda$.

In the general case, fix $a \in \mathcal{A}$ and $0 < \epsilon < 1$. We may and do assume that a is normal. Denote by $\mathcal{B} \subset \mathcal{A}$ the (commutative, not necessarily unital) C^*-algebra generated by a. By the previous paragraph applied to $\mathcal{B}$, there is $\alpha_0 \in \Lambda$ such that $\|a - e_{\alpha_0}a\| \le \epsilon$. If $\alpha_0 \le \alpha \in \Lambda$, then (calculating in $\mathcal{A}^\#$) $0 \le \mathbb{1} - e_\alpha \le \mathbb{1} - e_{\alpha_0} \le \mathbb{1}$, hence $0 \le a^*(\mathbb{1} - e_\alpha)a \le a^*(\mathbb{1} - e_{\alpha_0})a$, so

$$\begin{aligned}\|a - e_\alpha a\|^2 &= \|(\mathbb{1} - e_\alpha)a\|^2 = \left\|(\mathbb{1} - e_\alpha)^{1/2}(\mathbb{1} - e_\alpha)^{1/2}a\right\|^2 \le \left\|(\mathbb{1} - e_\alpha)^{1/2}a\right\|^2 \\ &= \|a^*(\mathbb{1} - e_\alpha)a\| \le \|a^*(\mathbb{1} - e_{\alpha_0})a\| \le \|a\|\,\|(\mathbb{1} - e_{\alpha_0})a\| \le \|a\|\,\epsilon.\end{aligned}$$

Thus $\lim_{\alpha \in \Lambda} e_\alpha a = a$. □

11.5 Ideals

The two main results of this section say that closed ideals of C^*-algebras, as well as quotients by them, are C^*-algebras. Remarkably, such ideals are automatically selfadjoint! This is established with the aid of approximate identities.

Lemma 11.11. *If L is a closed left ideal of a C^*-algebra $\mathcal{A}$, then L (as a Banach algebra) has a bounded right approximate identity $\{e_\alpha\}$, which is increasing, and all of its elements are positive and of norm at most 1.*

Proof. Set $\mathcal{B} := L \cap L^*$. Then $\mathcal{B}$ is a C^*-subalgebra of $\mathcal{A}$ and $L^*L \subset \mathcal{B}$. Let $\{e_\alpha\}_\alpha$ be an approximate identity for $\mathcal{B}$ (see Theorem 11.8). If $a \in L$ then $a^*a \in \mathcal{B}$, and thus $0 = \lim_\alpha (a^*a - a^*ae_\alpha)$. Consequently,

$$\|a - ae_\alpha\|^2 = \|(\mathbb{1} - e_\alpha)a^*a(\mathbb{1} - e_\alpha)\| \le \|a^*a(\mathbb{1} - e_\alpha)\| \underset{\alpha}{\to} 0$$

(calculating in $\mathcal{A}^\#$), so $ae_\alpha \underset{\alpha}{\to} a$. □

Theorem 11.12. *Let I be a closed ideal in a C^*-algebra $\mathcal{A}$. Then I is selfadjoint (i.e., closed under the involution operation), hence it is a C^*-subalgebra of $\mathcal{A}$.*

Proof. Let $\{e_\alpha\}_\alpha$ be as in the last lemma. Then for every $a \in I$, we have $a = \lim_\alpha ae_\alpha$, so $a^* = \lim_\alpha e_\alpha a^* \in I$ because $e_\alpha \in I$ for all α and I is a closed right ideal. □

Convention. Henceforth, unless otherwise stated, an *ideal of a C^*-algebra* will always be (two sided and) *closed.*

Theorem 11.13. *Let I be an ideal of a C^*-algebra $\mathcal{A}$. Put $(a + I)^* := a^* + I$ ($a \in \mathcal{A}$). Then this operation is as involution on $\mathcal{A}/I$, and $\mathcal{A}/I$ is a C^*-algebra.*

Proof. We already know that $\mathcal{A}/I$ is a Banach algebra; see Section 7.1. The function $a + I \to a^* + I$ ($a \in \mathcal{A}$) on $\mathcal{A}/I$ is well defined by the selfadjointness of I (Theorem 11.12). It is then easily seen to be an involution on $\mathcal{A}/I$.

It remains to prove the C^*-identity. Let $\{e_\alpha\}_\alpha$ be an approximate identity for I (cf. Theorems 11.12 and 11.8).

Claim: for all $a \in \mathcal{A}$ we have $\|a + I\| = \lim_\alpha \|a - ae_\alpha\|$. Indeed, for all α, we have $\|a + I\| \leq \|a - ae_\alpha\|$ as $e_\alpha \in I$ and hence $ae_\alpha \in I$. On the other hand, for every $\epsilon > 0$ there is $b \in I$ with $\|a - b\| \leq \|a + I\| + \epsilon$. Thus

$$\|a - ae_\alpha\| = \|(a-b)(\mathbb{1} - e_\alpha) + (b - be_\alpha)\| \leq \underbrace{\|a - b\|}_{\leq \|a+I\|+\epsilon} \underbrace{\|(\mathbb{1} - e_\alpha)\|}_{\leq 1}$$
$$+ \underbrace{\|b - be_\alpha\|}_{\underset{\alpha}{\rightarrow} 0},$$

so $\limsup_\alpha \|a - ae_\alpha\| \leq \|a + I\| + \epsilon$. This holds for all $\epsilon > 0$, proving the claim.

Take $a \in \mathcal{A}$. By the claim,

$$\begin{aligned} \|a + I\|^2 &= \lim_\alpha \|a - ae_\alpha\|^2 = \lim_\alpha \|(\mathbb{1} - e_\alpha)a^*a(\mathbb{1} - e_\alpha)\| \\ &\leq \lim_\alpha \|a^*a(\mathbb{1} - e_\alpha)\| = \|a^*a + I\|. \end{aligned}$$

The C^*-identity follows from this by Exercise 23 of Chapter 7. □

A beautiful application of this theory is the following result, which improves Theorem 11.2.

Theorem 11.14. *Let $\mathcal{A}, \mathcal{B}$ be C^*-algebras and $\varphi : \mathcal{A} \to \mathcal{B}$ a $*$-homomorphism. Then:*

(i) $\varphi(\mathcal{A})$ is a C^*-subalgebra of $\mathcal{B}$.

(ii) If $\varphi \neq 0$ then $\|\varphi\| = 1$.

Proof. (i) The ideal $\ker\varphi$ is closed in $\mathcal{A}$ because φ is continuous by Theorem 11.2. Consider the induced homomorphism $\tilde{\varphi} : \mathcal{A}/\ker\varphi \to \mathcal{B}$, $a + \ker\varphi \to \varphi(a)$. It is an injective $*$-homomorphism between two C^*-algebras by Theorem 11.13. Therefore, it is isometric by Theorem 11.2. Thus, its image $\tilde{\varphi}(\mathcal{A}/\ker\varphi) = \varphi(\mathcal{A})$ is a C^*-subalgebra of $\mathcal{B}$.

(ii) We already established that $\|\varphi\| \leq 1$ in Theorem 11.2. Let $\{e_\alpha\}_\alpha$ be an approximate identity for $\mathcal{A}$. Pick $a \in \mathcal{A}$ such that $\varphi(a) \neq 0$. Then as $ae_\alpha \underset{\alpha}{\to} a$, we have $\varphi(a)\varphi(e_\alpha) = \varphi(ae_\alpha) \underset{\alpha}{\to} \varphi(a)$. Thus

$$\|\varphi(a)\| \underset{\alpha}{\leftarrow} \|\varphi(a)\varphi(e_\alpha)\| \leq \|\varphi(a)\| \, \|\varphi\| \, \|e_\alpha\| \leq \|\varphi(a)\| \, \|\varphi\|.$$

Hence $\|\varphi\| \geq 1$. □

We can now prove an analogue of one of Noether's "isomorphism theorems":

Corollary 11.15. *Let $\mathcal{A}$ be a C^*-algebra, $\mathcal{B}$ be a C^*-subalgebra of $\mathcal{A}$, and I an ideal of $\mathcal{A}$. Then $\mathcal{B} + I$ is a C^*-subalgebra of $\mathcal{A}$, and there is a natural $*$-isomorphism from $(\mathcal{B} + I)/I$ onto $\mathcal{B}/(\mathcal{B} \cap I)$.*

Proof. Let $q : \mathcal{A} \to \mathcal{A}/I$ be the quotient mapping. Its restriction $q|_{\mathcal{B}} : \mathcal{B} \to \mathcal{A}/I$ has closed image by Theorems 11.13 and 11.14. Thus the $*$-subalgebra $\mathcal{B} + I = q^{-1}(q(\mathcal{B}))$ of $\mathcal{A}$ is closed, and hence it is a C^*-subalgebra of $\mathcal{A}$. Since $q|_{\mathcal{B}}$ is a $*$-homomorphism from $\mathcal{B}$ onto $(\mathcal{B} + I)/I$ whose kernel is $\mathcal{B} \cap I$, it induces the desired $*$-isomorphism. □

Example 11.16. Take $\mathcal{A} := C_0(\Omega)$, where Ω is a locally compact Hausdorff space. We know that $\Phi(\mathcal{A}) = \{\phi_t; t \in \Omega\}$, where ϕ_t is the "evaluation at t" character given by $f \to f(t)$ (Exercise 17 in Chapter 7). We now use this and Theorem 11.13 to characterize *all* (closed) ideals of $\mathcal{A}$.

If $S \subset \Omega$ is closed, then $I(S) := \{f \in \mathcal{A}; f|_S \equiv 0\}$ is an ideal of $\mathcal{A}$. Conversely, all ideals of $\mathcal{A}$ have this form. Indeed, if I is an ideal of $\mathcal{A}$, write $S(I) := \{t \in \Omega; f(t) = 0 \text{ for all } f \in I\}$. Then S is closed and evidently $I \subset I(S(I))$. The quotient $\mathcal{A}/I$ is a commutative C^*-algebra. Let $q : \mathcal{A} \to \mathcal{A}/I$ be the quotient map. Then $\{\varphi \circ q; \varphi \in \Phi(\mathcal{A}/I)\} = \{\phi \in \Phi(\mathcal{A}); \phi(I) = \{0\}\} = \{\phi_t; t \in S(I)\}$. Therefore, for $f \in \mathcal{A}$, we have $f \in I \iff q(f) = 0 \iff \varphi(q(f)) = 0$ for every $\varphi \in \Phi(\mathcal{A}/I) \iff f(t) = 0$ for all $t \in S(I) \iff f \in I(S(I))$. Hence $I = I(S(I))$.

The map $I \to S(I)$ is a order-reversing isomorphism between the poset of all ideals of $\mathcal{A}$ and the poset of all closed subsets of Ω, whose inverse is $S \to I(S)$. Indeed, this holds true because the latter map is injective by Urysohn's lemma.

Example 11.17. Let X be a Hilbert space. For $x, y \in X$, denote by $\theta_{x,y}$ the element $(\cdot, y)x$ of $K(X)$. Then span $\{\theta_{x,y}; x, y \in X\}$, which is the set of finite rank operators in $B(X)$, is dense in $K(X)$ (see Part (d) of Exercise 8 in Chapter 6).

We claim that every ideal $\mathcal{I}$ in $B(X)$ contains $K(X)$. Fix $0 \neq T \in \mathcal{I}$, and let $w \in X$ with $Tw \neq 0$. For every $x, y \in X$, we have

$$\theta_{x,y} = \frac{1}{\|Tw\|^2} \theta_{x,Tw} \circ T \circ \theta_{w,y} \in \mathcal{I}.$$

From the foregoing we conclude that $K(X) \subset \mathcal{I}$. In particular, $K(X)$ is a simple C^*-algebra. If X is an infinite-dimensional separable Hilbert space, then one can prove that the only non-trivial ideal in $B(X)$ is $K(X)$. As a result, the *Calkin algebra* $B(X)/K(X)$ is a simple C^*-algebra.

11.6 Positive linear functionals

Let $\mathcal{A}$ be a C^*-algebra. A linear functional on $\mathcal{A}$ is called *Hermitian* (*positive*) if it is *real-valued* on $\mathcal{A}_{\text{sa}}$ (*non-negative-valued* on $\mathcal{A}_+$, respectively).

Clearly, the linear functional ϕ is Hermitian iff $\phi(x^*) = \overline{\phi(x)}$ for all $x \in \mathcal{A}$ (this relation evidently implies that $\phi(x)$ is real for x selfadjoint; on the other hand, if ϕ is Hermitian, write $x = a + ib$ with $a, b \in \mathcal{A}_{\text{sa}}$; then $x^* = a - ib$, and therefore $\phi(x^*) = \phi(a) - i\phi(b)$ is the conjugate of $\phi(x) = \phi(a) + i\phi(b)$, since $\phi(a), \phi(b) \in \mathbb{R}$). In other words, ϕ is Hermitian iff it equals the linear functional ϕ^* on $\mathcal{A}$ given by $x \to \overline{\phi(x^*)}$. In this case we have $\Re(\phi(x)) = \phi(\Re x)$ for every $x \in \mathcal{A}$.

Every linear functional ϕ on $\mathcal{A}$ can be expressed uniquely as $\phi = \phi_1 + \mathrm{i}\phi_2$, where ϕ_1, ϕ_2 are Hermitian linear functionals on $\mathcal{A}$. Indeed, $\phi_1 := \frac{1}{2}(\phi + \phi^*)$ and $\phi_2 := \frac{1}{2\mathrm{i}}(\phi - \phi^*)$ do the job, and uniqueness is easy.

If ϕ is *positive*, it is necessarily Hermitian (write any selfadjoint x as $x^+ - x^-$; since $\phi(x^+), \phi(x^-) \in \mathbb{R}^+$, we have $\phi(x) = \phi(x^+) - \phi(x^-) \in \mathbb{R}$). It is also monotone on $\mathcal{A}_{\text{sa}}$: if $a, b \in \mathcal{A}_{\text{sa}}$ and $a \leq b$, then $\phi(a) \leq \phi(b)$.

Examples.

1. If Ω is a locally compact Hausdorff space and μ is a finite positive Borel measure on Ω, then $\omega(f) := \int_\Omega f\,d\mu$ defines a positive linear functional on $C_0(\Omega)$. By the Riesz–Markov representation theorem (3.18) or the Riesz representation theorem (4.9) and their proofs, all positive linear functional on $C_0(\Omega)$ have this form (and μ may be assumed regular).
2. For $n \in \mathbb{N}$ fixed, the trace $\operatorname{tr} : M_n \to \mathbb{C}$ (given by $A = (a_{ij}) \to \sum_{i=1}^n a_{ii}$) is positive, because if $A = B^*B$ for $B \in M_n$, then $\operatorname{tr}(a) = \sum_{i,j=1}^n \overline{b_{ji}} b_{ji}$. More generally, if $H \in (M_n)_+$, then $\operatorname{tr}_H : M_n \to \mathbb{C}$ given by $\operatorname{tr}_H(A) := \operatorname{tr}(HA)$ is positive. All positive linear functionals on M_n are of this form.
3. If X is a Hilbert space and $\mathcal{A}$ is a C^*-subalgebra of $B(X)$, then for each $x \in X$, the *vector functional* on $\mathcal{A}$ corresponding to x is $\omega_x : \mathcal{A} \to \mathbb{C}$ given by $\mathcal{A} \ni A \to (Ax, x)$. It is positive by Theorem 11.6.
4. If $\mathcal{A}$ is a C^*-algebra we define $\Phi(\mathcal{A})$ as we did in the commutative case: it is the set of all characters of $\mathcal{A}$, namely, the non-zero homomorphisms from $\mathcal{A}$ to $\mathbb{C}$. (When $\mathcal{A}$ is not commutative, it is possible that $\Phi(\mathcal{A})$ be empty!) The proof of Lemma 7.15 shows that each character is, in fact, a $*$-homomorphism from $\mathcal{A}$ to $\mathbb{C}$. Thus, every character $\phi \in \Phi(\mathcal{A})$ is positive, because $\phi(a^*a) = |\phi(a)|^2 \geq 0$ for all $a \in \mathcal{A}$.

We now connect positivity of a linear functional to its norm.

Theorem 11.18. *Let $\mathcal{A}$ be a C^*-algebra and $\omega : \mathcal{A} \to \mathbb{C}$ be a linear functional. Then the following are equivalent:*

(i) ω is positive;

(ii) ω is bounded, and for every approximate identity $\{e_\alpha\}$ of $\mathcal{A}$ we have $\|\omega\| = \lim_\alpha \omega(e_\alpha)$; and

(iii) ω is bounded, and for some approximate identity $\{e_\alpha\}$ of $\mathcal{A}$ we have $\|\omega\| = \lim_\alpha \omega(e_\alpha)$.

In particular, if $\mathcal{A}$ is unital, then ω is positive iff it is bounded and $\omega(\mathbb{1}) = \|\omega\|$.

Before proving the general case, it is constructive to first look at the case that $\mathcal{A}$ is unital. Let us prove that if ω is positive, then it is bounded and $\omega(\mathbb{1}) = \|\omega\|$.

Notice that $\mathbb{1} \in \mathcal{A}_+$, hence $\omega(\mathbb{1}) \geq 0$. If $x \in \mathcal{A}_{\text{sa}}$, then $-\|x\|\mathbb{1} \leq x \leq \|x\|\mathbb{1}$, thus $-\omega(\mathbb{1})\|x\| \leq \omega(x) \leq \omega(\mathbb{1})\|x\|$ by positivity of ω. Therefore $|\omega(x)| \leq \omega(\mathbb{1})\|x\|$

for all $x \in \mathcal{A}_{sa}$. Next, for $x \in \mathcal{A}$ arbitrary, write the complex number $\omega(x)$ in its polar form $|\omega(x)| e^{i\theta}$, $\theta \in \mathbb{R}$. Then

$$\begin{aligned}|\omega(x)| &= e^{-i\theta}\omega(x) = \omega(e^{-i\theta}x) \\ &= \Re\omega(e^{-i\theta}x) = \omega(\Re[e^{-i\theta}x]) \le \omega(\mathbb{1})\|\Re[e^{-i\theta}x]\| \le \omega(\mathbb{1})\|x\|.\end{aligned}$$

This shows that ω is bounded and $\|\omega\| \le \omega(\mathbb{1})$. On the other hand, $\omega(\mathbb{1}) \le \|\omega\|\|\mathbb{1}\| = \|\omega\|$. Therefore $\|\omega\| = \omega(\mathbb{1})$.

We turn to the proof of the general case. For a positive linear functional $\omega : \mathcal{A} \to \mathbb{C}$, put

$$(x, y)_\omega := \omega(y^*x) \qquad (x, y \in \mathcal{A}).$$

Since ω is positive and $x^*x \in \mathcal{A}_+$ for every $x \in \mathcal{A}$ by Theorem 11.6, the form $(\cdot, \cdot)_\omega : \mathcal{A} \times \mathcal{A} \to \mathbb{C}$ is a semi-inner product. Its induced semi-norm on $\mathcal{A}$ is

$$\|x\|_\omega := (x, x)_\omega^{1/2} = \omega(x^*x)^{1/2} \qquad (x \in \mathcal{A}).$$

By the Cauchy–Schwarz inequality for this semi-inner product,

$$|\omega(y^*x)| = |(x, y)_\omega| \le \|x\|_\omega \|y\|_\omega = \omega(x^*x)^{1/2}\omega(y^*y)^{1/2} \quad (\forall x, y \in \mathcal{A}). \qquad (1)$$

Proof

(i) $\implies$ (ii) Let $\omega : \mathcal{A} \to \mathbb{C}$ be a positive linear functional. We first show that $M := \sup\{\omega(a); a \in \mathcal{A}_+, \|a\| \le 1\}$ is finite. Else, there is a sequence $\{a_n\}_{n=1}^\infty$ in $\mathcal{A}_+$ with $\|a_n\| \le 1$ and $\omega(a_n) \ge 2^n$ for every $n \in \mathbb{N}$. The series $\sum_{n=1}^\infty \frac{1}{2^n} a_n$ converges absolutely, thus converges in $\mathcal{A}$ to an element a. For every $N \in \mathbb{N}$,

$$a - \sum_{n=1}^N \frac{1}{2^n} a_n = \sum_{n=N+1}^\infty \frac{1}{2^n} a_n \ge 0, \text{ so } \omega(a) \ge \omega(\sum_{n=1}^N \frac{1}{2^n} a_n) \ge N$$

by the positivity of ω; a contradiction. Thus $M < \infty$.

For general $a \in \mathcal{A}$, by decomposing $a = x + iy$, $x, y \in \mathcal{A}_{sa}$, $\|x\|, \|y\| \le \|a\|$, and further $x = x^+ - x^-$, $y = y^+ - y^-$, $x^\pm, y^\pm \in \mathcal{A}_+$, $\|x_\pm\|, \|y_\pm\| \le \|a\|$, we infer that $\|\omega\| \le 4M$. In particular, ω is bounded.

Assume for convenience that $\|\omega\| = 1$. Given an approximate identity $\{e_\alpha\}$ of $\mathcal{A}$, the net $\{\omega(e_\alpha)\}$ is increasing because $\{e_\alpha\}$ is increasing by definition and ω is positive, so it converges to its supremum m, which is at most $\|\omega\| = 1$. For every $a \in \mathcal{A}$ with $\|a\| \le 1$, we have by the Cauchy–Schwarz inequality (1)

$$|\omega(e_\alpha a)|^2 \le \omega(e_\alpha^2)\omega(a^*a) \le \omega(e_\alpha) \le m \qquad (\forall \alpha).$$

Since $\lim_\alpha e_\alpha a = a$ and ω is continuous, we get $|\omega(a)|^2 \le m$. Hence $1 \le m \le 1$, proving that $m = 1$.

(iii) $\implies$ (i) Assume that $\|\omega\| = \lim_\alpha \omega(e_\alpha) = 1$ for some approximate identity $\{e_\alpha\}$. We first show that ω is Hermitian. Let $a \in \mathcal{A}_{sa}$ with $\|a\| \le 1$. Write $\omega(a) = \zeta + i\eta$ where $\zeta, \eta \in \mathbb{R}$. For all $t \in \mathbb{R}$,

$$\begin{aligned}|\omega(a) + it\omega(e_\alpha)|^2 &= |\omega(a + ite_\alpha)|^2 \le \|a + ite_\alpha\|^2 = \|(a + ite_\alpha)^*(a + ite_\alpha)\| \\ &= \left\|a^2 + t^2e_\alpha^2 + it(ae_\alpha - e_\alpha a)\right\| \le 1 + t^2 + |t| \cdot \|ae_\alpha - e_\alpha a\|.\end{aligned}$$

Using that $\lim_\alpha \omega(e_\alpha) = 1$ and $\lim_\alpha(ae_\alpha - e_\alpha a) = 0$, we infer that

$$\zeta^2 + \eta^2 + 2\eta t + t^2 = |\zeta + \mathrm{i}\eta + \mathrm{i}t|^2 = |\omega(a) + \mathrm{i}t|^2 \le 1 + t^2.$$

Hence, $\zeta^2 + \eta^2 - 1 \le -2\eta t$, and since $t \in \mathbb{R}$ was arbitrary, we must have $\eta = 0$, namely $\omega(a) \in \mathbb{R}$. We proved that ω is Hermitian.

Finally, let $a \in \mathcal{A}_+$ be such that $\|a\| \le 1$. For every α, $e_\alpha - a \in \mathcal{A}_{\mathrm{sa}}$ and $-\mathbb{1} \le -a \le e_\alpha - a \le \mathbb{1} - a \le \mathbb{1}$ in $\mathcal{A}^\#$, so $\|e_\alpha - a\| \le 1$. Since, by the foregoing, $\omega(e_\alpha - a) \in \mathbb{R}$, we get $1 - \omega(a) \underset{\alpha}{\leftarrow} \omega(e_\alpha) - \omega(a) = \omega(e_\alpha - a) \le 1$, so $\omega(a) \ge 0$. This completes the proof. □

We write $\mathcal{A}^*_+$ for the set of all (bounded) positive linear functionals on $\mathcal{A}$. Evidently, $\mathcal{A}^*_+$ is a *weak**-closed positive cone in $\mathcal{A}^*$. We give the set of Hermitian functionals in $\mathcal{A}^*$ the partial order induced by this positive cone: for Hermitian $\phi, \psi \in \mathcal{A}^*$, $\phi \le \psi$ iff $\psi - \phi \in \mathcal{A}^*_+$.

Corollary 11.19. *If $\mathcal{A}$ is a C^*-algebra and $\omega_1, \omega_2 \in \mathcal{A}^*_+$, then $\|\omega_1 + \omega_2\| = \|\omega_1\| + \|\omega_2\|$.*

Proof. For an approximate identity $\{e_\alpha\}$ of $\mathcal{A}$, by Theorem 11.18 we have

$$\|\omega_1 + \omega_2\| = \lim_\alpha (\omega_1 + \omega_2)(e_\alpha) = \lim_\alpha \omega_1(e_\alpha) + \lim_\alpha \omega_2(e_\alpha) = \|\omega_1\| + \|\omega_2\|. \quad \square$$

The next result is concerned with the existence of positive linear functionals. It will be vital to the proof of the non-commutative Gelfand–Naimark theorem.

A positive linear functional ω on $\mathcal{A}$ that is *normalized*, that is, $\|\omega\| = 1$, is called a *state* of $\mathcal{A}$. The set of all states of $\mathcal{A}$ will be denoted by $\mathcal{S}(\mathcal{A})$. It will play in the non-commutative case a role as crucial as the role that $\Phi(\mathcal{A})$ played in the commutative Gelfand–Naimark theorem. Notice that $\Phi(\mathcal{A}) \subset \mathcal{S}(\mathcal{A})$.

Theorem 11.20. *Let $\mathcal{A}$ be a C^*-algebra and write $\mathcal{S} := \mathcal{S}(\mathcal{A})$. Then, for each $x \in \mathcal{A}$:*

(i) if $\mathcal{A}$ is unital then $\sigma(x) \subset \{\omega(x); \omega \in \mathcal{S}\}$;

(ii) if x is normal, then $\|x\| = \max_{\omega \in \mathcal{S}} |\omega(x)|$; for x arbitrary, $\|x\| = \max_{\omega \in \mathcal{S}} \|x\|_\omega$, where we recall that $\|x\|^2_\omega = \omega(x^*x)$;

(iii) if $\omega(x) = 0$ for all $\omega \in \mathcal{S}$, then $x = 0$;

(iv) if $\omega(x) \in \mathbb{R}$ for all $\omega \in \mathcal{S}$, then x is selfadjoint; and

(v) if $\omega(x) \in \mathbb{R}^+$ for all $\omega \in \mathcal{S}$, then $x \in \mathcal{A}_+$.

Also, if $\mathcal{B}$ is a C^*-subalgebra of $\mathcal{A}$, then

(vi) every $\omega \in \mathcal{B}^*_+$ extends to some $\tilde{\omega} \in \mathcal{A}^*_+$ with the same norm.

Proof. (i) Let $\lambda \in \sigma(x)$. Then for any $\alpha, \beta \in \mathbb{C}$, $\alpha\lambda + \beta \in \sigma(\alpha x + \beta\mathbb{1})$, and therefore

$$|\alpha\lambda + \beta| \le \|\alpha x + \beta\mathbb{1}\|. \tag{2}$$

Define $\omega_0 : Z := \text{span}\{x, \mathbb{1}\} \to \mathbb{C}$ by

$$\omega_0(\alpha x + \beta \mathbb{1}) := \alpha\lambda + \beta \qquad (\alpha, \beta \in \mathbb{C}).$$

If $\alpha x + \beta\mathbb{1} = \alpha' x + \beta'\mathbb{1}$, then by (2)

$$\begin{aligned}|(\alpha\lambda + \beta) - (\alpha'\lambda + \beta')| &= |(\alpha - \alpha')\lambda + (\beta - \beta')| \\ &\leq \|(\alpha - \alpha')x + (\beta - \beta')\mathbb{1}\| = 0.\end{aligned}$$

Therefore ω_0 is well defined. It is clearly linear and bounded, with norm at most 1 by (2). Since $\omega_0(\mathbb{1}) = 1$, we have $\|\omega_0\| = 1$. By the Hahn–Banach theorem, ω_0 has an extension ω as a bounded linear functional on $\mathcal{A}$ with norm $\|\omega\| = \|\omega_0\| = 1$. Since also $\omega(\mathbb{1}) = \omega_0(\mathbb{1}) = 1$, it follows from Theorem 11.18 that $\omega \in \mathcal{S}$, and $\lambda = \omega_0(x) = \omega(x)$.

(ii) Since all states are contractive, we have $\sup_{\omega \in \mathcal{S}} |\omega(x)| \leq \|x\|$ for any x. When $x \neq 0$ is normal, we have $r(x) = \|x\|$, and therefore there exists $\lambda_1 \in \sigma(x)$ such that $|\lambda_1| = \|x\|$. By (i) applied to $\mathcal{A}^\#$, $\lambda_1 = \omega_1^\#(x)$ for some $\omega_1^\# \in \mathcal{S}(\mathcal{A}^\#)$. If $\mathcal{A}$ is unital, we have finished: take $\omega_1 := \omega_1^\#$. If not, the restriction $\omega_1 := \omega_1^\#|_{\mathcal{A}}$ is positive, $\|\omega_1\| \leq \|\omega_1^\#\| = 1$, and $|\omega_1(\frac{1}{\|x\|}x)| = |\omega_1^\#(\frac{1}{\|x\|}x)| = \frac{|\lambda_1|}{\|x\|} = 1$, so $\|\omega_1\| = 1$. This shows that the supremum is a maximum, attained at $\omega_1 \in \mathcal{S}$, and is equal to $\|x\|$.

For x arbitrary, we apply the preceding identity to the selfadjoint (hence normal!) element x^*x:

$$\|x\|^2 = \|x^*x\| = \max_{\omega \in \mathcal{S}} |\omega(x^*x)| := \max_{\omega \in \mathcal{S}} \|x\|_\omega^2.$$

(iii) Suppose that $\omega(x) = 0$ for all $\omega \in \mathcal{S}$. Write $x = a + ib$ with $a, b \in \mathcal{A}$ selfadjoint. Since ω is real on selfadjoint elements (being a positive linear functional, hence Hermitian), the relation $0 = \omega(x) = \omega(a) + i\omega(b)$ implies that $\omega(a) = \omega(b) = 0$ for all $\omega \in \mathcal{S}$. By (ii), it follows that $a = b = 0$, hence $x = 0$.

(iv) If $\omega(x) \in \mathbb{R}$ for all ω, then (with notation as in (iii)) $\omega(b) = 0$ for all ω, and therefore $b = 0$ by (ii). Hence $x = a$ is selfadjoint.

(v) If $\omega(x) \in \mathbb{R}^+$ for all $\omega \in \mathcal{S}$, then x is selfadjoint by (iv). We also have $\omega(x) \in \mathbb{R}^+$ for all $\omega \in \mathcal{A}_+^*$, thus $\omega^\#(x) \in \mathbb{R}^+$ for all $\omega^\# \in (\mathcal{A}^\#)_+^*$ as $\omega^\#|_{\mathcal{A}} \in \mathcal{A}_+^*$. Consequently $\sigma_{\mathcal{A}}(x) = \sigma_{\mathcal{A}^\#}(x) \subset \mathbb{R}^+$ by (i); hence $x \in \mathcal{A}_+$.

(vi) Claim: suppose that $\mathcal{C}$ is a C^*-algebra, either unital or non-unital, and $\omega \in \mathcal{C}_+^*$. Let $\tilde{\mathcal{C}}$ be a unital C^*-algebra containing $\mathcal{C}$ and spanned by $\mathcal{C}$ and the unit $\mathbb{1}$ of $\tilde{\mathcal{C}}$, $\mathbb{1} \notin \mathcal{C}$. Then we can extend ω to $\tilde{\omega} \in \tilde{\mathcal{C}}_+^*$ with the same norm.

Proof of Claim: extend ω linearly to a functional $\tilde{\omega}$ on $\tilde{\mathcal{C}}$ by letting $\tilde{\omega}(\mathbb{1}) := \|\omega\|$. Let $\{e_\alpha\}$ be an approximate identity for $\mathcal{C}$. For every $x \in \mathcal{C}$ and $\zeta \in \mathbb{C}$, by Theorem 11.18,

$$|\tilde{\omega}(x + \zeta\mathbb{1})| = |\omega(x) + \zeta\,\|\omega\|| = \lim_\alpha |\omega(xe_\alpha) + \zeta\omega(e_\alpha)| = \lim_\alpha |\omega(xe_\alpha + \zeta e_\alpha)|\,.$$

But for every α we have

$$\begin{aligned}|\omega(xe_\alpha + \zeta e_\alpha)| &\leq \|\omega\|\,\|xe_\alpha + \zeta e_\alpha\| = \|\omega\|\,\|(x + \zeta\mathbb{1})e_\alpha\| \\ &\leq \|\omega\|\,\|x + \zeta\mathbb{1}\|\,\|e_\alpha\| \leq \|\omega\|\,\|x + \zeta\mathbb{1}\|\,.\end{aligned}$$

Therefore $|\tilde{\omega}(x+\zeta\mathbb{1})| \le \|\omega\|\,\|x+\zeta\mathbb{1}\|$. In other words, $\|\tilde{\omega}\| \le \|\omega\|$, whence $\|\tilde{\omega}\| = \|\omega\| = \tilde{\omega}(\mathbb{1})$, so $\tilde{\omega}$ is positive by Theorem 11.18.

By the claim, we may assume that $\mathcal{A}$ and $\mathcal{B}$ are unital with $\mathbb{1} := \mathbb{1}_{\mathcal{A}} \in \mathcal{B}$. Using the Hahn–Banach theorem, extend ω to $\tilde{\omega} \in \mathcal{A}^*$ with $\|\tilde{\omega}\| = \|\omega\|$. Thus $\|\tilde{\omega}\| = \|\omega\| = \omega(\mathbb{1}) = \tilde{\omega}(\mathbb{1})$, so $\tilde{\omega} \in \mathcal{A}^*_+$ by Theorem 11.18. □

11.7 Representations and the Gelfand–Naimark–Segal construction

In this section we establish the existence of "enough" representations, and prove the non-commutative Gelfand–Naimark theorem, saying that *every* C^*-algebra has a *faithful* representation on some Hilbert space. In other words, every C^*-algebra can be thought of as a C^*-subalgebra of some $B(X)$!

Let $\mathcal{A}$ be a C^*-algebra. A *(C^*-) representation of $\mathcal{A}$ on a Hilbert space X* is a $*$-homomorphism $\pi : \mathcal{A} \to B(X)$. We say that π is *faithful* if it is injective, or equivalently, if it is isometric (see Theorem 11.2). If $\mathcal{A}$ is unital and $\pi(\mathbb{1}) = I$, we say that π is *unital.*

If there exists a vector $\zeta_0 \in X$ such that $\pi(\mathcal{A})\zeta_0 := \{\pi(a)\zeta_0; a \in \mathcal{A}\}$ is dense in X, then π is called *cyclic* and ζ_0 is said to be a *cyclic vector* for π.

If $\pi(\mathcal{A})X := \operatorname{span}\{\pi(a)\zeta; a \in \mathcal{A}, \zeta \in X\}$ is dense in X, then π is called *non-degenerate.*

A cyclic representation is evidently non-degenerate. Also, if $\mathcal{A}$ is unital, then π is non-degenerate iff it is unital.

Example 11.21. Let Ω be a locally compact Hausdorff space and $\mathcal{A} := C_0(\Omega)$. Let μ be a positive Borel measure on Ω satisfying the assumptions of Lusin's theorem (3.20). Set $X := L^2(\Omega, \mu)$. For every $f \in C_0(\Omega)$, let $M_f \in B(X)$ be given by $g \to fg$ ($g \in X$). Then $\pi : \mathcal{A} \to B(X)$ given by $\pi(f) := M_f$ ($f \in \mathcal{A}$) is a representation of $\mathcal{A}$ on X. It is non-degenerate because $C_c(\Omega)$ ($\subset C_0(\Omega)$) is dense in X by Corollary 3.21 and because every function in $C_c(\Omega)$ is the product of two functions in $C_c(\Omega)$. If Ω is compact, then π is unital. If μ is finite, then the constant function 1 belongs to X and is cyclic for π, again because $C_c(\Omega)$, thus $C_0(\Omega)$, is dense in X. If the support of μ is all of Ω, that is, if $\mu(V) > 0$ for each non-empty open set $V \subset \Omega$ (see Section 3.5), then π is faithful, as if $0 \neq f \in \mathcal{A}$ then there is a non-empty open set V such that f does not vanish anywhere on V and $\mu(V) < \infty$; then the indicator $g := I_V$ of V belongs to X and $M_f g \neq 0$ in X.

Let $\omega \in \mathcal{S} := \mathcal{S}(\mathcal{A})$ be a state. Recall from Section 11.6 that ω induces on $\mathcal{A}$ the semi-inner product (s.i.p.) given by

$$(x, y)_\omega := \omega(y^* x) \qquad (x, y \in \mathcal{A}).$$

The induced semi-norm

$$\|x\|_\omega := (x, x)_\omega^{1/2} = \omega(x^* x)^{1/2} \qquad (x \in \mathcal{A})$$

is continuous on $\mathcal{A}$ (by continuity of ω, of the involution, and the multiplication). Also, if $\mathcal{A}$ is unital, then it is "normalized", that is, $\|\mathbb{1}\|_\omega = 1$ (because $\omega(\mathbb{1}) = 1$ by Theorem 11.18).

Let

$$J_\omega := \{x \in \mathcal{A}; \|x\|_\omega = 0\} = \|\cdot\|_\omega^{-1}(\{0\}).$$

The properties of the semi-norm imply that J_ω is a closed subspace of $\mathcal{A}$.

By the Cauchy–Schwarz inequality (see (1) of Section 11.6), if $x \in J_\omega$, then $(x, y)_\omega = (y, x)_\omega = 0$ for all $y \in \mathcal{A}$. This implies that J_ω is left $\mathcal{A}$-invariant (i.e., a left ideal), because for all $x \in J_\omega$ and $y \in \mathcal{A}$

$$\|yx\|_\omega^2 := \omega\Big((yx)^*(yx)\Big) = \omega\Big((y^*yx)^*x\Big) = (x, y^*yx)_\omega = 0.$$

Lemma 11.22. *Let $\omega \in \mathcal{S}(\mathcal{A})$. Then for all $x, y \in \mathcal{A}$,*

$$\|xy\|_\omega \le \|x\|\|y\|_\omega.$$

Proof. Since $x^*x \le \|x^*x\|\,\mathbb{1}$ (in $\mathcal{A}^\#$), we have $y^*x^*xy \le \|x^*x\|\,y^*y$ by Theorem 11.6, so by monotonicity,

$$\|xy\|_\omega^2 = \omega((xy)^*(xy)) = \omega(y^*x^*xy) \le \|x^*x\|\,\omega(y^*y) = \|x\|^2\,\|y\|_\omega^2. \qquad \square$$

The s.i.p. $(\cdot,\cdot)_\omega$ induces an inner product (same notation) on the quotient space $\mathcal{A}/J_\omega$:

$$(x + J_\omega, y + J_\omega)_\omega := (x, y)_\omega \quad (x, y \in \mathcal{A}).$$

This definition is independent on the cosets representatives, because if $x + J_\omega = x' + J_\omega$ and $y + J_\omega = y' + J_\omega$, then $x - x', y - y' \in J_\omega$, and therefore (dropping the subscript ω)

$$(x', y') = (x' - x, y') + (x, y) + (x, y' - y) = (x, y)$$

(the first and third summands vanish, because one of the factors of the s.i.p. is in J_ω).

If $\|x + J_\omega\|_\omega^2 := (x + J_\omega, x + J_\omega)_\omega = 0$, then $\|x\|_\omega^2 = (x, x)_\omega = 0$, that is, $x \in J_\omega$, hence $x + J_\omega$ is the zero coset J_ω. This means that $\|\cdot\|_\omega$ is a norm on $\mathcal{A}/J_\omega$. Let X_ω be the completion of $\mathcal{A}/J_\omega$ with respect to this norm. Then X_ω is a Hilbert space (its inner product is the unique continuous extension of $(\cdot,\cdot)_\omega$ from the dense subspace $\mathcal{A}/J_\omega \times \mathcal{A}/J_\omega$ to $X_\omega \times X_\omega$; the extension is also denoted by $(\cdot,\cdot)_\omega$).

For each $x \in \mathcal{A}$, consider the map

$$\pi(x) := \pi_\omega(x) : \mathcal{A}/J_\omega \to \mathcal{A}/J_\omega$$

defined by

$$\pi(x)(y + J_\omega) = xy + J_\omega \quad (y \in \mathcal{A}).$$

It is well defined, because if y, y' represent the same coset, then $\|y - y'\|_\omega = 0$, and therefore, by Lemma 11.22, $\|x(y - y')\|_\omega = 0$, which means that xy and xy' represent the same coset.

The map $\pi(x)$ is clearly linear. It is also *bounded*, with operator norm (on the normed space $\mathcal{A}/J_\omega$) $\|\pi(x)\| \le \|x\|$: indeed, by Lemma 11.22, for all $y \in \mathcal{A}$,

$$\|\pi(x)(y + J_\omega)\|_\omega = \|xy + J_\omega\|_\omega = \|xy\|_\omega \le \|x\|\|y\|_\omega = \|x\|\|y + J_\omega\|_\omega.$$

Therefore, $\pi(x)$ extends uniquely by continuity to a bounded operator on X_ω (also denoted $\pi(x)$), with operator norm $\|\pi(x)\| \le \|x\|$.

A routine calculation shows that $x \to \pi(x)$ is an algebra homomorphism of $\mathcal{A}$ into $B(\mathcal{A}/J_\omega)$, and a continuity argument implies that it is a homomorphism of $\mathcal{A}$ into $B(X_\omega)$. If $\mathcal{A}$ is unital, then since $\pi(\mathbb{1})$ is the identity operator on $\mathcal{A}/J_\omega$, we have $\pi(\mathbb{1}) = I$, the identity operator on X_ω.

For all $x, y, z \in \mathcal{A}$, we have (dropping the index ω)

$$\begin{aligned}(\pi(x)(y + J), z + J) &= (xy + J, z + J) = (xy, z) = \omega(z^*xy) = \omega\left((x^*z)^*y\right) \\ &= (y, x^*z) = (y + J, x^*z + J) = (y + J, \pi(x^*)(z + J)).\end{aligned}$$

By continuity, we obtain the identity

$$(\pi(x)\zeta, \eta) = (\zeta, \pi(x^*)\eta) \quad (\zeta, \eta \in X_\omega),$$

that is

$$(\pi(x))^* = \pi(x^*) \quad (x \in \mathcal{A}).$$

We conclude that $\pi : x \to \pi(x)$ is a (norm-decreasing) $*$-homomorphism of $\mathcal{A}$ into $B(X_\omega)$, namely, it is a *representation of $\mathcal{A}$ on the Hilbert space X_ω*, which is unital if $\mathcal{A}$ is. The construction of the "canonical" representation π is referred to as the *Gelfand–Naimark–Segal (GNS) construction*; accordingly, $\pi := \pi_\omega : x \to \pi_\omega(x)$ will be called the GNS representation associated with the given state ω on $\mathcal{A}$.

Assume for the moment that $\mathcal{A}$ is unital and consider the unit vector $\zeta_\omega := \mathbb{1} + J_\omega \in X_\omega$ (it is a unit vector because $\|\zeta_\omega\|_\omega = \|\mathbb{1}\|_\omega = \omega(\mathbb{1}^*\mathbb{1})^{1/2} = 1$). By definition of X_ω, the set

$$\{\pi(x)\zeta_\omega; x \in \mathcal{A}\} = \{x + J_\omega; x \in \mathcal{A}\} = \mathcal{A}/J_\omega$$

is dense in X_ω. Consequently, *the representation $\pi : x \to \pi(x)$ is cyclic, with cyclic vector ζ_ω*. Note also the identity (dropping the index ω)

$$\omega(x) = \omega(\mathbb{1}^*x) = (x, \mathbb{1}) = (x + J, \mathbb{1} + J) = (\pi(x)(\mathbb{1} + J), \mathbb{1} + J) = (\pi(x)\zeta, \zeta). \tag{1}$$

Thus, the state ω is realized through the representation π as the composition $\omega_\zeta \circ \pi$, where ω_ζ is the so-called *vector state* on $B(X_\omega)$ corresponding to the unit vector ζ defined by

$$\omega_\zeta(T) := (T\zeta, \zeta) \quad (T \in B(X_\omega)).$$

(see the examples in Section 11.6). One can show that a cyclic unit vector $\zeta = \zeta_\omega \in X_\omega$ for π with the same property $\omega = \omega_\zeta \circ \pi$ exists also when $\mathcal{A}$ is not unital; see Exercise 17.

The GNS representation is "universal" in a certain sense which we proceed to specify.

Suppose $\pi' : x \to \pi'(x)$ is *any* cyclic representation of $\mathcal{A}$ on a Hilbert space Z, with unit cyclic vector $\eta \in Z$ such that $\omega = \omega_\eta \circ \pi'$. Then by (1), for all $x \in \mathcal{A}$,

$$\begin{aligned}\|\pi'(x)\eta\|_Z^2 &= (\pi'(x)\eta, \pi'(x)\eta)_Z = (\pi'(x^*x)\eta, \eta)_Z \\ &= (\omega_\eta \circ \pi')(x^*x) = \omega(x^*x) = (\pi(x^*x)\zeta, \zeta)_\omega = \|\pi(x)\zeta\|_\omega^2. \end{aligned} \tag{2}$$

Since π and π' are *cyclic* representations on X_ω and Z with respective cyclic vectors ζ and η, the subspaces $\tilde{X}_\omega := \{\pi(x)\zeta; x \in \mathcal{A}\}$ and $\tilde{Z} := \{\pi'(x)\eta; x \in \mathcal{A}\}$ are dense in X_ω and Z, respectively. Define $U : \tilde{X}_\omega \to \tilde{Z}$ by

$$U(\pi(x)\zeta) := \pi'(x)\eta \quad (x \in \mathcal{A}).$$

It follows from (2) that U is a linear isometry of $\tilde{X}_\omega$ onto $\tilde{Z}$. It extends uniquely by continuity as a linear isometry of X_ω onto Z. Thus, U is a Hilbert space isomorphism of X_ω onto Z. For all $x, y \in \mathcal{A}$ we have

$$\begin{aligned}(U\pi(x))\,(\pi(y)\zeta) &= U\pi(xy)\zeta = \pi'(xy)\eta = \pi'(x)\pi'(y)\eta \\ &= \pi'(x)(U\pi(y)\zeta) = (\pi'(x)U)\,(\pi(y)\zeta).\end{aligned}$$

Thus, $U\pi(x) = \pi'(x)U$ on the dense subspace $\tilde{X}_\omega$ of X_ω; by continuity of the operators, it follows that $U\pi(x) = \pi'(x)U$, that is, $\pi'(x) = U\pi(x)U^*$ for all $x \in \mathcal{A}$ (one says that the representations π' and π are *unitarily equivalent*, through the *unitary equivalence* $U : X_\omega \to Z$). When $\mathcal{A}$ is unital, U carries ζ onto η because $U\zeta = UI\zeta = U\pi(\mathbb{1})\zeta := \pi'(\mathbb{1})\eta = I\eta = \eta$, where we use the notation I for the identity operator in both Hilbert spaces. This is also true when $\mathcal{A}$ is not unital; see again Exercise 17. This concludes the proof of the following

Theorem 11.23 (The Gelfand–Naimark–Segal theorem). *Let ω be a state of a C^*-algebra $\mathcal{A}$. Then the associated GNS representation $\pi := \pi_\omega$ is a cyclic representation of $\mathcal{A}$ on the Hilbert space $X := X_\omega$, with a unit cyclic vector $\zeta := \zeta_\omega$ such that $\omega = \omega_\zeta \circ \pi$. If π' is another cyclic representation of $\mathcal{A}$ on a Hilbert space Z with unit cyclic vector η such that $\omega = \omega_\eta \circ \pi'$, then π' is unitarily equivalent to π under a unitary equivalence $U : X \to Z$ such that $U\zeta = \eta$.*

Example 11.24 (The GNS construction of integration states of $C_0(\Omega)$). Let Ω be a locally compact Hausdorff space and μ be a *regular* probability Borel measure on Ω. The representation of $C_0(\Omega)$ on $L^2(\Omega, \mu)$ constructed in Example 11.21 ($\pi(f) = M_f$) is (unitarily equivalent to) the GNS representation of the state $\omega : C_0(\Omega) \to \mathbb{C}$, $f \to \int_\Omega f\, d\mu$. Indeed, the constant function 1 is cyclic for π, and for all $f \in C_0(\Omega)$, $(\pi(f)1, 1) = (f, 1) = \int_\Omega f \cdot \overline{1}\, d\mu = \omega(f)$.

Example 11.25 (The GNS construction of vector states). Let X be a Hilbert space, $\mathcal{A}$ be a C^*-subalgebra of $B(X)$, $\zeta \in X$ be a unit vector, and $\omega := \omega_\zeta|_{\mathcal{A}}$. Suppose that $\mathcal{A}$ is non-degenerate on X, i.e., $\overline{\operatorname{span}}\{x\eta; x \in \mathcal{A}, \eta \in X\} = X$.

Then the subspace $Y := \overline{\mathcal{A}\zeta}$ is $\mathcal{A}$-invariant (in fact, it is reducing for every element of $\mathcal{A}$), and we let the reader verify that $\zeta \in Y$. Obviously, the representation $\pi : x \to x|_Y$ of $\mathcal{A}$ on the Hilbert space Y has ζ as a cyclic vector and $\omega = \omega_\zeta \circ \pi$. Thus, the GNS representation of $\mathcal{A}$ associated to ω is unitarily equivalent to π.

Let X be the Hilbert space direct sum

$$X := \sum_{\omega \in \mathcal{S}} \oplus X_\omega$$

(recall the construction of Section 8.6). For each $x \in \mathcal{A}$, define $\pi(x) \in B(X)$ by

$$\pi(x) := \sum_{\omega \in \mathcal{S}} \oplus \pi_\omega(x).$$

This makes sense because $\|\pi_\omega(x)\| \leq \|x\|$ for each $\omega \in \mathcal{S}$, from which we also get $\|\pi(x)\| \leq \|x\|$. Actually, we have $\|\pi(x)\| = \|x\|$. Indeed, for each $\omega \in \mathcal{S}$, consider the vector $f_\omega \in X$ defined by

$$f_\omega(\omega) := \zeta_\omega; \qquad f_\omega(\rho) := 0 \quad (\rho \in \mathcal{S}, \rho \neq \omega). \tag{3}$$

Then $\|f_\omega\| = \|\zeta_\omega\|_\omega = 1$ and

$$\|\pi(x) f_\omega\| = \|\pi_\omega(x)\zeta_\omega\|_\omega = \|x\|_\omega.$$

Hence, $\|\pi(x)\| \geq \|x\|_\omega$ for all $\omega \in \mathcal{S}$, and therefore

$$\|\pi(x)\| \geq \max_{\omega \in \mathcal{S}} \|x\|_\omega = \|x\|$$

by Theorem 11.20 (ii). Together with the preceding inequality, we obtain $\|\pi(x)\| = \|x\|$ for all $x \in \mathcal{A}$.

An easy calculation shows that the map $\pi : x \to \pi(x)$ of $\mathcal{A}$ into $B(X)$ is an algebra homomorphism. Also $\pi(x^*) = (\pi(x))^*$ because for all $f, g \in X$,

$$\begin{aligned}(\pi(x^*)f, g) &= \sum_{\omega \in \mathcal{S}} \big(\pi_\omega(x^*) f(\omega), g(\omega)\big)_\omega = \sum_\omega \big((\pi_\omega(x))^* f(\omega), g(\omega)\big)_\omega \\ &= \sum_\omega \big(f(\omega), \pi_\omega(x) g(\omega)\big)_\omega = (f, \pi(x) g).\end{aligned}$$

Thus, π is an isometric $*$-isomorphism of the C^*-algebra $\mathcal{A}$ onto the C^*-subalgebra $\pi(\mathcal{A})$ of $B(X)$, namely a *faithful representation* of $\mathcal{A}$ on X. It is non-degenerate because each π_ω is non-degenerate. If $\mathcal{A}$ is unital, so is π, that is, $\pi(\mathbb{1}) = I$. The usual notation for π is $\sum_{\omega \in \mathcal{S}} \oplus \pi_\omega$; it is called *the direct sum of the representations* $\{\pi_\omega\}_{\omega \in \mathcal{S}}$. This particular representation of $\mathcal{A}$ is usually referred to as the *universal representation* of $\mathcal{A}$. We thus proved the following theorem, which can be considered the starting point of C^*-algebra theory.

Theorem 11.26 (The non-commutative Gelfand–Naimark theorem). *Every C^*-algebra $\mathcal{A}$ admits a* faithful *non-degenerate representation, that is, $\mathcal{A}$ is isometrically $*$-isomorphic to a non-degenerate C^*-subalgebra of $B(X)$ for some Hilbert space X.*

A special such representation (called the "universal representation") of the C^*-algebra $\mathcal{A}$ is the direct sum representation $\pi := \sum_{\omega\in\mathcal{S}} \oplus\pi_\omega$ on $X := \sum_{\omega\in\mathcal{S}} \oplus X_\omega$ of the GNS representations $\{\pi_\omega\}_{\omega\in\mathcal{S}}$, where $\mathcal{S} := \mathcal{S}(\mathcal{A})$.

The universality property of the universal representation π is that *every state of $\pi(\mathcal{A})$ ($\subset B(X)$) is a vector state.* Indeed, every state τ of $\pi(\mathcal{A})$ is of the form $\omega\circ\pi^{-1}$ for some state $\omega\in\mathcal{S}$. Consider the vector $f := f_\omega \in X$ of (3). Since $\omega = \omega_\zeta\circ\pi_\omega$ for $\zeta := \zeta_\omega$, it is clear that $\tau = \omega_f|_{\pi(\mathcal{A})}$. See also Exercise 24.

11.7.1 Irreducible representations

In this short section we introduce and characterize a special type of representations of C^*-algebras, namely *irreducible* ones. An elementary way to construct such representations appears in Section 11.8.

We start with several notions and a (very useful) lemma that connects them. For a Hilbert space X and a subset $\mathcal{R}\subset B(X)$, the *commutant* $\mathcal{R}'$ of $\mathcal{R}$ (in $B(X)$) is the set

$$\mathcal{R}' := \{S\in B(X); ST = TS \text{ for all } T\in\mathcal{R}\}.$$

Say that a closed subspace of X is *$\mathcal{R}$-invariant* if it is T-invariant for each $T\in\mathcal{R}$.

Lemma 11.27. *Let X be a Hilbert space and $\mathcal{R}$ be a $*$-subalgebra of $B(X)$. Consider the following conditions for a projection $P\in B(X)$:*

(i) $P\in\mathcal{R}'$;

(ii) the range of P is $\mathcal{R}$-invariant;

(iii) the range of P has the form $\overline{\mathcal{R}S}$, where $S\subset X$ and $\mathcal{R}S :=$ span$\{Tx; T\in\mathcal{R}, x\in S\}$.

Then (i) $\Longleftrightarrow$ (ii) $\Longleftarrow$ (iii), and all conditions are equivalent when $\mathcal{R}$ is non-degenerate, that is, $\overline{\mathcal{R}X} = X$.

Proof. (iii) $\Longrightarrow$ (ii) Immediate, because $\mathcal{R}$ is stable under multiplication.

(ii) $\Longrightarrow$ (i) For each $A\in\mathcal{R}$, the range of P is invariant for both A and A^* (because $A^*\in\mathcal{R}$, by selfadjointness of $\mathcal{R}$), so by Terminology 8.5, Point (c), we have $PA = AP$. That is, $P\in\mathcal{R}'$.

(i) $\Longrightarrow$ (ii) For each $x\in X$ and $T\in\mathcal{R}$, $TPx = PTx$ is in the range of P.

Finally, if $\mathcal{R}$ is non-degenerate then (i) $\Longrightarrow$ (iii), because the range PX of a projection $P\in\mathcal{R}'$ equals $\overline{\mathcal{R}(PX)}$. Indeed, $\overline{\mathcal{R}(PX)} = \overline{P(\mathcal{R}X)}$ since $P\in\mathcal{R}'$, and we have $PX = P(\overline{\mathcal{R}X}) \subset \overline{P(\mathcal{R}X)} \subset \overline{PX} = PX$. □

Let π be a representation of a C^*-algebra $\mathcal{A}$ on a Hilbert space X. If Y is a $\pi(\mathcal{A})$-invariant closed subspace of X, the function from $\mathcal{A}$ to $B(Y)$ given by $a \to \pi(a)|_Y$, $a \in \mathcal{A}$, is (well defined and) a representation of $\mathcal{A}$ on Y. Such a representation of $\mathcal{A}$ is called a *sub-representation of* π.

We say that π is *irreducible* if it has only trivial sub-representations, that is, if the only $\pi(\mathcal{A})$-invariant closed subspaces of X are $\{0\}$ and X. By Lemma 11.27, this means that the only projections in the commutant $\pi(\mathcal{A})'$ are 0 and the identity operator I. Since the projections in $\pi(\mathcal{A})'$ span a norm-dense subset of $\pi(\mathcal{A})'$ (see Section 12.1), this just means that the commutant is trivial: $\pi(\mathcal{A})' = \mathbb{C}I$.

Irreducibility of π has various implications. For instance, it is evidently non-degenerate. But more is true: if $\zeta \in X$ is non-zero, then the closed subspace $\overline{\pi(\mathcal{A})\zeta}$ is $\pi(\mathcal{A})$-invariant (and is also non-zero!) and thus equals X. In other words, every non-zero vector in X is cyclic for π. The converse is also true: if every non-zero vector in X is cyclic for π then π is irreducible, because every non-zero closed subspace of X invariant under $\mathcal{A}$ contains $\overline{\pi(\mathcal{A})\zeta} = X$ for some non-zero $\zeta \in X$.

11.8 Positive linear functionals and convexity

This section discusses notions and results on positive linear functionals that are related to convexity. Again, $\mathcal{A}$ is a C^*-algebra and $\mathcal{S}(\mathcal{A})$ is the set of all states of $\mathcal{A}$.

In this section the proofs in the non-unital case sometimes require a small technical change. The reader can safely assume that $\mathcal{A}$ is unital in first reading.

The set $\mathcal{S}(\mathcal{A}) = \{\omega \in \mathcal{A}^*_+; \|\omega\| = 1\}$ is convex in $\mathcal{A}^*$ by Corollary 11.19. The set $\mathcal{K}(\mathcal{A}) := \{\omega \in \mathcal{A}^*_+; \|\omega\| \leq 1\}$ is convex as the intersection of two convex sets: $\mathcal{A}^*_+$ and the norm-closed unit ball of $\mathcal{A}^*$. The first of these sets is *weak**-closed and the second is *weak**-compact by Alaoglu's theorem, so $\mathcal{K}(\mathcal{A})$ is also *weak**-compact. The set $\mathcal{S}(\mathcal{A})$ is *weak**-compact when $\mathcal{A}$ is unital by Theorem 11.18.

11.8.1 Pure states

A state ω of $\mathcal{A}$ is called *pure* if every $\rho \in \mathcal{A}^*_+$ dominated by ω, that is, $\rho \leq \omega$, is a multiple of ω, that is, $\rho = c\omega$ for some $c \in [0,1]$. Notice that $\rho \in \mathcal{A}^*_+$ is dominated by ω iff there exists $\rho' \in \mathcal{A}^*_+$ such that $\rho + \rho' = \omega$; and in this situation we have $\|\rho\| + \|\rho'\| = \|\omega\| = 1$ by Corollary 11.19. Consequently, the pure states of $\mathcal{A}$ are precisely the extremal points of the convex set $\mathcal{S}(\mathcal{A})$.

Denote the set of all pure states of $\mathcal{A}$ by $\mathcal{P}(\mathcal{A})$.

If $\mathcal{A}$ is unital, then since $\mathcal{S}(\mathcal{A})$ is *weak**-compact and convex in $\mathcal{A}^*$, the Krein–Milman theorem implies that $\mathcal{S}(\mathcal{A}) = \overline{\text{co}}^{weak^*}(\mathcal{P}(\mathcal{A}))$.

For general $\mathcal{A}$, we assert that $\mathcal{P}(\mathcal{A}) \cup \{0\}$ is the set of extremal points of the (*weak**-compact and convex) set $\mathcal{K}(\mathcal{A})$, so the Krein–Milman theorem implies that $\mathcal{K}(\mathcal{A}) = \overline{\text{co}}^{weak^*}(\mathcal{P}(\mathcal{A}) \cup \{0\})$. Indeed, if $\omega \in \mathcal{A}^*_+$ and $0 < \|\omega\| < 1$, then ω is not an extremal point of $\mathcal{K}(\mathcal{A})$ because $\omega = \frac{1}{2}\left(\frac{\|\omega\|+s}{\|\omega\|}\omega + \frac{\|\omega\|-s}{\|\omega\|}\omega\right)$ with

$s := \min(\|\omega\|, 1 - \|\omega\|)$. Conversely, the subsets $\mathcal{S}(\mathcal{A})$ and $\{0\}$ of $\mathcal{K}(\mathcal{A})$ are extremal in $\mathcal{K}(\mathcal{A})$, thus the elements of $\mathcal{P}(\mathcal{A}) \cup \{0\}$ are extremal points of $\mathcal{K}(\mathcal{A})$.

Remark that $\mathcal{P}(\mathcal{A})$ is generally not *weak**-closed in $\mathcal{A}^*$ even if $\mathcal{A}$ is unital.

We now establish that there are "enough" pure states with the following analogue of Theorem 11.20.

Theorem 11.28. *Let $\mathcal{A}$ be a C^*-algebra and denote $\mathcal{P} := \mathcal{P}(\mathcal{A})$. Then, for each $x \in \mathcal{A}$:*

(i) if x is normal, then $\|x\| = \max_{\omega \in \mathcal{P}} |\omega(x)|$; for x arbitrary, $\|x\| = \max_{\omega \in \mathcal{P}} \|x\|_\omega$;

(ii) if $\omega(x) = 0$ for all $\omega \in \mathcal{P}$, then $x = 0$;

(iii) if $\omega(x) \in \mathbb{R}$ for all $\omega \in \mathcal{P}$, then x is selfadjoint;

(iv) if $\omega(x) \in \mathbb{R}^+$ for all $\omega \in \mathcal{P}$, then $x \in \mathcal{A}_+$.

Proof. We explained that $\mathcal{K}(\mathcal{A}) = \overline{\text{co}}^{weak^*}(\mathcal{P}(\mathcal{A}) \cup \{0\})$. As $\mathcal{S}(\mathcal{A}) \subset \mathcal{K}(\mathcal{A})$, every state of $\mathcal{A}$ is the *weak**-limit of a net of convex combinations of elements of $\mathcal{P}(\mathcal{A}) \cup \{0\}$. Hence, Parts (ii)–(iv) follow from the analogous ones in Theorem 11.20.

(i) Suppose that x is normal and non-zero. By Theorem 11.20 there exists $\omega \in \mathcal{S}(\mathcal{A})$ such that $|\omega(x)| = \|x\|$. Denote $\lambda := \omega(x)$. The set $C := \{\rho \in \mathcal{K}(\mathcal{A}); \rho(x) = \lambda\}$ is non-empty because it contains ω. It is evidently *weak**-closed, thus *weak**-compact, in $\mathcal{A}^*$. It is also extremal in $\mathcal{K}(\mathcal{A})$, because if $\omega_1, \omega_2 \in \mathcal{K}(\mathcal{A})$ and $t \in [0,1]$ are such that $t\omega_1(x) + (1-t)\omega_2(x) = \lambda$, then as $|\omega_i(x)| \le \|x\| = |\lambda|$ $(i = 1,2)$ we must have $\omega_i(x) = \lambda$ $(i = 1,2)$. By the Krein–Milman theorem, C has an extremal point ω'. By the foregoing, ω' is also an extremal point of $\mathcal{K}(\mathcal{A})$, namely $\omega' \in \mathcal{P}(\mathcal{A}) \cup \{0\}$. But $\omega' \neq 0$ because $x \neq 0$ and thus $\lambda \neq 0$. Therefore ω' is a pure state. In conclusion, $\|x\| = \max_{\omega \in \mathcal{P}} |\omega(x)|$. The second assertion in (i) follows as in the proof of Theorem 11.20. □

Next, we characterize those GNS representations associated with pure states.

Theorem 11.29. *The GNS representation π_ω associated to a state ω of a C^*-algebra $\mathcal{A}$ is irreducible iff ω is pure.*

We require the following simple lemma, slightly extending Exercise 15 of Chapter 6. See also Lemma 10.12.

Lemma 11.30. *Let X be a Hilbert space and $\langle \cdot, \cdot \rangle : X \times X \to \mathbb{C}$ a function that is linear in the left variable and conjugate linear in the right variable. Assume that $\langle \cdot, \cdot \rangle$ is bounded, that is, there exists $M \in [0, \infty)$ such that*

$$|\langle x, y \rangle| \le M \|x\| \|y\| \qquad (\forall x, y \in X) \tag{1}$$

(M is then called a bound for $\langle \cdot, \cdot \rangle$). Then there is a unique $T \in B(X)$ such that

$$\langle x, y \rangle = (Tx, y) \qquad (\forall x, y \in X). \tag{2}$$

Moreover, $\|T\| \le M$. Furthermore, T is selfadjoint iff $\langle\cdot,\cdot\rangle$ is Hermitian, and T is positive iff $\langle\cdot,\cdot\rangle$ is positive semi-definite (i.e., a semi-inner product on X).

Proof. For a given $y \in X$, the linear functional $\langle\cdot, y\rangle$ on X is bounded by $M\,\|y\|$ according to (1). Therefore, by the "Little" Riesz representation theorem, there is a unique $Sy \in X$ such that

$$\langle x, y\rangle = (x, Sy) \tag{3}$$

for all $x \in X$. This construction yields a unique function $S : X \to X$ that satisfies (3) for all x, y, which is linear since $\langle\cdot,\cdot\rangle$ is conjugate linear in the right variable, and is bounded by M because $\|Sy\|_X = \|\langle\cdot, y\rangle\|_{X^*} \le M\,\|y\|$ for all $y \in X$. Thus, $T := S^*$ satisfies (2) (which uniquely determines it) and also has norm at most M. The other assertions are immediate. □

Proposition 11.31. *Let X be a Hilbert space and $\mathcal{A}$ be a C^*-subalgebra of $B(X)$. Let $\zeta \in X$ and let $\rho \in \mathcal{A}_+^*$ be dominated by the vector functional $\omega_\zeta|_{\mathcal{A}}$, that is, $\rho \le \omega_\zeta|_{\mathcal{A}}$. Then there exists $T \in \mathcal{A}' := \{S \in B(X); SA = AS \text{ for all } A \in \mathcal{A}\}$, $0 \le T \le I$, such that $\rho = \omega_{T^{1/2}\zeta}|_{\mathcal{A}}$. That is,*

$$\rho(A) = \left(AT^{1/2}\zeta, T^{1/2}\zeta\right) = (AT\zeta, \zeta) \qquad (\forall A \in \mathcal{A}). \tag{4}$$

Proof. Assume for convenience that ζ is cyclic for $\mathcal{A}$, namely that $\overline{\mathcal{A}\zeta} = X$. Consider the normed subspace $\mathcal{A}\zeta$ of X. The function $\langle\cdot,\cdot\rangle : (\mathcal{A}\zeta) \times (\mathcal{A}\zeta) \to \mathbb{C}$ given by

$$\langle A_1\zeta, A_2\zeta\rangle := \rho(A_2^*A_1) \qquad (A_1, A_2 \in \mathcal{A})$$

is a (well defined) semi-inner product on $\mathcal{A}\zeta$ bounded by 1. Indeed, on account of the positivity of ρ we have $\langle A\zeta, A\zeta\rangle = \rho(A^*A) \ge 0$ for all $A \in \mathcal{A}$ as $A^*A \in \mathcal{A}_+$, and for all $A_1, A_2 \in \mathcal{A}$, by the Cauchy–Schwarz inequality ((1) in Section 11.6) and by the assumption that $\rho \le \omega_\zeta|_{\mathcal{A}}$,

$$|\rho(A_2^*A_1)| \le \rho(A_2^*A_2)^{1/2}\rho(A_1^*A_1)^{1/2} \le \omega_\zeta(A_2^*A_2)^{1/2}\omega_\zeta(A_1^*A_1)^{1/2} = \|A_2\zeta\|\,\|A_1\zeta\|\,.$$

Consequently, $\langle\cdot,\cdot\rangle$ extends to a semi-inner product on the Hilbert space X bounded by 1. Therefore, by Lemma 11.30, there is an operator $T \in B(X)$, $0 \le T \le I$, such that $(T\xi, \eta) = \langle\xi, \eta\rangle$ for all $\xi, \eta \in X$. In particular, for every $A_1, A_2 \in \mathcal{A}$,

$$(TA_1\zeta, A_2\zeta) = \langle A_1\zeta, A_2\zeta\rangle = \rho(A_2^*A_1). \tag{5}$$

For all $A, A_1, A_2 \in \mathcal{A}$ we thus have

$$(TAA_1\zeta, A_2\zeta) = \rho(A_2^*AA_1) = \rho\left((A^*A_2)^*\,A_1\right) = (TA_1\zeta, A^*A_2\zeta) = (ATA_1\zeta, A_2\zeta)\,.$$

The fact that $\overline{\mathcal{A}\zeta} = X$ and the boundedness of T and A imply that $TA = AT$, proving that $T \in \mathcal{A}'$. In conclusion, (5) now says that for every $A_1, A_2 \in \mathcal{A}$,

$$(A_2^*A_1T\zeta, \zeta) = (TA_1\zeta, A_2\zeta) = \rho(A_2^*A_1). \tag{6}$$

Every element of $\mathcal{A}$ is the linear combination of positive elements, so span $\{A_2^* A_1; A_1, A_2 \in \mathcal{A}\} = \mathcal{A}$ (in fact, the "span" is not necessary by Exercise 6). Thus (6) proves (4). Alternatively, (6) implies (4) by Exercise 16. ☐

Proof of Theorem 11.29. Recall that π_ω is a representation of $\mathcal{A}$ on a Hilbert space X_ω with a unit cyclic vector $\zeta := \zeta_\omega$ for π_ω such that $\omega = \omega_\zeta \circ \pi_\omega$.

($\Longrightarrow$) Suppose that π_ω is irreducible and let $\rho \in \mathcal{A}_+^*$ be dominated by ω. Recall that $\pi_\omega(\mathcal{A})$ is a C^*-algebra (Theorem 11.14) and define on it a positive linear functional ρ' by $\rho'(\pi_\omega(a)) := \rho(a)$ for $a \in \mathcal{A}$. It is well defined because by the Cauchy–Schwarz inequality and the assumption $\rho \leq \omega$, for all $a \in \mathcal{A}$,

$$|\rho(a)| = |\rho(\mathbb{1}^* a)| \leq \rho(\mathbb{1}^*\mathbb{1})^{1/2} \rho(a^*a)^{1/2} \leq \rho(a^*a)^{1/2} \leq \omega(a^*a)^{1/2} = \|\pi_\omega(a)\zeta\|$$

(when $\mathcal{A}$ is unital; otherwise, use approximate identities). Now, the assumption $\rho \leq \omega$ implies that $\rho' \leq \omega_\zeta|_{\pi_\omega(\mathcal{A})}$. So by Proposition 11.31, there exists $T \in \pi_\omega(\mathcal{A})'$, $0 \leq T \leq I$, such that $\rho' = \omega_{T^{1/2}\zeta}|_{\pi_\omega(\mathcal{A})}$. But π_ω is irreducible, that is, $\pi_\omega(\mathcal{A})' = \mathbb{C}I$. If now $c \in [0,1]$ is such that $T = cI$, then for all $a \in \mathcal{A}$,

$$\rho(a) = \rho'(\pi_\omega(a)) = c\,(\pi_\omega(a)\zeta, \zeta) = c\omega(a).$$

In other words, $\rho = c\omega$. Thus, ω is pure.

($\Longleftarrow$) Suppose that ω is pure. Let $P' \in \pi_\omega(\mathcal{A})'$ be a projection. Consider the positive functional $\rho := \omega_{P'\zeta} \circ \pi_\omega$ on $\mathcal{A}$. For every $a \in \mathcal{A}$,

$$\begin{aligned}\rho(a^*a) &= (\pi_\omega(a^*a)P'\zeta, P'\zeta) = (\pi_\omega(a)P'\zeta, \pi_\omega(a)P'\zeta) = (P'\pi_\omega(a)\zeta, P'\pi_\omega(a)\zeta) \\ &\leq (\pi_\omega(a)\zeta, \pi_\omega(a)\zeta) = (\pi_\omega(a^*a)\zeta, \zeta) = \omega(a^*a).\end{aligned}$$

Hence $\rho \leq \omega$. By assumption, there is $c \in [0,1]$ such that $\rho = c\omega$. So for every $a, b \in \mathcal{A}$,

$$(P'\pi_\omega(a)\zeta, \pi_\omega(b)\zeta) = \rho(b^*a) = c\omega(b^*a) = (c\pi_\omega(a)\zeta, \pi_\omega(b)\zeta).$$

From ζ being cyclic for π_ω we infer that $P' = cI$. Thus, $P' \in \{0, I\}$, proving that π_ω is irreducible. ☐

11.8.2 Decompositions of functionals

Theorem 11.32 (The Jordan decomposition of bounded Hermitian functionals). *Let $\mathcal{A}$ be a C^*-algebra and $\omega \in \mathcal{A}^*$ be Hermitian. There exist unique $\omega^+, \omega^- \in \mathcal{A}_+^*$ such that $\omega = \omega^+ - \omega^-$ and $\|\omega\| = \|\omega^+\| + \|\omega^-\|$.*

Proof. We will only prove existence. Assume for simplicity that $\|\omega\| = 1$. The set

$$D := \{\rho^+ - \rho^-; \rho^+, \rho^- \in \mathcal{A}_+^*, \|\rho^+\| + \|\rho^-\| \leq 1\}$$

is convex in $\mathcal{A}^*$ by Corollary 11.19. Also, the set $\mathcal{K}(\mathcal{A})$ is *weak**-compact in $\mathcal{A}^*$ and the function $f : \mathcal{K}(\mathcal{A}) \times \mathcal{K}(\mathcal{A}) \times [0,1] \to \mathcal{A}_+^*$ given by

$$f(\rho_1, \rho_2, t) := t\rho_1 - (1-t)\rho_2 \qquad (\rho_1, \rho_2 \in \mathcal{K}(\mathcal{A}), t \in [0,1])$$

is *weak**-continuous. Thus, the image of f, namely D, is *weak**-compact in $\mathcal{A}^*$.

Suppose by contradiction that $\omega \notin D$. By Corollary 5.21 of the strict separation theorem, there exists $x \in \mathcal{A}$ that strictly separates ω and D, that is:

$$\sup_{\rho \in D} \Re\rho(x) < \Re\omega(x).$$

Since ω and all elements of D are Hermitian, one can replace x by $\frac{1}{2}(x + x^*)$, and thus assume that $x \in \mathcal{A}_{\mathrm{sa}}$ and

$$\sup_{\rho \in D} \rho(x) < \omega(x). \tag{7}$$

Since $\mathcal{S}(\mathcal{A}), -\mathcal{S}(\mathcal{A}) \subset D$ and x is selfadjoint, Theorem 11.20 implies that $\|x\| = \max_{\rho \in \mathcal{S}(\mathcal{A})} |\rho(x)| \leq \sup_{\rho \in D} \rho(x)$. This contradicts (7) because $\omega(x) \leq \|x\|$.

Hence there exist $\omega^+, \omega^- \in \mathcal{A}_+^*$ such that $\omega = \omega^+ - \omega^-$ and $\|\omega^+\| + \|\omega^-\| \leq 1$. By the triangle inequality,

$$1 = \|\omega\| \leq \|\omega^+\| + \|\omega^-\| \leq 1,$$

so we have the desired equality $\|\omega\| = \|\omega^+\| + \|\omega^-\|$. □

An immediate consequence is that every element of $\mathcal{A}^*$ is the linear combination of (at most) 4 states.

We close this section with the following theorem, which we give without proof:

Theorem 11.33 (The absolute value of bounded functionals). *Let $\omega \in \mathcal{A}^*$. There exists a unique positive functional $|\omega| \in \mathcal{A}_+^*$ such that $\||\omega|\| = \|\omega\|$ and*

$$|\omega(x)|^2 \leq \|\omega\| \cdot |\omega|(x^* x) \qquad (\forall x \in \mathcal{A}).$$

Exercises

Unless otherwise indicated, $\mathcal{A}, \mathcal{B}$ etc., denote C^*-algebras. Do not use the non-commutative Gelfand–Naimark theorem (11.26) unless asked, although it can simplify some of the solutions.

1. Let $\{\mathcal{A}_i\}_{i \in I}$ be a family of C^*-algebras. The *l^∞-direct sum of* $\{\mathcal{A}_i\}_{i \in I}$, denoted by $\sum_{i \in I}^{l^\infty} \oplus \mathcal{A}_i$, is the set of all elements $\{a_i\}_{i \in I}$ of the set-theoretic direct product $\prod_{i \in I} \mathcal{A}_i$ whose supremum norm $\|\{a_i\}_{i \in I}\| := \sup_{i \in I} \|a_i\|_{\mathcal{A}_i}$ is finite. The *c_0-direct sum of* $\{\mathcal{A}_i\}_{i \in I}$, denoted by $\sum_{i \in I}^{c_0} \oplus \mathcal{A}_i$, is the subset of the l^∞-direct sum consisting of all $\{a_i\}_{i \in I}$ that vanish at infinity in the sense that for each $\epsilon > 0$ there exists a finite set $F \subset I$ such that $\|a_i\|_{\mathcal{A}_i} < \epsilon$ for $i \in I \backslash F$.

   (a) Prove that $\sum_{i \in I}^{l^\infty} \oplus \mathcal{A}_i$ and $\sum_{i \in I}^{c_0} \oplus \mathcal{A}_i$ are C^*-algebras with respect to the pointwise $*$-algebra operations and the supremum norm, and the latter algebra is an ideal in the former.

   (b) For a set I, what are the C^*-algebras $\sum_{i \in I}^{l^\infty} \oplus \mathbb{C}$ and $\sum_{i \in I}^{c_0} \oplus \mathbb{C}$?

2. Let $\varphi : \mathcal{A} \to \mathcal{B}$ be a $*$-homomorphism. Show that φ maps $\mathcal{A}_+$ into $\mathcal{B}_+$ and the open unit ball of $\mathcal{A}$ into that of $\mathcal{B}$. Show that if $\varphi(\mathcal{A}) = \mathcal{B}$, then "into" can be replaced by "onto" in both places.

3. Suppose that $\mathcal{A}$ is unital and $u \in G(\mathcal{A})$. Prove that u is unitary iff $\|u\| = 1 = \|u^{-1}\|$. (Compare Exercise 24 of Chapter 7.)

4. The *absolute value* of $a \in \mathcal{A}$ is the element $|a| := (a^*a)^{\frac{1}{2}} \in \mathcal{A}_+$. Prove that $\||a|\| = \|a\|$, and that if a is selfadjoint, then $|a| = a^+ + a^-$.

5. Let $x \in \mathcal{A}$ and $a \in \mathcal{A}_+$ be such that $x^*x \leq a$. Let $0 < \alpha < \frac{1}{2}$. We will prove that there exists $v \in \overline{\mathcal{A}a} := \overline{\{ya; y \in \mathcal{A}\}} \subset \mathcal{A}$ such that $x = va^\alpha$.

   (a) As a "warm-up", explain why the assertion is clear if a is invertible.

   (b) For $n \in \mathbb{N}$, let $v_n := x(a + \frac{1}{n}\mathbb{1})^{-\frac{1}{2}}a^{\frac{1}{2}-\alpha}$ (the computations are in $\mathcal{A}^\#$). Verify that $v_n \in \overline{\mathcal{A}a}$.

   (c) Explain each of the following steps, where $b_n := (a + \frac{1}{n}\mathbb{1})^{-\frac{1}{2}}$ for $n \in \mathbb{N}$:

$$\begin{aligned} \|v_n - v_m\|^2 &= \left\| a^{\frac{1}{2}-\alpha}(b_n - b_m)x^*x(b_n - b_m)a^{\frac{1}{2}-\alpha} \right\| \\ &\leq \left\| a^{\frac{1}{2}-\alpha}(b_n - b_m)a(b_n - b_m)a^{\frac{1}{2}-\alpha} \right\| \\ &= \left\| a^{\frac{1}{2}}(b_n - b_m)a^{\frac{1}{2}-\alpha} \right\|^2 \xrightarrow[n,m\to\infty]{} 0. \end{aligned}$$

   Consequently, the limit $v := \lim_{n\to\infty} v_n$ exists in $\mathcal{A}$.

   (d) Explain each of the following steps:

$$\begin{aligned} \|x - v_n a^\alpha\|^2 &= \left\| (\mathbb{1} - b_n a^{\frac{1}{2}})x^*x(\mathbb{1} - b_n a^{\frac{1}{2}}) \right\| \\ &\leq \left\| (\mathbb{1} - b_n a^{\frac{1}{2}})a(\mathbb{1} - b_n a^{\frac{1}{2}}) \right\| \\ &= \left\| a^{\frac{1}{2}}(\mathbb{1} - b_n a^{\frac{1}{2}}) \right\|^2 \xrightarrow[n\to\infty]{} 0. \end{aligned}$$

   Conclude that $x = va^\alpha$, as desired.

6. Let $x \in \mathcal{A}$ and $0 < \alpha < 1$. Use Exercise 5 to prove that there exists $v \in \mathcal{A}$ such that $x = v|x|^\alpha$.

7. Prove that a separable C^*-algebra has an approximate identity that is a sequence.

8. Let I be an ideal of $\mathcal{A}$ and J an ideal of I. Prove that J is an ideal of $\mathcal{A}$.

9. A C^*-subalgebra $\mathcal{B}$ of a C^*-algebra $\mathcal{A}$ is called *hereditary* if for every $a \in \mathcal{A}$ and $b \in \mathcal{B}$, if $0 \leq a \leq b$ then $a \in \mathcal{B}$. Prove that every ideal I of a C^*-algebra $\mathcal{A}$ is hereditary. (Hint: let $a \in \mathcal{A}$ and $b \in I$ be such that $0 \leq a \leq b$.

For an approximate identity $\{e_\alpha\}$ for I, use the C^*-identity and positivity to show that $\lim_\alpha a^{1/2}e_\alpha = a^{1/2}$, which implies that $a \in I$.)

10. Let I be an ideal of $\mathcal{A}$. Let $a \in \mathcal{A}$ and $i \in I_+$ be such that $a^*a \leq i$. Prove that $a \in I$. (In particular, for $a \in \mathcal{A}$, this implies that $a \in I$ iff $a^*a \in I$. This also yields another proof that I is hereditary.) (Hint: Exercise 5.)

11. Let $\{\mathcal{B}_i\}$ be a family of C^*-subalgebras of $\mathcal{A}$ whose union $\mathcal{B} := \bigcup_i \mathcal{B}_i$ is a dense ($*$-) subalgebra of $\mathcal{A}$. We will prove that for every ideal I of $\mathcal{A}$, the intersection $I \cap \mathcal{B} = \bigcup_i (I \cap \mathcal{B}_i)$ is dense in I.
    (a) Let $J := \overline{I \cap \mathcal{B}}$. Prove that J is an ideal of $\mathcal{A}$ (contained in I).
    (b) Let $q : \mathcal{A} \to \mathcal{A}/J$ and $Q : q(\mathcal{A}) \to q(\mathcal{A})/q(I)$ be the quotient maps. Prove that for each i, $q(I) \cap q(\mathcal{B}_i) = \{0\}$, so that $Q|_{q(\mathcal{B}_i)}$ is injective. (Hint: $J \cap \mathcal{B}_i = I \cap \mathcal{B}_i$.)
    (c) Prove that Q is isometric and deduce that $J = I$.

12. It is easier to solve this question for unital $\mathcal{A}$ first.
    (a) Prove that for every $a \in \mathcal{A}$ and every state $\omega \in \mathcal{S}(\mathcal{A})$ we have $|\omega(a)| \leq \omega(|a|^2)^{\frac{1}{2}}$.
    (b) Let $a \in \mathcal{A}_{\text{sa}}$ and $\omega \in \mathcal{S}(\mathcal{A})$. Assume that $\omega(a^2) = \omega(a)^2$. Prove that $\omega(ab) = \omega(a)\omega(b) = \omega(ba)$ for all $b \in \mathcal{A}$.
    (c) Let $a \in \mathcal{A}_{\text{sa}}$. Assume that for all $0 \neq b \in \mathcal{A}_{\text{sa}}$ there is $\omega \in \mathcal{S}(\mathcal{A})$ such that $\omega(a^2) = \omega(a)^2$ and $\omega(b) \neq 0$. Prove that a belongs to the center $\{x \in \mathcal{A}; xy = yx \text{ for every } y \in \mathcal{A}\}$ of $\mathcal{A}$.

13. Suppose that $p \in \mathcal{A}$ is a projection (i.e., a selfadjoint idempotent: $p^2 = p = p^*$) that is not a unit for $\mathcal{A}$. Prove the existence of a state ω of $\mathcal{A}$ such that $\omega(p) = 0$. (Hint: there is $0 \neq a \in \mathcal{A}$ such that $ap = 0$. In fact, one can find such a that is positive and of norm 1. Then $a(\mathbb{1}-p) = a = (\mathbb{1}-p)a$ (where, as usual, $\mathbb{1}$ is the unit of $\mathcal{A}^\#$), thus $a = (\mathbb{1}-p)a(\mathbb{1}-p) \leq (\mathbb{1}-p)\mathbb{1}(\mathbb{1}-p) = (\mathbb{1}-p)$, and hence $0 \leq a+p \leq \mathbb{1}$. As a result, taking ω to be a state of $\mathcal{A}$ with $\omega(a) = 1$, we must have $\omega(p) = 0$.)

14. We know that if $\mathcal{A}$ is unital, then $\mathcal{S}(\mathcal{A})$ is *weak**-closed in $\mathcal{A}^*$. Let us prove the converse: if $\mathcal{A}$ is not unital, then $\mathcal{S}(\mathcal{A})$ is not *weak**-closed in $\mathcal{A}^*$. To this end, we show that $0 \in \overline{\mathcal{S}(\mathcal{A})}^{weak^*}$.
    (a) Prove that for each $x \in \mathcal{A}_+$ and $\epsilon > 0$ there is $\omega \in \mathcal{S}(\mathcal{A})$ such that $\omega(x) < \epsilon$ by considering the following complementary cases.

    Case 1: 0 is an isolated point in $\sigma(x)$. Use the operational calculus to show that x is dominated by a positive multiple of a projection in $\mathcal{A}$. Then use Exercise 13.

    Case 2: 0 is an accumulation point in $\sigma(x)$. By Theorem 11.20 (vi), it suffices to replace $\mathcal{A}$ by its C^*-subalgebra generated by x. Thus, by

Exercise 30 in Chapter 7, we can assume that for some compact subset $K \subset \mathbb{C}$ containing 0 as an accumulation point, $\mathcal{A}$ is the C^*-subalgebra $\{f \in C(K); f(0) = 0\}$ of $C(K)$ and x is the identity function.

(b) Use Part (a) to show that $0 \in \overline{\mathcal{S}(\mathcal{A})}^{weak^*}$, either by the bare definition of the *weak**-topology combined with positive functional techniques, or using separation.

15. Use Exercise 14 to show that if $\mathcal{A}$ is not unital, then $\overline{\text{co}}^{weak^*}(\mathcal{S}(\mathcal{A})) = \text{co}(\mathcal{S}(\mathcal{A}) \cup \{0\}) = \{\omega \in \mathcal{A}_+^*; \|\omega\| \leq 1\}$ $(= \mathcal{K}(\mathcal{A})$ of Section 11.8).

16. Let π be a representation of $\mathcal{A}$ on a Hilbert space X. Prove that the following conditions are equivalent:

 (i) π is non-degenerate;

 (ii) for every approximate identity $\{e_\alpha\}$ of $\mathcal{A}$ and every $\eta \in X$ we have $\lim_\alpha \pi(e_\alpha)\eta = \eta$; and

 (iii) for some approximate identity $\{e_\alpha\}$ of $\mathcal{A}$ and every η in some dense subspace of X we have $\lim_\alpha \pi(e_\alpha)\eta = \eta$.

17. In this exercise we fill the missing details in the proof of the GNS theorem (11.23) in the non-unital case. Let $\mathcal{A}$ be a non-unital C^*-algebra and ω a state of $\mathcal{A}$.

 (a) Suppose that π is a representation of $\mathcal{A}$ on a Hilbert space X and $\zeta \in X$ is a unit vector such that $\omega = \omega_\zeta \circ \pi$. Let $\{e_\alpha\}$ be an approximate identity for $\mathcal{A}$. Show that $\lim_\alpha \pi(e_\alpha)\zeta = \zeta$.

 (b) Let $\tilde{\omega}$ be the (unique) extension of ω to a state of $\mathcal{A}^\#$ (cf. Theorem 11.20). Apply the GNS construction to $(\mathcal{A}^\#, \tilde{\omega})$ to get a representation $\tilde{\pi}$ of $\mathcal{A}^\#$ on a Hilbert space X and a unit cyclic vector ζ for $\tilde{\pi}$ such that $\tilde{\omega} = \omega_\zeta \circ \tilde{\pi}$. Prove that $\pi := \tilde{\pi}|_{\mathcal{A}}$ is a representation of $\mathcal{A}$ on X with ζ as a cyclic vector (which clearly satisfies $\omega = \omega_\zeta \circ \pi$).

 (c) Let U be as in the uniqueness part of the statement of the GNS theorem. Show that indeed U maps ζ to η.

18. Recall that $\mathcal{P}(\mathcal{A})$ is the set of pure states of $\mathcal{A}$.

 (a) Prove that $\Phi(\mathcal{A}) \subset \mathcal{P}(\mathcal{A})$. (Hint: for $\phi \in \Phi(\mathcal{A})$, consider $\ker \phi$.)

 (b) Prove that if $\mathcal{A}$ is commutative then $\Phi(\mathcal{A}) = \mathcal{P}(\mathcal{A})$. (Hint: Exercise 17 of Chapter 7.)

19. A positive functional $\omega \in \mathcal{A}_+^*$ is called *faithful* if for every $a \in \mathcal{A}_+$, $\omega(a) = 0$ implies that $a = 0$. Prove that the GNS representation associated with a faithful state is faithful (i.e., injective). Remark that the converse is false.

20. Let $n \in \mathbb{N}$. Consider the matrix algebra $M_n(\mathcal{A})$ of $n \times n$-matrices with coefficients in $\mathcal{A}$. It becomes a $*$-algebra with the involution given by

$\left[(a_{ij})_{1\le i,j\le n}\right]^* := (a_{ji}^*)_{1\le i,j\le n}$. Prove that there is a (necessarily unique!) norm on $M_n(\mathcal{A})$ making it a C^*-algebra. (Hint: let X be a Hilbert space. Construct a natural $*$-isomorphism between $M_n(B(X))$ and $B(X^{\oplus n})$, where $X^{\oplus n}$ is the Hilbert space direct sum of n copies of X, and deduce that there is a norm on the $*$-algebra $M_n(B(X))$ making it a C^*-algebra. Finally, use the non-commutative Gelfand–Naimark theorem.)

21. We continue Exercise 20.

    (a) Let $n \in \mathbb{N}$, X be a Hilbert space and $(T_{ij})_{1\le i,j\le n} = T \in M_n(B(X))$. Prove that T is positive iff for every n vectors $x_1, \ldots, x_n$ in X we have $\sum_{i,j=1}^n (T_{ij}x_j, x_i) \ge 0$.

    (b) Assume that $\mathcal{A}$ is unital. Prove that for $a \in \mathcal{A}$ with $\|a\| \le 1$, the matrix $\left(\begin{smallmatrix} \mathbb{1} & a \\ a^* & \mathbb{1} \end{smallmatrix}\right)$ is positive in $M_2(\mathcal{A})$.

22. (a) Using the notation of Sections 11.7 and 11.8, prove that

    $$\sum_{\omega\in\mathcal{P}(\mathcal{A})} \oplus \pi_\omega$$

    is a *faithful* representation of $\mathcal{A}$. (Notice the difference between this and the non-commutative Gelfand–Naimark theorem.)

    (b) What form does this representation take if $\mathcal{A}$ is commutative?

23. Let π be a non-degenerate representation of $\mathcal{A}$ on a Hilbert space X.

    (a) Prove that there exists a subset $Z \subset X$ consisting of unit vectors such that $X = \sum_{\zeta\in Z} \oplus \overline{\pi(\mathcal{A})\zeta}$.

    (b) For every $\zeta \in Z$, let π_ζ be the representation of $\mathcal{A}$ on the Hilbert space $X_\zeta := \overline{\pi(\mathcal{A})\zeta}$ given by reducing $\pi(a)$ to X_ζ for each $a \in \mathcal{A}$ (notice that π_ζ is indeed a well-defined representation!). Prove that $\pi = \sum_{\zeta\in Z} \oplus \pi_\zeta$, and deduce that up to unitary equivalence, $\pi = \sum_{\zeta\in Z} \oplus \pi_{\omega_\zeta}$, where we used the notation of the GNS theorem. (Hint: Example 11.25.)

24. Prove that each non-degenerate representation of $\mathcal{A}$ is a sub-representation of the universal representation of $\mathcal{A}$. (This is the essence of its universality.)

25. Prove that for each $\omega \in \mathcal{A}^*$ there exist a representation π of $\mathcal{A}$ on a Hilbert space X and vectors $\zeta, \eta \in X$ such that $\omega(x) = (\pi(x)\zeta, \eta)$ for all $x \in \mathcal{A}$ and $\|\omega\| = \|\zeta\|\,\|\eta\|$. (Hint: for simplicity, assume that $\|\omega\| = 1$. Apply the GNS construction to the absolute value $|\omega|$ of ω (see Theorem 11.33) to obtain $(X := X_{|\omega|}, \pi := \pi_{|\omega|}, \zeta := \zeta_{|\omega|})$. Prove that the map from $\pi(\mathcal{A})\zeta$ to $\mathbb{C}$ given by $\pi(x)\zeta \to \omega(x)$, $x \in \mathcal{A}$, is a (well defined) linear functional of norm at most 1.)

12

Von Neumann algebras

In Chapter 11 we studied C^*-algebras. This chapter is devoted to a particular class of C^*-algebras called *von Neumann algebras* after John von Neumann who, together with Francis Joseph Murray, laid the foundations of this theory in a series of long papers of extraordinary depth and insight. A von Neumann algebra on a Hilbert space $\mathcal{H}$ is a $*$-subalgebra of $B(\mathcal{H})$ containing the identity I and closed in the weak/strong operator topology. This is more strict than (concrete) C^*-algebras acting on $\mathcal{H}$, which are only asked to be closed in the (operator) norm topology.

After introducing the chapter, we present the two great foundations of von Neumann algebra theory. The first is *von Neumann's double commutant theorem.* Published in 1930, it says that a $*$-subalgebra of $B(\mathcal{H})$ containing I is a von Neumann algebra—namely, it is weak operator closed—iff it equals its double commutant in $B(\mathcal{H})$. This connects an analytic property (w.o.-closedness) with an algebraic property (being equal to the double commutant). Kadison called this magnificent result "the first theorem of the subject of operator algebras", and practically the whole theory of von Neumann algebras depends on it.

The second is *Kaplansky's density theorem*, which states that for a von Neumann algebra $\mathcal{R}$ and a w.o.-dense $*$-subalgebra $\mathcal{A}$ of $\mathcal{R}$, every element $T \in \mathcal{R}$ can be approximated in the w.o.-topology by elements of $\mathcal{A}$ with norm at most $\|T\|$. Being able to control the norms of the approximating elements is a profoundly useful tool. G. K. Pedersen wrote: "The density theorem is Kaplansky's great gift to mankind".

It is noteworthy that the proofs of these two theorems use "*matrix tricks*". Over the years such elegant tricks have become common in operator algebras.

The next short topic is the *polar decomposition* of operators. In particular, we prove that if the original operator belongs to a von Neumann algebra, then so do the two factors of its decomposition. See Exercise 7 for the polar decomposition of unbounded operators.

The definition of von Neumann algebras is "concrete", that is, they consist of bounded operators on a Hilbert space. There is also an "abstract" definition

Introduction to Modern Analysis. Second Edition. Shmuel Kantorovitz and Ami Viselter, Oxford University Press.
 DOI: 10.1093/oso/9780192849540.003.0012

via Dixmier and Sakai: a C^*-algebra $\mathcal{R}$ is isomorphic to a von Neumann algebra iff it is dual to some Banach space $\mathcal{R}_*$, called a *predual* of $\mathcal{R}$. Such C^*-algebras are called *W^*-algebras.* Moreover, in this case, the Banach space $\mathcal{R}_*$ whose dual is $\mathcal{R}$ is unique. The elements of $\mathcal{R}_*$, viewed as functionals in $\mathcal{R}^*$, are called *normal.*

We then make a short detour through two classes of bounded operators on Hilbert spaces, namely *Hilbert–Schmidt* and *trace-class operators*, which are interesting in their own right. We use the latter class to provide another description of the preduals of von Neumann algebras.

In the introduction to Chapter 11 we described C^*-algebra theory as "non-commutative topology". Along this line, von Neumann algebra theory should be considered as "*non-commutative measure theory*". One reason for this is that the commutative von Neumann algebras "are" the L^∞ algebras of measure spaces; see the text for the precise statements.

For every C^*-algebra $\mathcal{A}$, the double dual $\mathcal{A}^{**}$ is shown to become a von Neumann algebra with a universality property related to the representations of $\mathcal{A}$. It is called the *enveloping von Neumann algebra* of $\mathcal{A}$.

Finally, note that projections have been omitted from this chapter except for places where they are essential. Projections are everywhere in von Neumann algebras and most of the theory revolves around them one way or the other. Doing them justice would require a book more focused on operator algebras. We thus mention them here only briefly.

12.1 Preliminaries

In contrast to previous chapters, here and in Chapter 13 we denote Hilbert spaces by $\mathcal{H}, \mathcal{K}$, etc., and reserve X, Y for other purposes.

Let $\mathcal{H}$ be a Hilbert space and $B(\mathcal{H})$ denote the algebra of all bounded operators on $\mathcal{H}$. Recall from Section 6.6 that the *strong* (respectively, *weak*) *operator topology* on $B(\mathcal{H})$, denoted by s.o.t. (w.o.t.), turns $B(\mathcal{H})$ into a locally convex t.v.s. where a net $\{T_\alpha\}_\alpha$ converges to T iff $T_\alpha x \underset{\alpha}{\to} Tx$ strongly (weakly) in $\mathcal{H}$ for all $x \in \mathcal{H}$ (the latter means that $(T_\alpha x, y) \underset{\alpha}{\to} (Tx, y)$ in $\mathbb{C}$ for all $x, y \in \mathcal{H}$).

A few basic facts are:

- The norm topology on $B(\mathcal{H})$ is finer than the s.o.t., which in turn is finer that the w.o.t.
- A convex subset of $B(\mathcal{H})$ is s.o.-closed (to be read: *strong operator closed*, i.e., closed in the s.o.t.) iff it is w.o.-closed (Corollary 6.20).
- The multiplication function $B(\mathcal{H}) \times B(\mathcal{H}) \ni (T, S) \to TS \in B(\mathcal{H})$ is not (jointly) s.o.-continuous, and is also not w.o.-continuous even when restricted to the unit ball of $B(\mathcal{H})$ cross itself, unless $\mathcal{H}$ is finite dimensional. However, it is s.o.-continuous when restricted to bounded sets (Exercise 17 in Chapter 8).

- The (Hilbert) adjoint function $T \to T^*$ on $B(\mathcal{H})$ is w.o.-continuous, but not s.o.-continuous, even when restricted to the unit ball of $B(\mathcal{H})$ unless $\mathcal{H}$ is finite dimensional.
- The (norm-) closed unit ball of $B(\mathcal{H})$ is w.o.-compact (Exercise 18 of Chapter 8).

A *von Neumann algebra on* $\mathcal{H}$ (or *acting on* $\mathcal{H}$) is a w.o.-closed $*$-subalgebra of $B(\mathcal{H})$ containing the identity I of $B(\mathcal{H})$. Note that every linear subspace, and in particular, every subalgebra, of $B(\mathcal{H})$ is convex, so it is w.o.-closed iff it is s.o.-closed.

Evidently, $B(\mathcal{H})$ itself is a von Neumann algebra on $\mathcal{H}$.

Every von Neumann algebra is a C^*-algebra, but the converse is not true. Observe that in contrast to C^*-algebras, von Neumann algebras are usually defined "concretely", as algebras of bounded operators on some Hilbert space. An equivalent "abstract" definition of von Neumann algebras, as the class of C^*-algebras having an additional property, is covered in Section 12.5.

Two von Neumann algebras $\mathcal{R}_1, \mathcal{R}_2$ on Hilbert spaces $\mathcal{H}_1, \mathcal{H}_2$, respectively, are *spatially isomorphic* if there exists a unitary equivalence $U : \mathcal{H}_1 \to \mathcal{H}_2$ such that $U\mathcal{R}_1U^* := \{UTU^*; T \in \mathcal{R}_1\}$ equals $\mathcal{R}_2$. This is evidently stronger than $\mathcal{R}_1, \mathcal{R}_2$ being isomorphic as C^*-algebras, namely $*$-isomorphic.

We begin with a few elementary results. Let $\mathcal{H}$ be a Hilbert space.

Proposition 12.1. *If $\{A_\alpha\}_\alpha$ is an increasing net in $B(\mathcal{H})_{\mathrm{sa}}$, bounded from above by some $B \in B(\mathcal{H})_{\mathrm{sa}}$, then $A := \lim_\alpha A_\alpha$ exists in the s.o.t., and it is the least upper bound of $\{A_\alpha\}$ in $B(\mathcal{H})_{\mathrm{sa}}$.*

Proof. We are given that $A_\alpha \le B$ for every α. We may assume without loss of generality that $C \le A_\alpha$ for some $C \in B(\mathcal{H})_{\mathrm{sa}}$, and by adding $-C$, we may assume that $A_\alpha \ge 0$ for every α. The net $\{A_\alpha\}_\alpha$ is thus uniformly bounded by $\|B\|$. For every $x \in \mathcal{H}$, the non-negative net $\{(A_\alpha x, x)\}_\alpha$ is increasing and bounded, so it is convergent. By the polarization identity, the limit $\langle x, y\rangle := \lim_\alpha (A_\alpha x, y)$ exists for every $x, y \in \mathcal{H}$. Now $\langle\cdot,\cdot\rangle : \mathcal{H} \times \mathcal{H} \to \mathbb{C}$ is a semi-inner product that is bounded by $\|B\|$, that is, $|\langle x, y\rangle| \le \|B\|\,\|x\|\,\|y\|$ for all $x, y \in \mathcal{H}$. By Lemma 11.30 there is $A \in B(\mathcal{H})_+$ such that $(Ax, y) = \lim_\alpha (A_\alpha x, y)$ for all $x, y \in \mathcal{H}$. As $(A_\alpha x, x) \nearrow_\alpha (Ax, x)$ for all $x \in \mathcal{H}$, we have $A_\alpha \le A$ for every α. Hence, for each $x \in \mathcal{H}$ we have

$$\left\|(A - A_\alpha)^{1/2}x\right\|^2 = ((A - A_\alpha)x, x) \underset{\alpha}{\to} 0,$$

that is, $(A - A_\alpha)^{1/2} \underset{\alpha}{\to} 0$ in the s.o.t. Since $\left\{(A - A_\alpha)^{1/2}\right\}_\alpha$ is uniformly bounded (why?), we deduce that $A - A_\alpha = \left[(A - A_\alpha)^{1/2}\right]^2 \underset{\alpha}{\to} 0$ in the s.o.t. If $A' \in B(\mathcal{H})_{\mathrm{sa}}$ is so that $A_\alpha \le A'$ for every α, namely, $(A_\alpha x, x) \le (A'x, x)$ for every $x \in \mathcal{H}$, then $(Ax, x) \le (A'x, x)$; that is, $A \le A'$. □

Recall from Chapter 11 that by a *projection* we mean "orthogonal projection" (Section 8.2).

Corollary 12.2. *Let $\mathcal{R}$ be a w.o.- (equivalently, s.o.-) closed $*$-subalgebra of $B(\mathcal{H})$. Then $\mathcal{R}$ has a unit, which is the projection of $\mathcal{H}$ onto $\overline{\text{span}}\{Tx; T \in \mathcal{R}, x \in \mathcal{H}\}$.*

Proof. Since $\mathcal{R}$ is a C^*-algebra, it has an approximate identity $\{P_\alpha\}_\alpha$ by Theorem 11.8. By Proposition 12.1, $\{P_\alpha\}_\alpha$ has an s.o.-limit, say $P \in \mathcal{R}$, which satisfies $0 \leq P \leq I$. For every $T \in \mathcal{R}$, PT is the s.o.-limit of the net $\{P_\alpha T\}_\alpha$, which converges in norm to T. Thus $PT = T$, and similarly $TP = T$. Hence, P is a unit for $\mathcal{R}$, and in particular a projection.

Since P is the s.o.-limit of a net in $\mathcal{R}$, its range is evidently contained in $\mathcal{H}_0 := \overline{\text{span}}\{Tx; T \in \mathcal{R}, x \in \mathcal{H}\}$. Conversely, the fact that $PT = T$ for all $T \in \mathcal{R}$ shows that $\mathcal{H}_0$ is contained in the range of P. □

Corollary 12.2 implies that we do not lose anything by requiring that a von Neumann algebra on $\mathcal{H}$ should contain the identity of $B(\mathcal{H})$ by definition. It also implies that for every $*$-subalgebra $\mathcal{R}$ of $B(\mathcal{H})$ that is non-degenerate, that is, $\overline{\text{span}}\{Tx; T \in \mathcal{R}, x \in \mathcal{H}\} = \mathcal{H}$, the closure $\overline{\mathcal{R}}^{\text{w.o.t.}} = \overline{\mathcal{R}}^{\text{s.o.t.}}$ is a von Neumann algebra on $\mathcal{H}$.

Theorem 12.3. *Let $\mathcal{R}$ be a von Neumann algebra on $\mathcal{H}$.*

(i) If $T \in \mathcal{R}$ is normal and E is the resolution of the identity for T, then $E(\sigma) \in \mathcal{R}$ for every Borel subset σ of $\mathbb{C}$.

(ii) For every $T \in \mathcal{R}$, the projections onto $\ker T$ and $\overline{T\mathcal{H}}$ belong to $\mathcal{R}$.

(iii) The set of all projections in $\mathcal{R}$ spans a norm-dense subspace of $\mathcal{R}$.

Proof. (i) Let $\sigma \subset \mathbb{C}$ be compact. It makes no difference to assume that $\sigma \subset \sigma(T)$. By Urysohn's lemma, there is a bounded sequence $\{f_n\}_{n=1}^\infty$ in $C(\sigma(T))$ converging pointwise to the indicator χ_σ of σ. Then $f_n(T) \in \mathcal{R}$ for every n, and $f_n(T) \to \chi_\sigma(T) = E(\sigma)$ in the s.o.t. by Theorem 9.6. Therefore $E(\sigma) \in \mathcal{R}$.

If $\sigma \subset \mathbb{C}$ is a general Borel set, consider the directed set $\mathbb{K}$ of all compact subsets K of σ ordered by inclusion. Then $E(K) \in \mathcal{R}$ for all $K \in \mathbb{K}$ by the foregoing, and $\lim_{K\in\mathbb{K}} E(K) = E(\sigma)$ in the w.o.t. by the regularity of E. In conclusion, $E(\sigma) \in \mathcal{R}$.

(ii) If $T \in \mathcal{R}$ is normal, then the projections onto $\ker T$ and $\overline{T\mathcal{H}}$ are $E(\{0\})$ and $E(\mathbb{C}\setminus\{0\})$, respectively (see Theorem 9.8 and Exercise 16 in Chapter 9), which belong to $\mathcal{R}$ by (i). If T is general, then $\ker T = \ker |T|$ and $\overline{T\mathcal{H}} = \overline{|T^*|\mathcal{H}}$ (see the text preceding Theorem 12.11) are in $\mathcal{R}$.

(iii) Every element in $\mathcal{R}$ is a linear combination of two selfadjoint elements in $\mathcal{R}$. By the spectral theorem, every selfadjoint operator is the (norm!) limit of linear combinations of its spectral projections, which belong to $\mathcal{R}$ by (i). Indeed, if $T \in \mathcal{R}_{sa}$ has resolution of the identity E, then the identity function on $\sigma(T)$ is the uniform limit of a sequence $\{f_n\}_{n=1}^\infty$ of simple measurable functions on $\sigma(T)$, and so every $f_n(T)$ is a linear combination of projections from $\{E(\sigma); \sigma \subset \mathbb{C} \text{ Borel}\}$ and $f_n(T) \to T$ in norm. □

Part (iii) of Theorem 12.3 shows that projections are plentiful in von Neumann algebras. This is in sharp contrast to C^*-algebras, in which projections may be scarce. For instance, the only projections in $C([0,1])$ are zero and the unit element; and there are even *simple* unital C^*-algebras with this property!

We conclude this section with direct sums of von Neumann algebras. Recall from Section 8.6 the construction of direct sums of Hilbert spaces and operators on them. Let $\{\mathcal{R}_\alpha\}_{\alpha\in A}$ be a family of von Neumann algebras acting on Hilbert spaces $\{\mathcal{H}_\alpha\}_{\alpha\in A}$, respectively. The set of all direct sums of operators of the form $\sum_{\alpha\in A}\oplus T_\alpha$, where $T_\alpha\in\mathcal{R}_\alpha$ for every $\alpha\in A$, is called *the direct sum of the von Neumann algebras* $\{\mathcal{R}_\alpha\}_{\alpha\in A}$ and is denoted by $\sum_{\alpha\in A}\oplus\mathcal{R}_\alpha$. It is not difficult to observe that $\sum_{\alpha\in A}\oplus\mathcal{R}_\alpha$ is a von Neumann algebra on the Hilbert space $\sum_{\alpha\in A}\oplus\mathcal{H}_\alpha$.

12.2 Commutants

This section presents the double commutant theorem of von Neumann. This spectacular result is the cornerstone of the entire deep theory of von Neumann algebras, and it cannot be overstated. It says that for particular subalgebras of $B(\mathcal{H})$ the algebraic property of being equal to the double commutant is equivalent to the analytic property of being w.o.-closed.

Let $\mathcal{H}$ be a Hilbert space and $\mathcal{R}\subset B(\mathcal{H})$. Recall that the *commutant* of $\mathcal{R}$ (in $B(\mathcal{H})$) is the set

$$\mathcal{R}' := \{T\in B(\mathcal{H}); TA = AT \text{ for all } A\in\mathcal{R}\}.$$

The commutant $\mathcal{R}'$ is a subalgebra of $B(\mathcal{H})$ containing I. It is also w.o.-closed. Indeed, if the net $\{T_j\}_{j\in J}\subset\mathcal{R}'$ converges to the operator $T\in B(\mathcal{H})$ in the w.o.t., then for all $x,y\in\mathcal{H}$ and $A\in\mathcal{R}$,

$$\begin{aligned}(TAx,y) &= \lim_j (T_jAx,y) = \lim_j (AT_jx,y)\\ &= \lim_j (T_jx,A^*y) = (Tx,A^*y) = (ATx,y),\end{aligned}$$

hence, $TA = AT$ for all $A\in\mathcal{R}$, that is, $T\in\mathcal{R}'$.

Also, if $\mathcal{R}$ is selfadjoint (i.e., $T^*\in\mathcal{R}$ whenever $T\in\mathcal{R}$), then so is $\mathcal{R}'$. In this case, $\mathcal{R}'$ is a von Neumann algebra on $\mathcal{H}$. It follows that the *second commutant*

$$\mathcal{R}'' := (\mathcal{R}')'$$

is a von Neumann algebra containing $\mathcal{R}$ (trivially). In particular, the relation $\mathcal{R}=\mathcal{R}''$ implies that $\mathcal{R}$ is w.o.-closed. We show below that the converse is also true provided that $\mathcal{R}$ is a $*$-*subalgebra*, that is, a *selfadjoint subalgebra*, of $B(\mathcal{H})$.

Theorem 12.4 (Von Neumann's double commutant theorem). *Let $\mathcal{R}$ be a $*$-subalgebra of $B(\mathcal{H})$ containing I. Then $\mathcal{R}$ is a von Neumann algebra (equivalently, is w.o.-closed) iff $\mathcal{R}=\mathcal{R}''$.*

We will make essential use of Lemma 11.27, by which when $\mathcal{R}$ is a $*$-*subalgebra* of $B(\mathcal{H})$, a projection $P\in B(\mathcal{H})$ whose range has the form $\overline{\mathcal{R}S}$, where $S\subset\mathcal{H}$ and $\mathcal{R}S := \operatorname{span}\{Tx; T\in\mathcal{R}, x\in S\}$, belongs to $\mathcal{R}'$.

Another ingredient in the proof is the following. Let $n \in \mathbb{N}$ and consider the Hilbert space $\mathcal{H}^{\oplus n}$, the direct sum of n copies of $\mathcal{H}$. The $*$-algebra $B(\mathcal{H}^{\oplus n})$ is $*$-isomorphic to the matrix $*$-algebra $M_n(B(\mathcal{H}))$ of $n \times n$-matrices with coefficients in $B(\mathcal{H})$ with the involution $\left[(T_{ij})_{1 \le i,j \le n}\right]^* := (T_{ji}^*)_{1 \le i,j \le n}$. In particular, two operators on $\mathcal{H}^{\oplus n}$ commute iff their associated matrices commute.

For $S \in B(\mathcal{H})$, let $S^{(n)}$ be the operator on $\mathcal{H}^{\oplus n}$ with matrix $\mathrm{diag}(S, \ldots, S)$ (the diagonal matrix with S on its diagonal). That is, $S^{(n)}$ is the direct sum of n copies of S (see Section 8.6). Then $S^{(n)} \in B(\mathcal{H}^{\oplus n})$ and clearly $(S^{(n)})^* = (S^*)^{(n)}$.

If $\mathcal{R}$ is as in the theorem, $\mathcal{R}^{(n)} := \{A^{(n)}; A \in \mathcal{R}\}$ is a $*$-subalgebra of $B(\mathcal{H}^{\oplus n})$ containing the identity operator. One verifies easily that for $S \in B(\mathcal{H}^{\oplus n}) \cong M_n(B(\mathcal{H}))$, $S \in (\mathcal{R}^{(n)})'$ iff its associated matrix has all entries S_{ij} in $\mathcal{R}'$.

Proof of Theorem 12.4. By the comments preceding the statement of Theorem 12.4, we must only show that if $T \in \mathcal{R}''$, then $T \in \overline{\mathcal{R}}^{\text{w.o.t.}} = \overline{\mathcal{R}}^{\text{s.o.t.}}$, that is, each s.o.-basic neighborhood $N(T, F, \epsilon) = \{S \in B(\mathcal{H}); \|(S - T)x\| < \epsilon \text{ for all } x \in F\}$ of T meets $\mathcal{R}$. Recall that F is an arbitrary finite subset of $\mathcal{H}$, say $F = \{x_1, \ldots, x_n\}$, and $\epsilon > 0$.

Consider first the special case when $n = 1$. Denote $\mathcal{R}x_1 := \mathcal{R}\{x_1\}$ and let P be the projection onto the closed subspace $\overline{\mathcal{R}x_1}$ of $\mathcal{H}$. By Lemma 11.27 we have $P \in \mathcal{R}'$. Therefore $TP = PT$ since $T \in \mathcal{R}''$. In particular, $\overline{\mathcal{R}x_1}$ is T-invariant. But $x_1 = Ix_1 \in \mathcal{R}x_1$, so $Tx_1 \in \overline{\mathcal{R}x_1}$. Thus, there exists $A \in \mathcal{R}$ such that $\|Tx_1 - Ax_1\| < \epsilon$, that is, $A \in N(T, \{x_1\}, \epsilon)$, as wanted.

The case of arbitrary n is reduced to the case $n = 1$ using $n \times n$-matrices. Let $T \in \mathcal{R}''$ and consider as shown the s.o.-neighborhood $N(T, \{x_1, \ldots, x_n\}, \epsilon)$. Set $\underline{x} := [x_1, \ldots, x_n] \in \mathcal{H}^{\oplus n}$. Then $T^{(n)} \in (\mathcal{R}^{(n)})''$ by the last paragraph before the proof, and therefore, by the case $n = 1$, the s.o.-neighborhood $N(T^{(n)}, \{\underline{x}\}, \epsilon)$ in $B(\mathcal{H}^{\oplus n})$ meets $\mathcal{R}^{(n)}$, that is, there exists $A \in \mathcal{R}$ such that

$$\|(T^{(n)} - A^{(n)})\underline{x}\| < \epsilon.$$

Hence, for all $k = 1, \ldots, n$,

$$\|(T - A)x_k\| \le \left(\sum_{j=1}^{n} \|(T - A)x_j\|^2 \right)^{1/2} = \|(T - A)^{(n)}\underline{x}\| < \epsilon,$$

that is, $A \in N(T, \{x_1, \ldots, x_n\}, \epsilon)$. □

An equivalent statement of the double commutant theorem is: if $\mathcal{R}$ is a non-degenerate $*$-subalgebra of $B(\mathcal{H})$, then $\overline{\mathcal{R}}^{\text{w.o.t.}} = \mathcal{R}''$.

We now show an application of the double commutant theorem involving cyclic and separating vectors. Let $\mathcal{R}$ be a von Neumann algebra on a Hilbert space $\mathcal{H}$. A vector $x \in \mathcal{H}$ is called *cyclic* for $\mathcal{R}$ if $\overline{\mathcal{R}x} = \mathcal{H}$, and it is called *separating* for $\mathcal{R}$ if for every $T \in \mathcal{R}$, $T = 0$ if (and only if) $Tx = 0$.

Suppose that x is cyclic for $\mathcal{R}$, and let $T' \in \mathcal{R}'$ be such that $T'x = 0$. Then for every $T \in \mathcal{R}$ we have $T'Tx = TT'x = 0$, and since x is cyclic for $\mathcal{R}$ and T' is bounded we infer that $T' = 0$. In other words, x is separating for $\mathcal{R}'$.

Conversely, if x is separating for $\mathcal{R}'$, let P' be the projection of $\mathcal{H}$ onto its closed subspace $\overline{\mathcal{R}x}$. Then $P' \in \mathcal{R}'$ by Lemma 11.27, and evidently $P'x = x$. So $I - P' \in \mathcal{R}'$ and $(I - P')x = 0$. Since x is separating for $\mathcal{R}'$, we conclude that $I = P'$, that is, $\overline{\mathcal{R}x} = \mathcal{H}$. That is, x is cyclic for $\mathcal{R}$.

We proved the first statement of the next proposition. The second one follows by applying the first to $\mathcal{R}'$ and using the double commutant theorem.

Proposition 12.5. *Let $\mathcal{R}$ be a von Neumann algebra. A vector is cyclic for $\mathcal{R}$ iff it is separating for $\mathcal{R}'$, and it is separating for $\mathcal{R}$ iff it is cyclic for $\mathcal{R}'$.*

To finish this section, we use commutants to give a simple example of *commutative* von Neumann algebras. We use the following observation.

Lemma 12.6. *Let $(X, \mathcal{A}, \mu)$ be a finite positive measure space, and $g \in L^2(X, \mathcal{A}, \mu)$ be such that there is $M < \infty$ with $\|fg\|_2 \le M \|f\|_2$ for every $f \in L^\infty(X, \mathcal{A}, \mu)$. Then actually $g \in L^\infty(X, \mathcal{A}, \mu)$ and $\|g\|_\infty \le M$.*

Proof. Let $M' > M$ and consider the set $E := [|g| \ge M'] \in \mathcal{A}$. Then the indicator function $f := \chi_E$ belongs to $L^\infty(X, \mathcal{A}, \mu)$, and by assumption,

$$M'\mu(E)^{\frac{1}{2}} < \Big(\int_E |g|^2 d\mu\Big)^{\frac{1}{2}} = \|fg\|_2 \le M \|f\|_2 = M\mu(E)^{\frac{1}{2}}.$$

This is possible only if $\mu(E) = 0$. Thus, $\|g\|_\infty \le M$. □

Example 12.7 (Compare Example 11.21). Let $(X, \mathcal{A}, \mu)$ be a positive measure space. For each $f \in L^\infty(X, \mathcal{A}, \mu)$, consider the multiplication operator $M_f \in B(L^2(X, \mathcal{A}, \mu))$ given by $g \to fg$, $g \in L^2(X, \mathcal{A}, \mu)$. The reader can verify easily that this operator is well defined and $\|M_f\| \le \|f\|_\infty$. Furthermore, the map $f \to M_f$ is an injective $*$-homomorphism from $L^\infty(X, \mathcal{A}, \mu)$ to $B(L^2(X, \mathcal{A}, \mu))$. We now prove that *if $(X, \mathcal{A}, \mu)$ is σ-finite*, then $\mathcal{R} := \{M_f; f \in L^\infty(X, \mathcal{A}, \mu)\}$ is a commutative von Neumann algebra on $L^2(X, \mathcal{A}, \mu)$. Since the C^*-algebras $L^\infty(X, \mathcal{A}, \mu)$ and $\mathcal{R}$ are ($*$-) isomorphic, it is colloquially customary to say that $L^\infty(X, \mathcal{A}, \mu)$ "is" a commutative von Neumann algebra in this case.

As $\mathcal{R}$ is commutative, we have $\mathcal{R} \subset \mathcal{R}'$. Let $T \in \mathcal{R}'$. Since $(X, \mathcal{A}, \mu)$ is σ-finite, we can write $X = \bigcup_{n=1}^\infty X_n$, where $\{X_n\}_{n=1}^\infty$ is a sequence of mutually disjoint sets in $\mathcal{A}$, each of which has finite measure. Let $g_n := T(\chi_{X_n}) \in L^2(X, \mathcal{A}, \mu)$. Then for every $f \in L^\infty(X, \mathcal{A}, \mu)$,

$$T(f\chi_{X_n}) = TM_f\chi_{X_n} = M_fT\chi_{X_n} = fg_n.$$

Taking $f := \chi_{X_n}$ we get $g_n = T(\chi_{X_n}) = \chi_{X_n}g_n$, so g_n is zero (a.e.) on X_n^c. Moreover, $\|fg_n\|_2 = \|T(f\chi_{X_n})\|_2 \le \|T\| \, \|f\chi_{X_n}\|_2$ for every $f \in L^\infty(X, \mathcal{A}, \mu)$. Consequently, by Lemma 12.6 applied to the "restriction" of $(X, \mathcal{A}, \mu)$ to X_n we get $g_n \in L^\infty(X, \mathcal{A}, \mu)$ and $\|g_n\|_\infty \le \|T\|$. Let $g \in L^\infty(X, \mathcal{A}, \mu)$ be defined by $g \equiv g_n$ on X_n, $n \in \mathbb{N}$. If $f \in L^\infty(X, \mathcal{A}, \mu)$ then for all $n \in \mathbb{N}$,

$$T(f\chi_{X_n}) = fg_n = f\chi_{X_n}g = M_g(f\chi_{X_n}).$$

Since $\operatorname{span}\{f\chi_{X_n}; f \in L^\infty(X, \mathcal{A}, \mu), n \in \mathbb{N}\}$ is dense in $L^2(X, \mathcal{A}, \mu)$ by Theorem 1.27, we infer that $T = M_g$. In conclusion, $\mathcal{R}' \subset \mathcal{R}$, hence $\mathcal{R} = \mathcal{R}'$. This means not only that $\mathcal{R}$ is a von Neumann algebra on $L^2(X, \mathcal{A}, \mu)$, but that it is a *maximal commutative* von Neumann subalgebra of $B(L^2(X, \mathcal{A}, \mu))$.

12.3 Density

If $\mathcal{R}$ is a normed space and $\mathcal{A}$ is a dense subspace, then evidently every element of the closed unit ball of $\mathcal{R}$ can be approximated in norm by elements of the closed unit ball of $\mathcal{A}$. One can ask whether this holds for algebras of operators when the operator norm topology is replaced by the weak/strong operator topologies. This turns out to be true when the algebras in question are *selfadjoint.*

The following theorem bears some resemblance to Goldstine's theorem (5.25), but their contexts and proofs are completely different.

Theorem 12.8 (Kaplansky's density theorem). *Let $\mathcal{R}$ be a $*$-subalgebra of $B(\mathcal{H})$ for a Hilbert space $\mathcal{H}$ and $\mathcal{A}$ an s.o.-dense $*$-subalgebra of $\mathcal{R}$. Then the (norm-) closed unit ball of $\mathcal{A}$ is s.o.-dense in the closed unit ball of $\mathcal{R}$.*

We first prove the theorem for selfadjoint operators.

Proposition 12.9. *The set of selfadjoint elements in the closed unit ball of $\mathcal{A}$ is s.o.-dense in the set of selfadjoint elements in the closed unit ball of $\mathcal{R}$.*

Proof. Since the closed unit ball of $\mathcal{R}$ is norm-dense in that of $\overline{\mathcal{R}}^{\|\cdot\|}$ and the same is true for $\mathcal{A}$, we may assume that $\mathcal{R}, \mathcal{A}$ are norm-closed, namely C^*-algebras. Moreover, note that $\mathcal{A}_{\text{sa}}$ is s.o.-dense in $\mathcal{R}_{\text{sa}}$. Indeed, for every $T \in \mathcal{R}_{\text{sa}}$, let $\{T_\alpha\}$ be a net in $\mathcal{A}$ that converges to T in the s.o.t. Then the net $\left\{\frac{1}{2}[T_\alpha + (T_\alpha)^*]\right\}$ in $\mathcal{A}_{\text{sa}}$ converges to T in the w.o.t., and therefore $T \in \overline{\mathcal{A}_{\text{sa}}}^{\text{w.o.t}}$. But $\overline{\mathcal{A}_{\text{sa}}}^{\text{w.o.t}} = \overline{\mathcal{A}_{\text{sa}}}^{\text{s.o.t}}$ since $\mathcal{A}_{\text{sa}}$ is convex, so that $T \in \overline{\mathcal{A}_{\text{sa}}}^{\text{s.o.t}}$.

Define $f : \mathbb{R} \to \mathbb{R}$ by $f(\lambda) := \frac{2\lambda}{1+\lambda^2}$, $\lambda \in \mathbb{R}$. Then f is continuous on $\mathbb{R}$ and $f(\mathbb{R}) = [-1, 1]$. We claim that f is "s.o.-continuous on $\mathcal{R}_{\text{sa}}$", namely, if a net $\{T_\alpha\}$ in $\mathcal{R}_{\text{sa}}$ converges to $T \in \mathcal{R}_{\text{sa}}$ in the s.o.t., then the net $\{f(T_\alpha)\}$ converges to $f(T)$ in the s.o.t. Indeed, for every α we have $\|f(T_\alpha)\| = \max_{\lambda \in \sigma(T_\alpha)} |f(\lambda)| \leq 1$ and

$$\begin{aligned}
\frac{1}{2}(f(T_\alpha) - f(T)) &= T_\alpha(I + T_\alpha^2)^{-1} - T(I + T^2)^{-1} \\
&= (I + T_\alpha^2)^{-1}\left[T_\alpha(I + T^2) - (I + T_\alpha^2)T\right](I + T^2)^{-1} \\
&= (I + T_\alpha^2)^{-1}\left[(T_\alpha - T) + (T_\alpha(T - T_\alpha)T)\right](I + T^2)^{-1} \\
&= (I + T_\alpha^2)^{-1}(T_\alpha - T)(I + T^2)^{-1} + \frac{1}{4}f(T_\alpha)(T - T_\alpha)f(T).
\end{aligned}$$

Now $T_\alpha \underset{\alpha}{\to} T$ in the s.o.t. and the nets $\{(I + T_\alpha^2)^{-1}\}$, $\{f(T_\alpha)\}$ are bounded (by 1), so $f(T_\alpha) \underset{\alpha}{\to} f(T)$ in the s.o.t.

Let $S \in \mathcal{R}_{\mathrm{sa}}$, $\|S\| \le 1$. The continuous function f is strictly increasing from -1 to 1 on $[-1,1]$. Thus, it has a (continuous) inverse there, say, $g : [-1,1] \to [-1,1]$. Let $T := g(S) \in \mathcal{R}_{\mathrm{sa}}$. Then $S = f(T)$. By assumption, there is a net $\{T_\alpha\}$ in $\mathcal{A}_{\mathrm{sa}}$ that converges to T in the s.o.t. By the foregoing, $\{f(T_\alpha)\}$ is a net in $\mathcal{A}_{\mathrm{sa}}$ that is bounded by 1 and converges to $f(T) = S$ in the s.o.t. □

We now use a 2×2-matrix trick to upgrade the selfadjoint case to the general case. Let $M_2(\mathcal{R})$ denote the $*$-subalgebra of $M_2(B(\mathcal{H})) \cong B(\mathcal{H}^{\oplus 2})$ consisting of all matrices $S = (S_{ij})_{1\le i,j\le 2}$ with $S_{ij} \in \mathcal{R}$ for each $1 \le i,j \le 2$. Doing the same for $\mathcal{A}$, we see easily that $M_2(\mathcal{A})$ is an s.o.-dense $*$-subalgebra of $M_2(\mathcal{R})$.

Proof of Theorem 12.8. Let S be in the closed unit ball of $\mathcal{R}$. The operator $\tilde{S} \in M_2(\mathcal{R})$ given by $\tilde{S} := \left(\begin{smallmatrix} 0 & S \\ S^* & 0 \end{smallmatrix}\right)$ is selfadjoint and $\|\tilde{S}\| = \|S\| \le 1$. By the proposition, there is a net $\{\tilde{T}_\alpha\}$ of selfadjoint elements of the closed unit ball of $M_2(\mathcal{A})$ converging to $\tilde{S}$ in the s.o.t. In particular, $\{(\tilde{T}_\alpha)_{12}\}$ converges to $(\tilde{S})_{12} = S$ in the s.o.t, and $\|(\tilde{T}_\alpha)_{12}\| \le \|\tilde{T}_\alpha\| \le 1$ for every α. □

12.4 The polar decomposition

Every $z \in \mathbb{C}$ has its (unique) polar decomposition $z = \lambda\,|z|$ for some λ in the complex unit circle. The polar decomposition of bounded operators on Hilbert spaces, which we prove in this section, is a far-reaching generalization of this. A further extension to unbounded operators is obtained in Exercise 7.

Let $\mathcal{H}$ denote a Hilbert space.

An operator $V \in B(\mathcal{H})$ is called a *partial isometry* if for some closed subspace M of $\mathcal{H}$, V acts isometrically on M (i.e., $\|Vx\| = \|x\|$ for every $x \in M$) and $V|_{M^\perp} = 0$. The closed subspace M is called the *initial space* of V, and the closed subspace $V(M)$ is called the *final space* of V.

Projections, unitaries, and isometries are examples of partial isometries.

The proof of the following lemma is left to the reader.

Lemma 12.10. *Let $V \in B(\mathcal{H})$.*

(i) V is a partial isometry iff V^* is a partial isometry. In this case, the initial and final spaces of V^* are the final and initial spaces of V, respectively.

(ii) V is a partial isometry iff V^*V is a projection iff VV^* is a projection iff $V^*VV^* = V^*$ iff $VV^*V = V$. In this case, V^*V, VV^* are the projections on the initial and final spaces of V, respectively.

Recall that the *absolute value* of an element a of a C^*-algebra $\mathcal{A}$ is $|a| := (a^*a)^{1/2} \in \mathcal{A}_+$.

Let $T \in B(\mathcal{H})$. The *support* of T is the closed subspace $(\ker T)^\perp = \overline{T^*\mathcal{H}}$ of $\mathcal{H}$. The support of a selfadjoint operator clearly equals the closure of its range, and the support of a partial isometry is its initial space.

For every $x \in \mathcal{H}$, we have

$$\|Tx\|^2 = (Tx,Tx) = (T^*Tx,x) = \left(|T|^2 x, x\right) = \||T|\,x\|^2 . \tag{1}$$

It follows that the operators T and $|T|$ share the same kernel. Thus, they also share the same support, which equals $\overline{|T|\,\mathcal{H}}$ as $|T|$ is selfadjoint (indeed, positive).

Theorem 12.11 (Polar decomposition). *Let $\mathcal{H}$ be a Hilbert space and $T \in B(\mathcal{H})$.*

(i) Existence: there exists a partial isometry $V \in B(\mathcal{H})$, whose initial space is the support of T and whose final space is $\overline{T\mathcal{H}}$, such that $T = V\,|T|$.

(ii) Uniqueness: if $A \in B(\mathcal{H})_+$ and $W \in B(\mathcal{H})$ is a partial isometry with initial space $\overline{A\mathcal{H}}$ such that $T = WA$, then $A = |T|$ and $W = V$.

(iii) The operator V of Part (i) also satisfies $|T^*| = V|T|V^*$ and $T = |T^*|V$.

(iv) If $\mathcal{R}$ is a von Neumann algebra on $\mathcal{H}$ and $T \in \mathcal{R}$, then $V, |T| \in \mathcal{R}$.

Proof. (i) For every $x \in \mathcal{H}$, define $V\,|T|\,x := Tx$. By (1), the resulting map $V : |T|\,\mathcal{H} \to T\mathcal{H}$ is well defined, isometric, and surjective, and so it extends to a surjective isometry $V : \overline{|T|\,\mathcal{H}} \to \overline{T\mathcal{H}}$. Extend it further to $V \in B(\mathcal{H})$ by letting V map $(\overline{|T|\,\mathcal{H}})^{\perp} = \ker T$ to 0. Then V is a partial isometry with initial and final spaces $\overline{|T|\,\mathcal{H}} = \overline{T^*\mathcal{H}}$ and $\overline{T\mathcal{H}}$, respectively, and we have $T = V\,|T|$.

(ii) If W, A are as above, then since the initial space of W is $\overline{A\mathcal{H}}$, we have $T^*T = AW^*WA = A^2$ by Lemma 12.10, thus $A = (T^*T)^{1/2} = |T|$ (see Section 11.3). So $V\,|T| = W\,|T|$. Hence, $W = V$, because the initial space of both partial isometries is the support of T, which equals $\overline{|T|\,\mathcal{H}}$.

(iii) We have $|T^*|^2 = TT^* = V|T|^2V^* = (V|T|V^*)^2$, where the last step holds because the initial space of V is the support of T (= that of $|T|$). Consequently, $|T^*| = V|T|V^*$, thus $|T^*|V = V|T|V^*V = V|T| = T$.

(iv) We already know that $|T| \in \mathcal{R}$. By von Neumann's double commutant theorem (12.4), it is left to show that $V \in \mathcal{R}''$. Every element in $\mathcal{R}'$ is the linear combination of at most 4 unitary elements in $\mathcal{R}'$ because $\mathcal{R}'$ is a unital C^*-algebra (see Section 11.2). Therefore, it suffices to show that V commutes with all unitaries in $\mathcal{R}'$. Let $U \in \mathcal{R}'$ be unitary. Since $T \in \mathcal{R}$,

$$T = UTU^* = UV\,|T|\,U^* = (UVU^*)(U\,|T|\,U^*) \tag{2}$$

(of course, we have $U\,|T|\,U^* = |T|$, but that will not be relied on). The operator $U\,|T|\,U^*$ is positive. Since V^*V is the projection on the initial space of V, which is $\overline{|T|\,\mathcal{H}}$, the operator $(UVU^*)^*(UVU^*) = UV^*U^*UVU^* = UV^*VU^*$ is the projection on $U(\overline{|T|\,\mathcal{H}}) = \overline{(U\,|T|\,U^*)\mathcal{H}}$. Equivalently UVU^* is a partial isometry with initial space $\overline{(U\,|T|\,U^*)\mathcal{H}}$. By (2) and the uniqueness of the polar decomposition, $UVU^* = V$; equivalently, U commutes with V, as desired. ☐

The decomposition $T = V|T|$ of Part (i) of the theorem is called the *polar decomposition* of T.

Remark 12.12. Let $\mathcal{A} \subset B(\mathcal{H})$ be a C^*-algebra and $T \in \mathcal{A}$. Let $T = V\,|T|$ be the polar decomposition of T. Then while $|T| \in \mathcal{A}$, normally $V \notin \mathcal{A}$ (even if $I \in \mathcal{A}$). Compare Exercise 6 in Chapter 11.

12.5 W^*-algebras

A *dual Banach space* is a Banach space Y for which there exists another Banach space X such that Y is isometrically isomorphic to X^*. Such X is called a *predual* of Y. Not every Banach space is dual: for instance, c_0 is not a dual Banach space (see Exercise 10). Also, if a predual exists, it is not necessarily unique up to isometric isomorphism (see Exercise 11).

Observe that if X is a predual of Y, then X can be viewed as a closed subspace of Y^*: indeed, if f is an isometric isomorphism of Y onto X^*, then the map $X \ni x \to (Y \ni y \to (f(y))(x))$ is an isometric isomorphism of X into Y^* (it equals $f^* \circ \kappa_X$, where κ_X is the canonical isometric isomorphism of X into X^{**} and f^* is the Banach adjoint of f). We will identify X with this subspace of Y^* tacitly.

A W^**-algebra* (or an *abstract* von Neumann algebra) is a C^*-algebra $\mathcal{R}$, which is a dual Banach space.

This section revolves around two facts due to Dixmier and Sakai:

(A) *The class of von Neumann algebras is, up to an isomorphism, the class of W^*-algebras.* In other words, every von Neumann algebra is a W^*-algebra (Theorem 12.14), and conversely, every W^*-algebra has a faithful representation as a von Neumann algebra (Theorem 12.18).

(B) *Every W^*-algebra $\mathcal{R}$ has a unique predual.* The uniqueness is not only up to isometric isomorphism: in fact, all preduals of $\mathcal{R}$ are *equal* as subspaces of $\mathcal{R}^*$ (Theorem 12.17); equivalently, they induce on $\mathcal{R}$ the same *weak** topology.

We start with one direction of Point (A), proving it first for $B(\mathcal{H})$.

Let $\mathcal{H}$ be a Hilbert space. For $x, y \in \mathcal{H}$ we define $\omega_{x,y} \in B(\mathcal{H})^*$ by $\omega_{x,y}(T) := (Tx, y)$, $T \in B(\mathcal{H})$. In particular, $\omega_x = \omega_{x,x}$. The subspace $B(\mathcal{H})_\sim := \operatorname{span}\{\omega_{x,y}; x, y \in \mathcal{H}\}$ of $B(\mathcal{H})^*$ consists precisely of all linear functionals on $B(\mathcal{H})$ that are w.o.- (equivalently, s.o.-) continuous (see Theorem 6.19). Denote by $B(\mathcal{H})_*$ the norm closure of $B(\mathcal{H})_\sim$ in $B(\mathcal{H})^*$. Then $B(\mathcal{H})_*$ is a Banach space

Proposition 12.13. *For every Hilbert space $\mathcal{H}$, $B(\mathcal{H})$ is a W^*-algebra. To elaborate, $B(\mathcal{H})$ is isometrically isomorphic to the dual of $B(\mathcal{H})_*$, the closure in $B(\mathcal{H})^*$ of the subspace of w.o.-continuous linear functionals on $B(\mathcal{H})$.*

Proof. There exists a natural linear map $\Psi : B(\mathcal{H}) \to (B(\mathcal{H})_*)^*$ given by $\Psi(A) := \hat{A}|_{B(\mathcal{H})_*}$, that is, $(\Psi(A))(\omega) := \omega(A)$ $(A \in B(\mathcal{H}), \omega \in B(\mathcal{H})_*)$. We shall prove that Ψ is an isometric isomorphism.

Isometricity. Let $A \in B(\mathcal{H})$. Since $|(\Psi(A))(\omega)| \le \|\omega\| \, \|A\|$ for each $\omega \in B(\mathcal{H})_*$, we have $\|\Psi(A)\| \le \|A\|$. On the other hand,

$$\begin{aligned} \|A\| &= \sup_{\|x\|=\|y\|=1} |(Ax, y)| = \sup_{\|x\|=\|y\|=1} |\omega_{x,y}(A)| \\ &= \sup_{\|x\|=\|y\|=1} |(\Psi(A))(\omega_{x,y})| \le \|\Psi(A)\| \,. \end{aligned}$$

Thus, $\|\Psi(A)\| = \|A\|$, proving that Ψ is isometric.

Surjectivity. Let $\varphi \in (B(\mathcal{H})_*)^*$. The map $\langle\cdot,\cdot\rangle : \mathcal{H}\times\mathcal{H} \to \mathbb{C}$ given by $\langle x, y\rangle := \varphi(\omega_{x,y})$, $x, y \in \mathcal{H}$, is linear in the left variable and conjugate linear in the right variable, and is bounded by $\|\varphi\|$ because $|\varphi(\omega_{x,y})| \leq \|\varphi\| \, \|\omega_{x,y}\| \leq \|\varphi\| \, \|x\| \, \|y\|$. Thus, by Lemma 11.30, there is $A \in B(\mathcal{H})$ with $(Ax, y) = \langle x, y\rangle = \varphi(\omega_{x,y})$ for every $x, y \in \mathcal{H}$. Hence, $\Psi(A) = \varphi$, because these two elements of $(B(\mathcal{H})_*)^*$ agree on $B(\mathcal{H})_\sim$, which is a dense subspace of $B(\mathcal{H})_*$. □

Recall that if X is a normed space and N is a *weak**-closed subspace of X^*, then N is a dual Banach space: it is isometrically isomorphic to the dual of the closed subspace $\{\kappa_X(x)|_N; x \in X\}$ of N^*; for this see Exercise 24 in Chapter 6.

Theorem 12.14. *Every von Neumann algebra is a* W^**-algebra. To elaborate, if* $\mathcal{R}$ *is a von Neumann algebra on a Hilbert space* $\mathcal{H}$*, then it is isometrically isomorphic to the dual of the Banach subspace* $\{\omega|_{\mathcal{R}}; \omega \in B(\mathcal{H})_*\}$ *of* $\mathcal{R}^*$.

Proof. Consider the predual $B(\mathcal{H})_*$ of $B(\mathcal{H})$ described in Proposition 12.13 (we still do not know that it is unique). Since $\omega_{x,y} \in B(\mathcal{H})_*$ for every $x, y \in \mathcal{H}$, the *weak** topology on $B(\mathcal{H})$ induced by $B(\mathcal{H})_*$ is stronger than the w.o.t. Since $\mathcal{R}$ is w.o.-closed in $B(\mathcal{H})$, it is also *weak**-closed there. The theorem's assertion thus follows from the paragraph preceding the theorem. □

The following result is required in the end of the next section.

Proposition 12.15. *Let* $\mathcal{R}$ *be a von Neumann algebra on a Hilbert space* $\mathcal{H}$*. The topologies induced on the closed unit ball of* $\mathcal{R}$ *by the w.o.t. and by the weak** *topology of* $B(\mathcal{H})$ *(given by the above predual) coincide.*

Proof. The w.o.t. is weaker than the *weak** topology (on $B(\mathcal{H})$ and thus on $\mathcal{R}$). Conversely, let $\{A_\alpha\}$ be a net in the closed unit ball of $\mathcal{R}$ that converges in the w.o.t. to an element A of the closed unit ball of $\mathcal{R}$. Let $\omega \in B(\mathcal{H})_*$ and $\epsilon > 0$ be given. By the definition of $B(\mathcal{H})_*$ in Proposition 12.13, there is a w.o.-continuous linear functional ω_0 on $B(\mathcal{H})$ such that $\|\omega - \omega_0\| \leq \epsilon$. Find α_0 so that $|\omega_0(A_\alpha - A)| \leq \epsilon$ for all $\alpha \geq \alpha_0$. Then for all such α,

$$\begin{aligned}|\omega(A_\alpha - A)| &\leq |\omega_0(A_\alpha - A)| + |(\omega - \omega_0)\,(A_\alpha - A)| \\ &\leq \epsilon + \|\omega - \omega_0\|\,(\|A_\alpha\| + \|A\|) \leq 3\epsilon.\end{aligned}$$

This proves that $A_a \underset{\alpha}{\to} A$ in the *weak** topology. □

We now turn to Point (B). We will not provide a full proof, but only indicate the steps required to prove it.

Definition 12.16. Let $\mathcal{R}$ be a W^*-algebra. A linear functional $\omega \in \mathcal{R}^*$ is called *normal* if for every bounded increasing net $\{T_\alpha\}$ in $\mathcal{R}_+$ with least upper bound $T \in \mathcal{R}_+$, one has $\omega(T) = \lim_\alpha \omega(T_\alpha)$.

Theorem 12.17. *Let* $\mathcal{R}$ *be a* W^**-algebra. Then* $\mathcal{R}$ *has a unique (up to isometric isomorphism) predual* $\mathcal{R}_*$.

In fact, viewed as a subspace of $\mathcal{R}^*$, the predual $\mathcal{R}_*$ consists precisely of the normal elements of $\mathcal{R}^*$. Also, the set $(\mathcal{R}_*)_+ := \mathcal{R}_* \cap \mathcal{R}^*_+$ linearly spans $\mathcal{R}_*$.

Outline of proof. Let $\mathcal{R}_*$ be a predual of $\mathcal{R}$. We view $\mathcal{R}_*$ as a closed subspace of $\mathcal{R}^*$.

1. $\mathcal{R}$ is unital. This relies on the fact that the closed unit ball of $\mathcal{R}$ is *weak**-compact by Alaoglu's theorem (5.24).
2. $\mathcal{R}_{\text{sa}}$ is *weak**-closed.
3. Abundance of elements of $\mathcal{R}_*$: if $T \in \mathcal{R}_{\text{sa}}$, then $T \geq 0$ iff $\omega(T) \geq 0$ for every $\omega \in (\mathcal{R}_*)_+$.
4. Every element of $\mathcal{R}_*$ decomposes as a linear combination of two Hermitian elements of $\mathcal{R}_*$; and for every Hermitian $\omega \in \mathcal{R}_*$ there exist $\omega^+, \omega^- \in (\mathcal{R}_*)_+$ with $\omega = \omega^+ - \omega^-$ and $\|\omega\| = \|\omega^+\| + \|\omega^-\|$ (compare the Jordan decomposition for C^*-algebras, Theorem 11.32).
5. Order completeness of $\mathcal{R}_{\text{sa}}$: if $\{T_\alpha\}$ is an increasing net in $\mathcal{R}_{\text{sa}}$ that is bounded from above, then it has a least upper bound in $\mathcal{R}_{\text{sa}}$.
   Let us prove this step. We can assume, without loss of generality, that $\{T_\alpha\}$ is norm bounded. Since the closed ball of any radius in $\mathcal{R}$ is *weak**-compact, we may assume by passing to a subnet that $\{T_\alpha\}$ *weak**-converges to some $T \in \mathcal{R}$. Since $\mathcal{R}_{\text{sa}}$ is *weak**-closed by Step 2, $T \in \mathcal{R}_{\text{sa}}$. Fix α_0. For every $\omega \in (\mathcal{R}_*)_+$, the net $\{\omega(T_\alpha)\}$ increases to $\omega(T)$; in particular, $\omega(T_{\alpha_0}) \leq \omega(T)$. By Step 3, $T_{\alpha_0} \leq T$. In conclusion, T is an upper bound of $\{T_\alpha\}$. If $S \in \mathcal{R}_{\text{sa}}$ is another upper bound of $\{T_\alpha\}$, then $\omega(T_\alpha) \leq \omega(S)$ for every $\omega \in (\mathcal{R}_*)_+$ and α, so $\omega(T) \leq \omega(S)$. Thus $T \leq S$ by Step 3.
6. An element ω of $\mathcal{R}^*$ belongs to $\mathcal{R}_*$ iff it is normal. This is proved first for positive ω and then generally. □

We finish with the second direction of Point (A).

Theorem 12.18. *Every W^*-algebra $\mathcal{R}$ has a faithful representation as a von Neumann algebra on some Hilbert space.*

Proof. The idea is to repeat the proof of the non-commutative Gelfand–Naimark theorem (11.26) but take only the *normal* states of $\mathcal{R}$ (i.e., those that belong to $\mathcal{R}_*$).

So, denote by $\mathcal{S}_{\text{n}}(\mathcal{R})$ the set of normal states of $\mathcal{R}$. Using the language of Section 11.7, for $\omega \in \mathcal{S}_{\text{n}}(\mathcal{R})$ let π_ω be the associated GNS representation of $\mathcal{R}$ on a Hilbert space $\mathcal{H}_\omega$ with cyclic vector ζ_ω such that $\omega = \omega_{\zeta_\omega} \circ \pi_\omega$ as in Theorem 11.23 (where $\mathcal{H}_\omega$ is denoted by X_ω). Let

$$\mathcal{H} := \sum_{\omega \in \mathcal{S}_{\text{n}}(\mathcal{R})} \oplus \mathcal{H}_\omega \quad \text{and} \quad \pi := \sum_{\omega \in \mathcal{S}_{\text{n}}(\mathcal{R})} \oplus \pi_\omega.$$

The representation π of $\mathcal{R}$ on $\mathcal{H}$ is injective, since if $T \in \mathcal{R}$ and $\pi(T) = 0$, then for each $\omega \in \mathcal{S}_{\text{n}}(\mathcal{R})$ we have $\pi_\omega(T) = 0$, hence $\omega(T) = (\omega_{\zeta_\omega} \circ \pi_\omega)(T) = 0$. From Theorem 12.17 we infer that $\omega(T) = 0$ for all $\omega \in \text{span}(\mathcal{R}_*)_+ = \mathcal{R}_*$, hence $T = 0$.

It remains to prove that $\pi(\mathcal{R})$ is a von Neumann algebra on $\mathcal{H}$. By Exercise 9, this is equivalent to the w.o.-closedness of the (norm-) closed unit ball of $\pi(\mathcal{R})$, which equals the image under π of the closed unit ball of $\mathcal{R}$ since π is faithful and thus isometric. To this end, let $\{T_\alpha\}_{\alpha \in A}$ be a net in the closed unit ball of $\mathcal{R}$ such that $\{\pi(T_\alpha)\}_{\alpha \in A}$ converges to some $S \in B(\mathcal{H})$ in the w.o.t. Then for every $\omega \in \mathcal{S}_{\mathrm{n}}(\mathcal{R})$ we have $\omega(T_\alpha) = (\pi(T_\alpha)\zeta_\omega, \zeta_\omega) \underset{\alpha}{\to} (S\zeta_\omega, \zeta_\omega)$. Hence, by Theorem 12.17, the (scalar) net $\{\omega(T_\alpha)\}_{\alpha \in A}$ converges for each $\omega \in \mathcal{R}_*$. Consequently, the map

$$\varphi : \mathcal{R}_* \ni \omega \to \lim_{\alpha \in A} \omega(T_\alpha)$$

is a well-defined linear functional on $\mathcal{R}_*$. This functional is bounded by 1, and in particular belongs to $(\mathcal{R}_*)^*$, since $\{T_\alpha\}_{\alpha \in A}$ is contained in the closed unit ball of $\mathcal{R}$. Let T be the element of the closed unit ball of $\mathcal{R}$ corresponding to $\varphi \in (\mathcal{R}_*)^*$ (recall that $\mathcal{R}_*$ is the predual of $\mathcal{R}$). This means that $\{T_\alpha\}_{\alpha \in A}$ converges in the *weak** topology to T. We leave to the reader to observe that $\{\pi(T_\alpha)\}_{\alpha \in A}$ converges to $\pi(T)$ in the w.o.t., and thus $S = \pi(T)$. □

The faithful representation of $\mathcal{R}$ constructed as shown is called the *universal normal representation* of $\mathcal{R}$.

Example 12.19. We can now revisit Example 12.7. Assume that $(X, \mathcal{A}, \mu)$ is a *σ-finite* positive measure space. The C^*-algebra $L^\infty(X, \mathcal{A}, \mu)$ is isometrically isomorphic to $L^1(X, \mathcal{A}, \mu)^*$ by Theorem 4.6, and hence it is a W^*-algebra.

12.6 Hilbert–Schmidt and trace-class operators

This section is devoted to two classes of operators: Hilbert–Schmidt and trace-class, and to the relations between them, compact operators and von Neumann algebras. Hilbert–Schmidt operators and their basic properties were introduced in Exercise 25 in Chapter 8.

Let $\mathcal{H}$ denote a Hilbert space.

For $T \in B(\mathcal{H})$, we would like to define the *trace* of T to be

$$\mathrm{Tr}(T) := \sum_{\alpha \in A} (Te_\alpha, e_\alpha), \tag{1}$$

where $\{e_\alpha\}_{\alpha \in A}$ is an orthonormal basis for $\mathcal{H}$ and the sum is defined as the limit of the net $\left\{\sum_{\alpha \in F} (Te_\alpha, e_\alpha)\right\}_{F \in \mathcal{F}}$, where $\mathcal{F}$ is the set of all finite subsets of A directed by inclusion. For this to make sense, this series has to converge (necessarily absolutely) for every choice of orthonormal basis and its sum has to be independent of this choice. These conditions do not hold true for every T; when they do, we say that *T has a trace*. If T has a trace and $U \in B(\mathcal{H})$ is unitary, then UTU^* also has a trace and $\mathrm{Tr}(UTU^*) = \mathrm{Tr}(T)$, because U maps each orthonormal basis of $\mathcal{H}$ onto another one.

For $T \in B(\mathcal{H})$, let

$$\|T\|_2 := \left(\sum_{\alpha \in A} \|Te_\alpha\|^2 \right)^{1/2} \in [0, \infty],$$

where $\{e_\alpha\}_A$ is an orthonormal basis for $\mathcal{H}$. This sum does not depend on the choice of orthonormal basis, because if $\{f_\alpha\}_A$ is another basis, then

$$\begin{aligned} \sum_{\alpha \in A} \|Te_\alpha\|^2 &= \sum_{\alpha,\beta \in A} |(Te_\alpha, f_\beta)|^2 = \sum_{\alpha,\beta \in A} |(e_\alpha, T^* f_\beta)|^2 \\ &= \sum_{\alpha,\beta \in A} |(T^* f_\beta, e_\alpha)|^2 = \sum_{\beta \in A} \|T^* f_\beta\|^2 . \end{aligned} \tag{2}$$

We say that T is a *Hilbert–Schmidt operator* if $\|T\|_2 < \infty$. The set of all such operators is denoted by $L^2(B(\mathcal{H}))$ or $L^2(\mathcal{H})$. Observe that

$$\begin{aligned} L^2(\mathcal{H}) &= \{T \in B(\mathcal{H}); \{\|Te_\alpha\|\}_{\alpha \in A} \in l^2(A)\} \\ \text{and } \|T\|_2 &= \big\|\{\|Te_\alpha\|\}_{\alpha \in A}\big\|_{l^2(A)} \text{ for all } T \in L^2(\mathcal{H}). \end{aligned}$$

As a result, $(L^2(\mathcal{H}), \|\cdot\|_2)$ is a normed space with respect to the pointwise operations.

Note that (2) also shows that $\|T\|_2 = \|T^*\|_2$. Moreover

$$\sum_{\alpha \in A} \|Te_\alpha\|^2 = \sum_{\alpha \in A} (Te_\alpha, Te_\alpha) = \sum_{\alpha \in A} (|T|^2 e_\alpha, e_\alpha).$$

Hence, $\|T\|_2 = \||T|\|_2$, and if $T \in L^2(\mathcal{H})$, then $|T|^2$ has a trace and $\|T\|_2 = (\mathrm{Tr}(|T|^2))^{1/2}$. In any case, since every positive operator has a (positive) square root, the sum in $[0, \infty]$ of the series (1) is independent of the choice of basis when T is positive.

Example 12.20. Let $\mathcal{H}$ be a Hilbert space with orthonormal basis $\{e_\alpha\}_{\alpha \in A}$ and let $\{\lambda_\alpha\}_{\alpha \in A} \in l^2(A)$. Then the diagonal operator $T \in B(\mathcal{H})$ given by $Tx := \sum_{\alpha \in A} \lambda_\alpha (x, e_\alpha) e_\alpha$, $x \in \mathcal{H}$, is Hilbert–Schmidt, and $\|T\|_2 = \left(\sum_{\alpha \in A} |\lambda_\alpha|^2\right)^{1/2} = \big\|\{\lambda_\alpha\}_{\alpha \in A}\big\|_{l^2(A)}$.

For $x, y \in \mathcal{H}$, let $\theta_{y,x} := (\cdot, x)y \in B(\mathcal{H})$. That is, $\theta_{y,x}$ is the operator $\mathcal{H} \ni w \to (w, x)y$. Then $\operatorname{span}\{\theta_{y,x}; x, y \in \mathcal{H}\}$ equals the set of finite rank operators in $B(\mathcal{H})$, denoted by $F(\mathcal{H})$.

Proposition 12.21 (Basic properties of Hilbert–Schmidt operators).

(i) *For every $T, S \in B(\mathcal{H})$ we have $\|T\|_2 = \|T^*\|_2$, $\|T\| \le \|T\|_2$ and $\|TS\|_2 \le \|T\| \|S\|_2, \|T\|_2 \|S\|$. Thus, $L^2(\mathcal{H})$ is a generally non-closed, selfadjoint ideal of $B(\mathcal{H})$.*

(ii) $F(\mathcal{H}) \subset L^2(\mathcal{H}) \subset K(\mathcal{H})$.

(iii) For every $S, T \in L^2(\mathcal{H})$, *the operator* ST *has a trace.*

(iv) The formula $(T, S)_2 := \mathrm{Tr}(S^*T)$, $S, T \in L^2(\mathcal{H})$ *defines an inner product on* $L^2(\mathcal{H})$ *(whose induced norm is* $\|\cdot\|_2$*), making it a Hilbert space.*

Proof. (i) Let $T, S \in B(\mathcal{H})$. We proved that $\|T\|_2 = \|T^*\|_2$. Every unit vector $x \in \mathcal{H}$ can be complemented to an orthonormal basis of $\mathcal{H}$. Thus, $\|Tx\| \leq \|T\|_2$, so $\|T\| \leq \|T\|_2$. Furthermore, if $\{e_\alpha\}_{\alpha\in A}$ is an orthonormal basis of $\mathcal{H}$, then

$$\|ST\|_2^2 = \sum_{\alpha\in A} \|STe_\alpha\|^2 \leq \|S\|^2 \sum_{\alpha\in A} \|Te_\alpha\|^2 = \|S\|^2 \|T\|_2^2$$

proving that $\|ST\|_2 \leq \|S\| \|T\|_2$, and

$$\|ST\|_2 = \|(ST)^*\|_2 = \|T^*S^*\|_2 \leq \|T^*\| \|S^*\|_2 = \|T\| \|S\|_2 .$$

(ii) For $x, y \in \mathcal{H}$, one sees easily that $\|\theta_{x,y}\|_2 = \|x\| \|y\|$, so $\theta_{x,y} \in L^2(\mathcal{H})$. Hence $F(\mathcal{H}) \subset L^2(\mathcal{H})$. The fact that $L^2(\mathcal{H}) \subset K(\mathcal{H})$, that is, every Hilbert–Schmidt operator is compact, was proved in Exercise 25 in Chapter 8.

(iii) Let $S, T \in L^2(\mathcal{H})$. By the polarization identity,

$$(STe_\alpha, e_\alpha) = (Te_\alpha, S^*e_\alpha) = \frac{1}{4}\sum_{k=0}^{3} \mathrm{i}^k \left\|Te_\alpha + \mathrm{i}^k S^* e_\alpha\right\|^2 \qquad (\forall \alpha \in A).$$

Summing over all $\alpha \in A$, the right-hand side converges to $\frac{1}{4}\sum_{k=0}^{3} \mathrm{i}^k \left\|T + \mathrm{i}^k S^*\right\|_2^2$. This proves that ST has a trace.

(iv) The fact that $(\cdot,\cdot)_2$, which is well defined by (iii), is an inner product on $L^2(\mathcal{H})$ is routine. The resulting inner product space is complete. Indeed, every Cauchy sequence $\{T_n\}_{n=1}^\infty$ in $(L^2(\mathcal{H}), \|\cdot\|_2)$ is also Cauchy in $B(\mathcal{H})$ as the $\|\cdot\|_2$-norm dominates the operator norm. Since $B(\mathcal{H})$ is a Banach space, $\{T_n\}_{n=1}^\infty$ converges in $B(\mathcal{H})$ to some T. An application of Fatou's lemma gives that $T \in L^2(\mathcal{H})$ and that $\{T_n\}_{n=1}^\infty$ also converges to T in $(L^2(\mathcal{H}), \|\cdot\|_2)$. □

Definition 12.22. For $T \in B(\mathcal{H})$, put

$$\|T\|_1 := \sum_{\alpha\in A} (|T| e_\alpha, e_\alpha) = \left\| |T|^{1/2} \right\|_2^2 \in [0, \infty],$$

where $\{e_\alpha\}_{\alpha\in A}$ is some orthonormal basis of $\mathcal{H}$ (since $|T|$ is positive, this sum is independent of the choice of basis as explained earlier). If $\|T\|_1 < \infty$, that is, if $|T|$ has a trace, we say that T is a *trace-class operator*, in which case $\|T\|_1 = \mathrm{Tr}(|T|)$. The set of all these operators is denoted by $L^1(B(\mathcal{H}))$ or $L^1(\mathcal{H})$. Notice that $T \in L^1(\mathcal{H}) \iff |T| \in L^1(\mathcal{H}) \iff |T|^{1/2} \in L^2(\mathcal{H})$.

Example 12.23. Let $\mathcal{H}$ be a Hilbert space with orthonormal basis $\{e_\alpha\}_{\alpha\in A}$, and let $\{\lambda_\alpha\}_{\alpha\in A} \in l^1(A)$. Then the diagonal operator $T \in B(\mathcal{H})$ given by

$Tx = \sum_{\alpha \in A} \lambda_\alpha (x, e_\alpha) e_\alpha$, $x \in \mathcal{H}$, is of trace class, and $\|T\|_1 = \sum_{\alpha \in A} |\lambda_\alpha| = \left\| \{\lambda_\alpha\}_{\alpha \in A} \right\|_{l^1(A)}$.

Proposition 12.24 (Basic properties of trace-class operators).

(i) *We have* $L^1(\mathcal{H}) = \{T_1 T_2; T_1, T_2 \in L^2(\mathcal{H})\}$, *and for every* $T_1, T_2 \in L^2(\mathcal{H})$ *we have* $\|T_1 T_2\|_1 \le \|T_1\|_2 \|T_2\|_2$.

(ii) *Every* $T \in L^1(\mathcal{H})$ *has a trace, and we have* $|\operatorname{Tr} T| \le \operatorname{Tr}(|T|) = \|T\|_1$.

(iii) $(L^1(\mathcal{H}), \|\cdot\|_1)$ *is a Banach space.*

(iv) *For every* $T, S \in B(\mathcal{H})$, *we have:* $\|T\|_1 = \|T^*\|_1$, $\|T\| \le \|T\|_1$ *and* $\|TS\|_1 \le \|T\| \|S\|_1, \|T\|_1 \|S\|$. *Thus,* $L^1(\mathcal{H})$ *is a generally non-closed, selfadjoint ideal of* $B(\mathcal{H})$.

(v) *For every* $T \in L^1(\mathcal{H})$ *and* $S \in B(\mathcal{H})$ *we have* $\operatorname{Tr}(TS) = \operatorname{Tr}(ST)$.

(vi) *We have* $F(\mathcal{H}) \subset L^1(\mathcal{H}) \subset L^2(\mathcal{H}) \subset K(\mathcal{H})$, *and for every* $x, y \in \mathcal{H}$ *we have* $\operatorname{Tr}(\theta_{x,y}) = (x, y)$.

Proof. We will use Proposition 12.21 freely.

(i)+(ii) Let $T_1, T_2 \in L^2(\mathcal{H})$ and let $T_1 T_2 = V |T_1 T_2|$ be the polar decomposition of $T_1 T_2$. Then $V^* T_1$ and T_2 are in $L^2(\mathcal{H})$, so that $(V^* T_1) T_2 = |T_1 T_2|$ has a trace, proving that $T_1 T_2 \in L^1(\mathcal{H})$. Also, the Cauchy–Schwarz inequality shows that

$$\begin{aligned} \|T_1 T_2\|_1 &= \operatorname{Tr} |T_1 T_2| = \operatorname{Tr}((V^* T_1) T_2) = (T_2, T_1^* V)_2 \le \|T_2\|_2 \|T_1^* V\|_2 \\ &\le \|T_2\|_2 \|T_1^*\|_2 \|V\| \le \|T_2\|_2 \|T_1\|_2 . \end{aligned}$$

On the other hand, if $T \in L^1(\mathcal{H})$ has polar decomposition $T = V|T|$, then $T = (V|T|^{1/2})|T|^{1/2}$ and $|T|^{1/2}$, thus also $V|T|^{1/2}$, belong to $L^2(\mathcal{H})$. This implies that T has a trace. Furthermore, as demonstrated, by the Cauchy–Schwarz inequality,

$$\begin{aligned} |\operatorname{Tr} T| &= \left|\operatorname{Tr}(V|T|^{1/2}|T|^{1/2})\right| = (|T|^{1/2}, |T|^{1/2} V^*)_2 \le \left\| |T|^{1/2} \right\|_2 \left\| |T|^{1/2} V^* \right\|_2 \\ &\le \left\| |T|^{1/2} \right\|_2^2 \|V^*\| \le \left\| |T|^{1/2} \right\|_2^2 = \operatorname{Tr}(|T|). \end{aligned}$$

(iii)+(iv) Let $T \in L^1(\mathcal{H})$. We have $\|T\|_1 = \left\| |T|^{1/2} \right\|_2^2 \ge \left\| |T|^{1/2} \right\|^2 = \| |T| \| = \|T\|$. Let again $T = V|T|$ be the polar decomposition of T. We have $T^* = |T| V^* = |T|^{1/2} (|T|^{1/2} V^*)$, so

$$\|T^*\|_1 \le \left\| |T|^{1/2} \right\|_2 \left\| |T|^{1/2} V^* \right\|_2 \le \left\| |T|^{1/2} \right\|_2^2 = \|T\|_1.$$

Hence, also $T^* \in L^1(\mathcal{H})$, and by symmetry, $\|T\|_1 = \|T^*\|_1$.

Very similarly, if $S \in B(\mathcal{H})$, then writing $TS = (V|T|^{1/2})(|T|^{1/2} S)$ proves that $TS \in L^1(\mathcal{H})$ and $\|TS\|_1 \le \|T\|_1 \|S\|$, and by the previous paragraph we also have $ST \in L^1(\mathcal{H})$ and $\|ST\|_1 \le \|T\|_1 \|S\|$.

Suppose that both $T, S \in L^1(\mathcal{H})$. Let $T + S = W|T + S|$ be the polar decomposition of $T+S$. Then $W^*T, W^*S \in L^1(\mathcal{H})$, hence, they have traces, and thus so does their sum, namely $|T + S|$. Also, $|\mathrm{Tr}(W^*T)| \leq \|W^*T\|_1 \leq \|T\|_1$ and the same for S. Therefore $\|T + S\|_1 = \mathrm{Tr}(|T+S|) = \mathrm{Tr}(W^*T)+\mathrm{Tr}(W^*S) \leq \|T\|_1 + \|S\|_1$.

Finally, we show that $(L^1(\mathcal{H}), \|\cdot\|_1)$ is complete. Every Cauchy sequence $\{T_n\}_{n=1}^\infty$ in $(L^1(\mathcal{H}), \|\cdot\|_1)$ is also Cauchy in $B(\mathcal{H})$, so it converges in $B(\mathcal{H})$ to some T. Now, continuity of the absolute value map $T \to |T|$ on $B(\mathcal{H})$ and Fatou's lemma yield that $T \in L^1(\mathcal{H})$ and that $\{T_n\}_{n=1}^\infty$ converges to T in $(L^1(\mathcal{H}), \|\cdot\|_1)$.

(v) By the text in the beginning of this section, for every $T \in L^1(\mathcal{H})$ and every unitary $U \in B(\mathcal{H})$ we have $\mathrm{Tr}(UTU^*) = \mathrm{Tr}(T)$. Equivalently for every $T \in L^1(\mathcal{H})$ and every unitary $U \in B(\mathcal{H})$ we have $\mathrm{Tr}(UT) = \mathrm{Tr}(TU)$. By linearity and the fact that every element of $B(\mathcal{H})$ is the linear combination of (at most four) unitary operators in $B(\mathcal{H})$, we conclude that $\mathrm{Tr}(ST) = \mathrm{Tr}(TS)$ for all $T \in L^1(\mathcal{H})$ and $S \in B(\mathcal{H})$.

(vi) If $x, y \in \mathcal{H}$, then $\|\theta_{x,y}\|_1 = \|x\| \|y\|$, so $\theta_{x,y} \in L^1(\mathcal{H})$. This proves that $F(\mathcal{H}) \subset L^1(\mathcal{H})$. Let $T \in B(\mathcal{H})$ have polar decomposition $T = V|T|$. Then $\|T\|_2 = \left\|V|T|^{1/2}|T|^{1/2}\right\|_2 \leq \left\|V|T|^{1/2}\right\| \left\||T|^{1/2}\right\|_2 = \left\|V|T|^{1/2}\right\| \|T\|_1^{1/2}$. Therefore $L^1(\mathcal{H}) \subset L^2(\mathcal{H})$. The equality $\mathrm{Tr}(\theta_{x,y}) = (x, y)$, $x, y \in \mathcal{H}$, is easy. □

Remark that the converse of Part (ii) is also true: every operator that has a trace is a trace-class operator; see Exercise 17.

Theorem 12.25. *For a Hilbert space $\mathcal{H}$, there are canonical isometric isomorphisms $K(\mathcal{H})^* \cong L^1(\mathcal{H})$ and $L^1(\mathcal{H})^* \cong B(\mathcal{H})$.*

Proof. Part I: $L^1(\mathcal{H})^* \cong B(\mathcal{H})$. Define a linear map $\Psi : B(\mathcal{H}) \to L^1(\mathcal{H})^*$ by $(\Psi(S))(T) := \mathrm{Tr}(ST)$ $(S \in B(\mathcal{H}), T \in L^1(\mathcal{H}))$. It is well defined and contractive $(\|\Psi(S)\| \leq \|S\|)$ because by Proposition 12.24,

$$ST \in L^1(\mathcal{H}) \text{ and } |\mathrm{Tr}(ST)| \leq \|TS\|_1 \leq \|S\| \|T\|_1 \quad (\forall S \in B(\mathcal{H}), T \in L^1(\mathcal{H})). \tag{3}$$

Note that

$$(\Psi(S))(\theta_{x,y}) = \mathrm{Tr}(S\theta_{x,y}) = \mathrm{Tr}(\theta_{Sx,y}) = (Sx, y) \quad (\forall S \in B(\mathcal{H}), x, y \in \mathcal{H}). \tag{4}$$

This implies that Ψ is injective. To show that Ψ is a surjective isometry, let $\varphi \in L^1(\mathcal{H})^*$. Define a map $\langle\cdot,\cdot\rangle : \mathcal{H} \times \mathcal{H} \to \mathbb{C}$ by $\langle x, y\rangle := \varphi(\theta_{x,y})$ $(x, y \in \mathcal{H})$. It is linear in the left variable and conjugate linear in the right variable. Since $|\langle x, y\rangle| \leq \|\varphi\| \|\theta_{x,y}\|_1 = \|\varphi\| \|x\| \|y\|$, $\langle\cdot,\cdot\rangle$ is bounded by $\|\varphi\|$, so we get from Lemma 11.30 an operator $S \in B(\mathcal{H})$ with $(Sx, y) = \langle x, y\rangle = \varphi(\theta_{x,y})$ for every $x, y \in \mathcal{H}$ and $\|S\| \leq \|\varphi\|$. Particularly, by (4), $\Psi(S)$ agrees with φ on $F(\mathcal{H})$, which is $\|\cdot\|_1$-dense in $L^1(\mathcal{H})$ by Theorem 12.26; therefore $\Psi(S) = \varphi$. Hence, $\|S\| \leq \|\varphi\| = \|\Psi(S)\| \leq \|S\|$, so $\|\Psi(S)\| = \|S\|$. This completes the proof.

Part II: $K(\mathcal{H})^* \cong L^1(\mathcal{H})$. Consider the linear map $\Psi : L^1(\mathcal{H}) \to K(\mathcal{H})^*$ given by $(\Psi(S))(K) := \mathrm{Tr}(SK)$ $(S \in L^1(\mathcal{H}), K \in K(\mathcal{H}))$. By (3), Ψ is a well-defined contraction, and as in Part I, Ψ is injective. To show that Ψ is a surjective isometry, let $\varphi \in K(\mathcal{H})^*$. Define a map $\langle\cdot,\cdot\rangle : \mathcal{H}\times\mathcal{H}\to\mathbb{C}$ by $\langle x,y\rangle := \varphi(\theta_{x,y})$ $(x,y\in\mathcal{H})$. Since $|\langle x,y\rangle| \le \|\varphi\|\,\|\theta_{x,y}\| = \|\varphi\|\,\|x\|\,\|y\|$, Lemma 11.30 yields an operator $S\in B(\mathcal{H})$ with $(Sx,y) = \langle x,y\rangle = \varphi(\theta_{x,y})$ for every $x,y\in\mathcal{H}$. To show that $S\in L^1(\mathcal{H})$, fix an orthonormal basis $\{e_\alpha\}_{\alpha\in A}$ of $\mathcal{H}$, and write the polar decomposition $S = V\,|S|$. Then $(|S|\,e_\alpha, e_\alpha) = (V^*Se_\alpha, e_\alpha) = (Se_\alpha, Ve_\alpha)$ for all $\alpha\in A$. For every finite $F\subset A$ we have

$$\sum_{\alpha\in F}(|S|\,e_\alpha,e_\alpha) = \sum_{\alpha\in F}(Se_\alpha,Ve_\alpha) = \sum_{\alpha\in F}\varphi(\theta_{e_\alpha,Ve_\alpha}) = \varphi\left(\sum_{\alpha\in F}\theta_{e_\alpha,Ve_\alpha}\right).$$

But we have $\left\|\sum_{\alpha\in F}\theta_{e_\alpha,Ve_\alpha}\right\| \le 1$ (why?), so $\sum_{\alpha\in F}(|S|\,e_\alpha,e_\alpha)\le\|\varphi\|$. This being true for all finite $F\subset A$, we conclude that $\|S\|_1 = \sum_{\alpha\in A}(|S|\,e_\alpha,e_\alpha)\le \|\varphi\|$, so that $S\in L^1(\mathcal{H})$. By construction, $\Psi(S)$ agrees with φ on $F(\mathcal{H})$, which is $\|\cdot\|$-dense in $K(\mathcal{H})$ by Exercise 8 in Chapter 6; therefore $\Psi(S)=\varphi$. Since Ψ is contractive we obtain $\|\Psi(S)\| = \|\varphi\| = \|S\|_1$. ☐

We saw that the predual $B(\mathcal{H})_*$ of $B(\mathcal{H})$, which equals the closure in $B(\mathcal{H})^*$ of the subspace of w.o.-continuous linear functionals (Proposition 12.13) as well as the subspace of normal linear functionals in $B(\mathcal{H})^*$ (Definition 12.16 and Theorem 12.17), is isometrically isomorphic to $L^1(\mathcal{H})$ (Theorem 12.25). This isomorphism is given by $\omega_{x,y}\longleftrightarrow\theta_{x,y}$, as $\mathrm{Tr}(S\theta_{x,y}) = (Sx,y) = \omega_{x,y}(S)$ for all $S\in B(\mathcal{H})$. We now find another description of $B(\mathcal{H})_*$.

Theorem 12.26. *Every $T\in L^1(\mathcal{H})$ can be written as*

$$T = \|\cdot\|_1\text{-}\sum_{n=1}^{\infty}\theta_{x_n,y_n}$$

("$\|\cdot\|_1$-" means that the series converges in the Banach space $(L^1(\mathcal{H}),\|\cdot\|_1)$), where $\{x_n\}_{n=1}^\infty$ and $\{y_n\}_{n=1}^\infty$ are sequences in $\mathcal{H}$ with $\sum_{n=1}^\infty\|x_n\|^2<\infty$, $\sum_{n=1}^\infty\|y_n\|^2<\infty$, and $\|T\|_1 = \sum_{n=1}^\infty\|x_n\|\,\|y_n\|$. If $T\ge 0$, it can be arranged so that $x_n = y_n$ for every $n\in\mathbb{N}$.

Proof. We have $T\in K(\mathcal{H})$ by Proposition 12.24. Suppose first that T is positive. By the spectral theorem for compact normal operators (Exercise 4 in Chapter 9), there exist an orthonormal sequence $\{e_n\}_{n=1}^\infty$ in $\mathcal{H}$ (or a finite orthonormal sequence if $\mathcal{H}$ is finite dimensional) and a non-negative sequence $\{\lambda_n\}_{n=1}^\infty\in c_0$ such that $T = \sum_{n=1}^\infty\lambda_n\theta_{e_n,e_n}$. This convergence holds in $(L^1(\mathcal{H}),\|\cdot\|_1)$ because $\sum_{n=1}^\infty\lambda_n = \|T\|_1<\infty$ and $\left\|T-\sum_{n=1}^N\lambda_n\theta_{e_n,e_n}\right\|_1 = \sum_{n=N+1}^\infty\lambda_n$ for all $N\in\mathbb{N}$. Thus, letting $x_n := \lambda_n^{1/2}e_n$ $(n\in\mathbb{N})$, we get a sequence with the desired properties.

For the general case, let $T = V|T|$ be the polar decomposition of T. Note that $V\theta_{x,y} = \theta_{Vx,y}$ for all $x,y\in\mathcal{H}$. Apply the previous paragraph to $|T|$ to express it as $\|\cdot\|_1\text{-}\sum_{n=1}^\infty\theta_{y_n,y_n}$ with $\{y_n\}_{n=1}^\infty$ in $\mathcal{H}$ satisfying $\|T\|_1 = \||T|\|_1 = \sum_{n=1}^\infty\|y_n\|^2<\infty$. Then with $x_n := Vy_n$ $(n\in\mathbb{N})$ we have $\sum_{n=1}^\infty\|x_n\|^2<\infty$

and $T = \|\cdot\|_1\text{-}\sum_{n=1}^{\infty} \theta_{x_n,y_n}$. Also, since $\|V\| \leq 1$, we have $\sum_{n=1}^{\infty} \|x_n\| \|y_n\| = \sum_{n=1}^{\infty} \|Vy_n\| \|y_n\| \leq \sum_{n=1}^{\infty} \|y_n\|^2 = \|T\|_1 \leq \sum_{n=1}^{\infty} \|\theta_{x_n,y_n}\| = \sum_{n=1}^{\infty} \|x_n\| \|y_n\|$, proving that $\|T\|_1 = \sum_{n=1}^{\infty} \|x_n\| \|y_n\|$. □

Corollary 12.27. *The predual $B(\mathcal{H})_*$ consists precisely of all $\omega \in B(\mathcal{H})^*$ of the form $\omega = \sum_{n=1}^{\infty} \omega_{x_n,y_n}$ (the convergence is in the norm topology of $B(\mathcal{H})^*$) where $\{x_n\}_{n=1}^{\infty}$ and $\{y_n\}_{n=1}^{\infty}$ are sequences in $\mathcal{H}$ with $\sum_{n=1}^{\infty} \|x_n\|^2 < \infty$ and $\sum_{n=1}^{\infty} \|y_n\|^2 < \infty$. It can be arranged that $\|\omega\| = \sum_{n=1}^{\infty} \|x_n\| \|y_n\|$, and when $\omega \in B(\mathcal{H})_*$ is positive, that also $x_n = y_n$ for every n.*

Proof. If $\{x_n\}_{n=1}^{\infty}$ and $\{y_n\}_{n=1}^{\infty}$ are sequences in $\mathcal{H}$ with $\sum_{n=1}^{\infty} \|x_n\|^2 < \infty$ and $\sum_{n=1}^{\infty} \|y_n\|^2 < \infty$, then the series $\sum_{n=1}^{\infty} \omega_{x_n,y_n}$ converges absolutely in $B(\mathcal{H})^*$ by the Cauchy–Schwarz inequality because $\|\omega_{x,y}\| = \|x\| \|y\|$ for each $x, y \in \mathcal{H}$. Its sum belongs to $B(\mathcal{H})_*$ by Proposition 12.13.

The converse, as well as the case of positive elements of $B(\mathcal{H})_*$, is just a restatement of Theorem 12.26. □

Definition 12.28. The *ultraweak* (or *σ-weak*) topology on $B(\mathcal{H})$ is the topology induced by all linear functionals in $B(\mathcal{H})^*$ of the form discussed in Corollary 12.27. That is, a net $\{A_\alpha\}_\alpha$ in $B(\mathcal{H})$ converges ultraweakly to $A \in B(\mathcal{H})$ iff for all sequences $\{x_n\}_{n=1}^{\infty}$ and $\{y_n\}_{n=1}^{\infty}$ in $\mathcal{H}$ with $\sum_{n=1}^{\infty} \|x_n\|^2 < \infty$ and $\sum_{n=1}^{\infty} \|y_n\|^2 < \infty$, we have $\omega(A_\alpha) \underset{\alpha}{\rightarrow} \omega(A)$ for $\omega := \sum_{n=1}^{\infty} \omega_{x_n,y_n}$.

Corollary 12.27 means that the ultraweak and the *weak** topologies on $B(\mathcal{H})$ coincide.

Finally, we summarize the characterizations of the predual of a von Neumann algebra $\mathcal{R}$ on $\mathcal{H}$ that we have found so far and add two more. Recall that $\mathcal{R}_*$ is identified with the restriction of elements of $B(\mathcal{H})_*$ to $\mathcal{R}$. By the previous paragraph, the *weak** topology of $\mathcal{R}$ coincides with the ultraweak topology induced from $B(\mathcal{H})$. Let $\omega \in \mathcal{R}^*$. Then $\omega \in \mathcal{R}_*$, that is, ω is *weak**-continuous (equivalently: normal) iff it is ultraweakly continuous. In addition, by the Krein–Šmulian theorem (5.52), ω is *weak**-continuous iff its restriction to the closed unit ball of $\mathcal{R}$ is *weak**-continuous, and by Proposition 12.15, this holds iff this restriction is w.o.-continuous. We proved:

Theorem 12.29. *Let $\mathcal{R}$ be a von Neumann algebra on a Hilbert space $\mathcal{H}$. For every $\omega \in \mathcal{R}^*$, the following conditions are equivalent:*

(i) ω is weak-continuous, that is, $\omega \in \mathcal{R}_*$;*

(ii) ω is ultraweakly continuous: there exist sequences $\{x_n\}_{n=1}^{\infty}$ and $\{y_n\}_{n=1}^{\infty}$ in $\mathcal{H}$ with $\sum_{n=1}^{\infty} \|x_n\|^2 < \infty$ and $\sum_{n=1}^{\infty} \|y_n\|^2 < \infty$ such that $\omega = \sum_{n=1}^{\infty} \omega_{x_n,y_n}|_{\mathcal{R}}$;

(iii) ω is normal;

(iv) the restriction of ω to the closed unit ball of $\mathcal{R}$ is weak-continuous (i.e., ultraweakly continuous); and*

(v) the restriction of ω to the closed unit ball of $\mathcal{R}$ is w.o.-continuous.

When ω satisfies these equivalent conditions and is positive, it can be arranged so that $x_n = y_n$ in (ii).

12.7 Commutative von Neumann algebras

This section presents some of the main results on the structure of commutative von Neumann algebras. In comparison to the commutative Gelfand–Naimark theorem, which says that a commutative C^*-algebra "is" the C_0 algebra of a topological space, we will see that a commutative von Neumann algebra "is", roughly, the L^∞ algebra of a measure space.

Recall from Section 12.1 the notion of *spatial equivalence* of von Neumann algebras and from Section 12.2 the notion of a *cyclic vector* of a von Neumann algebra. We use the notation M_f introduced in Example 12.7.

Theorem 12.30. *Let $\mathcal{R}$ be a commutative von Neumann algebra on a Hilbert space $\mathcal{H}$ admitting a cyclic vector. Then there exists a finite positive measure space $(X, \mathcal{A}, \mu)$ such that $\mathcal{R}$ is spatially isomorphic to the von Neumann algebra $\{M_f; f \in L^\infty(X, \mathcal{A}, \mu)\}$ acting on $L^2(X, \mathcal{A}, \mu)$.*

Proof. Since $\mathcal{R}$ is a unital commutative C^*-algebra, the commutative Gelfand–Naimark theorem (7.16) says that there exist a compact Hausdorff space X and a $*$-isomorphism Γ from $\mathcal{R}$ onto $C(X)$.

Let $\zeta \in \mathcal{H}$ be a unit cyclic vector for $\mathcal{R}$ and consider the state $\omega := \omega_\zeta|_{\mathcal{R}}$ of $\mathcal{R}$. Then $\omega \circ \Gamma^{-1}$ is a state of $C(X)$, so by the Riesz–Markov theorem (3.18) there are a σ-algebra $\mathcal{A}$ on X and a positive measure μ on $(X, \mathcal{A})$ such that

$$(\omega \circ \Gamma^{-1})(f) = \int_X f \, d\mu \qquad (\forall f \in C(X)). \tag{1}$$

In particular, taking $f := 1$ leads to $\mu(X) = 1$.

Define an operator $U_0 : \mathcal{R}\zeta \to C(X)$ by $T\zeta \to \Gamma(T)$, $T \in \mathcal{R}$. Then, viewing $C(X)$ as a normed subspace of $L^2(X, \mathcal{A}, \mu)$, U_0 is (well-defined and) isometric, because for each T, writing $f := \Gamma(T)$ we have $|f|^2 = \Gamma(T^*T)$ and

$$\|f\|_{L^2}^2 = \int_X |f|^2 \, d\mu = (\omega \circ \Gamma^{-1})(|f|^2) = \omega(T^*T) = \|T\zeta\|^2$$

by (1). Also $\overline{\mathcal{R}\zeta} = \mathcal{H}$ since ζ is cyclic for $\mathcal{R}$, and $C(X)$ is dense in $L^2(X, \mathcal{A}, \mu)$ by Corollary 3.21 of Lusin's theorem. All in all, U_0 extends to a unitary equivalence $U : \mathcal{H} \to L^2(X, \mathcal{A}, \mu)$.

Moreover for every $T \in \mathcal{R}$ we have $UTU^* = M_{\Gamma(T)}$. Indeed, for each $S \in \mathcal{R}$,

$$(UT)(S\zeta) = U(TS\zeta) = \Gamma(TS) = \Gamma(T) \cdot \Gamma(S) = \Gamma(T) \cdot (US\zeta) = M_{\Gamma(T)}(US\zeta).$$

In particular we have $U\mathcal{R}U^* = \{M_f; f \in C(X)\}$. But $U\mathcal{R}U^*$ is evidently a von Neumann algebra on $L^2(X, \mathcal{A}, \mu)$, and $\{M_f; f \in C(X)\}$ is s.o.-dense in the von Neumann algebra $\{M_f; f \in L^\infty(X, \mathcal{A}, \mu)\}$ by Exercise 18. Hence, the two algebras coincide. □

Remark 12.31. The map that takes $f \in L^\infty(X, \mathcal{A}, \mu)$ to M_f is injective. Thus, the end of the proof shows the surprising equality $C(X) = L^\infty(X, \mathcal{A}, \mu)$ *for this specific* $(X, \mathcal{A}, \mu)$; that is, for every $f \in L^\infty(X, \mathcal{A}, \mu)$ there exists $f' \in C(X)$ such that $f = f'$ a.e.! The reason for this seemingly unnatural equality has to do with the topology of X and the nature of the measure μ. We will not elaborate.

We conclude this section with two more theorems about commutative von Neumann algebra, which are given without proofs: the first deals with ones acting on separable Hilbert spaces, and the second discusses the general case.

Theorem 12.32. *Let $\mathcal{R}$ be a commutative von Neumann algebra on a separable Hilbert space. Then there exists a finite positive measure space $(X, \mathcal{A}, \mu)$ such that $\mathcal{R}$ is $*$-isomorphic to $L^\infty(X, \mathcal{A}, \mu)$.*

In comparison to Theorem 12.30, in which the existence of a cyclic vector guarantees a *spatial* isomorphism between $\mathcal{R}$ and $L^\infty(X, \mathcal{A}, \mu)$ when the latter acts canonically on $L^2(X, \mathcal{A}, \mu)$, in the situation of Theorem 12.32 we can only deduce that $\mathcal{R}$ and $L^\infty(X, \mathcal{A}, \mu)$ are isomorphic *as C^*-algebras.*

Theorem 12.33. *Let $\mathcal{R}$ be a commutative von Neumann algebra. Then $\mathcal{R}$ is spatially equivalent to the direct sum of a family $\{\mathcal{R}_\alpha\}_{\alpha \in A}$ of von Neumann algebras such that for each $\alpha \in A$ there exists a finite positive measure space $(X_\alpha, \mathcal{A}_\alpha, \mu_\alpha)$ such that $\mathcal{R}_\alpha$ is $*$-isomorphic to $L^\infty(X_\alpha, \mathcal{A}_\alpha, \mu_\alpha)$.*

12.8 The enveloping von Neumann algebra of a C^*-algebra

This section demonstrates how to obtain a canonical von Neumann algebra from an arbitrary C^*-algebra $\mathcal{A}$.

We begin with a non-degenerate representation π of $\mathcal{A}$ on a Hilbert space $\mathcal{H}$. Then $\overline{\pi(\mathcal{A})}^{\text{w.o.t.}}$ is a von Neumann algebra on $\mathcal{H}$. For $X \in \mathcal{A}^{**}$, consider the function $\langle \cdot, \cdot \rangle : \mathcal{H} \times \mathcal{H} \to \mathbb{C}$ given by

$$\langle x, y \rangle := X(\omega_{x,y} \circ \pi) \qquad (x, y \in \mathcal{H})$$

(which is well defined as $\omega_{x,y} \circ \pi$ belongs to $\mathcal{A}^*$). It is linear in the left variable and conjugate linear in the right variable, and for every $x, y \in \mathcal{H}$ we have

$$|\langle x, y \rangle| \leq \|X\| \, \|\omega_{x,y}\| \, \|\pi\| \leq \|X\| \, \|x\| \, \|y\| \, ,$$

so $\langle \cdot, \cdot \rangle$ is bounded by $\|X\|$. By Lemma 11.30, there exists a (unique) operator $\tilde{\pi}(X) \in B(\mathcal{H})$ such that for all $x, y \in \mathcal{H}$,

$$(\tilde{\pi}(X)x, y) = \langle x, y \rangle \, ,$$

that is,

$$(\tilde{\pi}(X)x, y) = X(\omega_{x,y} \circ \pi). \tag{1}$$

We claim that $\tilde{\pi}(X) \in \overline{\pi(\mathcal{A})}^{\text{w.o.t.}}$, and if $X \in \mathcal{A}$ (when $\mathcal{A}$ is viewed as embedded in $\mathcal{A}^{**}$) then $\tilde{\pi}(X) = \pi(X)$. The latter assertion follows from (1) since $(\pi(X)x, y) = (\omega_{x,y} \circ \pi)(X)$ for all $x, y \in \mathcal{H}$ when $X \in \mathcal{A}$. To prove the former assertion, let $T' \in (\pi(\mathcal{A}))'$. Then for all $x, y \in \mathcal{H}$ the functionals $\omega_{T'x,y}$ and $\omega_{x,(T')^*y}$ coincide on $\pi(\mathcal{A})$, namely, $\omega_{T'x,y} \circ \pi = \omega_{x,(T')^*y} \circ \pi$, and thus,

$$\begin{aligned}(\tilde{\pi}(X)T'x, y) &= X(\omega_{T'x,y} \circ \pi) = X(\omega_{x,(T')^*y} \circ \pi) \\ &= (\tilde{\pi}(X)x, (T')^*y) = (T'\tilde{\pi}(X)x, y),\end{aligned}$$

proving that $\tilde{\pi}(X)$ commutes with T', hence $\tilde{\pi}(X) \in (\pi(\mathcal{A}))'' = \overline{\pi(\mathcal{A})}^{\text{w.o.t.}}$ by von Neumann's double commutant theorem (12.4).

Since $X \in \mathcal{A}^{**}$ was arbitrary, we have constructed an extension

$$\tilde{\pi} : \mathcal{A}^{**} \to \overline{\pi(\mathcal{A})}^{\text{w.o.t.}}$$

of π, which is clearly linear and of norm at most 1. Also, by (1) and the fact that $\{\omega_{x,y}; x, y \in \mathcal{H}\}$ spans a dense subspace of $B(\mathcal{H})_*$ (Proposition 12.13), we have

$$\omega\left(\tilde{\pi}(X)\right) = X(\omega \circ \pi) \qquad (\forall X \in \mathcal{A}^{**}, \omega \in B(\mathcal{H})_*). \tag{2}$$

It is therefore clear that $\tilde{\pi}$ is *weak**-*weak**-continuous, or equivalently, *weak**-ultraweakly continuous. Notice that by (2), $\tilde{\pi}$ is nothing but the Banach adjoint of the bounded linear map $(\overline{\pi(\mathcal{A})}^{\text{w.o.t.}})_* \to \mathcal{A}^*$ given by $\omega \to \omega \circ \pi$.

(Remark that we could equally define $\tilde{\pi}$ by (2) or as in the previous sentence. This would be more "to the point". However, starting with (1) and essentially repeating part of the proof of Proposition 12.13 should be more illuminating to the reader.)

The closed unit ball of $\overline{\pi(\mathcal{A})}^{\text{w.o.t.}}$ is w.o.-compact and it contains the image under $\tilde{\pi}$ of the closed unit ball of $\mathcal{A}^{**}$. Furthermore, π maps the open unit ball of $\mathcal{A}$ onto that of $\pi(\mathcal{A})$ (see Exercise 2 in Chapter 11), and the latter is w.o.-dense in the closed unit ball of $\overline{\pi(\mathcal{A})}^{\text{w.o.t.}}$ by Kaplansky's density theorem (12.8). Finally, the closed unit ball of $\mathcal{A}^{**}$ is *weak**-compact by Alaoglu's theorem (5.24), and therefore it is mapped by the *weak**-ultraweakly continuous map $\tilde{\pi}$ to an ultraweakly compact subset of $\overline{\pi(\mathcal{A})}^{\text{w.o.t.}}$. All in all, $\tilde{\pi}$ maps the closed unit ball of $\mathcal{A}^{**}$ onto that of $\overline{\pi(\mathcal{A})}^{\text{w.o.t.}}$. In particular, this implies that $\tilde{\pi} : \mathcal{A}^{**} \to \overline{\pi(\mathcal{A})}^{\text{w.o.t.}}$ is surjective.

Recall from the non-commutative Gelfand–Naimark theorem (11.26) that the *universal representation* of $\mathcal{A}$, to be denoted here by π_{u}, is the direct sum representation $\sum_{\omega \in \mathcal{S}(\mathcal{A})} \oplus \pi_\omega$, where $\mathcal{S}(\mathcal{A})$ is the set of all states of $\mathcal{A}$ and for $\omega \in \mathcal{S}(\mathcal{A})$, π_ω is the cyclic representation of $\mathcal{A}$ on a Hilbert space $\mathcal{H}_\omega$ associated with ω by Theorem 11.23. The map π_{u} is a non-degenerate representation of $\mathcal{A}$ on the Hilbert space $\mathcal{H}_{\text{u}} := \sum_{\omega \in \mathcal{S}(\mathcal{A})} \oplus \mathcal{H}_\omega$.

Apply the shown construction to π_u to yield a surjective, *weak**-ultraweakly continuous, contractive linear map $\widetilde{\pi_u} : \mathcal{A}^{**} \to \overline{\pi_u(\mathcal{A})}^{\text{w.o.t.}}$. We prove in the next paragraph that $\widetilde{\pi_u}$ is injective. This has two implications, both relying on the fact that the restriction of $\widetilde{\pi_u}$ to the closed unit ball of $\mathcal{A}^{**}$ has the closed unit ball of $\overline{\pi_u(\mathcal{A})}^{\text{w.o.t.}}$ as its range. First, $\widetilde{\pi_u}$ is isometric. Second, this restriction of $\widetilde{\pi_u}$ is injective and *weak**-*weak**-continuous, its domain is compact and its range is Hausdorff in the respective topologies. Consequently, this restriction of $\widetilde{\pi_u}$ is a *weak**-*weak**-homeomorphism. By the Krein–Šmulian theorem (5.52), $\widetilde{\pi_u}$ itself is a *weak**-*weak**-homeomorphism.

It remains to prove that $\widetilde{\pi_u}$ is injective. As explained, $\widetilde{\pi_u}$ is the Banach adjoint of the bounded linear map $(\widetilde{\pi_u})_* : (\overline{\pi_u(\mathcal{A})}^{\text{w.o.t.}})_* \to \mathcal{A}^*$ given by $\omega \to \omega \circ \pi_u$. The map $(\widetilde{\pi_u})_*$ is surjective, because every $\rho \in \mathcal{A}^*$ is the linear combination of states of $\mathcal{A}$ by the Jordan decomposition (Theorem 11.32), and by the universality of π_u, every state ρ of $\mathcal{A}$ can be written as $\omega_x \circ \pi_u$ for a suitable $x \in \mathcal{H}_u$ (and the vector functional ω_x restricts to an element of $(\overline{\pi_u(\mathcal{A})}^{\text{w.o.t.}})_*$!). It follows from Part (e) of Exercise 9 in Chapter 6 that $\widetilde{\pi_u}$ is injective.

Theorem 12.34. *Let $\mathcal{A}$ be a C^*-algebra. Viewing $\mathcal{A}$ as a Banach subspace of $\mathcal{A}^{**}$, the universal representation π_u of $\mathcal{A}$ extends uniquely to a weak*-ultraweakly continuous function $\widetilde{\pi_u} : \mathcal{A}^{**} \to \overline{\pi_u(\mathcal{A})}^{\text{w.o.t.}}$. Furthermore:*

1. *$\widetilde{\pi_u}$ is linear, isometric, surjective, and a weak*-ultraweak-homeomorphism.*
2. *For every non-degenerate representation π of $\mathcal{A}$ on a Hilbert space $\mathcal{H}$ there exists a unique ultraweakly continuous function from $\overline{\pi_u(\mathcal{A})}^{\text{w.o.t.}}$ to $B(\mathcal{H})$ mapping $\pi_u(A)$ to $\pi(A)$ for each $A \in \mathcal{A}$. This map is a representation of $\overline{\pi_u(\mathcal{A})}^{\text{w.o.t.}}$ and its range is $\overline{\pi(\mathcal{A})}^{\text{w.o.t.}}$.*

Proof. Uniqueness is clear by the *weak**-density of $\mathcal{A}$ in $\mathcal{A}^{**}$. Existence and (i) were proved earlier.

(ii) Uniqueness is again clear by density. The linear map $\Pi := \widetilde{\pi} \circ (\widetilde{\pi_u})^{-1} : \overline{\pi_u(\mathcal{A})}^{\text{w.o.t.}} \to \overline{\pi(\mathcal{A})}^{\text{w.o.t.}}$, which is surjective and ultraweakly continuous, maps $\pi_u(A)$ to $\pi(A)$ for each $A \in \mathcal{A}$. It remains to prove that Π is a $*$-homomorphism.

Let $T, S \in \overline{\pi_u(\mathcal{A})}^{\text{w.o.t.}}$ and let $\{A_\alpha\}_{\alpha \in I}, \{B_\beta\}_{\beta \in J}$ be nets in $\mathcal{A}$ such that the nets $\{\pi_u(A_\alpha)\}_{\alpha \in I}, \{\pi_u(B_\beta)\}_{\beta \in J}$ converge ultraweakly to T, S, respectively. For every fixed $\beta \in J$, the net $\{\pi_u(A_\alpha B_\beta)\}_{\alpha \in I} = \{\pi_u(A_\alpha)\pi_u(B_\beta)\}_{\alpha \in I}$ converges ultraweakly to $T\pi_u(B_\beta)$, and thus the ultraweak continuity of Π implies that

$$\begin{aligned} \Pi\left(T\pi_u(B_\beta)\right) &= \lim_{\alpha \in I} \Pi\left(\pi_u(A_\alpha B_\beta)\right) = \lim_{\alpha \in I} \pi(A_\alpha B_\beta) = \lim_{\alpha \in I} \pi(A_\alpha)\pi(B_\beta) \\ &= \lim_{\alpha \in I} \Pi\left(\pi_u(A_\alpha)\right) \Pi\left(\pi_u(B_\beta)\right) = \Pi(T)\Pi\left(\pi_u(B_\beta)\right). \end{aligned} \tag{3}$$

Since also the net $\{\Pi(\pi_u(B_\beta))\}_{\beta \in J}$ converges ultraweakly to $\Pi(S)$, taking now the limit with respect to β in (3) gives $\Pi(TS) = \Pi(T)\Pi(S)$, proving that Π is multiplicative. A similar (and easier) argument shows that Π is $*$-preserving. ☐

The von Neumann algebra $\overline{\pi_u(\mathcal{A})}^{\text{w.o.t.}}$ is called the *enveloping von Neumann algebra* of the C^*-algebra $\mathcal{A}$. Part (ii) of the theorem is a universality property of this von Neumann algebra in terms of the representations of $\mathcal{A}$: every non-degenerate representation of $\mathcal{A}$ "extends" to an ultraweakly continuous representation of its enveloping von Neumann algebra.

Exercises

Let $\mathcal{H}$ be a Hilbert space and $\mathcal{R}$ be a von Neumann algebra on $\mathcal{H}$.

1. Let $\mathcal{I}$ be a w.o.-closed ideal in $\mathcal{R}$. Prove that there exists a projection P in the *center* $\mathcal{R} \cap \mathcal{R}'$ of $\mathcal{R}$ such that $\mathcal{I} = \mathcal{R}P := \{AP; A \in \mathcal{R}\}$.

2. Let $U \in \mathcal{R}$ be unitary. Prove that there is a selfadjoint $A \in \mathcal{R}$ such that $U = e^{iA}$. (Compare Exercise 28 in Chapter 7.)

3. Prove that a vector $x \in \mathcal{H}$ is separating for $\mathcal{R}$ iff the positive functional $\omega_x|_{\mathcal{R}}$ is faithful (see Exercise 19 in Chapter 11).

4. Prove Lemma 12.10.

5. Prove that if $T \in B(\mathcal{H})$ is normal, then its support is $\overline{T\mathcal{H}}$.

6. (a) Let $A \in B(\mathcal{H})_+$. Prove that $\ker A = \ker A^{1/2}$. In fact, with a little more thought, prove that $\ker A = \ker A^\alpha$ for every $\alpha > 0$.

   (b) Let $T \in B(\mathcal{H})$. Prove that the kernels of $T, T^*T, |T|$ are equal, and thus so are their supports.

7. (The polar decomposition of unbounded operators.) In this exercise we extend Theorem 12.11. We use freely the terminology and results of Chapter 10 and Exercises 15 and 17 therein. Let T be a (generally unbounded) closed densely defined operator on $\mathcal{H}$.

   (a) The support of T is defined to be the closed subspace $(\ker T)^\perp$ of $\mathcal{H}$. Prove that it equals $\overline{T^*\mathcal{H}}$.

   (b) Imitating Exercise 6, show that the kernels of $T, T^*T, |T|$ are equal, and thus so are their supports, and that the latters are equal to $\overline{|T|\,\mathcal{H}}$.

   (c) Existence: using Exercise 17 in Chapter 10, verify that the proof of Part (i) of Theorem 12.11 carries verbatim, proving the existence of a partial isometry $V \in B(\mathcal{H})$, whose initial space is the support of T and whose final space is $\overline{T\mathcal{H}}$, such that $T = V\,|T|$.

   (d) Uniqueness: using Exercise 15 in Chapter 10, verify that the proof of Part (ii) of Theorem 12.11 carries verbatim, proving that if A is a positive selfadjoint operator on $\mathcal{H}$ and $W \in B(\mathcal{H})$ is a partial isometry with initial space $\overline{A\mathcal{H}}$ such that $T = WA$, then $A = |T|$ and $W = V$.

(e) Verify that the proof of Part (iii) of Theorem 12.11 also carries.

8. Under the conditions of Kaplansky's density theorem, show that for every $T \in \mathcal{R}$ there is a net $\{T_\alpha\}$ in $\mathcal{A}$ that converges to T in the s.o.t. and satisfies $\|T_\alpha\| = \|T\|$ for each α. (Hint: Exercise 25 of Chapter 6.)

9. Prove that a non-degenerate $*$-subalgebra of $B(\mathcal{H})$ is a von Neumann algebra on $\mathcal{H}$ iff its (norm-) closed unit ball is w.o.-compact.

10. Prove that $c_0 := C_0(\mathbb{N})$ is not a dual Banach space.

11. Prove that the Banach space l^1 admits (at least) two non-isometrically isomorphic preduals. (Hint: use Exercises 8 and 9 in Chapter 4.)

12. Let X be a predual of a Banach space Y. View X as a closed subspace of Y^* as explained in Section 12.5. Then:

    (i) $X \subset Y^*$ is *norming* for Y, that is, for every $y \in Y$, $\|y\| = \sup\{x(y); x \in X, \|x\| \leq 1\}$.

    (ii) The closed unit ball of Y is compact in the X-topology.

    Indeed, (i) follows since Y is isometrically isomorphic to X^* and we view X as embedded in Y^* canonically, and (ii) holds by Alaoglu's theorem.

    We show that properties (i) and (ii) *characterize* X being a predual of Y. Thus, let again Y be a Banach space and X be a closed subspace of Y^* such that (i) and (ii) hold.

    (a) For $y \in Y$, consider the map $f(y) := \hat{y}|_X : x \to x(y)$ on X. It is obviously a bounded linear functional, thus an element of X^*. Explain why $f : Y \to X^*$ is a linear isometry.

    (b) Prove that the range of f is dense in X^* in the X-topology.

    (c) Explain why the image under f of the closed unit ball of Y is compact in the X-topology, and deduce that the range of f is closed in the X-topology. In conclusion, f is onto X^*.

13. Let $\mathcal{R}_1, \mathcal{R}_2$ be von Neumann algebras.

    (a) Let $\Phi : \mathcal{R}_1 \to \mathcal{R}_2$ be a linear map. Prove that the following conditions are equivalent:

        (i) Φ is *weak**-continuous (i.e., continuous when both $\mathcal{R}_1$ and $\mathcal{R}_2$ are equipped with their respective *weak** topologies);

        (ii) Φ is ultraweakly continuous;

        (iii) the restriction of Φ to the closed unit ball of $\mathcal{R}_1$ is *weak**-continuous; and

        (iv) the restriction of Φ to the closed unit ball of $\mathcal{R}_1$ is w.o.-continuous.

Prove that if Φ is a $*$-homomorphism, then the conditions shown are also equivalent to:

(v) Φ is normal: for every bounded increasing net $\{T_\alpha\}_\alpha$ in $(\mathcal{R}_1)_+$ with least upper bound T, the least upper bound of $\{\Phi(T_\alpha)\}_\alpha$ in $(\mathcal{R}_2)_+$ is $\Phi(T)$; equivalently, $\Phi(T) = \lim_\alpha \Phi(T_\alpha)$ in the w.o.t.

(b) Prove that every $*$-isomorphism from $\mathcal{R}_1$ onto $\mathcal{R}_2$ is *weak**-continuous.

(c) Prove that if $\Phi : \mathcal{R}_1 \to \mathcal{R}_2$ is a unital $*$-homomorphism satisfying the equivalent conditions of Part (a), then its image $\Phi(\mathcal{R}_1)$ is a von Neumann subalgebra of $\mathcal{R}_2$.

14. Recall from Section 11.8 that the set $\mathcal{S}(\mathcal{R})$ of states of $\mathcal{R}$ is convex and *weak**-compact, that is, compact with respect to the *weak** topology on $\mathcal{R}^*$. Prove that the set $\mathcal{S}_n(\mathcal{R})$ of normal states of $\mathcal{R}$ is *weak**-dense in $\mathcal{S}(\mathcal{R})$.

15. Prove that a $*$-subalgebra of $B(\mathcal{H})$ is ultraweakly closed in $B(\mathcal{H})$ iff it is w.o.-closed in $B(\mathcal{H})$ (iff it is s.o.-closed in $B(\mathcal{H})$).

16. Prove that $\mathrm{Tr}(TS) = \mathrm{Tr}(ST)$ for all $T, S \in L^2(\mathcal{H})$.

17. We proved in Proposition 12.24 that every element of $L^1(\mathcal{H})$ has a trace. In this exercise we prove the converse. In fact, we show that the following a priori weaker condition on $T \in B(\mathcal{H})$ implies that $T \in L^1(\mathcal{H})$: for each orthonormal basis $\{e_\alpha\}_{\alpha\in A}$ of $\mathcal{H}$ we have $\sum_{\alpha\in A} |(Te_\alpha, e_\alpha)| < \infty$.

(a) Recall that we denote $\Re T := \frac{1}{2}(T+T^*)$ and $\Im T := \frac{1}{2\mathrm{i}}(T-T^*)$. Show that $|((\Re T)e,e)|, |((\Im T)e,e)| \le |(Te,e)|$ for all $e \in \mathcal{H}$.

(b) Use Part (a) to reduce the problem to selfadjoint operators T.

(c) Assuming that T is selfadjoint, let E be the resolution of the identity for T. Let $\{e_\alpha\}_{\alpha\in B}$ and $\{e_\alpha\}_{\alpha\in C}$ be orthonormal bases of $E([0,\infty))\mathcal{H}$ and $E((-\infty,0])\mathcal{H}$, respectively, with $B \cap C = \emptyset$. Then, with $A := B \cup C$, $\{e_\alpha\}_{\alpha\in A}$ is an orthonormal basis of $\mathcal{H}$. Prove that $\|T\|_1 = \sum_{\alpha\in A} |(Te_\alpha, e_\alpha)|$.

18. Let $(X, \mathcal{A}, \mu)$ be a positive measure space as in Lusin's theorem (3.20) and assume additionally that it is σ-finite. Prove that $\{M_f; f \in C_c(X)\}$ is s.o.-dense in the von Neumann algebra $\{M_f; f \in L^\infty(X, \mathcal{A}, \mu)\}$ (cf. Corollary 3.21).

19. (a) Prove that $\mathcal{R}$ is commutative iff $\mathcal{R} \subset \mathcal{R}'$.

(b) Prove that $\mathcal{R}$ is a *maximal* commutative von Neumann subalgebra of $B(\mathcal{H})$ iff $\mathcal{R} = \mathcal{R}'$.

Henceforth we assume that $\mathcal{R}$ is commutative.

(c) Prove that every cyclic vector for $\mathcal{R}$ is also separating for $\mathcal{R}$.

(d) Prove that if $\mathcal{R}$ has a cyclic vector, then it is a maximal commutative von Neumann subalgebra of $B(\mathcal{H})$.

(e) Prove that if $\mathcal{H}$ is separable then the converse of Part (d) holds: if $\mathcal{R}$ is a maximal commutative von Neumann subalgebra of $B(\mathcal{H})$, then $\mathcal{R}$ has a cyclic vector. (Hint: use Zorn's lemma to obtain a (possibly finite) sequence $\{x_n\}$ in $\mathcal{H}$ such that the family $\{\overline{\mathcal{R}x_n}\}$ consists of mutually orthogonal closed subspaces of $\mathcal{H}$ whose direct sum is $\mathcal{H}$ (notice that the projection P_n of $\mathcal{H}$ onto $\overline{\mathcal{R}x_n}$ belongs to $\mathcal{R}' = \mathcal{R}$). Set $x := \sum_n \frac{1}{n} x_n$. Show that for each n we have $P_n x = \frac{1}{n} x_n$ and therefore $\mathcal{R}x$ contains $\mathcal{R}x_n$.)

20. What is the relationship between the universal representation of a C^*-algebra $\mathcal{A}$ and the universal normal representation of the enveloping von Neumann algebra of $\mathcal{A}$?

21. Let $\mathcal{A}$ be a C^*-algebra, and let $\widetilde{\pi_{\mathrm{u}}} : \mathcal{A}^{**} \to \overline{\pi_{\mathrm{u}}(\mathcal{A})}^{\text{w.o.t.}}$ be the isomorphism from Theorem 12.34. Define a product on $\mathcal{A}^{**}$ by "pulling back" the product on $\overline{\pi_{\mathrm{u}}(\mathcal{A})}^{\text{w.o.t.}}$, that is, letting $X \cdot Y := (\widetilde{\pi_{\mathrm{u}}})^{-1}(\widetilde{\pi_{\mathrm{u}}}(X)\widetilde{\pi_{\mathrm{u}}}(Y))$ for $X, Y \in \mathcal{A}^{**}$. Prove that this product is the same as the first and second Arens products on $\mathcal{A}^{**}$ (see Section 7.5). In particular, $\mathcal{A}$ is Arens regular.

13

Constructions of C^*-algebras

This chapter provides two examples of constructions of C^*-algebras, namely, *tensor products of C^*-algebras* and *group C^*-algebras*, which are fundamental and prevalent in C^*-algebra theory. Both constructions have von Neumann algebraic analogues, which in fact preceded the C^*-algebraic ones and go back to Murray and von Neumann. Tensor products of von Neumann algebras are discussed in the exercises.

The theory of *tensor products* (initially called direct products) *of C^*-algebras* started with the work of Turumaru in 1952 and received a big push from Takesaki. To define a tensor product of two C^*-algebras $\mathcal{A}$ and $\mathcal{B}$ we need a norm on the algebraic tensor product $\mathcal{A} \otimes_{\text{alg}} \mathcal{B}$ with respect to which the completion is a C^*-algebra. Surprisingly, there is usually more than one such norm, but there is always a smallest and a largest. Section 13.1 identifies these two norms and characterizes the associated tensor products.

For a locally compact topological group G, the commutative C^*-algebra $C_0(G)$ takes into account only the topology on G and ignores the group structure of G. Section 13.2 constructs the *group C^*-algebra* of G, denoted $C^*(G)$. This is a C^*-algebra reflecting both the topology and the group structure of G. The C^*-algebra $C^*(G)$ is constructed in such a way that its non-degenerate representations are in bijective correspondence with the (s.o.-continuous) unitary representations of G. When G is discrete, $C^*(G)$ is a C^*-algebraic analogue of the complex group algebra $\mathbb{C}[G]$.

13.1 Tensor products of C^*-algebras

This section shows how one can tensor C^*-algebras. Tensor products are fundamental in operator algebras and form a rich source of examples.

Introduction to Modern Analysis. Second Edition. Shmuel Kantorovitz and Ami Viselter, Oxford University Press.
 DOI: 10.1093/oso/9780192849540.003.0013

Roughly speaking, the tensor product of two algebras is to multiplication what the direct sum of two vector spaces is to summation and what the direct product of two groups is to multiplication. The direct sum $U \oplus V$ of two vector spaces U, V is a new vector space that contains copies of U and V in a way that they span $U \oplus V$ and do not "interact" linearly, that is, $U \cap V = \{0\}$. Similarly, the direct product $H \times K$ of two groups H, K is a new group containing *commuting* copies of the groups H and K that generate $H \times K$ as a group and do not interact algebraically apart from commutation. Likewise, the (algebraic) tensor product $\mathcal{A} \otimes \mathcal{B}$ of two (say unital) algebras $\mathcal{A}, \mathcal{B}$ will be a new algebra containing *commuting* copies of $\mathcal{A}$ and $\mathcal{B}$ that generate $\mathcal{A} \otimes \mathcal{B}$ as an algebra and do not interact algebraically apart from commutation. (Remark that removing commutativity of the copies of $\mathcal{A}$ and $\mathcal{B}$ from this last description leads to *free products* of algebras, which are not discussed in this book.)

A C^*-algebraic tensor product of two C^*-algebras $\mathcal{A}$ and $\mathcal{B}$ is a C^*-algebra containing the algebraic tensor product of $\mathcal{A}$ and $\mathcal{B}$ as a dense subalgebra. Thus, one should come up with a "suitable" norm on the algebraic tensor product and then complete it. However, it turns out that finding such a norm is not at all straightforward. In fact, usually there is more than one such norm!

We assume that the reader is familiar with Section 8.8.

13.1.1 Tensor products of algebras

We begin with the algebraic setting. Let $\mathcal{A}, \mathcal{B}$ be algebras over a field $\mathbb{F}$. Consider the algebraic tensor product $\mathcal{A} \otimes \mathcal{B}$ of $\mathcal{A}$ and $\mathcal{B}$ *as vector spaces*. Recall that $\mathcal{A} \otimes \mathcal{B}$ is spanned by the "simple tensors" $a \otimes b$ ($a \in \mathcal{A}$, $b \in \mathcal{B}$) and that the map $(a, b) \to a \otimes b$ is bilinear.

Endow the vector space $\mathcal{A} \otimes \mathcal{B}$ with the product defined on simple tensors by

$$(a_1 \otimes b_1) \cdot (a_2 \otimes b_2) := (a_1 a_2) \otimes (b_1 b_2) \qquad (a_1, a_2 \in \mathcal{A}, b_1, b_2 \in \mathcal{B})$$

and extended linearly. It is immediate from the definition of the vector space $\mathcal{A} \otimes \mathcal{B}$ that this product is well defined and that it makes $\mathcal{A} \otimes \mathcal{B}$ into an algebra over $\mathbb{F}$. The latter is called the *tensor product* of the algebras $\mathcal{A}$ and $\mathcal{B}$, and is also denoted by $\mathcal{A} \otimes \mathcal{B}$.

If $\mathcal{B}$ is unital with unit $\mathbb{1}_{\mathcal{B}}$, then the map $a \to a \otimes \mathbb{1}_{\mathcal{B}}$ is an injective homomorphism from $\mathcal{A}$ into $\mathcal{A} \otimes \mathcal{B}$ (injectivity follows from Part (vi) of Theorem 8.16), giving a "copy" $\mathcal{A} \otimes \mathbb{1}_{\mathcal{B}} := \{a \otimes \mathbb{1}_{\mathcal{B}}; a \in \mathcal{A}\}$ of the algebra $\mathcal{A}$ inside the algebra $\mathcal{A} \otimes \mathcal{B}$. A similar statement holds true when $\mathcal{A}$ is unital. Hence, when both $\mathcal{A}, \mathcal{B}$ are unital, there are canonical commuting copies of $\mathcal{A}$ and $\mathcal{B}$ inside $\mathcal{A} \otimes \mathcal{B}$, which generate $\mathcal{A} \otimes \mathcal{B}$ as an algebra because $a \otimes b = (a \otimes \mathbb{1}_{\mathcal{B}})(\mathbb{1}_{\mathcal{A}} \otimes b) = (\mathbb{1}_{\mathcal{A}} \otimes b)(a \otimes \mathbb{1}_{\mathcal{B}})$ for all $a \in \mathcal{A}$ and $b \in \mathcal{B}$.

Parts (i)–(iv) of Theorem 8.16 have analogues for tensor products of algebras, in which the linear maps are algebra homomorphisms.

Example 13.1. If $\mathcal{A}$ is an algebra over a field $\mathbb{F}$ and $n \in \mathbb{N}$, then the algebras $M_n(\mathbb{F}) \otimes \mathcal{A}$ and $M_n(\mathcal{A})$ are canonically isomorphic. Indeed, writing $(e_{ij})_{1 \le i,j \le n}$ for the matrix units of $M_n(\mathbb{F})$, the map that sends a matrix $(a_{ij})_{1 \le i,j \le n} \in M_n(\mathcal{A})$

to $\sum_{i,j=1}^{n} e_{ij} \otimes a_{ij} \in M_n(\mathbb{F}) \otimes \mathcal{A}$ is a surjective algebra isomorphism. Indeed, it is a surjective linear map, which is injective by Part (vi) of Theorem 8.16, and is also multiplicative.

For the next example, recall that for a group G and a field $\mathbb{F}$, the *group algebra* $\mathbb{F}[G]$ is the set of all formal sums of the form $\sum_{g\in G} \alpha_g g$, where $\alpha_g \in \mathbb{F}$ for all $g \in G$ and $\{g \in G; \alpha_g \neq 0\}$ is finite. It is a unital algebra with respect to the operations

$$\sum_{g\in G} \alpha_g g + \sum_{g\in G} \beta_g g := \sum_{g\in G} (\alpha_g + \beta_g)\, g$$

$$\alpha \sum_{g\in G} \alpha_g g := \sum_{g\in G} (\alpha\alpha_g)\, g$$

$$\left(\sum_{g\in G} \alpha_g g\right)\left(\sum_{g\in G} \beta_g g\right) := \sum_{g_1,g_2\in G} (\alpha_{g_1}\beta_{g_2})\,(g_1 g_2)$$

for $\sum_{g\in G} \alpha_g g, \sum_{g\in G} \beta_g g \in \mathbb{F}[G]$ and $\alpha \in \mathbb{F}$, and the unit $1_{\mathbb{F}} e$. Write g for $1_{\mathbb{F}} g$. (As a vector space, $\mathbb{F}[G]$ is the direct sum of $|G|$ copies of $\mathbb{F}$.)

Example 13.2. If H, K are groups and $\mathbb{F}$ is a field, then the algebras $\mathbb{F}[H] \otimes \mathbb{F}[K]$ and $\mathbb{F}[H \times K]$ are canonically isomorphic through the map that sends $\sum_{(h,k)\in H\times K} \alpha_{(h,k)}(h,k) \in \mathbb{F}[H\times K]$ to $\sum_{(h,k)\in H\times K} \alpha_{(h,k)} h \otimes 1_{\mathbb{F}} k \in \mathbb{F}[H] \otimes \mathbb{F}[K]$.

13.1.2 Tensor products of C^*-algebras through representations

We return to the analytic setting. Henceforth, we denote the tensor product of two Hilbert spaces $\mathcal{H}, \mathcal{K}$ by $\mathcal{H} \otimes \mathcal{K}$, and recall that this Hilbert space is the completion of the algebraic tensor product $\mathcal{H} \otimes_{\text{alg}} \mathcal{K}$ with respect to a canonical inner product. We also denote the tensor product of two algebras $\mathcal{A}, \mathcal{B}$, as defined earlier, by $\mathcal{A} \otimes_{\text{alg}} \mathcal{B}$. If $\mathcal{A}, \mathcal{B}$ are $*$-algebras, then $\mathcal{A} \otimes_{\text{alg}} \mathcal{B}$ becomes a $*$-algebra with the involution $(a \otimes b)^* := a^* \otimes b^*$.

A norm $\|\cdot\|$ on a $*$-algebra $\mathcal{C}$ is called a *C^*-norm* if it is submultiplicative ($\|xy\| \leq \|x\|\,\|y\|$ for all $x, y \in \mathcal{C}$) and satisfies the C^*-identity ($\|x^*x\| = \|x\|^2$ for all $x \in \mathcal{C}$). A semi-norm with the same additional properties is called a C^*-semi-norm. The completion of $\mathcal{C}$ with respect to a C^*-norm becomes a C^*-algebra upon continuously extending the products and the involution. Such a C^*-algebra is called a *C^*-completion of $\mathcal{C}$*.

Let $\mathcal{A}, \mathcal{B}$ be C^*-algebras. A *C^*-algebraic tensor product of $\mathcal{A}$ and $\mathcal{B}$* is a C^*-completion of $\mathcal{A} \otimes_{\text{alg}} \mathcal{B}$. It is natural to ask whether we should "accept" any C^*-norm on $\mathcal{A} \otimes_{\text{alg}} \mathcal{B}$ or be more restrictive. In particular, it would be desirable to have the additional property

$$\|a \otimes b\| = \|a\|_{\mathcal{A}}\, \|b\|_{\mathcal{B}} \qquad (\forall a \in \mathcal{A}, b \in \mathcal{B}). \tag{1}$$

A norm on $\mathcal{A} \otimes_{\mathrm{alg}} \mathcal{B}$ satisfying (1) is called a *cross norm*. This notion makes sense whenever $\mathcal{A}, \mathcal{B}$ are Banach spaces, and it is indeed taken from the theory of tensor products of Banach spaces, not discussed in this book. Notice that the norm on the tensor product of Hilbert spaces is a cross norm; and moreover, if $\mathcal{H}_1, \mathcal{K}_1, \mathcal{H}_2, \mathcal{K}_2$ are Hilbert spaces and $T \in B(\mathcal{H}_1, \mathcal{H}_2), S \in B(\mathcal{K}_1, \mathcal{K}_2)$, then $\|T \otimes S\| = \|T\| \, \|S\|$ by Part (vi) of Proposition 8.18.

It is a surprising fact that *all* C^*-norms on the algebraic tensor product $\mathcal{A} \otimes_{\mathrm{alg}} \mathcal{B}$ of two C^*-algebras are cross norms! The proof of this in Corollary 13.17 requires several stages. As an appetizer, we show that if $\mathcal{A}, \mathcal{B}$ are *unital* C^*-algebras, then every C^*-norm on $\mathcal{A} \otimes_{\mathrm{alg}} \mathcal{B}$ satisfies $\|a \otimes b\| \leq \|a\|_{\mathcal{A}} \|b\|_{\mathcal{B}}$ for all $a \in \mathcal{A}$ and $b \in \mathcal{B}$. Indeed, the maps

$$a \to a \otimes \mathbb{1}_{\mathcal{B}} \text{ and } b \to \mathbb{1}_{\mathcal{A}} \otimes b \tag{2}$$

are injective $*$-homomorphisms from $\mathcal{A}$ and $\mathcal{B}$, respectively, to the C^*-completion $\mathcal{A} \otimes \mathcal{B}$ of $\mathcal{A} \otimes_{\mathrm{alg}} \mathcal{B}$ associated with the given C^*-norm. These maps are thus isometric by Theorem 11.2. As a result, for all $a \in \mathcal{A}$ and $b \in \mathcal{B}$ we have

$$\|a \otimes b\| = \|(a \otimes \mathbb{1}_{\mathcal{B}})(\mathbb{1}_{\mathcal{A}} \otimes b)\| \leq \|a \otimes \mathbb{1}_{\mathcal{B}}\| \, \|\mathbb{1}_{\mathcal{A}} \otimes b\| = \|a\|_{\mathcal{A}} \|b\|_{\mathcal{B}}.$$

In order to proceed we need to examine the *representations* of $\mathcal{A} \otimes_{\mathrm{alg}} \mathcal{B}$. This occurs in Propositions 13.3, 13.4, and 13.6. By a representation of a $*$-algebra $\mathcal{C}$ on a Hilbert space $\mathcal{H}$ we mean a $*$-homomorphism π from $\mathcal{C}$ to $B(\mathcal{H})$. Such π is called *faithful* if it is injective, and it is called *non-degenerate* if $\pi(\mathcal{C})\mathcal{H} := \operatorname{span}\{\pi(X)\zeta; X \in \mathcal{C}, \zeta \in \mathcal{H}\}$ is dense in $\mathcal{H}$.

Representations matter because they are the source of C^*-norms.

Proposition 13.3. *Let $\mathcal{C}$ be a $*$-algebra. If π is a representation of $\mathcal{C}$ on a Hilbert space $\mathcal{H}$, then the formula*

$$\|X\|_\pi := \|\pi(X)\|_{B(\mathcal{H})} \qquad (X \in \mathcal{C})$$

defines a C^*-semi-norm $\|\cdot\|_\pi$ on $\mathcal{C}$, which is a C^*-norm (i.e., definite) iff π is faithful. Conversely, every C^*-norm on $\mathcal{C}$ equals $\|\cdot\|_\pi$ for some faithful non-degenerate representation π of $\mathcal{C}$.

Proof. The first assertion follows readily from π being a $*$-homomorphism and $B(\mathcal{H})$ being a C^*-algebra.

Conversely, let $\|\cdot\|$ be a C^*-norm on $\mathcal{C}$ and write $\overline{\mathcal{C}}^{\|\cdot\|}$ for the associated C^*-completion. Let π be a faithful (hence isometric, by Theorem 11.2) non-degenerate representation of $\overline{\mathcal{C}}^{\|\cdot\|}$; such π exists by the non-commutative Gelfand–Naimark theorem (11.26). Then $\pi|_{\mathcal{C}}$ is a faithful non-degenerate representation of $\mathcal{C}$, and $\|X\| = \|X\|_{\overline{\mathcal{C}}^{\|\cdot\|}} = \|\pi(X)\|_{B(\mathcal{H})}$ for all $X \in \mathcal{C}$ by isometricity of π. □

The C^*-completion of $\mathcal{C}$ associated with a C^*-norm $\|\cdot\|_\pi$ as shown (when π is faithful) is naturally identified with the (norm-) closure of $\pi(\mathcal{C})$ in $B(\mathcal{H})$.

The next proposition analyzes the general structure of the representations of $\mathcal{A} \otimes_{\mathrm{alg}} \mathcal{B}$.

Proposition 13.4. *Let $\mathcal{A}, \mathcal{B}$ be C^*-algebras.*

(i) *If $\pi_\mathcal{A}, \pi_\mathcal{B}$ are representations with commuting ranges of $\mathcal{A}, \mathcal{B}$, respectively, on the same Hilbert space $\mathcal{H}$, then there exists a unique representation $\pi_\mathcal{A} \cdot \pi_\mathcal{B}$ of $\mathcal{A} \otimes_{\text{alg}} \mathcal{B}$ on $\mathcal{H}$ satisfying*

$$(\pi_\mathcal{A} \cdot \pi_\mathcal{B})(a \otimes b) = \pi_\mathcal{A}(a)\pi_\mathcal{B}(b) \qquad (\forall a \in \mathcal{A}, b \in \mathcal{B}). \tag{3}$$

Also, $\pi_\mathcal{A} \cdot \pi_\mathcal{B}$ is non-degenerate iff both $\pi_\mathcal{A}$ and $\pi_\mathcal{B}$ are non-degenerate.

(ii) *Every non-degenerate representation π of $\mathcal{A} \otimes_{\text{alg}} \mathcal{B}$ is equal to $\pi_\mathcal{A} \cdot \pi_\mathcal{B}$ for unique representations with commuting ranges $\pi_\mathcal{A}, \pi_\mathcal{B}$ of $\mathcal{A}, \mathcal{B}$, respectively.*

The representations $\pi_\mathcal{A}, \pi_\mathcal{B}$ of Part (ii) are called the *restrictions* of π to $\mathcal{A}, \mathcal{B}$, respectively. The reason for this is that when both algebras are unital, $\pi_\mathcal{A}, \pi_\mathcal{B}$ are obtained by composing π with the embeddings (2).

Proof. (i) Since the map $(a, b) \to \pi_\mathcal{A}(a)\pi_\mathcal{B}(b)$ from $\mathcal{A} \times \mathcal{B}$ to $B(\mathcal{H})$ is bilinear, a unique linear map $\pi_\mathcal{A} \cdot \pi_\mathcal{B} : \mathcal{A} \otimes_{\text{alg}} \mathcal{B} \to B(\mathcal{H})$ satisfying (3) exists. It is multiplicative and $*$-preserving because so are $\pi_\mathcal{A}, \pi_\mathcal{B}$ and because their ranges commute.

The assertion on non-degeneracy holds since $(\pi_\mathcal{A} \cdot \pi_\mathcal{B})(\mathcal{A} \otimes_{\text{alg}} \mathcal{B})\mathcal{H} = \pi_\mathcal{A}(\mathcal{A})\,(\pi_\mathcal{B}(\mathcal{B})\mathcal{H}) = \pi_\mathcal{B}(\mathcal{B})\,(\pi_\mathcal{A}(\mathcal{A})\mathcal{H})$.

(ii) Let π be a representation of $\mathcal{A} \otimes_{\text{alg}} \mathcal{B}$ on a Hilbert space $\mathcal{H}$.

It is quite clear how $\pi_\mathcal{A}, \pi_\mathcal{B}$ need be defined: for example, we should have $\pi_\mathcal{A}(a)(\pi(x \otimes y)\zeta) := \pi(ax \otimes y)\zeta$. We will show that this makes sense and gives representations with the desired properties. The trick is to use *positive functionals*.

For each $a \in \mathcal{A}$ define a linear operator $\pi_\mathcal{A}(a)$ on $\pi(\mathcal{A} \otimes_{\text{alg}} \mathcal{B})\mathcal{H}$ by

$$\pi_\mathcal{A}(a) \sum_{i=1}^n \pi(x_i \otimes y_i)\zeta_i := \sum_{i=1}^n \pi(ax_i \otimes y_i)\zeta_i$$
$$(n \in \mathbb{N} \text{ and } x_i \in \mathcal{A}, y_i \in \mathcal{B}, \zeta_i \in \mathcal{H} \text{ for } 1 \le i \le n). \tag{4}$$

To show that these operators are well defined and bounded, fix $n \in \mathbb{N}$ and x_i, y_i, ζ_i $(1 \le i \le n)$ as in (4) and consider the linear functional ω on *the unitization* $\mathcal{A}^\#$ of $\mathcal{A}$ given by

$$\omega(x) := \left(\sum_{i=1}^n \pi(xx_i \otimes y_i)\zeta_i, \sum_{j=1}^n \pi(x_j \otimes y_j)\zeta_j \right) \qquad (x \in \mathcal{A}^\#)$$

(if we already knew that $\pi_\mathcal{A}$ was well defined, we would simply have $\omega(x) = (\pi_\mathcal{A}(x)\xi, \xi)$ for $\xi := \sum_{i=1}^n \pi(x_i \otimes y_i)\zeta_i$). We assert that ω is positive. Indeed, for all $a \in \mathcal{A}^\#$ we have

$$\begin{aligned}
\omega(a^*a) &= \sum_{i,j=1}^{n} (\pi(a^*ax_i \otimes y_i)\zeta_i, \pi(x_j \otimes y_j)\zeta_j) \\
&= \sum_{i,j=1}^{n} (\pi(x_j \otimes y_j)^*\pi(a^*ax_i \otimes y_i)\zeta_i, \zeta_j) \\
&= \sum_{i,j=1}^{n} (\pi(x_j^*a^*ax_i \otimes y_j^*y_i)\zeta_i, \zeta_j) = \sum_{i,j=1}^{n} (\pi(ax_j \otimes y_j)^*\pi(ax_i \otimes y_i)\zeta_i, \zeta_j) \\
&= \left(\sum_{i=1}^{n} \pi(ax_i \otimes y_i)\zeta_i, \sum_{j=1}^{n} \pi(ax_j \otimes y_j)\zeta_j\right) \geq 0.
\end{aligned}$$

As a result, $\|\omega\| = \omega(\mathbb{1}_{\mathcal{A}^\#})$ by Theorem 11.18. Hence, for all $a \in \mathcal{A}^\#$ we have $\omega(a^*a) \leq \|a^*a\|\, \omega(\mathbb{1}_{\mathcal{A}^\#})$, which exactly means that

$$\left\|\sum_{i=1}^{n} \pi(ax_i \otimes y_i)\zeta_i\right\|^2 \leq \|a\|^2 \left\|\sum_{i=1}^{n} \pi(x_i \otimes y_i)\zeta_i\right\|^2.$$

This being true for all $a \in \mathcal{A}^\#$ and all x_i, y_i, ζ_i as above proves that for each $a \in \mathcal{A}$, the formula (4) gives rise to an operator $\pi_{\mathcal{A}}(a)$ on $\pi(\mathcal{A} \otimes_{\text{alg}} \mathcal{B})\mathcal{H}$ of norm at most $\|a\|$. The subspace $\pi(\mathcal{A} \otimes_{\text{alg}} \mathcal{B})\mathcal{H}$ is dense in $\mathcal{H}$ as π is non-degenerate, thus, $\pi_{\mathcal{A}}(a)$ extends to an element of $B(\mathcal{H})$ of the same norm, also denoted $\pi_{\mathcal{A}}(a)$. Straightforward calculations show that $\pi_{\mathcal{A}}$ is a $*$-homomorphism, thus a representation of $\mathcal{A}$ on $\mathcal{H}$.

One does the same for $\pi_{\mathcal{B}}$. Evidently, $\pi_{\mathcal{A}}$ and $\pi_{\mathcal{B}}$ have commuting ranges and $\pi = \pi_{\mathcal{A}} \cdot \pi_{\mathcal{B}}$.

Lastly, suppose that also $\pi = \pi'_{\mathcal{A}} \cdot \pi'_{\mathcal{B}}$ for non-degenerate representations $\pi'_{\mathcal{A}}, \pi'_{\mathcal{B}}$ as in Part (i). Fix an approximate identity $\{e_\alpha\}$ of $\mathcal{A}$ and let $b \in \mathcal{B}$. For every α we have $\pi_{\mathcal{A}}(e_\alpha)\pi_{\mathcal{B}}(b) = \pi(e_\alpha \otimes b) = \pi'_{\mathcal{A}}(e_\alpha)\pi'_{\mathcal{B}}(b)$. By Exercise 16 in Chapter 11, taking the s.o.-limit in α yields that $\pi_{\mathcal{B}}(b) = \pi'_{\mathcal{B}}(b)$. Similarly, $\pi_{\mathcal{A}} = \pi'_{\mathcal{A}}$. □

An immediate result is that every C^*-norm on $\mathcal{A} \otimes_{\text{alg}} \mathcal{B}$ is at least "half" a cross norm:

Corollary 13.5. *Let $\mathcal{A}, \mathcal{B}$ be C^*-algebras.*

(i) Every representation π of $\mathcal{A} \otimes_{\text{alg}} \mathcal{B}$ satisfies

$$\|\pi(a \otimes b)\| \leq \|a\|_{\mathcal{A}} \|b\|_{\mathcal{B}} \qquad (\forall a \in \mathcal{A}, b \in \mathcal{B}).$$

(ii) Every C^*-norm $\|\cdot\|$ on $\mathcal{A} \otimes_{\text{alg}} \mathcal{B}$ satisfies

$$\|a \otimes b\| \leq \|a\|_{\mathcal{A}} \|b\|_{\mathcal{B}} \qquad (\forall a \in \mathcal{A}, b \in \mathcal{B})$$

(i.e., this inequality is satisfied in every C^*-completion of $\mathcal{A} \otimes_{\text{alg}} \mathcal{B}$).

Proof. (i) We can assume that π is non-degenerate, because otherwise it can be replaced by its restriction to a non-degenerate representation on $\overline{\pi(\mathcal{A} \otimes_{\text{alg}} \mathcal{B})\mathcal{H}}$, where $\mathcal{H}$ is the Hilbert space on which π represents.

With $\pi_{\mathcal{A}}, \pi_{\mathcal{B}}$ as in Part (ii) of Proposition 13.4 we get $\|\pi(a \otimes b)\| = \|\pi_{\mathcal{A}}(a)\pi_{\mathcal{B}}(b)\| \leq \|a\|_{\mathcal{A}} \|b\|_{\mathcal{B}}$ because representations of C^*-algebras are contractive.

(ii) This follows from Part (i) and Proposition 13.3. □

We have been discussing C^*-norms on $\mathcal{A} \otimes_{\text{alg}} \mathcal{B}$, but we still have not proved the existence of such norms. By Proposition 13.3, this amounts to exhibiting faithful representations of $\mathcal{A} \otimes_{\text{alg}} \mathcal{B}$. This is what we do next.

Let ρ, σ be representations of the C^*-algebras $\mathcal{A}, \mathcal{B}$ on Hilbert spaces $\mathcal{H}, \mathcal{K}$, respectively. The linear map

$$\rho \otimes \sigma : \mathcal{A} \otimes_{\text{alg}} \mathcal{B} \to B(\mathcal{H}) \otimes_{\text{alg}} B(\mathcal{K})$$

sending $a \otimes b$ to $\rho(a) \otimes \sigma(b)$ for $a \in \mathcal{A}, b \in \mathcal{B}$ (see Part (iv) of Proposition 8.16) is a $*$-homomorphism of $*$-algebras. It is injective if so are ρ and σ by Part (ix) of the same proposition. Furthermore, by Exercise 27 in Chapter 8, we can identify $B(\mathcal{H}) \otimes_{\text{alg}} B(\mathcal{K})$ as a $*$-subalgebra of $B(\mathcal{H} \otimes \mathcal{K})$ by identifying $T \otimes S \in B(\mathcal{H}) \otimes_{\text{alg}} B(\mathcal{K})$ with the element of $B(\mathcal{H} \otimes \mathcal{K})$ having the same notation given by $(T \otimes S)(\zeta \otimes \eta) = T\zeta \otimes S\eta$ for all $\zeta \in \mathcal{H}, \eta \in \mathcal{K}$. Hence, in the sequel we view the $*$-homomorphism $\rho \otimes \sigma$ as a map

$$\rho \otimes \sigma : \mathcal{A} \otimes_{\text{alg}} \mathcal{B} \to B(\mathcal{H} \otimes \mathcal{K}),$$

which is thus a representation of $\mathcal{A} \otimes_{\text{alg}} \mathcal{B}$ on $\mathcal{H} \otimes \mathcal{K}$, faithful if so are ρ and σ. From Proposition 13.3 we infer the following.

Proposition 13.6. *Let ρ, σ be representations of C^*-algebras $\mathcal{A}, \mathcal{B}$ on Hilbert spaces $\mathcal{H}, \mathcal{K}$, respectively. The formula*

$$\|X\|_{\rho\otimes\sigma} := \|(\rho \otimes \sigma)(X)\|_{B(\mathcal{H}\otimes\mathcal{K})} \qquad (X \in \mathcal{A} \otimes_{\text{alg}} \mathcal{B})$$

defines a C^*-semi-norm $\|\cdot\|_{\rho\otimes\sigma}$ on $\mathcal{A} \otimes_{\text{alg}} \mathcal{B}$, which is a C^*-norm if ρ and σ are faithful.

Faithful representations of $\mathcal{A}$ and $\mathcal{B}$ do exist by the non-commutative Gelfand–Naimark theorem. As a result, we can state the following.

Corollary 13.7. *For C^*-algebras $\mathcal{A}$ and $\mathcal{B}$ there is at least one C^*-norm on $\mathcal{A} \otimes_{\text{alg}} \mathcal{B}$.*

We denote by $\mathcal{A} \otimes_{\rho,\sigma} \mathcal{B}$ the C^*-completion of $(\mathcal{A} \otimes_{\text{alg}} \mathcal{B}, \|\cdot\|_{\rho\otimes\sigma})$ for faithful ρ, σ as in Proposition 13.6. Then $\mathcal{A} \otimes_{\rho,\sigma} \mathcal{B}$ is naturally identified with the closure in $B(\mathcal{H} \otimes \mathcal{K})$ of the algebraic tensor product $\rho(\mathcal{A}) \otimes_{\text{alg}} \sigma(\mathcal{B})$.

Proposition 13.6 raises the following question: does the C^*-norm $\|\cdot\|_{\rho\otimes\sigma}$ really depend on ρ and σ? The answer is negative. We prove this in Section 13.1.5.

13.1.3 The maximal tensor product

The results of Section 13.1.2 allow a clear identification of the *largest* C^*-norm on $\mathcal{A} \otimes_{\mathrm{alg}} \mathcal{B}$.

Theorem 13.8. *Let $\mathcal{A}, \mathcal{B}$ be C^*-algebras. The formula*

$$\|X\|_{\max} := \sup \{ \|\pi(X)\| \,; \pi \text{ is a representation of } \mathcal{A} \otimes_{\mathrm{alg}} \mathcal{B}\} \tag{5}$$

for $X \in \mathcal{A} \otimes_{\mathrm{alg}} \mathcal{B}$ defines a C^*-norm $\|\cdot\|_{\max}$ on $\mathcal{A} \otimes_{\mathrm{alg}} \mathcal{B}$, which is larger than every other C^*-norm $\|\cdot\|$ on $\mathcal{A} \otimes_{\mathrm{alg}} \mathcal{B}$: $\|X\| \leq \|X\|_{\max}$ for all $X \in \mathcal{A} \otimes_{\mathrm{alg}} \mathcal{B}$.

Proof. First, for $X \in \mathcal{A} \otimes_{\mathrm{alg}} \mathcal{B}$, the supremum in (5) is finite, because writing $X = \sum_{i=1}^{n} a_i \otimes b_i$, for every representation π of $\mathcal{A} \otimes_{\mathrm{alg}} \mathcal{B}$ we have $\|\pi(X)\| \leq \sum_{i=1}^{n} \|a_i\|_{\mathcal{A}} \|b_i\|_{\mathcal{B}}$ by Corollary 13.5.

The function $\|\cdot\|_{\max}$ dominates every C^*-norm $\|\cdot\|$ on $\mathcal{A} \otimes_{\mathrm{alg}} \mathcal{B}$ by Proposition 13.3. And since there exists at least one C^*-norm on $\mathcal{A} \otimes_{\mathrm{alg}} \mathcal{B}$ by Corollary 13.7, we deduce that $\|X\|_{\max} > 0$ for all $0 \neq X \in \mathcal{A} \otimes_{\mathrm{alg}} \mathcal{B}$.

The rest of the conditions for $\|\cdot\|_{\max}$ to be a submultiplicative norm on $\mathcal{A} \otimes_{\mathrm{alg}} \mathcal{B}$ that satisfies the C^*-identity are easily checked. □

Remark 13.9. In (5) it suffices to take just non-degenerate representations π.

The C^*-norm $\|\cdot\|_{\max}$ is called the *maximal C^*-norm*, and the associated C^*-completion of $\mathcal{A} \otimes_{\mathrm{alg}} \mathcal{B}$ is called the *maximal tensor product of $\mathcal{A}$ and $\mathcal{B}$* and is denoted by $\mathcal{A} \otimes_{\max} \mathcal{B}$.

The maximal tensor product is "universal with respect to the representations of $\mathcal{A} \otimes_{\mathrm{alg}} \mathcal{B}$".

Corollary 13.10. *Let $\mathcal{A}, \mathcal{B}$ be C^*-algebras. Every representation of $\mathcal{A} \otimes_{\mathrm{alg}} \mathcal{B}$ extends to a representation of $\mathcal{A} \otimes_{\max} \mathcal{B}$. Conversely, every representation of $\mathcal{A} \otimes_{\max} \mathcal{B}$ arises this way from a unique representation of $\mathcal{A} \otimes_{\mathrm{alg}} \mathcal{B}$.*

Proof. If π is a representation of $\mathcal{A} \otimes_{\mathrm{alg}} \mathcal{B}$ on a Hilbert space $\mathcal{H}$, then $\|\pi(X)\| \leq \|X\|_{\max}$ by the definition of $\|\cdot\|_{\max}$. In other words, π is contractive as a map from $(\mathcal{A} \otimes_{\mathrm{alg}} \mathcal{B}, \|\cdot\|_{\max})$ to $B(\mathcal{H})$. It thus extends uniquely to a bounded map from $\mathcal{A} \otimes_{\max} \mathcal{B}$ to $B(\mathcal{H})$, which is a representation of $\mathcal{A} \otimes_{\max} \mathcal{B}$. The converse statement is obvious. □

13.1.4 Tensor products of bounded linear functionals

We prove several technical results on the existence of "tensor products" of bounded linear functionals, which are used in Section 13.1.5.

The reader should not be confused by $\omega \otimes \tau$ being given several meanings depending on the context: an element of $B(\mathcal{H} \otimes \mathcal{K})_*$ (Lemma 13.11), a bounded linear functional on the (norm-) closure of $B(\mathcal{H}) \otimes_{\mathrm{alg}} B(\mathcal{K}) \subset B(\mathcal{H} \otimes \mathcal{K})$ (Lemma 13.12) and on $\mathcal{A} \otimes_{\rho,\sigma} \mathcal{B}$ (Proposition 13.13), and of course a functional on various algebraic tensor products. These are related to one another in obvious ways. For instance, if $\omega \in B(\mathcal{H})_*$ and $\tau \in B(\mathcal{K})_*$, then the functional $\omega \otimes \tau$ of

Lemma 13.11 extends the functional $\omega \bar{\otimes} \tau$ of Lemma 13.12, which in turn extends the functional $\omega \otimes \tau$ on $B(\mathcal{H}) \otimes_{\mathrm{alg}} B(\mathcal{K})$.

Lemma 13.11. *Let $\mathcal{H}, \mathcal{K}$ be Hilbert spaces. Let $\omega \in B(\mathcal{H})_*$ and $\tau \in B(\mathcal{K})_*$. There exists a unique linear functional $\omega \otimes \tau \in B(\mathcal{H} \otimes \mathcal{K})_*$ satisfying*

$$(\omega \otimes \tau)(a \otimes b) = \omega(a)\tau(b) \qquad (\forall a \in B(\mathcal{H}), b \in B(\mathcal{K})). \tag{6}$$

Additionally, $\|\omega \otimes \tau\| = \|\omega\| \, \|\tau\|$, and $\omega \otimes \tau$ is positive if ω, τ are positive.

Proof. By Corollary 12.27, we can write $\omega = \sum_{n=1}^{\infty} \omega_{\zeta_n, \eta_n}$ and $\tau = \sum_{m=1}^{\infty} \omega_{\zeta'_m, \eta'_m}$ (with convergence in the norm topology of $B(\mathcal{H})^*$ and $B(\mathcal{K})^*$, respectively) for sequences $\{\zeta_n\}_{n=1}^{\infty}$ and $\{\eta_n\}_{n=1}^{\infty}$ in $\mathcal{H}$ and $\{\zeta'_m\}_{m=1}^{\infty}$ and $\{\eta'_m\}_{m=1}^{\infty}$ in $\mathcal{K}$ satisfying

$$\sum_{n=1}^{\infty} \|\zeta_n\|^2, \sum_{n=1}^{\infty} \|\eta_n\|^2, \sum_{m=1}^{\infty} \|\zeta'_m\|^2, \sum_{m=1}^{\infty} \|\eta'_m\|^2 < \infty$$

and

$$\|\omega\| = \sum_{n=1}^{\infty} \|\zeta_n\| \, \|\eta_n\|, \quad \|\tau\| = \sum_{n=1}^{\infty} \|\zeta'_n\| \, \|\eta'_n\|,$$

and moreover, if ω and τ are positive, we can arrange that $\zeta_n = \eta_n$ and $\zeta'_m = \eta'_m$ for all $n, m \in \mathbb{N}$. For $n, m \in \mathbb{N}$ consider the vectors $\zeta_n \otimes \zeta'_m$ and $\eta_n \otimes \eta'_m$ in $\mathcal{H} \otimes \mathcal{K}$. We have

$$\sum_{n,m=1}^{\infty} \|\zeta_n \otimes \zeta'_m\|^2 = \sum_{n,m=1}^{\infty} \|\zeta_n\|^2 \, \|\zeta'_m\|^2 = \left(\sum_{n=1}^{\infty} \|\zeta_n\|^2 \right) \left(\sum_{m=1}^{\infty} \|\zeta'_m\|^2 \right) < \infty,$$

and similarly $\sum_{n,m=1}^{\infty} \|\eta_n \otimes \eta'_m\|^2 < \infty$. As a result, the series $\sum_{n,m=1}^{\infty} \omega_{\zeta_n \otimes \zeta'_m, \eta_n \otimes \eta'_m}$ converges (absolutely) in $B(\mathcal{H} \otimes \mathcal{K})^*$ to an element of $\omega \otimes \tau$ of $B(\mathcal{H} \otimes \mathcal{K})_*$. For $a \in B(\mathcal{H})$ and $b \in B(\mathcal{K})$ we have

$$\begin{aligned}
(\omega \otimes \tau)(a \otimes b) &= \sum_{n,m=1}^{\infty} ((a \otimes b)(\zeta_n \otimes \zeta'_m), \eta_n \otimes \eta'_m) \\
&= \sum_{n,m=1}^{\infty} (a\zeta_n \otimes b\zeta'_m, \eta_n \otimes \eta'_m) \\
&= \sum_{n,m=1}^{\infty} (a\zeta_n, \eta_n) \, (b\zeta'_m, \eta'_m) \\
&= \left(\sum_{n=1}^{\infty} \omega_{\zeta_n, \eta_n}(a) \right) \left(\sum_{m=1}^{\infty} \omega_{\zeta'_m, \eta'_m}(b) \right) = \omega(a)\tau(b).
\end{aligned}$$

Also

$$\begin{aligned}\|\omega \otimes \tau\| &\le \sum_{n,m=1}^{\infty} \|\omega_{\zeta_n\otimes\zeta'_m,\eta_n\otimes\eta'_m}\| = \sum_{n,m=1}^{\infty} \|\zeta_n \otimes \zeta'_m\| \, \|\eta_n \otimes \eta'_m\| \\ &= \sum_{n,m=1}^{\infty} \|\zeta_n\| \, \|\zeta'_m\| \, \|\eta_n\| \, \|\eta'_m\| \\ &= \left(\sum_{n=1}^{\infty} \|\zeta_n\| \, \|\eta_n\|\right)\left(\sum_{m=1}^{\infty} \|\zeta'_m\| \, \|\eta'_m\|\right) = \|\omega\| \, \|\tau\| \,.\end{aligned}$$

On the other hand, it follows from (6) that $\|\omega\| \, \|\tau\| \le \|\omega \otimes \tau\|$, because if $a \in B(\mathcal{H})$ and $b \in B(\mathcal{K})$ are of norm 1, then so is $a \otimes b \in B(\mathcal{H} \otimes \mathcal{K})$.

If $\zeta_n = \eta_n$ and $\zeta'_m = \eta'_m$ for all $n, m \in \mathbb{N}$, then $\zeta_n \otimes \zeta'_m = \eta_n \otimes \eta'_m$ for all $n, m \in \mathbb{N}$, and hence, $\omega \otimes \tau$ is positive.

The functional $\omega \otimes \tau$ is unique due to (6) because it is ultraweakly continuous and $B(\mathcal{H}) \otimes_{\mathrm{alg}} B(\mathcal{K}) \subset B(\mathcal{H} \otimes \mathcal{K})$ is ultraweakly dense in $B(\mathcal{H} \otimes \mathcal{K})$ by Exercise 1. □

Lemma 13.12. *Let $\mathcal{H}, \mathcal{K}$ be Hilbert spaces. Let $\omega \in B(\mathcal{H})^*$ and $\tau \in B(\mathcal{K})^*$. There exists a unique bounded linear functional $\omega \otimes \tau$ on the (norm-) closure $\mathcal{C}$ of $B(\mathcal{H}) \otimes_{\mathrm{alg}} B(\mathcal{K}) \subset B(\mathcal{H} \otimes \mathcal{K})$ in $B(\mathcal{H} \otimes \mathcal{K})$ satisfying*

$$(\omega \otimes \tau)(a \otimes b) = \omega(a)\tau(b) \qquad (\forall a \in B(\mathcal{H}), b \in B(\mathcal{K})). \tag{7}$$

Additionally, $\|\omega \otimes \tau\| = \|\omega\| \, \|\tau\|$, and $\omega \otimes \tau$ is positive if ω, τ are positive.

Proof. By Goldstine's theorem (5.25), there are nets $\{\omega_\alpha\}_{\alpha \in A}$ in $B(\mathcal{H})_*$ and $\{\tau_\beta\}_{\beta \in B}$ in $B(\mathcal{K})_*$, bounded by $\|\omega\|$ and $\|\tau\|$, and *weak**-converging to ω and τ, respectively. Moreover, if ω and τ are positive, we can choose these nets to consist of positive functionals by Exercise 14 in Chapter 12. Consider the net $\{\omega_\alpha \otimes \tau_\beta\}_{(\alpha,\beta) \in A \times B}$ in $B(\mathcal{H} \otimes \mathcal{K})_*$ defined by Lemma 13.11. It is bounded by $\|\omega\| \, \|\tau\|$, and thus has a *weak**-cluster point φ in $B(\mathcal{H} \otimes \mathcal{K})^*$. If ω and τ are positive, so is φ.

Let $\omega \otimes \tau := \varphi|_{\mathcal{C}}$. It is evident that (7) holds. Also $\|\omega \otimes \tau\| \le \|\varphi\| \le \|\omega\| \, \|\tau\|$. On the other hand, it follows from (7) that $\|\omega\| \, \|\tau\| \le \|\omega \otimes \tau\|$.

Finally, uniqueness of $\omega \otimes \tau$ is a consequence of (7), the continuity of $\omega \otimes \tau$ and $B(\mathcal{H}) \otimes_{\mathrm{alg}} B(\mathcal{K}) \subset B(\mathcal{H} \otimes \mathcal{K})$ being (norm-) dense in $\mathcal{C}$. □

We are ready to prove the existence of certain bounded linear functionals on $\mathcal{A} \otimes_{\rho,\sigma} \mathcal{B}$.

Proposition 13.13. *Let $\mathcal{A}, \mathcal{B}$ be C^*-algebras, ρ, σ be faithful representations of $\mathcal{A}, \mathcal{B}$ on Hilbert spaces $\mathcal{H}, \mathcal{K}$, respectively, and $\omega \in \mathcal{A}^*, \tau \in \mathcal{B}^*$. Then there exists a (unique) functional $\omega \otimes \tau \in (\mathcal{A} \otimes_{\rho,\sigma} \mathcal{B})^*$ satisfying*

$$(\omega \otimes \tau)(a \otimes b) = \omega(a)\tau(b) \qquad (\forall a \in \mathcal{A}, b \in \mathcal{B}). \tag{8}$$

Furthermore, $\|\omega \otimes \tau\| = \|\omega\| \, \|\tau\|$, *and* $\omega \otimes \tau$ *is positive if* ω, τ *are positive.*

Proof. Identify $\mathcal{A} \otimes_{\rho,\sigma} \mathcal{B}$ with the closure of $(\rho \otimes \sigma)(\mathcal{A} \otimes_{\text{alg}} \mathcal{B}) = \rho(\mathcal{A}) \otimes_{\text{alg}} \sigma(\mathcal{B})$ in $B(\mathcal{H} \otimes \mathcal{K})$ and write $\mathcal{C}$ for the closure of $B(\mathcal{H}) \otimes_{\text{alg}} B(\mathcal{K})$ in $B(\mathcal{H} \otimes \mathcal{K})$. Evidently, $\mathcal{A} \otimes_{\rho,\sigma} \mathcal{B} \subset \mathcal{C}$.

Use the Hahn–Banach theorem to extend $\omega \circ \rho^{-1} \in \rho(\mathcal{A})^*$ and $\tau \circ \sigma^{-1} \in \sigma(\mathcal{B})^*$ to elements $\tilde{\omega} \in B(\mathcal{H})^*$ and $\tilde{\tau} \in B(\mathcal{K})^*$ with the same norms as ω and τ, respectively, and being positive if ω and τ are positive (the last part is by virtue of Part (vi) of Theorem 11.20).

Let $\omega \otimes \tau$ be the restriction of $\tilde{\omega} \otimes \tilde{\tau} \in \mathcal{C}^*$ of Lemma 13.12 to $\mathcal{A} \otimes_{\rho,\sigma} \mathcal{B}$. Then $\omega \otimes \tau \in (\mathcal{A} \otimes_{\rho,\sigma} \mathcal{B})^*$, $\|\omega \otimes \tau\| \leq \|\omega\| \, \|\tau\|$, and $\omega \otimes \tau$ is positive if ω, τ are positive. As before, it follows from (8) that $\|\omega\| \, \|\tau\| \leq \|\omega \otimes \tau\|$, thus $\|\omega \otimes \tau\| = \|\omega\| \, \|\tau\|$. Uniqueness is clear. □

13.1.5 The minimal tensor product

We discussed in Proposition 13.6 the C^*-norm $\|\cdot\|_{\rho \otimes \sigma}$ on $\mathcal{A} \otimes_{\text{alg}} \mathcal{B}$ constructed from faithful representations ρ, σ of $\mathcal{A}, \mathcal{B}$, respectively. If ω, τ are states of $\mathcal{A}, \mathcal{B}$, respectively, then the functional $\omega \otimes \tau$ constructed in Proposition 13.13 is a state of $\mathcal{A} \otimes_{\rho,\sigma} \mathcal{B}$. Theorem 13.14 shows that states of this form are enough for the computation of norms in $\mathcal{A} \otimes_{\rho,\sigma} \mathcal{B}$, leading to the conclusion that $\mathcal{A} \otimes_{\rho,\sigma} \mathcal{B}$ *is independent of the faithful representations* ρ, σ.

Theorem 13.14. *Let* $\mathcal{A}, \mathcal{B}$ *be* C^**-algebras.*

(i) Let ρ, σ *be faithful representations of* $\mathcal{A}, \mathcal{B}$ *on Hilbert spaces* $\mathcal{H}, \mathcal{K}$, *respectively. Then for each* $X \in \mathcal{A} \otimes_{\rho,\sigma} \mathcal{B}$ *we have*

$$\|X\|^2_{\rho \otimes \sigma} = \sup \left\{ \frac{(\omega \otimes \tau)(Y^* X^* X Y)}{(\omega \otimes \tau)(Y^* Y)} ; \omega \in \mathcal{S}(\mathcal{A}), \tau \in \mathcal{S}(\mathcal{B}), \right.$$

$$\left. Y \in \mathcal{A} \otimes_{\text{alg}} \mathcal{B}, (\omega \otimes \tau)(Y^* Y) > 0 \right\}. \quad (9)$$

(ii) The C^**-norms* $\|\cdot\|_{\rho \otimes \sigma}$ *on* $\mathcal{A} \otimes_{\text{alg}} \mathcal{B}$ *are independent of the faithful representations* ρ, σ. *In other words, the tensor product* C^**-algebras* $\mathcal{A} \otimes_{\rho,\sigma} \mathcal{B}$ *are independent of* ρ, σ.

Proof. (i) Let $X \in \mathcal{A} \otimes_{\rho,\sigma} \mathcal{B}$. For every $Y \in \mathcal{A} \otimes_{\rho,\sigma} \mathcal{B}$ we have $Y^* X^* X Y \leq \|X\|^2_{\rho \otimes \sigma} Y^* Y$, so by positivity of $\omega \otimes \tau \in (\mathcal{A} \otimes_{\rho,\sigma} \mathcal{B})^*$ we get $(\omega \otimes \tau)(Y^* X^* X Y) \leq \|X\|^2_{\rho \otimes \sigma} (\omega \otimes \tau)(Y^* Y)$. This proves that $\|X\|^2_{\rho \otimes \sigma} \geq \sup\{\cdots\}$ in (9).

To show the converse, we can and will assume that ρ and σ are non-degenerate. Recall from Exercise 23 in Chapter 11 that every non-degenerate representation of a C^*-algebra is the direct sum of cyclic ones (up to unitary equivalence). So, let $\{\rho_\alpha\}_{\alpha \in I}$ and $\{\sigma_\beta\}_{\beta \in J}$ be families of cyclic (not necessarily faithful) representations of $\mathcal{A}$ and $\mathcal{B}$, respectively, such that $\rho = \sum_{\alpha \in I} \oplus \rho_\alpha$ and

$\sigma = \sum_{\beta \in J} \oplus \sigma_\beta$. Hence the representation $\rho \otimes \sigma$ of $\mathcal{A} \otimes_{\text{alg}} \mathcal{B}$ is the direct sum of the representations $\{\rho_\alpha \otimes \sigma_\beta\}_{(\alpha,\beta) \in I \times J}$. Consequently

$$\|X\|^2_{\rho \otimes \sigma} = \sup_{(\alpha,\beta) \in I \times J} \|X\|^2_{\rho_\alpha \otimes \sigma_\beta}. \tag{10}$$

Let $(\alpha, \beta) \in I \times J$. The representations ρ_α and σ_β of $\mathcal{A}$ and $\mathcal{B}$ on Hilbert spaces $\mathcal{H}_\alpha \subset \mathcal{H}$ and $\mathcal{K}_\beta \subset \mathcal{K}$ have cyclic unit vectors ζ_α and η_β, thus the subspaces $\rho_\alpha(\mathcal{A})\zeta_\alpha = \rho(\mathcal{A})\zeta_\alpha$ and $\sigma_\beta(\mathcal{B})\eta_\beta = \sigma(\mathcal{B})\eta_\beta$ are dense in $\mathcal{H}_\alpha$ and $\mathcal{K}_\beta$, respectively. Consider the states $\omega_\alpha := \omega_{\zeta_\alpha} \circ \rho$ and $\tau_\beta := \omega_{\eta_\beta} \circ \sigma$ of $\mathcal{A}$ and $\mathcal{B}$, respectively. Then $\omega_\alpha \otimes \tau_\beta = \omega_{\zeta_\alpha \otimes \eta_\beta} \circ (\rho \otimes \sigma)$ on $\mathcal{A} \otimes_{\text{alg}} \mathcal{B}$. Hence

$$(\omega_\alpha \otimes \tau_\beta)(Z^*Z) = \|((\rho \otimes \sigma)(Z))\,(\zeta_\alpha \otimes \eta_\beta)\|^2 \qquad (\forall Z \in \mathcal{A} \otimes_{\text{alg}} \mathcal{B}). \tag{11}$$

Notice that

$$\{((\rho \otimes \sigma)(Y))\,(\zeta_\alpha \otimes \eta_\beta); Y \in \mathcal{A} \otimes_{\text{alg}} \mathcal{B}\} \\ = \rho(\mathcal{A})\zeta_\alpha \otimes_{\text{alg}} \sigma(\mathcal{B})\eta_\beta \text{ is dense in } \mathcal{H}_\alpha \otimes \mathcal{K}_\beta. \tag{12}$$

Thus, by the definition of the operator norm,

$$\|X\|^2_{\rho_\alpha \otimes \sigma_\beta} = \|(\rho_\alpha \otimes \sigma_\beta)(X)\|^2 = \sup\left\{\frac{\|((\rho \otimes \sigma)(X))\,\xi\|^2}{\|\xi\|^2}; 0 \neq \xi \in \mathcal{H}_\alpha \otimes \mathcal{K}_\beta\right\}$$

$$\overset{\text{by (12)}}{=} \sup\left\{\frac{\|((\rho \otimes \sigma)(X))\,((\rho \otimes \sigma)(Y))\,(\zeta_\alpha \otimes \eta_\beta)\|^2}{\|((\rho \otimes \sigma)(Y))\,(\zeta_\alpha \otimes \eta_\beta)\|^2};\right.$$

$$\left. Y \in \mathcal{A} \otimes_{\text{alg}} \mathcal{B}, ((\rho \otimes \sigma)(Y))\,(\zeta_\alpha \otimes \eta_\beta) \neq 0\right\}$$

$$\overset{\text{by (11)}}{=} \sup\left\{\frac{(\omega_\alpha \otimes \tau_\beta)(Y^*X^*XY)}{(\omega_\alpha \otimes \tau_\beta)(Y^*Y)}; Y \in \mathcal{A} \otimes_{\text{alg}} \mathcal{B}, (\omega_\alpha \otimes \tau_\beta)(Y^*Y) > 0\right\}.$$

In combination with (10), this shows that $\|X\|^2_{\rho \otimes \sigma} \leq \sup\{\cdots\}$ in (9).

(ii) For $X \in \mathcal{A} \otimes_{\text{alg}} \mathcal{B}$, the right-hand side of (9), which equals $\|X\|^2_{\rho \otimes \sigma}$ by Part (i), is independent of the faithful representations ρ, σ. □

The C^*-norm discussed in Theorem 13.14 is called the *minimal* or the *spatial C^*-norm of $\mathcal{A} \otimes_{\text{alg}} \mathcal{B}$* and it is denoted by $\|\cdot\|_{\min}$. The associated C^*-completion of $\mathcal{A} \otimes_{\text{alg}} \mathcal{B}$ is called the *minimal* or the *spatial tensor product of $\mathcal{A}$ and $\mathcal{B}$* and is denoted by $\mathcal{A} \otimes_{\min} \mathcal{B}$. The word "spatial" comes from (Hilbert) space: indeed, this C^*-norm is obtained by embedding $\mathcal{A}$ and $\mathcal{B}$ in $B(\mathcal{H})$ and $B(\mathcal{K})$ for Hilbert spaces $\mathcal{H}$ and $\mathcal{K}$, respectively (in any way we choose!), and then naturally embedding $\mathcal{A} \otimes_{\text{alg}} \mathcal{B}$ in $B(\mathcal{H} \otimes \mathcal{K})$ from which we take the norm.

Corollary 13.15. *If ρ, σ are representations of C^*-algebras $\mathcal{A}, \mathcal{B}$ on Hilbert spaces $\mathcal{H}, \mathcal{K}$, then $\rho \otimes \sigma : \mathcal{A} \otimes_{\text{alg}} \mathcal{B} \to B(\mathcal{H} \otimes \mathcal{K})$ extends to a $*$-homomorphism* *from $\mathcal{A} \otimes_{\min} \mathcal{B}$ to $B(\mathcal{H} \otimes \mathcal{K})$, also denoted by $\rho \otimes \sigma$, which is faithful if ρ, σ are faithful.*

Proof. This is clear when ρ, σ are faithful. The non-faithful case is left to the reader. □

The reason for the adjective "minimal" is the following fundamental theorem, which we will not prove.

Theorem 13.16 (Takesaki's theorem). *Let $\mathcal{A}, \mathcal{B}$ be C^*-algebras.*

(i) The minimal C^*-norm on $\mathcal{A} \otimes_{\text{alg}} \mathcal{B}$ is the smallest of all C^*-norms on $\mathcal{A} \otimes_{\text{alg}} \mathcal{B}$.

(ii) If either $\mathcal{A}$ or $\mathcal{B}$ is commutative, then there is only one C^*-norm on $\mathcal{A} \otimes_{\text{alg}} \mathcal{B}$.

Corollary 13.17. *Let $\mathcal{A}, \mathcal{B}$ be C^*-algebras. Every C^*-norm on $\mathcal{A} \otimes_{\text{alg}} \mathcal{B}$ is a cross norm.*

Proof. One inequality comes from Corollary 13.5 (which was used to define the maximal tensor product). The other comes from Part (i) of Theorem 13.16, because the minimal C^*-norm is a cross norm. □

Theorem 13.8 and Part (i) of Theorem 13.16 imply that every C^*-norm $\|\cdot\|$ on $\mathcal{A} \otimes_{\text{alg}} \mathcal{B}$ lies between the minimal and the maximal ones: $\|\cdot\|_{\min} \leq \|\cdot\| \leq \|\cdot\|_{\max}$.

Observe that if $\mathcal{C}$ is a $*$-algebra with C^*-norms $\|\cdot\|_1, \|\cdot\|_2$ and associated C^*-completions $\mathcal{C}_1, \mathcal{C}_2$, and if $\|\cdot\|_1 \leq \|\cdot\|_2$, then the identity map of $\mathcal{C}$ extends to a (contractive) $*$-homomorphism from $\mathcal{C}_2$ *onto* $\mathcal{C}_1$, for the image of a $*$-homomorphism between two C^*-algebras is always a C^*-algebra. As a result, we have a canonical surjective $*$-homomorphism $\mathcal{A} \otimes_{\max} \mathcal{B} \to \mathcal{A} \otimes_{\min} \mathcal{B}$, and for every C^*-completion $\mathcal{A} \otimes \mathcal{B}$ of $\mathcal{A} \otimes_{\text{alg}} \mathcal{B}$ there are canonical surjective $*$-homomorphism $\mathcal{A} \otimes_{\max} \mathcal{B} \to \mathcal{A} \otimes \mathcal{B}$ and $\mathcal{A} \otimes \mathcal{B} \to \mathcal{A} \otimes_{\min} \mathcal{B}$ such that the following diagram commutes:

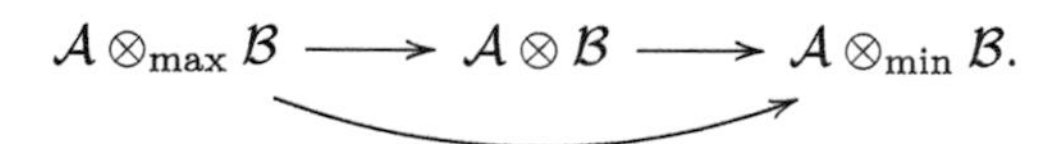

From this and Corollary 13.15 we infer the following.

Corollary 13.18. *Let $\mathcal{A}, \mathcal{B}$ be C^*-algebras and $\mathcal{A} \otimes \mathcal{B}$ be a C^*-completion of $\mathcal{A} \otimes_{\text{alg}} \mathcal{B}$. If ρ, σ are representations of $\mathcal{A}, \mathcal{B}$ on Hilbert spaces $\mathcal{H}, \mathcal{K}$, then $\rho \otimes \sigma : \mathcal{A} \otimes_{\text{alg}} \mathcal{B} \to B(\mathcal{H} \otimes \mathcal{K})$ extends to a $*$-homomorphism from $\mathcal{A} \otimes \mathcal{B}$ to $B(\mathcal{H} \otimes \mathcal{K})$.*

A C^*-algebra $\mathcal{A}$ such that for *every* C^*-algebra $\mathcal{B}$ there is a unique C^*-norm on $\mathcal{A} \otimes_{\text{alg}} \mathcal{B}$ is said to be *nuclear*. Part (ii) of Theorem 13.16 says that every commutative C^*-algebra is nuclear; Section 13.1.6 provides an explicit realization of the tensor product in this case.

13.1.6 Tensor products by commutative C^*-algebras

Let Ω be a locally compact Hausdorff space and $\mathcal{A}$ be a C^*-algebra. Let

$$C_0(\Omega, \mathcal{A}) := \{F : \Omega \to \mathcal{A}; F \text{ is continuous and vanishes at infinity}\}$$

(the vanishing at infinity condition means that for every $\epsilon > 0$ there is a compact $K \subset \Omega$ such that $\|F(x)\| < \epsilon$ for all $x \in \Omega \backslash K$). Then $C_0(\Omega, \mathcal{A})$ is a C^*-algebra with respect to the pointwise operations and the norm

$$\|F\| := \max_{x \in \Omega} \|F(x)\|_{\mathcal{A}} \qquad (F \in C_0(\Omega, \mathcal{A})).$$

This section shows that the unique C^*-algebraic tensor product of $C_0(\Omega)$ and $\mathcal{A}$ is naturally isomorphic to $C_0(\Omega, \mathcal{A})$ (cf. Exercise 29 in Chapter 8).

For $f \in C_0(\Omega)$ and $a \in \mathcal{A}$ let $fa : \Omega \to \mathcal{A}$ be given by $x \to f(x)a$ $(x \in \Omega)$. Evidently $fa \in C_0(\Omega, \mathcal{A})$. The map from $C_0(\Omega) \times \mathcal{A}$ to $C_0(\Omega, \mathcal{A})$ given by $[f, a] \to fa$ is bilinear, so there is a (unique) linear map $\Psi : C_0(\Omega) \otimes_{\text{alg}} \mathcal{A} \to C_0(\Omega, \mathcal{A})$ mapping $f \otimes a$ to fa for all $f \in C_0(\Omega)$ and $a \in \mathcal{A}$ (see Exercise 22 in Chapter 8). In fact, Ψ is a $*$-homomorphism. It is injective, because if $\sum_{i=1}^n f_i \otimes a_i \in C_0(\Omega) \otimes_{\text{alg}} \mathcal{A}$ and $\sum_{i=1}^n f_i a_i = 0$ in $C_0(\Omega, \mathcal{A})$ then assuming, as we may, that $a_1, \ldots, a_n$ are linearly independent, we get $f_i(x) = 0$ for all $1 \leq i \leq n$ and $x \in \Omega$, hence surely $\sum_{i=1}^n f_i \otimes a_i = 0$.

Let us show that Ψ has dense range. Fix $F \in C_0(\Omega, \mathcal{A})$. Given $\epsilon > 0$, let $K \subset \Omega$ be a compact set such that $\|F(x)\| < \epsilon$ for each $x \in \Omega \backslash K$. For every $x \in K$ let $U_x \subset \Omega$ be an open neighborhood of x such that $\|F(y) - F(x)\| < \epsilon$ for all $y \in U_x$. By compactness of K there are $x_1, \ldots, x_n \in K$ such that $V_1 := U_{x_1}, \ldots, V_n := U_{x_n}$ is a finite open cover of K. By Theorem 3.3 there is an associated partition of unity, namely, $h_1, \ldots, h_n \in C_c(\Omega)$ with values in $[0, 1]$ and supports contained in $V_1, \ldots, V_n$, respectively, such that $h_1 + \ldots + h_n = 1$ on K and $0 \leq h_1 + \ldots + h_n \leq 1$ everywhere. Consider the function $G := \Psi(\sum_{i=1}^n h_i \otimes F(x_i)) = \sum_{i=1}^n h_i F(x_i)$. We will show that $\|F - G\| \leq 3\epsilon$.

Indeed, let $x \in \Omega$ and $I := \{1 \leq i \leq n; x \in V_i\}$. Then for each $i \in I$ we have $\|F(x) - F(x_i)\| < \epsilon$ and for each $i \notin I$ we have $h_i(x) = 0$. If $x \in K$, then $\sum_{i \in I} h_i(x) = \sum_{i=1}^n h_i(x) = 1$, thus

$$\begin{aligned} \|F(x) - G(x)\| &= \left\| \sum_{i \in I} h_i(x)\,(F(x) - F(x_i)) \right\| \leq \sum_{i \in I} h_i(x)\,\|F(x) - F(x_i)\| \\ &< \epsilon \sum_{i \in I} h_i(x) = \epsilon. \end{aligned}$$

If $x \in \Omega \backslash K$, then $\|F(x)\| < \epsilon$, and either $I = \emptyset$ and hence $G(x) = 0$, or $I \neq \emptyset$, in which case for each $i \in I$ we have $\|F(x_i)\| \leq \|F(x)\| + \epsilon < 2\epsilon$, and therefore,

$$\|F(x) - G(x)\| \leq \|F(x)\| + \|G(x)\| < \epsilon + \sum_{i \in I} h_i(x)\,\|F(x_i)\| \leq 3\epsilon.$$

The foregoing proves that $\|F - G\| \leq 3\epsilon$, as desired.

In conclusion, the formula

$$\|C\| := \|\Psi(C)\|_{C_0(\Omega,\mathcal{A})} \qquad (C \in C_0(\Omega) \otimes_{\mathrm{alg}} \mathcal{A})$$

defines on $C_0(\Omega) \otimes_{\mathrm{alg}} \mathcal{A}$ a C^*-norm, whose associated C^*-completion can be identified with $C_0(\Omega, \mathcal{A})$. By Theorem 13.16, this is the *unique* C^*-norm on $C_0(\Omega) \otimes_{\mathrm{alg}} \mathcal{A}$. As a result, *both* $C_0(\Omega) \otimes_{\max} \mathcal{A}$ *and* $C_0(\Omega) \otimes_{\min} \mathcal{A}$ *are canonically* $*$*-isomorphic to* $C_0(\Omega, \mathcal{A})$.

We give another proof of the isomorphism between $C_0(\Omega) \otimes_{\min} \mathcal{A}$ and $C_0(\Omega, \mathcal{A})$ in Exercise 6.

13.2 Group C^*-algebras

Throughout this section let G be a locally compact topological group with identity e. We construct a C^*-algebra $C^*(G)$ whose structure reflects not only the topology on G but also its algebraic structure. This is accomplished similarly to the construction of the *maximal* tensor product. We begin with $C_c(G)$ endowed with a suitable $*$-algebra structure in a way that its $L^1(G)$-contractive representations are in bijection with the unitary representations of G. Such representations are used to produce a C^*-norm on $C_c(G)$ whose completion $C^*(G)$ has a universal property.

Denote by μ a left Haar measure on G and by $\delta : G \to (0, \infty)$ the modular function of G (see Section 4.5), and write $L^p(G)$ for $L^p(G, \mu)$.

For $f : G \to \mathbb{C}$ and $t \in G$, the *left* t*-translate of* f is $f_t : G \to \mathbb{C}$ defined by $f_t(s) := f(ts)$ $(s \in G)$. By the definition of μ, if $f \in L^1(G)$ then $f_t \in L^1(G)$ and $\int_G f\, d\mu = \int_G f_t\, d\mu$ for all $t \in G$. It follows that for $p \in [1, \infty)$ and $t \in G$, the map $f \to f_t$ is a linear isometry of $L^p(G)$ onto itself, and in the particular case when $p = 2$, this map is unitary.

Also, for all $f \in L^1(G)$, the function $G \to \mathbb{C}$ given by $t \to \delta(t^{-1})f(t^{-1})$ belongs to $L^1(G)$ and its integral equals $\int_G f\, d\mu$.

Moreover, if $p \in [1, \infty)$ and $f \in L^p(G)$, then the function $G \to L^p(G)$ given by $t \to f_t$ is continuous. This is proved precisely like Exercise 1 in Chapter 3 using the left translation invariance of μ.

For $f, g \in C_c(G)$ define functions $f * g$ (the *convolution of* f *and* g) and f^* in $C_c(G)$ by

$$\left.\begin{aligned} (f * g)(t) &:= \int_G f(ts)g(s^{-1})\, d\mu(s) = \int_G f(s)g(s^{-1}t)\, d\mu(s), \\ f^*(t) &:= \delta(t)^{-1}\overline{f(t^{-1})} \end{aligned}\right. \qquad (t \in G).$$

The reader can verify that these operations turn $C_c(G)$ into a $*$-algebra.

Example 13.19. Assume that G is discrete. Then the complex group algebra $\mathbb{C}[G]$ (see Section 13.1.1) is isomorphic to $C_c(G)$ as algebras when for $g \in G$, the element $g \in \mathbb{C}[G]$ is identified with the indicator function of $\{g\}$. Under this identification we have $g^* = g^{-1}$ for $g \in G$ ($\delta(\cdot) = 1$ as G is discrete).

Denote by $\mathcal{V}$ the directed set of all open neighborhoods of e ordered by reverse inclusion. For $V \in \mathcal{V}$ choose a function $f_V \in C_c^+(G)$ that is supported in V and such that $\int_G f_V d\mu = 1$. We obtain a net $\{f_V\}_{V \in \mathcal{V}}$. Such a net is called an *approximate identity for* $C_c(G)$, because for each $g \in C_c(G)$ we have

$$(f_V * g - g)(t) = \int_G f_V(s)\left(g(s^{-1}t) - g(t)\right) d\mu(s) \qquad (\forall t \in G),$$

so Fubini's theorem implies that

$$\begin{aligned} \|f_V * g - g\|_{L^1(G)} &\leq \int_G \int_G \left|f_V(s)\left(g(s^{-1}t) - g(t)\right)\right| d\mu(s) d\mu(t) \\ &= \int_G |f_V(s)| \int_G \left|g(s^{-1}t) - g(t)\right| d\mu(t) d\mu(s) \qquad (1) \\ &\leq \sup_{s \in V} \|g_{s^{-1}} - g\|_{L^1(G)} \xrightarrow[V \in \mathcal{V}]{} 0 \end{aligned}$$

by the continuity of the map $G \to L^1(G)$ given by $s \to g_{s^{-1}}$.

Remark 13.20. We could have replaced $C_c(G)$ by $L^1(G)$ throughout this section. We use $C_c(G)$ because at times it is technically more convenient.

13.2.1 Unitary representations

For a Hilbert space $\mathcal{H}$ write $U(\mathcal{H})$ for the group of unitary operators in $B(\mathcal{H})$. A *unitary representation of G on $\mathcal{H}$* is a group homomorphism $\rho : G \to U(\mathcal{H})$ that is s.o.-continuous, in other words, continuous when $U(\mathcal{H})$ is equipped with the relative s.o.t. inherited from $B(\mathcal{H})$. Note that $\rho(t^{-1}) = \rho(t)^*$ for all $t \in G$ since $\rho(e) = I$.

Example 13.21. The unitary representation of G on $\mathbb{C}$ given by $t \to 1$ is called the *trivial representation of G*.

Example 13.22. For $t \in G$ let λ_t be the operator on $L^2(G)$ given by $\lambda_t f := f_{t^{-1}}$, that is, $(\lambda_t f)(s) := f(t^{-1}s)$ $(f \in L^2(G), s \in G)$. As explained previously, λ_t is well defined and unitary. The map $\lambda : G \to U(L^2(G))$ given by $\lambda(t) := \lambda_t$ $(t \in G)$ is a group homomorphism, as a simple calculation shows. It is also s.o.-continuous by the continuity of the inverse map on G and the map $t \to f_t$ for each fixed $f \in L^2(G)$. Thus, λ is a unitary representation of G on $L^2(G)$. It is called the *left regular representation of G*.

Let ρ be a unitary representation of G on a Hilbert space $\mathcal{H}$ and let $f \in C_c(G)$. Set

$$\tilde{\rho}(f) := \int_G f(t)\rho(t)\, d\mu(t),$$

where the integral convergence in the w.o.t. This means that the function from $\mathcal{H} \times \mathcal{H}$ to $\mathbb{C}$ that maps the pair $[\zeta, \eta]$ to $\int_G f(t)\,(\rho(t)\zeta, \eta)\, d\mu(t)$ (the integrand belongs to $C_c(G)$!) is linear in the first variable, conjugate linear in the second variable, and bounded by $\|f\|_{L^1(G)}$ (as $\|\rho(t)\| = 1$ for all t because $\rho(t)$ is unitary). Thus, there exists a (unique) operator $\tilde{\rho}(f) = \int_G f(t)\rho(t)\,d\mu(t) \in B(\mathcal{H})$ such that $\left(\left(\int_G f(t)\rho(t)\,d\mu(t)\right)\zeta, \eta\right) = \int_G f(t)\,(\rho(t)\zeta, \eta)\,d\mu(t)$ for all $\zeta, \eta \in \mathcal{H}$, and we have $\|\tilde{\rho}(f)\| \leq \|f\|_{L^1(G)}$.

The function $f \to \tilde{\rho}(f)$ is evidently linear. For all $\zeta, \eta \in \mathcal{H}$ we have

$$\begin{aligned}(\tilde{\rho}(f)\tilde{\rho}(g)\zeta, \eta) &= \int_G f(s)\,(\rho(s)\tilde{\rho}(g)\zeta, \eta)\,d\mu(s) \\ &= \int_G f(s)\,(\tilde{\rho}(g)\zeta, \rho(s)^*\eta)\,d\mu(s) \\ &= \int_G f(s)\left(\int_G g(t)\,(\rho(t)\zeta, \rho(t)^*\eta)\,d\mu(t)\right)d\mu(s) \\ &= \int_G f(s)\left(\int_G g(t)\,(\rho(st)\zeta, \eta)\,d\mu(t)\right)d\mu(s).\end{aligned}$$

On the other hand,

$$\begin{aligned}(\tilde{\rho}(f * g)\zeta, \eta) &= \int_G (f * g)(t)\,(\rho(t)\zeta, \eta)\,d\mu(t) \\ &= \int_G \left(\int_G f(s)g(s^{-1}t)\,d\mu(s)\right)(\rho(t)\zeta, \eta)\,d\mu(t) \\ &= \int_G f(s)\left(\int_G g(s^{-1}t)\,(\rho(t)\zeta, \eta)\,d\mu(t)\right)d\mu(s),\end{aligned}$$

where in the last step we used Fubini's theorem. By the left invariance of μ,

$$(\tilde{\rho}(f * g)\zeta, \eta) = \int_G f(s)\left(\int_G g(t)\,(\rho(st)\zeta, \eta)\,d\mu(t)\right)d\mu(s) = (\tilde{\rho}(f)\tilde{\rho}(g)\zeta, \eta)\,,$$

proving that $\tilde{\rho}(f * g) = \tilde{\rho}(f)\tilde{\rho}(g)$. Similarly, for all $\zeta, \eta \in \mathcal{H}$ we have

$$\begin{aligned}(\tilde{\rho}(f^*)\zeta, \eta) &= \int_G f^*(t)\,(\rho(t)\zeta, \eta)\,d\mu(t) = \int_G \delta(t)^{-1}\overline{f(t^{-1})}\,(\rho(t)\zeta, \eta)\,d\mu(t) \\ &= \int_G \overline{f(t)}\,(\rho(t^{-1})\zeta, \eta)\,d\mu(t) = \int_G \overline{f(t)}\,(\zeta, \rho(t)\eta)\,d\mu(t) \\ &= \overline{(\tilde{\rho}(f)\eta, \zeta)} = (\tilde{\rho}(f)^*\zeta, \eta)\,,\end{aligned}$$

thus, $\tilde{\rho}(f^*) = \tilde{\rho}(f)^*$. This proves that $\tilde{\rho}$ is a representation of $C_c(G)$ on $\mathcal{H}$.

Furthermore, if $\{f_V\}_{V \in \mathcal{V}}$ is an approximate identity for $C_c(G)$, then $\lim_{V \in \mathcal{V}} \tilde{\rho}(f_V) = I$ in the w.o.t. because

$$\begin{aligned}|(\tilde{\rho}(f_V)\zeta, \eta) - (\zeta, \eta)| &= \left|\int_G f_V(t)\,((\rho(t) - I)\zeta, \eta)\,d\mu(t)\right| \\ &\leq \sup_{t \in V} |((\rho(t) - I)\zeta, \eta)| \xrightarrow[V \in \mathcal{V}]{} 0\end{aligned} \qquad (\forall \zeta, \eta \in \mathcal{H})$$

since ρ is s.o.-continuous. This implies that $\tilde{\rho}$ is non-degenerate. The same reasoning shows that $\lim_{V\in\mathcal{V}} \tilde{\rho}\left((f_V)_{t^{-1}}\right) = \rho(t)$ in the w.o.t. for each $t \in G$, so that the unitary representation ρ of G can be reconstructed from the representation $\tilde{\rho}$ of $C_c(G)$ it induces. We proved the following.

Proposition 13.23. *For a unitary representation ρ of G on a Hilbert space $\mathcal{H}$, the function $\tilde{\rho} : C_c(G) \to B(\mathcal{H})$ is a non-degenerate representation of the $*$-algebra $C_c(G)$ on $\mathcal{H}$ satisfying $\|\tilde{\rho}(f)\| \leq \|f\|_{L^1(G)}$ for all $f \in C_c(G)$.*

Conversely, suppose that $\varrho : C_c(G) \to B(\mathcal{H})$ is a non-degenerate representation of the $*$-algebra $C_c(G)$ on a Hilbert space $\mathcal{H}$ satisfying $\|\varrho(f)\| \leq \|f\|_{L^1(G)}$ for all $f \in C_c(G)$. We will demonstrate that it is induced as shown by a unitary representation of G.

Pick an approximate identity $\{f_V\}_{V\in\mathcal{V}}$ for $C_c(G)$. Let $t \in G$. We claim that the net $\{\varrho\left((f_V)_t\right)\}_{V\in\mathcal{V}}$ converges in the s.o.t. of $B(\mathcal{H})$. To prove the claim, note that $\|\varrho\left((f_V)_t\right)\| \leq \|(f_V)_t\|_{L^1(G)} = \|f_V\|_{L^1(G)} = 1$ for each V, thus the net is bounded. Also, for each $g \in C_c(G)$ we have

$$\varrho\left((f_V)_t\right)\varrho(g) = \varrho\left((f_V)_t * g\right) = \varrho\left((f_V * g)_t\right) \xrightarrow[V\in\mathcal{V}]{} \varrho(g_t)$$

(in the operator norm!) because $f_V * g \xrightarrow[V\in\mathcal{V}]{} g$, thus $(f_V * g)_t \xrightarrow[V\in\mathcal{V}]{} g_t$, in $L^1(G)$. In particular, $\varrho\left((f_V)_t\right)\varrho(g) \xrightarrow[V\in\mathcal{V}]{} \varrho(g_t)$ in the s.o.t. Since ϱ is non-degenerate, $\mathcal{H}$ is a Hilbert space, and $\{\varrho\left((f_V)_t\right)\}_{V\in\mathcal{V}}$ is bounded, we infer that $\{\varrho\left((f_V)_t\right)\}_{V\in\mathcal{V}}$ converges in the s.o.t. (and the limit does not depend on the choice of $\{f_V\}_{V\in\mathcal{V}}$).

For $t \in G$ write $\underset{\sim}{\varrho}(t) \in B(\mathcal{H})$ for the s.o.-limit of $\{\varrho\left((f_V)_{t^{-1}}\right)\}_{V\in\mathcal{V}}$. It is a technical exercise, which we skip to show that $t \to \underset{\sim}{\varrho}(t)$ is a unitary representation of G on $\mathcal{H}$ such that $\left(\underset{\sim}{\varrho}\right)^{\sim} = \varrho$. To summarize:

Proposition 13.24. *There is a one-to-one correspondence between the unitary representations ρ of G on a Hilbert space $\mathcal{H}$ and the non-degenerate representations ϱ of the $*$-algebra $C_c(G)$ on $\mathcal{H}$ satisfying $\|\varrho(f)\| \leq \|f\|_{L^1(G)}$ for all $f \in C_c(G)$. It is given by the maps $\rho \to \tilde{\rho}$ and $\varrho \to \underset{\sim}{\varrho}$, which are inverse to one another.*

13.2.2 The definition and representations of the group C^*-algebra

We assume familiarity with C^*-norms and C^*-completions of $*$-algebras (see Section 13.1.2).

For each $f \in C_c(G)$ let

$$\|f\| := \sup\left\{\|\tilde{\rho}(f)\| \,;\, \rho \text{ is a unitary representation of } G\right\}.$$

By Proposition 13.23, $\|f\| \leq \|f\|_{L^1(G)}$, so in particular the sup is finite. By Proposition 13.24, $\|f\|$ equals the sup of all numbers $\|\varrho(f)\|$ where ϱ goes over the non-degenerate representations of the $*$-algebra $C_c(G)$ satisfying $\|\varrho(g)\| \leq \|g\|_{L^1(G)}$ for each $g \in C_c(G)$.

It is immediate that $\|\cdot\|$ is a C^*-semi-norm on $C_c(G)$. To prove that it is a C^*-norm, we show that if $0 \neq f \in C_c(G)$ then $\tilde{\lambda}(f) \neq 0$, where λ is the left regular representation of G on $L^2(G)$ (Example 13.22), thus $\|f\| \geq \|\tilde{\lambda}(f)\| > 0$. Indeed, for $g, h \in C_c(G) \subset L^2(G)$ we have by Fubini's theorem

$$\begin{aligned}
(\tilde{\lambda}(f)g, h)_{L^2(G)} &= \int_G f(t)\,(\lambda(t)g, h)\,d\mu(t) = \int_G f(t)\,(g_{t^{-1}}, h)\,d\mu(t) \\
&= \int_G f(t) \left(\int_G g(t^{-1}s)\overline{h(s)}\,d\mu(s) \right) d\mu(t) \\
&= \int_G \left(\int_G f(t) g(t^{-1}s)\,d\mu(t) \right) \overline{h(s)}\,d\mu(s) = (f * g, h)_{L^2(G)},
\end{aligned}$$

proving that $\tilde{\lambda}(f)g = f * g$ by the density of $C_c(G)$ in $L^2(G)$. Defining $g \in C_c(G)$ by $g(t) := \overline{f(t^{-1})}$, $t \in G$, we have $\big(\tilde{\lambda}(f)g\big)(e) = (f * g)(e) = \|f\|^2_{L^2(G)} > 0$. This proves that $\tilde{\lambda}(f) \neq 0$.

Definition 13.25. The C^*-completion of $C_c(G)$ with respect to the C^*-norm $\|\cdot\|$ above is called the (*universal* or *full*) *group C^*-algebra of G*. It is denoted by $C^*(G)$.

If ρ is a unitary representation of G on a Hilbert space $\mathcal{H}$, then the inequality $\|\tilde{\rho}(\cdot)\| \leq \|\cdot\|$ on $C_c(G)$ proves that $\tilde{\rho} : C_c(G) \to B(\mathcal{H})$ extends uniquely to a bounded linear map from $C^*(G)$ to $B(\mathcal{H})$, also denoted by $\tilde{\rho}$, which is clearly a non-degenerate representation of $C^*(G)$. Conversely, given a non-degenerate representation Φ of $C^*(G)$ on a Hilbert space $\mathcal{H}$, the restriction $\Phi|_{C_c(G)}$ is a non-degenerate representation of $C_c(G)$ and $\|\Phi(f)\| \leq \|f\| \leq \|f\|_{L^1(G)}$ for all $f \in C_c(G)$, so by Proposition 13.24, $\Phi|_{C_c(G)}$ equals $\tilde{\rho}$ for some unitary representation ρ of G on $\mathcal{H}$.

So, continuing Proposition 13.24, there is a one-to-one correspondence between the unitary representations of G and the non-degenerate representations of the C^*-algebra $C^*(G)$. We can loosely say that $C^*(G)$ is "universal with respect to the unitary representations of G".

13.2.3 Properties of the group C^*-algebra

We list several properties of group C^*-algebras, which are important in *quantum group* theory.

Theorem 13.26.

(i) $C^*(G)$ is commutative iff G is abelian.

(ii) $C^*(G)$ is unital iff G is discrete.

(iii) $C^*(G)$ admits a character, that is, a non-zero homomorphism from $C^*(G)$ to $\mathbb{C}$.

(iv) *Assume that G is discrete. There exists a unique bounded linear map $\Delta : C^*(G) \to C^*(G) \otimes_{\max} C^*(G)$ that, using the identification of $C_c(G)$ with $\mathbb{C}[G]$ as in Example 13.19, maps $t \in G$ to $t \otimes t$. It is an isometric $*$-homomorphism.*

Proof. (i) The C^*-algebra $C^*(G)$ is commutative iff the $*$-algebra $C_c(G)$ is commutative. If G is abelian, then in particular it is unimodular, and for all $f, g \in C_c(G)$ we have

$$\begin{aligned}(f * g)(t) &= \int_G f(ts)g(s^{-1})\,d\mu(s) = \int_G f(ts^{-1})g(s)\,d\mu(s) \\ &= \int_G g(s)f(s^{-1}t)\,d\mu(s) = (g * f)(t)\end{aligned} \qquad (\forall t \in G),$$

proving that $f * g = g * f$. Thus $C_c(G)$ is commutative. The converse is left as an exercise.

(ii) If G is discrete, then identifying $C_c(G)$ with $\mathbb{C}[G]$, the element $e \in \mathbb{C}[G]$ is evidently a unit for $\mathbb{C}[G]$, thus for $C^*(G)$. For the converse, see Exercise 21.

(iii) Denote by 1 the trivial representation of G (Example 13.21). Then $\tilde{1}$ is a non-degenerate (in particular, non-zero) representation of $C^*(G)$ on $\mathbb{C}$, that is, a character of $C^*(G)$.

(iv) Uniqueness follows from the density of $\mathbb{C}[G]$ in $C^*(G)$.

To prove the existence of the bounded linear map Δ we should show that for each $\sum_{t\in G} \alpha_t t$ in $\mathbb{C}[G]$ (identified with $C_c(G)$) we have

$$\left\| \sum_{t\in G} \alpha_t (t \otimes t) \right\|_{C^*(G)\otimes_{\max} C^*(G)} \leq \left\| \sum_{t\in G} \alpha_t t \right\|_{C^*(G)}. \tag{2}$$

Let π be a non-degenerate representation of $C^*(G) \otimes_{\text{alg}} C^*(G)$ on a Hilbert space $\mathcal{H}$. Let π_1, π_2 be the (unique) non-degenerate representations of $C^*(G)$ on $\mathcal{H}$ with commuting ranges such that $\pi = \pi_1 \cdot \pi_2$ (Proposition 13.4), and for $i = 1, 2$ let ρ_i be the (unique) unitary representation of G on $\mathcal{H}$ such that $\pi_i = \tilde{\rho}_i$ (see the paragraph after Definition 13.25). It follows from the proof of Proposition 13.24 that the ranges of ρ_1, ρ_2 commute. Consequently, the function $\rho_1 \cdot \rho_2 : G \to U(\mathcal{H})$ given by $(\rho_1 \cdot \rho_2)(t) := \rho_1(t)\rho_2(t)$, $t \in G$, is a unitary representation of G on $\mathcal{H}$.

For each $\sum_{t\in G} \alpha_t t \in \mathbb{C}[G]$ we have

$$\begin{aligned}\pi\left(\sum_{t\in G} \alpha_t (t \otimes t)\right) &= \sum_{t\in G} \alpha_t \pi_1(t)\pi_2(t) = \sum_{t\in G} \alpha_t \tilde{\rho}_1(t)\tilde{\rho}_2(t) = \sum_{t\in G} \alpha_t \rho_1(t)\rho_2(t) \\ &= \sum_{t\in G} \alpha_t (\rho_1 \cdot \rho_2)(t) = \widetilde{(\rho_1 \cdot \rho_2)}\left(\sum_{t\in G} \alpha_t t\right).\end{aligned}$$

Therefore, by the definition of the norm on $C^*(G)$,

$$\left\| \pi\left(\sum_{t\in G} \alpha_t (t \otimes t)\right) \right\| = \left\| \widetilde{(\rho_1 \cdot \rho_2)}\left(\sum_{t\in G} \alpha_t t\right) \right\| \leq \left\| \sum_{t\in G} \alpha_t t \right\|_{C^*(G)}.$$

This holds true for each non-degenerate representation π of $C^*(G) \otimes_{\text{alg}} C^*(G)$, thus verifying (2) by the definition of the maximal C^*-norm (Theorem 13.8).

The bounded linear map Δ is easily seen to be a $*$-homomorphism. It remains to show that Δ is injective (thus isometric). The character $\tilde{1} \in C^*(G)^*$ of $C^*(G)$ discussed in Part (iii) satisfies $\tilde{1}(t) = 1$ for all $t \in G$ (we still identify $\mathbb{C}[G]$ with $C_c(G)$ and embed it in $C^*(G)$). Consider the slice map $\tilde{1} \otimes \mathrm{id}_{C^*(G)} \in B(C^*(G) \otimes_{\max} C^*(G), C^*(G))$, see Exercise 5. For all $t \in G$ we have

$$\left((\tilde{1} \otimes \mathrm{id}_{C^*(G)}) \circ \Delta\right)(t) = (\tilde{1} \otimes \mathrm{id}_{C^*(G)})(t \otimes t) = \left(\tilde{1}(t)\right) t = t.$$

By the density of $\mathbb{C}[G]$ in $C^*(G)$ and by the linearity and boundedness of the operators in both sides we deduce that $(\tilde{1} \otimes \mathrm{id}_{C^*(G)}) \circ \Delta = \mathrm{id}_{C^*(G)}$. In particular, Δ is injective. □

Remark 13.27. Since $C^*(G)$ is, in general, not commutative, the existence of a character on $C^*(G)$ is not clear. There are "many" C^*-algebras without characters!

Remark 13.28. In the proof of Part (iv) the discreteness of G was required for the definition of Δ (by the formula $\sum_{t \in G} \alpha_t t \to \sum_{t \in G} \alpha_t (t \otimes t)$). However, notice that the paragraph starting with "Let π be a non-degenerate representation of $C^*(G) \otimes_{\text{alg}} C^*(G)$" remains valid when G is not discrete. And indeed, there is an extension of Part (iv) to general locally compact topological groups G with the co-domain being a canonical C^*-algebra strictly larger than $C^*(G) \otimes_{\max} C^*(G)$.

Exercises

Tensor products of C^*-algebras

Unless otherwise stated, $\mathcal{A}, \mathcal{B}$ are C^*-algebras.

1. Let $\mathcal{H}, \mathcal{K}$ be Hilbert spaces. Recall that for $x, y \in \mathcal{H}$ we write $\theta_{y,x}$ (here $\theta^{\mathcal{H}}_{y,x}$ for clarity) for the element $(\cdot, x)\, y$ of $B(\mathcal{H})$.
   (a) Let $x, y \in \mathcal{H}$ and $w, z \in \mathcal{K}$. Prove that $\theta^{\mathcal{H}}_{y,x} \otimes \theta^{\mathcal{K}}_{z,w} = \theta^{\mathcal{H} \otimes \mathcal{K}}_{y \otimes z, x \otimes w}$ (where the tensor in the left-hand side is given by Part (vi) of Proposition 8.18).
   (b) Let $A \in B(\mathcal{H} \otimes \mathcal{K})$. Assume that $\mathrm{Tr}((\theta^{\mathcal{H}}_{y,x} \otimes \theta^{\mathcal{K}}_{z,w}) A) = 0$ for all $x, y \in \mathcal{H}$ and $w, z \in \mathcal{K}$. Prove that $A = 0$. (See Section 12.6 for the Tr notation.)
   (c) Use Theorem 12.25 and separation to deduce that $B(\mathcal{H}) \otimes_{\text{alg}} B(\mathcal{K}) \subset B(\mathcal{H} \otimes \mathcal{K})$ is ultraweakly dense in $B(\mathcal{H} \otimes \mathcal{K})$ (hence also w.o.-dense and s.o.-dense by Exercise 15 in Chapter 12).
2. (a) Let $\mathcal{C}$ be a $*$-algebra. Assume that $\|\cdot\|_1$ is a norm on $\mathcal{C}$ making it a C^*-algebra and $\|\cdot\|_2$ is a C^*-norm on $\mathcal{C}$. Prove that $\|\cdot\|_1 = \|\cdot\|_2$.
   (b) Prove that for all $n \in \mathbb{N}$, $\mathbb{C}^n$ and $M_n(\mathbb{C})$ are nuclear C^*-algebras.

3. The minimal and the maximal tensor products of C^*-algebras are commutative, associative, and distributive. Provide precise statements of this and prove them.

4. (Tensor products of $*$-homomorphisms.) Let $\mathcal{A}_1, \mathcal{A}_2, \mathcal{B}_1, \mathcal{B}_2$ be C^*-algebras. Suppose that $\varphi : \mathcal{A}_1 \to \mathcal{A}_2$ and $\psi : \mathcal{B}_1 \to \mathcal{B}_2$ are $*$-homomorphisms.
   (a) Prove that the $*$-homomorphism $\varphi \otimes \psi : \mathcal{A}_1 \otimes_{\text{alg}} \mathcal{B}_1 \to \mathcal{A}_2 \otimes_{\text{alg}} \mathcal{B}_2$ extends uniquely both to a bounded linear map $\mathcal{A}_1 \otimes_{\min} \mathcal{B}_1 \to \mathcal{A}_2 \otimes_{\min} \mathcal{B}_2$ and to a bounded linear map $\mathcal{A}_1 \otimes_{\max} \mathcal{B}_1 \to \mathcal{A}_2 \otimes_{\max} \mathcal{B}_2$, also denoted by $\varphi \otimes \psi$, that these map are $*$-homomorphisms, and that they are surjective if φ, ψ are surjective.
   (b) Prove that if φ, ψ are injective, then so is the map $\varphi \otimes \psi : \mathcal{A}_1 \otimes_{\min} \mathcal{B}_1 \to \mathcal{A}_2 \otimes_{\min} \mathcal{B}_2$. (This is *not* true for the maximal tensor product.)

5. (Slice maps.) Let $\omega \in \mathcal{A}^*$. Denote by $\mathrm{id}_{\mathcal{B}}$ the identity map on $\mathcal{B}$. Prove that the linear map $\omega \otimes \mathrm{id}_{\mathcal{B}} : \mathcal{A} \otimes_{\text{alg}} \mathcal{B} \to \mathcal{B}$ (sending $a \otimes b$ to $\omega(a)b$) extends uniquely both to a bounded linear map $\mathcal{A} \otimes_{\min} \mathcal{B} \to \mathcal{B}$ and to a bounded linear map $\mathcal{A} \otimes_{\max} \mathcal{B} \to \mathcal{B}$, also denoted by $\omega \otimes \mathrm{id}_{\mathcal{B}}$. Prove that these map have norm $\|\omega\|$ and that they are positive (i.e., send positive elements to positive elements) if so is ω.
   (Hint: it suffices to prove the assertion for the minimal tensor product. In this case there are several similar ways to show existence, either by assuming first that ω is positive and using the GNS construction, or by adapting the proof of Proposition 13.13 using Exercise 16.)

6. (Tensor products by commutative C^*-algebras revisited.)
   As in Section 13.1.6, let Ω be a locally compact Hausdorff space and $\mathcal{A}$ be a C^*-algebra. In this exercise we show explicitly that $C_0(\Omega) \otimes_{\min} \mathcal{A}$ is $*$-isomorphic to $C_0(\Omega, \mathcal{A})$ canonically.
   Let ρ be the natural faithful representation of $C_0(\Omega)$ on $l^2(\Omega) = \sum_{x\in\Omega} \oplus \mathbb{C}$ given by "multiplication": for $f \in C_0(\Omega)$, $\rho(f)$ takes $\sum_{x\in\Omega} \oplus \lambda_x \in l^2(\Omega)$ to $\sum_{x\in\Omega} \oplus f(x)\lambda_x \in l^2(\Omega)$. Let σ be a faithful representation of $\mathcal{A}$ on a Hilbert space $\mathcal{H}$. Consider the faithful representation π of $C_0(\Omega, \mathcal{A})$ on $\sum_{x\in\Omega} \oplus \mathcal{H}$ given by $C_0(\Omega, \mathcal{A}) \ni F \to \sum_{x\in\Omega} \oplus \sigma(F(x)) \in B(\sum_{x\in\Omega} \oplus \mathcal{H})$. Identifying $l^2(\Omega) \otimes \mathcal{H}$ with $\sum_{x\in\Omega} \oplus \mathcal{H}$ canonically, prove that $(\rho \otimes \sigma)(f \otimes a) = \pi(fa)$ for all $f \in C_0(\Omega)$ and $a \in \mathcal{A}$. Deduce that the images of the faithful representations $\rho \otimes \sigma$ and π of $C_0(\Omega) \otimes_{\min} \mathcal{A}$ and $C_0(\Omega, \mathcal{A})$, respectively, on $l^2(\Omega) \otimes \mathcal{H}$, are equal.

7. (Compare Exercise 28 in Chapter 8 as well as Exercise 17 here.) Let Ω_1, Ω_2 be locally compact Hausdorff spaces. In this exercise we prove that $C_0(\Omega_1) \otimes C_0(\Omega_2)$ "is" $C_0(\Omega_1 \times \Omega_2)$, where $C_0(\Omega_1) \otimes C_0(\Omega_2)$ is the unique C^*-algebraic tensor product of $C_0(\Omega_1)$ and $C_0(\Omega_2)$; see Takesaki's theorem.
   Precisely: for $f \in C_0(\Omega_1)$ and $g \in C_0(\Omega_2)$, define $f \times g : \Omega_1 \times \Omega_2 \to \mathbb{C}$ by $[x, y] \to f(x)g(y)$ $(x \in \Omega_1, y \in \Omega_2)$, and notice that $f \times g \in C_0(\Omega_1 \times \Omega_2)$.

Prove that there exists a unique bounded linear map from $C_0(\Omega_1)\otimes C_0(\Omega_2)$ to $C_0(\Omega_1\times\Omega_2)$ that sends $f\otimes g$ to $f\times g$ for all $f\in C_0(\Omega_1)$ and $g\in C_0(\Omega_2)$, and that this map is a surjective $*$-isomorphism.

8. (Compare Proposition 13.4 and Corollary 13.10.) Let $\mathcal{C}$ be a C^*-algebra and let $\pi_{\mathcal{A}}:\mathcal{A}\to\mathcal{C}$ and $\pi_{\mathcal{B}}:\mathcal{B}\to\mathcal{C}$ be $*$-homomorphisms with commuting ranges. Prove that there exists a unique bounded linear map from $\mathcal{A}\otimes_{\max}\mathcal{B}$ to $\mathcal{C}$ mapping $a\otimes b$ to $\pi_{\mathcal{A}}(a)\pi_{\mathcal{B}}(b)$, and that this map is a $*$-homomorphism.

Tensor products of von Neumann algebras

Unless otherwise stated, $\mathcal{R},\mathcal{S}$ are von Neumann algebras acting on Hilbert spaces $\mathcal{H},\mathcal{K}$, respectively. We define the (normal spatial) *von Neumann algebraic tensor product of* $\mathcal{R}$ *and* $\mathcal{S}$, denoted by $\mathcal{R}\mathbin{\overline{\otimes}}\mathcal{S}$, to be the w.o.-(equivalently, ultraweak or s.o.-) closure of $\mathcal{R}\otimes_{\mathrm{alg}}\mathcal{S}$ in $B(\mathcal{H}\otimes\mathcal{K})$; for this we view $\mathcal{R}\otimes_{\mathrm{alg}}\mathcal{S}$ as a $*$-subalgebra of $B(\mathcal{H})\otimes_{\mathrm{alg}}B(\mathcal{K})$, which in turn is viewed as a $*$-subalgebra of $B(\mathcal{H}\otimes\mathcal{K})$. Then $\mathcal{R}\mathbin{\overline{\otimes}}\mathcal{S}$ is a von Neumann algebra on $\mathcal{H}\otimes\mathcal{K}$. Notice that $\mathcal{R}\mathbin{\overline{\otimes}}\mathcal{S}$ is the w.o.-closure of $\mathcal{R}\otimes_{\min}\mathcal{S}$, when the latter is viewed as sitting in $B(\mathcal{H}\otimes\mathcal{K})$ using the given representations of $\mathcal{R}$ and $\mathcal{S}$ on $\mathcal{H}$ and $\mathcal{K}$, respectively.

9. (a) Observe that $B(\mathcal{H})\mathbin{\overline{\otimes}}B(\mathcal{K})=B(\mathcal{H}\otimes\mathcal{K})$ trivially by Exercise 1.

   (b) (Matrix representation of elements of $B(\mathcal{H})\mathbin{\overline{\otimes}}B(\mathcal{K})=B(\mathcal{H}\otimes\mathcal{K})$.) Let $\{f_j\}_{j\in J}$ be an orthonormal basis of $\mathcal{K}$. To every $T\in B(\mathcal{H}\otimes\mathcal{K})$ there is an associated matrix $(T_{ij})_{i,j\in J}$ with entries in $B(\mathcal{H})$ obtained by the identification of $\mathcal{H}\otimes\mathcal{K}$ with $\sum_{j\in J}\oplus\mathcal{H}$ (see Part (iv) of Proposition 8.18), that is, $T_{ij}\zeta=(I_{\mathcal{H}}\otimes\langle\cdot,f_i\rangle)T(\zeta\otimes f_j)$ for $i,j\in J$ and $\zeta\in\mathcal{H}$, where the map $I_{\mathcal{H}}\otimes\langle\cdot,f_i\rangle:\mathcal{H}\otimes\mathcal{K}\to\mathcal{H}$ is given by Part (vi) of Proposition 8.18. Prove that $\|T_{ij}\|\le\|T\|$ for all $i,j\in J$ and that $T\left(\sum_{j\in J}\zeta_j\otimes f_j\right)=\sum_{i\in J}\left(\sum_{j\in J}T_{ij}\zeta_j\right)\otimes f_i$ for all $\sum_{j\in J}\oplus\zeta_j\in\sum_{j\in J}\oplus\mathcal{H}$. Prove that the map $T\to(T_{ij})_{i,j\in J}$ is an injective $*$-homomorphism from $B(\mathcal{H}\otimes\mathcal{K})$ to the $*$-algebra $M_J(B(\mathcal{H}))$.

   (c) Prove that $T\in\mathcal{R}\mathbin{\overline{\otimes}}B(\mathcal{K})$ iff $T_{ij}\in\mathcal{R}$ for all $i,j\in J$, and that $T\in\mathcal{R}\mathbin{\overline{\otimes}}\mathbb{C}I_{\mathcal{K}}$ iff there is $S\in\mathcal{R}$ such that $T_{ij}=\delta_{i,j}S$ for all $i,j\in J$ (where δ is Kronecker's delta).

10. For $n\in\mathbb{N}$, describe $\mathcal{R}\mathbin{\overline{\otimes}}\mathbb{C}^n$ and $\mathcal{R}\mathbin{\overline{\otimes}}M_n(\mathbb{C})$.

11. Prove that the commutants of $\mathcal{R}\mathbin{\overline{\otimes}}\mathbb{C}I_{\mathcal{K}}$ and $\mathcal{R}\mathbin{\overline{\otimes}}B(\mathcal{K})$ inside $B(\mathcal{H}\otimes\mathcal{K})$ are $\mathcal{R}'\mathbin{\overline{\otimes}}B(\mathcal{K})$ and $\mathcal{R}'\mathbin{\overline{\otimes}}\mathbb{C}I_{\mathcal{K}}$, respectively.

12. The von Neumann algebraic tensor product is commutative, associative, and distributive. Provide precise statements of this and prove them.

13. (Tensor product of normal functionals.) Let $\omega\in\mathcal{R}_*$ and $\tau\in\mathcal{S}_*$. Use Lemma 13.11 to prove that there exists a unique linear functional $\omega\otimes\tau\in$

$(\mathcal{R} \mathbin{\overline{\otimes}} \mathcal{S})_*$ satisfying $(\omega \otimes \tau)(A \otimes B) = \omega(A)\tau(B)$ for all $A \in \mathcal{R}$ and $B \in \mathcal{S}$. Also prove that $\|\omega \otimes \tau\| = \|\omega\| \, \|\tau\|$ and that $\omega \otimes \tau$ is positive if so are ω, τ.

14. Prove that $\operatorname{span}\{\omega \otimes \tau; \omega \in \mathcal{R}_*, \tau \in \mathcal{S}_*\}$ is norm-dense in $(\mathcal{R} \mathbin{\overline{\otimes}} \mathcal{S})_*$.

15. (Tensor products of normal $*$-homomorphisms; compare Exercise 4.) Recall Exercise 13 in Chapter 12. Let $\mathcal{R}_1, \mathcal{R}_2, \mathcal{S}_1, \mathcal{S}_2$ be von Neumann algebras. Suppose that $\varphi : \mathcal{R}_1 \to \mathcal{R}_2$ and $\psi : \mathcal{S}_1 \to \mathcal{S}_2$ are ultraweakly continuous $*$-homomorphisms. Prove that the $*$-homomorphism $\varphi \otimes \psi : \mathcal{R}_1 \otimes_{\mathrm{alg}} \mathcal{S}_1 \to \mathcal{R}_2 \otimes_{\mathrm{alg}} \mathcal{S}_2$ extends uniquely to an ultraweakly continuous linear map $\mathcal{R}_1 \mathbin{\overline{\otimes}} \mathcal{S}_1 \to \mathcal{R}_2 \mathbin{\overline{\otimes}} \mathcal{S}_2$, also denoted by $\varphi \otimes \psi$, that this map is a $*$-homomorphism, and that it is injective/surjective if so are φ, ψ.
(Hint: for existence: let $X \in \mathcal{R}_1 \mathbin{\overline{\otimes}} \mathcal{S}_1$. Take a bounded net $\{X_\alpha\}_\alpha$ in $\mathcal{R}_1 \otimes_{\mathrm{alg}} \mathcal{S}_1$ that converges ultraweakly to X in $\mathcal{R}_1 \mathbin{\overline{\otimes}} \mathcal{S}_1$ (why does such a net exist?), and for $\omega \in (\mathcal{R}_2)_*$ and $\tau \in (\mathcal{S}_2)_*$, prove that the scalar net $\{(\omega \otimes \tau)((\varphi \otimes \psi)(X_\alpha))\}_\alpha$ converges. Now apply Exercises 4 and 14.)

16. (Slice maps; compare Exercise 5.) Fix $\omega \in \mathcal{R}_*$. Prove that the linear map $\omega \otimes \mathrm{id}_{\mathcal{S}} : \mathcal{R} \otimes_{\mathrm{alg}} \mathcal{S} \to \mathcal{S}$ (sending $A \otimes B$ to $\omega(A)B$) extends uniquely to an ultraweakly continuous linear map $\mathcal{R} \mathbin{\overline{\otimes}} \mathcal{S} \to \mathcal{S}$, also denoted by $\omega \otimes \mathrm{id}_{\mathcal{S}}$. Observe that $\tau \circ (\omega \otimes \mathrm{id}_{\mathcal{S}}) = \omega \otimes \tau$, and prove that $\|\omega \otimes \mathrm{id}_{\mathcal{S}}\| = \|\omega\|$ and that $\omega \otimes \mathrm{id}_{\mathcal{S}}$ is positive (i.e., sends positive elements to positive elements) if so is ω.
(Hint: for existence: consider the Banach adjoint of the linear map from $\mathcal{S}_*$ to $(\mathcal{R} \mathbin{\overline{\otimes}} \mathcal{S})_*$ given by $\tau \to \omega \otimes \tau$.)

17. (Compare Exercise 28 in Chapter 8 as well as Exercise 7 here.) Let $(\Omega, \mathcal{A}, \mu)$ and $(\Theta, \mathcal{B}, \nu)$ be complete σ-finite positive measure spaces. In this exercise we prove that, essentially, $L^\infty(\Omega, \mathcal{A}, \mu) \mathbin{\overline{\otimes}} L^\infty(\Theta, \mathcal{B}, \nu)$ "is" $L^\infty(\Omega \times \Theta, \mathcal{A} \times \mathcal{B}, \mu \times \nu)$. Precisely: use Exercise 28 in Chapter 8 to prove that $\{M_f; f \in L^\infty(\Omega, \mathcal{A}, \mu)\} \mathbin{\overline{\otimes}} \{M_g; g \in L^\infty(\Theta, \mathcal{B}, \nu)\}$ (acting on $L^2(\Omega, \mathcal{A}, \mu) \otimes L^2(\Theta, \mathcal{B}, \nu)$) is spatially isomorphic to $\{M_F; F \in L^\infty(\Omega \times \Theta, \mathcal{A} \times \mathcal{B}, \mu \times \nu)\}$ (acting on $L^2(\Omega \times \Theta, \mathcal{A} \times \mathcal{B}, \mu \times \nu)$).

Group C^*-algebras

18. (a) Show that in the definition of a unitary representation one can replace the s.o.t. by the w.o.t., yielding an equivalent definition.
    (b) Prove that the left regular representation is continuous in the operator norm topology iff the group is discrete.

19. This exercise explores $C^*(\mathbb{R})$, where the additive group $\mathbb{R}$ is considered with its usual topology. Recall from Exercise 7 in Chapter 2 or from Section II.3.2 that the Fourier transform of $f \in L^1(\mathbb{R})$ is the function $\hat{f} \in C_0(\mathbb{R})$ given by $\hat{f}(t) := \int_{\mathbb{R}} f(x)e^{-ixt}\,dx$ $(t \in \mathbb{R})$. For every $f \in L^1(\mathbb{R}) \cap L^2(\mathbb{R})$ we have $\hat{f} \in L^2(\mathbb{R})$ and $\|f\|_{L^2(\mathbb{R})} = \|\hat{f}\|_{L^2(\mathbb{R})}$. Thus, the

function from $L^1(\mathbb{R}) \cap L^2(\mathbb{R})$ to $L^2(\mathbb{R})$ given by $f \to \hat{f}$ extends (uniquely) to a unitary operator $\mathcal{F} \in B(L^2(\mathbb{R}))$.

(a) Prove that for each $f \in C_c(\mathbb{R})$ we have $\tilde{\lambda}(f) = \mathcal{F}^* M_{\hat{f}} \mathcal{F}$, where λ is the left regular representation of $\mathbb{R}$ and $M_{\hat{f}} \in B(L^2(\mathbb{R}))$ is the multiplication operator on $L^2(\mathbb{R})$ given by $g \to \hat{f}g$. Deduce that $\|\tilde{\lambda}(f)\| = \|\hat{f}\|_u := \sup_{\mathbb{R}} |\hat{f}|$.

(b) Prove that for every unitary representation ρ of $\mathbb{R}$ and every $f \in C_c(\mathbb{R})$, we have $\|\tilde{\rho}(f)\| \leq \|\hat{f}\|_u$.
(Hint: Stone's theorem (Exercise 9 in Chapter 10).)

(c) Deduce that the norm in $C^*(\mathbb{R})$ of each $f \in C_c(\mathbb{R})$ is equal to $\|\hat{f}\|_u$.

(d) Prove that there is a natural (isometric) isomorphism between $C^*(\mathbb{R})$ and $C_0(\mathbb{R})$.

20. Prove, similarly to Exercise 19, that there are natural isomorphisms between $C^*(\mathbb{Z})$ and $C(\mathbb{T})$ as well as between $C^*(\mathbb{T})$ and $C_0(\mathbb{Z})$. This requires introducing the Fourier transform on the appropriate groups.

21. Let G be a locally compact topological group and λ be its left regular representation.

(a) The (generally non-unital) C^*-subalgebra of $B(L^2(G))$ generated by $\{\tilde{\lambda}(f); f \in C_c(G)\}$ is called the *reduced group C^*-algebra of G* and is denoted by $C^*_{\mathrm{r}}(G)$. Explain why the map $C_c(G) \ni f \to \tilde{\lambda}(f)$ extends (uniquely) to a $*$-homomorphism from $C^*(G)$ *onto* $C^*_{\mathrm{r}}(G)$.

(b) Let $\{f_V\}_{V \in \mathcal{V}}$ be an approximate identity for $C_c(G)$ (see Section 13.2). For $V \in \mathcal{V}$, notice that $\|f_V^{1/2}\|_{L^2(G)} = 1$, and consider the restriction ω_V of the vector state $\omega_{f_V^{1/2}}$ of $B(L^2(G))$ to $C^*_{\mathrm{r}}(G)$. Explain why the net $\{\omega_V\}_{V \in \mathcal{V}}$ consists of states of $C^*_{\mathrm{r}}(G)$.

(c) Prove that if G is not discrete then $\omega_V \xrightarrow[V \in \mathcal{V}]{} 0$ in the *weak** topology of $C^*_{\mathrm{r}}(G)^*$. (Hint: $\mu(V) \xrightarrow[V \in \mathcal{V}]{} 0$ when G is not discrete.)

(d) Prove that if $C^*_{\mathrm{r}}(G)$ is unital then G is discrete.

(e) Conclude that $C^*(G)$ is unital iff $C^*_{\mathrm{r}}(G)$ is unital iff G is discrete.

22. Let G_1, G_2 be locally compact topological groups. Prove that there exists a canonical $*$-isomorphism from $C^*(G_1 \times G_2)$ onto $C^*(G_1) \otimes_{\max} C^*(G_2)$. For technical reasons you may assume that G_1, G_2 are σ-compact. The case that G_1, G_2 are discrete is technically easier, and it suffices for conveying the general case's idea (cf. Example 13.2).

Application I

Probability

I.1 Heuristics

A fundamental concept in probability theory is that of an *event*. The "real world" content of the "event" plays no role in the mathematical analysis. What matters is only the event's *occurrence* or non-occurrence.

Two "extreme" events are the *empty event* $\emptyset$ (which cannot occur), and the *sure event* Ω (which occurs always).

To each event A, one associates *the complementary event* A^c, which occurs iff A does not occur.

If the occurrence of the event A forces the occurrence of the event B, one says that A *implies* B, and one writes $A \subset B$. One has trivially $\emptyset \subset A \subset \Omega$ for any event A.

The events A, B are *equivalent* (notation: $A = B$) if they imply each other. Such events are identified.

The *intersection* $A \cap B$ of the events A and B occurs iff A and B both occur. If $A \cap B = \emptyset$ (i.e., if A and B cannot occur together), one says that the events are *mutually disjoint*; for example, for any event A, the events A and A^c are mutually disjoint.

The *union* $A \cup B$ of the events A, B is the event that occurs iff at least one of the events A, B occurs. The operations $\cap$ and $\cup$ are trivially commutative, and satisfy the following relations:

$$A \cup A^c = \Omega;$$

$$A \cap B \subset A \subset A \cup B.$$

One verifies that the algebra of events satisfies the usual associative and distributive laws for the family $\mathbb{P}(\Omega)$ of all subsets of a set Ω, with standard operations between subsets, as well as the DeMorgan (dual) laws:

$$\left(\bigcup_k A_k\right)^c = \bigcap_k A_k^c; \qquad \left(\bigcap_k A_k\right)^c = \bigcup_k A_k^c,$$

for any sequence of events $\{A_k\}$. Mathematically, we may then view the sure event Ω as a given set (called the *sample space*), and the set of all events (for a particular probability problem) as an algebra of subsets of Ω.

Since limiting processes are central in probability theory, countable unions of events should also be events. Therefore, in the set-theoretical model, the algebra of events is required to be a σ-algebra $\mathcal{A}$.

The second fundamental concept of probability theory is that of a *probability*. Each event $A \in \mathcal{A}$ is assigned a probability $P(A)$ (also denoted PA), such that:

(1) $0 \le P(A) \le 1$ for all $A \in \mathcal{A}$;

(2) $P(\bigcup A_k) = \sum P(A_k)$ for any sequence of mutually disjoint events A_k; and

(3) $P(\Omega) = 1$.

In other words, P is a "normalized" finite positive measure on the *measurable space* $(\Omega, \mathcal{A})$. The measure space $(\Omega, \mathcal{A}, P)$ is called a *probability space*. Note that $P(A^c) = 1 - P(A)$.

Examples.

(1) The *trivial* probability space $(\Omega, \mathcal{A}, P)$ has Ω arbitrary, $\mathcal{A} = \{\emptyset, \Omega\}, P(\emptyset) = 0$, and $P(\Omega) = 1$.

(2) *Discrete probability space.* Ω is the union of finitely many mutually disjoint events $A_1, \ldots, A_n$, with probabilities $P(A_k) = p_k, p_k \ge 0$, and $\sum p_k = 1$. The family $\mathcal{A}$ consists of $\emptyset$ and all finite unions $A = \bigcup_{k \in J} A_k$, where $J \subset \{1, \ldots, n\}$. One lets $P(\emptyset) = 0$ and $P(A) = \sum_{k \in J} p_k$.

This probability space is the (finite) discrete probability space. When $p_k = p$ for all k (so that $p = 1/n$), one gets the (finite) uniform probability space. The formula for the probability reduces in this special case to

$$P(A) = \frac{|A|}{n},$$

where $|A|$ denotes the number of points in the index set J (i.e., the number of "elementary events" A_k contained in A).

(3) *Random sampling.* A sample of size s from a population $\mathcal{P}$ of $N \ge s$ objects is a subset $S \subset \mathcal{P}$ with s elements ($|S| = s$). The sampling is random if all $\binom{N}{s}$ samples of size s are assigned the same probability (i.e., the corresponding probability space is a uniform probability space, where Ω is the set of all samples of given size s; this is actually the origin of the name "sample space" given to Ω). The elementary event of getting any particular sample of size s has probability $1/\binom{N}{s}$.

Suppose the population $\mathcal{P}$ is the disjoint union of m sub-populations ("layers") $\mathcal{P}_i$ of size $N_i (\sum N_i = N)$. The number of size s samples with s_i objects from $\mathcal{P}_i (i = 1, \ldots, m; \sum s_i = s)$ is the product of the binomial coefficients $\binom{N_i}{s_i}$. Therefore, if $A_{s_1, \ldots, s_m}$ denotes the event of getting s_i objects from $\mathcal{P}_i (i = 1, \ldots, m)$ in a random sampling of s objects from the multi-layered population $\mathcal{P}$, then

$$P(A_{s_1,\ldots,s_m}) = \frac{\binom{N_1}{s_1} \cdots \binom{N_m}{s_m}}{\binom{N}{s}}.$$

An *ordered sample* of size s is an ordered s-tuple $(x_1, \ldots, x_s) \subset \mathcal{P}$ (we may think of x_i as the object drawn at the ith drawing from the population). The number of such samples is clearly $N(N-1)\cdots(N-s+1)$ (since there are N possible outcomes of the first drawing, $N-1$ for the second, etc.). Fixing one specific object, let A denote the event of getting that object in some specific drawing. Since the procedure is equivalent to (ordered) sampling of size $s-1$ from a population of size $N-1$, we have $|A| = (N-1)\cdots[(N-1)-(s-1)+1]$, and therefore, for random sampling (the uniform model!),

$$P(A) = |A|/|\Omega| = \frac{(N-1)\cdots(N-s+1)}{N(N-1)\cdots(N-s+1)} = 1/N.$$

This probability is *independent* of the drawing considered! This fact is referred to as the "equivalence law of ordered sampling".

I.2 Probability space

Let $(\Omega, \mathcal{A}, P)$ be a probability space, that is, a normalized finite positive measure space. Following our terminology, the "measurable sets" $A \in \mathcal{A}$ are called the *events*; Ω is the *sure event*; $\emptyset$ is the *empty event*; the measure P is called the *probability*. One says *almost surely* (a.s.) instead of "almost everywhere" (or "with probability one", since the complement of the exceptional set has probability one).

If f is a real valued function on Ω, it is desirable that the sets $[f > c]$ be events for any real c, that is, that f be measurable. Such functions will be called real *random variables* (r.v.). Similarly, a complex r.v. is a complex measurable function on Ω.

The simplest r.v. is the indicator I_A of an event $A \in \mathcal{A}$. We clearly have

$$I_{A^c} = 1 - I_A; \tag{1}$$

$$I_{A\cap B} = I_A I_B; \tag{2}$$

$$I_{A\cup B} = I_A + I_B - I_{A\cap B} \tag{3}$$

for any events A, B, and

$$I_{\bigcup_k A_k} = \sum_k I_{A_k} \tag{4}$$

for any sequence $\{A_k\}$ of mutually disjoint events.

A finite linear combination of indicators is a "simple random variable"; $L^1(P)$ is the space of "integrable r.v.'s" (real or complex, as needed); the integral over

Ω of an integrable r.v. X is called its *expectation*, and is denoted by $E(X)$ or EX:

$$E(X) := \int_{\Omega} X\,dP, \quad X \in L^1(P).$$

The functional E on $L^1(P)$ is linear, positive, bounded (with norm 1), and $E1 = 1$. For any $A \in \mathcal{A}$,

$$E(I_A) = P(A).$$

For a simple r.v. X, EX is then the weighted arithmetical average of its values, with weights equal to the probabilities that X assume these values.

The obvious relations

$$P(A^c) = 1 - P(A); \qquad P(A \cup B) = PA + PB - P(A \cap B),$$

parallel (1) and (3); however, the probability analogue of (2), namely $P(A \cap B) = P(A)P(B)$, is *not* true in general. *One says that the events A and B are (stochastically) independent* if

$$P(A \cap B) = P(A)P(B).$$

More generally, a family $\mathcal{F} \subset \mathcal{A}$ of events is (stochastically) independent if

$$P\left(\bigcap_{k \in J} A_k\right) = \prod_{k \in J} P(A_k)$$

for any *finite subset* $\{A_k; k \in J\} \subset \mathcal{F}$.

The random variables $X_1, \ldots, X_n$ are (stochastically) independent if for any choice of Borel sets $B_1, \ldots, B_n$ in $\mathbb{R}$ (or $\mathbb{C}$), the events $X_1^{-1}(B_1), \ldots, X_n^{-1}(B_n)$ are independent.

Theorem I.2.1. *If $X_1, \ldots, X_n$ are (real) independent r.v.'s, and $f_1, \ldots, f_n$ are (real or complex) Borel functions on $\mathbb{R}$, then $f_1(X_1), \ldots, f_n(X_n)$ are independent r.v.'s.*

Proof. For simplicity of notation, we take $n = 2$ (the general case is analogous). Thus, X, Y are independent r.v.'s, and f, g are real (or complex) Borel functions on $\mathbb{R}$. Let A, B be Borel subsets of $\mathbb{R}$ (or $\mathbb{C}$). Then

$$\begin{aligned} P(f(X)^{-1}(A) \cap g(Y)^{-1}(B)) &= P(X^{-1}[f^{-1}(A)] \cap Y^{-1}[g^{-1}(B)]) \\ &= P(X^{-1}[f^{-1}(A)])P(Y^{-1}[g^{-1}(B)]) = P(f(X)^{-1}(A))P(g(Y)^{-1}(B)). \end{aligned}$$

□

In particular, when X, Y are independent r.v.'s, the random variables $aX + b$ and $cY + d$ are independent for any constants a, b, c, d. For example, if X, Y are independent integrable r.v.'s, then $X - EX$ and $Y - EY$ are independent *central* (integrable) r.v.'s, where "central" means "with expectation zero".

Theorem I.2.2 (Multiplicativity of E on independent r.v.'s). *If $X_1, \ldots, X_n$ are independent integrable r.v.'s, then $\prod X_k$ is integrable and*

$$E\left(\prod X_k\right) = \prod E(X_k).$$

Proof. The proof is by induction on n. It suffices therefore to prove the theorem for two independent integrable r.v.'s X, Y.

Case 1. Simple r.v.'s:

$$X = \sum x_j I_{A_j}, \qquad Y = \sum y_k I_{B_k},$$

with all x_j distinct, and all y_k distinct. Thus $A_j = X^{-1}(\{x_j\})$ and $B_k = Y^{-1}(\{y_k\})$ are independent events. Hence

$$\begin{aligned} E(XY) = E\left(\sum_{j,k} x_j y_k I_{A_j} I_{B_k}\right) &= \sum x_j y_k E(I_{A_j \cap B_k}) \\ &= \sum x_j y_k P(A_j \cap B_k) = \sum x_j y_k P(A_j) P(B_k) \\ &= \sum_j x_j P(A_j) \sum_k y_k P(B_k) = E(X)E(Y). \end{aligned}$$

Case 2. Non-negative (integrable) r.v.'s X, Y:
For $n = 1, 2, \ldots$, let

$$A_{n,j} := X^{-1}\left(\left[\frac{j-1}{2^n}, \frac{j}{2^n}\right)\right)$$

and

$$B_{n,k} := Y^{-1}\left(\left[\frac{k-1}{2^n}, \frac{k}{2^n}\right)\right),$$

with $j, k = 1, \ldots, n2^n$. Consider the simple r.v's

$$X_n = \sum_{j=1}^{n2^n} \frac{j-1}{2^n} I_{A_{n,j}},$$

$$Y_n = \sum_{k=1}^{n2^n} \frac{k-1}{2^n} I_{B_{n,k}}.$$

For each n, X_n, Y_n are independent, so that by Case 1, $E(X_n Y_n) = E(X_n)E(Y_n)$. Since the non-decreasing sequences $\{X_n\}, \{Y_n\}$, and $\{X_n Y_n\}$ converge to X, Y, and XY, respectively, it follows from the Lebesgue monotone convergence theorem that

$$E(XY) = \lim E(X_n Y_n) = \lim E(X_n)E(Y_n) = E(X)E(Y)$$

(and in particular, XY is integrable).

Case 3. X, Y real independent integrable r.v.'s:
In this case, $|X|, |Y|$ are independent (by Theorem I.2.1), and by Case 2,

$$E(|XY|) = E(|X|)E(|Y|) < \infty,$$

so that XY is integrable. Also by Theorem I.2.1, X', Y' are independent r.v.'s, where the prime stands for either $+$ or $-$. Therefore, by Case 2,

$$\begin{aligned} E(XY) &= E((X^+ - X^-)(Y^+ - Y^-)) \\ &= E(X^+)E(Y^+) - E(X^-)E(Y^+) - E(X^+)E(Y^-) + E(X^-)E(Y^-) \\ &= [E(X^+) - E(X^-)][E(Y^+) - E(Y^-)] = E(X)E(Y). \end{aligned}$$

The case of complex X, Y follows from Case 3 in a similar fashion. □

Definition I.2.3. If X is a *real* r.v., its *characteristic function* (ch.f.) is defined by

$$f_X(u) := E(\mathrm{e}^{\mathrm{i}uX}) \quad (u \in \mathbb{R}).$$

Clearly f_X is a well-defined complex valued function, $|f_X| \le 1, f_X(0) = 1$, and one verifies easily that it is uniformly continuous on $\mathbb{R}$.

Corollary I.2.4. *The ch.f. of the sum of independent real r.v.'s is the product of their ch.f.'s.*

Proof. If $X_1, \ldots, X_n$ are independent real r.v.'s, it follows from Theorem I.2.1 that $\mathrm{e}^{\mathrm{i}uX_1}, \ldots, \mathrm{e}^{\mathrm{i}uX_n}$ are independent (complex) integrable r.v.'s, and therefore, if $X := \sum X_k$ and $u \in \mathbb{R}$,

$$f_X(u) := E(\mathrm{e}^{\mathrm{i}uX}) = E\left(\prod_k \mathrm{e}^{\mathrm{i}uX_k}\right) = \prod_k E(\mathrm{e}^{\mathrm{i}uX_k}) = \prod_k f_{X_k}(u)$$

by Theorem I.2.2. □

I.2.1 L^2-random variables

Terminology I.2.5. If $X \in L^2(P)$, Schwarz's inequality shows that

$$E(|X|) = E(1.|X|) \le \|1\|_2 \|X\|_2 = \|X\|_2,$$

that is, X is integrable, and

$$\sigma(X) := \|X - EX\|_2 < \infty$$

is called the *standard deviation* (s.d.) of X (this is the L^2-distance from X to its expectation). The square of the s.d. is the *variance* of X.

If X, Y are real L^2-r.v.'s, the product $(X - EX)(Y - EY)$ is integrable (by Schwarz's inequality). One defines the *covariance* of X and Y by

$$\mathrm{cov}(X, Y) := E((X - EX)(Y - EY)).$$

In particular,

$$\operatorname{cov}(X,X)=\sigma^2(X).$$

By Schwarz's inequality,

$$|\operatorname{cov}(X,Y)|\le\sigma(X)\sigma(Y).$$

The linearity of E implies that

$$\operatorname{cov}(X,Y)=E(XY)-E(X)E(Y), \tag{5}$$

and in particular (for $Y=X$),

$$\sigma^2(X)=E(X^2)-(EX)^2. \tag{6}$$

By (5), $\operatorname{cov}(X,Y)=0$ if X,Y are independent (cf. Theorem I.2.2). The converse is false in general, as can be seen by simple counter-examples.

The L^2-r.v.'s X,Y are said to be *uncorrelated* if $\operatorname{cov}(X,Y)=0$.
If $X=I_A$ and $Y=I_B(A,B\in\mathcal{A})$, then by (5),

$$\operatorname{cov}(I_A,I_B)=E(I_AI_B)-E(I_A)E(I_B)=P(A\cap B)-P(A)P(B). \tag{7}$$

Thus, indicators are uncorrelated iff they are independent! Taking $B=A$ (with $PA=p$, so that $P(A^c)=1-p:=q$), we see from (6) that

$$\sigma^2(I_A)=E(I_A)-E(I_A)^2=p-p^2=pq. \tag{8}$$

Lemma I.2.6. *Let $X_1,\ldots,X_n$ be real L^2-r.v.'s. Then*

$$\sigma^2\left(\sum_k X_k\right)=\sum_k\sigma^2(X_k)+2\sum_{1\le j<k\le n}\operatorname{cov}(X_j,X_k).$$

In particular, if X_j are pairwise uncorrelated, then

$$\sigma^2\left(\sum_k X_k\right)=\sum_k\sigma^2(X_k)$$

(BienAyme's identity).

Proof.

$$\begin{aligned}\sigma^2\left(\sum X_k\right)&=E\left(\sum X_k-\sum EX_k\right)^2=E\left(\sum[X_k-EX_k]\right)^2\\&=E\left[\sum(X_k-EX_k)^2+2\sum_{j<k}(X_j-EX_j)(X_k-EX_k)\right]\\&=\sum\sigma^2(X_k)+2\sum_{j<k}\operatorname{cov}(X_j,X_k).\end{aligned}$$

□

Example I.2.7. Let $\{A_k\} \subset \mathcal{A}$ be a sequence of pairwise independent events. Let

$$S_n := \sum_{k=1}^{n} I_{A_k}, \quad n = 1, 2, \ldots$$

Then

$$ES_n = \sum_{k=1}^{n} PA_k; \qquad \sigma^2(S_n) = \sum_{k=1}^{n} PA_k(1 - PA_k).$$

In particular, when $PA_k = p$ for all k (the "Bernoulli case"), we have

$$ES_n = np; \qquad \sigma^2(S_n) = npq.$$

Note that for each $\omega \in \Omega$, $S_n(\omega)$ is the number of events A_k with $k \leq n$ for which $\omega \in A_k$ ("the number of successes in the first n trials").

For $0 \leq j \leq n$, $S_n(\omega) = j$ iff there are precisely j events $A_k, 1 \leq k \leq n$, such that $\omega \in A_k$ (and $\omega \in A_k^c$ for the remaining $n - j$ events). Since there are $\binom{n}{j}$ possibilities to choose j indices k from the set $\{1, \ldots, n\}$ (for which $\omega \in A_k$), and these choices define mutually disjoint events, we have in the Bernoulli case

$$P[S_n = j] = \binom{n}{j} p^j q^{n-j}. \tag{*}$$

One calls S_n the "Bernoulli random variable", and (*) is its *distribution*.

Example I.2.8. Consider random sampling from a two-layered population (see Section I.1, Example 3). Let B_k be the event of getting an object from the layer $\mathcal{P}_1$ in the kth drawing, and let $D_s = \sum_{k=1}^{s} I_{B_k}$. In our previous notations (with $m = 2$),

$$P[D_s = s_1] = \frac{\binom{N_1}{s_1}\binom{N_2}{s_2}}{\binom{N}{s}},$$

where $s_1 + s_2 = s$ and $N_1 + N_2 = N$.

By the equivalence principle of ordered sampling (cf. Section I.1), $PB_k = N_1/N$ for all k, and therefore

$$ED_s = \sum PB_k = s\frac{N_1}{N}.$$

Note that the events B_k are *dependent* (drawing without return!). In the case of drawings with returns, the events B_k are independent, and D_s is the Bernoulli r.v., for which we saw that $ED_s = sp = s(N_1/N)$ (since $p := PB_k$). Note that the expectation is the same in both cases (of drawings with or without returns).

For $1 \leq j < k \leq s$, one has

$$P(B_k \cap B_j) = \frac{N_1}{N}\frac{N_1 - 1}{N - 1}$$

by the equivalence principle of ordered sampling. Therefore, by (7),

$$\operatorname{cov}(I_{B_k}, I_{B_j}) = \frac{N_1}{N}\frac{N_1-1}{N-1} - \left(\frac{N_1}{N}\right)^2,$$

independently of k, j. By Lemma I.2.6,

$$\begin{aligned}\sigma^2(D_s) &= s\frac{N_1}{N}\left(1-\frac{N_1}{N}\right) + s(s-1)\left[\frac{N_1}{N}\frac{N_1-1}{N-1} - \left(\frac{N_1}{N}\right)^2\right] \\ &= \frac{N-s}{N-1}s\frac{N_1}{N}\left(1-\frac{N_1}{N}\right).\end{aligned}$$

Thus, the difference between the variances for the methods of drawing with or without returns appears in the *correcting factor* $(N-s)/(N-1)$, which is close to 1 when the sample size s is small relative to the population size N.

One calls D_s the *hypergeometric random variable.*

Example I.2.9. Suppose we mark $N \geq 1$ objects with numbers $1, \dots, N$. In drawings without returns from this population of objects, let A_k denote the event of drawing precisely the kth object in the kth drawing ("matching" in the kth drawing). In this case, the r.v.

$$S = \sum_{k=1}^{N} I_{A_k}$$

"is" the number of matchings in N drawings.

By the equivalence principle of ordered sampling, $PA_k = 1/N$ and $P(A_k \cap A_j) = (1/N)(1/(N-1))$, independently of k and $j < k$. Hence,

$$ES = \sum_{k=1}^{N} PA_k = 1,$$

$$\operatorname{cov}(I_{A_k}, I_{A_j}) = \frac{1}{N(N-1)} - \frac{1}{N^2},$$

and consequently, by Lemma I.2.6,

$$\sigma^2(S) = N\frac{1}{N}\left(1-\frac{1}{N}\right) + 2\binom{N}{2}\left[\frac{1}{N(N-1)} - \frac{1}{N^2}\right] = 1.$$

Lemma I.2.10. *Let X be any r.v. and $\epsilon > 0$.*

(1) If $X \in L^2(P)$, then

$$P[|X - EX| \geq \epsilon] \leq \frac{\sigma^2(X)}{\epsilon^2}.$$

(Tchebichev's inequality).

(2) If $|X| \le 1$, then

$$P[|X| \ge \epsilon] \ge E(|X|^2) - \epsilon^2.$$

(Kolmogorov's inequality).

Proof. Denote $A = [|X - EX| \ge \epsilon]$. Since $|X - EX|^2 \ge |X - EX|^2 I_A \ge \epsilon^2 I_A$, the monotonicity of E implies that

$$\sigma^2(X) := E(|X - EX|^2) \ge \epsilon^2 E(I_A) = \epsilon^2 P(A),$$

and Part (1) is verified.

In case $|X| \le 1$, denote $A = [|X| \ge \epsilon]$. Then

$$|X|^2 = |X|^2 I_A + |X|^2 I_{A^c} \le I_A + \epsilon^2,$$

and Part (2) follows by applying E. □

Corollary I.2.11. *Let X be an integrable r.v. with $|X - EX| \le 1$. Then for any $\epsilon > 0$,*

$$\sigma^2(X) - \epsilon^2 \le P[|X - EX| \ge \epsilon] \le \frac{\sigma^2(X)}{\epsilon^2}.$$

(Note that X is bounded, hence in $L^2(P)$, so that we may apply Part (1) to X and Part (2) to $X - EX$.)

Corollary I.2.12. *Let $\{A_k\}$ be a sequence of events, and let $X_n = (1/n)\sum_{k=1}^n I_{A_k}$ (the occurrence frequency of the first n events). Then $X_n - EX_n$ converge to zero in probability (i.e. $P[|X_n - EX_n| \ge \epsilon] \to 0$ as $n \to \infty$, for any $\epsilon > 0$) if and only if $\sigma^2(X_n) \to 0$.*

Proof. Since $0 \le I_{A_k}, PA_k \le 1$, we clearly have $|I_{A_k} - PA_k| \le 1$, and therefore

$$|X_n - EX_n| \le \frac{1}{n}\sum_{k=1}^n |I_{A_k} - PA_k| \le 1.$$

The result follows then by applying Corollary I.2.11. □

Example I.2.13. Suppose the events A_k are pairwise independent and $PA_k = p$ for all k. By Example I.2.7, $EX_n = p$ and $\sigma^2(X_n) = pq/n \to 0$, and consequently, by Corollary I.2.12, $X_n \to p$ in probability. This is the *Bernoulli Law of Large Numbers* (the "success frequencies" converge in probability to the probability of success when the number of trials tends to ∞).

Example I.2.14. Let $\{X_k\}$ be a sequence of pairwise uncorrelated L^2-random variables, with

$$EX_k = \mu, \quad \sigma(X_k) = \sigma \qquad (k = 1, 2, \ldots)$$

(e.g., X_k is the outcome of the kth random drawing from an infinite population, or from a finite population with returns). Let

$$M_n := \frac{1}{n}\sum_{k=1}^{n} X_k$$

(the "sample mean" for a sample of size n). Then, by Lemma I.2.6,

$$E(M_n) = \mu; \qquad \sigma^2(M_n) = \sigma^2/n.$$

By Lemma I.2.10,

$$P[|M_n - \mu| \geq \epsilon] \leq \frac{\sigma^2}{n\epsilon^2}$$

for any $\epsilon > 0$, and therefore $M_n \to \mu$ in probability (when $n \to \infty$). This is the so-called *weak law of large numbers* (the sample means converge to the expectation μ when the sample size tends to infinity). The special case $X_k = I_{A_k}$ for pairwise independent events A_k with $PA_k = p$ is precisely the Bernoulli law of large numbers of Example I.2.13.

The Bernoulli law is generalized to *dependent* events in the next section.

Theorem I.2.15 (Generalized Bernoulli law of large numbers). *Let $\{A_k\}$ be a sequence of events. Set*

$$p_1(n) := \frac{1}{n}\sum_{k=1}^{n} PA_k,$$

$$p_2(n) := \frac{1}{\binom{n}{2}}\sum_{1\leq j<k\leq n} P(A_k \cap A_j),$$

and

$$d_n := p_2(n) - p_1^2(n).$$

Let X_n be as in Corollary I.2.12 [the occurrence frequency of the first n events].

Then $X_n - EX_n \to 0$ in probability if and only if $d_n \to 0$.

Proof. By Lemma I.2.6 and relations (7) and (8) preceding it, we obtain by a straightforward calculation

$$\sigma^2(X_n) = \frac{1}{n^2}\sum_{k=1}^{n}[PA_k - (PA_k)^2] + \frac{2}{n^2}\sum_{1\leq j<k\leq n}[P(A_k \cap A_j) - PA_k \cdot PA_j]$$
$$= d_n + (1/n)[p_1(n) - p_2(n)]. \tag{9}$$

Therefore

$$|\sigma^2(X_n) - d_n| \leq (1/n)|p_1(n) - p_2(n)|.$$

However, $p_i(n)$ are arithmetical means of numbers in the interval $[0,1]$, hence they belong to $[0,1]$; therefore $|p_1 - p_2| \leq 1$ and so

$$|\sigma^2(X_n) - d_n| \leq \frac{1}{n}. \tag{10}$$

In particular, $\sigma^2(X_n) \to 0$ iff $d_n \to 0$, and the theorem follows then from Corollary I.2.12. □

Remark I.2.16. Note that when the events A_k are pairwise independent,

$$|d_n| = \left| \frac{2}{n(n-1)} \sum_{1 \le j < k \le n} P(A_k)P(A_j) - \frac{2}{n^2} \sum_{j<k} P(A_k)P(A_j) - \frac{1}{n^2} \sum_{k=1}^{n} P(A_k)^2 \right|$$

$$= (1/n) \left| \frac{1}{\binom{n}{2}} \sum_{j<k} P(A_k)P(A_j) - \frac{1}{n} \sum_{k=1}^{n} P(A_k)^2 \right|.$$

Both arithmetical means between the absolute value signs are means of numbers in [0,1], and are therefore in [0,1]. The distance between them is thus ≤ 1, hence $|d_n| \le 1/n$, and the condition of the theorem is satisfied. Hence, $X_n - EX_n$ converge in probability to zero, even *without* the assumption $PA_k = p$ (for all k) of the Bernoulli case.

We consider next the stronger property of *almost sure convergence to zero of* $X_n - EX_n$.

Theorem I.2.17 (Borel's strong law of large numbers). *With notations as in Theorem I.2.15, suppose that* $d_n = O(1/n)$. *Then* $X_n - EX_n$ *converge to zero almost surely.*

This happens in particular when the events A_k are pairwise independent, hence in the Bernoulli case.

Proof. We first prove the following.

Lemma. *Let* $\{X_n\}$ *be any sequence of r.v.'s such that*

$$\sum_n P[|X_n| \ge 1/m] < \infty$$

for all $m = 1, 2, \ldots$.

Then $X_n \to 0$ *almost surely.*

Proof of lemma. Observe that by definition of convergence to 0,

$$[X_n \to 0] = \bigcap_m \bigcup_n \bigcap_k [|X_{n+k}| < 1/m],$$

where all indices run from 1 to ∞. By DeMorgan's laws, we then have

$$[X_n \to 0]^c = \bigcup_m \bigcap_n \bigcup_k [|X_{n+k}| \ge 1/m]. \tag{11}$$

Denote the "innermost" union in (11) by B_{nm}, and let $B_m := \bigcap_n B_{nm}$. We have (by the σ-subadditivity of P):

$$PB_{nm} \le \sum_k P[|X_{n+k}| \ge 1/m] = \sum_{r=n+1}^{\infty} P[|X_r| \ge 1/m] \to_{n\to\infty} 0$$

by the lemma's hypothesis, for all m. Therefore, since $PB_m \le PB_{nm}$ for all n, we have $PB_m = 0$ (for all m), and consequently

$$P([X_n \to 0]^c) = P(\bigcup_m B_m) = 0.$$

□

Back to the proof of the theorem, recall (10) from Section I.2.15:

$$|\sigma^2(X_n) - d_n| \le 1/n.$$

Since $|d_n| \le c/n$ by hypothesis, we have $\sigma^2(X_n) \le (c+1)/n$. By Tchebichev's inequality,

$$\sum_k P[|X_{k^2} - EX_{k^2}| \ge 1/m] \le m^2 \sum_k \sigma^2(X_{k^2})$$
$$\le (c+1)m^2 \sum_k (1/k^2) < \infty.$$

By the lemma, we then have almost surely

$$X_{k^2} - EX_{k^2} \to 0.$$

For each $n \in \mathbb{N}$, let k be the unique $k \in \mathbb{N}$ such that

$$k^2 \le n < (k+1)^2.$$

Necessarily $n - k^2 \le 2k$ and $k \to \infty$ when $n \to \infty$. We have

$$|X_n - X_{k^2}| = \left| \left(\frac{1}{n} - \frac{1}{k^2} \right) \sum_{j=1}^{k^2} I_{A_j} + \frac{1}{n} \sum_{j=k^2+1}^{n} I_{A_j} \right|$$
$$\le \frac{n-k^2}{nk^2} k^2 + \frac{n-k^2}{n} = 2\frac{n-k^2}{n} \le \frac{4k}{k^2} = 4/k.$$

Hence, also

$$|EX_{k^2} - EX_n| = |E(X_{k^2} - X_n)| \le 4/k,$$

and therefore,

$$|X_n - EX_n| \le |X_n - X_{k^2}| + |X_{k^2} - EX_{k^2}| + |EX_{k^2} - EX_n|$$
$$\le 8/k + |X_{k^2} - EX_{k^2}| \to 0$$

almost surely, when $n \to \infty$. □

I.2.18. Let $\{A_n\}$ be a sequence of events. Recall the notation

$$\limsup A_n := \bigcap_{k=1}^{\infty} \bigcup_{j=k}^{\infty} A_j.$$

This event occurs iff for each $k \in \mathbb{N}$, there exists $j \geq k$ such that A_j occurs, that is, iff *infinitely many A_n occur.*

Lemma I.2.19 (The Borel–Cantelli lemma). *If*

$$\sum P(A_n) < \infty, \tag{*}$$

then

$$P(\limsup A_n) = 0.$$

Proof.

$$P(\limsup A_n) \leq P\Big(\bigcup_{j=k}^{\infty} A_j\Big) \leq \sum_{j=k}^{\infty} P(A_j)$$

for all k, and the conclusion follows from $(*)$ by letting $k \to \infty$. □

Example I.2.20 (the mouse problem). Consider a row of three connected chambers, denoted L (left), M (middle), and R (right). Chamber R has also a right exit to "freedom" (F). Chamber L has also a left exit to a "death" trap D. A mouse, located originally in M, moves to a chamber to its right (left) with a fixed probability p $(q := 1 - p)$. The moves between chambers are independent. Thus, after 2^1 moves, we have

$$P(F_1) = p^2; \quad P(D_1) = q^2; \quad P(M_1) = 2pq,$$

where F_1, D_1, M_1 denote, respectively, the events that the mouse reaches F, D, or M after precisely 2^1 moves. In general, let M_k denote the event that the mouse reaches back M (for the first time) after precisely 2^k moves. Clearly

$$P(M_k) = (2pq)^k.$$

Since $\sum P(M_k) < \infty$, the Borel–Cantelli lemma implies that

$$P(\limsup M_n) = 0,$$

that is, with probability 1, there exists $k \in \mathbb{N} \cup \{0\}$ such that the mouse moves either to F or to D at its 2^{k+1}-th move. The probability of these events is, $(2pq)^k p^2$ and $(2pq)^k q^2$, respectively. Denoting also by F (or D) the event that the mouse reaches freedom (or death) in *some* move, then F is the disjoint union of the F_k (and similarly for D). Thus

$$PF = \sum_k (2pq)^k p^2 = \frac{p^2}{1 - 2pq}$$

and similarly

$$PD = \frac{q^2}{1 - 2pq}.$$

This is coherent with the preceding observation (that with probability 1, either F or D occurs), since the sum of these two probabilities is clearly 1.

The case of events with $\sum P(A_n) = \infty$ is considered in Theorem I.2.21. Notation is as in Theorem I.2.15.

Theorem I.2.21 (Erdos–Renyi). *Let $\{A_n\}$ be a sequence of events such that*

$$\sum P(A_n) = \infty \tag{12}$$

and

$$\liminf \frac{p_2(n)}{p_1^2(n)} = 1. \tag{13}$$

Then

$$P(\limsup A_n) = 1.$$

Proof. Let $X_n = (1/n)\sum_{k=1}^{n} I_{A_k}$. Then

$$EX_n = p_1(n)$$

and

$$\sigma^2(X_n) = p_2(n) - p_1^2(n) + (1/n)[p_1(n) - p_2(n)]$$

(cf. (9) in Theorem I.2.15). By Tchebichev's inequality,

$$\begin{aligned} P[|X_n - p_1(n)| \geq p_1(n)/2] &\leq \frac{\sigma^2(X_n)}{[p_1(n)/2]^2} \\ &= 4\left[\left(1 - \frac{1}{n}\right)\frac{p_2(n)}{p_1^2(n)} - 1 + \frac{1}{np_1(n)}\right]. \end{aligned}$$

By (12),

$$np_1(n) = \sum_{k=1}^{n} PA_k \to \infty. \tag{14}$$

Hence by (13), the $\liminf$ of the right-hand side is 0. Thus

$$\liminf P[|X_n - p_1(n)| \geq p_1(n)/2] = 0. \tag{15}$$

Clearly

$$[X_n < p_1(n)/2] \subset [|X_n - p_1(n)| \geq p_1(n)/2],$$

and therefore

$$\liminf P[X_n < p_1(n)/2] = 0. \tag{16}$$

We may then choose a sequence of integers $1 \leq n_1 < n_2 < \cdots$ such that

$$\sum_k P[X_{n_k} < p_1(n_k)/2] < \infty.$$

By the Borel–Cantelli lemma,

$$P(\limsup[X_{n_k} < p_1(n_k)/2]) = 0,$$

that is, with probability 1, $X_{n_k} < p_1(n_k)/2$ *for only finitely many* ks. Thus, with probability 1, there exists k_0 such that $X_{n_k} \geq p_1(n_k)/2$ for *all* $k > k_0$, that is,

$$\sum_{j=1}^{n_k} I_{A_j} \geq n_k p_1(n_k)/2$$

for all $k > k_0$, and since the right-hand side diverges to ∞ by (14), we have

$$\sum_{j=1}^{\infty} I_{A_j} = \infty$$

with probability 1, that is, infinitely many A_js occur with probability 1. □

Corollary I.2.22. *Let $\{A_n\}$ be a sequence of pairwise independent events such that*

$$\sum PA_n = \infty.$$

Then

$$P(\limsup A_n) = 1.$$

Proof. We show that (13) of the Erdos–Renyi theorem is satisfied. We have

$$\begin{aligned}\binom{n}{2} p_2(n) &= \sum_{1\leq j<k\leq n} P(A_k \cap A_j) = (1/2) \sum_{j\neq k, 1\leq j,k\leq n} P(A_j)P(A_k) \\ &= (1/2)\left[\sum_{j,k=1}^{n} P(A_j)P(A_k) - \sum_{k=1}^{n} P(A_k)^2\right] \\ &= (1/2)\left[n^2 p_1^2(n) - \sum_{k=1}^{n} P(A_k)^2\right].\end{aligned}$$

Therefore

$$\frac{p_2(n)}{p_1^2(n)} = \frac{n}{n-1} - \frac{\sum_{k=1}^{n} P(A_k)^2}{n(n-1)p_1^2(n)}.$$

However, since $PA_k \leq 1$, the sum shown is $\leq \sum_{k=1}^{n} PA_k := np_1(n)$. Therefore, the second (non-negative) term on the right-hand side is $\leq 1/[(n-1)p_1(n)] \to 0$ by (14) (consequence of the divergence hypothesis). Hence, $\lim p_2(n)/p_1^2(n) = 1$. □

Corollary I.2.23 (Zero-one law). *Let $\{A_n\}$ be a sequence of pairwise independent events. Then the event* $\limsup A_n$ *(that infinitely many A_n's occur) has probability 0 or 1, according to whether the series $\sum PA_n$ converges or diverges (respectively).*

I.3 Probability distributions

Let $(\Omega, \mathcal{A}, P)$ be a probability space, let X be a real- (or complex-, or $\mathbb{R}^n$-) valued random variable on Ω, and let $\mathcal{B}$ denote the Borel σ-algebra of the range space ($\mathbb{R}$, etc.). We set

$$P_X(B) := P[X \in B] = P(X^{-1}(B)) \quad (B \in \mathcal{B}).$$

The set function P_X is called the *distribution* of X.

Theorem I.3.1 (stated for the case of a real r.v.).

(1) $(\mathbb{R}, \mathcal{B}, P_X)$ *is a probability space.*

(2) For any finite Borel function g *on* $\mathbb{R}$*, the distribution of the r.v.* $g(X)$ *is given by*

$$P_{g(X)}(B) = P_X(g^{-1}(B)) \quad (B \in \mathcal{B}).$$

(3) If the Borel function g *is integrable with respect to* P_X*, then*

$$E(g(X)) = \int_{\mathbb{R}} g\, dP_X.$$

Proof.

(1) Clearly, $0 \le P_X \le 1$, and $P_X(\mathbb{R}) = P(X^{-1}(\mathbb{R})) = P(\Omega) = 1$. If $\{B_k\} \subset \mathcal{B}$ is a sequence of mutually disjoint sets, then the sets $X^{-1}(B_k)$ are mutually disjoint sets in $\mathcal{A}$, and therefore

$$P_X\Big(\bigcup_k B_k\Big) := P\Big(X^{-1}\Big(\bigcup_k B_k\Big)\Big) = P\Big(\bigcup_k X^{-1}(B_k)\Big)$$
$$= \sum_k P(X^{-1}(B_k)) = \sum P_X(B_k).$$

(2) We have for all $B \in \mathcal{B}$:

$$P_{g(X)}(B) := P(g(X)^{-1}(B)) = P(X^{-1}(g^{-1}(B))) := P_X(g^{-1}(B)).$$

(3) If $g = I_B$ for some $B \in \mathcal{B}$, then $g(X) = I_{X^{-1}(B)}$, and therefore

$$E(g(X)) = P(X^{-1}(B)) := P_X(B) = \int_{\mathbb{R}} g\, dP_X.$$

By linearity, Statement (3) is then valid for simple Borel functions g. If g is a non-negative Borel function, there exists an increasing sequence of non-negative simple Borel functions converging pointwise to g, and (3) follows from the monotone convergence theorem (applied to the measures P and P_X). If g is any (real) Borel function in $L^1(P_X)$, then, by the preceding case, $E(|g(X)|) = \int_{\mathbb{R}} |g|\, dP_X < \infty$, that is, $g(X) \in L^1(P)$, and $E(g(X)) = E(g^+(X) - g^-(X)) = \int_{\mathbb{R}} g^+ dP_X - \int_{\mathbb{R}} g^- dP_X = \int_{\mathbb{R}} g\, dP_X$.

The routine extension to complex g is omitted. □

Definition I.3.2. The *distribution function* of the *real* r.v. X is the function

$$F_X(x) := P_X((-\infty, x)) = P[X < x] \quad (x \in \mathbb{R}).$$

The integral $\int_{\mathbb{R}} g\, dP_X$ is also denoted $\int_{\mathbb{R}} g\, dF_X$.

Proposition I.3.3. F_X *is a non-decreasing, left-continuous function with range in* $[0,1]$*, such that*

$$F_X(-\infty) = 0; \qquad F_X(\infty) = 1. \tag{*}$$

Proof. Exercise.

Definition I.3.4. Any function F with the properties listed in Proposition I.3.3 is called a *distribution function*. If Property (*) is omitted, the function F is called a *quasi-distribution function*.

Any (quasi-) distribution function induces a unique finite positive Lebesgue–Stieltjes measure (it is a probability measure on $\mathbb{R}$ if (*) is satisfied), and integration with respect to that measure is denoted by $\int_{\mathbb{R}} g\,dF$. In case g is a bounded continuous function on $\mathbb{R}$, this integral coincides with the (improper) Riemann–Stieltjes integral $\int_{-\infty}^{\infty} g(x)\,dF(x)$.

The *characteristic function* of the (quasi-) distribution function F is defined by

$$f(u) := \int_{\mathbb{R}} e^{iux}\,dF(x) \quad (x \in \mathbb{R}).$$

By Theorem I.3.1, if $F = F_X$ for a real r.v. X, then f coincides with the ch.f. f_X of Definition I.2.3.

In general, the ch.f. f is a uniformly continuous function on $\mathbb{R}$, $|f| \le 1$, and $f(0) = 1$ in case F satisfies (*).

Proposition I.3.5. *Let X be a real r.v., $b > 0$, and $a \in \mathbb{R}$. Then*

$$f_{a+bX}(u) = e^{iua} f_X(bu) \quad (u \in \mathbb{R}).$$

Proof. Write $Y = a + bX$ and $y = a + bx$ $(x \in \mathbb{R})$. Since $b > 0$,

$$F_X(x) = P[X < x] = P[Y < y] = F_Y(y),$$

and therefore,

$$f_Y(u) = \int_{-\infty}^{\infty} e^{iuy}\,dF_Y(y) = \int_{-\infty}^{\infty} e^{iu(a+bx)}\,dF_X(x) = e^{iua} f_X(bu).$$

□

We consider r.v.'s of class $L^r(P)$ *for* $r \ge 0$. The L^r-"norm", denoted $\|X\|_r$, satisfies (by Theorem I.3.1)

$$\|X\|_r^r := E(|X|^r) = \int_{-\infty}^{\infty} |x|^r dF_X(x).$$

This expression is called the rth *absolute central moment* of X (or of F_X). It always exists, but could be infinite (unless $X \in L^r$).

The rth central moment of X (or F_X) is

$$m_r := E(X^r) = \int_{-\infty}^{\infty} x^r\,dF_X(x).$$

These concepts are used with any quasi-distribution function F, whenever they make sense.

Lemma I.3.6.

(1) The function $\phi(r) := \log E(|X|^r)$ *is convex on* $[0, \infty)$*, and* $\phi(0) = 0$.

(2) $\|X\|_r$ *is a non-decreasing function of* r*. In particular, if* X *is of class* L^r *for some* $r > 0$*, then it is of class* L^s *for all* $0 \le s \le r$.

Proof. For any r.v.'s Y, Z, Schwarz's inequality gives

$$E(|YZ|) \le \|Y\|_2 \|Z\|_2.$$

For $r \ge s \ge 0$, choose $Y = |X|^{(r-s)/2}$ and $Z = |X|^{(r+s)/2}$. Then

$$E(|X|^r) \le [E(|X|^{r-s})E(|X|^{r+s}]^{1/2},$$

so that

$$\phi(r) \le (1/2)[\phi(r-s) + \phi(r+s)],$$

and (1) follows.

The slope of the chord joining the points (0,0) and $(r, \phi(r))$ on the graph of ϕ is $\phi(r)/r$, and it increases with r, by convexity of ϕ. Therefore $\|X\|_r = e^{\phi(r)/r}$ increases with r. □

Theorem I.3.7. *Let* X *be an* L^r *r.v. for some* $r \ge 1$*, and let* $f = f_X$*. Then for all integers* $1 \le k \le r$*, the derivative* $f^{(k)}$ *exists, is uniformly continuous and bounded by* $E(|X|^k)(< \infty)$*, and is given by*

$$f^{(k)}(u) = i^k \int_{-\infty}^{\infty} e^{iux} x^k dF(x), \qquad (*)$$

(where $F := F_X$*). In particular, the moment* m_k *exists and is given by*

$$m_k = f^{(k)}(0)/i^k \quad (k = 1, \ldots, [r]).$$

Proof. By Lemma I.3.6, $E(|X|^k) < \infty$ for $k \le r$, and therefore the integral in (*) converges absolutely and defines a continuous function $g_k(u)$. Also, by Fubini's theorem,

$$\int_0^t g_k(u)du = \int_{-\infty}^{\infty} \int_0^t i^k e^{iux} du x^k dF(x)$$
$$= g_{k-1}(t) - g_{k-1}(0).$$

Assuming (*) for $k-1$, the last expression is equal to $f^{(k-1)}(t) - f^{(k-1)}(0)$. Since the left-hand side is differentiable, with derivative $g_k(t)$, it follows that $f^{(k)}$ exists and equals g_k. Since (*) reduces to the definition of f for $k = 0$, Relation (*) for general $k \le r$ follows by induction. □

Corollary I.3.8. *If the r.v.* X *is in* L^k *for all* $k = 1, 2, \ldots$*, and if* $f := f_X$ *is analytic in some real interval* $|u| < R$*, then*

$$f(u) = \sum_{k=0}^{\infty} i^k m_k u^k / k! \quad (|u| < R).$$

Example I.3.9 (discrete r.v.). Let X be a *discrete* real r.v., that is, its range is the set $\{x_k\}$, with $x_k \in \mathbb{R}$ distinct, and let $P[X = x_k] = p_k$ ($\sum p_k = 1$). Then

$$f_X(u) = \sum_k e^{iux_k} p_k. \tag{1}$$

For the Bernoulli r.v. with parameters n, p, we have $x_k = k \quad (0 \le k \le n)$ and $p_k = \binom{n}{k} p^k q^{n-k}$ ($q := 1 - p$). A short calculation starting from (1) gives

$$f_X(u) = (pe^{iu} + q)^n.$$

By Theorem I.3.7,

$$m_1 = f'_X(0)/i = np,$$
$$m_2 = f''_X(0)/i^2 = (np)^2 + npq,$$

and therefore,

$$\sigma^2(X) = m_2 - m_1^2 = npq$$

(cf. Example I.2.7).

The *Poisson r.v. X with parameter* $\lambda > 0$ assumes exclusively the values k ($k = 0, 1, 2, \ldots$) with

$$P[X = k] = e^{-\lambda}\lambda^k/k!.$$

Then $F_X(x) = 0$ for $x \le 0$, and $= e^{-\lambda}\sum_{k<x}\lambda^k/k!$ for $x > 0$. Clearly, $F_X(\infty) = 1$, and by (1),

$$f_X(u) = e^{-\lambda}\sum_k e^{iuk}\lambda^k/k! = e^{\lambda(e^{iu}-1)}.$$

Note that $E(|X|^n) = e^{-\lambda}\sum_k k^n\lambda^k/k! < \infty$ for all n, and Corollary I.3.8 implies then that $m_k = f^{(k)}(0)/i^k$ for all k. Thus, for example, one calculates that

$$m_1 = \lambda, \qquad m_2 = \lambda(\lambda + 1),$$

and therefore

$$\sigma^2(X) = \lambda.$$

(This can be reached of course directly from the definitions.)

The Poisson distribution is the limit of Bernoulli distributions in the following sense.

Proposition I.3.10. *Let $\{A_{k,n}; k = 0, \ldots, n; n = 1, 2, \ldots\}$ be a "triangular array" of events such that, for each $n = 1, 2, \ldots$, $\{A_{k,n}; k = 0, \ldots, n\}$ is a Bernoulli system with parameter $p_n = \lambda/n$ ($\lambda > 0$ fixed) (cf. Example I.2.7). Let $X_n := \sum_{k=0}^n I_{A_{k,n}}$ be the Bernoulli r.v. corresponding to the nth system. Then the distribution of X_n converges pointwise to the Poisson distribution when $n \to \infty$:*

$$P[X_n = k] \to_{n\to\infty} e^{-\lambda}\lambda^k/k! \quad (k = 0, 1, 2, \ldots).$$

Proof.

$$
\begin{aligned}
P[X_n = k] &= \binom{n}{k} (\lambda/n)^k (1-\lambda/n)^{n-k} \\
&= \left(1-\frac{1}{n}\right)\cdots\left(1-\frac{k-1}{n}\right)\frac{\lambda^k}{k!}\left(1-\frac{\lambda}{n}\right)^n\left(1-\frac{\lambda}{n}\right)^{-k} \\
&\to_{n\to\infty} \frac{\lambda^k}{k!}\mathrm{e}^{-\lambda}.
\end{aligned}
$$

□

Example I.3.11 (Distributions with density). If there exists an $L^1(dx)$-function h such that the distribution function F has the form

$$F(x) = \int_{-\infty}^{x} h(t)\,dt \quad (x \in \mathbb{R}),$$

then h is uniquely determined a.e. (with respect to Lebesgue measure dx on $\mathbb{R}$), and one has a.e. $F'(x) = h(x)$. In particular, $h \geq 0$ a.e., and since it is only determined a.e., one assumes that $h \geq 0$ everywhere, and one calls h (or F') the *density* of F. For any $g \in L^1(F)$,

$$\int_B g\,dF = \int_B gF'\,dx \quad (B \in \mathcal{B}).$$

We consider a few common densities.

I.3.12. *The normal density.*
The "standard" normal (or Gaussian) density is

$$F'(x) = (2\pi)^{-1/2}\mathrm{e}^{-x^2/2}.$$

To verify that the corresponding F is a distribution function, we need only to show that $F(\infty) = 1$. We write

$$F(\infty)^2 = (1/2\pi)\iint_{\mathbb{R}^2} \mathrm{e}^{-(t^2+s^2)/2}\,dt\,ds = (1/2\pi)\int_0^{2\pi}\int_0^{\infty} \mathrm{e}^{-r^2/2} r\,dr\,d\theta = 1,$$

where we used polar coordinates. The ch.f. is

$$
\begin{aligned}
f(u) &= (2\pi)^{-1/2}\int_{\mathbb{R}} \mathrm{e}^{iux}\mathrm{e}^{-x^2/2}\,dx = (2\pi)^{-1/2}\int_{\mathbb{R}} \mathrm{e}^{-[(x-iu)^2+u^2]/2}\,dx \\
&= c(u)\mathrm{e}^{-u^2/2},
\end{aligned}
$$

where $c(u) = (2\pi)^{-1/2}\int_{\mathbb{R}} \mathrm{e}^{-(x-iu)^2/2}\,dx$. The Cauchy integral theorem is applied to the entire function $\mathrm{e}^{-z^2/2}$, with the rectangular path having vertices at $-M$, N, $N - iu$, and $-M - iu$; letting then $M, N \to \infty$, one sees that

$c(u) = (2\pi)^{-1/2} \int_{-\infty}^{\infty} e^{-x^2/2}\,dx = F(\infty) = 1$. Thus the ch.f. of the standard normal distribution is $e^{-u^2/2}$.

If the r.v. X has the standard normal distribution, and $Y = a + bX$ with $a \in \mathbb{R}$ and $b > 0$, then for $y = a + bx$, we have (through the change of variable $s = a + bt$):

$$F_Y(y) = F_X(x) = (2\pi)^{-1/2} \int_{-\infty}^{x} e^{-t^2/2}\,dt = (2\pi b^2)^{-1/2} \int_{-\infty}^{y} e^{-(s-a)^2/2b^2}\,ds.$$

This distribution is called the normal (or Gaussian) distribution with parameters a, b^2 (or briefly, the $N(a, b^2)$ distribution). By Proposition I.3.5, its ch.f. is

$$f_Y(u) = e^{iua} e^{-(bu)^2/2}.$$

Since $E(|Y|^n) < \infty$ for all n, Corollary I.3.8 applies. In particular, one calculates that

$$m_1 = f'_Y(0)/i = a; \qquad m_2 = -f''_Y(0) = a^2 + b^2,$$

so that

$$\sigma^2(Y) = m_2 - m_1^2 = b^2.$$

Thus, the parameters of the $N(a, b^2)$ distribution are its expectation and its variance.

For the $N(0, 1)$ distribution, we write the power series for $f(u) = e^{-u^2/2}$ and deduce from Corollary I.3.8 that $m_{2j+1} = 0$ and $m_{2j} = (2j)!/(2^j j!)$, $j = 0, 1, 2, \ldots$.

Proposition. *The sum of independent normally distributed r.v.'s is normally distributed.*

Proof. Let X_k be $N(a_k, b_k^2)$ distributed independent r.v.'s ($k = 1, \ldots, n$), and let $X = X_1 + \cdots + X_n$. By Corollary I.2.4,

$$f_X(u) = \prod_k f_{X_k}(u) = \prod_k e^{iua_k} e^{-b_k^2 u^2/2} = e^{iua} e^{-b^2 u^2/2}$$

with $a = \sum_k a_k$ and $b^2 = \sum_k b_k^2$. By the uniqueness theorem for ch.f.'s (see Theorem I.4.2), X is $N(a, b^2)$ distributed. □

I.3.13. *The Laplace density.*

$$F'(x) = (1/2b)e^{-|x-a|/b},$$

where $a \in \mathbb{R}$ and $b > 0$ are its "parameters". One calculates that

$$f(u) = \frac{e^{iua}}{1 + b^2 u^2}.$$

In particular, $F(\infty) = f(0) = 1$, so that F is indeed a distribution function. One verifies that

$$m_1 = f'(0)/i = a; \qquad m_2 = f''(0)/i^2 = a^2 + b^2,$$

so that $\sigma^2 = b^2$.

I.3.14. *The Cauchy density.*

$$F'(x) = (1/\pi)\frac{b}{b^2 + (x-a)^2},$$

where $a \in \mathbb{R}$ and $b > 0$ are its "parameters". To calculate its ch.f., one uses the residues theorem with positively oriented rectangles in $\Re z \geq 0$ and in $\Re z \leq 0$ for $u \geq 0$ and $u \leq 0$, respectively, for the function $\mathrm{e}^{\mathrm{i}uz}/(b^2 + (z-a)^2)$. One gets

$$f(u) = \mathrm{e}^{\mathrm{i}ua} e^{-b|u|}. \tag{2}$$

In particular, $f(0) = 1$, so that $F(\infty) = 1$ as needed.

Note that f is not differentiable at 0. Also m_k do not exist for $k \geq 1$.

As in the case of normal r.v.'s, it follows from (2) that the sum of independent Cauchy-distributed r.v.'s is Cauchy distributed, with parameters equal to the sum of the corresponding parameters. This property is not true however for Laplace-distributed r.v.'s.

I.3.15. *The Gamma density.*

$$F'(x) = \frac{b^p}{\Gamma(p)} x^{p-1} e^{-bx} \quad (x > 0),$$

and $F(x) = 0$ for $x \leq 0$. The "parameters" b, p are positive. The special case $p = 1$ gives the *exponential distribution density.*

The function F is trivially non-decreasing, continuous, $F(-\infty) = 0$, and

$$F(\infty) = \Gamma(p)^{-1} \int_0^\infty \mathrm{e}^{-bx} (bx)^{p-1}\, d(bx) = 1,$$

so that F is indeed a distribution function. We have

$$\begin{aligned} f(u) &= \frac{b^p}{\Gamma(p)} \int_0^\infty x^{p-1} \mathrm{e}^{-(b-\mathrm{i}u)x}\, dx \\ &= (b - \mathrm{i}u)^{-p} \frac{b^p}{\Gamma(p)} \int_0^\infty [(b-\mathrm{i}u)x]^{p-1} \mathrm{e}^{-(b-\mathrm{i}u)x}\, d(b-\mathrm{i}u)x. \end{aligned}$$

By Cauchy's integral theorem, integration along the ray $\{(b - \mathrm{i}u)x; x \geq 0\}$ can be replaced by integration along the ray $[0, \infty)$, and therefore the integral equals $\Gamma(p)$, and

$$f(u) = (1 - \mathrm{i}u/b)^{-p}. \tag{3}$$

Thus

$$f^{(k)}(u) = p(p+1)\cdots(p+k-1)(\mathrm{i}/b)^k (1 - \mathrm{i}u/b)^{-(p+k)}$$

and

$$m_k = f^{(k)}(0)/\mathrm{i}^k = p(p+1)\cdots(p+k-1)b^{-k}.$$

In particular,

$$m_1 = p/b, \quad m_2 = p(p+1)/b^2, \quad \sigma^2 = m_2 - m_1^2 = p/b^2.$$

As in the case of the normal distribution, (3) implies the following.

Proposition 1. *The sum of independent Gamma-distributed r.v.'s with parameters $b, p_k, k = 1, \ldots, n$, is Gamma-distributed with parameters $b, \sum p_k$.*

Note the special case of the exponential distribution ($p = 1$):

$$f(u) = (1 - \mathrm{i}u/b)^{-1}; \quad m_1 = 1/b; \quad \sigma^2 = 1/b^2.$$

Another important special case has $p = b = 1/2$. The Gamma distribution with these parameters is called the *standard χ^2 distribution with one degree of freedom.* Its density equals 0 for $x \le 0$, and since $\Gamma(1/2) = \pi^{1/2}$,

$$F'(x) = (2\pi)^{-1/2} x^{-1/2} \mathrm{e}^{-x/2}$$

for $x > 0$.

By Proposition 1, the sum of n independent random variables with the standard χ^2-distribution has the *standard χ^2-distribution with n degrees of freedom*, that is, the Gamma distribution with $p = n/2$ and $b = 1/2$. Its density for $x > 0$ is

$$F'(x) = \frac{1}{2^{n/2}\Gamma(n/2)} x^{n/2-1} \mathrm{e}^{-x/2}.$$

Note that $m_1 = p/b = n$ and $\sigma^2 = p/b^2 = 2n$.

The χ^2 distribution arises naturally as follows. Suppose X is a real r.v. with continuous distribution F_X, and let $Y = X^2$. Then $F_Y(x) := P[Y < x] = 0$ for $x \le 0$, and for $x > 0$,

$$F_Y(x) = P[-x^{1/2} < X < x^{1/2}] = F_X(x^{1/2}) - F_X(-x^{1/2}).$$

If F'_X exists and is continuous on $\mathbb{R}$, then

$$F'_Y(x) = \frac{1}{2x^{1/2}} [F'_X(x^{1/2}) + F'_X(-x^{1/2})]$$

for $x > 0$ (and trivially 0 for $x < 0$). In particular, if X is $N(0,1)$-distributed, then $Y = X^2$ has the density $(2\pi)^{-1/2} x^{-1/2} \mathrm{e}^{-x/2}$ for $x > 0$ (and 0 for $x < 0$), which is precisely the standard χ^2 density for one degree of freedom. Consequently, we have the following.

Proposition 2. *Let $X_1, \ldots, X_n$ be $N(0,1)$-distributed independent r.v.'s and let*

$$\chi^2 := \sum_{k=1}^{n} X_k^2.$$

Then F_{χ^2} is the standard χ^2 distribution with n degrees of freedom (denoted $F_{\chi^2,n}$).

If we start with $N(\mu, \sigma^2)$ independent r.v.'s X_k, the *standardized r.v.'s*

$$Z_k := \frac{X_k - \mu}{\sigma}$$

are independent $N(0,1)$ variables, and therefore the sum $V := \sum_{k=1}^{n} Z_k^2$ has the $F_{\chi^2,n}$ distribution. Hence, if we let

$$\chi_\sigma^2 := \sum_{k=1}^{n} (X_k - \mu)^2,$$

then, for $x > 0$,

$$F_{\chi_\sigma^2}(x) = P[V < x/\sigma^2] = F_{\chi^2,n}(x/\sigma^2).$$

This is clearly the Gamma distribution with parameters $p = n/2$ and $b = 1/2\sigma^2$. In particular, we have then $m_1 = p/b = n\sigma^2$ (not surprisingly!), and $\sigma^2(\chi_\sigma^2) = p/b^2 = 2n\sigma^4$.

I.4 Characteristic functions

Let F be a distribution function on $\mathbb{R}$. Its *normalization* is the distribution function

$$F^*(x) := (1/2)[F(x-0) + F(x+0)] = (1/2)[F(x) + F(x+0)]$$

(since F is left-continuous). Of course, $F^*(x) = F(x)$ at all continuity points x of F.

Theorem I.4.1 (The inversion theorem). *Let f be the ch.f. of the distribution function F. Then*

$$F^*(b) - F^*(a) = \lim_{U\to\infty} \frac{1}{2\pi} \int_{-U}^{U} \frac{e^{-iua} - e^{-iub}}{iu} f(u)du$$

for $-\infty < a < b < \infty$.

Proof. Let J_U denote the integral on the right-hand side. By Fubini's theorem,

$$\begin{aligned} J_U &= (1/2\pi) \int_{-U}^{U} \frac{e^{-iua} - e^{-iub}}{iu} \int_{-\infty}^{\infty} e^{iux}\, dF(x)\, du \\ &= \int_{-\infty}^{\infty} K_U(x)\, dF(x), \end{aligned}$$

where

$$\begin{aligned} K_U(x) :&= (1/2\pi) \int_{-U}^{U} \frac{e^{iu(x-a)} - e^{iu(x-b)}}{iu}\, du \\ &= (1/\pi) \int_{U(x-b)}^{U(x-a)} \frac{\sin t}{t}\, dt. \end{aligned}$$

The convergence of the Dirichlet integral $\int_{-\infty}^{\infty} (\sin t/t) dt$ (to the value π) implies that $|K_U(x)| \le M < \infty$ for all $x \in \mathbb{R}$ and $U > 0$, and as $U \to \infty$,

$$K_U(x) \to \phi(x) := I_{(a,b)}(x) + (1/2)[I_{\{a\}} + I_{\{b\}}](x)$$

pointwise. Therefore, by dominated convergence,

$$\lim_{U\to\infty} J_U = \int_{\mathbb{R}} \phi\, dF = F^*(b) - F^*(a).$$

□

Theorem I.4.2 (The uniqueness theorem). *A distribution function is uniquely determined by its ch.f.*

Proof. Let F, G be distribution functions with ch.f.'s f, g, and suppose that $f = g$. By the inversion theorem,

$$F^*(b) - F^*(a) = G^*(b) - G^*(a)$$

for all real $a < b$. Letting $a \to -\infty$, we get $F^* = G^*$, and therefore $F = G$ at all points where F, G are both continuous. Since these points are dense on $\mathbb{R}$ and F, G are left-continuous, it follows that $F = G$. □

Definition I.4.3. Let C_F denote the set of all continuity points of the quasi-distribution function F (its complement in $\mathbb{R}$ is finite or countable). A sequence $\{F_n\}$ of quasi-distribution functions converges *weakly* to F if $F_n(x) \to F(x)$ for all $x \in C_F$. One writes then $F_n \to_w F$. In case F_n, F are distributions of r.v.'s X_n, X respectively, we also write $X_n \to {}_w X$.

Lemma I.4.4 (Helly–Bray). *Let F_n, F be quasi-distribution functions, $F_n \to_w F$, and suppose $a < b$ are such that $F_n(a) \to F(a)$ and $F_n(b) \to F(b)$. Then*

$$\int_a^b g\, dF_n \to \int_a^b g\, dF$$

for any continuous function g on $[a, b]$.

Proof. Consider partitions

$$a = x_{m,1} < \cdots < x_{m,k_m+1} = b, \quad x_{m,j} \in C_F,$$

such that

$$\delta_m := \sup_j (x_{m,j+1} - x_{m,j}) \to 0$$

as $m \to \infty$. Let

$$g_m = \sum_{j=1}^{k_m} g(x_{m,j}) I_{[x_{m,j}, x_{m,j+1})}.$$

Then

$$\sup_{[a,b]} |g - g_m| \to_{m\to\infty} 0. \tag{1}$$

The hypothesis implies that

$$F_n(x_{m,j+1}) - F_n(x_{m,j}) \to F(x_{m,j+1}) - F(x_{m,j})$$

when $n \to \infty$, for all $j = 1, \ldots, k_m$; $m = 1, 2, \ldots$ Therefore

$$\begin{aligned}\int_a^b g_m\, dF_n &= \sum_{j=1}^{k_m} g(x_{m,j})[F_n(x_{m,j+1}) - F_n(x_{m,j})] \\ &\to_{n\to\infty} \sum_{j=1}^{k_m} g(x_{m,j})[F(x_{m,j+1}) - F(x_{m,j})] \\ &= \int_a^b g_m\, dF \quad (m = 1, 2, \ldots). \qquad (2)\end{aligned}$$

Write

$$\begin{aligned}&\left|\int_a^b g\, dF_n - \int_a^b g\, dF\right| \\ &\le \int_a^b |g - g_m|\, dF_n + \left|\int_a^b g_m\, dF_n - \int_a^b g_m\, dF\right| + \int_a^b |g_m - g|\, dF \\ &\le 2 \sup_{[a,b]} |g - g_m| + \left|\int_a^b g_m\, dF_n - \int_a^b g_m\, dF\right|\end{aligned}$$

If $\epsilon > 0$, we may fix m such that $\sup_{[a,b]} |g - g_m| < \epsilon/4$ (by (1)); for this m, it follows from (2) that there exists n_0 such that the second summand shown is $<\epsilon/2$ for all $n > n_0$. Hence

$$\left|\int_a^b g\, dF_n - \int_a^b g\, dF\right| < \epsilon \quad (n > n_0).$$

□

We consider next integration over $\mathbb{R}$.

Theorem I.4.5 (Helly–Bray). *Let F_n, F be quasi-distribution functions such that $F_n \to_w F$. Then for every $g \in C_0(\mathbb{R})$ (the continuous functions vanishing at ∞),*

$$\int_{\mathbb{R}} g\, dF_n \to \int_{\mathbb{R}} g\, dF.$$

In case F_n, F are distribution functions, the conclusion is valid for all $g \in C_b(\mathbb{R})$ (the bounded continuous functions on $\mathbb{R}$).

Proof. Let $\epsilon > 0$. For $a < b$ in C_F and $g \in C_b(\mathbb{R})$, write

$$\left|\int_{\mathbb{R}} g\, dF_n - \int_{\mathbb{R}} g\, dF\right| \le \int_{[a,b]^c} |g|\, d(F_n + F) + \left|\int_a^b g\, dF_n - \int_a^b g\, dF\right|.$$

In case of quasi-distribution functions and $g \in C_0$, we may choose $[a, b]$ such that $|g| < \epsilon/4$ on $[a, b]^c$; the first term on the right-hand side is then $<\epsilon/2$ for all

n. The second term is $<\epsilon/2$ for all $n > n_0$, by Lemma I.4.4, and the conclusion follows.

In the case of distribution functions and $g \in C_b$, let $M = \sup_{\mathbb{R}} |g|$. Then the first term on the right-hand side is

$$\leq M[F_n(a) + 1 - F_n(b) + F(a) + 1 - F(b)].$$

Letting $n \to \infty$, we have by Lemma I.4.4

$$\limsup \left| \int_{\mathbb{R}} g\, dF_n - \int_{\mathbb{R}} g\, dF \right| \leq 2M[F(a) + 1 - F(b)],$$

for any $a < b$ in C_F. The right-hand side is arbitrarily small, since $F(-\infty) = 0$ and $F(\infty) = 1$. □

Corollary I.4.6. *Let F_n, F be distribution functions such that $F_n \to_w F$, and let f_n, f be their respective ch.f.'s. Then $f_n \to f$ pointwise on $\mathbb{R}$.*

In order to prove a converse to this corollary, we need the following:

Lemma I.4.7 (Helly). *Every sequence of quasi-distribution functions contains a subsequence converging weakly to a quasi-distribution function ("weak sequential compactness of the space of quasi-distribution functions").*

Proof. Let $\{F_n\}$ be a sequence of quasi-distribution functions, and let $\mathbb{Q} = \{x_n\}$ be the sequence of all rational points on $\mathbb{R}$.

Since $\{F_n(x_1)\} \subset [0,1]$, the Bolzano–Weierstrass theorem asserts the existence of a convergent subsequence $\{F_{n1}(x_1)\}$. Again, $\{F_{n1}(x_2)\} \subset [0,1]$, and has therefore a convergent subsequence $\{F_{n2}(x_2)\}$, etc. Inductively, we obtain subsequences $\{F_{nk}\}$ such that the kth subsequence is a subsequence of the $(k-1)$th subsequence, and converges at the points $x_1, \ldots, x_k$. The diagonal subsequence $\{F_{nn}\}$ converges therefore at all the rational points. Let $F_{\mathbb{Q}} := \lim_n F_{nn}$, defined pointwise on $\mathbb{Q}$. For arbitrary $x \in \mathbb{R}$, define

$$F(x) := \sup_{r \in \mathbb{Q}; r \leq x} F_{\mathbb{Q}}(r).$$

Clearly, F is non-decreasing, has range in $[0,1]$, and coincides with $F_{\mathbb{Q}}$ on $\mathbb{Q}$. Its left-continuity is verified as follows: given $\epsilon > 0$, there exists a rational $r < x$ (for $x \in \mathbb{R}$ given) such that $F_{\mathbb{Q}}(r) > F(x) - \epsilon$. If $t \in (r, x)$,

$$F(x) \geq F(t) \geq F(r) = F_{\mathbb{Q}}(r) > F(x) - \epsilon,$$

so that $0 \leq F(x) - F(t) < \epsilon$.

Thus F is a quasi-distribution function.

Given $x \in C_F$, if $r, s \in \mathbb{Q}$ satisfy $r < x < s$, then

$$F_{nn}(r) \leq F_{nn}(x) \leq F_{nn}(s).$$

Therefore

$$F(r) = F_{\mathbb{Q}}(r) \leq \liminf F_{nn}(x) \leq \limsup F_{nn}(x) \leq F_{\mathbb{Q}}(s) = F(s).$$

Hence,

$$F(x) := \sup_{r\in\mathbb{Q};r\leq x} F(r) \leq \liminf F_{nn}(x) \leq \limsup F_{nn}(x) \leq F(s),$$

and since $x \in C_F$, letting $s \to x+$, we conclude that $F_{nn}(x) \to F(x)$. □

Theorem I.4.8 (Paul Levy continuity theorem). *Let F_n be* distribution functions *such that their ch.f.'s f_n converge pointwise to a function g continuous at zero. Then there exists a distribution function F (with ch.f. f) such that $F_n \to_w F$ and $f = g$.*

Proof. Since $|f_n| \leq 1$ (ch.f.'s!) and $f_n \to g$ pointwise, it follows by dominated convergence that

$$\int_0^u f_n(t)dt \to \int_0^u g(t)dt \quad (u \in \mathbb{R}). \tag{3}$$

By Lemma I.4.7, there exists a subsequence $\{F_{n_k}\}$ converging weakly to a quasi-distribution function F. Let f be its ch.f. By Fubini's theorem and Theorem I.4.5,

$$\begin{aligned}\int_0^u f_{n_k}(t)\,dt &= \int_{\mathbb{R}} \frac{\mathrm{e}^{\mathrm{i}ux}-1}{\mathrm{i}x}\,dF_{n_k}(x)\\ &\to_{k\to\infty} \int_{\mathbb{R}} \frac{\mathrm{e}^{\mathrm{i}ux}-1}{\mathrm{i}x}\,dF(x) = \int_0^u f(t)\,dt.\end{aligned}$$

By (3), it follows that $\int_0^u g(t)\,dt = \int_0^u f(t)\,dt$ for all real u, and since both g and f are continuous at zero, it follows that $f(0) = g(0) := \lim f_n(0) = 1$, that is, $F(\infty) - F(-\infty) = 1$, hence necessarily $F(-\infty) = 0$ and $F(\infty) = 1$. Thus F is a distribution function. By Corollary I.4.6, any distribution function that is the weak limit of some subsequence of $\{F_n\}$ has the ch.f. g, and therefore, by Theorem I.4.2, the full sequence $\{F_n\}$ converges weakly to F. □

We proceed now to prove Lyapounov's central limit theorem.

Lemma I.4.9. *Let X be a real r.v. of class L^r $(r \geq 2)$, and let $f := f_X$. Then for any non-negative integer $n \leq r-1$,*

$$f(u) = \sum_{k=0}^{n} m_k(\mathrm{i}u)^k/k! + R_n(u),$$

with

$$|R_n(u)| \leq E(|X|^{n+1})|u|^{n+1}/(n+1)!,$$

for all $u \in \mathbb{R}$.

In particular, if X is a central L^3-r.v., then

$$f(u) = 1 - \sigma^2u^2/2 + R_2(u),$$

where $\sigma^2 := \sigma^2(X)$ and

$$|R_2(u)| \leq E(|X|^3)|u|^3/3! \quad (u \in \mathbb{R}).$$

Proof. Apply Theorem I.3.7 and Taylor's formula. □

Consider next a sequence of *independent central real r.v.'s* $X_k, k = 1, 2, \ldots$ of class L^3. Denote $\sigma_k := \sigma(X_k)$. We assume that $\sigma_k \neq 0$ (i.e., X_k is "non-degenerate", which means that X_k is not a.s. zero) for all k. We fix the following notation:

$$f_k := f_{X_k}; \quad S_n := \sum_{k=1}^{n} X_k; \quad s_n := \sigma(S_n).$$

Of course, $s_n^2 = \sum_{k=1}^{n} \sigma_k^2$. In particular, $s_n \neq 0$ for all n, and we may consider the "standardized" sums S_n/s_n. We denote their ch.f.'s by ϕ_n:

$$\phi_n(u) := f_{S_n/s_n}(u) = \prod_{k=1}^{n} f_k(u/s_n). \tag{4}$$

Finally, we let

$$M_n = s_n^{-3} \sum_{k=1}^{n} E(|X_k|^3) \quad (n = 1, 2, \ldots).$$

Lemma I.4.10. *Let $\{X_k\}$ be as in the previous paragraph, and suppose that $M_n \to 0$ ("Lyapounov's condition"). Then $\phi_n(u) \to \mathrm{e}^{-u^2/2}$ pointwise everywhere on $\mathbb{R}$.*

Proof. By Lemma I.3.6, for all k,

$$\sigma_k \leq \|X_k\|_3. \tag{5}$$

Therefore,

$$\frac{\sigma_k}{s_n} \leq \left[\frac{E(|X_k|^3)}{s_n^3}\right]^{1/3} \leq M_n^{1/3}, \quad k = 1, \ldots, n; \;\; n = 1, 2, \ldots. \tag{6}$$

By (4) and Lemma I.4.9,

$$\log \phi_n(u) = \sum_{k=1}^{n} \log[1 - \frac{\sigma_k^2}{s_n^2} u^2/2 + R_2^{[k]}(u/s_n)]$$

and

$$|R_2^{[k]}(u/s_n)| \leq \frac{E(|X_k|^3)}{s_n^3} |u|^3/3!. \tag{7}$$

Write

$$|\log \phi_n(u) - (-u^2/2)| = \left| \sum_{k=1}^{n} \log\left[1 - \frac{\sigma_k^2}{s_n^2}u^2/2 + R_2^{[k]}(u/s_n)\right] \right.$$
$$\left. - \sum_{k=1}^{n}\left[-\frac{\sigma_k^2}{s_n^2}u^2/2 + R_2^{[k]}(u/s_n)\right] + \sum_{k=1}^{n} R_2^{[k]}(u/s_n)\right|$$
$$\leq \sum_{k=1}^{n}\left|\log(1+z_{k,n}) - z_{k,n}\right| + \sum_{k=1}^{n}\left|R_2^{[k]}(u/s_n)\right|, \tag{8}$$

where

$$z_{k,n} := -\frac{\sigma_k^2}{s_n^2}u^2/2 + R_2^{[k]}(u/s_n).$$

By (6) and (7),

$$|z_{k,n}| \leq M_n^{2/3}u^2/2 + M_n|u|^3/3! \quad (k \leq n). \tag{9}$$

Since $M_n \to 0$ by hypothesis, there exists n_0 such that the right-hand side of (9) is $<1/2$ for all $n > n_0$. Thus

$$|z_{k,n}| < 1/2 \quad (k = 1, \ldots, n; n > n_0). \tag{10}$$

By Taylor's formula, for $|z| < 1$,

$$|\log(1+z) - z| \leq \frac{|z|^2/2}{(1-|z|)^2}.$$

Hence, by (7) and (10),

$$|\log(1+z_{k,n}) - z_{k,n}| \leq 2|z_{k,n}|^2$$
$$\leq \left(\frac{\sigma_k}{s_n}\right)^4 u^4/2 + \left(\frac{\sigma_k}{s_n}\right)^2 \frac{E(|X_k|^3)}{s_n^3}|u|^5/3 + \left(\frac{E(|X_k|^3)}{s_n^3}\right)^2 u^6/18,$$

for $k \leq n$ and $n > n_0$. By (6), the first summand is

$$\leq M_n^{1/3}\frac{E(|X_k|^3)}{s_n^3}u^4/2.$$

The second summand is

$$\leq M_n^{2/3}\frac{E(|X_k|^3)}{s_n^3}|u|^5/3.$$

The third summand is

$$\leq M_n\frac{E(|X_k|^3)}{s_n^3}u^6/18.$$

Therefore (by (8)),

$$\begin{aligned}
&|\log \phi_n(u) - (-u^2/2)| \\
&\leq [M_n^{1/3}u^4/2 + M_n^{2/3}|u|^5/3 + M_n u^6/18 + |u|^3/6] \sum_{k=1}^{n} \frac{E(|X_k|^3)}{s_n^3} \\
&= M_n^{4/3}u^4/2 + M_n^{5/3}|u|^5/3 + M_n^2 u^6/18 + M_n|u|^3/6 \to 0
\end{aligned}$$

as $n \to \infty$. □

Theorem I.4.11 (The Lyapounov central limit theorem). *Let $\{X_k\}$ be a sequence of non-degenerate, real, central, independent, L^3-r.v.'s, such that*

$$\lim_{n\to\infty} s_n^{-3} \sum_{k=1}^{n} E(|X_k|^3) = 0.$$

Then the distribution function of

$$\frac{\sum_{k=1}^{n} X_k}{s_n} \left(:= \frac{S_n}{\sigma(S_n)}\right)$$

converges pointwise to the standard normal distribution as $n \to \infty$.

Proof. By Lemma I.4.10, the ch.f. of S_n/s_n converges pointwise to the ch.f. $e^{-u^2/2}$ of the standard normal distribution (cf. section I.3.12). By Theorem I.4.8 and Theorem I.4.2, the distribution function of S_n/s_n converges pointwise to the standard normal distribution. □

Corollary I.4.12 (Central limit theorem for uniformly bounded r.v.'s). *Let $\{X_k\}$ be a sequence of non-degenerate, real, central, independent r.v.'s such that $|X_k| \leq K$ for all $k \in \mathbb{N}$ and $s_n (:= \sigma(S_n)) \to \infty$. Then the distribution functions of S_n/s_n converge pointwise to the standard normal distribution.*

Proof. We have $E(|X_k|^3) \leq K\sigma_k^2$. Therefore

$$s_n^{-3} \sum_{k=1}^{n} E(|X_k|^3) \leq K/s_n \to 0,$$

and Theorem I.4.11 applies. □

Corollary I.4.13 (Laplace central limit theorem). *Let $\{A_k\}$ be a sequence of independent events with $PA_k = p, 0 < p < 1, k = 1, 2, \ldots$. Let B_n be the (Bernoulli) r.v., whose value is the number of occurrences of the first n events ("number of successes"). Then the distribution function of the "standardized Bernoulli r.v."*

$$B_n^* := \frac{B_n - np}{(npq)^{1/2}}$$

converges pointwise to the standard normal distribution.

Proof. Let $X_k = I_{A_k} - p$. Then X_k are non-degenerate (since $\sigma^2(X_k) = pq > 0$ when $0 < p < 1$), real, central, independent r.v.'s, and $|X_k| < 1$. Also $s_n = (npq)^{1/2} \to \infty$ (since $pq > 0$). By Corollary I.4.12, the distribution function of $S_n/s_n = B_n^*$ converges pointwise to the standard normal distribution. □

Corollary I.4.14 (Central limit theorem for equidistributed r.v.'s). *Let $\{X_k\}$ be a sequence of non-degenerate, real, central, independent, equidistributed, L^3-r.v.'s. Then the distribution function of S_n/s_n converges pointwise to the standard normal distribution.*

Proof. Denote (independently of k, since the r.v.'s are equidistributed):

$$E(|X_k|^3) = \alpha; \qquad \sigma^2(X_k) = \sigma^2(>0).$$

Since X_k are independent, $s_n^2 = n\sigma^2$ by BienAyme's identity, and therefore, as $n \to \infty$,

$$M_n = \frac{n\alpha}{(n\sigma^2)^{3/2}} = (\alpha/\sigma^3)n^{-1/2} \to 0.$$

The result follows now from Theorem I.4.11. □

I.5 Vector-valued random variables

Let $X = (X_1, \ldots, X_n)$ be an $\mathbb{R}^n$-valued r.v. on the probability space $(\Omega, \mathcal{A}, P)$. We say that X has a density if there exists a non-negative Borel function h on $\mathbb{R}^n$ (called a *density of* X, or a *joint density of* $X_1, \ldots, X_n$), such that

$$P[X \in B] = \int_B h\,dx \quad (B \in \mathcal{B}(\mathbb{R}^n)),$$

where $dx = dx_1 \ldots dx_n$ is Lebesgue measure on $\mathbb{R}^n$.

When X has a density h, the later is uniquely determined on $\mathbb{R}^n$ almost everywhere with respect to Lebesgue measure dx (we may then refer to *the* density of X).

Suppose the density of X is of the form

$$h(x) = u(x_1, \ldots, x_k)v(x_{k+1}, \ldots, x_n), \tag{*}$$

for some $1 \le k < n$, where u, v are the densities of $(X_1, \ldots, X_k)$ and $(X_{k+1}, \ldots, X_n)$, respectively. Then for any $A \in \mathcal{B}(\mathbb{R}^k)$ and $B \in \mathcal{B}(\mathbb{R}^{n-k})$, we have by Fubini's theorem

$$\begin{aligned}
&P[(X_1, \ldots, X_k) \in A] \cdot P[(X_{k+1}, \ldots, X_n) \in B] \\
&= \int_A u\,dx_1 \ldots dx_k \cdot \int_B v\,dx_{k+1} \ldots dx_n \\
&= \int_{A \times B} h(x_1, \ldots, x_n) dx_1 \ldots dx_n = P[(X_1, \ldots, X_n) \in A \times B] \\
&= P([(X_1, \ldots, X_k) \in A] \cap [(X_{k+1}, \ldots, X_n) \in B]).
\end{aligned}$$

Thus $(X_1, \ldots, X_k)$ and $(X_{k+1}, \ldots, X_n)$ are independent. Conversely, if $(X_1, \ldots, X_k)$ and $(X_{k+1}, \ldots, X_n)$ are independent with respective densities u, v, then for all "measurable rectangles" $A \times B$ with A, B Borel sets in $\mathbb{R}^k$ and $\mathbb{R}^{n-k}$, respectively,

$$\begin{aligned}
P[(X_1, \ldots, X_n) \in A \times B] &= P([(X_1, \ldots, X_k) \in A] \cap [(X_{k+1}, \ldots, X_n) \in B]) \\
&= P[(X_1, \ldots, X_k) \in A] \cdot P[(X_{k+1}, \ldots, X_n) \in B] \\
&= \int_A u \, dx_1 \ldots dx_k \cdot \int_B v \, dx_{k+1} \ldots dx_n \\
&= \int_{A \times B} uv \, dx_1 \ldots dx_n,
\end{aligned}$$

and therefore

$$P[X \in H] = \int_H uv \, dx$$

for all $H \in \mathcal{B}(\mathbb{R}^n)$, that is, X has a density of the form (*).

We proved the following.

Proposition I.5.1. *Let X be an $\mathbb{R}^n$-valued r.v., and suppose that for some $k \in \{1, \ldots, n-1\}$, X has a density of the form $h = uv$, where $u = u(x_1, \ldots, x_k)$ and $v = v(x_{k+1}, \ldots, x_n)$ are densities for the r.v.'s $(X_1, \ldots, X_k)$ and $(X_{k+1}, \ldots, X_n)$, respectively. Then $(X_1, \ldots, X_k)$ and $(X_{k+1}, \ldots, X_n)$ are independent. Conversely, if, for some k as indicated, $(X_1, \ldots, X_k)$ and $(X_{k+1}, \ldots, X_n)$ are independent with densities u, v, respectively, then $h := uv$ is a density for $(X_1, \ldots, X_n)$.*

If the $\mathbb{R}^n$-r.v. X has the density h and $x = x(y)$ is a C^1-transformation of $\mathbb{R}^n$ with inverse $y = y(x)$ and Jacobian $J \neq 0$, we have

$$P[X \in B] = \int_{x^{-1}(B)} h(x(y)) |J(y)| dy$$

for all $B \in \mathcal{B}(\mathbb{R}^n)$.

Example I.5.2. *The distribution of a sum.*

Suppose (X, Y) has the density $h : \mathbb{R}^2 \to [0, \infty)$. Consider the transformation

$$x = u - v; \quad y = v$$

with Jacobian identically 1 and inverse

$$u = x + y; \quad v = y.$$

The set

$$B = \{(x, y) \in \mathbb{R}^2; x + y < c, y \in \mathbb{R}\}$$

corresponds to the set

$$B' = \{(u, v) \in \mathbb{R}^2; u < c, v \in \mathbb{R}\}.$$

Therefore, by Tonelli's theorem,

$$F_{X+Y}(c) : = P[X+Y<c] = P[(X,Y) \in B] = \int_{B'} h(u-v,v)\,du\,dv$$
$$= \int_{-\infty}^{c} \left(\int_{-\infty}^{\infty} h(u-v,v)\,dv \right) du.$$

This shows that the r.v. $X+Y$ has the density

$$h_{X+Y}(u) = \int_{-\infty}^{\infty} h(u-v,v)\,dv.$$

In particular, when X, Y are independent with respective densities h_X and h_Y, then $X+Y$ has the density

$$h_{X+Y}(u) = \int_{\mathbb{R}} h_X(u-v) h_Y(v)\,dv := (h_X * h_Y)(u),$$

(the *convolution* of h_X and h_Y).

Example I.5.8. *The distribution of a ratio.*

Let Y be a positive r.v., and X any real r.v. We assume that (X,Y) has a density h. The transformation

$$x = uv; \quad y = v$$

has the inverse

$$u = x/y; \quad v = y,$$

and Jacobian $J = v > 0$. Therefore,

$$F_{X/Y}(c) := P[X/Y < c] = P[(X,Y) \in B],$$

where

$$B = \{(x,y); -\infty < x < cy, y > 0\}$$

corresponds to

$$B' = \{(u,v); -\infty < u < c, v > 0\}.$$

Therefore, by Tonelli's theorem,

$$F_{X/Y}(c) = \int_{B'} h(uv,v) v\,du\,dv = \int_{-\infty}^{c} \left(\int_{0}^{\infty} h(uv,v) v\,dv \right) du$$

for all real c. This shows that X/Y has the density

$$h_{X/Y}(u) = \int_{0}^{\infty} h(uv,v) v\,dv \quad (u \in \mathbb{R}).$$

When X, Y are independent, this formula becomes

$$h_{X/Y}(u) = \int_0^\infty h_X(uv)h_Y(v)v\,dv \quad (u \in \mathbb{R}).$$

Let X be an $\mathbb{R}^n$-valued r.v., and let g be a real Borel function on $\mathbb{R}^n$. The r.v. $g(X)$ is called a *statistic*. For example,

$$\bar{X} := (1/n)\sum_{k=1}^n X_k$$

and

$$S^2 := (1/n)\sum_{k=1}^n (X_k - \bar{X})^2$$

are the statistics corresponding to the Borel functions

$$\bar{x}(x_1, \ldots, x_n) := (1/n)\sum_{k=1}^n x_k$$

and

$$s^2(x_1, \ldots, x_n) := (1/n)\sum_{k=1}^n [x_k - \bar{x}(x_1, \ldots, x_n)]^2,$$

respectively.

The statistics $\bar{X}$ and S^2 are called the *sample mean* and the *sample variance*, respectively.

Theorem I.5.4 (Fisher). *Let $X_1, \ldots, X_n$ be independent $N(0, \sigma^2)$-distributed r.v.'s. Let*

$$Z_k := \frac{X_k - \bar{X}}{S}, \quad k = 1, \ldots, n.$$

Then

(1) $(Z_1, \ldots, Z_{n-2}), \bar{X}$, and S are independent;

(2) $(Z_1, \ldots, Z_{n-2})$ has density independent of σ;

(3) $\bar{X}$ is $N(0, \sigma^2/n)$-distributed; and

(4) nS^2 is χ^2_σ-distributed, with $n-1$ degrees of freedom.

Proof. The map sending $X_1, \ldots, X_n$ to the statistics $Z_1, \ldots, Z_{n-2}, \bar{X}$, and S^2 is given by the following equations:

$$\begin{aligned} z_k &= \frac{x_k - \bar{x}}{s}, \quad k = 1, \ldots, n-2; \\ \bar{x} &= (1/n)\sum_{k=1}^n x_k; \\ s^2 &= (1/n)\sum_{k=1}^n (x_k - \bar{x})^2. \end{aligned} \tag{1}$$

Note the relations (for z_k defined as in (1) for *all* $k = 1, \ldots, n$).

$$\sum_{k=1}^{n} z_k = 0; \qquad \sum_{k=1}^{n} z_k^2 = n; \tag{2}$$

$$\sum_{k=1}^{n} x_k^2 = n(\bar{x})^2 + ns^2. \tag{3}$$

By (2),

$$z_{n-1} + z_n = u \left(:= -\sum_{k=1}^{n-2} z_k \right)$$

and

$$z_{n-1}^2 + z_n^2 = w \left(:= n - \sum_{k=1}^{n-2} z_k^2 \right).$$

Thus,

$$(u - z_n)^2 + z_n^2 = w,$$

that is,

$$z_n^2 - uz_n + (u^2 - w)/2 = 0.$$

Therefore,

$$z_n = (u + v)/2; \qquad z_{n-1} = (u - v)/2,$$

where $v := \sqrt{2w - u^2}$; a second solution has v replaced by $-v$. Note that

$$2w - u^2 = 2z_{n-1}^2 + 2z_n^2 - (z_{n-1} + z_n)^2 = (z_{n-1} - z_n)^2 \geq 0,$$

so that v is real.

The inverse transformations are then

$$\begin{aligned} x_k &= \bar{x} + sz_k, k = 1, \ldots, n-2; \\ x_{n-1} &= \bar{x} + s(u - v)/2; \\ x_n &= \bar{x} + s(u + v)/2, \end{aligned}$$

with v replaced by $-v$ in the second inverse; u, v are themselves functions of $z_1, \ldots, z_{n-2}$.

The corresponding Jacobian

$$J := \frac{\partial(x_1, \ldots, x_n)}{\partial(z_1, \ldots, z_{n-2}, \bar{x}, s)}$$

has the form

$$s^{n-2} g\ (z_1, \ldots, z_{n-2}),$$

where g is a function of $z_1, \ldots, z_{n-2}$ only, and does not depend on the parameter σ^2 of the given normal distribution of the $X_k (k = 1, \ldots, n)$.

Replacing v by $-v$ only interchanges the last two columns of the determinant, so that $|J|$ remains unchanged. Therefore, using (3),

$$\begin{aligned} h_X(x)\,dx &= (2\pi\sigma^2)^{-n/2}\mathrm{e}^{-\sum_{k=1}^n x_k^2/2\sigma^2}\,dx_1\ldots dx_n \\ &= 2(2\pi\sigma^2)^{-n/2}\mathrm{e}^{-((n(\bar{x})^2+ns^2)/2\sigma^2)}s^{n-2}g(z_1,\ldots,z_{n-2}) \\ &\quad\times dz_1\ldots dz_{n-2}\,d\bar{x}\,ds, \end{aligned}$$

where the factor 2 comes from the fact that the inverse is bivalued, with the same value of $|J|$ for both possible choices, hence doubling the "mass element".

The last expression can be written in the form

$$h_1(\bar{x})\,d\bar{x}\,h_2(s)\,ds\,h_3(z_1,\ldots,z_{n-2})\,dz_1\ldots dz_{n-2},$$

where

$$h_1(\bar{x}) = \frac{1}{\sqrt{2\pi}\sigma/\sqrt{n}}\,\mathrm{e}^{-((\bar{x})^2/2(\sigma/\sqrt{n})^2)}$$

is the $N(0,\sigma^2/n)$-density;

$$h_2(s) = \frac{n^{(n-1)/2}s^{n-2}}{2^{(n-3)/2}\Gamma((n-1)/2)\sigma^{n-1}}\mathrm{e}^{-ns^2/2\sigma^2} \quad (s>0)$$

(and $h_2(s)=0$ for $s\le 0$) is seen to be a density (i.e., has integral $=1$); and

$$h_3(z_1,\ldots,z_{n-2}) = \frac{\Gamma((n-1)/2)}{n^{n/2}\pi^{(n-1)/2}}g(z_1,\ldots,z_{n-2})$$

is necessarily the density for $(Z_1,\ldots,Z_{n-2})$, and clearly does not depend on σ^2.

The shown decomposition implies by Proposition I.5.1 that Statements 1–3 of the theorem are correct. Moreover, for $x>0$, we have

$$F_{nS^2}(x) = P[nS^2<x] = P[S<\sqrt{x/n}] = \int_0^{\sqrt{x/n}} h_2(s)\,ds.$$

Therefore

$$\begin{aligned} h_{nS^2}(x) &:= \frac{d}{dx}F_{nS^2}(x) = h_2(\sqrt{x/n})(1/2)(x/n)^{-1/2}(1/n) \\ &= \frac{x^{(n-1)/2-1}\mathrm{e}^{-x/2\sigma^2}}{(2\sigma^2)^{(n-1)/2}\Gamma((n-1)/2)}. \end{aligned}$$

This is precisely the χ^2_σ density with $n-1$ degrees of freedom. □

Theorem I.5.4 will be applied in the sequel to obtain the distributions of some important statistics.

Theorem I.5.5. *Let U,V be independent r.v.'s, with U normal $N(0,1)$ and V χ^2_1-distributed with ν degrees of freedom. Then the statistic*

$$T := \frac{U}{\sqrt{V/\nu}}$$

has the density

$$h_\nu(t) = \nu^{-1/2} B(1/2, \nu/2)^{-1} (1 + t^2/\nu)^{-(\nu+1)/2} \quad (t \in \mathbb{R}),$$

called the "t-density" or the "Student density with ν degrees of freedom".

In the previous formula, $B(\cdot\,,\cdot)$ denotes the beta function:

$$B(s,t) = \frac{\Gamma(s)\Gamma(t)}{\Gamma(s+t)},$$

$(s,t > 0)$.

Proof. We apply Example I.5.3 with the independent r.v.'s $X = U$ and $Y = \sqrt{V/\nu}$. The distribution F_Y (for $y > 0$) is given by

$$\begin{aligned} F_Y(y) &:= P[\sqrt{V/\nu} < y] = P[V < \nu y^2] \\ &= 2^{-\nu/2}\Gamma(\nu/2)^{-1} \int_0^{\nu y^2} s^{(\nu/2)-1} e^{-s/2}\, ds \end{aligned}$$

(cf. Section I.3.15). The corresponding density is (for $y > 0$)

$$h_Y(y) := \frac{d}{dy} F_Y(y) = \frac{\nu^{\nu/2}}{2^{\nu/2-1}\Gamma(\nu/2)} y^{\nu-1} e^{-\nu y^2/2},$$

and of course $h_Y(y) = 0$ for $y \le 0$.

By hypothesis, $h_X(x) = e^{-x^2/2}/\sqrt{2\pi}$. By Example I.5.3, the density of T is

$$\begin{aligned} h_T(t) &= \int_0^\infty h_X(vt) h_Y(v) v\, dv \\ &= \frac{\nu^{\nu/2}}{\sqrt{2\pi} 2^{\nu/2-1}\Gamma(\nu/2)} \int_0^\infty e^{-v^2(t^2+\nu)/2} v^\nu\, dv. \end{aligned}$$

Write $s = v^2(t^2+\nu)/2$:

$$h_T(t) = \frac{\nu^{\nu/2}}{\sqrt{\pi}\Gamma(\nu/2)(t^2+\nu)^{(\nu+1)/2}} \int_0^\infty e^{-s} s^{(\nu+1)/2-1}\, ds.$$

Since $\Gamma(1/2) = \sqrt{\pi}$, the last expression coincides with $h_\nu(t)$. □

Corollary I.5.6. *Let $X_1, \ldots, X_n$ be independent $N(\mu, \sigma^2)$-r.v.'s. Then the statistic*

$$T := \frac{\bar{X} - \mu}{S} \sqrt{n-1} \tag{*}$$

has the Student distribution with $\nu = n-1$ degrees of freedom.

Proof. Take $U = (\bar{X} - \mu)/(\sigma/\sqrt{n})$ and $V = nS^2/\sigma^2$. By Fisher's theorem, U, V satisfy the hypothesis of Theorem I.5.5, and the conclusion follows (since the statistic T in Theorem I.5.5 coincides in the present case with (*)). □

Corollary I.5.7. *Let $X_1,\dots,X_n$ and $Y_1,\dots,Y_m$ be independent $N(\mu,\sigma^2)$-r.v.'s. Let $\bar X$, $\bar Y$, S_X^2, and S_Y^2 be the "sample means" and "sample variances" (for the "samples" $X_1,\dots,X_n$ and $Y_1,\dots,Y_m$). Then the statistic*

$$W := \frac{\bar X - \bar Y}{\sqrt{nS_X^2 + mS_Y^2}}\sqrt{(n+m-2)nm/(n+m)}$$

has the Student distribution with $\nu = n+m-2$ degrees of freedom.

Proof. The independence hypothesis implies that $\bar X$ and $\bar Y$ are independent normal r.v.'s with parameters $(\mu,\sigma^2/n)$ and $(\mu,\sigma^2/m)$, respectively. Therefore $\bar X - \bar Y$ is $N(0,\sigma^2(n+m)/nm)$-distributed, and

$$U := \frac{\bar X - \bar Y}{\sigma\sqrt{(n+m)/nm}}$$

is $N(0,1)$-distributed.

By Fisher's theorem, the r.v.'s nS_X^2/σ^2 and mS_Y^2/σ^2 are χ_1^2-distributed with $n-1$ and $m-1$ degrees of freedom, respectively, and are *independent* (as Borel functions of the independent r.v.'s $X_1,\dots,X_n$ and $Y_1,\dots,Y_m$, resp.). Since the χ^2-distribution with r degrees of freedom is the Gamma distribution with $p=r/2$ and $b=1/2$, it follows from Proposition 1 in Section I.3.15 that the r.v.

$$V := \frac{nS_X^2 + mS_Y^2}{\sigma^2}$$

is χ_1^2-distributed with $\nu = (n-1)+(m-1)$ degrees of freedom.

Also, by Fisher's theorem, U,V are independent. We may then apply Theorem I.5.5 to the present choice of U,V. An easy calculation shows that for this choice $T=W$, and the conclusion follows. □

Remark I.5.8. The statistic T is used in "testing hypothesis" about the value of the mean μ of a normal "population", using the "sample outcomes" $X_1,\dots,X_n$. The statistic W is used in testing the "zero-hypothesis" that two normal populations have the same mean (using the outcomes of samples taken from the respective populations). Its efficiency is enhanced by the fact that it is independent of the unknown parameters (μ,σ^2) of the normal population (cf. Section I.6).

Tables of the Student distribution are usually available for $\nu < 30$. For $\nu \ge 30$, the normal distribution is a good approximation. The following theorem supports this fact:

Theorem I.5.9. *Let h_ν be the Student distribution density with ν degrees of freedom. Then as $\nu\to\infty$, h_ν converges pointwise to the $N(0,1)$-density, and*

$$\lim_{\nu\to\infty} P[a \le T < b] = (1/\sqrt{2\pi})\int_a^b \mathrm{e}^{-t^2/2}\,dt \tag{*}$$

for all real $a<b$.

Proof. By Stirling's formula, $\Gamma(n)$ is asymptotically equal to $(n/e)^n$. Therefore (since $\Gamma(1/2) = \sqrt{\pi}$)

$$\lim_{\nu \to \infty} \frac{\nu^{1/2} B(1/2, \nu/2)}{\sqrt{2\pi}}$$

$$= \lim_{\nu} \frac{\Gamma(\nu/2)}{(\nu/2e)^{\nu/2}} \cdot \frac{((\nu+1)/2e)^{(\nu+1)/2}}{\Gamma((\nu+1)/2)} \cdot \frac{e^{1/2}}{(1+1/\nu)^{\nu/2+1/2}} = 1.$$

Hence, as $\nu \to \infty$,

$$h_\nu(t) := [\nu^{1/2} B(1/2, \nu/2)]^{-1} (1 + t^2/\nu)^{-\nu/2-1/2} \to (1/\sqrt{2\pi}) e^{-t^2/2}.$$

For real $a < b$,

$$P[a \le T < b] = [\nu^{1/2} B(1/2, \nu/2)]^{-1} \int_a^b \frac{dt}{(1+t^2/\nu)^{(\nu+1)/2}}.$$

The coefficient before the integral was seen to converge to $1/\sqrt{2\pi}$; the integrand converges pointwise to $e^{-t^2/2}$ and is bounded by 1; therefore (*) follows by dominated convergence. ☐

Theorem I.5.10. *Let U_i be independent χ_1^2-distributed r.v.'s with ν_i degrees of freedom ($i = 1, 2$). Assume $U_2 > 0$, and consider the statistic*

$$F := \frac{U_1/\nu_1}{U_2/\nu_2}.$$

Then F has the distribution density

$$h(u; \nu_1, \nu_2) = \frac{\nu_1^{\nu_1/2} \nu_2^{\nu_2/2}}{B(\nu_1/2, \nu_2/2)} \frac{u^{\nu_1/2-1}}{(\nu_1 u + \nu_2)^{(\nu_1+\nu_2)/2}}$$

for $u > 0$ (and $= 0$ for $u \le 0$).

The discussed density is called the "F-density" or "Snedecor density" with (ν_1, ν_2) degrees of freedom.

Proof. We take in Example I.5.3 the independent r.v.'s $X = U_1/\nu_1$ and $Y = U_2/\nu_2$. We have for $x > 0$:

$$F_X(x) = P[U_1 < \nu_1 x] = [2^{\nu_1/2} \Gamma(\nu_1/2)]^{-1} \int_0^{\nu_1 x} t^{\nu_1/2-1} e^{-t/2} \, dt,$$

and $F_X(x) = 0$ for $x \le 0$. Therefore, for $x > 0$,

$$h_X(x) = \frac{d}{dx} F_X(x) = \frac{(\nu_1/2)^{\nu_1/2}}{\Gamma(\nu_1/2)} x^{\nu_1/2-1} e^{-\nu_1 x/2},$$

and $h_X(x) = 0$ for $x \leq 0$. A similar formula is valid for Y, with ν_2 replacing ν_1. By Example I.5.3, the density of $F := X/Y$ is 0 for $u \leq 0$, and for $u > 0$,

$$\begin{aligned} h_F(u) &= \int_0^\infty h_X(uv)h_Y(v)v\,dv \\ &= \frac{(\nu_1/2)^{\nu_1/2}(\nu_2/2)^{\nu_2/2}}{\Gamma(\nu_1/2)\Gamma(\nu_2/2)} \int_0^\infty (uv)^{\nu_1/2-1}v^{\nu_2/2}e^{-(\nu_1 uv+\nu_2 v)/2}\,dv. \qquad (*) \end{aligned}$$

The integral is

$$u^{\nu_1/2-1}\int_0^\infty \mathrm{e}^{-v(\nu_1 u+\nu_2)/2}v^{(\nu_1+\nu_2)/2-1}\,dv.$$

Making the substitution $v(\nu_1 u + \nu_2)/2 = s$, the integral takes the form

$$\frac{u^{\nu_1/2-1}}{[(\nu_1 u+\nu_2)/2]^{(\nu_1+\nu_2)/2}}\int_0^\infty \mathrm{e}^{-s}s^{(\nu_1+\nu_2)/2-1}\,ds.$$

Since the last integral is $\Gamma((\nu_1+\nu_2)/2)$, it follows from (*) that $h_F(u) = h(u;\nu_1,\nu_2)$, for h as in the theorem's statement. □

Corollary I.5.11. *Let $X_1,\ldots,X_n$ and $Y_1,\ldots,Y_m$ be independent $N(0,\sigma^2)$-distributed r.v.'s. Then the statistic*

$$F := \frac{S_X^2}{S_Y^2}(1-1/m)/(1-1/n)$$

has Snedecor's density with $\nu_1 = n-1$ and $\nu_2 = m-1$ degrees of freedom.

Proof. Let $U_1 := nS_X^2/\sigma^2$ and $U_2 := mS_Y^2/\sigma^2$. By Fisher's theorem, U_i are χ_1^2-distributed with ν_i degrees of freedom ($i = 1,2$). Since they are independent, the r.v. $F = (U_1/\nu_1)/(U_2/\nu_2)$ has the Snedecor density with (ν_1,ν_2) degrees of freedom, by Theorem I.5.10. □

I.5.12. The statistic F of Corollary I.5.11 is used, for example, to test the "zero hypothesis" that two normal "populations" have the same variance (see Section I.6). Statistical tables give u_α, defined by

$$P[F \geq u_\alpha]\left(= \int_{u_\alpha}^\infty h(u;\nu_1,\nu_2)\,du\right) = \alpha,$$

for various values of α and of the degrees of freedom ν_i.

I.6 Estimation and decision

We consider random sampling of size n from a given population. The n outcomes are a value of a $\mathbb{R}^n$-valued r.v. $X = (X_1,\ldots,X_n)$, where X_k have the same distribution function (the "population distribution") $F(.;\theta)$; the "parameter vector" θ is usually unknown. For example, a normal population has the parameter vector (μ, σ^2), etc.

Estimation is concerned with the problem of "estimating" θ by using the sample outcomes $X_1, \ldots, X_n$, say, by means of some Borel function of $X_1, \ldots, X_n$:

$$\theta^* := g(X_1, \ldots, X_n).$$

This statistic is called an *estimator* of θ.

Consider the case of a single real parameter θ.

A measure of the estimator's precision is its mean square deviation from θ,

$$E(\theta^* - \theta)^2,$$

called the *risk function* of the estimator.

We have

$$\begin{aligned} E(\theta^* - \theta)^2 &= E[(\theta^* - E\theta^*) + (E\theta^* - \theta)]^2 \\ &= E(\theta^* - E\theta^*)^2 + 2E(\theta^* - E\theta^*) \cdot (E\theta^* - \theta) + (E\theta^* - \theta)^2. \end{aligned}$$

The middle term vanishes, so that the risk function is equal to

$$\sigma^2(\theta^*) + (E\theta^* - \theta)^2. \tag{*}$$

The difference $\theta - E\theta^*$ is called the *bias* of the estimator θ^*; the estimator is *unbiased* if the bias is zero, that is, if $E\theta^* = \theta$. In this case the risk function is equal to the variance $\sigma^2(\theta^*)$.

Example 1.6.1. We wish to estimate the expectation μ of the population distribution. Any weighted average

$$\mu^* = \sum_{k=1}^{n} a_k X_k \quad \left(a_k > 0, \sum a_k = 1\right)$$

is a reasonable unbiased estimator of μ:

$$E\mu^* = \sum a_k E X_k = \sum a_k \mu = \mu.$$

By BienAyme's identity (since the X_k are independent in random sampling), the risk function is given by

$$\sigma^2(\mu^*) = \sum_{k=1}^{n} a_k^2 \sigma^2(X_k) = \sum a_k^2 \sigma^2,$$

where σ^2 is the population variance (assuming that it exists). However, since $\sum a_k = 1$,

$$\sum a_k^2 = \sum (a_k - 1/n)^2 + 1/n \geq 1/n,$$

and the minimum $1/n$ is attained when $a_k = 1/n$ for all k. Thus, among all estimators of μ that are weighted averages of the sample outcomes, the estimator with minimal risk function is the arithmetical mean $\mu^* = \bar{X}$; its risk function is σ^2/n.

Example I.6.2. As an estimator of the parameter p of a binomial population we may choose the "successes frequency" $p^* := S_n/n$ (cf. Example I.2.7). The r.v. p^* takes on the values k/n ($k = 0, \ldots, n$), and

$$P[p^* = k/n] = P[S_n = k] = \binom{n}{k} p^k q^{n-k}.$$

The estimator p^* is unbiased, since

$$Ep^* = ES_n/n = np/n = p.$$

Its risk function is

$$\sigma^2(p^*) = npq/n^2 = pq/n \leq 1/4n.$$

By Corollary I.4.13, $(p^* - p)/\sqrt{pq/n}$ is approximately $N(0,1)$-distributed for n "large". Thus, for example,

$$P[|p^* - p| < 2\sqrt{pq/n}] > 0.95,$$

and since $pq \leq 1/4$, we surely have

$$P[|p^* - p| < 1/\sqrt{n}] > 0.95.$$

Thus, the estimated parameter p lies in the interval $(p^* - 1/\sqrt{n}, p^* + 1/\sqrt{n})$ with "confidence" 0.95 (at least). We return to this idea later.

In comparing two binomial populations (e.g., in quality control problems), we may wish to estimate the difference $p_1 - p_2$ of their parameters, using samples of sizes n and m from the respective populations. A reasonable estimator is the difference $V = S_n/n - S'_m/m$ of the success frequencies in the two samples. The estimator V is clearly unbiased, and its risk function is (by BienAyme's identity)

$$\sigma^2(V) = \sigma^2(S_n/n) + \sigma^2(S'_m/m) = p_1q_1/n + p_2q_2/m \leq 1/4n + 1/4m.$$

For large samples, we may use the normal approximation (cf. Corollary I.4.13) to test the "zero hypothesis" that $p_1 - p_2 = 0$ (i.e., that the two binomial populations are equidistributed) by using the statistic V.

Example I.6.3. Consider the two-layered population of Example I.2.8. An estimator of the proportion N_1/N of the layer $\mathcal{P}_1$ in the population, could be the sample frequency $U := D_s/s$ of $\mathcal{P}_1$-objects in a random sample of size s.

We have $EU = (1/s)(sN_1/N) = N_1/N$, so that U is unbiased.

The risk function is (cf. Example I.2.8)

$$\sigma^2(U) = (1/s)\frac{N-s}{N-1}\frac{N_1}{N}\left(1 - \frac{N_1}{N}\right) < 1/4s.$$

Example I.6.4. The sample average $\bar{X}$ is a natural unbiased estimator λ^* of the parameter λ of a Poissonian population. Its risk function is $\sigma^2(\lambda^*) = \lambda/n$ (cf. Example I.3.9).

Example I.6.5. Let $X_1, \ldots, X_n$ be independent $N(\mu, \sigma^2)$-r.v.'s. For any weights $a_1, \ldots, a_n$, the statistic

$$V := \frac{n}{n-1}\left[\sum_{k=1}^{n} a_k X_k^2 - \bar{X}^2\right]$$

is an unbiased estimator of σ^2. Indeed,

$$E(\bar{X})^2 = \sigma^2(\bar{X}) + [E\bar{X}]^2 = \sigma^2/n + \mu^2,$$

and therefore,

$$EV = \frac{n}{n-1}\left[\sum_k a_k(\sigma^2 + \mu^2) - (\sigma^2/n + \mu^2)\right] = \frac{n}{n-1}(1 - 1/n)\sigma^2 = \sigma^2.$$

When $a_k = 1/n$ for all k, the estimator V is the "sample error"

$$V = nS^2/(n-1) = \frac{1}{n-1}\sum_{k=1}^{n}(X_k - \bar{X})^2.$$

Example I.6.6. Let $X_1, X_2, \ldots$ be independent r.v.'s with the same distribution. Assume that the moment μ_{2r} of that distribution exists for some $r \geq 1$. For each n, the arithmetical means

$$m_{r,n} := \frac{1}{n}\sum_{k=1}^{n} X_k^r$$

are unbiased estimators of the rth moment μ_r of the distribution. By Example I.2.14 applied to $Y_k = X_k^r$, $m_{r,n} \to \mu_r$ in probability (as $n \to \infty$). We say that the sequence of estimators $\{m_{r,n}\}_n$ of the parameter μ_r is *consistent* (the general definition of consistency is the same, *mutatis mutandis*).

If $\{\theta_n^*\}$ is a consistent sequence of estimators for θ and $\{a_n\}$ is a real sequence converging to 1, then $\{a_n\theta_n^*\}$ is clearly consistent as well. Biased estimators could be consistent (start with any consistent sequence of unbiased estimators θ_n^*; then $((n-1)/n)\theta_n^*$ are still consistent estimators, but their bias is $\theta/n \neq 0$ (unless $\theta = 0$)).

I.6.7. Maximum likelihood estimators (MLEs).

The distribution density $f(x_1, \ldots, x_n; \theta)$ of $(X_1, \ldots, X_n)$, considered as a function $L(\theta)$, is called the *likelihood function* (for $X_1, \ldots, X_n$). The MLE (for $X_1, \ldots, X_n$) is $\theta^*(X_1, \ldots, X_n)$, where $\theta^* = \theta^*(x_1, \ldots, x_n)$ is the value of θ for which $L(\theta)$ is maximal (if such a value exists), that is,

$$f(x_1, \ldots, x_n; \theta^*) \geq f(x_1, \ldots, x_n; \theta)$$

for all $(x_1,\ldots,x_n)\in\mathbb{R}^n$ and θ in the relevant range. Hence,

$$P_{\theta^*}[X\in B]\geq P_\theta[X\in B]$$

for all $B\in\mathcal{B}(\mathbb{R}^n)$ and θ, where the subscript of P means that the distribution of X is taken with the parameter indicated.

We consider the case when $X_1,\ldots,X_n$ are independent $N(\mu,\sigma^2)$-r.v.'s. Thus

$$L(\theta)=(2\pi\sigma^2)^{-n/2}\mathrm{e}^{-\sum(x_k-\mu)^2/2\sigma^2}\quad \theta=(\mu,\sigma^2).$$

Maximizing L is equivalent to maximizing the function

$$\phi(\theta):=\log L(\theta)=-(n/2)\log(2\pi\sigma^2)-\sum(x_k-\mu)^2/2\sigma^2.$$

Case 1. MLE for μ when σ is given. The necessary condition for $\mu=\mu^*$,

$$\frac{\partial\phi}{\partial\mu}=\sum(x_k-\mu)/\sigma^2=0, \tag{1}$$

implies that $\mu^*=\mu^*(x_1,\ldots,x_n)=\bar{x}$. This is indeed a maximum point, since

$$\frac{\partial^2\phi}{\partial\mu^2}=-n/\sigma^2<0.$$

Thus the MLE for μ (when σ is given) is

$$\mu^*(X_1,\ldots,X_n)=\bar{X}.$$

Case 2. MLE for σ^2 when μ is given. The solution of the equation

$$\frac{\partial\phi}{\partial(\sigma^2)}=-n/2\sigma^2+(1/2\sigma^4)\sum(x_k-\mu)^2=0 \tag{2}$$

is

$$(\sigma^2)^*=(1/n)\sum(x_k-\mu)^2.$$

Since the second derivative of ϕ at $(\sigma^2)^*$ is equal to $-n/2[(\sigma^2)^*]^2<0$, we obtained indeed a maximum point of ϕ, and the corresponding MLE for σ^2 is

$$(\sigma^2)^*(X_1,\ldots,X_n)=(1/n)\sum(X_k-\mu)^2.$$

Case 3. MLE for $\theta:=(\mu,\sigma^2)$ (as an unknown vector parameter). We need to solve the equations (1) and (2) simultaneously. From (1), we get $\mu^*=\bar{x}$; from (2) we get

$$(\sigma^2)^*=(1/n)\sum(x_k-\mu^*)^2=(1/n)\sum(x_k-\bar{x})^2:=s^2.$$

The solution $(\bar{x},s^2)$ is indeed a maximum point for ϕ, since the Hessian for ϕ at this point equals $n^2/2s^6>0$, and the second partial derivative of ϕ with respect to μ (at this point) equals $-n/s^2<0$. Thus the MLE for θ is

$$\theta^*(X_1,\ldots,X_n)=(\bar{X},S^2).$$

Note that the estimator S^2 is *biased*, since

$$ES^2 = \frac{n-1}{n}\sigma^2,$$

but *consistent*: indeed,

$$S^2 = (1/n)\sum_{k=1}^{n} X_k^2 - \left[(1/n)\sum_{k=1}^{n} X_k\right]^2;$$

by the weak law of large numbers, the first average on the right-hand side converges in probability to the second moment μ_2, while the second average converges to the first moment $\mu_1 = \mu$; hence S^2 converges in probability to $\mu_2 - \mu^2 = \sigma^2$ (when $n \to \infty$).

I.6.1 Confidence intervals

I.6.8. Together with the estimator θ^* of a real parameter θ, it is useful to have an interval around θ^* that contains θ with some high probability $1-\alpha$ (called the *confidence* of the interval). In fact, θ *is not a random variable*, and the rigorous approach is to find an interval $(a(\theta), b(\theta))$ such that

$$P[\theta^* \in (a(\theta), b(\theta))] = 1 - \alpha. \tag{3}$$

The corresponding interval for θ is a $(1-\alpha)$-confidence interval for θ.

Example 1. Consider an $N(\mu, \sigma^2)$-population with known variance. Let

$$Z := \frac{\bar{X} - \mu}{\sigma/\sqrt{n}}.$$

Then Z is $N(0,1)$-distributed. In our case, (3) for the estimator $\bar{X}$ of μ takes the equivalent form

$$P\left[\frac{a(\mu)-\mu}{\sigma/\sqrt{n}} < Z < \frac{b(\mu)-\mu}{\sigma/\sqrt{n}}\right] = 1 - \alpha. \tag{4}$$

For simplicity, take a symmetric Z-interval,

$$\frac{b(\mu)-\mu}{\sigma/\sqrt{n}} = c; \qquad \frac{a(\mu)-\mu}{\sigma/\sqrt{n}} = -c,$$

that is,

$$a(\mu) = \mu - c\sigma/\sqrt{n}; \qquad b(\mu) = \mu + c\sigma/\sqrt{n}.$$

Let Φ denote the $N(0,1)$-distribution function. By symmetry of the normal density,

$$\Phi(-c) = 1 - \Phi(c),$$

and therefore (4) takes the form

$$1 - \alpha = \Phi(c) - \Phi(-c) = 2\Phi(c) - 1,$$

that is,

$$\Phi(c) = 1 - \alpha/2,$$

and we get the unique solution for c:

$$c = \Phi^{-1}(1 - \alpha/2) := z_{1-\alpha/2}. \tag{5}$$

By symmetry of the normal density, $-c = z_{\alpha/2}$. The interval for the estimator $\bar{X}$ is then

$$\mu + z_{\alpha/2}\sigma/\sqrt{n} := a(\mu) < \bar{X} < b(\mu) := \mu + z_{1-\alpha/2}\sigma/\sqrt{n},$$

and the corresponding $(1 - \alpha)$-confidence interval for μ is

$$\bar{X} - z_{1-\alpha/2}\sigma/\sqrt{n} < \mu < \bar{X} - z_{\alpha/2}\sigma/\sqrt{n}.$$

Example 2. Consider an $N(\mu, \sigma^2)$-population with *both parameters unknown.* We still use the MLE $\mu^* = \bar{X}$. By Corollary I.5.6, the statistic

$$T := \frac{\bar{X} - \mu}{S}\sqrt{n-1}$$

has the Student distribution with $n - 1$ degrees of freedom. By symmetry of the Student density, the argument in Example 1 applies in this case by replacing Φ with $F_{T,n-1}$, the Student distribution function for $n - 1$ degrees of freedom, and $\sigma/\sqrt{n}$ by $S/\sqrt{n-1}$. Let

$$t_{\gamma,n-1} := F_{T,n-1}^{-1}(\gamma).$$

Then a $(1 - \alpha)$-confidence interval for μ is

$$\bar{X} - t_{1-\alpha/2,n-1}S/\sqrt{n-1} < \mu < \bar{X} - t_{\alpha/2,n-1}S/\sqrt{n-1}.$$

Example 3. In the context of Example 2, we look for a $(1 - \alpha)$-confidence interval for σ^2. By Fisher's theorem, the statistic $V := nS^2/\sigma^2$ has the χ_1^2 distribution with $n - 1$ degrees of freedom (denoted for simplicity by F_{n-1}). Denote

$$\chi^2_{\gamma,n-1} = F_{n-1}^{-1}(\gamma) \quad (\gamma \in \mathbb{R}).$$

Choosing

$$a = \chi^2_{\alpha/2,n-1}; \qquad b = \chi^2_{1-\alpha/2,n-1},$$

we get

$$P[a < V < b] = F_{n-1}(b) - F_{n-1}(a) = (1 - \alpha/2) - \alpha/2 = 1 - \alpha,$$

which is equivalent to

$$P\left[\frac{nS^2}{\chi^2_{1-\alpha/2,n-1}} < \sigma^2 < \frac{nS^2}{\chi^2_{\alpha/2,n-1}}\right] = 1 - \alpha,$$

from which we read off the wanted $(1 - \alpha)$-confidence interval for σ^2.

I.6.2 Testing of hypothesis and decision

I.6.9. Let $X_1, \ldots, X_n$ be independent r.v.'s with common distribution $F(.;\theta)$, with an unknown parameter θ. A *simple hypothesis* is a hypothesis of the form

$$H_0 : \theta = \theta_0.$$

This is the so-called "zero hypothesis".

We may consider an "alternative hypothesis" that is also simple, that is,

$$H_1 : \theta = \theta_1.$$

Let $P_{\theta_i} (i = 0, 1)$ denote the probability of any event "involving" $X_1, \ldots, X_n$, under the assumption that their common distribution is $F(.;\theta_i)$.

The set $C \in \mathcal{B}(\mathbb{R}^n)$ is called the *rejection region* of a statistical test if the zero hypothesis is rejected when $X \in C$, where $X = (X_1, \ldots, X_n)$.

The *significance* of the test is the probability α of rejecting H_0 when H_0 is true.

For the simple hypothesis H_0,

$$\alpha = \alpha(C) := P_{\theta_0}[X \in C].$$

Similarly, the probability

$$\beta = \beta(C) := P_{\theta_1}[X \in C]$$

is called the *power* of the test. It is the probability of rejecting H_0 when the alternative hypothesis H_1 is true.

It is clearly desirable to choose C such that α is minimal and β is maximal. The following result goes in this direction.

Lemma I.6.10 (Neyman–Pearson). *Suppose the population distribution has the density $h(.;\theta)$. For $k \in \mathbb{R}$, let*

$$C_k := \left\{(x_1, \ldots, x_n) \in \mathbb{R}^n; \prod_{j=1}^{n} h(x_j;\theta_1) > k \prod_{j=1}^{n} h(x_j;\theta_0)\right\}.$$

Then among all $C \in \mathcal{B}(\mathbb{R}^n)$ with $\alpha(C) \le \alpha(C_k)$, the set C_k has maximal power.

In symbols, $\beta(C) \le \beta(C_k)$ for all C with $\alpha(C) \le \alpha(C_k)$.

Proof. Let $C \in \mathcal{B}(\mathbb{R}^n)$ be such that $\alpha(C) \le \alpha(C_k)$. Denote $D = C \cap C_k$. Since

$$C - D = C \cap C_k^c \subset C_k^c,$$

we have

$$
\begin{aligned}
\beta(C-D) &= \beta(C\cap C_k^c) := P_{\theta_1}[X\in C\cap C_k^c] \\
&= \int_{C\cap C_k^c}\prod_j h(x_j;\theta_1)\,dx_1\cdots dx_n \le k\int_{C\cap C_k^c}\prod_j h(x_j;\theta_0)\,dx_1\cdots dx_n \\
&= kP_{\theta_0}[X\in C-D] = kP_{\theta_0}[X\in C]-kP_{\theta_0}[X\in D] \\
&\le kP_{\theta_0}[X\in C_k]-kP_{\theta_0}[X\in D] = kP_{\theta_0}[X\in C_k-D] \\
&= k\int_{C_k-D}\prod_j h(x_j;\theta_0)dx_1\cdots dx_n \\
&\le \int_{C_k-D}\prod_j h(x_j;\theta_1)\,dx_1\cdots dx_n = P_{\theta_1}[X\in C_k-D] = \beta(C_k-D).
\end{aligned}
$$

Hence

$$\beta(C) = \beta(C-D)+\beta(D) \le \beta(C_k-D)+\beta(D) = \beta(C_k).$$

□

Note that the proof does not depend on the special form of the joint density of $(X_1,\ldots,X_n)$. Thus, if C_k is defined using the *joint density* (with the values θ_i of the parameter), the Neyman–Pearson lemma is valid *without the independence assumption* on $X_1,\ldots,X_n$.

Application I.6.11. Suppose F is the $N(\mu,\sigma^2)$ distribution with σ^2 known, and consider simple hypothesis

$$H_i : \mu = \mu_i, \quad i=0,1.$$

For $k>0$, we have (by taking logarithms):

$$
\begin{aligned}
C_k &= \Big\{(x_1,\ldots,x_n)\in\mathbb{R}^n; (-1/2\sigma^2)\sum_{j=1}^n[(x_j-\mu_1)^2-(x_j-\mu_0)^2] > \log k\Big\} \\
&= \Big\{(x_1,\ldots,x_n); (\mu_1-\mu_0)\Big[\sum x_j - n(\mu_1+\mu_0)/2\Big] > \sigma^2\log k\Big\}.
\end{aligned}
$$

Denote

$$k^* := \frac{\mu_1+\mu_0}{2} + (\sigma^2/n)\frac{\log k}{\mu_1-\mu_0}.$$

Then if $\mu_1>\mu_0$,

$$C_k = \{(x_1,\ldots,x_n); \bar{x} > k^*\},$$

and if $\mu_1<\mu_0$,

$$C_k = \{(x_1,\ldots,x_n); \bar{x} < k^*\}.$$

We choose the rejection region $C=C_k$ for maximal power (by the Neyman–Pearson lemma). Note that it is determined by the statistic $\bar{X}$: H_0 is rejected if $\bar{X} > k^*$ (in case $\mu_1 > \mu_0$). The *critical value* k^* is found by means of the requirement that *the significance be equal to some given* α:

$$\alpha(C_k) = \alpha,$$

that is, when $\mu_1 > \mu_0$,

$$P_{\mu_0}[\bar{X} > k^*] = \alpha.$$

Since $\bar{X}$ is $N(\mu_0, \sigma^2/n)$-distributed under the hypothesis H_0, the statistic

$$Z := \frac{\bar{X} - \mu_0}{\sigma/\sqrt{n}}$$

is $N(0,1)$-distributed. Using the notation of Example 1 in Section I.6.8, we get

$$\alpha = P_{\mu_0}\left[Z > \frac{k^* - \mu_0}{\sigma/\sqrt{n}}\right] = 1 - \Phi\left(\frac{k^* - \mu_0}{\sigma/\sqrt{n}}\right),$$

so that

$$\frac{k^* - \mu_0}{\sigma/\sqrt{n}} = z_{1-\alpha},$$

and

$$k^* = \mu_0 + z_{1-\alpha}\sigma/\sqrt{n}.$$

We thus arrive to the "optimal" rejection region

$$C_k = \{(x_1, \ldots, x_n) \in \mathbb{R}^n; \bar{x} > \mu_0 + z_{1-\alpha}\sigma/\sqrt{n}\}$$

for significance level α (in case $\mu_1 > \mu_0$).

An analogous calculation for the case $\mu_1 < \mu_0$ gives the "optimal" rejection region at significance level α

$$C_k = \{(x_1, \ldots, x_n); \bar{x} < \mu_0 + z_\alpha\sigma/\sqrt{n}\}.$$

Application I.6.12. Suppose again that F is the $N(\mu, \sigma^2)$ distribution, this time with μ known. Consider the simple hypothesis

$$H_i : \sigma = \sigma_i, \quad i = 0, 1.$$

We deal with the case $\sigma_1 > \sigma_0$ (the other case is analogous).

For $k > 0$ given, the Neyman–Pearson rejection region is (after taking logarithms)

$$C_k = \left\{x = (x_1, \ldots, x_n) \in \mathbb{R}^n; \sum_{j=1}^{n}(x_j - \mu)^2 > k^*\right\},$$

where

$$k^* := \frac{2\log k + n\log(\sigma_1/\sigma_0)}{\sigma_0^{-2} - \sigma_1^{-2}}.$$

We require the significance level α, that is,

$$\alpha = \alpha(C_k) = P_{\sigma_0}\left[\sum (X_j - \mu)^2 > k^*\right].$$

Since $(X_j - \mu)/\sigma_0^2$ are independent $N(0,1)$-distributed r.v.'s (under the hypothesis H_0), the statistic

$$\chi^2 := (1/\sigma_0^2)\sum_{j=1}^{n}(X_j - \mu)^2$$

has the standard χ_1^2 distribution with n degrees of freedom (cf. Section I.3.15, Proposition 2). Thus

$$\alpha = P_{\sigma_0}[\chi^2 > k^*/\sigma_0^2] = 1 - F_{\chi_1^2}(k^*/\sigma_0^2).$$

Denote

$$c_\gamma = F_{\chi_1^2}^{-1}(\gamma) \quad (\gamma > 0)$$

(for n degrees of freedom). Then $k^* = \sigma_0^2 c_{1-\alpha}$, and the Neyman–Pearson rejection region for H_0 at significance level α is

$$C_k = \left\{x \in \mathbb{R}^n; \sum (x_j - \mu)^2 > \sigma_0^2 c_{1-\alpha}\right\}.$$

I.6.3 Tests based on a statistic

I.6.13. Suppose we wish to test the hypothesis

$$H_0 : \theta = \theta_0$$

against the alternative hypothesis

$$H_1 : \theta \neq \theta_0$$

about the parameter θ of the population distribution $F = F(.;\theta)$.

Let $X_1, \ldots, X_n$ be independent F-distributed r.v.'s (i.e., a random sample from the population), and suppose that the distribution $F_{g(X)}$ of some statistic $g(X)$ (where $g : \mathbb{R}^n \to \mathbb{R}$ is a Borel function) is known explicitely. Denote this distribution, under the hypothesis H_0, by F_0. It is reasonable to reject H_0 when $g(X) > c$ ("one-sided test") or when either $g(X) < a$ or $g(X) > b$ ("two-sided test"), where c and $a < b$ are some "critical values". The corresponding rejection regions are

$$C = \{x \in \mathbb{R}^n; g(x) > c\} \tag{6}$$

and

$$C = \{x \in \mathbb{R}^n; g(x) < a \text{ or } g(x) > b\}. \tag{7}$$

For the one-sided test, the significance α requirement is (assuming that F_0 is continuous):

$$\alpha = \alpha(C) := P_{\theta_0}[g(X) > c] = 1 - F_0(c),$$

and the corresponding *critical value* of c is

$$c_\alpha = F_0^{-1}(1-\alpha).$$

In case (7) (which is more adequate for the "decision problem" with the alternative hypothesis H_1), it is convenient to choose the values a, b by requiring

$$P_{\theta_0}[g(X) < a] = P_{\theta_0}[g(X) > b] = \alpha/2,$$

which is sufficient for having $\alpha(C) = \alpha$. The *critical values* of a, b are then

$$a_\alpha = F_0^{-1}(\alpha/2); \qquad b_\alpha = F_0^{-1}(1-\alpha/2). \tag{8}$$

Example I.6.14. *The z-test*

Suppose F is the normal distribution with known variance. We wish to test the hypothesis

$$H_0 : \mu = \mu_0$$

against

$$H_1 : \mu \neq \mu_0.$$

Using the $N(0,1)$-distributed statistic Z as in Application I.6.11, the two-sided critical values at significance level α are

$$a_\alpha = z_{\alpha/2}; \qquad b_\alpha = z_{1-\alpha/2}.$$

By symmetry of the normal density, $a_\alpha = -b_\alpha$. The zero hypothesis is rejected (at significance level α) if either $Z < z_{\alpha/2}$ or $Z > z_{1-\alpha/2}$, that is, if $|Z| > z_{1-\alpha/2}$.

Example I.6.15. *The t-test*

Suppose *both parameters* of the *normal* distribution F are unknown, and consider the hypothesis H_i of Example I.6.14. By Corollary I.5.6, the statistic

$$T := \frac{\bar{X} - \mu}{S/\sqrt{n-1}}$$

has the Student distribution with $n-1$ degrees of freedom. With notation as in Section I.6.8, Example 2, the critical values (at significance level α, for the test based on the statistic T) are

$$a_\alpha = t_{\alpha/2,n-1}; \qquad b_\alpha = t_{1-\alpha/2,n-1}.$$

By symmetry of the Student density, the zero hypothesis is rejected (at significance level α) if $|T| > t_{1-\alpha/2,n-1}$.

Example I.6.16. *Comparing the means of two normal populations.*

The zero hypothesis is that the two populations have the same normal distribution.

Two random samples $X_1, \ldots, X_n$ and $Y_1, \ldots, Y_m$ are taken from the respective populations. Under the zero hypothesis H_0, the statistic W of

Corollary I.5.7 has the Student distribution with $\nu = n + m - 2$ degrees of freedom. By symmetry of this distribution, the two-sided test at significance level α rejects H_0 if $|W| > t_{1-\alpha/2,\nu}$.

Example I.6.17. *Comparing the variances of two normal populations.*
With H_0, X, and Y as in Example I.6.16, the statistic F of Corollary I.5.11 has the Snedecor distribution with $(n-1, m-1)$ degrees of freedom. If F_0 is this distribution, the critical values at significance level α are given by (8), Section I.6.13.

I.7 Conditional probability

I.7.1 Heuristics

Let $(\Omega, \mathcal{A}, P)$ be a probability space, and $A_i, B_j \in \mathcal{A}$. Consider a two-stage experiment, with possible outcomes $A_1, \ldots, A_m$ in Stage 1 and $B_1, \ldots, B_n$ in Stage 2.

On the basis of the "counting principle", it is intuitively acceptable that

$$P(A_i \cap B_j) = P(A_i)P(B_j|A_i), \tag{1}$$

where $P(B_j|A_i)$ denotes the so-called *conditional probability* that B_j will occur in Stage 2, *when it is given that* A_i *occurred in Stage 1.* We take (1) as the *definition* of $P(B_j|A_i)$ (whenever $P(A_i) \neq 0$).

Definition I.7.1. If $A \in \mathcal{A}$ has $PA \neq 0$, the conditional probability of $B \in \mathcal{A}$ given A is

$$P(B|A) := \frac{P(A \cap B)}{PA}.$$

It is clear that $P(.|A)$ is a probability measure on $\mathcal{A}$. For any $L^1(P)$ real r.v. X, the expectation of X relative to $P(.|A)$ makes sense. It is called the *conditional expectation* of X *given* A, and is denoted by $E(X|A)$:

$$E(X|A) := \int_\Omega X\, dP(\cdot|A) = (1/PA) \int_A X\, dP. \tag{2}$$

Equivalently,

$$E(X|A)PA = \int_A X\, dP \quad (A \in \mathcal{A}, PA \neq 0). \tag{3}$$

Since $P(B|A) = E(I_B|A)$, we may take the conditional expectation as the basic concept, and view the conditional probability as a derived concept.

Let $\{A_i\} \subset \mathcal{A}$ be a partition of Ω (with $PA_i \neq 0$), and let $\mathcal{A}_0$ be the σ-algebra generated by $\{A_i\}$. Denote

$$E(X|\mathcal{A}_0) := \sum_i E(X|A_i) I_{A_i}. \tag{4}$$

This is an $\mathcal{A}_0$-measurable function, which takes the constant value $E(X|A_i)$ on A_i $(i = 1, 2, \ldots)$. Any $A \in \mathcal{A}_0$ has the form $A = \bigcup_{i \in J} A_i$, where $J \subset \mathbb{N}$. By (3),

$$\int_A E(X|\mathcal{A}_0)\, dP = \sum_i E(X|A_i) P(A_i \cap A) = \sum_{i \in J} E(X|A_i) PA_i$$
$$= \sum_{i \in J} \int_{A_i} X\, dP = \int_A X\, dP \quad (A \in \mathcal{A}_0). \tag{5}$$

Relation (5) may be used to *define* the conditional expectation of X, *given the (arbitrary)* σ*-subalgebra* $\mathcal{A}_0$ *of* $\mathcal{A}$.

Definition I.7.2. Let $\mathcal{A}_0$ be a σ-subalgebra of $\mathcal{A}$, and let X be an $L^1(P)$-real r.v. The conditional expectation of X given $\mathcal{A}_0$ is the (P-a.s. determined) $\mathcal{A}_0$-measurable function $E(X|\mathcal{A}_0)$ satisfying the identity

$$\int_A E(X|\mathcal{A}_0) dP = \int_A X\, dP \quad (A \in \mathcal{A}_0). \tag{6}$$

Note that the right-hand side of (6) defines a real-valued measure ν on $\mathcal{A}_0$, absolutely continuous with respect to P (restricted to $\mathcal{A}_0$). By the Radon–Nikodym theorem, there exists a P-a.s. determined $\mathcal{A}_0$-measurable function, integrable on $(\Omega, \mathcal{A}_0, P)$, such that (6) is valid. Actually, $E(X|\mathcal{A}_0)$ is the Radon–Nikodym derivative of ν with respect to (the restriction of) P.

The conditional probability of $B \in \mathcal{A}$ given $\mathcal{A}_0$ is then defined by

$$P(B|\mathcal{A}_0) := E(I_B|\mathcal{A}_0).$$

By (6), it is the P-a.s. determined $\mathcal{A}_0$-measurable function satisfying

$$\int_A P(B|\mathcal{A}_0)\, dP = P(A \cap B) \quad (A \in \mathcal{A}_0). \tag{7}$$

We show that $E(X|\mathcal{A}_0)$ defined by (6) coincides with the function defined before for the special case of a σ-subalgebra generated by a sequence of mutually disjoint *atoms*. The idea is included in the following.

Theorem I.7.3. *The conditional expectation $E(X|\mathcal{A}_0)$ has a.s. the constant value $E(X|A)$ on each P-atom $A \in \mathcal{A}_0$. ($A \in \mathcal{A}_0$ is a P-atom if $PA > 0$, and A is not the disjoint union of two $\mathcal{A}_0$-measurable sets with positive P-measure.)*

Proof. Suppose $f : \Omega \to [-\infty, \infty]$ is $\mathcal{A}_0$-measurable, and let $A \in \mathcal{A}_0$ be a P-atom. We show that f is a.s. constant on A.

Denote $A_x := \{\omega \in A; f(\omega) < x\}$, for $x > -\infty$.

By monotonicity of P, if $-\infty \le y \le x \le \infty$ and $PA_x = 0$, then also $PA_y = 0$. Let

$$h = \sup\{x; PA_x = 0\}.$$

Then $PA_x = 0$ for all $x < h$. Since

$$A_h = \bigcup_{r < h; r \in \mathbb{Q}} A_r,$$

we have

$$PA_h = 0. \tag{8}$$

By definition of h, we have $PA_x > 0$ for $x > h$, and since A is a P-atom and $A_x \in \mathcal{A}_0$ (because f is $\mathcal{A}_0$-measurable) is a subset of A, it follows that $P\{\omega \in A; f(\omega) \geq x\} = 0$ for all $x > h$. Writing

$$\{\omega \in A; f(\omega) > h\} = \bigcup_{r>h;r\in\mathbb{Q}} \{\omega \in A; f(\omega) \geq r\},$$

we see that $P\{\omega \in A; f(\omega) > h\} = 0$. Together with (8), this proves that

$$P\{\omega \in A; f(\omega) \neq h\} = 0,$$

that is, $f(\omega) = h$ P-a.s. on A.

Applying the conclusion to $f = E(X|\mathcal{A}_0)$, we see from (6) that this constant value is necessarily $E(X|A)$ (cf. (2)). □

We collect some elementary properties of the conditional expectation in the following.

Theorem I.7.4.

(1) $E(E(X|\mathcal{A}_0)) = EX$.

(2) If X is $\mathcal{A}_0$-measurable, then $E(X|\mathcal{A}_0) = X$ a.s. (this is true in particular for X constant, and for any r.v. X if $\mathcal{A}_0 = \mathcal{A}$).

(3) Monotonicity: for real r.v.'s $X, Y \in L^1(P)$ such that $X \leq Y$ a.s., $E(X|\mathcal{A}_0) \leq E(Y|\mathcal{A}_0)$ a.s. (in particular, since $-|X| \leq X \leq |X|$, $|E(X|\mathcal{A}_0)| \leq E(|X||\mathcal{A}_0)$ a.s.).

(4) Linearity: for $X, Y \in L^1(P)$ and $\alpha, \beta \in \mathbb{C}$,

$$E(\alpha X + \beta Y|\mathcal{A}_0) = \alpha E(X|\mathcal{A}_0) + \beta E(Y|\mathcal{A}_0) \quad \text{a.s.}$$

Proof.

(1) Take $A = \Omega$ in (6).

(2) X is $\mathcal{A}_0$-measurable (hypothesis!) and satisfies trivially (6).

(3) The right-hand side of (6) is monotonic; therefore

$$\int_A E(X|\mathcal{A}_0)\, dP \leq \int_A E(Y|\mathcal{A}_0)\, dP$$

for all $A \in \mathcal{A}_0$, and the conclusion follows for example from the "averages lemma".

(4) The right-hand side of the equation in property (4) is $\mathcal{A}_0$-measurable and its integral over A equals $\int_A(\alpha X + \beta Y)\, dP$ for all $A \in \mathcal{A}_0$. By (6), it coincides a.s. with $E(\alpha X + \beta Y|\mathcal{A}_0)$.

□

We show next that conditional expectations behave like "projections" in an appropriate sense.

Theorem I.7.5. *Let $\mathcal{A}_0 \subset \mathcal{A}_1$ be σ-subalgebras of $\mathcal{A}$. Then for all $X \in L^1(P)$,*

$$E(E(X|\mathcal{A}_0)|\mathcal{A}_1) = E(E(X|\mathcal{A}_1)|\mathcal{A}_0) = E(X|\mathcal{A}_0) \quad \text{a.s.} \tag{9}$$

Proof. $E(X|\mathcal{A}_0)$ is $\mathcal{A}_0$-measurable, hence it is also $\mathcal{A}_1$-measurable (since $\mathcal{A}_0 \subset \mathcal{A}_1$), and therefore, by Theorem I.7.4, Part 2, the far left and far right in (9) coincide a.s.

Next, for all $A \in \mathcal{A}_0 (\subset \mathcal{A}_1)$, we have by (6)

$$\int_A E(E(X|\mathcal{A}_1)|\mathcal{A}_0)dP = \int_A E(X|\mathcal{A}_1)dP = \int_A X\,dP = \int_A E(X|\mathcal{A}_0)dP,$$

so that the middle and far right expressions in (9) coincide a.s. □

"Almost sure" versions of the usual convergence theorems for integrals are valid for the conditional expectation.

Theorem I.7.6 (Monotone convergence theorem for $E(.|\mathcal{A}_0)$). *Let $0 \le X_1 \le X_2 \le \cdots$ (a.s.) be r.v.'s such that $\lim X_n := X \in L^1(P)$. Then*

$$E(X|\mathcal{A}_0) = \lim E(X_n|\mathcal{A}_0) \quad \text{a.s.}$$

Proof. By Part (3) in Theorem I.7.4,

$$0 \le E(X_1|\mathcal{A}_0) \le E(X_2|\mathcal{A}_0) \le \cdots \quad \text{a.s.}$$

Therefore, $h := \lim_n E(X_n|\mathcal{A}_0)$ exists a.s., and is $\mathcal{A}_0$-measurable (after being extended as 0 on some P-null set). By the usual monotone convergence theorem, we have for all $A \in \mathcal{A}_0$

$$\int_A h\,dP = \lim_n \int_A E(X_n|\mathcal{A}_0)\,dP = \lim_n \int_A X_n\,dP = \int_A X\,dP,$$

hence $h = E(X|\mathcal{A}_0)$ a.s. □

Corollary I.7.7 (Beppo Levi theorem for conditional expectation). *Let $X_n \ge 0$ (a.s.) be r.v.'s such that $\sum_n X_n \in L^1(P)$. Then*

$$E\Big(\sum_n X_n|\mathcal{A}_0\Big) = \sum_n E(X_n|\mathcal{A}_0) \quad \text{a.s.}$$

Taking in particular $X_n = I_{B_n}$ with mutually disjoint $B_n \in \mathcal{A}$, we obtain the a.s. σ-additivity of $P(.|\mathcal{A}_0)$:

$$P\Big(\bigcup_{n=1}^{\infty} B_n|\mathcal{A}_0\Big) = \sum_{n=1}^{\infty} P(B_n|\mathcal{A}_0) \quad \text{a.s.} \tag{10}$$

Theorem I.7.8 (Dominated convergence theorem for conditional expectation). *Let $\{X_n\}$ be a sequence of r.v.'s such that*

$$X_n \to X \quad \text{a.s.}$$

and

$$|X_n| \le Y \in L^1(P).$$

Then

$$E(X|\mathcal{A}_0) = \lim_n E(X_n|\mathcal{A}_0) \quad \text{a.s.}$$

Proof. By Properties (1) and (3) in Theorem I.7.4, $E(|E(X_n|\mathcal{A}_0)|) \le E(Y) < \infty$, and therefore $E(X_n|\mathcal{A}_0)$ is finite a.s., and similarly $E(X|\mathcal{A}_0)$. Hence $E(X_n|\mathcal{A}_0) - E(X|\mathcal{A}_0)$ is well-defined and finite a.s., and has absolute value equal a.s. to

$$|E(X_n - X|\mathcal{A}_0)| \le E(|X_n - X||\mathcal{A}_0) \le E(Z_n|\mathcal{A}_0), \tag{11}$$

where

$$Z_n := \sup_{k \ge n} |X_k - X| (\in L^1(P)).$$

Since Z_n is a non-increasing sequence (with limit 0 a.s.), Property (3) in Theorem I.7.4 implies that $E(Z_n|\mathcal{A}_0)$ is a non-increasing sequence a.s. Let h be its (a.s.) limit. After proper extension on a P-null set, h is a non-negative $\mathcal{A}_0$-measurable function. Since $0 \le Z_n \le 2Y \in L^1(P)$, the usual dominated convergence theorem gives

$$0 \le \int_\Omega h\,dP \le \int_\Omega E(Z_n|\mathcal{A}_0)\,dP = \int_\Omega Z_n\,dP \to_n 0,$$

hence $h = 0$ a.s. By (11), this gives the conclusion of the theorem. □

Property (2) in Theorem I.7.4 means that $\mathcal{A}_0$-measurable functions behave like constants relative to the operation $E(.|\mathcal{A}_0)$. This "constant-like" behavior is a special case of the following.

Theorem I.7.9. *Let X, Y be r.v.'s such that $X, Y, XY \in L^1(P)$. If X is $\mathcal{A}_0$-measurable, then*

$$E(XY|\mathcal{A}_0) = X\,E(Y|\mathcal{A}_0) \quad \text{a.s.} \tag{12}$$

Proof. If $B \in \mathcal{A}_0$ and $X = I_B$, then for all $A \in \mathcal{A}_0$,

$$\int_A XE(Y|\mathcal{A}_0)\,dP = \int_{A\cap B} E(Y|\mathcal{A}_0)\,dP = \int_{A\cap B} Y\,dP = \int_A XY\,dP,$$

so that (12) is valid for $\mathcal{A}_0$-measurable indicators, and by linearity, for all $\mathcal{A}_0$-measurable simple functions. For an arbitrary $\mathcal{A}_0$-measurable r.v. $X \in L^1(P)$, there exists a sequence $\{X_n\}$ of $\mathcal{A}_0$-measurable simple functions such that $X_n \to X$ and $|X_n| \le |X|$. We have

$$E(X_nY|\mathcal{A}_0) = X_nE(Y|\mathcal{A}_0) \quad \text{a.s.}$$

Since $E(Y|\mathcal{A}_0)$ is P-integrable, it is a.s. finite, and therefore the right-hand side converges a.s. to $X\,E(Y|\mathcal{A}_0)$.

Since $X_nY \to XY$ and $|X_nY| \le |XY| \in L^1(P)$, the left-hand side converges a.s. to $E(XY|\mathcal{A}_0)$ by Theorem I.7.8, and the result follows. □

I.7.2 Conditioning by an r.v.

I.7.10. Given an r.v. X, it induces a σ-subalgebra $\mathcal{A}_X$ of $\mathcal{A}$, where

$$\mathcal{A}_X := \{X^{-1}(B); B \in \mathcal{B}\},$$

and $\mathcal{B}$ is the Borel algebra of $\mathbb{R}$ (or $\mathbb{C}$). It is then "natural" to define

$$E(Y|X) := E(Y|\mathcal{A}_X) \tag{13}$$

for any integrable r.v. Y.

Thus $E(Y|X)$ is the a.s. uniquely determined ($\mathcal{A}_X$)-measurable function such that

$$\int_{X^{-1}(B)} E(Y|X)\,dP = \int_{X^{-1}(B)} Y\,dP \tag{14}$$

for all $B \in \mathcal{B}$.

As a function of B, the right-hand side of (14) is a real (or complex) measure on $\mathcal{B}$, absolutely continuous with respect to the probability measure P_X [$P_X(B) = 0$ means that $P(X^{-1}(B)) = 0$, which implies that the right-hand side of (14) is zero]. By the Radon–Nikodym theorem, there exists a unique (up to P_X-equivalence) Borel $L^1(P_X)$-function h such that

$$\int_B h\,dP_X = \int_{X^{-1}(B)} Y\,dP \quad (B \in \mathcal{B}). \tag{15}$$

We shall denote (for X real valued)

$$h(x) := E(Y|X = x) \quad (x \in \mathbb{R}), \tag{16}$$

and call this function the "conditional expectation of Y, given $X = x$". Thus, by definition,

$$\int_B E(Y|X = x)\,dP_X(x) = \int_{X^{-1}(B)} Y\,dP \quad (B \in \mathcal{B}). \tag{17}$$

Taking $B = \mathbb{R}$ in (17), we see that

$$E_{P_X}(E(Y|X = x)) = E(Y), \tag{18}$$

where E_{P_X} denotes the expectation operator on $L^1(\mathbb{R}, \mathcal{B}, P_X)$.

The proof of Theorem I.7.3 shows that $E(Y|X=x)$ is P_X-a.s. constant on each P_X-atom $B \in \mathcal{B}$. By (17), we have

$$E(Y|X=x) = \frac{1}{P(X^{-1}(B))} \int_{X^{-1}(B)} Y\,dP \tag{19}$$

P_X-a.s. on B, for each P_X-atom $B \in \mathcal{B}$.

As before, the "conditional probability of $A \in \mathcal{A}$ given $X = x$" is defined by

$$P(A|X=x) := E(I_A|X=x) \quad (x \in \mathbb{R}),$$

or directly by (17) for the special case $Y = I_A$:

$$\int_B P(A|X=x)\,dP_X(x) = P(A \cap X^{-1}(B)) \quad (B \in \mathcal{B}). \tag{20}$$

If $B \in \mathcal{B}$ is a P_X-atom, we have by (19)

$$P(A|X=x) = \frac{P(A \cap X^{-1}(B))}{P(X^{-1}(B))} = P(A|[X \in B]) \tag{21}$$

P_X-almost surely on B, where the right-hand side of (21) is the "elementary" conditional probability of the event $A \in \mathcal{A}$, given the event $[X \in B]$. In particular, if $B = \{x\}$ is a P_X-atom (i.e., if $P_X(\{x\}) > 0$; i.e. if $P[X=x] > 0$), then

$$P(A|X=x) = P(A|[X=x]) \quad (A \in \mathcal{A}), \tag{22}$$

so that the notation is "consistent".

The relation between the $\mathcal{A}_X$-measurable function $E(Y|X)$ and the Borel function $h(x) := E(Y|X=x)$ is stated next.

Theorem I.7.11. *Let $(\Omega, \mathcal{A}, P)$ be a probability space, and let X, Y be (real) r.v.'s, with Y integrable. Then, P-almost surely,*

$$E(Y|X) = h(X),$$

where $h(x) := E(Y|X=x)$.

Proof. By (14) and (15),

$$\int_{X^{-1}(B)} E(Y|X)\,dP = \int_B h\,dP_X \quad (B \in \mathcal{B}). \tag{23}$$

We claim that

$$\int_B h\,dP_X = \int_{X^{-1}(B)} h(X)\,dP \quad (B \in \mathcal{B}) \tag{24}$$

for any real Borel P_X-integrable function h on $\mathbb{R}$. If $h = I_C$ for $C \in \mathcal{B}$, (24) is valid, since

$$\begin{aligned}\int_B h\,dP_X &= P_X(B \cap C) = P(X^{-1}(B \cap C)) \\ &= P(X^{-1}(B) \cap X^{-1}(C)) = \int_{X^{-1}(B)} I_{X^{-1}(C)}\,dP = \int_{X^{-1}(B)} h(X)\,dP.\end{aligned}$$

By linearity, (24) is then valid for simple Borel functions h. If h is a non-negative Borel function, let $\{h_n\}$ be a sequence of simple Borel functions such that $0 \leq h_1 \leq h_2 \leq \cdots$, and $\lim h_n = h$. Then $\{h_n(X)\}$ is a sequence of $\mathcal{A}_X$-measurable functions such that

$$0 \leq h_1(X) \leq h_2(X) \leq \cdots$$

and $\lim h_n(X) = h(X)$. By the monotone convergence theorem applied in the measure spaces $(\mathbb{R}, \mathcal{B}, P_X)$ and $(\Omega, \mathcal{A}, P)$, we have for all $B \in \mathcal{B}$:

$$\int_B h\, dP_X = \lim_n \int_B h_n\, dP_X = \lim_n \int_{X^{-1}(B)} h_n(X)\, dP = \int_{X^{-1}(B)} h(X)\, dP.$$

For any real P_X-integrable Borel function h, write $h = h^+ - h^-$; then

$$\begin{aligned} \int_B h\, dP_X &:= \int_B h^+\, dP_X - \int_B h^- dP_X \\ &= \int_{X^{-1}(B)} h^+(X) dP - \int_{X^{-1}(B)} h^-(X)\, dP \\ &= \int_{X^{-1}(B)} h(X)\, dP \quad (B \in \mathcal{B}). \end{aligned}$$

Thus (24) is verified, and by (23), we have

$$\int_A E(Y|X)\, dP = \int_A h(X)\, dP$$

for all $A \in \mathcal{A}_X (:= \{X^{-1}(B); B \in \mathcal{B}\})$.

Since both integrands are in $L^1(\Omega, \mathcal{A}_X, P)$, it follows that they coincide P-almost surely. □

Theorem I.7.12. *Let X, Y be (real) r.v.'s, with $Y \in L^2(P)$. Then $Z = E(Y|X)$ is the (real) $\mathcal{A}_X$-measurable solution in $L^2(P)$ of the extremal problem*

$$\|Y - Z\|_2 = \min.$$

(Geometrically, Z is the orthogonal projection of Y onto $L^2(\Omega, \mathcal{A}_X, P)$.)

Proof. Write (for $Y, Z \in L^2(P)$):

$$\begin{aligned} (Y - Z)^2 = [Y - E(Y|X)]^2 + [E(Y|X) - Z]^2 \\ + 2[E(Y|X) - Z][Y - E(Y|X)]. \end{aligned} \tag{25}$$

In particular, the third term is $\leq (Y - Z)^2$. Similarly, we see that the negative of the third term is also $\leq (Y - Z)^2$. Hence this term has absolute value $\leq (Y - Z)^2 \in L^1(P)$. Since $E(Y|X) \in L^1(P)$, the functions $U := E(Y|X) - Z$, $V := Y - E(Y|X)$, and UV are all in $L^1(P)$, and U is $\mathcal{A}_X$-measurable whenever Z is. By Theorem I.7.9 with $\mathcal{A}_0 = \mathcal{A}_X$,

$$\begin{aligned} E(UV|X) &:= E(UV|\mathcal{A}_X) = UE(V|\mathcal{A}_X) = UE(V|X) \\ &= U[E(Y|X) - E(Y|X)] = 0. \end{aligned}$$

Hence by (25)

$$E([Y - Z]^2|X) = E(U^2|X) + E(V^2|X).$$

Applying E, we obtain

$$E((Y - Z)^2) = E(U^2) + E(V^2) \geq E(V^2),$$

that is, $\|Y - Z\|_2 \geq \|Y - E(Y|X)\|_2$, with the minimum attained when $U = 0$ (P-a.s.), that is, when $Z = E(Y|X)$ a.s. □

Applying Theorem I.7.11, we obtain the following extremal property of $h = E(Y|X = \cdot)$.

Corollary I.7.13. *Let X, Y be (real) r.v.'s, with $Y \in L^2(P)$. Then the extremal problem for (real) Borel functions g on $\mathbb{R}$ with $g(X) \in L^2(P)$*

$$\|Y - g(X)\|_2 = \min$$

has the solution

$$g = h := E(Y|X = \cdot) \quad a.s.$$

Thus $h(X)$ gives the best "mean square approximation" of Y by "functions of X". The graph of the equation

$$y = h(x)(:= E(Y|X = x))$$

is called the *regression curve of Y on X*.

I.7.14. Linear regression. We consider the extremal problem of Corollary I.7.13 with the stronger restriction that g be *linear*. Thus we wish to find values of the real parameters a, b such that

$$\|Y - (aX + b)\|_2 = \min,$$

where X, Y are given non-degenerate $L^2(P)$-r.v.'s. Necessarily, X, Y have finite expectations μ_k and standard deviations $\sigma_k > 0$, and we may define the so-called *correlation coefficient of X and Y*

$$\rho = \rho(X, Y) := \frac{\operatorname{cov}(X, Y)}{\sigma_1 \sigma_2}.$$

By I.2.5, $|\rho| \leq 1$.

We have

$$\begin{aligned}
&\|Y - (aX + b)\|_2^2 \\
&\quad = E([Y - \mu_2] - a[X - \mu_1] + [\mu_2 - (a\mu_1 + b)])^2 \\
&\quad = E(Y - \mu_2)^2 + a^2 E(X - \mu_1)^2 + [\mu_2 - (a\mu_1 + b)]^2 - 2aE((X - \mu_1)(Y - \mu_2)) \\
&\quad = \sigma_2^2 + a^2\sigma_1^2 + [\mu_2 - (a\mu_1 + b)]^2 - 2a\rho\sigma_1\sigma_2 \\
&\quad = (a\sigma_1 - \rho\sigma_2)^2 + (1 - \rho^2)\sigma_2^2 + [\mu_2 - (a\mu_1 + b)]^2 \geq (1 - \rho^2)\sigma_2^2,
\end{aligned}$$

with equality (giving the minimal L^2-distance $\sigma_2\sqrt{1-\rho^2}$) attained when

$$a\sigma_1 - \rho\sigma_2 = 0, \qquad \mu_2 - (a\mu_1 + b) = 0,$$

that is, when

$$a = a^* := \rho\sigma_2/\sigma_1; \qquad b = b^* := \mu_2 - a^*\mu_1.$$

In conclusion, the *linear solution* of our extremum problem (the so-called *linear regression of* Y *on* X) has the equation

$$y = \mu_2 + \rho\frac{\sigma_2}{\sigma_1}(x - \mu_1).$$

Note that the minimal L^2-distance vanishes iff $|\rho| = 1$; in that case $Y = a^*X + b^*$ a.s.

I.7.15. Conditional distribution; discrete case.

Let X, Y be discrete real r.v.'s, with respective ranges $\{x_j\}$ and $\{y_k\}$. The vector-valued r.v. (X, Y) assumes the value (x_j, y_k) with the positive probability p_{jk}, where

$$\sum_{j,k} p_{jk} = 1.$$

The *joint distribution function* of (X, Y) is defined (in general, for any real r.v.'s) by

$$F(x, y) := P[X < x, Y < y] \quad (x, y \in \mathbb{R}).$$

In discrete case I.7.15,

$$F(x, y) = \sum_{x_j < x, y_k < y} p_{jk}.$$

The *marginal distributions* are defined in general by

$$F_X(x) := P[X < x] = F(x, \infty); \qquad F_Y(y) := P[Y < y] = F(\infty, y).$$

In our case,

$$F_X(x) = \sum_{x_j < x} p_{j.}; \qquad F_Y(y) = \sum_{y_k < y} p_{.k},$$

where

$$p_{j.} := \sum_k p_{jk} = P[X = x_j]; \qquad p_{.k} := \sum_j p_{jk} = P[Y = y_k].$$

Each singleton $\{x_j\}$ is a P_X-atom (because $P[X = x_j] = p_{j.} > 0$). By (22) in Section I.7.10 (with $A = [Y = y_k]$), we have

$$P(Y = y_k | X = x_j) = \frac{p_{jk}}{p_{j.}} \quad (j, k = 1, 2, \ldots)$$

and similarly

$$P(X = x_j | Y = y_k) = \frac{p_{jk}}{p_{.k}}.$$

Note that

$$\sum_k P(Y = y_k | X = x_j) = 1,$$

and therefore the function of y given by

$$F(y|X = x_j) := \sum_{y_k < y} P(Y = y_k | X = x_j) = (1/p_{j.}) \sum_{y_k < y} p_{jk} \quad (j = 1, 2, \ldots)$$

is a distribution function. It is called the *conditional distribution of Y, given $X = x_j$.*

I.7.16. Conditional distribution; Continuous case. Consider now the case where the vector-valued r.v. (X, Y) has a (joint) density h (cf. Section I.5). Then the distribution function of (X, Y) is given by

$$F(x, y) = \int_{-\infty}^{x} \int_{-\infty}^{y} h(s, t)\, ds\, dt \quad (x, y \in \mathbb{R}).$$

By Tonelli's theorem, the order of integration is irrelevant. At all continuity points (x, y) of h, one has $h(x, y) = \partial^2 F / \partial x\, \partial y$. The *marginal density functions* are defined by

$$h_X(x) := \int_{\mathbb{R}} h(x, y) dy; \quad h_Y(y) := \int_{\mathbb{R}} h(x, y) dx \qquad (x, y \in \mathbb{R}).$$

These are densities for the distribution function F_X and F_Y, respectively.

If $S := \{(x, y) \in \mathbb{R}^2; h_X(x) = 0\} (= h_X^{-1}(\{0\}) \times \mathbb{R})$, then

$$P[(X, Y) \in S] = P[X \in h_X^{-1}(\{0\})] = \int_{h_X^{-1}(\{0\})} h_X(x)\, dx = 0,$$

so that S may be disregarded. On $\mathbb{R}^2 - S$, define

$$h(y|x) := \frac{h(x, y)}{h_X(x)}. \tag{26}$$

This function is called the *conditional distribution density of Y, given $X = x$.* The terminology is motivated by the following.

Proposition I.7.17. *In the setting shown, we have P_X-almost surely*

$$E(Y|X = x) = \int_{\mathbb{R}} y\, h(y|x)\, dy.$$

Proof. For all $B \in \mathcal{B}(\mathbb{R})$, we have by Fubini's theorem:

$$\int_{X^{-1}(B)} Y\, dP = \iint_{B \times \mathbb{R}} y h(x, y)\, dx\, dy = \int_B h_X(x) \int_{\mathbb{R}} y h(y|x)\, dy\, dx$$
$$= \int_B \left(\int_{\mathbb{R}} y h(y|x)\, dy \right) dP_X(x),$$

and the conclusion follows from (17) in Section I.7.10. □

If h is *continuous* on $\mathbb{R}^2$, we also have the following.

Proposition I.7.18. *Suppose the joint distribution density h of (X,Y) is continuous on $\mathbb{R}^2$. Then for all $x \in \mathbb{R}$ for which $h_X(x) \neq 0$ and for all $B \in \mathcal{B}(\mathbb{R})$, we have*

$$\int_B h(y|x)\,dy = \lim_{\delta\to 0+} P(Y \in B \big| x-\delta < X < x+\delta).$$

Proof. For $\delta > 0$,

$$P(Y \in B \mid x-\delta < X < x+\delta) = \frac{P([Y \in B] \cap [x-\delta < X < x+\delta])}{P[x-\delta < X < x+\delta]}$$

$$= \frac{\int_{x-\delta}^{x+\delta} \int_B h(s,y)\,dy\,ds}{\int_{x-\delta}^{x+\delta} h_X(s)\,ds}.$$

Divide both numerator and denominator by 2δ and let $\delta \to 0$. The continuity assumption implies that, for all x for which $h_X(x) \neq 0$, the last expression has the limit

$$\frac{\int_B h(x,y)\,dy}{h_X(x)} = \int_B h(y|x)\,dy.$$

□

It follows from (26) that $\int_{\mathbb{R}} h(y|x)dy = 1$, so that $h(y|x)$ (defined for all x such that $h_X(x) \neq 0$) is the density of a distribution function:

$$F(y|x) := \int_{-\infty}^{y} h(t|x)\,dt,$$

called the *conditional distribution of Y given $X = x$.*

Example I.7.19 (The binormal distribution). We say that X, Y are *binormally distributed* if they have the joint density function (called the *binormal density*) given by

$$h(x,y) = (1/c)\exp\Big(-Q\Big(\frac{x-\mu_1}{\sigma_1}, \frac{y-\mu_2}{\sigma_2}\Big)\Big),$$

where Q is the positive definite quadratic form

$$Q(s,t) := \frac{s^2 - 2\rho st + t^2}{2(1-\rho^2)},$$

and

$$c = 2\pi\sigma_1\sigma_2\sqrt{1-\rho^2}$$

($\mu_k \in \mathbb{R}$, $\sigma_k > 0$, $-1 < \rho < 1$ are the parameters of the distribution).

Note that

$$Q(s,t) = [(s-\rho t)^2 + (1-\rho^2)t^2]/[2(1-\rho^2)] \ge 0$$

for all real s, t, with equality holding iff $s = t = 0$. Therefore h attains its absolute maximum $1/c$ at the unique point $(x,y) = (\mu_1, \mu_2)$.

The sections of the surface $z = h(x,y)$ with the planes $z = a$ are empty for $a > 1/c$ (and $a \le 0$); a single point for $a = 1/c$; and ellipses for $0 < a < 1/c$ (the surface is "bell-shaped").

In order to calculate the integral $\int_{\mathbb{R}^2} h(x,y)\,dx\,dy$, we make the transformation

$$x = \mu_1 + \sigma_1 s = \mu_1 + \sigma_1(u + \rho t); \qquad y = \mu_2 + \sigma_2 t,$$

where $u := s - \rho t$. Then (u,t) ranges over $\mathbb{R}^2$ when (x,y) does, and

$$\frac{\partial(x,y)}{\partial(u,t)} = \sigma_1\sigma_2 > 0.$$

Therefore, the previous integral is equal to

$$\begin{aligned}(1/c)&\iint_{\mathbb{R}^2} \mathrm{e}^{-((u^2+(1-\rho^2)t^2)/(2(1-\rho^2))}\sigma_1\sigma_2\,du\,dt \\ &= (1/\sqrt{2\pi(1-\rho^2)})\int_{\mathbb{R}} \mathrm{e}^{-u^2/2(1-\rho^2)}\,du = 1,\end{aligned}$$

since the last integral is that of the $N(0, 1-\rho^2)$-density.

Thus h is indeed the density of a two-dimensional distribution function.

Since $Q(s,t) = s^2/2 + (t-\rho s)^2/2(1-\rho^2)$, we get (for $x \in \mathbb{R}$ fixed, with $s = (x-\mu_1)/\sigma_1$ and $t = (y-\mu_2)/\sigma_2$, so that $dy = \sigma_2\,dt$):

$$h_X(x) = (1/c)\mathrm{e}^{-s^2/2}\int_{\mathbb{R}} \mathrm{e}^{-(t-\rho s)^2/2(1-\rho^2)}\sigma_2\,dt = \frac{1}{\sqrt{2\pi}\sigma_1}\mathrm{e}^{-(x-\mu_1)^2/2\sigma_1^2}.$$

Thus h_X is the $N(\mu_1, \sigma_1^2)$-density. By symmetry, the marginal density h_Y is the $N(\mu_2, \sigma_2^2)$-density. In particular, the meaning of the parameters μ_k and σ_k^2 has been clarified (as the expectations and variance of X and Y).

We have (with s, t related to x, y as before and $c' = \sqrt{2\pi\sigma_2^2(1-\rho^2)}$):

$$\begin{aligned}h(y|x) := \frac{h(x,y)}{h_X(x)} &= (1/c')\exp\left\{-\frac{\rho^2 s^2 - 2\rho st + t^2}{2(1-\rho^2)}\right\} \\ &= (1/c')\exp\{-(t-\rho s)^2/2(1-\rho^2)\} \\ &= (1/c')\exp\left\{-\frac{[y-(\mu_2+\rho(\sigma_2/\sigma_1)(x-\mu_1)]^2}{2\sigma_2^2(1-\rho^2)}\right\}.\end{aligned}$$

Thus $h(y|x)$ is the $N(\mu_2 + \rho(\sigma_2/\sigma_1)(x-\mu_1), \sigma_2^2(1-\rho^2))$ density.

By Proposition I.7.17, for all real x,

$$E(Y|X=x)=\mu_2+\rho(\sigma_2/\sigma_1)(x-\mu_1),$$

with an analogous formula for $E(X|Y=y)$.

Thus, for binormally distributed X, Y, *the regression curves* $y=E(Y|X=x)$ *and* $x=E(X|Y=y)$ *coincide with the linear regression curves* (cf. Section I.7.14). They intersect at (μ_1,μ_2), and the coefficient ρ here coincides with the correlation coefficient $\rho(X,Y)$ (cf. Section I.7.14). Indeed (with previous notations),

$$\begin{aligned}\rho(X,Y)&=E\Big(\frac{X-\mu_1}{\sigma_1}\cdot\frac{Y-\mu_2}{\sigma_2}\Big)\\&=(1/c)\iint_{\mathbb{R}^2}(u+\rho t)t\exp\Big\{-\frac{u^2+(1-\rho^2)t^2}{2(1-\rho^2)}\Big\}\sigma_1\sigma_2\,du\,dt.\end{aligned}$$

The integrand splits as the sum of two terms. The term with the factor ut is odd in each variable; by Fubini's theorem, its integral vanishes. The remaining integral is

$$\rho\frac{1}{\sqrt{2\pi(1-\rho^2)}}\int_{\mathbb{R}}e^{-u^2/2(1-\rho^2)}\,du\,\frac{1}{\sqrt{2\pi}}\int_{\mathbb{R}}t^2e^{-t^2/2}\,dt=\rho.$$

Note in particular that if the binormally distributed r.v.'s X, Y are uncorrelated, then $\rho=\rho(X,Y)=0$, and therefore $h(x,y)=h_X(x)h_Y(y)$. By Proposition I.5.1, it follows that X, Y are independent. Since the converse is generally true (cf. Section I.2.5), we have the following.

Proposition I.7.20. *If the r.v.'s X, Y are binormally distributed, then they are independent iff they are uncorrelated.*

I.8 Series of L^2 random variables

This section considers the a.s. convergence of series of independent r.v.'s.

We fix the following notation: $\{X_k\}_{k=1}^{\infty}$ is a sequence of real independent central $L^2(P)$ random variables; for $n=1,2,\ldots$, we let

$$S_n=\sum_{k=1}^{n}X_k;\quad s_n^2=\sigma^2(S_n)=\sum_{k=1}^{n}\sigma_k^2;\quad \sigma_k^2=\sigma^2(X_k);$$

and

$$T_n=\max_{1\le m\le n}|S_m|.$$

Lemma I.8.1 (Kolmogorov). *For each $\epsilon>0$ and $n=1,2,\ldots$,*

$$P[T_n\ge\epsilon]\le s_n^2/\epsilon^2.$$

Proof. Write

$$[T_n \geq \epsilon] = [|S_1| \geq \epsilon] \cup \{[|S_2| \geq \epsilon] \cap [|S_1| < \epsilon]\} \cup \cdots \\ \cup \{[|S_n| \geq \epsilon] \cap [|S_k| < \epsilon; k = 1, \ldots, n-1]\},$$

and denote the kth set in this union by A_k.

By independence and centrality of X_k,

$$\int_{A_k} S_n^2 \, dP = \sum_{j=1}^{n} \int_{A_k} X_j^2 \, dP \geq \sum_{j=1}^{k} \int_{A_k} X_j^2 \, dP \\ = \int_{A_k} S_k^2 \, dP \geq \epsilon^2 P(A_k).$$

Since the sets A_k are mutually disjoint, we get

$$\epsilon^2 P[T_n \geq \epsilon] = \sum_{k=1}^{n} \epsilon^2 P(A_k) \leq \sum_{k=1}^{n} \int_{A_k} S_n^2 \, dP \\ = \int_{[T_n \geq \epsilon]} S_n^2 \, dP \leq E(S_n^2) = s_n^2.$$

□

Theorem I.8.2. (For X_k as previously.) *If $\sum_k \sigma_k^2 < \infty$, then $\sum_k X_k$ converges a.s.*

Proof. Fix $\epsilon > 0$. For $n, m \in \mathbb{N}$, denote

$$A_{nm} = [\max_{1 \leq k \leq n} |\sum_{j=m+1}^{m+k} X_j| > \epsilon]$$

and

$$A_m = [\sup_{1 \leq k < \infty} |\sum_{j=m+1}^{m+k} X_j| > \epsilon].$$

Then A_m is the union of the increasing sequence $\{A_{nm}\}_n$, so that

$$P(A_m) = \lim_n P(A_{nm}).$$

By Lemma I.8.1,

$$P(A_{nm}) \leq (1/\epsilon^2) \sum_{k=m+1}^{m+n} \sigma_k^2 \leq (1/\epsilon^2) \sum_{k=m+1}^{\infty} \sigma_k^2.$$

Hence, for all m

$$P(A_m) \leq (1/\epsilon^2) \sum_{k=m+1}^{\infty} \sigma_k^2,$$

and therefore

$$P[\inf_m \sup_k |\sum_{j=m+1}^{m+k} X_j| > \epsilon] \leq (1/\epsilon^2) \sum_{k=m+1}^{\infty} \sigma_k^2.$$

The right-hand side tends to zero when $m \to \infty$ (by hypothesis), and therefore the left-hand side equals 0. Thus,

$$\inf_m \sup_k |\sum_{j=m+1}^{m+k} X_j| \leq \epsilon \quad \text{a.s.},$$

hence, there exists m, such that for all k, one has $|\sum_{j=m+1}^{m+k} X_j| < 2\epsilon$ (a.s.). □

For an a.s. *bounded* sequence of r.v.'s, a converse result is:

Theorem I.8.3. *Let $\{X_k\}$ be a sequence of independent central r.v.'s such that*

(i) $|X_k| \leq c$ a.s.; and

(ii) $P[\sum_{k=1}^{\infty} X_k$ converges $] > 0$.

Then $\sum_k \sigma_k^2 < \infty$.

Proof. By (i), $|S_n| \leq nc$ a.s.

Let A be the set on which $\{S_n\}$ converges. Since $PA > 0$ by (ii), it follows from Theorem 1.57 that $\{S_n\}$ converges uniformly on some measurable subset $B \subset A$ with $PB > 0$. Hence $|S_n| \leq d$ for all n on some measurable subset $E \subset B$ with $PE > 0$. Let

$$E_n = [|S_k| \leq d; 1 \leq k \leq n] \quad (n \in \mathbb{N}).$$

The sequence $\{E_n\}$ is decreasing, with intersection E. Let $\alpha_0 = 0$ and

$$\alpha_n := \int_{E_n} S_n^2 \, dP \quad (n \in \mathbb{N}).$$

Write

$$F_n = E_{n-1} - E_n (\subset E_{n-1}); \qquad E_n = E_{n-1} - F_n,$$

so that

$$\alpha_n - \alpha_{n-1} = \int_{E_{n-1}} S_n^2 \, dP - \int_{F_n} S_n^2 \, dP - \int_{E_{n-1}} S_{n-1}^2 \, dP.$$

Since X_n and S_{n-1} are central and independent, we have by BienAyme's identity

$$\int_{E_{n-1}} S_n^2 \, dP = \int_{E_{n-1}} X_n^2 \, dP + \int_{E_{n-1}} S_{n-1}^2 \, dP,$$

and therefore

$$\alpha_n - \alpha_{n-1} = \int_{E_{n-1}} X_n^2 \, dP - \int_{F_n} S_n^2 \, dP.$$

On $F_n(\subset E_{n-1})$, we have

$$|S_n| \leq |X_n| + |S_{n-1}| \leq c + d \quad \text{a.s.}$$

Therefore

$$\alpha_n - \alpha_{n-1} \geq \int_{E_{n-1}} X_n^2 \, dP - (c+d)^2 P(F_n). \tag{1}$$

Since $I_{E_{n-1}}$ (which is defined exclusively by means of $X_1, \ldots, X_{n-1}$) and X_n^2 are independent, it follows from Theorem I.2.2 that

$$\int_{E_{n-1}} X_n^2 \, dP = E(I_{E_{n-1}} X_n^2) = P(E_{n-1})\sigma^2(X_n) \geq P(E)\sigma_n^2.$$

Hence, by (1), and summing all the inequalities for $n = 1, \ldots, k$, we obtain

$$\alpha_k \geq P(E) \sum_{n=1}^{k} \sigma_n^2 - (c+d)^2 \sum_{n=1}^{k} P(F_n).$$

However, the sets F_n are disjoint, so that

$$\sum_{n=1}^{k} P(F_n) = P\left(\bigcup_{n=1}^{k} F_n \right) \leq 1,$$

hence,

$$\alpha_k \geq P(E) \sum_{n=1}^{k} \sigma_n^2 - (c+d)^2.$$

Since $P(E) > 0$, we obtain for all $k \in \mathbb{N}$

$$\sum_{n=1}^{k} \sigma_n^2 \leq \frac{\alpha_k + (c+d)^2}{P(E)} \leq \frac{d^2 + (c+d)^2}{P(E)},$$

so that $\sum_n \sigma_n^2 < \infty$. □

We consider next the non-central case.

Theorem I.8.4. *Let $\{X_k\}$ be a sequence of independent r.v.'s, such that $|X_k| \leq c, k = 1, 2, \ldots$ a.s. Then $\sum_k X_k$ converges a.s. iff the two (numerical) series $\sum_k E(X_k)$ and $\sum_k \sigma_k^2$ converge.*

Proof. Suppose the two "numerical" series converge (for this part of the proof, the hypothesis $|X_k| \leq c$ a.s. is *not* needed, and X_k are only assumed in L^2, our standing hypothesis). Let $Y_k = X_k - E(X_k)$. Then Y_k are independent central L^2 random variables, and $\sum_k \sigma^2(Y_k) = \sum_k \sigma^2(X_k) < \infty$. By Theorem I.8.2, $\sum_k Y_k$ converges a.s., and therefore, $\sum_k X_k = \sum_k [Y_k + E(X_k)]$ converges a.s., since $\sum E(X_k)$ converges by hypothesis.

Conversely, suppose that $\sum X_k$ converges a.s.

Define on the product probability space

$$(\Omega, \mathcal{A}, P) \times (\Omega, \mathcal{A}, P)$$

the random variables

$$Z_n(\omega_1, \omega_2) := X_n(\omega_1) - X_n(\omega_2).$$

Then Z_n are independent. They are central, since

$$\begin{aligned} E(Z_n) &= \int_{\Omega\times\Omega} [X_n(\omega_1) - X_n(\omega_2)]\, d(P \times P) \\ &= \int_\Omega X_n(\omega_1)\, dP(\omega_1) - \int_\Omega X_n(\omega_2)\, dP(\omega_2) = E(X_n) - E(X_n) = 0. \end{aligned}$$

Also $|Z_n| \le 2c$.

Furthermore, $\sum Z_n$ converges almost surely, because

$$\begin{aligned} &\Big\{(\omega_1, \omega_2) \in \Omega \times \Omega; \sum Z_n \text{ diverges}\Big\} \\ &\quad \subset \Big\{(\omega_1, \omega_2); \sum X_n(\omega_1) \text{ diverges}\} \cup \{(\omega_1, \omega_2); \sum_n X_n(\omega_2) \text{ diverges}\Big\}, \end{aligned}$$

and both sets in the union have $P \times P$-measure zero (by our a.s. convergence hypothesis on $\sum X_n$).

By Theorem I.8.3, it follows that $\sum \sigma^2(Z_n) < \infty$. However,

$$\sigma^2(Z_n) = \int_{\Omega\times\Omega} [X_n(\omega_1) - X_n(\omega_2)]^2\, d(P \times P).$$

Expanding the square and integrating, we see that

$$\sigma^2(Z_n) = 2[E(X_n^2) - E(X_n)^2] = 2\sigma^2(X_n).$$

Therefore, $\sum \sigma^2(X_n) < \infty$, and since Y_n are central, and $\sum \sigma^2(Y_n) = \sum \sigma^2(X_n) < \infty$, we conclude from Theorem I.8.2 that $\sum Y_n$ converges a.s.; but then $\sum E(X_n) = \sum (X_n - Y_n)$ converges as well. □

We consider finally the general case of a series of independent L^2 random variables.

Theorem I.8.5 (Kolmogorov's "three series theorem"). *Let $\{X_n\}$ be a sequence of real independent $L^2(P)$ random variables. For any real $k > 0$ and $n \in \mathbb{N}$, denote*

$$E_n := [|X_n| \le k]; \qquad X_n' = I_{E_n} X_n.$$

Then the series $\sum X_n$ converges a.s. iff the following numerical series (a), (b), and (c) converge:

(a) $\sum_n P(E_n^c)$;

(b) $\sum_n E(X_n')$;

(c) $\sum_n \sigma^2(X_n')$.

Proof. Consider the "truncated" r.v.'s

$$Y_n^+ := I_{E_n} X_n + k I_{E_n^c},$$

and Y_n^- defined similarly with $-k$ instead of k.

If $\sum X_n(\omega)$ converges, then $X_n(\omega) \to 0$, so that $\omega \in E_n$ for $n > n_0$, and therefore, $Y_n^+(\omega) = Y_n^-(\omega) = X_n(\omega)$ for all $n > n_0$; hence, both series $\sum Y_n^+(\omega)$ and $\sum Y_n^-(\omega)$ converge. Conversely, if one of these two series converge (say the first), then $Y_n^+(\omega) \to 0$, so that $Y_n^+(\omega) \neq k$ for $n > n_0$, hence, necessarily $Y_n^+(\omega) = X_n(\omega)$ for $n > n_0$, and so $\sum X_n(\omega)$ converges.

We showed that $\sum X_n(\omega)$ converges iff the series of Y_n^+ and Y_n^- both converge at ω; therefore, $\sum X_n$ converges a.s. iff both Y-series converge a.s. Since Y_n^+ and Y_n^- satisfy the hypothesis of Theorem I.8.4, we conclude from that theorem that $\sum X_n$ converges a.s. iff the numerical series $\sum E(Y_n^+)$, $\sum \sigma^2(Y_n^+)$, and the corresponding series for Y_n^- converge. It then remains to show that the convergence of these four series is equivalent to the convergence of the three series (a)–(c).

Since

$$E(Y_n^+) = E(X_n') + kP(E_n^c)$$

(and a similar formula for Y_n^-), we see by addition and subtraction that the convergence of the series (a) and (b) is equivalent to the convergence of the two series

$$\sum E(Y_n^+), \qquad \sum E(Y_n^-).$$

Next

$$\begin{aligned}
\sigma^2(Y_n^+) &= E\{(Y_n^+)^2\} - \{E(Y_n^+)\}^2 \\
&= E\{(X_n')^2\} + k^2 P(E_n^c) - [E(X_n') + kP(E_n^c)]^2 \\
&= \sigma^2(X_n') + k^2 P(E_n)P(E_n^c) - 2kE(X_n')P(E_n^c),
\end{aligned} \tag{2}$$

and similarly for Y_n^- (with $-k$ replacing k).

If the series (a)–(c) converge, then we already know that $\sum E(Y_n^+)$ and $\sum E(Y_n^-)$ converge. The convergence of (b) also implies that $|E(X_n')| \leq M$ for all n, and therefore,

$$|E(X_n')P(E_n^c)| \leq MP(E_n^c),$$

so that $\sum E(X_n')P(E_n^c)$ converges (by convergence of (a)).

Since $0 \leq P(E_n)P(E_n^c) \leq P(E_n^c)$, the series $\sum P(E_n)P(E_n^c)$ converges (by convergence of (a)).

Relation (2) and the convergence of (c) imply therefore that $\sum \sigma^2(Y_n^+)$ converges (and similarly for Y_n^-), as wanted.

Conversely, if the "four series" mentioned converge, we saw already that the series (a) and (b) converge, and this in turn implies the convergence of the series

$$\sum E(X_n')P(E_n^c) \quad \text{and} \quad \sum P(E_n)P(E_n^c).$$

By Relation (2), the convergence of $\sum \sigma^2(Y_n^+)$ implies therefore the convergence of the series (c) as well. □

I.9 Infinite divisibility

Definition I.9.1. A random variable X (or its distribution function F, or its characteristic function f) is *infinitely divisible* (i.d.) if, for each $n \in \mathbb{N}$, F is the distribution function of a sum of n independent r.v.'s with the same distribution function F_n.

Equivalently, X is i.d. if there exists a *triangular array* of r.v.'s

$$\{X_{nk}; 1 \leq k \leq n, n = 1, 2, \ldots\} \tag{1}$$

such that, for each n, the r.v.'s of the nth row are independent and equidistributed, and their sum $T_n =_d X$ (if X, Y are r.v.'s, we write $X =_d Y$ when they have the same distribution).

By the uniqueness theorem for ch.f.'s, X is i.d. iff, for each n, there exists a ch.f. f_n such that

$$f = f_n^n. \tag{2}$$

In terms of distribution functions, infinite divisibility of F means the existence, for each n, of a distribution function F_n, such that

$$F = F_n^{(n)}, \tag{3}$$

where $G^{(n)} := G * \cdots * G$ (n times) for any distribution function G. The *convolution* $F * G$ of two distribution functions is defined by

$$(F * G)(x) := \int_{\mathbb{R}} F(x - y)\, dG(y) \quad (x \in \mathbb{R}).$$

It is clearly a distribution function. An application of Fubini's theorem shows that its ch.f. is precisely fg (where f, g are the ch.f.'s of F, G, respectively). It then follows from the uniqueness theorem for ch.f.'s (or directly!) that convolution of distribution functions is commutative and associative. We may then omit parenthesis and write $F_1 * F_2 * \cdots * F_n$ for the convolution of finitely many distribution functions. In particular, the repeated convolutions $G^{(n)}$ mentioned earlier make sense, and criterion (3) is clearly equivalent to (2).

Example I.9.2.

(1) The Poisson distribution is i.d.: take F_n to be Poisson with parameter λ/n (where λ is the parameter of F). Then indeed (cf. Section I.3.9):

$$f_n^n(u) = [e^{(\lambda/n)(e^{iu}-1)}]^n = f(u).$$

(2) The normal distribution (parameters μ, σ^2) is i.d.: take F_n to be the normal distribution with parameters $\mu/n, \sigma^2/n$. Then (cf. Section I.3.12)

$$f_n^n(u) = [e^{iu\mu/n} e^{-(\sigma u)^2/2n}]^n = f(u).$$

(3) The Gamma distribution (parameters p, b) is i.d.: take F_n to be the *Gamma* distribution with parameters $p/n, b$. Then (cf. Section I.3.15):

$$f_n^n(u) = \left[\left(1 - \frac{iu}{b}\right)^{-p/n}\right]^n = f(u).$$

It is also clear that the Cauchy distribution is i.d. (cf. Section I.3.14), while the Laplace distribution is not.

We have the following criterion for infinite divisibility (its necessity is obvious, by (1); we omit the proof of its sufficiency).

Theorem I.9.3. *A random variable X is i.d. iff there exists a triangular array (1) such that (cf. Definition I.4.3)*

$$T_n \to_w X. \tag{4}$$

Some elementary properties of i.d. random variables (or ch.f.'s) are stated in the next theorem.

Theorem I.9.4.

(a) If X is i.d., so is $Y := a + bX$.

(b) If f, g are i.d. characteristic functions, so is fg.

(c) If f is an i.d. characteristic function, so are $\bar{f}$ and $|f|^2$.

(d) If f is an i.d. characteristic function, then $f \neq 0$ everywhere.

(e) If $\{f_k\}$ is a sequence of i.d. characteristic functions converging pointwise to a function g continuous at 0, then g is an i.d. ch.f.

(f) If f is an i.d. characteristic function, then its representation (2) is unique (for each n).

Proof.

(a) By Proposition I.3.5 and (2), for all $n \in \mathbb{N}$,

$$f_Y(u) = e^{iua} f(bu) = [e^{iua/n} f_n(bu)]^n = g_n^n,$$

where g_n is clearly a ch.f.

(b) Represent f, g as in (2), for each n. Then $fg = [f_n g_n]^n$. By Corollary I.2.4, fg and $f_n g_n$ are ch.f.'s, and they satisfy (2) as needed.

(c) First, $\bar{f}$ is a ch.f., since

$$\bar{f}(u) = \overline{\int_{\mathbb{R}} e^{iux}\, dF(x)} = \int_{\mathbb{R}} e^{-iux}\, dF(x)$$
$$= \int_{\mathbb{R}} e^{iux}\, d[1 - F(-x)] = \int_{\mathbb{R}} e^{iux}\, dF_{-X}(x) = f_{-X}(u).$$

If f is i.d., then by (2), $\bar{f} = (\overline{f_n})^n$, where $\overline{f_n}$ is a ch.f., as needed. The conclusion about $|f|^2$ follows then from (b).

(d) Since it suffices to prove that $|f|^2 \neq 0$, and $|f|^2$ is a non-negative i.d. ch.f. (by (c)), we may assume without loss of generality that $f \geq 0$. Let then $f_n := f^{1/n}$ be the unique non-negative nth root of f. Then $g := \lim f_n$ (pointwise) exists:

$$g = I_E, \quad E = f^{-1}(\mathbb{R}^+).$$

Since $f(0) = 1$, the point 0 belongs to the open set E, and so $g = 1$ in a neighborhood of 0; in particular, g is continuous at 0. By the Paul Levy continuity theorem (Theorem I.4.8), g is a ch.f., and is therefore continuous everywhere. In particular, its range is connected; since it is a subset of $\{0, 1\}$ containing 1, it must be precisely $\{1\}$, that is, $g = 1$ everywhere. This means that $f > 0$ everywhere, as claimed.

(e) By the Paul Levy continuity theorem, g is a ch.f. By (d), $f_k \neq 0$ everywhere (for each k), and has therefore a continuous logarithm $\log f_k$, uniquely determined by the condition $\log f_k(0) = 0$. Since f_k is i.d., $f_k = f_{k,n}^n$, with $f_{k,n}$ ch.f.'s (by (2)). We have

$$e^{(1/n)\log f_k} = e^{(1/n)\log f_{k,n}^n} = f_{k,n}.$$

The left-hand side converges pointwise (as $k \to \infty$) to $e^{(1/n)\log g} := g_n$. Since g is a ch.f., $g(0) = 1$, and therefore $\log g$ is continuous at 0, and the same is true of $g_n (= \lim_k f_{k,n})$. By Paul Levy's theorem, g_n is a ch.f., and clearly $g_n^n = g$. Hence g is i.d.

(f) Fix n, and suppose g, h are ch.f.'s such that

$$g^n = h^n = f. \qquad (*)$$

By (d), $h \neq 0$ everywhere, and therefore g/h is continuous, and $(g/h)^n = 1$ everywhere. The continuity implies that g/h has a connected range, which is a subset of the finite set of nth roots of unity. Since $g(0) = h(0) = 1$ (these are ch.f.'s!), the range contains 1, and coincides therefore with the singleton $\{1\}$, that is, $g/h = 1$ identically. □

By Example I.9.2(1) and Theorem I.9.4 (Part (a)), if $Y = a + bX$ with X Poisson-distributed, then Y is i.d. We call the distribution F_Y of such an r.v. Y a *Poisson-type distribution.* By Proposition I.3.5 and Section I.3.9,

$$f_Y(u) = \mathrm{e}^{iua+\lambda(\mathrm{e}^{iub}-1)}. \tag{5}$$

The Poisson-type distributions "generate" all the i.d. distributions in the following sense (compare with Theorem I.9.3).

Theorem I.9.5. *A random variable X is infinitely divisible iff there exists an array*

$$\{X_{nk}; 1 \le k \le r(n),\, n = 1, 2, \ldots\}$$

such that, for each n, the r.v.'s in the nth row are independent Poisson-type, and

$$T_n := \sum_{k=1}^{r(n)} X_{nk} \to_w X.$$

Proof. *Sufficiency.* As we just observed, each X_{nk} is i.d., hence T_n are i.d. by Theorem I.9.4, Part (b), and therefore X is i.d. by Part (e) of Theorem I.9.4 and Corollary I.4.6.

Necessity. Let X be i.d. By (2), there exist ch.f.'s f_n such that $f := f_X = f_n^n$, $n = 1, 2, \ldots$. By Theorem I.9.4, Part (f) and the proof of Part (e), the f_n are uniquely determined, and can be written as

$$f_n = \mathrm{e}^{(1/n)\log f},$$

where $\log f$ is continuous and uniquely determined by the condition $\log f(0) = 0$.

Fix $u \in \mathbb{R}$. Then

$$\begin{aligned} n[f_n(u) - 1] &= n[\mathrm{e}^{(1/n)\log f(u)} - 1] = n[(1/n)\log f(u) + o(1/n)] \\ &\to_{n\to\infty} \log f(u), \end{aligned}$$

that is, if F_n denotes the distribution function with ch.f. f_n, then

$$\log f(u) = \lim_n n \int_{\mathbb{R}} (\mathrm{e}^{iux} - 1)\, dF_n(x). \tag{6}$$

For each n, let $m = m(n)$ be such that

$$1 - F_n(m) + F_n(-m) < \frac{1}{2n^2}.$$

Then for all $u \in \mathbb{R}$,

$$\left| \int_{\mathbb{R}} (\mathrm{e}^{iux} - 1)\, dF_n(x) - \int_{-m(n)}^{m(n)} (\mathrm{e}^{iux} - 1)\, dF_n(x) \right| < \frac{1}{n^2}.$$

Approximate the Stieltjes integral over $[-m(n), m(n)]$ by Riemann–Stieltjes sums, such that

$$\left| \int_{-m(n)}^{m(n)} (\cdots) - \sum_{k=1}^{r(n)} (e^{iux_k} - 1)[F_n(x_k) - F_n(x_{k-1})] \right| < \frac{1}{n^2},$$

where $x_k = x_k(n) = -m(n) + 2m(n)k/r(n)$, $k = 1, \ldots, r(n)$. By (6), f is the pointwise limit (as $n \to \infty$) of the products

$$\prod_{k=1}^{r(n)} \exp\{\lambda_{nk}(e^{ia_{nk}u} - 1)\}, \tag{7}$$

where $\lambda_{nk} := n[F_n(x_k) - F_n(x_{k-1})]$ and $a_{nk} := x_k (= x_k(n))$.

The products in (7) are the ch.f.'s of sums of $r(n)$ independent Poisson-type r.v.'s. □

I.10 More on sequences of random variables

Let $\{X_n\}$ be a sequence of (complex) random variables on the probability space $(\Omega, \mathcal{A}, P)$. For $c > 0$, set

$$m(c) := \sup_n \int_{[|X_n| \geq c]} |X_n| dP.$$

Definition I.10.1. $\{X_n\}$ is *uniformly integrable* (u.i.) if

$$\lim_{c \to \infty} m(c) = 0.$$

For example, if $|X_n| \leq g \in L^1(P)$ for all n, then $[|X_n| \geq c] \subset [g \geq c]$, so that

$$m(c) \leq \int_{[g \geq c]} g \, dP \to 0$$

as $c \to \infty$.

A less trivial example is given in the following.

Proposition I.10.2. *Let $\mathcal{A}_n$ be sub-σ-algebras of $\mathcal{A}$, let $Y \in L^1(P)$, and*

$$X_n = E(Y|\mathcal{A}_n) \quad n = 1, 2, \ldots.$$

Then X_n are u.i.

Proof. Since $|X_n| \leq E(|Y||\mathcal{A}_n)$, and X_n are $\mathcal{A}_n$-measurable (so that $A_n := [|X_n| \geq c] \in \mathcal{A}_n$),

$$\int_{A_n} |X_n| \, dP \leq \int_{A_n} E(|Y||\mathcal{A}_n) \, dP = \int_{A_n} |Y| \, dP.$$

We have

$$P(A_n) \leq E(|X_n|)/c \leq E(|Y|)/c.$$

Given $\epsilon > 0$, choose $K > 0$ such that $\int_{[|Y|>K]} |Y|\, dP < \epsilon$. Then

$$\int_{A_n} |Y|\, dP = \Big(\int_{A_n \cap [|Y| \leq K]} + \int_{A_n \cap [|Y|>K]} \Big) |Y|\, dP$$
$$\leq KP(A_n) + \epsilon \leq K\|Y\|_1/c + \epsilon,$$

hence

$$m(c) \leq K\|Y\|_1/c + \epsilon.$$

Since ϵ was arbitrary, it follows that $\lim_{c\to\infty} m(c) = 0$. □

Theorem I.10.3. *$\{X_n\}$ is u.i. iff*

(1) $\sup_n \|X_n\|_1 < \infty$ ("norm-boundedness") and

(2) $\sup_n \int_A |X_n|\, dP \to 0$ when $PA \to 0$ ("uniform absolute continuity").

Proof. If the sequence is u.i., let

$$M := \sup_{c>0} m(c) \quad (< \infty).$$

Then for any $c > 0$,

$$\|X_n\|_1 = \Big(\int_{[|X_n|<c]} + \int_{[|X_n|\geq c]} \Big) |X_n|\, dP \leq c + m(c) \leq c + M,$$

so that (1) is valid.

Also, for any $A \in \mathcal{A}$,

$$\int_A |X_n|\, dP = \Big(\int_{A\cap[|X_n|<c]} + \int_{A\cap[|X_n|\geq c]} \Big) |X_n|\, dP \leq cPA + m(c).$$

Given $\epsilon > 0$, fix $c > 0$ such that $m(c) < \epsilon/2$. For this c, if $PA < \epsilon/2c$, then $\int_A |X_n|\, dP < \epsilon$ for all n, proving (2).

Conversely, if (1) and (2) hold, then by (1), for all n

$$cP[|X_n| \geq c] \leq \int_{[|X_n|\geq c]} |X_n|\, dP \leq \sup_n \|X_n\|_1 := R < \infty,$$

so that

$$P[|X_n| \geq c] \leq R/c.$$

Given $\epsilon > 0$, there exists $\delta > 0$ such that $\int_A |X_n|\, dP < \epsilon$ for all n whenever $A \in \mathcal{A}$ has $PA < \delta$ (by (2)). Therefore, if $c > R/\delta$, surely $P[|X_n| \geq c] < \delta$, and consequently $m(c) < \epsilon$. □

Fatou's lemma extends as follows to u.i. *real* r.v.'s:

Theorem I.10.4. *Let X_n be u.i. real r.v.'s. Then*

$$E(\liminf X_n) \le \liminf E(X_n) \le \limsup E(X_n) \le E(\limsup X_n).$$

Proof. Since $[X_n < -c] \subset [|X_n| > c]$, we have

$$\left| \int_{[X_n<-c]} X_n \, dP \right| \le m(c)$$

for any $c > 0$. Given $\epsilon > 0$, we may fix $c > 0$ such that $m(c) < \epsilon$ (since X_n are u.i.).

Denote $A_n = [X_n \ge -c]$.

We apply Fatou's lemma to the *non-negative* measurable functions $c + X_n I_{A_n}$:

$$E(c + \liminf X_n I_{A_n}) \le \liminf E(c + X_n I_{A_n}),$$

hence,

$$E(\liminf X_n I_{A_n}) \le \liminf E(X_n I_{A_n}). \tag{*}$$

However,

$$E(X_n I_{A_n}) = E(X_n) - \int_{A_n^c} X_n \, dP < E(X_n) + \epsilon,$$

and therefore the right-hand side of (*) is

$$\le \liminf E(X_n) + \epsilon.$$

Since $X_n I_{A_n} \ge X_n$, the left-hand side of (*) is $\ge E(\liminf X_n)$, and the left inequality of the theorem follows. The right inequality is then obtained by replacing X_n by $-X_n$. □

Corollary I.10.5. *Let X_n be u.i. (complex) r.v.'s, such that $X_n \to X$ a.s. or in probability. Then $X_n \to X$ in L^1 (and in particular, $E(X_n) \to E(X)$).*

Proof. Let $K := \sup_n \|X_n\|_1 (< \infty$, by Theorem I.10.3(1)). By Fatou's lemma, if $X_n \to X$ a.s., then

$$\|X\|_1 \le \liminf_m \|X_m\|_1 \le K,$$

so $X \in L^1$ and

$$\sup_n \|X_n - X\|_1 \le 2K.$$

Also, for $A \in \mathcal{A}$, again by Fatou's lemma,

$$\int_A |X_n - X| \, dP \le \liminf_m \int_A |X_n - X_m| \, dP,$$

hence,

$$\sup_n \int_A |X_n - X| \, dP \le 2 \sup_n \int_A |X_n| \, dP \to 0$$

as $PA \to 0$, by Theorem I.10.3(2).

Consequently (by the same theorem) $|X_n - X|$ are u.i., and since $|X_n - X| \to 0$ a.s., Theorem I.10.4 applied to $|X_n - X|$ shows that $\|X_n - X\|_1 \to 0$.

In case $X_n \to X$ in probability, there exists a subsequence X_{n_k} converging a.s. to X. By the first part of the proof, $X \in L^1$, and $\|X_{n_k} - X\|_1 \to 0$. The previous argument with $m = n_k$ shows that $|X_{n_k} - X|$ are u.i. Therefore *any* subsequence of X_n has a subsequence converging to X in L^1. If we assume that $\{X_n\}$ itself does not converge to X in L^1, then given $\epsilon > 0$, there exists a subsequence X_{n_k} such that $\|X_{n_k} - X\|_1 \geq \epsilon$ for all k, a contradiction. □

Definition I.10.6. A *submartingale* is a sequence of ordered pairs $(X_n, \mathcal{A}_n)$, where

(1) $\{\mathcal{A}_n\}$ is an increasing sequence of sub-σ-algebras of $\mathcal{A}$;

(2) the real r.v. $X_n \in L^1$ is $\mathcal{A}_n$-measurable ($n = 1, 2, \ldots$); and

(3) $X_n \leq E(X_{n+1}|\mathcal{A}_n)$ a.s. ($n = 1, 2, \ldots$).

If equality holds in (3), the sequence is called a *martingale*. The sequence is a *supermartingale* if the inequality (3) is reversed.

By definition of conditional expectation, a sequence $(X_n, \mathcal{A}_n)$ satisfying (1) and (2) is a submartingale iff

$$\int_A X_n \, dP \leq \int_A X_{n+1} \, dP \quad (A \in \mathcal{A}_n).$$

For example, if $\mathcal{A}_n$ are as in (1) and Y is an L^1- r.v., then setting

$$X_n := E(Y|\mathcal{A}_n), \quad n = 1, 2, \ldots,$$

the sequence $(X_n, \mathcal{A}_n)$ is a martingale: indeed (2) is clear by definition, and equality in (3) follows from Theorem I.7.5.

An important example is given in the following.

Proposition I.10.7. *Let $\{Y_n\}$ be a sequence of L^1, central, independent r.v.'s. Let $\mathcal{A}_n$ be the smallest σ-algebra for which $Y_1, \ldots, Y_n$ are measurable, and let $X_n = Y_1 + \cdots + Y_n$. Then $(X_n, \mathcal{A}_n)$ is a martingale.*

Proof. The requirements (1) and (2) are clear.

If $Y_{n+1} = I_B$ with $B \in \mathcal{A}$ independent of all $A \in \mathcal{A}_n$, then for all $A \in \mathcal{A}_n$,

$$\int_A Y_{n+1} \, dP = P(A \cap B) = P(A)P(B) = \int_A E(Y_{n+1}) \, dP,$$

and this identity between the extreme terms remains true, by linearity, for all simple r.v.'s Y_{n+1} independent of $Y_1, \ldots, Y_n$. By monotone convergence, the identity is true for all $Y_{n+1} \geq 0$, and finally for all L^1-r.v.'s Y_{n+1} independent of $Y_1, \ldots, Y_n$. Therefore,

$$E(Y_{n+1}|\mathcal{A}_n) = E(Y_{n+1}) \quad \text{a.s.},$$

and in the central case,

$$E(Y_{n+1}|\mathcal{A}_n) = 0 \quad \text{a.s.}$$

Since X_n is $\mathcal{A}_n$-measurable,

$$E(X_n|\mathcal{A}_n) = X_n \quad \text{a.s.}$$

Adding these equations, we obtain $E(X_{n+1}|\mathcal{A}_n) = X_n$ a.s. □

Proposition I.10.8. *If $(X_n, \mathcal{A}_n)$ is a submartingale, and $g : \mathbb{R} \to \mathbb{R}$ is a convex, increasing function such that $g(X_n) \in L^1$ for all n, then $(g(X_n), \mathcal{A}_n)$ is a submartingale. If $(X_n, \mathcal{A}_n)$ is a martingale, the preceding conclusion is valid without assuming that g is increasing.*

Proof. Since g is increasing, and $X_n \leq E(X_{n+1}|\mathcal{A}_n)$ a.s., we have a.s.

$$g(X_n) \leq g(E(X_{n+1}|\mathcal{A}_n)).$$

By Jensen's inequality for the convex function g, the right-hand side is $\leq E(g(X_{n+1})|\mathcal{A}_n)$ a.s., proving the first statement. In the martingale case, since $X_n = E(X_{n+1}|\mathcal{A}_n)$, we get a.s.

$$g(X_n) = g(E(X_{n+1}|\mathcal{A}_n)) \leq E(g(X_{n+1})|\mathcal{A}_n).$$ □

We omit the proof of the following important theorem.

Theorem I.10.9 (submartingale convergence theorem). *If $(X_n, \mathcal{A}_n)$ is a submartingale such that*

$$\sup_n E(X_n^+) < \infty,$$

then there exists an L^1-r.v. X such that $X_n \to X$ a.s.

By the proposition following Definition I.10.1 and the comments following Definition I.10.6, if $\mathcal{A}_n$ are increasing sub-σ-algebras, Y is an L^1-r.v., and $X_n := E(Y|\mathcal{A}_n)$, then $(X_n, \mathcal{A}_n)$ is a u.i. martingale. The converse is also true:

Theorem I.10.10. *$(X_n, \mathcal{A}_n)$ is a u.i. martingale iff there exists an L^1-r.v. Y such that $X_n = E(Y|\mathcal{A}_n)$ (a.s.) for all n. When this is the case, $X_n \to Y = E(Y|\mathcal{A}_\infty)$ a.s. and in L^1, where $\mathcal{A}_\infty$ is the σ-algebra generated by the algebra $\bigcup_n \mathcal{A}_n$.*

Proof. We just observed that if $X_n = E(Y|\mathcal{A}_n)$, then $(X_n, \mathcal{A}_n)$ is a u.i. martingale. Conversely, let $(X_n, \mathcal{A}_n)$ be a u.i. martingale. By Theorem I.10.3 (1), $\sup_n \|X_n\|_1 < \infty$, hence by Theorem I.10.9, there exists an L^1-r.v. Y such that $X_n \to Y$ a.s. By Corollary I.10.5, $X_n \to Y$ in L^1 as well. Hence, for all $A \in \mathcal{A}_n$ and $m \geq n$,

$$\int_A X_n \, dP = \int_A X_m \, dP \to_m \int_A Y \, dP = \int_A E(Y|\mathcal{A}_n) \, dP$$

and it follows that $X_n = E(Y|\mathcal{A}_n)$ a.s.

For any Borel set $B \subset \mathbb{R}$, we have $X_n^{-1}(B) \in \mathcal{A}_n \subset \mathcal{A}_\infty$. Hence X_n is $\mathcal{A}_\infty$-measurable for all n. If we give Y some arbitrary value on the null set where it is not determined, Y is $\mathcal{A}_\infty$ measurable, and therefore $E(Y|\mathcal{A}_\infty) = Y$. □

Application II

Distributions

II.1 Preliminaries

II.1.1. Let Ω be an open subset of $\mathbb{R}^n$. For $0 \le k < \infty$, let $C^k(\Omega)$ denote the space of all complex functions on Ω with continuous (mixed) partial derivatives of order $\le k$. The intersection of all these spaces is denoted by $C^\infty(\Omega)$; it is the space of all (complex) functions with continuous partial derivatives of *all* orders in Ω. For $0 \le k \le \infty$, $C_c^k(\Omega)$ stands for the space of all $f \in C^k(\Omega)$ with compact support (in Ω). We shall also use the notation $C_c^k(\Delta)$ for the space of all functions in $C^k(\mathbb{R}^n)$ with compact support in the *arbitrary* set $\Delta \subset \mathbb{R}^n$. (The latter notation is consistent with the preceding one when Δ is *open*, since in that case any C^k-function in Δ with compact support in Δ extends trivially to a C^k-function on $\mathbb{R}^n$.)

Define $f : \mathbb{R} \to [0, \infty)$ by $f(t) = 0$ for $t \ge 0$, and $f(t) = \mathrm{e}^{1/t}$ for $t < 0$. Then $f \in C^\infty(\mathbb{R})$, and therefore, with a suitable choice of the constant γ, the function $\phi(x) := \gamma f(|x|^2 - 1)$ on $\mathbb{R}^n$ has the following properties:

(1) $\phi \in C_c^\infty(\mathbb{R}^n)$;

(2) $\operatorname{supp} \phi = \{x \in \mathbb{R}^n; |x| \le 1\}$; and

(3) $\phi \ge 0$ and $\int \phi \, dx = 1$.

(For $x \in \mathbb{R}^n, |x|$ denotes the usual Euclidean norm; $\int \cdot \, dx$ denotes integration over $\mathbb{R}^n$ with respect to the n-dimensional Lebesgue measure dx.)

In the following, ϕ denotes *any* fixed function with Properties (1)–(3).

If $u : \mathbb{R}^n \to \mathbb{C}$ is locally integrable, then for any $r > 0$, we consider its *regularization*

$$u_r(x) := \int u(x - ry)\phi(y) \, dy = r^{-n} \int u(y)\phi\left(\frac{x-y}{r}\right) dy, \tag{1}$$

(i.e., u_r is the convolution of u with $\phi_r := r^{-n}\phi(\cdot/r)$; note that the subscript r has different meanings when assigned to u and to ϕ.)

Theorem II.1.2. *Let K be a compact subset of (the open set) $\Omega \subset \mathbb{R}^n$, and let $u \in L^1(\mathbb{R}^n)$ vanish outside K. Then*

(1) $u_r \in C_c^\infty(\Omega)$ for all $r < \delta := dist(K, \Omega^c)$;

(2) $\lim_{r\to 0} u_r = u$ in L^p-norm if $u \in L^p (1 \le p < \infty)$, and uniformly if u is continuous.

Proof. Since $u_r = u * \phi_r$, one sees easily that (mixed) differentiation of any order can be performed on u_r by performing the operation on ϕ_r; the resulting convolution is clearly continuous. Hence $u_r \in C^\infty(\Omega)$.

Let

$$K_r := \{x \in \mathbb{R}^n;\ \operatorname{dist}(x, K) \le r\}. \tag{2}$$

It is a compact set, contained in Ω for $r < \delta$. If y is in the closed unit ball S of $\mathbb{R}^n$ and $x - ry \in K$, then $\operatorname{dist}(x, K) \le |x - (x - ry)| = |ry| \le r$, that is, $x \in K_r$. Hence, for $x \notin K_r$, $x - ry \notin K$ for all $y \in S$. Since u and ϕ vanish outside K and S, respectively, it follows from (1) that $u_r = 0$ outside K_r. Therefore $\operatorname{supp} u_r \subset K_r \subset \Omega$ (and so $u_r \in C_c^\infty(\Omega)$) for $r < \delta$.

By Property (3) of ϕ,

$$u_r(x) - u(x) = \int_S [u(x - ry) - u(x)]\phi(y)\, dy.$$

If u is continuous (hence uniformly continuous, since its support is in K), then for given $\epsilon > 0$, there exists $\eta > 0$ such that $|u(x - ry) - u(x)| < \epsilon$ for all $x \in \mathbb{R}^n$ and all $y \in S$ if $r < \eta$. Hence $\|u_r - u\|_\infty \le \epsilon$ for $r < \eta$.

In case $u \in L^p$, we have

$$\|u_r\|_p = \|u * \phi_r\|_p \le \|u\|_p \|\phi_r\|_1 = \|u\|_p.$$

Fix $v \in C_c(\Omega)$ such that $\|u - v\|_p < \epsilon$ (by density of $C_c(K)$ in $L^p(K)$). Let M be a bound for the (Lebesgue) measure of $(\operatorname{supp} v)_r$ for all $r < 1$. Then for $r < 1$,

$$\|u_r - u\|_p \le \|(u - v)_r\|_p + \|v_r - v\|_p + \|v - u\|_p < 2\epsilon + \|v_r - v\|_\infty M^{1/p} < 3\epsilon$$

(by the preceding case), for r small enough. □

The inequality

$$\|f * g\|_p \le \|f\|_p \|g\|_1 \quad (f \in L^p; g \in L^1) \tag{3}$$

used in the preceding proof, can be verified as follows:

$$\begin{aligned}
\|f * g\|_p &= \sup\left\{\left|\int (f * g) h\, dx\right|; h \in L^q, \|h\|_q \le 1\right\} \\
&\le \sup_h \iint |f(x - y)||g(y)|\, dy |h(x)|\, dx \\
&= \sup_h \iint |f(x - y)|\, |h(x)|\, dx |g(y)|\, dy \\
&\le \sup_h \int \|f(\cdot - y)\|_p \|h\|_q |g(y)|\, dy = \sup_h \|f\|_p \|h\|_q \|g\|_1 = \|f\|_p \|g\|_1,
\end{aligned}$$

where the suprema are taken over all h in the unit ball of L^q. (We used Theorems 4.6, 2.18, and 1.33, and the translation invariance of Lebesgue measure.)

Corollary II.1.3.

(1) $C_c^\infty(\Omega)$ *is dense in* $L^p(\Omega)$ $(1 \le p < \infty)$.

(2) A regular complex Borel measure μ *on* Ω *is uniquely determined by the integrals* $\int_\Omega f\,d\mu$ *with* $f \in C_c^\infty(\Omega)$.

Corollary II.1.4. *Let* K *be a compact subset of the open set* $\Omega \subset \mathbb{R}^n$. *Then there exists* $\psi \in C_c^\infty(\Omega)$ *such that* $0 \le \psi \le 1$ *and* $\psi = 1$ *in a neighborhood of* K.

Proof. Let δ be as in Theorem II.1.2, $r < \delta/3$, and $\psi = u_r$, where u is the indicator of K_{2r} (cf. (2)). Then $\operatorname{supp}\psi \subset (K_{2r})_r = K_{3r} \subset \Omega$, $\psi \in C_c^\infty(\Omega), 0 \le \psi \le 1$, and $\psi = 1$ on K_r. □

Corollary II.1.4 is the special case $k = 1$ of the following.

Theorem II.1.5 (Partitions of unity in $C_c^\infty(\Omega)$). *Let* $\Omega_1, \ldots, \Omega_k$ *be an open covering of the compact set* K *in* $\mathbb{R}^n$. *Then there exist* $\phi_j \in C_c^\infty(\Omega_j)$ $(j = 1, \ldots, k)$ *such that* $\phi_j \ge 0, \sum_j \phi_j \le 1$, *and* $\sum_j \phi_j = 1$ *in a neighborhood of* K.

The set $\{\phi_j\}$ is called a *partition of unity subordinate to the covering* $\{\Omega_j\}$.

Proof. There exist open sets with compact closures $K_j \subset \Omega_j$ such that $K \subset \bigcup_j K_j$. Let ψ_j be associated with K_j and Ω_j as in Corollary II.1.4. Define $\phi_1 = \psi_1$ and $\phi_j = \psi_j(1-\psi_{j-1})\ldots(1-\psi_1)$ for $j = 2, \ldots, k$ (as in the proof of Theorem 3.3). Then

$$\sum_j \phi_j = 1 - \prod_j \psi_j,$$

from which we read off the desired properties of ϕ_j. □

II.2 Distributions

II.2.1. *Topology on* $C_c^\infty(\Omega)$.

Let $D_j := -i\partial/\partial x_j$ $(j = 1, \ldots, n)$. For any "multi-index" $\alpha = (\alpha_1, \ldots, \alpha_n)$ $(\alpha_j = 0, 1, 2, \ldots)$ and $x = (x_1, \ldots, x_n) \in \mathbb{R}^n$, set

$$D^\alpha := D_1^{\alpha_1} \cdots D_n^{\alpha_n},$$

$$\alpha! := \alpha_1! \cdots \alpha_n!, |\alpha| := \sum \alpha_j, \text{ and } x^\alpha := x_1^{\alpha_1} \cdots x_n^{\alpha_n}.$$

We denote also

$$f^{(\alpha)} = \frac{\partial^{|\alpha|} f}{\partial x_1^{\alpha_1} \cdots \partial x_n^{\alpha_n}}$$

for any function f for which the shown derivative makes sense.

Let K be a compact subset of the open set $\Omega \subset \mathbb{R}^n$. The sequence of semi-norms on $C_c^\infty(K)$

$$\|\phi\|_k := \sum_{|\alpha|\le k} \sup |D^\alpha \phi|, \quad k = 0, 1, \ldots,$$

induces a (locally convex) topology on the vector space $C_c^\infty(K)$: it is the weakest topology on the space for which all these semi-norms are continuous. Basic neighborhoods of 0 in this topology have the form

$$\{\phi \in C_c^\infty(K); \|\phi\|_k < \epsilon\}, \tag{1}$$

with $\epsilon > 0$ and $k = 0, 1, \ldots$.

A sequence $\{\phi_j\}$ converges to ϕ in $C_c^\infty(K)$ iff $D^\alpha \phi_j \to D^\alpha \phi$ uniformly, for all α.

A linear functional u on $C_c^\infty(K)$ is continuous iff there exist a constant $C > 0$ and a non-negative integer k such that

$$|u(\phi)| \le C\|\phi\|_k \quad (\phi \in C_c^\infty(K)) \tag{2}$$

(cf. Theorem 4.2).

Let $\{\Omega_j\}$ be a sequence of open subsets of Ω, with union Ω, such that, for each j, Ω_j has compact closure K_j contained in Ω_{j+1}.

Since $C_c^\infty(\Omega) = \bigcup_j C_c^\infty(K_j)$, we may topologize $C_c^\infty(\Omega)$ in a natural way so that any linear functional u on $C_c^\infty(\Omega)$ is continuous iff its restriction to $C_c^\infty(K)$ is continuous for all compact $K \subset \Omega$. (This so-called "inductive limit topology" will not be described here systematically.) We note the following facts:

(i) a linear functional u on $C_c^\infty(\Omega)$ is continuous iff for each compact $K \subset \Omega$, there exist $C > 0$ and $k \in \mathbb{N} \cup \{0\}$ such that (2) holds;

(ii) a sequence $\{\phi_j\} \subset C_c^\infty(\Omega)$ converges to 0 iff $\{\phi_j\} \subset C_c^\infty(K)$ for some compact $K \subset \Omega$, and $\phi_j \to 0$ in $C_c^\infty(K)$.

(iii) a linear functional u on $C_c^\infty(\Omega)$ is continuous iff $u(\phi_j) \to 0$ for any *sequence* $\{\phi_j\} \subset C_c^\infty(\Omega)$ converging to 0.

Definition II.2.2. The space $C_c^\infty(\Omega)$ with the topology described earlier is denoted $\mathcal{D}(\Omega)$ and is called *the space of test functions on* Ω. The elements of its dual $\mathcal{D}'(\Omega)$ (= the space of all continuous linear functionals on $\mathcal{D}(\Omega)$) are called *distributions* in Ω.

The topology on $\mathcal{D}'(\Omega)$ is the "weak*" topology: the net $\{u_\nu\}$ of distributions converges to 0 iff $u_\nu(\phi) \to 0$ for all $\phi \in \mathcal{D}(\Omega)$.

II.2.3. Measures and functions.

If μ is a regular complex Borel measure on Ω, it may be identified with a continuous linear functional on $C_c(\Omega)$ (through the Riesz representation theorem); since it is uniquely determined by its restriction to $\mathcal{D}(\Omega)$ (cf. Corollary II.1.3), and this restriction *is continuous* (with respect to the stronger topology of $\mathcal{D}(\Omega)$), the measure μ can (and will) be identified with the *distribution* $u := \mu|_{\mathcal{D}(\Omega)}$. We say in this case that the distribution u *is a measure.*

In the special case where $d\mu = f\,dx$ with $f \in L^1_{loc}(\Omega)$ (i.e., f "locally integrable", that is, integrable on compact subsets), the function f is usually identified with the distribution it induces (as shown) through the formula

$$f(\phi) := \int_\Omega \phi f\,dx \quad (\phi \in \mathcal{D}(\Omega)).$$

In such event, we say that the distribution *is a function.*

If Ω' is an open subset of Ω, the restriction of a distribution u in Ω to $\mathcal{D}(\Omega')$ is a distribution in Ω', denoted $u|_{\Omega'}$ (and called *the restriction of* u *to* Ω'). If the distributions u_1, u_2 have equal restrictions to some open neighborhood of x, one says that they are *equal in a neighborhood of* x.

Proposition II.2.4. *If two distributions in* Ω *are equal in a neighborhood of each point of* Ω*, then they are equal.*

Proof. Fix $\phi \in \mathcal{D}(\Omega)$, and let $K = \operatorname{supp}\phi$. Each $x \in K$ has an open neighborhood $\Omega_x \subset \Omega$ in which the given distributions u_1, u_2 are equal. By compactness of K, there exist open sets $\Omega_j := \Omega_{x_j}$ $(j = 1, \dots, m)$ such that $K \subset \bigcup_j \Omega_j$. Let $\{\phi_j\}$ be a partition of unity subordinate to the open covering $\{\Omega_j\}$ of K. Then $\phi = \sum_j \phi\phi_j$ and $\phi\phi_j \in \mathcal{D}(\Omega_j)$. Hence $u_1(\phi\phi_j) = u_2(\phi\phi_j)$ by hypothesis, and therefore $u_1(\phi) = u_2(\phi)$. □

II.2.5. The support.

For any distribution u in Ω, the set

$$Z(u) := \{x \in \Omega;\, u = 0 \text{ in a neighborhood of } x\}$$

is open, and $u|_{Z(u)} = 0$ by Proposition II.2.4; furthermore, $Z(u)$ is *the largest open subset* Ω' *of* Ω *such that* $u|_{\Omega'} = 0$ (if $x \in \Omega'$ for such a set Ω', then $u = 0$ in the neighborhood Ω' of x, that is, $x \in Z(u)$; hence $\Omega' \subset Z(u)$). The *support* of u, denoted $\operatorname{supp} u$, is the set $\Omega - Z(u)$ (relatively *closed* in Ω). The previous statement may be rephrased as follows in terms of the support: $\operatorname{supp} u$ is the smallest relatively closed subset S of Ω such that $u(\phi) = 0$ for all $\phi \in \mathcal{D}(\Omega)$ such that $\operatorname{supp}\phi \cap S = \emptyset$.

If the distribution u is a measure or a function, its support as a distribution coincides with its support as a measure or a function, respectively (exercise).

II.2.6. Differentiation.

Fix the open set $\Omega \subset \mathbb{R}^n$.

For $j = 1, \dots, n$ and $u \in \mathcal{D}' := \mathcal{D}'(\Omega)$, we set

$$(D_j u)\phi = -u(D_j\phi) \quad (\phi \in \mathcal{D} := \mathcal{D}(\Omega)).$$

Then $D_j u \in \mathcal{D}'$ and the map $u \to D_j u$ is a continuous linear map of $\mathcal{D}'$ into itself. Furthermore, $D_k D_j = D_j D_k$ for all $j, k \in \{1, \ldots, n\}$, and

$$(D^\alpha u)\phi = (-1)^{|\alpha|} u(D^\alpha \phi) \quad (\phi \in \mathcal{D}).$$

For example, if δ denotes the Borel measure

$$\delta(E) = I_E(0) \quad (E \in \mathcal{B}(\mathbb{R}^n))$$

(the so-called *delta measure at* 0), then for all $\phi \in \mathcal{D}$,

$$(D^\alpha \delta)\phi = (-1)^{|\alpha|}\delta(D^\alpha \phi) = (-1)^{|\alpha|}(D^\alpha \phi)(0).$$

If u is a function such that $\partial u/\partial x_j$ exists and is locally integrable in Ω, then for all $\phi \in \mathcal{D}$, an integration by parts shows that

$$(D_j u)\phi := -u(D_j\phi) = \mathrm{i}\int u(\partial\phi/\partial x_j)\,dx = -\mathrm{i}\int (\partial u/\partial x_j)\phi\,dx := \left(\frac{1}{\mathrm{i}}\frac{\partial u}{\partial x_j}\right)(\phi),$$

so that $D_j u$ *is the function* $(1/\mathrm{i})\partial u/\partial x_j$ in this case, as desired.

Proposition II.2.7 (du Bois–Reymond). *Let $u, f \in C(\Omega)$, and suppose $D_j u = f$ (in the distribution sense). Then $D_j u = f$ in the classical sense.*

Proof. *Case $u \in C_c(\Omega)$.* Let u_r, f_r be regularizations of u, f (using the same ϕ as in II.1). Then

$$\begin{aligned} r^n D_j u_r(x) &= \int u(y) D_{x_j}\phi((x-y)/r)\,dy = -\int u(y) D_{y_j}\phi((x-y)/r)\,dy \\ &= \int (D_j u)\phi((x-y)/r)\,dy = r^n f_r. \end{aligned}$$

By Theorem II.1.2, $u_r \to u$ and $D_j u_r = f_r \to f$ *uniformly* as $r \to 0$. Therefore, $D_j u = f$ in the classical sense.

$u \in C(\Omega)$ arbitrary. Let $\psi \in \mathcal{D}(\Omega)$. Then $v := \psi u \in C_c(\Omega)$, and $D_j v = (D_j\psi)u + \psi f := g \in C(\Omega)$, so by the first case, $D_j v = g$ in the classical sense. For any point $x \in \Omega$, we may choose ψ not vanishing at x. Then $u = v/\psi$ is differentiable with respect to x_j at x, and $D_j u = f$ at x (in the classical sense). □

Let ω be an open set with compact closure contained in Ω (this relation between ω and Ω is denoted $\omega \subset\subset \Omega$), and let $\rho = \mathrm{diam}(\omega) := \sup\{|x-y|; x, y \in \omega\} (<\infty)$. Denote the unit vectors in the x_j-direction by e_j. Given $x \in \omega$, let t_j be the smallest positive number t such that $x + te_j \in \partial\omega$. If $\phi \in \mathcal{D}(\omega)$, then $\phi(x + t_j e_j) = 0$, and by the mean value theorem, there exists $0 < \tau_j < t_j$ such that

$$|\phi(x)| = |\phi(x) - \phi(x + t_j e_j)| = t_j |D_j \phi(x + \tau_j e_j)| \le \rho \sup |D_j \phi|.$$

Hence

$$\sup|\phi| \le \rho \sup|D_j\phi| \quad (\phi \in \mathcal{D}(\omega)).$$

Therefore, for any multi-index α with $|\alpha| \le k$,

$$\sup|D^\alpha\phi| \le \rho^{nk-|\alpha|} \sup|D_1^k \cdots D_n^k\phi|,$$

and consequently

$$\|\phi\|_k \le C' \sup|D_1^k \cdots D_n^k\phi| \quad (\phi \in \mathcal{D}(\omega)),$$

where $C' = C'(\rho, k, n)$ is a positive constant. Let $u \in \mathcal{D}'(\Omega)$. By (2), there exist a constant $C > 0$ and a non-negative integer k such that

$$|u(\phi)| \le CC' \sup|D_1^k \cdots D_n^k\phi| \quad (\phi \in \mathcal{D}(\omega)).$$

Write

$$(D_1^k \cdots D_n^k\phi)(x) = \mathrm{i}^n \int_{[y<x]} D_1^{k+1} \cdots D_n^{k+1}\phi\, dy,$$

where $[y < x] := \{y \in \mathbb{R}^n; y_j < x_j \ (j = 1, \ldots, n)\}$ and ϕ is extended to $\mathbb{R}^n$ in the usual way ($\phi = 0$ on ω^c). Writing $s := k + 1$, we have therefore

$$\sup|D_1^k \cdots D_n^k\phi| \le \int_\omega |D_1^s \cdots D_n^s\phi|\, dy,$$

and consequently,

$$|u(\phi)| \le CC'\|D_1^s \cdots D_n^s\phi\|_{L^1(\omega)} \quad (\phi \in \mathcal{D}(\omega)).$$

This means that the linear functional

$$(-1)^{ns} D_1^s \cdots D_n^s\phi \to u(\phi)$$

is continuous with norm $\le CC'$ on the subspace $D_1^s \cdots D_n^s \mathcal{D}(\omega)$ of $L^1(\omega)$. By the Hahn–Banach theorem, it has an extension as a continuous linear functional on $L^1(\omega)$ (with the same norm). Therefore there exists $f \in L^\infty(\omega)$ such that

$$u(\phi) = (-1)^{ns} \int_\omega f D_1^s \cdots D_n^s\phi\, dx \quad (\phi \in \mathcal{D}(\omega)).$$

This means that

$$u|_\omega = D_1^s \cdots D_n^s f.$$

We may also define

$$g(x) = \mathrm{i}^n \int_{[y<x]} I_\omega(y) f(y)\, dy;$$

Then g is continuous, and one verifies easily that $f = D_1 \cdots D_n g$ (in the distribution sense). Hence

$$u|_\omega = D_1^{s+1} \cdots D_n^{s+1} g$$

(with g continuous in ω). We proved the following.

Theorem II.2.8. *Let $u \in \mathcal{D}'(\Omega)$. Then for any $\omega \subset\subset \Omega$, there exist a non-negative integer s and a function $f \in L^\infty(\omega)$ such that $u|_\omega = D_1^s \cdots D_n^s f$. Moreover, f may be chosen to be continuous.*

II.2.9. *Leibnitz's formula.*

If $\phi, \psi \in C^\infty(\Omega)$, then for any multi-index α

$$D^\alpha(\phi\psi) = \sum_{\beta \leq \alpha} \frac{\alpha!}{\beta!(\alpha-\beta)!} D^\beta \phi D^{\alpha-\beta}\psi \tag{3}$$

(the sum goes over all multi-indices β with $\beta_j \leq \alpha_j$ for all j). This general Leibnitz formula follows by repeated application of the usual one-variable formula.

Multiplication of a distribution by a function.

Let $u \in \mathcal{D}'(\Omega)$ and $\psi \in C^\infty(\Omega)$. Since $\psi\phi \in \mathcal{D}(\Omega)$ for all $\phi \in \mathcal{D}(\Omega)$, the map $\phi \to u(\psi\phi)$ is well defined on $\mathcal{D}(\Omega)$. It is trivially linear, and for any compact $K \subset \Omega$, if C and k are as in (2), then for all $\phi \in C_c^\infty(K)$, we have by (3)

$$|u(\psi\phi)| \leq C\|\psi\phi\|_k \leq C'\|\phi\|_k,$$

where C' is a constant depending on n, k, K, and ψ. This means that the map defined above is a distribution in Ω; it is denoted ψu (and called the product of u by ψ). Thus

$$(\psi u)(\phi) = u(\psi\phi) \quad (\phi \in \mathcal{D}(\Omega)) \tag{4}$$

for all $\psi \in C^\infty(\Omega)$.

This definition clearly coincides with the usual pointwise multiplication when u is a function or a measure.

We verify easily the inclusion $Z(\psi) \cup Z(u) \subset Z(\psi u)$, that is,

$$\operatorname{supp}(\psi u) \subset \operatorname{supp}\psi \cap \operatorname{supp} u.$$

It follows from the definitions that

$$D_j(\psi u) = (D_j\psi)u + \psi(D_j u)$$

($\psi \in C^\infty(\Omega)$, $u \in \mathcal{D}'(\Omega)$, $j = 1, \ldots, n$), and therefore, by the same formal arguments as in the classical case, Leibnitz's formula (3) is valid in the present situation.

If P is a polynomial on $\mathbb{R}^n$, we denote by $P(D)$ the differential operator obtained by substituting formally D for the variable $x \in \mathbb{R}^n$. Let $P_\alpha(x) := x^\alpha$ for any multi-index α. Since $P_\alpha^{(\beta)}(x) = (\alpha!/(\alpha-\beta)!)x^{\alpha-\beta}$ for $\beta \leq \alpha$ (and equals zero otherwise), we can rewrite (3) in the form

$$P(D)(\psi u) = \sum_\beta (1/\beta!)(D^\beta\psi)P^{(\beta)}(D)u \tag{5}$$

for the special polynomials $P = P_\alpha$, hence by linearity, for *all* polynomials P. This is referred to as the "general Leibnitz formula".

II.2.10. The space $\mathcal{E}(\Omega)$ and its dual.

The space $\mathcal{E}(\Omega)$ is the space $C^\infty(\Omega)$ as a (locally convex) t.v.s. with the topology induced by the family of semi-norms

$$\phi \to \|\phi\|_{k,K} := \sum_{|\alpha|\le k} \sup_K |D^\alpha \phi|, \tag{6}$$

with $k = 0, 1, 2, \ldots$, and K varying over all compact subsets of Ω.

A sequence $\{\phi_k\}$ converges to 0 in $\mathcal{E}(\Omega)$ iff $D^\alpha \phi_k \to 0$ uniformly on every compact subset of Ω, for all multi-indices α.

A linear functional u on $\mathcal{E}(\Omega)$ is continuous iff there exist constants $k \in \mathbb{N}\cup\{0\}$ and $C > 0$ and a compact set K such that

$$|u(\phi)| \le C\|\phi\|_{k,K} \quad (\phi \in \mathcal{E}(\Omega)). \tag{7}$$

The dual space $\mathcal{E}'(\Omega)$ of $\mathcal{E}(\Omega)$ consists of all these continuous linear functionals, with the weak*-topology: the net u_ν converges to u in $\mathcal{E}'(\Omega)$ if $u_\nu(\phi) \to u(\phi)$ for all $\phi \in \mathcal{E}(\Omega)$.

If $u \in \mathcal{E}'(\Omega)$, then by (7), $u(\phi) = 0$ whenever $\phi \in \mathcal{E}(\Omega)$ vanishes in a neighborhood of the compact set K (appearing in (7)). For all $\phi \in \mathcal{D}(\Omega)$, $|u(\phi)| \le C\|\phi\|_{k,K} \le C\|\phi\|_k$, that is, $\tilde{u} := u|_{\mathcal{D}(\Omega)} \in \mathcal{D}'(\Omega)$. Also, if $x \in \Omega - K$ and ω is an open neighborhood of x contained in $\Omega - K$, then for all $\phi \in \mathcal{D}(\omega)$, $\|\phi\|_{k,K} = 0$, and (7) implies that $u(\phi) = 0$. This shows that $\Omega - K \subset Z(\tilde{u})$, that is, $\operatorname{supp} \tilde{u} \subset K$, that is, $\tilde{u}$ is a *distribution with compact support* in Ω.

Conversely, let v be a distribution with compact support in Ω, and let K be any compact subset of Ω containing this support. Fix $\psi \in \mathcal{D}(\Omega)$ such that $\psi = 1$ in a neighborhood of K. For any $\phi \in \mathcal{E}(\Omega)$, define $u(\phi) = v(\phi\psi)$. Then u is a well-defined linear functional on $\mathcal{E}(\Omega)$, $u(\phi) = 0$ whenever $\phi = 0$ in a neighborhood of K, and $u(\phi) = v(\phi)$ for $\phi \in \mathcal{D}(\Omega)$. On the other hand, if w is a linear functional on $\mathcal{E}(\Omega)$ with these properties, then for all $\phi \in \mathcal{E}(\Omega)$, $\phi\psi \in \mathcal{D}(\Omega)$ and $\phi(1-\psi) = 0$ in a neighborhood of K, and consequently,

$$w(\phi) = w(\phi\psi) + w(\phi(1-\psi)) = v(\phi\psi) = u(\phi).$$

This shows that v has a *unique* extension as a linear functional on $\mathcal{E}(\Omega)$ such that $v(\phi) = 0$ whenever ϕ vanishes in a neighborhood of K.

Let $Q = \operatorname{supp}\psi$. By (2) applied to the compact set Q, there exist C and k such that

$$|u(\phi)| = |v(\phi\psi)| \le C\|\phi\psi\|_k$$

for all $\phi \in \mathcal{E}(\Omega)$ (because $\phi\psi \in \mathcal{D}(Q)$). Hence, by Leibnitz's formula, $|u(\phi)| \le C'\|\phi\|_{k,Q}$ for some constant C', that is, $u \in \mathcal{E}'(\Omega)$.

We have established, therefore, that each distribution v with compact support has a unique extension as an element $u \in \mathcal{E}'(\Omega)$, and conversely, each $u \in \mathcal{E}'(\Omega)$ restricted to $\mathcal{D}(\Omega)$ is a distribution with compact support. This relationship allows us to *identify $\mathcal{E}'(\Omega)$ with the space of all distributions with compact support in Ω.*

II.2.11. Convolution.

Let $u \in \mathcal{D}' := \mathcal{D}'(\mathbb{R}^n)$ and $\phi \in \mathcal{D} := \mathcal{D}(\mathbb{R}^n)$. For $x \in \mathbb{R}^n$ *fixed*, let

$$(u * \phi)(x) := u(\phi(x - \cdot)).$$

The *function* $u * \phi$ is called the *convolution of* u *and* ϕ.

For x fixed, $h \neq 0$ real, and $j = 1, \ldots, n$, the functions $(ih)^{-1}[\phi(x+he_j-\cdot)-\phi(x-\cdot)]$ converge to $(D_j\phi)(x-\cdot)$ (as $h \to 0$) in $\mathcal{D}$. Therefore, $(ih)^{-1}[(u*\phi)(x+he_j)-(u*\phi)(x)] \to (u*(D_j\phi))(x)$, that is, $D_j(u*\phi) = u*(D_j\phi)$ *in the classical sense.* Also $u*(D_j\phi) = (D_ju)*\phi$ by definition of the derivative of a distribution. Iterating, we obtain that $u * \phi \in \mathcal{E} := \mathcal{E}(\mathbb{R}^n)$ and for any multi-index α,

$$D^\alpha(u * \phi) = u * (D^\alpha\phi) = (D^\alpha u) * \phi. \tag{8}$$

If $\operatorname{supp} u \cap \operatorname{supp} \phi(x - \cdot) = \emptyset$, then $(u * \phi)(x) = 0$. Equivalently, if $(u * \phi)(x) \neq 0$ then $\operatorname{supp} u$ meets $\operatorname{supp} \phi(x - \cdot)$ at some point y, that is, $x - y \in \operatorname{supp} \phi$ and $y \in \operatorname{supp} u$, that is, $x \in \operatorname{supp} u + \operatorname{supp} \phi$. This shows that

$$\operatorname{supp}(u * \phi) \subset \operatorname{supp} u + \operatorname{supp} \phi. \tag{9}$$

Hence

$$\mathcal{E}' * \mathcal{D} \subset \mathcal{D} \tag{10}$$

and in particular

$$\mathcal{D} * \mathcal{D} \subset \mathcal{D}. \tag{11}$$

Let $\phi_m \to 0$ in $\mathcal{D}$. There exits then a compact set K containing $\operatorname{supp} \phi_m$ for all m, and $D^\alpha\phi_m \to 0$ uniformly for all α. Let Q be any compact set in $\mathbb{R}^n$. It follows that $\operatorname{supp} \phi_m(x - \cdot) \subset Q - K := \{x - y; x \in Q, y \in K\}$ for all $x \in Q$. By (2) with the compact set $Q - K$ and the distribution $D^\alpha u$, there exist C, k such that

$$|D^\alpha(u * \phi_m)(x)| = |(D^\alpha u)(\phi_m(x - \cdot))| \leq C\|\phi_m(x - \cdot)\|_k = C\|\phi_m\|_k \to 0$$

for all $x \in Q$. Hence, $D^\alpha(u * \phi_m) \to 0$ uniformly on Q, and we conclude that $u*\phi_m \to 0$ in the topological space $\mathcal{E}$. In other words, *the (linear) map* $\phi \to u*\phi$ *is sequentially continuous from* $\mathcal{D}$ *to* $\mathcal{E}$. If $u \in \mathcal{E}'$, the map is (sequentially) continuous from $\mathcal{D}$ into itself (cf. (10)). In this case, *the definition of* $u*\phi$ *makes sense for all* $\phi \in \mathcal{E}$, *and the (linear) map* $\phi \to u*\phi$ *from* $\mathcal{E}$ *into itself is continuous* (note that $\mathcal{E}(\Omega)$ is metrizable, so there is no need to qualify the continuity; the metrizability follows from the fact that the topology of $\mathcal{E}(\Omega)$ is induced by the *countable* family of semi-norms $\{\|\cdot\|_{k,K_m}; k, m = 0, 1, 2, \ldots\}$, where $\{K_m\}$ is a suitable sequence of compact sets with union equal to Ω).

If $\phi, \psi \in \mathcal{D}$ and $u \in \mathcal{D}'$, it follows from (11) and the fact that $u * \phi \in \mathcal{E}$ that both $u * (\phi * \psi)$ and $(u * \phi) * \psi$ make sense. In order to show that they coincide, we approximate $(\phi * \psi)(x) = \int_Q \phi(x - y)\psi(y)\,dy$ (where Q is an n-dimensional cube containing the support of ψ) by (finite) Riemann sums of the form

$$\chi_m(x) = m^{-n} \sum_{y \in \mathbb{Z}^n} \phi(x - y/m)\psi(y/m).$$

If $\chi_m(x) \neq 0$ for some x and m, then there exists $y \in \mathbb{Z}^n$ such that $y/m \in \operatorname{supp}\psi$ and $x - y/m \in \operatorname{supp}\phi$, that is, $x \in y/m + \operatorname{supp}\phi \subset \operatorname{supp}\psi + \operatorname{supp}\phi$. This shows that for all m, χ_m have support in the fixed compact set $\operatorname{supp}\psi + \operatorname{supp}\phi$. Also for all multi-indices α,

$$(D^\alpha \chi_m)(x) = m^{-n} \sum_y D^\alpha \phi(x - y/m)\psi(y/m) \to ((D^\alpha \phi) * \psi)(x) = (D^\alpha(\phi * \psi))(x)$$

uniformly (in x). This means that $\chi_m \to \phi * \psi$ in $\mathcal{D}$. By continuity of u on $\mathcal{D}$,

$$\begin{aligned} [u * (\phi * \psi)](x) &:= u((\phi * \psi)(x - \cdot)) = \lim_m u(\chi_m(x - \cdot)) \\ &= \lim_m (u * \chi_m)(x) = \lim_m m^{-n} \sum_m (u * \phi)(x - y/m)\psi(y/m) \\ &= [(u * \phi) * \psi](x), \end{aligned}$$

for all x, that is,

$$u * (\phi * \psi) = (u * \phi) * \psi \quad (\phi, \psi \in \mathcal{D}; u \in \mathcal{D}'). \tag{12}$$

Fix ϕ as in Section II.1, consider ϕ_r as before, and define $u_r := u * \phi_r$ for any $u \in \mathcal{D}'$.

Proposition II.2.12 (Regularization of distributions). *For any distribution u in $\mathbb{R}^n$,*

(i) $u_r \in \mathcal{E}$ for all $r > 0$;

(ii) $\operatorname{supp} u_r \subset \operatorname{supp} u + \{x; |x| \le r\}$;

(iii) $u_r \to u$ in the space $\mathcal{D}'$.

Proof. Since $\operatorname{supp}\phi_r = \{x; |x| \le r\}$, (i) and (ii) are special cases of properties of the convolution discussed in II.2.11.

Denote $J : \psi(x) \in \mathcal{D} \to \tilde{\psi}(x) := \psi(-x)$. Then

$$u(\psi) = (u * \tilde{\psi})(0). \tag{13}$$

By Theorem II.1.2 applied to $\tilde{\psi}$, $\phi_r * \tilde{\psi} \to \tilde{\psi}$ in $\mathcal{D}$. Therefore, by (13),

$$\begin{aligned} u_r(\psi) &= [(u * \phi_r) * \tilde{\psi}](0) = [u * (\phi_r * \tilde{\psi})](0) \\ &= u(J(\phi_r * \tilde{\psi})) \to u(\psi) \end{aligned}$$

for all $\psi \in \mathcal{D}$, that is, $u_r \to u$ in $\mathcal{D}'$. □

In particular, *$\mathcal{E}$ is sequentially dense in $\mathcal{D}'$.*

Note also that if $u * \mathcal{D} = \{0\}$, then $u_r = 0$ for all $r > 0$; letting $r \to 0$, it follows that $u = 0$.

II.2.13. Commutation with translations.

Consider the *translation operators* (for $h \in \mathbb{R}^n$)

$$\tau_h : \phi(x) \to \phi(x-h)$$

from $\mathcal{D}$ into itself. For any $u \in \mathcal{D}'$, it follows from the definitions that

$$\tau_h(u * \phi) = u * (\tau_h \phi),$$

that is, *convolution with* u *commutes with translations.* This commutation property, together with the previously observed fact that convolution with u is a (sequentially) continuous linear map from $\mathcal{D}$ to $\mathcal{E}$, *characterizes convolution with distributions.* Indeed, let $U : \mathcal{D} \to \mathcal{E}$ be linear, sequentially continuous, and commuting with translations. Define $u(\phi) = (U(\tilde{\phi}))(0), (\phi \in \mathcal{D})$. Then u is a linear functional on $\mathcal{D}$, and if $\phi_k \to 0$ in $\mathcal{D}$, the sequential continuity of U on $\mathcal{D}$ implies that $u(\phi_k) \to 0$. Hence $u \in \mathcal{D}'$. For any $x \in \mathbb{R}^n$ and $\phi \in \mathcal{D}$,

$$\begin{aligned}(U\phi)(x) &= [\tau_{-x} U\phi](0) = [U(\tau_{-x}\phi)](0) \\ &= u(J(\tau_{-x}\phi)) = [u * (\tau_{-x}\phi)](0) = [\tau_{-x}(u * \phi)](0) = (u * \phi)(x),\end{aligned}$$

that is, $U\phi = u * \phi$, as wanted.

II.2.14. *Convolution of distributions.*

Let u, v be distributions in $\mathbb{R}^n$, one of which has compact support. The map

$$W : \phi \in \mathcal{D} \to u * (v * \phi) \in \mathcal{E}$$

is linear, continuous, and commutes with translations. By II.2.13, there exists a distribution w such that $W\phi = w * \phi$. By the final observation in II.2.12, w is uniquely determined; we call it *the convolution of* u *and* v and denote it by $u * v$; thus, *by definition,*

$$(u * v) * \phi = u * (v * \phi) \quad (\phi \in \mathcal{D}). \tag{14}$$

If $v = \psi \in \mathcal{D}$, the right-hand side of (14) equals $(u * \psi) * \phi$ by (12) (where $u * \psi$ is the "usual" convolution of the distribution u with ψ). Again by the final observation of II.2.12, it follows that the convolution of the two distributions u, v coincides with the previous definition when v is a function in $\mathcal{D}$ (the same is true if $u \in \mathcal{E}'$ and $v = \psi \in \mathcal{E}$).

One verifies easily that

$$\operatorname{supp}(u * v) \subset \operatorname{supp} u + \operatorname{supp} v. \tag{15}$$

(With ϕ_r as in Section II.1, it follows from (9) and Proposition II.2.12 that

$$\begin{aligned}\operatorname{supp}[(u * v) * \phi_r] = \operatorname{supp}[u * (v * \phi_r)] &\subset \operatorname{supp} u + \operatorname{supp} v_r \\ &\subset \operatorname{supp} u + \operatorname{supp} v + \{x; |x| \le r\},\end{aligned}$$

and we obtain (15) by letting $r \to 0$.)

If u, v, w are distributions, two of which (at least) having compact support, then by (15) both convolutions $(u * v) * w$ and $u * (v * w)$ are well-defined distributions. Since their convolutions with any given $\phi \in \mathcal{D}$ coincide, the "associative law" for distributions follows (cf. end of Proposition II.2.12).

Convolution of *functions* in $\mathcal{E}$ (one of which at least having compact support) is seen to be commutative by a change of variable in the defining integral. If $u \in \mathcal{D}'$ and $\psi \in \mathcal{D}$, we have for all $\phi \in \mathcal{D}$ (by definition and the associative law we just verified!):

$$\begin{aligned}(u * \psi) * \phi &:= u * (\psi * \phi) = u * (\phi * \psi) \\ &:= (u * \phi) * \psi = \psi * (u * \phi) = (\psi * u) * \phi,\end{aligned}$$

and therefore $u * \psi = \psi * u$. The same is valid (with a similar proof) when $u \in \mathcal{E}'$ and $\psi \in \mathcal{E}$.

For any two distributions u, v (one of which at least having compact support), the *commutative law of convolution* follows now from the same formal calculation with ψ replaced by v.

Let δ be the "delta measure at 0" (cf. II.2.6). Then for any multi-index α and $u \in \mathcal{D}'$,

$$(D^\alpha \delta) * u = D^\alpha u. \tag{16}$$

Indeed, observe first that for all $\phi \in \mathcal{D}$, $(\delta * \phi)(x) = \int \phi(x - y)\, d\delta(y) = \phi(x)$, that is, $\delta * \phi = \phi$. Therefore, for any $v \in \mathcal{D}'$, $(v * \delta) * \phi = v * (\delta * \phi) = v * \phi$, and consequently

$$v * \delta = v \quad (v \in \mathcal{D}'). \tag{17}$$

Now for all $\phi \in \mathcal{D}$,

$$\begin{aligned}(u * D^\alpha \delta) * \phi &= u * (D^\alpha \delta * \phi) = u * (D^\alpha \phi * \delta) \\ &= u * D^\alpha \phi = (D^\alpha u) * \phi,\end{aligned}$$

and (16) follows (cf. end of II.2.12 and (8)).

Next, for any distributions u, v with one at least having compact support, we have by (16) and the associative and commutative laws for convolution:

$$\begin{aligned}D^\alpha (u * v) &= (D^\alpha \delta) * (u * v) = (D^\alpha \delta * u) * v = (D^\alpha u) * v \\ &= D^\alpha (v * u) = (D^\alpha v) * u = u * (D^\alpha v).\end{aligned}$$

This generalizes (8) to the case when both factors in the convolution are distributions.

II.3 Temperate distributions

II.3.1. The Schwartz space.

The Schwartz space $\mathcal{S} = \mathcal{S}(\mathbb{R}^n)$ of *rapidly decreasing functions* consists of all $\phi \in \mathcal{E} = \mathcal{E}(\mathbb{R}^n)$ such that

$$\|\phi\|_{\alpha,\beta} := \sup_x |x^\beta D^\alpha \phi(x)| < \infty.$$

The topology induced on $\mathcal{S}$ by the family of semi-norms $\|\cdot\|_{\alpha,\beta}$ (where α, β range over all multi-indices) makes $\mathcal{S}$ into a locally convex (metrizable) topological vector space. It follows from Leibnitz's formula that $\mathcal{S}$ is a topological algebra for pointwise multiplication. It is also closed under multiplication by polynomials and application of any operator D^α, with both operations continuous from $\mathcal{S}$ into itself.

We have the topological inclusions

$$\mathcal{D} \subset \mathcal{S} \subset \mathcal{E}; \quad \mathcal{S} \subset L^1 := L^1(\mathbb{R}^n).$$

(Let $p_n(x) = \prod_{j=1}^n (1 + x_j^2)$; since $\|1/p_n\|_{L^1} = \pi^n$, we have

$$\|\phi\|_{L^1} \le \pi^n \sup_x |p_n(x)\phi(x)|. \tag{1}$$

If $\phi_k \to 0$ in $\mathcal{S}$, it follows from (1) that $\phi_k \to 0$ in L^1.)

Fix ϕ as in II.1.1, and let χ be the indicator of the ball $\{x; |x| \le 2\}$. The function $\psi := \chi * \phi$ belongs to $\mathcal{D}$ (cf. Theorem II.1.2), and $\psi = 1$ on the closed unit ball (because for $|x| \le 1, \{y; |y| \le 1$ and $|x-y| \le 2\} = \{y; |y| \le 1\} = \operatorname{supp}\phi$, and therefore $\psi(x) = \int_{\operatorname{supp}\phi} \chi(x-y)\phi(y)\,dy = \int \phi(y)\,dy = 1$).

Now, for *any* $\phi \in \mathcal{S}$ and the function ψ defined earlier, consider the functions $\phi_r(x) := \phi(x)\psi(rx), (r > 0)$ (not to be confused with the functions defined in II.1.1). Then $\phi_r \in \mathcal{D}$ and $\phi - \phi_r = 0$ for $|x| \le 1/r$. Therefore $\|\phi - \phi_r\|_{\alpha,\beta} = \sup_{|x|>1/r} |x^\beta D^\alpha(\phi - \phi_r)|$. We may choose M such that $\sup_x |x^\beta |x|^2 D^\alpha(\phi - \phi_r)| < M$ for all $0 < r \le 1$. Then, $\|\phi - \phi_r\|_{\alpha,\beta} \le \sup_{|x|>1/r} \frac{M}{|x|^2} < Mr^2 \to 0$ when $r \to 0$. Thus, $\phi_r \to \phi$ in $\mathcal{S}$, and we conclude that $\mathcal{D}$ *is dense in* $\mathcal{S}$. A similar argument shows that $\phi_r \to \phi$ in $\mathcal{E}$ for *any* $\phi \in \mathcal{E}$; hence $\mathcal{D}$ (and therefore $\mathcal{S}$) is dense in $\mathcal{E}$.

II.3.2. The Fourier transform on $\mathcal{S}$.

Denote the inner product in $\mathbb{R}^n$ by $x \cdot y$ ($x \cdot y := \sum_j x_j y_j$), and let $F : f \to \hat{f}$ be the Fourier transform on L^1:

$$\hat{f}(y) = \int_{\mathbb{R}^n} \mathrm{e}^{-\mathrm{i}x\cdot y} f(x)\,dx \quad (f \in L^1). \tag{2}$$

(All integrals in the sequel are over $\mathbb{R}^n$, unless specified otherwise.)

If $\phi \in \mathcal{S}, x^\alpha \phi(x) \in \mathcal{S} \subset L^1$ for all multi-indices α, and therefore, the integral $\int \mathrm{e}^{-\mathrm{i}x\cdot y}(-x)^\alpha \phi(x)\,dx$ converges *uniformly* in y. Since this integral is the result of applying D^α to the integrand of (2), we have $\hat{\phi} \in \mathcal{E}$ and

$$D^\alpha \hat{\phi} = F[(-x)^\alpha \phi(x)]. \tag{3}$$

For any multi-index β, it follows from (3) (by integration by parts) that

$$y^\beta (D^\alpha \hat{\phi})(y) = F D^\beta [(-x)^\alpha \phi(x)]. \tag{4}$$

In particular ($\alpha = 0$)

$$y^\beta \hat{\phi}(y) = [F D^\beta \phi](y). \tag{5}$$

Since $D^\beta[(-x)^\alpha\phi(x)] \in \mathcal{S} \subset L^1$, it follows from (4) that $y^\beta(D^\alpha\hat{\phi})(y)$ is a bounded function of y, that is, $\hat{\phi} \in \mathcal{S}$. Moreover, by (1) and (4),

$$\sup_y \left| y^\beta (D^\alpha \hat{\phi})(y) \right| \le \|D^\beta[(-x)^\alpha\phi(x)]\|_{L^1}$$
$$\le \pi^n \sup_x \left| p_n(x) D^\beta[(-x)^\alpha \phi(x)] \right|.$$

This inequality shows that if $\phi_k \to 0$ in $\mathcal{S}$, then $\hat{\phi}_k \to 0$ in $\mathcal{S}$, that is, the map $F : \phi \in \mathcal{S} \to \hat{\phi} \in \mathcal{S}$ is a *continuous* (linear) operator. Denote

$$M^\beta : \phi(x) \in \mathcal{S} \to x^\beta \phi(x) \in \mathcal{S}.$$

Then (4) can be written as the operator identity on $\mathcal{S}$:

$$M^\beta D^\alpha F = F D^\beta (-M)^\alpha. \tag{6}$$

A change of variables shows that

$$(F\psi(ry))(s) = r^{-n}\hat{\psi}(s/r) \tag{7}$$

for any $\psi \in L^1$ and $r > 0$.

If $\phi, \psi \in L^1$, an application of Fubini's theorem gives

$$\int \hat{\phi}(y)\psi(y)\mathrm{e}^{\mathrm{i}x\cdot y}\, dy = \int \hat{\psi}(t-x)\phi(t)\, dt = \int \hat{\psi}(s)\phi(x+s)\, ds. \tag{8}$$

Replacing $\psi(y)$ by $\psi(ry)$ in (8), we obtain by (7) and the change of variable $s = rt$

$$\int \hat{\phi}(y)\psi(ry)\mathrm{e}^{\mathrm{i}x\cdot y}\, dy = r^{-n}\int \hat{\psi}(s/r)\phi(x+s)\, ds = \int \hat{\psi}(t)\phi(x+rt)\, dt. \tag{9}$$

In case $\phi, \psi \in \mathcal{S}(\subset L^1)$, we have $\hat{\phi}, \hat{\psi} \in \mathcal{S} \subset L^1$, and ϕ, ψ are bounded and continuous. By Lebesgue's dominated convergence theorem, letting $r \to 0$ in (9) gives

$$\psi(0)\int \hat{\phi}(y)\mathrm{e}^{\mathrm{i}x\cdot y}\, dy = \phi(x)\int \hat{\psi}(t)\, dt \tag{10}$$

for all $\phi, \psi \in \mathcal{S}$ and $x \in \mathbb{R}^n$.

Choose, for example, $\psi(x) = (2\pi)^{-n/2}\,\mathrm{e}^{-|x|^2/2}$. Then $\hat{\psi}(t) = \mathrm{e}^{-|t|^2/2}$ (cf. I.3.12) and $\int \hat{\psi}(t)\, dt = (2\pi)^{n/2}$. Substituting these values in (10), we obtain (for all $\phi \in \mathcal{S}$)

$$\phi(x) = (2\pi)^{-n}\int \hat{\phi}(y)\mathrm{e}^{\mathrm{i}x\cdot y}\, dy. \tag{11}$$

This is the *inversion formula for the Fourier transform* F on $\mathcal{S}$. It shows that F is an automorphism of $\mathcal{S}$, whose inverse is given by

$$F^{-1} = (2\pi)^{-n} J F, \tag{12}$$

where $J : \phi \to \tilde{\phi}$.

Note that $JF = FJ$ and $F^2 = (2\pi)^n J$.

Also, by definition and the inversion formula,

$$\hat{\bar{\psi}}(y) = \int e^{ix\cdot y}\overline{\psi(x)}\,dx = (2\pi)^n (F^{-1}\bar{\psi})(y),$$

that is,

$$F(\hat{\bar{\psi}}) = (2\pi)^n \bar{\psi} \quad (\psi \in \mathcal{S}). \tag{13}$$

(It is sometimes advantageous to define the Fourier transform by $\mathcal{F} = (2\pi)^{-n/2}F$; the inversion formula for $\mathcal{F} : \mathcal{S} \to \mathcal{S}$ is then $\mathcal{F}^{-1} = J\mathcal{F}$, and the last identities become $\mathcal{F}^2 = J$ and $\mathcal{F}C\mathcal{F} = C$, where $C : \psi \to \bar{\psi}$ is the conjugation operator.)

An application of Fubini's theorem shows that

$$F(\phi * \psi) = \hat{\phi}\hat{\psi} \quad (\phi, \psi \in \mathcal{S}). \tag{14}$$

(This is true actually for all $\phi, \psi \in L^1$.)

Replacing ϕ, ψ by $\hat{\phi}, \hat{\psi} (\in \mathcal{S})$, respectively, we get

$$\begin{aligned} F(\hat{\phi} * \hat{\psi}) &= (F^2\phi)(F^2\psi) = (2\pi)^{2n}(J\phi)(J\psi) \\ &= (2\pi)^{2n} J(\phi\psi) = (2\pi)^n F^2(\phi\psi). \end{aligned}$$

Hence

$$F(\phi\psi) = (2\pi)^{-n}\hat{\phi} * \hat{\psi} \quad (\phi, \psi \in \mathcal{S}). \tag{15}$$

For $x = 0$, the identity (8) becomes

$$\int \hat{\phi}\psi\,dx = \int \phi\hat{\psi}\,dx \quad (\phi, \psi \in L^1). \tag{16}$$

In case $\psi \in \mathcal{S}$ (so that $\hat{\psi} \in \mathcal{S} \subset L^1$), we replace ψ by $\hat{\bar{\psi}}$ in (16); using (13), we get

$$\int \hat{\phi}\bar{\hat{\psi}}\,dx = (2\pi)^n \int \phi\bar{\psi}\,dx \quad (\phi, \psi \in \mathcal{S}). \tag{17}$$

This is *Parseval's formula* for the Fourier transform.

In terms of the operator $\mathcal{F}$, the formula takes the form

$$(\mathcal{F}\phi, \mathcal{F}\psi) = (\phi, \psi) \quad (\phi, \psi \in \mathcal{S}), \tag{18}$$

where $(\cdot, \cdot)$ is the L^2 inner product.

In particular,

$$\|\mathcal{F}\phi\|_2 = \|\phi\|_2, \tag{19}$$

where $\|\cdot\|_2$ denotes here the L^2-norm.

Thus, $\mathcal{F}$ is a (linear) isometry of $\mathcal{S}$ onto itself, with respect to the L^2-norm on $\mathcal{S}$. Since $\mathcal{S}$ is dense in L^2 (recall that $\mathcal{D} \subset \mathcal{S}$, and $\mathcal{D}$ is dense in L^2), the operator $\mathcal{F}$ extends uniquely as a linear isometry of L^2 onto itself. This operator, also denoted $\mathcal{F}$, is called *the L^2-Fourier transform.*

Example. Consider the orthonormal sequence $\{f_k;\ k \in \mathbb{Z}\}$ in $L^2(\mathbb{R})$ defined in the second example of Section 8.11. Since $\mathcal{F}$ is a Hilbert automorphism of $L^2(\mathbb{R})$, *the sequence* $\{g_k := \mathcal{F} f_k;\ k \in \mathbb{Z}\}$ *is orthonormal in* $L^2(\mathbb{R})$. The fact that f_k are also in $L^1(\mathbb{R})$ allows us to calculate as follows

$$g_k(y) = (1/\sqrt{2\pi}) \int_{\mathbb{R}} \mathrm{e}^{-ixy} f_k(x)\, dx = (1/2\pi) \int_{-\pi}^{\pi} \mathrm{e}^{-ix(y-k)}\, dx$$
$$= (-1)^k \frac{\sin \pi y}{\pi(y-k)}.$$

Note in particular that

$$\int_{\mathbb{R}} \frac{\sin^2(\pi y)}{(\pi y)^2}\, dy = \|g_0\|_2^2 = 1,$$

that is, $\int_{\mathbb{R}} \sin^2 t/t^2\, dt = \pi$. Integrating by parts, we have for all $a < b$ real

$$\int_a^b \sin^2 t/t^2\, dt = \int_a^b \sin^2 t\, d(-1/t) = \sin^2 a/a - \sin^2 b/b + \int_a^b \sin(2t)/t\, dt.$$

Letting $a \to -\infty$ and $b \to \infty$, we see that the integral $\int_{\mathbb{R}} \sin(2t)/t\, dt$ converges and has the value π, that is,

$$\int_{\mathbb{R}} \frac{\sin t}{t}\, dt = \pi.$$

This is the so-called *Dirichlet integral.*

If $g = \mathcal{F} f$ is in the closure of the span of $\{g_k\}$ in $L^2(\mathbb{R})$, f is necessarily in the closure of the span of $\{f_k\}$, hence, vanishing on $(-\pi, \pi)^{\mathrm{c}}$; in particular, f is also in $L^1(\mathbb{R})$, and therefore g is continuous. Also g has the unique $L^2(\mathbb{R})$-convergent generalized Fourier expansion $g = \sum_{k\in\mathbb{Z}} a_k g_k$ (equality in L^2). Since both $\{a_k\}$ and $\{\|g_k\|_\infty\}$ are in $l^2(\mathbb{Z})$, it follows (by Schwarz's inequality for $l^2(\mathbb{Z})$) that the shown series for g converges (absolutely and) uniformly on $\mathbb{R}$; in particular, $\sum a_k g_k$ is continuous. Since g is continuous as well, $g = \sum a_k g_k$ everywhere, that is,

$$g(y) = \frac{\sin \pi y}{\pi} \sum_k (-1)^k a_k/(y-k).$$

Letting $y \to n$ for any given $n \in \mathbb{Z}$, we get $g(n) = a_n$ Thus

$$g(y) = \frac{\sin \pi y}{\pi} \sum_k (-1)^k g(k)/(y-k).$$

II.3.3. The dual space $\mathcal{S}'$.

If $u \in \mathcal{S}'$, then $u|_{\mathcal{D}} \in \mathcal{D}'$ (because the inclusion $\mathcal{D} \subset \mathcal{S}$ is topological). Moreover, since $\mathcal{D}$ is dense in $\mathcal{S}, u$ is uniquely determined by its restriction to $\mathcal{D}$.

The one-to-one map $u \in \mathcal{S}' \to u|_{\mathcal{D}} \in \mathcal{D}'$ allows us to identify $\mathcal{S}'$ as a subspace of $\mathcal{D}'$; its elements are called *temperate distributions.* We also have $\mathcal{E}' \subset \mathcal{S}'$:

$$\mathcal{E}' \subset \mathcal{S}' \subset \mathcal{D}',$$

topologically.

Note that $L^p \subset \mathcal{S}'$ for any $p \in [1, \infty]$.

The *Fourier transform of the temperate distribution* u is defined by

$$\hat{u}(\phi) = u(\hat{\phi}) \quad (\phi \in \mathcal{S}). \tag{20}$$

If u *is a function* in L^1, it follows from (16) and the density of $\mathcal{S}$ in L^1 that its Fourier transform as a temperate distribution "is the function $\hat{u}$", the usual L^1-Fourier transform. Similarly, if $u \in M(\mathbb{R}^n)$, that is, if u *is a (regular Borel) complex measure*, $\hat{u}$ "is" the usual *Fourier–Stieltjes transform* of u, defined by

$$\hat{u}(y) = \int_{\mathbb{R}^n} \mathrm{e}^{-\mathrm{i}x \cdot y} du(x) \quad (y \in \mathbb{R}^n).$$

In general, (20) defines $\hat{u}$ as a temperate distribution, since the map $F : \phi \to \hat{\phi}$ is a continuous linear map of $\mathcal{S}$ into itself and $\hat{u} := u \circ F$. We shall write Fu for $\hat{u}$ (using the same notation for the "extended" operator); F is trivially a continuous linear operator on $\mathcal{S}'$.

Similarly, we define the (continuous linear) operators J and $\mathcal{F}$ on $\mathcal{S}'$ by

$$(Ju)(\phi) = u(J\phi) \quad (\phi \in \mathcal{S})$$

and $\mathcal{F} = (2\pi)^{-n/2} F$.

We have for all $\phi \in \mathcal{S}$

$$(F^2 u)(\phi) = u(F^2 \phi) = (2\pi)^n u(J\phi) = (2\pi)^n (Ju)(\phi),$$

that is

$$F^2 = (2\pi)^n J$$

on $\mathcal{S}'$. It follows that F is a continuous automorphism of $\mathcal{S}'$; its inverse is given by the Fourier inversion formula $F^{-1} = (2\pi)^{-n} JF$ (equivalently, $\mathcal{F}^{-1} = J\mathcal{F}$) on $\mathcal{S}'$.

It follows in particular that the restrictions of F to L^1 and to $M := M(\mathbb{R}^n)$ are one-to-one. This is the so-called *uniqueness property* of the L^1-Fourier transform and of the Fourier–Stieltjes transform, respectively.

If $u \in L^2 (\subset \mathcal{S}')$, then for all $\phi \in \mathcal{S}$

$$\begin{aligned} |(\mathcal{F}u)(\phi)| = |u(\mathcal{F}\phi)| &= \left| \int u \cdot \mathcal{F}\phi \, dx \right| \\ &\leq \|u\|_2 \|\mathcal{F}\phi\|_2 = \|u\|_2 \|\phi\|_2. \end{aligned}$$

Thus, $\mathcal{F}u$ is a continuous linear functional on the dense subspace $\mathcal{S}$ of L^2 with norm $\leq \|u\|_2$. It extends uniquely as a continuous linear functional on L^2 with the same norm. By the ("Little") Riesz representation theorem, there exists a unique $g \in L^2$ such that $\|g\|_2 \le \|u\|_2$ and

$$(\mathcal{F}u)(\phi) = \int \phi g \, dx \quad (\phi \in \mathcal{S}).$$

This shows that $\mathcal{F}u$ "is" the L^2-function g, that is, $\mathcal{F}$ maps L^2 into itself. The identity $u = \mathcal{F}^2 Ju$ shows that $\mathcal{F}L^2 = L^2$. Also $\|\mathcal{F}u\|_2 = \|g\|_2 \le \|u\|_2$, so that

$$\|u\|_2 = \|Ju\|_2 = \|\mathcal{F}^2 u\|_2 \le \|\mathcal{F}u\|_2,$$

and the equality $\|\mathcal{F}u\|_2 = \|u\|_2$ follows. This proves that $\mathcal{F}|_{L^2}$ is a (linear) isometry of L^2 onto itself. Its restriction to the dense subspace $\mathcal{S}$ of L^2 is the operator $\mathcal{F}$ originally defined on $\mathcal{S}$; therefore $\mathcal{F}|_{L^2}$ coincides with the L^2 Fourier transform defined at the end of II.3.2.

The formulae relating the operators F, D^α, and M^β on $\mathcal{S}$ extend easily to $\mathcal{S}'$: if $u \in \mathcal{S}'$, then for all $\phi \in \mathcal{S}$,

$$\begin{aligned}[FD^\beta u]\phi &= (D^\beta u)(F\phi) = u\left((-D)^\beta F\phi\right) \\ &= u\left(FM^\beta \phi\right) = (M^\beta F u)(\phi),\end{aligned}$$

that is, $FD^\beta = M^\beta F$ on $\mathcal{S}'$. By linearity of F, it follows that for any polynomial P on $\mathbb{R}^n, FP(D) = P(M)F$ on $\mathcal{S}'$.

Theorem II.3.4. *If $u \in \mathcal{E}'$, $\hat{u}$ "is the function" $\psi(y) := u(\mathrm{e}^{-\mathrm{i}x\cdot y})$ (y is a parameter on the right of the equation), and extends to $\mathbb{C}^n$ as the entire function $\psi(z) := u(\mathrm{e}^{-\mathrm{i}x\cdot z})$, ($z \in \mathbb{C}^n$). In particular, $\hat{u} \in \mathcal{E}$.*

Proof.

(i) *Case $u \in \mathcal{D}(\subset \mathcal{E}')$.* Since $\psi(z) = \int_{\operatorname{supp} u} \mathrm{e}^{-\mathrm{i}x\cdot z} u(x)\, dx$, it is clear that ψ is entire, and coincides with $\hat{u}$ on $\mathbb{R}^n$.

(ii) *General case.* Let ϕ be as in II.1.1, and consider the regularizations $u_r := u * \phi_r \in \mathcal{D}$. By Proposition II.2.12, $u_r \to u$ in $\mathcal{D}'$ and $\operatorname{supp} u_r \subset \operatorname{supp} u + \{x; |x| \le 1\} := K$ for all $0 < r \le 1$. Given $\phi \in \mathcal{E}$, choose $\phi' \in \mathcal{D}$ that coincides with ϕ in a neighborhood of K. Then for $0 < r \le 1, u_r(\phi) = u_r(\phi') \to u(\phi') = u(\phi)$ as $r \to 0$, that is, $u_r \to u$ in $\mathcal{E}'$, hence also in $\mathcal{S}'$. Therefore $\hat{u}_r \to \hat{u}$ in $\mathcal{S}'$ (by continuity of the Fourier transform on $\mathcal{S}'$). By Case (i), $\hat{u}_r(y) = u_r(\mathrm{e}^{-\mathrm{i}x\cdot y}) \to u(\mathrm{e}^{-\mathrm{i}x\cdot y})$, since $u_r \to u$ in $\mathcal{E}'$. More precisely, for any $z \in \mathbb{C}^n$,

$$u_r(\mathrm{e}^{-\mathrm{i}x\cdot z}) = [(u * \phi_r) * \mathrm{e}^{\mathrm{i}x\cdot z}](0) = [u * (\phi_r * \mathrm{e}^{\mathrm{i}x\cdot z})](0).$$

However,

$$\begin{aligned}\phi_r * \mathrm{e}^{\mathrm{i}x\cdot z} &= r^{-n} \int \mathrm{e}^{\mathrm{i}(x-y)\cdot z} \phi(y/r)\, dy \\ &= \mathrm{e}^{\mathrm{i}x\cdot z} \int \mathrm{e}^{-\mathrm{i}t\cdot rz} \phi(t)\, dt = \mathrm{e}^{\mathrm{i}x\cdot z} \hat{\phi}(rz).\end{aligned}$$

Therefore,

$$u_r(\mathrm{e}^{-\mathrm{i}x\cdot z}) = \hat{\phi}(rz)(u * \mathrm{e}^{\mathrm{i}x\cdot z})(0) = \hat{\phi}(rz)u(\mathrm{e}^{-\mathrm{i}x\cdot z}),$$

and so, for all $0 < r \le 1$,

$$\begin{aligned}|u_r(\mathrm{e}^{-\mathrm{i}x\cdot z}) - u(\mathrm{e}^{-\mathrm{i}x\cdot z})| &= |\hat{\phi}(rz) - 1|\,|u(\mathrm{e}^{-\mathrm{i}x\cdot z})| \\ &= \left|\int_{|x|\le 1} (\mathrm{e}^{-\mathrm{i}x\cdot rz} - 1)\phi(x)\,dx\right|\,|u(\mathrm{e}^{-\mathrm{i}x\cdot z})| \\ &\le C\|\mathrm{e}^{-\mathrm{i}x\cdot z}\|_{k,K}|z|e^{|\Im z|}r,\end{aligned}$$

with the constants C, k and the compact set K independent of r. Consequently $u_r(\mathrm{e}^{-\mathrm{i}x\cdot z}) \to u(\mathrm{e}^{-\mathrm{i}x\cdot z})$ as $r \to 0$, *uniformly* with respect to z on compact subsets of $\mathbb{C}^n$. Since $u_r(\mathrm{e}^{-\mathrm{i}x\cdot z})$ are entire (cf. Case (i)), it follows that $u(\mathrm{e}^{-\mathrm{i}x\cdot z})$ is entire.

In order to verify that $\hat{u}$ is the function $u(\mathrm{e}^{-\mathrm{i}x\cdot y})$, it suffices to show that

$$\hat{u}(\phi) = \int \phi(y)u(\mathrm{e}^{-\mathrm{i}x\cdot y})\,dy$$

for arbitrary $\phi \in \mathcal{D}$, since $\mathcal{D}$ is dense in $\mathcal{S}$. Let K be the compact support of ϕ, and $0 < r \le 1$. Since $\phi(y)\hat{u}_r(y) \to \phi(y)u(\mathrm{e}^{-\mathrm{i}x\cdot y})$ uniformly on K as $r \to 0$, we get

$$\hat{u}(\phi) = \lim_{r\to 0} \hat{u}_r(\phi) = \lim_r \int_K \phi(y)\hat{u}_r(y)\,dy = \int_K \phi(y)u(\mathrm{e}^{-\mathrm{i}x\cdot y})\,dy,$$

as desired. □

The entire function $u(\mathrm{e}^{-\mathrm{i}x\cdot z})$ will be denoted $\hat{u}(z)$; since the distribution $\hat{u}$ "is" this function restricted to $\mathbb{R}^n$ (by Theorem II.3.4), the notation is justified. The function $\hat{u}(z)$ is called the *Fourier–Laplace transform* of $u \in \mathcal{E}'$.

Theorem II.3.5. *If $u \in \mathcal{E}'$ and $v \in \mathcal{S}'$, then $u * v \in \mathcal{S}'$ and $F(u * v) = \hat{u}\,\hat{v}$.*

Proof. By Theorem II.3.4, $\hat{u} \in \mathcal{E}$, and therefore the product $\hat{u}\,\hat{v}$ makes sense as a distribution (cf. (4), II.2.9). We prove next that $u * v \in \mathcal{S}'$; then $F(u * v)$ will make sense as well (and belong to $\mathcal{S}'$), and it will remain to verify the identity.

For all $\phi \in \mathcal{D}$,

$$\begin{aligned}(u * v)(\phi) &= [(u * v) * J\phi](0) = [u * (v * J\phi)](0) \\ &= u(J(v * J\phi)) = (Ju)(v * J\phi).\end{aligned} \tag{21}$$

Since $Ju \in \mathcal{E}'$, there exist $K \subset \mathbb{R}^n$ compact and constants $C > 0$ and $k \in \mathbb{N}\cup\{0\}$ (all independent of ϕ) such that

$$|(u * v)(\phi)| \le C\|v * J\phi\|_{k,K} \tag{22}$$

(cf. (7) in II.2.10).

For each multi-index α, $D^\alpha v \in \mathcal{S}'$. Therefore, there exist a constant $C' > 0$ and multi-indices β, γ (all independent of ϕ and x), such that

$$|(D^\alpha v * J\phi)(x)| = |(D^\alpha v)(\phi(y - x))| \le C' \sup_y |y^\beta (D^\gamma \phi)(y)|. \tag{23}$$

By (22) and (23), $|(u * v)(\phi)|$ can be estimated by semi-norms of ϕ in $\mathcal{S}$ (for all $\phi \in \mathcal{D}$). Since $\mathcal{D}$ is dense in $\mathcal{S}$, $u * v$ extends uniquely as a continuous linear functional on $\mathcal{S}$. Thus, $u * v \in \mathcal{S}'$.

We now verify the identity of the theorem, first in the special case $u \in \mathcal{D}$. Then $\hat{u} \in \mathcal{S}$, and therefore $\hat{u}\hat{v} \in \mathcal{S}'$ (by a simple application of Leibnitz's formula).

By (21), if $\psi \in \mathcal{S}$ is such that $\hat{\psi} \in \mathcal{D}$,

$$[F(u * v)](\psi) = (u * v)(\hat{\psi}) = (v * u)(\hat{\psi}) = v(J(u * J\hat{\psi})).$$

For $f, g \in \mathcal{D}$, it follows from the integral definition of the convolution that $J(f * g) = (Jf) * (Jg)$. Then, by (12) and (15) in II.3.2,

$$\begin{aligned}[F(u * v)](\psi) &= v((Ju) * \hat{\psi}) = v[(2\pi)^{-n}(F^2u) * F\psi)] \\ &= v(F(\hat{u}\psi)) = \hat{v}(\hat{u}\psi) = (\hat{u}\hat{v})(\psi),\end{aligned}$$

where the last equality follows from the definition of the product of the distribution $\hat{v}$ by the function $\hat{u} \in \mathcal{S} \subset \mathcal{E}$. Hence $F(u * v) = \hat{u}\hat{v}$ on the set $\mathcal{S}_0 = \{\psi \in \mathcal{S}; \hat{\psi} \in \mathcal{D}\}$. For $\psi \in \mathcal{S}$ arbitrary, since $\hat{\psi} \in \mathcal{S}$ and $\mathcal{D}$ is dense in $\mathcal{S}$, there exists a sequence $\phi_k \in \mathcal{D}$ such that $\phi_k \to \hat{\psi}$ in $\mathcal{S}$. Let $\psi_k = F^{-1}\phi_k$. Then $\psi_k \in \mathcal{S}, F\psi_k = \phi_k \in \mathcal{D}$ (that is, $\psi_k \in \mathcal{S}_0$), and $\psi_k \to \psi$ by continuity of F^{-1} on $\mathcal{S}$ (i.e., $\mathcal{S}_0$ is dense in $\mathcal{S}$).

Since $F(u * v)$ and $\hat{u}\hat{v}$ are in $\mathcal{S}'$ (as observed) and coincide on $\mathcal{S}_0$, they are indeed equal.

Consider next the general case $u \in \mathcal{E}'$. If $\phi \in \mathcal{D}$, $\phi * u \in \mathcal{D}$ and $v \in \mathcal{S}'$, and also $\phi \in \mathcal{D}$ and $u * v \in \mathcal{S}'$. Applying the special case to these two pairs, we get

$$\begin{aligned}(\hat{\phi}\hat{u})\hat{v} &= [F(\phi * u)]\hat{v} = F[(\phi * u) * v] \\ &= F[\phi * (u * v)] = \hat{\phi}F(u * v).\end{aligned}$$

For each point y, we can choose $\phi \in \mathcal{D}$ such that $\hat{\phi}(y) \neq 0$; it follows that $\hat{u}\hat{v} = F(u * v)$. □

The next theorem characterizes Fourier–Laplace transforms of distributions with compact support (cf. Theorem II.3.4).

Theorem II.3.6 (Paley–Wiener–Schwartz).

(i) The entire function f on $\mathbb{C}^n$ is the Fourier–Laplace transform of a distribution u with support in the ball $S_A := \{x \in \mathbb{R}^n; |x| \leq A\}$ iff there exist a constant $C > 0$ and a non-negative integer m such that

$$|f(z)| \leq C(1 + |z|)^m \, e^{A|\Im z|} \quad (z \in \mathbb{C}^n).$$

(ii) The entire function f is the Fourier–Laplace transform of a function $u \in \mathcal{D}(S_A)$ iff for each m, there exists a positive constant $C = C(m)$ such that

$$|f(z)| \leq C(m)(1 + |z|)^{-m} \, e^{A|\Im z|} \quad (z \in \mathbb{C}^n).$$

Proof. Let $r > 0$, and suppose $u \in \mathcal{E}'$ with $\operatorname{supp} u \subset S_A$. By Theorem II.2.8 with $\omega = \{x; |x| < A + r\}$, there exists $g \in L^\infty(\omega)$ such that $u|_\omega = D_1^s \cdots D_n^s g$. We may take g and s independent of r for all $0 < r \leq 1$. Extend g to $\mathbb{R}^n$ by setting $g = 0$ for $|x| \geq A + 1$ (then of course $\|g\|_{L^1} < \infty$). Since $\operatorname{supp} u \subset \omega$, we have $u = D_1^s \cdots D_n^s g$, and therefore,

$$|\hat{u}(z)| = \left| \int_\omega \mathrm{e}^{-\mathrm{i}x\cdot z} D_1^s \cdots D_n^s g(x)]\, dx \right| = \left| \int_\omega D_1^s \cdots D_n^s (\mathrm{e}^{-\mathrm{i}x\cdot z}) g(x)\, dx \right|$$

$$= \left| \int_\omega (-1)^{sn} (z_1 \cdots z_n)^s \mathrm{e}^{-\mathrm{i}x\cdot z} g(x)\, dx \right| \leq (|z_1| \cdots |z_n|)^s \|g\|_{L^1} \sup_\omega \mathrm{e}^{x\cdot \Im z}$$

$$\leq C(1 + |z|)^m \mathrm{e}^{(A+r)|\Im z|},$$

for any constant $C > \|g\|_{L^1}, m = sn$, and $r \leq 1$ (since $|x \cdot \Im z| \leq |x||\Im z| \leq (A+r)|\Im z|$ on ω, by Schwarz's inequality). Letting $r \to 0$, we obtain the necessity of the estimate in (i).

If $u \in \mathcal{D}$ with support in S_A, we have for any multi-index β and $z \in \mathbb{C}^n$

$$|z^\beta \hat{u}(z)| = |(FD^\beta u)(z)| = |\int_{S_A} \mathrm{e}^{-\mathrm{i}x\cdot z} (D^\beta u)(x)\, dx| \leq \|D^\beta u\|_{L^1} \mathrm{e}^{A|\Im z|},$$

and the necessity of the estimates in (ii) follows.

Suppose next that the entire function f satisfies the estimates in (ii). In particular, its restriction to $\mathbb{R}^n$ is in L^1, and we may *define* $u = (2\pi)^{-n/2} J\mathcal{F} f|_{\mathbb{R}^n}$. The estimates in (ii) show that $y^\alpha f(y) \in L^1(\mathbb{R}^n)$ for all multi-indices α, and therefore D^α may be applied to the integral defining u under the integration sign; in particular, $u \in \mathcal{E}$.

The estimates in (ii) show also that the integral defining u can be shifted (by Cauchy's integral theorem) to $\mathbb{R}^n + \mathrm{i}t$, with $t \in \mathbb{R}^n$ fixed (but arbitrary). Therefore

$$|u(x)| \leq C(m) \exp[A|t| - x \cdot t] \int (1 + |y|)^{-m}\, dy$$

for all $x \in \mathbb{R}^n$ and $m \in \mathbb{N}$. Fix m so that the last integral converges, and choose $t = \lambda x (\lambda > 0)$. Then for a suitable constant C' and $|x| > A, |u(x)| \leq C' \exp[-\lambda|x|(|x| - A)] \to 0$ as $\lambda \to \infty$. This shows that $\operatorname{supp} u \subset S_A$, and so $u \in \mathcal{D}(S_A)$. But then its Fourier–Laplace transform is entire, and coincides with the entire function f on $\mathbb{R}^n$ (hence on $\mathbb{C}^n$), by the Fourier inversion formula.

Finally, suppose the estimate in (i) is satisfied. Then $f|_{\mathbb{R}^n} \in \mathcal{S}'$, and therefore $f|_{\mathbb{R}^n} = \hat{u}$ for a unique $u \in \mathcal{S}'$. It remains to show that $\operatorname{supp} u \subset S_A$. Let ϕ be as in II.1.1, and let $u_r := u * \phi_r$ be the corresponding regularization of u. By Theorem II.3.5, $\hat{u}_r = \hat{u}\hat{\phi}_r$. Since $\phi_r \in \mathcal{D}(S_r)$, it follows from the necessity part of (ii) that $|\hat{\phi}_r(z)| \leq C(m)(1 + |z|)^{-m} \mathrm{e}^{r|\Im z|}$ for all m. Therefore, by the estimate in (i),

$$|f(z)\hat{\phi}_r(z)| \leq C'(k)(1 + |z|)^{-k} \mathrm{e}^{(A+r)|\Im z|}$$

for all integers k. By the sufficiency part of (ii), the entire function $f(z)\hat{\phi}_r(z)$ is the Fourier–Laplace transform of some $\psi_r \in \mathcal{D}(S_{A+r})$. Since F is injective on $\mathcal{S}'$, we conclude (by restricting to $\mathbb{R}^n$) that the distribution u_r "is the function" ψ_r. In particular, $\operatorname{supp} u_r \subset S_{A+r}$ for all $r > 0$. If $\chi \in \mathcal{D}$ has support in (the open set) S_A^c, there exists $r_0 > 0$ such that $\operatorname{supp}\chi \subset S_{A+r_0}^c$; then for all $0 < r \le r_0, u_r(\chi) = 0$ because the supports of u_r and χ are contained in the disjoint sets S_{A+r} and $S_{A+r_0}^c$ (respectively). Letting $r \to 0$, it follows that $u(\chi) = 0$, and we conclude that u has support in S_A. □

Example. Consider the orthonormal sequences $\{f_k; k \in \mathbb{Z}\}$ and $\{g_k; k \in \mathbb{Z}\}$ in $L^2(\mathbb{R})$ defined in the example at the end of II.3.2. We saw that $f \in L^2(\mathbb{R})$ belongs to the closure of the span of $\{f_k\}$ in $L^2(\mathbb{R})$ iff it vanishes in $(-\pi,\pi)^c$. Since $\mathcal{F}$ is a Hilbert isomorphism of $L^2(\mathbb{R})$ onto itself, $g := \mathcal{F}f \in L^2(\mathbb{R})$ belongs to the closure of the span of $\{g_k\}$ iff it extends to $\mathbb{C}$ *as an entire function of exponential type* $\le \pi$ (by Theorem II.3.6). The expansion we found for g in that example extends to $\mathbb{C}$:

$$g(z) = (1/\pi)\sin \pi z \sum_k (-1)^k g(k)/(z-k) \quad (z \in \mathbb{C}), \tag{24}$$

where the series converges uniformly in $|\Re z| \le r$, for each r. We proved that *any entire function g of exponential type $\le \pi$, whose restriction to $\mathbb{R}$ belongs to $L^2(\mathbb{R})$, admits the expansion* (24).

Suppose h is an entire function of exponential type $\le \pi$, whose restriction to $\mathbb{R}$ is *bounded.* Let $g(z) := [h(z) - h(0)]/z$ for $z \ne 0$ and $g(0) = h'(0)$. Then g is entire of exponential type $\le \pi$, and $g|_{\mathbb{R}} \in L^2(\mathbb{R})$. Applying (24) to g, we obtain

$$h(z) = h(0) + (1/\pi)\sin \pi z\Big(h'(0) + \sum_{k\ne 0}(-1)^k[h(k)-h(0)][1/k + 1/(z-k)]\Big). \tag{25}$$

The series $s(z)$ in (25) can be differentiated term by term (because the series thus obtained converges uniformly in any strip $|\Re z| \le r$). We then obtain

$$h'(z) = \cos\pi z(h'(0) + s(z)) + (1/\pi)\sin\pi z\sum_{k\ne 0}(-1)^{k-1}[h(k) - h(0)]/(z-k)^2.$$

In particular,

$$\begin{aligned} h'(1/2) &= (1/\pi)\sum_{k\ne 0}(-1)^{k-1}[h(k) - h(0)]/(k-1/2)^2 \\ &= (4/\pi)\sum_{k\in\mathbb{Z}}(-1)^{k-1}\frac{h(k)-h(0)}{(2k-1)^2}. \end{aligned}$$

Since $\sum_{k\in\mathbb{Z}}(-1)^{k-1}/(2k-1)^2 = 0$, we can rewrite the last formula in the form

$$h'(1/2) = (4/\pi)\sum_{k\in\mathbb{Z}}(-1)^{k-1}\frac{h(k)}{(2k-1)^2}. \tag{26}$$

For $t \in \mathbb{R}$ fixed, the function $\tilde{h}(z) := h(z + t - 1/2)$ is entire of exponential type $\leq \pi, \sup |\tilde{h}\Big|_{\mathbb{R}}| = \sup |h\Big|_{\mathbb{R}}| := M < \infty$, and $\tilde{h}'(1/2) = h'(t)$. Therefore, by (26) applied to $\tilde{h}$,

$$h'(t) = (4/\pi) \sum_{k \in \mathbb{Z}} (-1)^{k-1} \frac{h(t + k - 1/2)}{(2k-1)^2}. \tag{27}$$

Hence (cf. example at the end of Terminology 8.11)

$$|h'(t)| \leq (4/\pi) M \sum_{k \in \mathbb{Z}} \frac{1}{(2k-1)^2} = (4/\pi) M (\pi^2/4) = \pi M. \tag{28}$$

Thus, considering h *restricted to* $\mathbb{R}$,

$$\|h'\|_\infty \leq \pi \|h\|_\infty. \tag{29}$$

Let $1 \leq p < \infty$, and let q be its conjugate exponent. For any simple measurable function ϕ on $\mathbb{R}$ with $\|\phi\|_q = 1$, we have by (27) and Holder's inequality

$$\left| \int_{\mathbb{R}} h' \phi \, dt \right| = (4/\pi) \left| \sum_{k \in \mathbb{Z}} \frac{(-1)^{k-1}}{(2k-1)^2} \int h(t + k - 1/2) \phi(t) \, dt \right|$$

$$\leq (4/\pi) \sum_{k \in \mathbb{Z}} \frac{1}{(2k-1)^2} \|h\|_p \|\phi\|_q = \pi \|h\|_p.$$

Taking the supremum over all such functions ϕ, it follows that

$$\|h'\|_p \leq \pi \|h\|_p. \tag{30}$$

(For $p = 1$, (30) follows directly from (27): for any real numbers $a < b$,

$$\int_a^b |h'(t)| \, dt \leq (4/\pi) \sum_{k \in \mathbb{Z}} \frac{1}{(2k-1)^2} \|h\|_1 = \pi \|h\|_1,$$

and (30) for $p = 1$ is obtained by letting $a \to -\infty$ and $b \to \infty$.)

If f is an entire function of exponential type $\leq \nu > 0$ and is bounded on $\mathbb{R}$, the function $h(z) := f(\pi z/\nu)$ is entire of exponential type $\leq \pi$; $\|h\|_\infty = \|f\|_\infty$ (norms in $L^\infty(\mathbb{R})$); and $h'(t) = (\pi/\nu) f'(\pi t/\nu)$. A simple calculation starting from (30) for h shows that

$$\|f'\|_p \leq \nu \|f\|_p \tag{31}$$

for all $p \in [1, \infty]$. This is *Bernstein's inequality* (for f entire of exponential type $\leq \nu$, that is bounded on $\mathbb{R}$).

II.3.1 The spaces $\mathcal{W}_{p,k}$

II.3.7. Temperate weights.

A (temperate) weight on $\mathbb{R}^n$ is a *positive* function k on $\mathbb{R}^n$ such that

$$\frac{k(x+y)}{k(y)} \leq (1 + C|x|)^m \quad (x, y \in \mathbb{R}^n) \tag{32}$$

for some constants $C, m > 0$.

By (32),

$$(1 + C|x|)^{-m} \leq \frac{k(x+y)}{k(y)} \leq (1 + C|x|)^m \quad (x, y \in \mathbb{R}^n), \tag{33}$$

and it follows that k is *continuous*, satisfies the estimate

$$k(0)(1 + C|x|)^{-m} \leq k(x) \leq k(0)(1 + C|x|)^m \quad (x \in \mathbb{R}^n), \tag{34}$$

and $1/k$ is also a weight.

For any real s, k^s is a weight (trivial for $s > 0$, and since $1/k$ is a weight, the conclusion follows for $s < 0$ as well).

An elementary calculation shows that $1 + |x|^2$ is a weight; therefore, $k_s(x) := (1 + |x|^2)^{s/2}$ is a weight for any real s.

Sums and product of weights are weights. For any weight k, set

$$\underline{k}(x) := \sup_y \frac{k(x+y)}{k(y)}, \tag{35}$$

so that, by definition,

$$\underline{k}(x) \leq (1 + C|x|)^m \quad \text{and} \quad k(x+y) \leq \underline{k}(x)k(y). \tag{36}$$

Also

$$\underline{k}(x+y) \leq \underline{k}(x)\underline{k}(y). \tag{37}$$

By (36) and (37), $\underline{k}$ is a weight with the additional "normal" properties (37) and

$$1 = \underline{k}(0) \leq \underline{k}(x). \tag{38}$$

($\underline{k}(0) = 1$ by (35); then by (37) and (36), for all $r = 1, 2, \ldots$,

$$1 = \underline{k}(0) = \underline{k}(rx - rx) \leq \underline{k}(x)^r \underline{k}(-rx) \leq \underline{k}(x)^r (1 + Cr|x|)^m;$$

taking the rth root and letting $r \to \infty$, we get $1 \leq \underline{k}(x)$.)

Given a weight k and $t > 0$, define

$$k^t(x) := \sup_y k(x-y) \exp(-t|y|) = \sup_y \exp(-t|x-y|)k(y).$$

(x, y range in $\mathbb{R}^n$.)

We have

$$k(x) \leq k^t(x) \leq \sup_y (1 + C|y|)^m k(x) \exp(-t|y|) \leq C_t k(x),$$

that is,

$$1 \leq \frac{k^t(x)}{k(x)} \leq C_t \quad (x \in \mathbb{R}^n),$$

where C_t is a constant depending on t.

Also

$$k^t(x+x') = \sup_y k(x+x'-y)\exp(-t|y|)$$
$$\leq \sup_y (1+C|x'|)^m k(x-y)\exp(-t|y|) = (1+C|x'|)^m k^t(x),$$

that is, k^t is a weight (with the constants C, m of k, whence independent of t). By the last inequality,

$$(1\leq)\underline{k^t}(x) \leq (1+C|x|)^m.$$

Since

$$k^t(x+x') = \sup_y \exp(-t|x+x'-y|)k(y) \leq e^{t|x'|} \sup_y \exp(-t|x-y|)k(y) = e^{t|x'|}k^t(x),$$

therefore,

$$(1\leq)\underline{k^t}(x) \leq e^{t|x|}.$$

In particular, $\underline{k^t} \to 1$ as $t \to 0+$, uniformly on compact subsets of $\mathbb{R}^n$.

A weight associated with the differential operator $P(D)$ (for any polynomial P on $\mathbb{R}^n$) is defined by

$$k_P := \left[\sum_\alpha |P^{(\alpha)}|^2\right]^{1/2}, \tag{39}$$

where the (finite) sum extends over all multi-indices α. The estimate (32) follows from Taylor's formula and Schwarz's inequality:

$$k_P^2(x+y) = \sum_\alpha \left|\sum_{|\beta|\leq m} P^{(\alpha+\beta)}(y)x^\beta/\beta!\right|^2$$
$$\leq \sum_\alpha \sum_\beta |P^{(\alpha+\beta)}(y)|^2 \sum_{|\beta|\leq m} |x^\beta/\beta!|^2$$
$$\leq k_P^2(y)(1+C|x|)^{2m},$$

where $m = \deg P$.

Extending the sum in (39) over multi-indices $\alpha \neq 0$ only, we get a weight k'_P (same verification!), that will also play a role in the sequel.

II.3.8. Weighted L^p-spaces.

For any (temperate) weight k and $p \in [1, \infty]$, consider the normed space

$$L_{p,k} := (1/k)L^p = \{f; kf \in L^p\}$$

(where $L^p := L^p(\mathbb{R}^n)$), with the natural norm $\|f\|_{L_{p,k}} = \|kf\|_p(\|\cdot\|_p$ denotes the L^p-norm). One verifies easily that $L_{p,k}$ is a Banach space for all $p \in [0, \infty]$, and $(L_{p,k})^*$ is isomorphic and isometric to $L_{q,1/k}$ for $1 \leq p < \infty$ (q denotes the conjugate exponent of p): if $\Lambda \in (L_{p,k})^*$, there exists a unique $g \in L_{q,1/k}$ such that

$$\Lambda f = \int fg\,dy \quad (f \in L_{p,k})$$

and

$$\|\Lambda\| = \|g\|_{L_{q,1/k}}.$$

By (34), $\mathcal{S} \subset L_{p,k}$ topologically.

Given $f \in L_{p,k}$, Holder's inequality shows that

$$\left|\int \phi f\,dx\right| \le \|f\|_{L_{p,k}} \|\phi\|_{L_{q,1/k}} \quad (\phi \in \mathcal{S}). \tag{40}$$

Since $\mathcal{S} \subset L_{q,1/k}$ topologically, it follows from (40) that the map $\phi \to \int \phi f\,dx$ is continuous on $\mathcal{S}$, and belongs therefore to $\mathcal{S}'$. With the usual identification, this means that $L_{p,k} \subset \mathcal{S}'$, and it follows also from (40) that the inclusion is topological (if $f_j \to 0$ in $L_{p,k}$, then $\int \phi f_j, dx \to 0$ by (40), for all $\phi \in \mathcal{S}$, that is, $f_j \to 0$ in $\mathcal{S}'$). We showed therefore that

$$\mathcal{S} \subset L_{p,k} \subset \mathcal{S}' \tag{41}$$

topologically.

II.3.9. The spaces $\mathcal{W}_{p,k}$.

Let

$$\mathcal{F} : u \in \mathcal{S}' \to \mathcal{F}u := (2\pi)^{-n/2}\hat{u} \in \mathcal{S}',$$

and consider the normed space (for p, k given as before)

$$\mathcal{W}_{p,k} := \mathcal{F}^{-1}L_{p,k} = \{u \in \mathcal{S}'; \mathcal{F}u \in L_{p,k}\} \tag{42}$$

with the norm

$$\|u\|_{p,k} := \|\mathcal{F}u\|_{L_{p,k}} = \|k\mathcal{F}u\|_p \quad (u \in \mathcal{W}_{p,k}). \tag{43}$$

Note that for any $t > 0$,

$$\mathcal{W}_{p,k} = \mathcal{W}_{p,k^t}$$

(because of the inequality $1 \le k^t/k \le C_t$, cf. II.3.7).

By definition, $\mathcal{F} : \mathcal{W}_{p,k} \to L_{p,k}$ is a (linear) surjective isometry, and therefore $\mathcal{W}_{p,k}$ is a Banach space. Since $\mathcal{F}^{-1}$ is a continuous automorphism of both $\mathcal{S}$ and $\mathcal{S}'$, it follows from (41) in II.3.8 (and the said isometry) that

$$\mathcal{S} \subset \mathcal{W}_{p,k} \subset \mathcal{S}' \tag{44}$$

topologically.

Fix ϕ as in II.1.1, and consider the regularizations $u_r = u * \phi_r \in \mathcal{E}$ of $u \in \mathcal{W}_{p,k}, p < \infty$. As $r \to 0+$, $k(x)\hat{u}_r(x) = k(x)\hat{u}(x)\hat{\phi}(rx) \to k(x)\hat{u}(x)$ pointwise, and $|k\hat{u}_r| \le |k\hat{u}| \in L^p$; therefore, $k\hat{u}_r \to k\hat{u}$ in L^p, that is, $u_r \to u$ in $\mathcal{W}_{p,k}$. One verifies easily that $u_r \in \mathcal{S}$, and consequently, $\mathcal{S}$ is dense in $\mathcal{W}_{p,k}$. Since $\mathcal{D}$ is dense in $\mathcal{S}$, and $\mathcal{S}$ is topologically included in $\mathcal{W}_{p,k}$, we conclude that $\mathcal{D}$ *is dense in*

$\mathcal{W}_{p,k}$ (and so $\mathcal{W}_{p,k}$ *is the completion of* $\mathcal{D}$ *with respect to the norm* $\|\cdot\|_{p,k}$) for any $1 \le p < \infty$ and any weight k. The special space $\mathcal{W}_{2,k_s}$ is called *Sobolev's space*, and is usually denoted $\mathcal{H}^s$.

Let $L \in \mathcal{W}^*_{p,k}$ (for some $1 \le p < \infty$). Since $\mathcal{F}$ is a linear isometry of $\mathcal{W}_{p,k}$ onto $L_{p,k}$, the map $\Lambda = L\mathcal{F}^{-1}$ is a continuous linear functional on $L_{p,k}$ with norm $\|L\|$. By the preceding characterization of $L^*_{p,k}$, there exists a unique $g \in L_{q,1/k}$ such that $\|g\|_{L_{q,1/k}} = \|L\|$ and $\Lambda f = \int fg\,dx$ for all $f \in L_{p,k}$. Define $v = \mathcal{F}^{-1}g$. For all $u \in \mathcal{W}_{p,k}$, denoting $f = \mathcal{F}u (\in L_{p,k})$, we have $\|v\|_{q,1/k} = \|L\|$ and

$$Lu = L\mathcal{F}^{-1}f = \Lambda f = \int (\mathcal{F}u)(\mathcal{F}v)\,dx. \tag{45}$$

The continuous functional L is uniquely determined by its restriction to the dense subspace $\mathcal{S}$ of $\mathcal{W}_{p,k}$. For $u \in \mathcal{S}$, we may write (45) in the form

$$Lu = v(\mathcal{F}^2 u) = v(Ju) = (Jv)(u) \quad (u \in \mathcal{S}), \tag{46}$$

that is, $L|_{\mathcal{S}} = Jv$. Conversely, any $v \in \mathcal{W}_{q,1/k}$ determines through (45) an element $L \in \mathcal{W}^*_{p,k}$ such that $\|L\| = \|v\|_{q,1/k}$ (and $L|_{\mathcal{S}} = Jv$). We conclude that $\mathcal{W}^*_{p,k}$ *is isometrically isomorphic with* $\mathcal{W}_{q,1/k}$. In particular, $(\mathcal{H}^s)^*$ is isometrically isomorphic with $\mathcal{H}^{-s}$.

If the distribution $u \in \mathcal{W}_{p,k}$ has compact support, and $v \in \mathcal{W}_{\infty,k'}$, then by Theorem II.3.5, $u * v \in \mathcal{S}'$ and

$$\begin{aligned}\|u * v\|_{p,kk'} &= \|kk'\mathcal{F}(u*v)\|_p = (2\pi)^{n/2}\|(k\mathcal{F}u)(k'\mathcal{F}v)\|_p \\ &\le (2\pi)^{n/2}\|k\mathcal{F}u\|_p\|k'\mathcal{F}v\|_\infty = (2\pi)^{n/2}\|u\|_{p,k}\|v\|_{\infty,k'}.\end{aligned}$$

In particular,

$$(\mathcal{W}_{p,k} \cap \mathcal{E}') * \mathcal{W}_{\infty,k'} \subset \mathcal{W}_{p,kk'}. \tag{47}$$

Let P be any polynomial on $\mathbb{R}^n$. We have $P(D)\delta \in \mathcal{E}' \subset \mathcal{S}'$, and

$$\mathcal{F}P(D)\delta = P(M)\mathcal{F}\delta = (2\pi)^{-n/2}P(M)1 = (2\pi)^{-n/2}P.$$

Thus

$$\|P(D)\delta\|_{\infty,k'} = (2\pi)^{-n/2}\|k'P\|_\infty < \infty,$$

for any weight k' such that $k'P$ is bounded (we may take for example $k' = 1/k_P$, or $k' = k_s$ with $s \le -m$, where $m = \deg P$).

Thus, $P(D)\delta \in \mathcal{W}_{\infty,k'}$ (for such k'), and for any $u \in \mathcal{W}_{p,k}, P(D)u = (P(D)\delta) * u \in \mathcal{W}_{p,kk'}$.

Formally stated, *for any weight* k' *such that* $k'P$ *is bounded,*

$$P(D)\mathcal{W}_{p,k} \subset \mathcal{W}_{p,kk'}. \tag{48}$$

The shown calculations also demonstrate that

$$\|P(D)u\|_{p,kk'} \le \|k'P\|_\infty \|u\|_{p,k} \qquad (u \in \mathcal{W}_{p,k}), \tag{49}$$

that is, $P(D)$ is a *continuous* (linear) map of $\mathcal{W}_{p,k}$ into $\mathcal{W}_{p,kk'}$.

If $u \in \mathcal{W}_{p,k}$ and $\phi \in \mathcal{D}$, then $\hat{u} \in L_{p,k}$ and $\hat{\phi} \in \mathcal{S} \subset L_{q,1/k}$, so that the convolution $\hat{\phi} * \hat{u}$ makes sense as a usual integral. On the other hand, ϕu is well defined and belongs to $\mathcal{S}'$. Using Theorem II.3.5 and the Fourier inversion formula on $\mathcal{S}'$, we see that

$$(2\pi)^{n/2}\mathcal{F}(\phi u) = (\mathcal{F}\phi) * (\mathcal{F}u).$$

Hence (since $k(x) \leq \underline{k}(x-y)k(y)$),

$$\begin{aligned}(2\pi)^{n/2}\|\phi u\|_{p,k} &= \|k(\mathcal{F}\phi) * (\mathcal{F}u)\|_p \\ &\leq \|(\underline{k}|\mathcal{F}\phi|) * (k|\mathcal{F}u|)\|_p \leq \|\underline{k}\mathcal{F}\phi\|_1 \|k\mathcal{F}u\|_p,\end{aligned}$$

that is,

$$\|\phi u\|_{p,k} \leq (2\pi)^{-n/2}\|\phi\|_{1,\underline{k}}\|u\|_{p,k}. \tag{50}$$

Since $\mathcal{D}$ is dense in $\mathcal{S}$ and $\mathcal{S} \subset \mathcal{W}_{1,\underline{k}}$, it follows from (50) that the multiplication operator $\phi \in \mathcal{D} \to \phi u \in \mathcal{W}_{p,k}$ extends uniquely as an operator from $\mathcal{S}$ to $\mathcal{W}_{p,k}$ (same notation!), that is, $\mathcal{S}W_{p,k} \subset \mathcal{W}_{p,k}$, and (50) is valid for all $\phi \in \mathcal{S}$ and $u \in \mathcal{W}_{p,k}$.

Apply (50) to the weights k^t associated with k (cf. II.3.7). We have (for any $\phi \in \mathcal{S}$)

$$\|\phi\|_{1,\underline{k^t}} = \int |\underline{k}^t\mathcal{F}\phi|\, dx.$$

As $t \to 0+$, the integrand converges pointwise to $\mathcal{F}\phi$, and are dominated by $(1 + C|x|)^m|\mathcal{F}\phi| \in L^1$ (with C, m independent of t). By Lebesgue's dominated convergence theorem, the integral tends to $\|\mathcal{F}\phi\|_1 := \|\phi\|_{1,1}$. There exists therefore $t_0 > 0$ (depending on ϕ) such that $\|\phi\|_{1,\underline{k^t}} \leq 2\|\phi\|_{1,1}$ for all $t < t_0$. Hence by (50)

$$\|\phi u\|_{p,k^t} \leq 2(2\pi)^{-n/2}\|\phi\|_{1,1}\|u\|_{p,k^t} \tag{51}$$

for all $0 < t < t_0, \phi \in \mathcal{S}$, and $u \in \mathcal{W}_{p,k} = \mathcal{W}_{p,k^t}$, with t_0 depending on ϕ. This inequality is used in the proof of Theorem II.7.2.

Let j be a non-negative integer. If $|y|^j \in L_{q,1/k}$ for some weight k and some $1 \leq q \leq \infty$, then $y^\alpha \in L_{q,1/k}$ for all multi-indices α with $|\alpha| \leq j$. Consequently, for any $u \in \mathcal{W}_{p,k}$ (with p conjugate to q), $y^\alpha \hat{u}(y) \in L^1$. By the Fourier inversion formula, $u(x) = (2\pi)^{-n}\int e^{ix\cdot y}\hat{u}(y)\, dy$, and the integrals obtained by formal differentiations under the integral sign up to the order j converge absolutely and uniformly, and are equal therefore to the classical derivatives of u. In particular, $u \in C^j$. This shows that

$$\mathcal{W}_{p,k} \subset C^j \tag{52}$$

if $|y|^j \in L_{q,1/k}$. This is a *regularity property* of the distributions in $\mathcal{W}_{p,k}$.

II.4 Fundamental solutions

II.4.1. Let P be a polynomial on $\mathbb{R}^n$. A *fundamental solution* for the (partial) differential operator $P(D)$ is a distribution v on $\mathbb{R}^n$ such that

$$P(D)v = \delta. \tag{1}$$

For any $f \in \mathcal{E}'$, the (well-defined) distribution $u := v * f$ is then a solution of the (partial) differential equation

$$P(D)u = f. \tag{2}$$

(Indeed, $P(D)u = (P(D)v) * f = \delta * f = f$.) The identity

$$P(D)(v * u) = v * (P(D)u) = u \quad (u \in \mathcal{E}') \tag{3}$$

means that the map $V : u \in \mathcal{E}' \to v * u$ is the inverse of the map $P(D) : \mathcal{E}' \to \mathcal{E}'$.

Theorem II.4.2 (Ehrenpreis–Malgrange–Hormander). *Let P be a polynomial on $\mathbb{R}^n$, and $\epsilon > 0$. Then there exists a fundamental solution v for $P(D)$ such that*

$$sech(\epsilon|x|)v \in \mathcal{W}_{\infty,k_P},$$

and $\|sech(\epsilon|x|)v\|_{\infty,k_P}$ is bounded by a constant depending only on ϵ, n, and $m = \deg P$.

Note that $\operatorname{sech}(\epsilon|x|) \in \mathcal{E}$ (since $\cosh(\epsilon|x|) = \sum_k \epsilon^{2k}(x_1^2 + \cdots + x_n^2)^k/(2k)!$), and therefore its product with the distribution v is well defined.

For any $\psi \in \mathcal{D}$ (and v as in the theorem), write

$$\psi v = [\psi \cosh(\epsilon|x|)][\operatorname{sech}(\epsilon|x|)v].$$

The function in the first square brackets belongs to $\mathcal{D}$; the distribution in the second square brackets belongs to $\mathcal{W}_{\infty,k_P}$. Hence $\psi v \in \mathcal{W}_{\infty,k_P}$ by (50) in Section II.3.9. Denoting

$$\mathcal{W}_{p,k}^{\mathrm{loc}} = \{u \in \mathcal{D}'; \psi u \in \mathcal{W}_{p,k} \text{ for all } \psi \in \mathcal{D}\},$$

the above observation means that *the operator $P(D)$ has a fundamental solution in $\mathcal{W}_{\infty,k_P}^{\mathrm{loc}}$*.

The basic estimate needed for the proof of the theorem is stated in the following.

Lemma II.4.3 (notation as in Theorem II.4.2.). *There exists a constant $C > 0$ (depending only on ϵ, n, and m) such that, for all $u \in \mathcal{D}$,*

$$|u(0)| \le C\|\cosh(\epsilon|x|)P(D)u\|_{1,1/k_P}.$$

Proof of Theorem II.4.2. Assuming the lemma, we proceed with the proof of the theorem. Consider the linear functional

$$w : P(D)u \to u(0) \quad (u \in \mathcal{D}). \tag{4}$$

By the lemma and the Hahn–Banach theorem, w extends as a continuous linear functional on $\mathcal{D}$ such that

$$|w(\phi)| \leq C\|\cosh(\epsilon|x|)\phi\|_{1,1/k_P} \quad (\phi \in D). \tag{5}$$

Since $\mathcal{D} \subset \mathcal{W}_{1,1/k_P}$ topologically, it follows that w is continuous on $\mathcal{D}$, that is, $w \in \mathcal{D}'$. By (5),

$$|[\operatorname{sech}(\epsilon|x|)w](\phi)| = |w(\operatorname{sech}(\epsilon|x|)\phi)| \leq C\|\phi\|_{1,1/k_P}$$

for all $\phi \in \mathcal{D}$. Since $\mathcal{D}$ is dense in $\mathcal{W}_{1,1/k_P}$, the distribution $\operatorname{sech}(\epsilon|x|)w$ extends uniquely to a continuous linear functional on $\mathcal{W}_{1,1/k_P}$ with norm $\leq C$. Therefore

$$\operatorname{sech}(\epsilon|x|)w \in \mathcal{W}_{\infty,k_P} \tag{6}$$

and

$$\|\operatorname{sech}(\epsilon|x|)w\|_{\infty,k_P} \leq C. \tag{7}$$

Define $v = Jw := \tilde{w}$. Then the distribution $\operatorname{sech}(\epsilon|x|)v \in \mathcal{W}_{\infty,k_P}$ has $\|\cdot\|_{\infty,k_P}$-norm $\leq C$ (the constant in the lemma), and for all $\phi \in \mathcal{D}$, we have by (4)

$$\begin{aligned}(P(D)v)(\phi) &= [(P(D)v) * \tilde{\phi}](0) = (v * P(D)\tilde{\phi})(0) \\ &= [\tilde{w} * P(D)\tilde{\phi}](0) = \tilde{w}(J[P(D)\tilde{\phi}]) = w(P(D)\tilde{\phi}) \\ &= \tilde{\phi}(0) = \phi(0) = \delta(\phi),\end{aligned}$$

that is, $P(D)v = \delta$. □

Proof of Lemma II.4.3. (1) Let p be a monic polynomial of degree m in one complex variable, say $p(z) = \sum_{j=0}^{m} a_j z^j, a_m = 1$. The polynomial $q(z) = \sum_{j=0}^{m} \overline{a_j} z^{m-j}$ satisfies $q(0) = 1$ and

$$|q(\mathrm{e}^{\mathrm{i}t})| = \left|\mathrm{e}^{\mathrm{i}mt}\overline{\sum_j a_j\, \mathrm{e}^{\mathrm{i}jt}}\right| = |p(\mathrm{e}^{\mathrm{i}t})|.$$

If f is analytic on the closed unit disc, it follows from Cauchy's formula applied to the function fq that

$$|f(0)| = |f(0)q(0)| \leq \frac{1}{2\pi}\int_0^{2\pi} |f(\mathrm{e}^{\mathrm{i}t})q(\mathrm{e}^{\mathrm{i}t})|\,dt = \frac{1}{2\pi}\int_0^{2\pi} |f(\mathrm{e}^{\mathrm{i}t})p(\mathrm{e}^{\mathrm{i}t})|\,dt. \tag{8}$$

Writing $p(z) = \prod_{j=1}^{m}(z + z_j)$, we have for $k \leq m$

$$p^{(k)}(z) = \sum_{n_1} \sum_{n_2 \notin \{n_1\}} \cdots \sum_{n_k \notin \{n_1,\dots,n_{k-1}\}} \prod_{j \notin \{n_1,\dots,n_k\}} (z + z_j), \tag{9}$$

where all indices range in $\{1, \dots, m\}$.

Using (8) with the analytic function

$$f(z) \prod_{j \notin \{n_1,\dots,n_k\}} (z+z_j)$$

and the polynomial

$$p(z) = \prod_{j \in \{n_1,\dots,n_k\}} (z+z_j),$$

we obtain

$$\Big| f(0) \prod_{j \notin \{n_1,\dots,n_k\}} z_j \Big| \le \frac{1}{2\pi} \int_0^{2\pi} |f(\mathrm{e}^{\mathrm{i}t}) p(\mathrm{e}^{\mathrm{i}t})| \, dt.$$

Since the number of summands in (9) is $m(m-1)\cdots(m-k+1) = m!/(m-k)!$, it follows that

$$|f(0)p^{(k)}(0)| \le \frac{m!}{(m-k)!} \frac{1}{2\pi} \int_0^{2\pi} |f(\mathrm{e}^{\mathrm{i}t}) p(\mathrm{e}^{\mathrm{i}t})| \, dt. \tag{10}$$

Since (10) remains valid when p is replaced by cp with $c \neq 0$ complex, the inequality (10) is true for *any* polynomial p and any function f analytic on the closed unit disc.

(2) If f is entire, we apply (10) to the function $f(rz)$ and the polynomial $p(rz)$ for each $r > 0$. Then

$$|f(0)p^{(k)}(0)| 2\pi r^k \le \frac{m!}{(m-k)!} \int_0^{2\pi} |f(r\,\mathrm{e}^{\mathrm{i}t}) p(r\,\mathrm{e}^{\mathrm{i}t})| \, dt. \tag{11}$$

Let g be a non-negative function with compact support, integrable with respect to Lebesgue measure on $\mathbb{C}$, and depending only on $|z|$. We multiply (11) by $rg(r\,\mathrm{e}^{\mathrm{i}t})$ and integrate with respect to r over $[0,\infty)$. Thus,

$$|f(0)p^{(k)}(0) \int_{\mathbb{C}} |z^k| g(z) \, dz \le \frac{m!}{(m-k)!} \int_{\mathbb{C}} |f(z)p(z)| g(z) \, dz, \tag{12}$$

where $dz = r\,dr\,dt$ is the area measure in $\mathbb{C}$.

(3) The n-dimensional version of (12) is obtained by applying (12) "one variable at a time": let f be an entire function on $\mathbb{C}^n$, p a polynomial on $\mathbb{C}^n$, and g a non-negative function with compact support, integrable with respect to Lebesgue measure dz on $\mathbb{C}^n$, and depending only on $|z_1|, \dots, |z_n|$. Then

$$|f(0)p^{(\alpha)}(0)| \int_{\mathbb{C}^n} |z^\alpha| g(z) \, dz \le \frac{m!}{(m-|\alpha|)!} \int_{\mathbb{C}^n} |f(z)p(z)| g(z) \, dz. \tag{13}$$

(4) Let $u \in \mathcal{D}$, fix $y \in \mathbb{R}^n$, and apply (13) to the entire function $f(z) = \hat{u}(y+z)$, the polynomial $p(z) = P(y+z)$, and the function g equal to the indicator of the ball $B := \{z \in \mathbb{C}^n; |z| < \epsilon/2\}$. Then

$$\Big| \hat{u}(y) P^{(\alpha)}(y) \Big| \int_B |z^\alpha| \, dz \le \frac{m!}{(m-|\alpha|)!} \int_B |\hat{u}(y+z) P(y+z)| \, dz.$$

Therefore,

$$\begin{aligned}
k_P(y)|\hat{u}(y)| &\le \sum_{|\alpha|\le m} |P^{(\alpha)}(y)||\hat{u}(y)| \\
&\le C_1 \int_B |\hat{u}(y+z)P(y+z)|\,dz = C_1(2\pi)^{n/2}\int_B |(\mathcal{F}P(D)u)(y+z)|\,dz \\
&= C_1(2\pi)^{n/2}\int_B |\mathcal{F}[\mathrm{e}^{-\mathrm{i}x\cdot z}P(D)u](y)|\,dz,
\end{aligned}$$

where C_1 is a constant depending only on m and n. Denote $k := 1/k_P$. Then

$$\begin{aligned}
(2\pi)^{n/2}|u(0)| &= (2\pi)^{-n/2}\left|\int_{\mathbb{R}^n}\hat{u}(y)\,dy\right| \\
&\le C_1\int_{\mathbb{R}^n}\int_B k(y)|\mathcal{F}[\mathrm{e}^{-\mathrm{i}x.z}P(D)u](y)|\,dz\,dy \\
&= C_1\int_B \|\mathrm{e}^{-\mathrm{i}x.z}P(D)u\|_{1,k}\,dz \le C_1|B|\sup_{z\in B}\|\mathrm{e}^{-\mathrm{i}x\cdot.z}P(D)u\|_{1,k},
\end{aligned}$$

where $|B|$ denotes the $\mathbb{C}^n$-Lebesgue measure of B (depends only on n and ϵ).

For $z \in B$, all derivatives of the function $\phi_z(x) := \mathrm{e}^{-\mathrm{i}x.z}/\cosh(\epsilon|x|)$ are $O(\mathrm{e}^{-\epsilon|x|/2})$, and therefore the family $\phi_B := \{\phi_z; z \in B\}$ is *bounded* in $\mathcal{S}$ (this means that given any zero neighborhood U in $\mathcal{S}$, there exists $\eta > 0$ such that $\lambda\phi_B \subset U$ for all scalars λ with modulus $< \eta$. For a topological vector space with topology induced by a family of semi-norms, the above condition is equivalent to the boundedness of the semi-norms of the family on the set ϕ_B; thus, in the special case of $\mathcal{S}$, the boundedness of ϕ_B means that $\sup_{z\in B}\|\phi_z\|_{\alpha,\beta} < \infty$). Since $\mathcal{S} \subset \mathcal{W}_{p,k}$ topologically (for any p,k), it follows that the set ϕ_B is bounded in $\mathcal{W}_{p,k}$ for any p,k. In particular, $M_B := \sup_{z\in B}\|\phi_z\|_{1,\underline{k}} < \infty$. ($M_B$ depends only on n and ϵ.) By (50) in II.3.9,

$$\begin{aligned}
|u(0)| &\le C_1|B|\sup_{z\in B}(2\pi)^{-n/2}\|\phi_z[\cosh(\epsilon|x|)P(D)u]\|_{1,k} \\
&\le (2\pi)^{-n}C_1|B|\sup_{z\in B}\|\phi_z\|_{1,\underline{k}}\|\cosh(\epsilon|x|)P(D)u\|_{1,k} = C\|\cosh(\epsilon|x|)P(D)u\|_{1,k},
\end{aligned}$$

where $C = (2\pi)^{-n}C_1|B|M_B$ depends only on n, m and ϵ. □

II.5 Solution in $\mathcal{E}'$

II.5.1. Consider the operator $P(D)$ *restricted to* $\mathcal{E}'$. We look for necessary and sufficient conditions on $f \in \mathcal{E}'$ such that the equation $P(D)u = f$ has a solution $u \in \mathcal{E}'$. A necessary condition is immediate. For any solution $\phi \in \mathcal{E}$ of the so-called homogeneous "adjoint" equation $P(-D)\phi = 0$, we have (for u as shown)

$$f(\phi) = (P(D)u)(\phi) = u(P(-D)\phi) = u(0) = 0,$$

that is, f annihilates the null space of $P(-D)|_{\mathcal{E}}$. In particular, f annihilates elements of the null space of the special form $\phi(x) = q(x)e^{ix\cdot z}$, where $z \in \mathbb{C}^n$ and q is a polynomial on $\mathbb{R}^n$ (let us call such elements "exponential solutions" of the homogeneous adjoint equation). Theorem II.5.2 establishes that the later condition is also sufficient.

Theorem II.5.2. *The following statements are equivalent for $f \in \mathcal{E}'$:*

(1) The equation $P(D)u = f$ has a solution $u \in \mathcal{E}'$.

(2) f annihilates every exponential solution of the homogeneous adjoint equation.

(3) The function $F(z) := \hat{f}(z)/P(z)$ is entire on $\mathbb{C}^n$.

Proof. $1 \Longrightarrow 2$. See II.5.1.

$2 \Longrightarrow 3$. In order to make one-complex-variable arguments, we consider the function

$$F(t; z, w) := \frac{\hat{f}(tw + z)}{P(tw + z)} \quad (t \in \mathbb{C}; z, w \in \mathbb{C}^n). \tag{1}$$

Let P_m be the principal part of P, and *fix* $w \in \mathbb{C}^n$ such that $P_m(w) \neq 0$. Since

$$\begin{aligned} P(tw + z) &= P_m(tw + z) + \text{terms of lower degree} \\ &= \sum_\alpha P_m^{(\alpha)}(tw) z^\alpha / \alpha! + \text{terms of lower degree} \\ &= P_m(tw) + \text{terms of lower degree} = t^m P_m(w) + \text{ terms of lower degree}, \end{aligned}$$

and $P_m(w) \neq 0$, $P(tw + z)$ is a polynomial of degree m in t (for each given z). Fix $z = z_0$, and let t_0 be a zero of order k of $P(tw + z_0)$. For $j < k$, set

$$\phi_j(x, t) := (x \cdot w)^j \exp(-ix \cdot (tw + z_0)). \tag{2}$$

Then

$$\begin{aligned} P(-D)\phi_j(x, t) &= P(-D) \left(i\frac{\partial}{\partial t} \right)^j \exp\left(-ix \cdot (tw + z_0)\right) \\ &= \left(i\frac{\partial}{\partial t} \right)^j P(-D) \exp(-ix \cdot (tw + z_0)) \\ &= \left(i\frac{\partial}{\partial t} \right)^j P(tw + z_0) \exp(-ix \cdot (tw + z_0)), \end{aligned}$$

and therefore $P(-D)\phi_j(x, t_0) = 0$, that is, $\phi_j(\cdot, t_0)$ are exponential solutions of the homogeneous adjoint equation. By hypothesis, we then have $f(\phi_j(\cdot, t_0)) = 0$ for all $j < k$. This means that, for all $j < k$,

$$\frac{\partial^j}{\partial t^j} |_{t=t_0} \hat{f}(tw + z_0) = 0,$$

that is, $\hat{f}(tw+z_0)$ has a zero of order $\geq k$ at $t = t_0$, and $F(\cdot; z_0, w)$ is consequently entire (cf. (1)).

Choose $r > 0$ such that $P(tw + z_0) \neq 0$ on the circle $\Gamma := \{t; |t| = r\}$ (this is possible since, as a polynomial in t, $P(tw + z_0)$ has finitely many zeros). By continuity, $P(tw + z) \neq 0$ for all t on Γ and z in a neighborhood U of z_0. Therefore, the function $G(z) := 1/2\pi i \int_\Gamma F(t; z, w)\, dt/t$ is analytic in U, that is, *G is entire* (by the arbitrariness of z_0). However, by Cauchy's integral theorem for the entire function $F(\cdot; z, w)$, we have $G(z) = F(0; z, w) = \hat{f}(z)/P(z)$, and Statement (3). is proved.

$3 \Longrightarrow 1$. By Theorem II.3.6, it suffices to show that F satisfies the estimate in Part (i) of Theorem II.3.6 (for then F is the Fourier–Laplace transform of some $u \in \mathcal{E}'$; restricting to $\mathbb{R}^n$, we have therefore $\mathcal{F}f = P\mathcal{F}u = \mathcal{F}P(D)u$, hence $P(D)u = f$).

Fix $\zeta \in \mathbb{C}^n$ and apply (13) in the proof of Lemma II.4.3 to the entire function $f(z) = F(\zeta + z)$, the polynomial $p(z) = P(\zeta + z)$, and g the indicator of the unit ball B of $\mathbb{C}^n$. Then

$$|F(\zeta)P^{(\alpha)}(\zeta)| \leq C \int_B |\hat{f}(\zeta + z)|\, dz \leq C|B| \sup_{z\in B} |\hat{f}(\zeta + z)|,$$

where C is a constant depending only on n and $m = \deg P$. Choose α such that $P^{(\alpha)}$ is a non-zero constant. Then, by the necessity of the estimate in Part (i) of Theorem II.3.6 (applied to the distribution $f \in \mathcal{E}'$), we have

$$|F(\zeta)| \leq C_1 \sup_{z\in B} (1 + |\zeta + z|)^k e^{A|\Im(\zeta+z)|} \leq C_2(1 + |\zeta|)^k e^{A|\Im\zeta|},$$

as desired. □

II.6 Regularity of solutions

II.6.1. Let Ω be an open subset of $\mathbb{R}^n$. If $\phi \in \mathcal{D}(\Omega)$ and $u \in \mathcal{D}'(\Omega)$, the product ϕu is a distribution with compact support in Ω, and may then be considered as an element of $\mathcal{E}' := \mathcal{E}'(\mathbb{R}^n) \subset \mathcal{S}' := \mathcal{S}'(\mathbb{R}^n)$. Set (for any weight k and $p \in [1, \infty]$)

$$\mathcal{W}_{p,k}^{\text{loc}}(\Omega) := \{u \in \mathcal{D}'(\Omega); \phi u \in \mathcal{W}_{p,k} \text{ for all } \phi \in \mathcal{D}(\Omega)\}.$$

(cf. comments following Theorem II.4.2.) Note that if $u \in \mathcal{E}(\Omega)$, then $\phi u \in \mathcal{D}(\Omega) \subset \mathcal{S} \subset \mathcal{W}_{p,k}$ for all $\phi \in \mathcal{D}(\Omega)$, that is, $\mathcal{E}(\Omega) \subset \mathcal{W}_{p,k}^{\text{loc}}(\Omega)$ for all p, k. Conversely, if $u \in \mathcal{W}_{p,k}^{\text{loc}}(\Omega)$ for all p, k (or even for some p and all weights k_s), it follows from (52) in Section II.3.9 that $u \in \mathcal{E}(\Omega)$. This observation gives an approach for proving *regularity* of distribution solutions of the equation $P(D)u = f$ in Ω (for suitable f): it would suffice to prove that the solutions u belong to all the spaces $\mathcal{W}_{p,k}^{\text{loc}}(\Omega)$ (since then $u \in \mathcal{E}(\Omega)$).

II.6.2. Hypoellipticity.

The polynomial P (or the differential operator $P(D)$) is *hypoelliptic* if there exist constants $C, c > 0$ such that

$$\left|\frac{P(x)}{P^{(\alpha)}(x)}\right| \geq C|x|^{c|\alpha|} \tag{1}$$

as $x \in \mathbb{R}^n \to \infty$, for all multi-indices $\alpha \neq 0$.

Conditions equivalent to (1) are *any one* of the following conditions (2)–(4):

$$\lim_{|x|\to\infty} \frac{P^{(\alpha)}(x)}{P(x)} = 0 \tag{2}$$

for all $\alpha \neq 0$;

$$\lim_{|x|\to\infty} \operatorname{dist}(x, N(P)) = \infty, \tag{3}$$

where $N(P) := \{z \in \mathbb{C}^n; P(z) = 0\}$;

$$\operatorname{dist}(x, N(P)) \geq C|x|^c \tag{4}$$

as $x \in \mathbb{R}^n \to \infty$, for suitable positive constants C, c (these equivalent descriptions of hypoellipticity are not used in the sequel). For example, if the principal part P_m of P does not vanish for $0 \neq x \in \mathbb{R}^n$ (in this case, P and $P(D)$ are said to be *elliptic*), Condition (2) is clearly satisfied; thus elliptic differential operators are hypoelliptic.

Theorem II.6.3. *Let P be a hypoelliptic polynomial, and let Ω be an open subset of $\mathbb{R}^n$. If $u \in \mathcal{D}'(\Omega)$ is a solution of the equation $P(D)u = f$ with $f \in \mathcal{W}^{\mathrm{loc}}_{p,k}(\Omega)$, then $u \in \mathcal{W}^{\mathrm{loc}}_{p,kk_P}(\Omega)$. In particular, if $f \in \mathcal{E}(\Omega)$, then $u \in \mathcal{E}(\Omega)$.*

(The following converse is also true (proof omitted): suppose that for some Ω, some $p \in [1, \infty]$, and some weight k, every solution of the equation $P(D)u = 0$ in $\mathcal{W}^{\mathrm{loc}}_{p,k}(\Omega)$ is in $\mathcal{E}(\Omega)$. Then P is hypoelliptic.)

Proof. Fix $\omega \subset\subset \Omega$. For any $u \in \mathcal{D}'(\Omega)$ and $\phi \in \mathcal{D}(\omega)$, we view ϕu as an element of $\mathcal{E}'$ with support in ω (cf. II.6.1). By the necessity of the estimate in Part (i) of Theorem II.3.6), $|\mathcal{F}(\phi u)(x)| \leq M(1 + |x|)^r$ for some constants M, r independent of ϕ. Hence, for any given p, there exists s (independent of ϕ) such that $k_{-s}\mathcal{F}(\phi u) \in L^p$. Denote $k' = k_{-s}$ for such an s (fixed from now on). Thus, $\phi u \in \mathcal{W}_{p,k'}$ for all $\phi \in \mathcal{D}(\omega)$, that is, $u \in \mathcal{W}^{\mathrm{loc}}_{p,k'}(\omega)$.

The hypoellipticity condition (1) (Section II.6.2) implies the existence of a constant $C' > 0$ such that $|P^{(\alpha)}/P| \leq (1/C')|x|^{-c|\alpha|}$ for all $\alpha \neq 0$. Summing over all $\alpha \neq 0$ with $|\alpha| \leq m$, we get that $k'_P/|P| \leq (1/C'')(1 + |x|)^{-c}$ for some constant $C'' > 0$ (cf. notation at the end of II.3.7). Hence

$$\frac{k_P}{k'_P} \geq \frac{|P|}{k'_P} \geq C''(1 + |x|)^c. \tag{5}$$

Given the weight k, kk_P/k' is a weight, and therefore it is $O((1+|x|)^\nu)$ for some ν. Consequently there exists a positive integer r (depending only on the ratio k/k') such that $kk_P/k' \le \text{const.}(1+|x|)^{cr}$. By (5), it then follows that

$$kk_P \le Ck'\left(\frac{k_P}{k'_P}\right)^r \tag{6}$$

for some constant C.

Claim. *If k is any weight such that $f \in \mathcal{W}^{\mathrm{loc}}_{p,k}(\omega)$ and (6) with $r=1$ is valid (that is, $k \le C(k'/k'_P)$), then any solution $u \in \mathcal{W}^{\mathrm{loc}}_{p,k'}(\omega)$ of the equation $P(D)u = f$ is necessarily in $\mathcal{W}^{\mathrm{loc}}_{p,kk_P}(\omega)$.*

Proof of claim. We first observe that

$$P(D)\mathcal{W}^{\mathrm{loc}}_{p,k}(\omega) \subset \mathcal{W}^{\mathrm{loc}}_{p,k/k_P}(\omega) \tag{7}$$

(cf. II.3.9, Relation (48)). Therefore, $P^{(\alpha)}(D)u \in \mathcal{W}^{\mathrm{loc}}_{p,k'/k'_P}(\omega) \subset \mathcal{W}^{\mathrm{loc}}_{p,k}(\omega)$ for all $\alpha \neq 0$ (for u as in the claim, because $k \le C(k'/k'_P)$).

If $\phi \in \mathcal{D}(\omega)$, we have by Leibnitz's formula (II.2.9, (5))

$$P(D)(\phi u) = \phi f + \sum_{\alpha \neq 0} D^\alpha \phi P^{(\alpha)}(D)u/\alpha!.$$

The first term is in $\mathcal{W}_{p,k}$ by hypothesis. The sum over $\alpha \neq 0$ is in $\mathcal{W}_{p,k}$ by the preceding observation. Hence $P(D)(\phi u) \in \mathcal{W}_{p,k}$ (and has compact support).

Let $v \in \mathcal{W}^{\mathrm{loc}}_{\infty,k_P}(\mathbb{R}^n)$ be a fundamental solution for $P(D)$ (by Theorem II.4.2 and the observation following its statement). Then

$$\phi u = v * [P(D)(\phi u)] \in \mathcal{W}_{p,kk_P}$$

since for any weights k, k_1 (cf. II.3.9, (47))

$$\mathcal{W}^{\mathrm{loc}}_{\infty,k_1}(\mathbb{R}^n) * [\mathcal{W}_{p,k} \cap \mathcal{E}'] \subset \mathcal{W}_{p,kk_1}.$$

This concludes the proof of the claim.

Suppose now that $r > 1$. Consider the weights

$$k_j = k\left(\frac{k'_P}{k_P}\right)^j \qquad j = 0, \ldots, r-1.$$

Since

$$k = k_0 \ge k_1 \ge \cdots \ge k_{r-1},$$

we have $f \in \mathcal{W}^{\mathrm{loc}}_{p,k_j}(\omega)$ for all $j = 0, \ldots, r-1$. Also by (6)

$$k_P k_{r-1} \left(= k_P k \left(\frac{k'_P}{k_P}\right)^{r-1}\right) \le Ck'\frac{k_P}{k'_P}.$$

We may then apply the claim with the weight k_{r-1} replacing k. Then $u \in \mathcal{W}^{\mathrm{loc}}_{p,k_P k_{r-1}}(\omega)$, $f \in \mathcal{W}^{\mathrm{loc}}_{p,k_{r-2}}(\omega)$, and $k_{r-2} = k_P k_{r-1}/k'_P$. By the claim with the weights k, k' replaced by $k_{r-2}, k_P k_{r-1}$ (respectively), it follows that $u \in \mathcal{W}^{\mathrm{loc}}_{p,k_P k_{r-2}}(\omega)$. Repeating this argument, we obtain finally (since $k_0 = k$) that $u \in \mathcal{W}^{\mathrm{loc}}_{p,k_P k}(\omega)$. This being true for any $\omega \subset\subset \Omega$, we conclude that $u \in \mathcal{W}^{\mathrm{loc}}_{p,k_P k}(\Omega)$. □

II.7 Variable coefficients

II.7.1. The constant coefficients theory of II.4.1 and Theorem II.4.2 can be applied "locally" to linear differential operators $P(x, D)$ with (locally) C^∞-coefficients. (This means that $P(x, y)$ is a polynomial in $y \in \mathbb{R}^n$, with coefficients that are C^∞-functions of x in some neighborhood $\Omega \subset \mathbb{R}^n$ of x^0.) Denote $P_0 = P(x^0, \cdot)$. *We shall assume that there exist* $\epsilon > 0$ *and* $0 < M < \infty$ *such that the* ϵ*-neighborhood* V *of* x^0 *is contained in* Ω *and for all* $x \in V$

$$\frac{k_{P(x,\cdot)}}{k_{P_0}} \leq M. \tag{1}$$

The method described next regards the operator $P(x, D)$ as a "perturbation" of the operator $P_0(D)$ for x in a "small" neighborhood of x^0.

Let $r + 1$ be the (finite!) dimension of the space of polynomials Q such that k_Q/k_{P_0} is bounded, and choose a basis $P_0, P_1, \ldots, P_r$ for this space. By (1), we have a unique representation

$$P(x, \cdot) = P_0 + \sum_{j=1}^{r} c_j(x) P_j \tag{2}$$

for all $x \in V$. Necessarily $c_j(x^0) = 0$ (take $x = x^0$) and $c_j \in C^\infty(V)$.

By Theorem II.4.2, we may choose a fundamental solution $v \in \mathcal{W}^{\mathrm{loc}}_{\infty,k_{P_0}}$ for the operator $P_0(D)$. Fix $\chi \in \mathcal{D}$ such that $\chi = 1$ in a 3ϵ-neighborhood of x^0. Then

$$w := \chi v \in \mathcal{W}_{\infty,k_{P_0}}, \tag{3}$$

and for all $h \in \mathcal{E}'(V)$,

$$P_0(D)(w * h) = w * (P_0(D)h) = v * P_0(D)h = h. \tag{4}$$

(The second equality follows from the fact that $\operatorname{supp} P_0(D)h \subset V$ and $w * g = v * g$ for all $g \in \mathcal{E}'(V)$.) By (2) and (4)

$$P(x, D)(w * h) = h + \sum_{j=1}^{r} c_j(x) P_j(D)(w * h) \tag{5}$$

for all $h \in \mathcal{E}'(V)$.

We localize to a suitable δ-neighborhood of x^0 by fixing some function $\phi \in \mathcal{D}$ such that $\phi = 1$ for $|x| \leq 1$ and $\operatorname{supp}\phi \subset \{x; |x| < 2\}$, and letting $\phi_\delta(x) = \phi((x - x^0)/\delta)$. (Thus $\phi_\delta = 1$ for $|x - x^0| \leq \delta$ and $\operatorname{supp}\phi_\delta \subset \{x; |x - x^0| < 2\delta$.)

By (5), whenever $\delta < \epsilon$ and $h \in \mathcal{E}'(V)$,

$$P(\cdot, D)(w * h) = h + \sum_{j=1}^{r} \phi_\delta c_j P_j(D)(w * h) \tag{6}$$

in $|x - x^0| < \delta$.

Claim. *There exists $\delta_0 < \epsilon/2$ such that, for $\delta < \delta_0$, the equation*

$$h + \sum_{j=1}^{r} \phi_\delta c_j P_j(D)(w * h) = \phi_\delta f \tag{7}$$

(for any $f \in \mathcal{E}'$) has a unique solution $h \in \mathcal{E}'$.

Assuming the claim, the solution h of (7) satisfies (by (6))

$$P(\cdot, D)(w * h) = \phi_\delta f = f \tag{8}$$

in $V_\delta := \{x; |x - x^0| < \delta\}$. (Since $2\delta < \epsilon$, $\operatorname{supp}\phi_\delta \subset V$, and therefore, $\operatorname{supp} h \subset V$ by (7), and (6) applies.)

In other words, $u = w * h \in \mathcal{E}'$ solves the equation $P(\cdot, D)u = f$ in V_δ. Equivalently, the map

$$T : f \in \mathcal{E}' \to w * h \in \mathcal{E}' \tag{9}$$

(with h as in the "claim") is "locally" a right inverse of the operator $P(\cdot, D)$, that is,

$$P(x, D)Tf = f \quad (f \in \mathcal{E}'; x \in V_\delta). \tag{10}$$

The operator T is also a left inverse of $P(\cdot, D)$ (in the shown local sense). Indeed, given $u \in \mathcal{E}'(V_\delta)$, we take $f := P(\cdot, D)u$ and $h := P_0(D)u$. By (4), $w * h = w * P_0(D)u = u$ (since $u \in \mathcal{E}'(V)$). Therefore, the left-hand side of (7) equals

$$P_0(D)u + \sum_j \phi_\delta c_j P_j(D)u = P(\cdot, D)u = f = \phi_\delta f$$

in V_δ (since $\phi_\delta = 1$ in V_δ). Thus "our" h is the (unique) solution of (7) (for "our" f) in V_δ. Consequently

$$TP(\cdot, D)u := w * h = u \quad (u \in \mathcal{E}'(V_\delta)) \tag{11}$$

in V_δ. Since $2\delta < \epsilon$, (11) is true in V_δ for all $u \in \mathcal{E}'(\mathbb{R}^n)$. Modulo the "claim", we proved the first part of the following.

Theorem II.7.2. *Let $P(\cdot, D)$ have C^∞-coefficients and satisfy*

$$k_{P(x,\cdot)} \leq M\, k_{P_0} \tag{*}$$

in an ϵ-neighborhood of x^0 (where M is a constant and $P_0 := P(x^0, \cdot)$). Then there exists a δ-neighborhood V_δ of x^0 (with $\delta < \epsilon$) and a linear map $T : \mathcal{E}' \to \mathcal{E}'$ such that

$$P(\cdot, D)Tg = TP(\cdot, D)g = g \quad \text{in } V_\delta \qquad (g \in \mathcal{E}').$$

Moreover, the restriction of T to the subspace $\mathcal{W}_{p,k} \cap \mathcal{E}'$ of $\mathcal{W}_{p,k}$ is a bounded operator into $\mathcal{W}_{p,kk_{P_0}}$, for any weight k.

Proof. We first prove the "claim" (this will complete the proof of the first part of the theorem.)

For any $\delta < \epsilon$, consider the map

$$S_\delta : h \in \mathcal{S}' \to \sum_{j=1}^{r} \phi_\delta c_j P_j(D)(w * h).$$

Since $w \in \mathcal{W}_{\infty, k_{P_0}}$ and k_{P_j}/k_{P_0} are bounded (by definition of P_j), we have

$$|P_j \mathcal{F} w| \le k_{P_j} |\mathcal{F} w| \le \text{ const } \cdot k_{P_0} |\mathcal{F} w| \le C < \infty \quad (j = 0, \ldots, r). \tag{12}$$

Let k be any given weight. By (51) in Section II.3.9, there exists $t_0 > 0$ such that, for $0 < t < t_0$,

$$\|S_\delta h\|_{p,k^t} \le 2(2\pi)^{-n/2} \sum_{j=1}^{r} \|\phi_\delta c_j\|_{1,1} \|P_j(D)(w * h)\|_{p,k^t}. \tag{13}$$

By the inequality preceding (47) in II.3.9,

$$\|P_j(D)(w * h)\|_{p,k^t} = \|[P_j(D)w] * h\|_{p,k^t} \le (2\pi)^{n/2} \|P_j(D)w\|_{\infty,1} \|h\|_{p,k^t}$$

for all $h \in \mathcal{W}_{p,k} = \mathcal{W}_{p,k^t}$. Since by (12)

$$\|P_j(D)w\|_{\infty,1} = \|\mathcal{F}[P_j(D)w]\|_\infty = \|P_j \mathcal{F} w\|_\infty \le C$$

(for $j = 1, \ldots, r$), we obtain from (13)

$$\|S_\delta h\|_{p,k^t} \le 2C \sum_{j=1}^{r} \|\phi_\delta c_j\|_{1,1} \|h\|_{p,k^t} \tag{14}$$

for all $h \in \mathcal{W}_{p,k}$.

Since $c_j(x^0) = 0$, $c_j = O(\delta)$ on $\operatorname{supp} \phi_\delta$ by the mean value inequality. Using the definition of ϕ_δ, it follows that $c_j D^\alpha \phi_\delta = O(\delta^{1-|\alpha|})$. Hence by Leibnitz's formula, $D^\alpha(\phi_\delta c_j) = O(\delta^{1-|\alpha|})$. Therefore, since the measure of $\operatorname{supp}(\phi_\delta c_j)$ is $O(\delta^n)$, we have

$$x^\alpha \mathcal{F}(\phi_\delta c_j) = \mathcal{F}[D^\alpha(\phi_\delta c_j)] = O(\delta^{n+1-|\alpha|}).$$

Hence

$$(1 + \delta|x|)^{n+1} \mathcal{F}(\phi_\delta c_j) = O(\delta^{n+1}).$$

Consequently

$$\|\phi_\delta c_j\|_{1,1} := \int |\mathcal{F}(\phi_\delta c_j)|\,dx \le \text{const } \cdot \delta^{n+1} \int \frac{dx}{(1+\delta|x|)^{n+1}} = O(\delta).$$

We may then choose $0 < \delta_0 < \epsilon/2$ such that

$$\sum_{j=1}^{r} \|\phi_\delta c_j\|_{1,1} < \frac{1}{4C} \tag{15}$$

for $0 < \delta < \delta_0$. By (14), we then have

$$\|S_\delta h\|_{p,k^t} \le (1/2)\|h\|_{p,k^t}$$

for all $h \in \mathcal{W}_{p,k} = \mathcal{W}_{p,k^t}$. This means that for $\delta < \delta_0$, the operator S_δ on the Banach space $\mathcal{W}_{p,k^t}$ has norm $\le 1/2$, and therefore $I + S_\delta$ has a bounded inverse (I is the identity operator). Thus, (7) (in II.7.1) has a unique solution $h \in \mathcal{W}_{p,k}$ for each $f \in \mathcal{W}_{p,k}$. By the equation, h is necessarily in $\mathcal{E}'$ (since $\phi_\delta \in \mathcal{D}$). If f is an arbitrary distribution in $\mathcal{E}'$, the trivial part of the Paley–Wiener–Schwartz theorem (II.3.6) shows that $\hat{f}(x) = O(1+|x|)^N$ for some N, and therefore, $f \in \mathcal{W}_{p,k}$ for suitable weight k (e.g., $k = k_{-s}$ with s large enough). Therefore (for $\delta < \delta_0$) there exists a solution h of (7) in $\mathcal{W}_{p,k} \cap \mathcal{E}'$. The solution is unique (in $\mathcal{E}'$), because if $h, h' \in \mathcal{E}'$ are solutions, there exists a weight k such that $f, h, h' \in \mathcal{W}_{p,k}$, and therefore, $h = h'$ by the uniqueness of the solution in $\mathcal{W}_{p,k}$. This completes the proof of the claim.

Since $\|S_\delta\| \le 1/2$ (the norm is the $B(\mathcal{W}_{p,k^t})$-norm!), we have $\|(I+S_\delta)^{-1}\| \le 2$ (by the Neumann expansion of the resolvent!), and therefore, $\|h\|_{p,k^t} \le 2\|\phi_\delta f\|_{p,k^t}$. Consequently (with h related to f as in the "claim", and t small enough), we have by the inequality preceding (47) and by (51) in II.3.9:

$$\begin{aligned}\|Tf\|_{p,k_{P_0}k^t} &= \|w * h\|_{p,k_{P_0}k^t} \le (2\pi)^{n/2}\|w\|_{\infty,k_{P_0}}\|h\|_{p,k^t}\\ &\le 2(2\pi)^{n/2}\|w\|_{\infty,k_{P_0}}\|\phi_\delta f\|_{p,k^t} \le 4\|w\|_{\infty,k_{P_0}}\|\phi_\delta\|_{1,1}\|f\|_{p,k^t}.\end{aligned}$$

This proves the second part of the theorem, since the norms $\|\cdot\|_{p,k}$ ($\|\cdot\|_{p,k_{P_0}k}$) and $\|\cdot\|_{p,k^t}$ ($\|\cdot\|_{p,k_{P_0}k^t}$, respectively) are equivalent. □

Corollary II.7.3. *For $P(\cdot, D)$ and V_δ as in Theorem II.7.2, the equation $P(\cdot, D)u = f$ has a solution $u \in C^\infty(V_\delta)$ for each $f \in C^\infty(\mathbb{R}^n)$.*

Proof. Fix $\phi \in \mathcal{D}$ such that $\phi = 1$ in a neighborhood of $\overline{V_\delta}$. For $f \in C^\infty$, $\phi f \in \mathcal{W}_{p,k} \cap \mathcal{E}'$ for all k; therefore, $u := T(\phi f)$ is a solution of $P(\cdot, D)u = f$ in V_δ (because $\phi f = f$ in V_δ), which belongs to $\mathcal{W}_{p,kk_{P_0}}$ for all weights k, hence $u \in C^\infty(V_\delta)$. □

II.8 Convolution operators

Let $h : \mathbb{R}^n \to \mathbb{C}$ be locally Lebesgue integrable, and consider the *convolution operator*

$$T : u \to h * u,$$

originally defined on the space L^1_c of integrable functions u on $\mathbb{R}^n$ with compact support.

We set $h^t(x) := t^n h(tx)$, and make the following.

Hypothesis I.

$$\int_{|x|\geq 2} |h^t(x-y) - h^t(x)|\,dx \leq K < \infty \quad (|y| \leq 1;\ t > 0). \tag{1}$$

Lemma II.8.1. *If $u \in L^1_c$ vanishes outside the ball $B(a,t)$ and $\int u\,dx = 0$, then*

$$\int_{B(a,2t)^c} |Tu|\,dx \leq K\|u\|_1.$$

Proof. Denote $u_a(x) = u(x+a)$. Since $(Tu)_a = Tu_a$, we have

$$\begin{aligned}\int_{B(a,2t)^c} |Tu|\,dx &= \int_{B(0,2t)^c} |(Tu)_a|\,dx = \int_{B(0,2t)^c} |h * u_a|\,dx \\ &= \int_{B(0,2)^c} |h^t * (u_a)^t|\,dx.\end{aligned}$$

Since $(u_a)^t(y) = t^n u(ty+a) = 0$ for $|y| \geq 1$ and $\int (u_a)^t(y)\,dy = \int u(x)\,dx = 0$, the last integral is equal to

$$\begin{aligned}&\int_{B(0,2)^c} \left| \int_{|y|<1} [h^t(x-y) - h^t(x)](u_a)^t(y)\,dy \right| dx \\ &\quad\leq \int_{B(0,2)^c} \int_{|y|<1} |h^t(x-y) - h^t(x)|\,|(u_a)^t(y)|\,dy\,dx \\ &\quad= \int_{|y|<1} \left(\int_{B(0,2)^c} |h^t(x-y) - h^t(x)|\,dx \right) |(u_a)^t(y)|\,dy \leq K\|(u_a)^t\|_1 = K\|u\|_1.\end{aligned}$$

□

We shall need the following version of the *Calderon–Zygmund decomposition lemma.*

Lemma II.8.2. *Fix $s > 0$, and let $u \in L^1(\mathbb{R}^n)$. Then there exist disjoint open (hyper)cubes I_k and functions $u_k, v \in L^1(\mathbb{R}^n)$ $(k \in \mathbb{N})$, such that*

(1) u_k vanishes outside I_k and $\int u_k\,dx = 0$ for all $k \in \mathbb{N}$;

(2) $|v| \leq 2^n s$ a.e.;

(3) $u = v + \sum_k u_k$;

(4) $\|v\|_1 + \sum_k \|u_k\|_1 \leq 3\|u\|_1$; and

(5) $\sum_k |I_k| \leq \|u\|_1/s$ (where $|I_k|$ denotes the volume of I_k).

Proof. We first partition $\mathbb{R}^n$ into cubes of volume $>\|u\|_1/s$. For any such cube Q, the average on Q of $|u|$,

$$A_Q(|u|) := |Q|^{-1}\int_Q |u|\,dx,$$

satisfies

$$A_Q(|u|) \le |Q|^{-1}\|u\|_1 < s. \tag{2}$$

Subdivide Q into 2^n congruent subcubes Q_i (by dividing each side of Q into two equal intervals). If $A_{Q_i}(|u|) \ge s$ for all i, then

$$A_Q(|u|) = |Q|^{-1}\sum_i \int_{Q_i} |u|\,dx \ge |Q|^{-1}s\sum_i |Q_i| = s,$$

contradicting (2). Let $Q_{1,j}$ be the *open* subcubes of Q on which the averages of $|u|$ are $\ge s$, and let $Q'_{1,l}$ be the remaining subcubes (there is at least one subcube of the latter kind). We have

$$s|Q_{1,j}| \le \int_{Q_{1,j}} |u|\,dx \le \int_Q |u|\,dx < s|Q| = 2^n s|Q_{1,j}|. \tag{3}$$

We define v on the the cubes $Q_{1,j}$ as the *constant* $A_{Q_{1,j}}(u)$ (for each j), and we let $u_{1,j}$ be equal to $u - v$ on $Q_{1,j}$ and to zero on $Q^c_{1,j}$.

For each cube $Q'_{1,l}$ we repeat the construction we did with Q (since the average of $|u|$ over such cubes is $< s$, as it was over Q). We obtain the *open* subcubes $Q_{2,j}$ (of the cubes $Q'_{1,l}$) on which the average of $|u|$ is $\ge s$, and the remaining subcubes $Q'_{2,l}$ on which the average is $<s$. We then extend the definition of v to the subcubes $Q_{2,j}$ by assigning to v the constant value $A_{Q_{2,j}}(u)$ on $Q_{2,j}$ (for each j). The functions $u_{2,j}$ are then defined in the same manner as $u_{1,j}$, with $Q_{2,j}$ replacing $Q_{1,j}$.

Continuing this process (and renaming) we obtain a sequence of mutually disjoint open cubes I_k, a sequence of measurable functions u_k defined on $\mathbb{R}^n$, and a measurable function v defined on $\Omega := \bigcup I_k$ (which we extend to $\mathbb{R}^n$ by setting $v = u$ on Ω^c). By construction, Property 3 is satisfied.

Since the average of $|u|$ on each I_k is $\ge s$ (by definition), we have

$$s\sum |I_k| \le \sum \int_{I_k} |u|\,dx = \int_\Omega |u|\,dx \le \|u\|_1,$$

and Property 5 is satisfied.

If $x \in \Omega$, then $x \in I_k$ for precisely one k, and therefore

$$|v(x)| = |A_{I_k}(u)| \le A_{I_k}(|u|) \le 2^n s$$

by (3) (which is true for all the cubes I_k, by construction). If $x \notin \Omega$, there is a sequence of open cubes J_k containing x, over which the average of $|u|$ is $< s$, such that $|J_k| \to 0$. This implies that $|u(x)| \le s$ a.e. on Ω^c, and since $v = u$ on Ω^c, we conclude that v has Property 2.

By construction, u_k vanishes outside I_k and $\int u_k\,dx = \int_{I_k} u\,dx - \int_{I_k} v\,dx = 0$ for all $k \in \mathbb{N}$ (Property 1).

Since I_k are mutually disjoint and $\operatorname{supp} u_k \subset I_k$, we have

$$\|v\|_1 + \sum \|u_k\|_1 = \int_{\Omega^c} |v|\,dx + \sum \int_{I_k} (|v| + |u_k|)\,dx.$$

However $v = u$ on Ω^c and $u_k = u - v$ on I_k; therefore, the right-hand side is

$$\leq \int_{\Omega^c} |u|\,dx + \sum_k \Big(2\int_{I_k} |v|\,dx + \int_{I_k} |u|\,dx\Big).$$

Since v has the constant value $A_{I_k}(u)$ on I_k,

$$\int_{I_k} |v|\,dx = |A_{I_k}(u)|\,|I_k| \leq \int_{I_k} |u|\,dx,$$

and Property 4 follows. □

Consider now $u \in L^1(\mathbb{R}^n)$ *with compact support and* $\|u\|_1 = 1$. It follows from the construction in the last proof that v has compact support; by Property 1 in Lemma II.8.2, u_k have compact support as well, for all k. Therefore, Tv and Tu_k are well defined (for all k), and

$$Tu = Tv + \sum_k Tu_k. \tag{4}$$

For any $r > 0$, we then have

$$[|Tu| > r] \subset [|Tv| > r/2] \cup \Big[\sum |Tu_k| > r/2\Big]. \tag{5}$$

Denote the sets in the last union by F_r and G_r.

Let $B(a_k, t_k)$ be the smallest ball containing the cube I_k and let c_n be the ratio of their volumes (depends only on the dimension n of $\mathbb{R}^n$). Since u_k vanishes outside $B(a_k, t_k)$ and $\int u_k\,dx = 0$, we have by Lemma II.8.1

$$\int_{B(a_k,2t_k)^c} |Tu_k|\,dx \leq K\|u_k\|_1. \tag{6}$$

Let

$$E := \bigcup_{k=1}^{\infty} B(a_k, 2t_k).$$

Then (for $s > 0$ given as in Lemma II.8.2)

$$\begin{aligned} |E| &\leq \sum_k |B(a_k, 2t_k)| = 2^n \sum_k |B(a_k, t_k)| \\ &= 2^n c_n \sum_k |I_k| \leq 2^n c_n/s. \end{aligned} \tag{7}$$

Therefore

$$|G_r| = |G_r \cap E^c| + |G_r \cap E| \le |G_r \cap E^c| + |E|$$
$$\le |G_r \cap E^c| + 2^n c_n/s. \tag{8}$$

Since $E^c \subset B(a_k, 2t_k)^c$ for all k, we have by (6)

$$\int_{E^c} |Tu_k|\, dx \le K\|u_k\|_1 \quad (k = 1, 2, \ldots),$$

and therefore

$$|G_r \cap E^c| \le (2/r) \int_{E^c} \sum |Tu_k|\, dx = (2/r) \sum \int_{E^c} |Tu_k|\, dx$$
$$\le (2/r)K \sum \|u_k\|_1 \le (6/r)K$$

by Property 4 of the functions u_k (cf. Lemma II.8.2). We then conclude from (8) that

$$|G_r| \le 6K/r + 2^n c_n/s. \tag{9}$$

In order to get an estimate for $|F_r|$, we make the following.

Hypothesis II.

$$\|T\phi\|_2 \le C\|\phi\|_2 \quad (\phi \in \mathcal{D}), \tag{10}$$

for some finite constant $C > 0$.

Since v is bounded a.e. (Property 2 in Lemma II.8.2) with compact support, it belongs to L^2, and it follows from (10) and the density of $\mathcal{D}$ in L^2 that

$$\|Tv\|_2^2 \le C^2\|v\|_2^2 \le C^2\|v\|_\infty\|v\|_1 \le 3C^2 2^n s, \tag{11}$$

where Properties 2 and 4 in Lemma II.8.2 were used. Therefore,

$$|F_r| \le (4/r^2)\|Tv\|_2^2 \le 12\, C^2 2^n s/r^2,$$

and we conclude from (5) and (9) that

$$|[|Tu| > r]| \le 6K/r + 2^n c_n/s + 12\, C^2 2^n s/r^2.$$

The left-hand side being independent of $s > 0$, we may minimize the right-hand side with respect to s; hence,

$$|[|Tu| > r]| \le C'/r, \tag{12}$$

where $C' := 6K + 2^{n+2}\sqrt{3c_n}C$ depends linearly on the constants K and C of the hypothesis (1) and (10) (and on the dimension n).

If $u \in L^1$ (with compact support) is not necessarily normalized, we consider $w = u/\|u\|_1$ (when $\|u\|_1 > 0$). Then by (12) for w,

$$|[|Tu| > r]| = |[|Tw| > r/\|u\|_1]| \le C'\|u\|_1/r. \tag{13}$$

Since (13) is trivial when $\|u\|_1 = 0$, we proved the following.

Lemma II.8.3. *Let $h : \mathbb{R}^n \to \mathbb{C}$ be locally Lebesgue integrable and satisfy Hypotheses I and II (where T denotes the convolution operator $T : u \to h * u$, originally defined on L^1_c). Then there exists a positive constant C' depending linearly on the constants K and C of the hypothesis (and on the dimension n) such that*

$$\|[|Tu| > r]\| \le C'\|u\|_1/r \quad (r > 0)$$

for all $u \in L^1_c$.

Under the hypothesis of Lemma II.8.3, the linear operator T is of weak type $(1,1)$ with weak $(1,1)$-norm $\le C'$, and of strong (hence weak) type $(2,2)$ with strong (hence weak) $(2,2)$-norm $\le C < C'$. By Theorem 5.41, it follows that T is of strong type (p,p) with strong (p,p)-norm $\le A_p C'$, for any p in the interval $1 < p \le 2$, where A_p depends only on p (and is bounded when p is bounded away from 1).

Let $\tilde{h}(x) := h(-x)$, and let $\tilde{T}$ be the corresponding convolution operator. Since $\tilde{h}$ satisfies hypotheses (1) and (10) (when h does) with the same constants K and C, the operator $\tilde{T}$ is of strong type (p,p) with strong (p,p)-norm $\le A_p C'$ for all $p \in (1,2]$. Let q be the conjugate exponent of p (for a given $p \in (1,2]$). Then for all $u \in \mathcal{D}$,

$$\begin{aligned}
\|Tu\|_q &= \sup\left\{\left|\int (Tu)v\,dx\right|; v \in \mathcal{D}, \|v\|_p = 1\right\} \\
&= \sup_v \left|\iint h(x-y)u(y)\,dy\; v(x)\,dx\right| \\
&= \sup_v \left|\iint \tilde{h}(y-x)v(x)\,dx\; u(y)\,dy\right| = \sup_v \left|\int (\tilde{T}v)u\,dy\right| \\
&\le \sup_v \|\tilde{T}v\|_p \, \|u\|_q \le A_p C' \, \|u\|_q.
\end{aligned}$$

Thus, T is of strong type (q,q) with strong (q,q)-norm $\le A_p C'$. Since q varies over the interval $[2,\infty)$ when p varies in $(1,2]$, we conclude that T is of strong type (p,p) with strong (p,p)-norm $\le A'_p C'$ for all $p \in (1,\infty)$ ($A'_p = A_p$ for $p \in (1,2]$ and $A'_p = A_{p'}$ for $p \in [2,\infty)$, where p' is the conjugate exponent of p). Observe that A'_p is a bounded function of p in any compact subset of $(1,\infty)$. We proved the following.

Theorem II.8.4 (Hormander). *Let $h : \mathbb{R}^n \to \mathbb{C}$ be locally integrable, and let $T : u \to h * u$ be the corresponding convolution operator (originally defined on L^1_c). Assume Hypotheses I and II. Then for all $p \in (1,\infty)$, T is of strong type (p,p) with strong (p,p)-norm $\le A_p C'$, where the constant A_p depends only on p and is a bounded function of p in any compact subset of $(1,\infty)$, and C' depends linearly on the constants K and C of the hypothesis (and on the dimension n).*

In order to apply Theorem II.8.4 to some special convolution operators, we need the following.

Lemma II.8.5. *Let* $S := \{y \in \mathbb{R}^n; 1/2 < |y| < 2\}$. *There exists* $\phi \in \mathcal{D}(S)$ *with range in* $[0,1]$ *such that*

$$\sum_{k\in\mathbb{Z}} \phi(2^{-k}y) = 1 \quad (y \in \mathbb{R}^n - \{0\}).$$

(For each $0 \neq y \in \mathbb{R}^n$, *at most two summands of the series are* $\neq 0$.*)*

Proof. Fix $\psi \in \mathcal{D}(S)$ such that $\psi(x) = 1$ for $3/4 \le |x| \le 3/2$. Let $y \in \mathbb{R}^n - \{0\}$ and $k \in \mathbb{Z}$. If $\psi(2^{-k}y) \neq 0$, then $2^{-k}y \in S$, that is, $\log_2|y| - 1 < k < \log_2|y| + 1$; there are at most two values of the integer k in that range. Moreover, $\psi(2^{-k}y) = 1$ if $3/4 \le 2^{-k}|y| \le 3/2$, that is, if $\log_2(|y|/3) + 1 \le k \le \log_2(|y|/3) + 2$, there is at least one value of k in this range. It follows that

$$1 \le \sum_{k\in\mathbb{Z}} \psi(2^{-k}y) < \infty$$

(there are at most two non-zero terms in the series, all terms are ≥ 0, and at least one term equals 1).

Define

$$\phi(y) = \frac{\psi(y)}{\sum_{j\in\mathbb{Z}} \psi(2^{-j}y)}.$$

Then $\phi \in \mathcal{D}(S)$ has range in $[0,1]$ and for each $y \neq 0$

$$\phi(2^{-k}y) = \frac{\psi(2^{-k}y)}{\sum_{j\in\mathbb{Z}} \psi(2^{-j-k}y)} = \frac{\psi(2^{-k}y)}{\sum_{j\in\mathbb{Z}} \psi(2^{-j}y)},$$

hence, $\sum_{k\in\mathbb{Z}} \phi(2^{-k}y) = 1$. □

The following discussion is restricted to the case $n = 1$ for simplicity (a similar analysis can be done in the general case). Fix ϕ as in Lemma II.8.5; let $c := \max(1, \sup|\phi'|)$, and denote $\phi_k(y) := \phi(2^{-k}y)$ for $k \in \mathbb{Z}$.

Let f be a measurable complex function, locally square integrable on $\mathbb{R}$. Denote $f_k := f\phi_k$. Then

$$\operatorname{supp} f_k \subset 2^k \operatorname{supp} \phi \subset 2^k S,$$

$$f = \sum_{k\in\mathbb{Z}} f_k \text{ on } \mathbb{R} - \{0\},$$

and $|f_k| \le |f|$.

Let I_k denote the indicator of the set $2^k S$. Then $|f_k| = |f_k I_k| \le |f I_k|$, and therefore,

$$\|f_k\|_2 \le \|f I_k\|_2 < \infty. \tag{14}$$

Consider f and f_k as distributions, and suppose that Df (in distribution sense) is a locally square integrable (measurable) function. Since

$$|Df_k| \le |Df|\phi_k + 2^{-k}\sup|\phi'|\,|f| \le c(|Df| + 2^{-k}|f|),$$

it follows that

$$\begin{aligned}\|Df_k\|_2 &= \|(Df_k)I_k\|_2 \le c(\|(Df)I_k\|_2 + 2^{-k}\|fI_k\|_2)\\ &= c2^{-k/2}[2^{-k/2}\|fI_k\|_2 + 2^{k/2}\|(Df)I_k\|_2]. \qquad (15)\end{aligned}$$

Notation. We denote by $\mathcal{H}$ the space of all measurable complex functions f on $\mathbb{R}$, locally square integrable on $\mathbb{R}$, for which

$$\|f\|_{\mathcal{H}} := \sup_{k\in\mathbb{Z}}[2^{-k/2}\|fI_k\|_2 + 2^{k/2}\|(Df)I_k\|_2] < \infty.$$

Assume $f \in \mathcal{H}$. It follows from (14) and (15) that

$$\|f_k\|_2 \le 2^{k/2}\|f\|_{\mathcal{H}} \text{ and } \|Df_k\|_2 \le c2^{-k/2}\|f\|_{\mathcal{H}} \qquad (16)$$

for all $k \in \mathbb{Z}$.

Let $g_k = \mathcal{F}^{-1}f_k$. Since $\mathcal{F}^{-1}$ is isometric on L^2 and $\mathcal{F}^{-1}D = -M\mathcal{F}^{-1}$ (where M denotes the operator of multiplication by the independent variable), we have by (16):

$$\begin{aligned}\int_{\mathbb{R}}(1+2^{2k}x^2)|g_k(x)|^2\,dx &= \|g_k\|_2^2 + 2^{2k}\|(-M)g_k\|_2^2\\ &= \|f_k\|_2^2 + 2^{2k}\|Df_k\|_2^2 \le 2^k(1+c^2)\|f\|_{\mathcal{H}}^2. \qquad (17)\end{aligned}$$

Therefore, by Schwarz's inequality

$$\begin{aligned}\|g_k\|_1 &= \int_{\mathbb{R}}(1+2^{2k}x^2)^{-1/2}[(1+2^{2k}x^2)^{1/2}|g_k(x)|]\,dx\\ &\le \left(\int_{\mathbb{R}}\frac{dx}{1+2^{2k}x^2}\right)^{1/2}\left(\int_{\mathbb{R}}(1+2^{2k}x^2)|g_k(x)|^2\,dx\right)^{1/2}\\ &\le \sqrt{1+c^2}\|f\|_{\mathcal{H}}\left(\int_{\mathbb{R}}\frac{2^k\,dx}{1+(2^kx)^2}\right)^{1/2} = c'\|f\|_{\mathcal{H}},\end{aligned}$$

where $c' = \sqrt{\pi(1+c^2)}$. Hence

$$|f_k| = |\mathcal{F}g_k| \le \|g_k\|_1 \le c'\|f\|_{\mathcal{H}} \quad (k \in \mathbb{Z}). \qquad (18)$$

Consider now the "partial sums"

$$s_m = \sum_{|k|\le m} f_k \quad (m = 1, 2, \ldots),$$

and let $h_m := \mathcal{F}^{-1}s_m = \sum_{|k|\le m} g_k$.

Since at most two summands f_k are $\neq 0$ at each point $y \neq 0$ (and $s_m(0) = 0$), we have by (18)

$$|s_m| \leq 2c' \|f\|_{\mathcal{H}} \quad (m = 1, 2, \ldots). \tag{19}$$

Therefore, for all $\psi \in \mathcal{S} := \mathcal{S}(\mathbb{R})$ and $m \in \mathbb{N}$,

$$\begin{aligned}\|h_m * \psi\|_2 &= \|\mathcal{F}(h_m * \psi)\|_2 = \sqrt{2\pi}\|\mathcal{F}h_m \mathcal{F}\psi\|_2 \\ &= \sqrt{2\pi}\|s_m \mathcal{F}\psi\|_2 \leq c''\|f\|_{\mathcal{H}}\|\mathcal{F}\psi\|_2 = c''\|f\|_{\mathcal{H}}\|\psi\|_2,\end{aligned}$$

where $c'' := 2\sqrt{2\pi}c'$. Thus, h_m satisfy Hypothesis II of Theorem II.8.4, with $C = c''\|f\|_{\mathcal{H}}$ *independent of* m.

Claim. *h_m satisfies Hypothesis I of Theorem II.8.4 with $K = K'\|f\|_{\mathcal{H}}$, where K' is a constant independent of m.*

Assuming the claim, it follows from Theorem II.8.4 that

$$\|h_m * \psi\|_p \leq C_p \|f\|_{\mathcal{H}} \|\psi\|_p \quad (m \in \mathbb{N}) \tag{20}$$

for all $p \in (1, \infty)$, where C_p is a constant *depending only on* p.

Let $\psi, \chi \in \mathcal{S}$. By (19)

$$|s_m \mathcal{F}\psi\, \overline{\mathcal{F}\chi}| \leq 2c'\, \|f\|_{\mathcal{H}} |\mathcal{F}\psi|\, |\mathcal{F}\chi| \in \mathcal{S} \subset L^1.$$

Since $s_m \to f$ pointwise a.e., it follows from the Lebesgue dominated convergence theorem that

$$\lim_m \int_{\mathbb{R}} s_m \mathcal{F}\psi\, \overline{\mathcal{F}\chi}\, dx = \int_{\mathbb{R}} f \mathcal{F}\psi\, \overline{\mathcal{F}\chi}\, dx. \tag{21}$$

On the other hand, by Parseval's identity and (20) (with $p' = p/(p-1)$)

$$\begin{aligned}\sqrt{2\pi}\left|\int_{\mathbb{R}} s_m \mathcal{F}\psi\, \overline{\mathcal{F}\chi}\, dx\right| &= \left|\int_{\mathbb{R}} \mathcal{F}(h_m * \psi)\, \overline{\mathcal{F}\chi}\, dx\right| = \left|\int_{\mathbb{R}} (h_m * \psi)\bar{\chi}\, dx\right| \\ &\leq \|h_m * \psi\|_p \|\chi\|_{p'} \leq C_p\, \|f\|_{\mathcal{H}} \|\psi\|_p\, \|\chi\|_{p'}\end{aligned}$$

for all $m \in \mathbb{N}$. Hence

$$\left|\int_{\mathbb{R}} f \mathcal{F}\psi\, \overline{\mathcal{F}\chi}\, dx\right| \leq (2\pi)^{-1/2} C_p\, \|f\|_{\mathcal{H}}\, \|\psi\|_p\, \|\chi\|_{p'}. \tag{22}$$

Let $u \in \mathcal{S}'$ be such that $\mathcal{F}u \in \mathcal{H}$. We may then apply (22) to $f = \mathcal{F}u$. Since $f\, \mathcal{F}\psi = \mathcal{F}u\, \mathcal{F}\psi = (2\pi)^{-1/2}\mathcal{F}(u * \psi)$, the integral on the left-hand side of (22) is equal to $(2\pi)^{-1/2} \int_{\mathbb{R}} (u * \psi)\bar{\chi}\, dx$ (by Parseval's identity). Hence

$$\left|\int_{\mathbb{R}} (u * \psi)\bar{\chi}\, dx\right| \leq C_p\, \|f\|_{\mathcal{H}}\, \|\psi\|_p\, \|\chi\|_{p'} \quad (\psi, \chi \in \mathcal{S}).$$

Therefore

$$\|u * \psi\|_p \leq C_p\, \|f\|_{\mathcal{H}}\, \|\psi\|_p \quad (\psi \in \mathcal{S}). \tag{23}$$

This proves the following result (once the "claim" is verified).

Theorem II.8.6. *Let $u \in \mathcal{S}'(\mathbb{R})$ be such that $\mathcal{F}u \in \mathcal{H}$. Then for each $p \in (1, \infty)$, the map $T : \psi \in \mathcal{S} \to u * \psi$ is of strong type (p,p), with strong (p,p)-norm $\leq C_p \|\mathcal{F}u\|_{\mathcal{H}}$, where the constant C_p depends only on p, and is a bounded function of p in any compact subset of $(1, \infty)$.*

Proof of the "claim". The change of variables $tx \to x$ and $ty \to y$ shows that we must prove the estimate

$$\int_{|x|\geq 2t} |h_m(x-y) - h_m(x)|\, dx \leq K \tag{24}$$

for all $t > 0$ and $|y| \leq t$. Since $h_m = \sum_{|k|\leq m} g_k$, we consider the corresponding integrals for g_k.

$$\int_{|x|\geq 2t} |g_k(x-y) - g_k(x)|\, dx \leq \int_{|x|\geq 2t} |g_k(x-y)|\, dx + \int_{|x|\geq 2t} |g_k(x)|\, dx. \tag{25}$$

The change of variable $x' = x - y$ in the first integral on the right-hand side tranforms it to $\int_{\{x';|x'+y|\geq 2t\}} |g_k(x')|\, dx'$. However, for $|y| \leq t$, we have $\{x'; |x'+y| \geq 2t\} \subset \{x'; |x'| \geq t\}$; therefore the first integral (and trivially, the second as well) is $\leq \int_{|x|\geq t} |g_k(x)|\, dx$.

By the Cauchy–Schwarz's inequality and (17),

$$\begin{aligned}
\int_{|x|\geq t} |g_k(x)|\, dx &\leq \left(\int_{\mathbb{R}} (1 + 2^{2k}x^2)|g_k(x)|^2\, dx\right)^{1/2} \left(\int_{|x|\geq t} \frac{dx}{2^{2k}x^2}\right)^{1/2} \\
&\leq (1+c^2)^{1/2}\|f\|_{\mathcal{H}} 2^{-k/2} \left(\int_{|x|\geq t} x^{-2}\, dx\right)^{1/2} \\
&= \sqrt{2(1+c^2)}\|f\|_{\mathcal{H}}(2^k t)^{-1/2}.
\end{aligned}$$

Therefore, for all $t > 0$ and $|y| \leq t$,

$$\int_{|x|\geq 2t} |g_k(x-y) - g_k(x)|\, dx \leq 2\sqrt{2(1+c^2)}\|f\|_{\mathcal{H}}(2^k t)^{-1/2}. \tag{26}$$

Another estimate of the integral on the left-hand side of (26) is obtained by writing it in the form

$$\int_{|x|\geq 2t} \left| \int_x^{x-y} g_k'(s)\, ds \right| dx. \tag{27}$$

The integrand of the outer integral is $\leq \int_x^{x-y} |g_k'(s)|\, ds$ for $y \leq 0$ ($\leq \int_{x-y}^x |g_k'(s)|\, ds$ for $y > 0$). Therefore, by Tonelli's theorem, the expression in (27) is $\leq \int_{\mathbb{R}} (\int_{s+y}^s dx)|g_k'(s)|\, ds$ for $y \leq 0$ ($\leq \int_{\mathbb{R}} (\int_s^{s+y} dx)|g_k'(s)|\, ds$ for $y > 0$, respectively) $= |y|\, \|g_k'\|_1 \leq t\|g_k'\|_1$.

Since $\operatorname{supp} f_k \subset 2^k S \subset B(0, 2^{k+1})$ and $g_k = \mathcal{F}^{-1} f_k = \mathcal{F}\tilde{f}_k$, Theorem II.3.6) implies that g_k extends to $\mathbb{C}$ as an entire function of exponential type $\le 2^{k+1}$, and is bounded on $\mathbb{R}$. By Bernstein's inequality (cf. Example in Section II.3.6, (31)) and (18)

$$\|g_k'\|_1 \le 2^{k+1} \|g_k\|_1 \le c' \|f\|_{\mathcal{H}} 2^{k+1},$$

and it follows that the integral on the left-hand side of (26) is $\le 2c'\|f\|_{\mathcal{H}} 2^k t$. Hence (for all $t > 0$ and $|y| \le t$),

$$\int_{|x|\ge 2t} |g_k(x-y) - g_k(x)|\, dx \le C\, \|f\|_{\mathcal{H}} \min(2^k t, (2^k t)^{-1/2}),$$

where $C = 2\sqrt{\pi(1+c^2)}$. It follows that

$$\int_{|x|\ge 2t} |h_m(x-y) - h_m(x)|\, dx \le C\, \|f\|_{\mathcal{H}} \sum_{k\in\mathbb{Z}} \min(2^k t, (2^k t)^{-1/2}). \tag{28}$$

We split the sum as $\sum_{k\in J} + \sum_{k\in J^c}$, where

$$J := \{k \in \mathbb{Z};\, 2^k t \le (2^k t)^{-1/2}\} = \{k; 2^k t \le 1\} = \{k; k = -j, j \ge \log_2 t\},$$

and $J^c := \mathbb{Z} - J$. We have

$$\sum_{k\in J} = t \sum_{j \ge \log_2 t} \frac{1}{2^j} = \sum_{j - \log_2 t \ge 0} \frac{1}{2^{j - \log_2 t}} \le \sum_{j - \log_2 t \ge 0} \frac{1}{2^{[j - \log_2 t]}} \le 2.$$

Similarly

$$\sum_{k\in J^c} = t^{-1/2} \sum_{k > -\log_2 t} (1/\sqrt{2})^k = \sum_{k + \log_2 t > 0} (1/\sqrt{2})^{k + \log_2 t}$$

$$\le \sum_{k+\log_2 t > 0} (1/\sqrt{2})^{[k + \log_2 t]} \le \frac{1}{1 - (1/\sqrt{2})}.$$

We then conclude from (28) that h_m satisfies Hypothesis I of Theorem II.8.4 with $K = K'\|f\|_{\mathcal{H}}$, where $K' = C(2 + (\sqrt{2}/\sqrt{2} - 1))$ is independent of m. □

Notation. Let

$$\mathcal{K} := \{f \in L^\infty;\ MDf \in L^\infty\},$$

where $L^\infty := L^\infty(\mathbb{R})$, M denotes the multiplication by x operator, and D is understood in the distribution sense.

The norm on $\mathcal{K}$ is

$$\|f\|_{\mathcal{K}} = \|f\|_\infty + \|MDf\|_\infty.$$

If $f \in \mathcal{K}$, we have for all $k \in \mathbb{Z}$

$$2^{-k/2} \|f I_k\|_2 \le 2^{-k/2} \|f\|_\infty |2^k S|^{1/2} = \sqrt{3} \|f\|_\infty,$$

and

$$2^{k/2}\|(Df)I_k\|_2 = 2^{k/2}\|(MDf)(1/x)I_k\|_2$$
$$\leq \|MDf\|_\infty \left(2^k \int_{2^k S} x^{-2}\, dx\right)^{1/2} = \sqrt{3}\|MDf\|_\infty.$$

Therefore

$$\|f\|_{\mathcal{H}} \leq \sqrt{3}\|f\|_{\mathcal{K}}$$

and $\mathcal{K} \subset \mathcal{H}$. We then have the following.

Corollary II.8.7. *Let $u \in \mathcal{S}'$ be such that $\mathcal{F}u \in \mathcal{K}$. Then for each $p \in (1, \infty)$, the map $T : \psi \in \mathcal{S} \to u * \psi$ is of strong type (p, p), with strong (p, p)-norm $\leq C_p\|\mathcal{F}u\|_{\mathcal{K}}$.*

(The constant C_p here may be taken as $\sqrt{3}$ times the constant C_p in Theorem II.8.6.)

II.9 Some holomorphic semigroups

II.9.1. We shall apply Corollary II.8.7 to the study of some holomorphic semigroups of operators and their boundary groups.

Let $\mathbb{C}^+ = \{z \in \mathbb{C}; \Re z > 0\}$. For any $z \in \mathbb{C}^+$ and $\epsilon > 0$, consider the function

$$K_{z,\epsilon}(x) = \Gamma(z)^{-1}\, \mathrm{e}^{-\epsilon x} x^{z-1} \quad (x > 0) \tag{1}$$

and $K_{z,\epsilon}(x) = 0$ for $x \leq 0$.

Clearly, $K_{z,\epsilon} \in L^1 \subset \mathcal{S}'$, and a calculation using residues shows that

$$(\mathcal{F}K_{z,\epsilon})(y) = (1/\sqrt{2\pi})(\epsilon^2 + y^2)^{-z/2}\, \mathrm{e}^{-\mathrm{i}z \arctan(y/\epsilon)}. \tag{2}$$

We get easily from (2) that

$$(MD\mathcal{F}K_{z,\epsilon})(y) = -\frac{zy}{\epsilon + \mathrm{i}y}(\mathcal{F}K_{z,\epsilon})(y). \tag{3}$$

Hence

$$|MD\mathcal{F}K_{z,\epsilon}| \leq |z|\, |\mathcal{F}K_{z,\epsilon}|. \tag{4}$$

Therefore, for all $z = s + \mathrm{i}t,\ s \in \mathbb{R}^+,\ t \in \mathbb{R}$,

$$\|\mathcal{F}K_{z,\epsilon}\|_{\mathcal{K}} \leq (1 + |z|)\|\mathcal{F}K_{z,\epsilon}\|_\infty \leq (1/\sqrt{2\pi})\epsilon^{-s}\, \mathrm{e}^{\pi|t|/2}(1 + |z|). \tag{5}$$

By Corollary II.8.7, it follows from (5) that the operator

$$T_{z,\epsilon} : f \to K_{z,\epsilon} * f$$

acting on $L^p(\mathbb{R})$ $(1 < p < \infty)$ has $B(L^p(\mathbb{R}))$-norm $\leq C\,\epsilon^{-s}\,\mathrm{e}^{\pi|t|/2}(1+|z|)$, where C is a constant depending only on p. In the special case $p = 2$, the factor $C(1+|z|)$ can be omitted from the estimate, since

$$\begin{aligned}\|T_{z,\epsilon}f\|_2 &= \|K_{z,\epsilon} * f\|_2 = \|\mathcal{F}(K_{z,\epsilon} * f)\|_2 = \sqrt{2\pi}\|(\mathcal{F}K_{z,\epsilon})(\mathcal{F}f)\|_2 \\ &\leq \sqrt{2\pi}\|\mathcal{F}K_{z,\epsilon}\|_\infty \|f\|_2 \leq \epsilon^{-s}\,\mathrm{e}^{\pi|t|/2}\|f\|_2,\end{aligned}$$

by (2) and the fact that $\mathcal{F}$ is isometric on $L^2(\mathbb{R})$.

Consider $L^p(\mathbb{R}^+)$ as the closed subspace of $L^p(\mathbb{R})$ consisting of all (equivalence classes of) functions in $L^p(\mathbb{R})$ vanishing a.e. on $(-\infty, 0)$. This is an *invariant subspace* for $T_{z,\epsilon}$, and

$$(T_{z,\epsilon}f)(x) = \Gamma(z)^{-1}\int_0^x (x-y)^{z-1}\,\mathrm{e}^{-\epsilon(x-y)}f(y)\,dy \quad (f \in L^p(\mathbb{R}^+)). \tag{6}$$

We have for all $f \in L^p(\mathbb{R}^+)$

$$\begin{aligned}\|T_{z,\epsilon}f\|_{L^p(\mathbb{R}^+)} &= \|T_{z,\epsilon}f\|_{L^p(\mathbb{R})} \leq \|T_{z,\epsilon}\|_{B(L^p(\mathbb{R}))}\|f\|_{L^p(\mathbb{R})} \\ &= \|T_{z,\epsilon}\|_{B(L^p(\mathbb{R}))}\|f\|_{L^p(\mathbb{R}^+)}\end{aligned}$$

hence

$$\|T_{z,\epsilon}\|_{B(L^p(\mathbb{R}^+))} \leq \|T_{z,\epsilon}\|_{B(L^p(\mathbb{R}))} \leq C\epsilon^{-s}\,\mathrm{e}^{\pi|t|/2}(1+|z|) \tag{7}$$

for all $z = s+it \in \mathbb{C}^+$. (Again, the factor $C(1+|z|)$ can be omitted in (7) in the special case $p = 2$.)

For $z, w \in \mathbb{C}^+$ and $y \in \mathbb{R}$, we have by (14) in II.3.2 and (2)

$$\begin{aligned}[\mathcal{F}(K_{z,\epsilon} * K_{w,\epsilon})](y) &= \sqrt{2\pi}[(\mathcal{F}K_{z,\epsilon})(\mathcal{F}K_{w,\epsilon})](y) \\ &= (1/\sqrt{2\pi})(\epsilon^2+y^2)^{-(z+w)/2}\,\mathrm{e}^{-\mathrm{i}(z+w)\arctan(y/\epsilon)} \\ &= [\mathcal{F}K_{z+w,\epsilon}](y).\end{aligned}$$

By the uniqueness property of the L^1-Fourier transform (cf. II.3.3), it follows that

$$K_{z,\epsilon} * K_{w,\epsilon} = K_{z+w,\epsilon}. \tag{8}$$

Therefore, for all $f \in L^p$,

$$\begin{aligned}(T_{z,\epsilon}T_{w,\epsilon})f &= T_{z,\epsilon}(T_{w,\epsilon}f) = K_{z,\epsilon} * (K_{w,\epsilon} * f) \\ &= (K_{z,\epsilon} * K_{w,\epsilon}) * f = K_{z+w,\epsilon} * f = T_{z+w,\epsilon}f.\end{aligned}$$

Thus, $z \to T_{z,\epsilon}$ is *a semigroup of operators* on $\mathbb{C}^+$, i.e.,

$$T_{z,\epsilon}T_{w,\epsilon} = T_{z+w,\epsilon} \quad (z, w \in \mathbb{C}^+). \tag{9}$$

For any $N > 0$, consider the space $L^p(0, N)$ (with Lebesgue measure) and the classical Riemann–Liouville *fractional integration* operators

$$(J^z f)(x) = \Gamma(z)^{-1}\int_0^x (x-y)^{z-1}f(y)\,dy \quad (f \in L^p(0,N)).$$

It is known that $z \to J^z \in B(L^p(0,N))$ is strongly continuous in $\mathbb{C}^+$. Since

$$\|(T_{z,\epsilon} - T_{w,\epsilon})f\|_{L^p(0,N)} \le \|(J^z - J^w)(\mathrm{e}^{\epsilon x} f)\|_{L^p(0,N)} \tag{10}$$

the function $z \to T_{z,\epsilon} \in B(L^p(0,N))$ is strongly continuous as well. For the space $L^p(\mathbb{R}^+)$, we have by (10)

$$\begin{aligned}\|T_{z,\epsilon}f - T_{w,\epsilon}f\|^p_{L^p(\mathbb{R}^+)} \le \|(J^z - J^w)(\mathrm{e}^{\epsilon x} f)\|^p_{L^p(0,N)} \\ + \int_N^\infty \mathrm{e}^{-\epsilon p x/2} |T_{z,\epsilon/2}(\mathrm{e}^{\epsilon x/2} f)|^p \, dx \\ + \int_N^\infty \mathrm{e}^{-\epsilon p x/2} |T_{w,\epsilon/2}(\mathrm{e}^{\epsilon x/2} f)|^p \, dx.\end{aligned}$$

For $f \in C_c(\mathbb{R}^+)$, $\mathrm{e}^{\epsilon x/2} f \in L^p(\mathbb{R}^+)$, and therefore, for all $z = s + it$ and $w = u + iv$ in $\mathbb{C}^+$, it follows from (7) that the sum of the integrals over (N, ∞) can be estimated by

$$\begin{aligned}\mathrm{e}^{-\epsilon p N/2} C^p \Big[(\epsilon/2)^{-ps} \, \mathrm{e}^{\pi p|t|/2} (1 + |z|)^p \\ + (\epsilon/2)^{-pu} \; \mathrm{e}^{\pi p |v|/2} (1 + |w|)^p\Big] \, \|\mathrm{e}^{\epsilon x/2} f\|^p_{L^p(\mathbb{R}^+)}.\end{aligned}$$

Fix $z \in \mathbb{C}^+$, and let M be a bound for the expression in square brackets when w belongs to some closed disc $\bar{B}(z, r) \subset \mathbb{C}^+$. Given $\delta > 0$, we may choose N such that

$$\mathrm{e}^{-\epsilon p N/2} C^p M \, \|\mathrm{e}^{\epsilon x/2} f\|^p_{L^p(\mathbb{R}^+)} < \delta^p.$$

For this N, it then follows from the strong continuity of J^z on $L^p(0,N)$ that

$$\limsup_{w \to z} \|(T_{z,\epsilon} - T_{w,\epsilon})f\|_{L^p(\mathbb{R}^+)} \le \delta.$$

Thus, $T_{w,\epsilon} f \to T_{z,\epsilon} f$ as $w \to z$ for all $f \in C_c(\mathbb{R}^+)$. Since $\|T_{w,\epsilon}\|_{B(L^p(\mathbb{R}^+))}$ is bounded for w in compact subsets of $\mathbb{C}^+$ (by (7)) and $C_c(\mathbb{R}^+)$ is dense in $L^p(\mathbb{R}^+)$, it follows that $T_{w,\epsilon} \to T_{z,\epsilon}$ in the strong operator topology (as $w \to z$). Thus the function $z \to T_{z,\epsilon}$ is strongly continuous in $\mathbb{C}^+$. A similar argument (based on the corresponding known property of J^z) shows that $T_{z,\epsilon} \to I$ strongly, as $z \in \mathbb{C}^+ \to 0$. (Another way to prove this is to rely on the strong continuity of $T_{z,\epsilon}$ at $z = 1$, which was proved earlier; by the semigroup property, it follows that $T_{z,\epsilon} f \to f$ in $L^p(\mathbb{R}^+)$-norm for all f in the range of $T_{1,\epsilon}$, which is easily seen to be dense in $L^p(\mathbb{R}^+)$. Since the $B(L^p(\mathbb{R}^+))$-norms of $T_{z,\epsilon}$ are uniformly bounded in the rectangle $Q := \{z = s + it;\ 0 < s \le 1,\ |t| \le 1\}$, the result follows.) An application of Morera's theorem shows now that $T_{z,\epsilon}$ is an analytic function of z in $\mathbb{C}^+$. ($T_{z,\epsilon}$ is said to be a *holomorphic semigroup on* $\mathbb{C}^+$.)

Fix $t \in \mathbb{R}$ and $\delta > 0$. For $z \in B^+(\mathrm{i}t, \delta) := \{z \in \mathbb{C}^+;\ |z - \mathrm{i}t| < \delta\}$, we have by (7)

$$\|T_{z,\epsilon}\| \le C \max(\epsilon^{-\delta}, 1) e^{(\pi/2)(|t|+\delta)} (1 + |t| + \delta). \tag{11}$$

If $z_n \in B^+(it, \delta)$ converge to it and $f = T_{1,\epsilon} g$ for some $g \in L^p(\mathbb{R}^+)$, then

$$T_{z_n,\epsilon} f = T_{z_n+1,\epsilon} g \to T_{\mathrm{i}t+1,\epsilon} g$$

in $L^p(\mathbb{R}^+)$, by strong continuity of $T_{z,\epsilon}$ at the point $it + 1$. Thus, $\{T_{z_n,\epsilon} f\}$ is Cauchy in $L^p(\mathbb{R}^+)$ for each f in the *dense* range of $T_{1,\epsilon}$, and therefore, by (11), it is Cauchy for all $f \in L^p(\mathbb{R}^+)$. If $z'_n \in B^+(\mathrm{i}t, \delta)$ also converge to it, then $T_{z_n,\epsilon} f - T_{z'_n,\epsilon} f \to 0$ in $L^p(\mathbb{R}^+)$ for all f in the range of $T_{1,\epsilon}$ (by strong continuity of $T_{z,\epsilon}$ at $z = 1 + \mathrm{i}t$), hence, for all $f \in L^p(\mathbb{R}^+)$, by (11) and the density of the said range. Therefore the L^p-limit of the Cauchy sequence $\{T_{z_n,\epsilon} f\}$ (for each $f \in L^p(\mathbb{R}^+)$) exists and is independent of the particular sequence $\{z_n\}$ in $B^+(\mathrm{i}t, \delta)$. This limit (denoted as usual $\lim_{z \to \mathrm{i}t} T_{z,\epsilon} f$) defines a linear operator denoted by $T_{\mathrm{i}t,\epsilon}$. By (11)

$$\|T_{\mathrm{i}t,\epsilon} f\|_p \le C \max(\epsilon^{-\delta}, 1) \mathrm{e}^{(\pi/2)(|t|+\delta)} (1 + |t| + \delta) \|f\|_p,$$

where the norms are $L^p(\mathbb{R}^+)$-norms. Since $\delta > 0$ is arbitrary, we conclude that

$$\|T_{\mathrm{i}t,\epsilon}\| \le C\, \mathrm{e}^{\pi|t|/2} (1 + |t|) \quad (t \in \mathbb{R}) \tag{12}$$

where the norm is the $B(L^p(\mathbb{R}^+))$-norm and C depends only on p. (The factor $C(1 + |t|)$ can be omitted in case $p = 2$.)

Since $T_{w,\epsilon}$ is a bounded operator on $L^p(\mathbb{R}^+)$ (cf. (7)), we have for each $w \in \mathbb{C}^+$, $t \in \mathbb{R}$, and $f \in L^p(\mathbb{R}^+)$,

$$T_{w,\epsilon} T_{\mathrm{i}t,\epsilon} f = \lim_{z \to \mathrm{i}t} T_{w,\epsilon} T_{z,\epsilon} f = \lim_{z \to \mathrm{i}t} T_{w+z,\epsilon} f = T_{w+\mathrm{i}t,\epsilon} f.$$

Also, by definition,

$$T_{\mathrm{i}t,\epsilon} T_{w,\epsilon} f = \lim_{z \to \mathrm{i}t} T_{z,\epsilon} T_{w,\epsilon} f = \lim_{z \to \mathrm{i}t} T_{z+w,\epsilon} f = T_{w+\mathrm{i}t,\epsilon} f.$$

Thus

$$T_{w,\epsilon} T_{\mathrm{i}t,\epsilon} f = T_{\mathrm{i}t,\epsilon} T_{w,\epsilon} f = T_{w+\mathrm{i}t,\epsilon} f \tag{13}$$

for all $w \in \mathbb{C}^+$ and $t \in \mathbb{R}$. (In particular, for all $s \in \mathbb{R}^+$ and $t \in \mathbb{R}$,

$$T_{s+\mathrm{i}t,\epsilon} = T_{s,\epsilon} T_{\mathrm{i}t,\epsilon} = T_{\mathrm{i}t,\epsilon} T_{s,\epsilon}.)$$

Letting $w \to is$ in (13) (for any $s \in \mathbb{R}$), it follows from the definition of the operators $T_{\mathrm{i}t,\epsilon}$ and their boundedness over $L^p(\mathbb{R}^+)$ that

$$T_{\mathrm{i}s,\epsilon} T_{\mathrm{i}t,\epsilon} f = T_{\mathrm{i}(s+t),\epsilon} f \quad (s, t \in \mathbb{R}; f \in L^p(\mathbb{R}^+)). \tag{14}$$

Thus, $\{T_{\mathrm{i}t,\epsilon};\ t \in \mathbb{R}\}$ is a *group of operators*, called *the boundary group* of the holomorphic semigroup $\{T_{z,\epsilon};\ z \in \mathbb{C}^+\}$. The boundary group is strongly continuous. (We use the preceding argument: for $f = T_{1,\epsilon} g$ for some $g \in L^p$,

$$T_{\mathrm{i}s,\epsilon} f = T_{1+\mathrm{i}s,\epsilon} g \to T_{1+\mathrm{i}t,\epsilon} g = T_{\mathrm{i}t,\epsilon} f$$

as $s \to t$, by strong continuity of $T_{z,\epsilon}$ at $z = 1 + \mathrm{i}t$. By (12), the same is true for all $f \in L^p$, since $T_{1,\epsilon}$ has dense range in L^p.) We formalize the previous results as follows.

Theorem II.9.2. *For each $\epsilon > 0$ and $p \in (1, \infty)$, the family of operators $\{T_{z,\epsilon}; z \in \mathbb{C}^+\}$ defined by (6) has the following properties:*

(1) It is a holomorphic semigroup of operators in $L^p(\mathbb{R}^+)$;

(2) $\lim_{z \in \mathbb{C}^+ \to 0} T_{z,\epsilon} = I$ in the s.o.t.;

(3) $\|T_{z,\epsilon}\| \le C(1 + |z|)\epsilon^{-s}\,\mathrm{e}^{\pi|t|/2}$ for all $z = s + \mathrm{i}t \in \mathbb{C}^+$, where the norm is the operator norm on $L^p(\mathbb{R}^+)$ and C is a constant depending only on p (in case $p = 2$, the factor $C(1 + |z|)$ can be omitted).

(4) The boundary group

$$T_{\mathrm{i}t,\epsilon} := (\textit{strong}) \lim_{z \in \mathbb{C}^+ \to \mathrm{i}t} T_{z,\epsilon} \quad (t \in \mathbb{R})$$

exists, and is a strongly continuous group of operators in $L^p(\mathbb{R}^+)$ with Properties (12) and (13).

We may apply the theorem to the classical Riemann–Liouville semigroup J^z on $L^p(0, N)$ $(N > 0)$. Elements of $L^p(0, N)$ are regarded as elements of $L^p(\mathbb{R}^+)$ vanishing outside the interval $(0, N)$. All p-norms below are $L^p(0, N)$-norms. The inequality (7) takes the form

$$\|T_{z,\epsilon}\|_{B(L^p(0,N))} \le C(1 + |z|)\epsilon^{-s}\,\mathrm{e}^{\pi|t|/2} \quad (z = s + \mathrm{i}t \in \mathbb{C}^+). \tag{15}$$

For all $f \in L^p(0, N)$,

$$\|J^z f\|_p \le \|K_z\|_1 \|f\|_p,$$

where $K_z(x) = \Gamma(z)^{-1} x^{z-1}$ for $x \in (0, N)$. Calculating the L^1-norm above we then have

$$\|J^z\| \le \frac{N^s}{s|\Gamma(z)|} \quad (z = s + \mathrm{i}t \in \mathbb{C}^+). \tag{16}$$

Since $0 \le 1 - \mathrm{e}^{-\epsilon(x-y)} \le \epsilon(x - y)$ for $0 \le y \le x \le N$, we get for $z = s + \mathrm{i}t \in \mathbb{C}^+$

$$|J^z f - T_{z,\epsilon} f| \le \epsilon \frac{\Gamma(s+1)}{|\Gamma(z)|} J^{s+1}|f|,$$

and therefore by (16)

$$\|J^z f - T_{z,\epsilon} f\|_p \le \epsilon \frac{|z|\, N^{s+1}}{(s+1)|\Gamma(z+1)|} \|f\|_p. \tag{17}$$

By (15) and (17), we have for all $z \in Q := \{z = s + \mathrm{i}t \in \mathbb{C}^+; s \le 1, |t| \le 1\}$

$$\|J^z f\|_p \le \|J^z f - T_{z,1} f\|_p + \|T_{z,1} f\|_p \le M\,\|f\|_p, \tag{18}$$

where

$$M = \sup_{0<s\le 1;|t|\le 1} \frac{N^{s+1}\sqrt{2}}{|\Gamma(s+1+\mathrm{i}t)|} + C(1+\sqrt{2})\mathrm{e}^{\pi/2}.$$

Given $t \in \mathbb{R}$, let $n = [|t|] + 1$. Then $z = s + \mathrm{i}t \in nQ$ for $s \le 1$, and therefore, by the semigroup property (with $w = z/n \in Q$),

$$\|J^z\| = \|J^{nw}\| = \|(J^w)^n\| \le \|J^w\|^n \le M^n = M^{[|t|]+1}.$$

This inequality is surely valid in a neighborhood $B^+(it, \delta)$, and the argument of Section II.8 implies the existence of the boundary group $J^{\mathrm{i}t}$, defined as the strong limit of J^z as $z \to \mathrm{i}t$. By (17) and (15), we also have (for $z = s+\mathrm{i}t \in \mathbb{C}^+$)

$$\begin{aligned}\|J^z f\|_p &\le \|J^z f - T_{z,\epsilon} f\|_p + \|T_{z,\epsilon} f\|_p \\ &\le \epsilon \frac{|z|\, N^{s+1}}{(s+1)|\Gamma(z+1)|}\|f\|_p + C(1+|z|)\epsilon^{-s}\mathrm{e}^{\pi|t|/2}\|f\|_p.\end{aligned}$$

Letting $s \to 0$, we get for all $f \in L^p(0, N)$

$$\|J^{\mathrm{i}t} f\|_p \le \epsilon \frac{|t|\, N}{|\Gamma(\mathrm{i}t+1)|}\|f\|_p + C(1+|t|)\mathrm{e}^{\pi|t|/2}\|f\|_p.$$

Since ϵ is arbitrary, we conclude that

$$\|J^{\mathrm{i}t}\| \le C(1+|t|)\mathrm{e}^{\pi|t|/2} \quad (t \in \mathbb{R}). \tag{19}$$

As before, the factor $C(1 + |t|)$ can be omitted in case $p = 2$. Formally:

Corollary II.9.3. *For each $p \in (1, \infty)$ and $N > 0$, the Riemann–Liouville semigroup J^z on $L^p(0, N)$ has a boundary group $J^{\mathrm{i}t}$ (defined as the strong limit of J^z as $z \in \mathbb{C}^+ \to \mathrm{i}t$). The boundary group is strongly continuous, satisfies and it the identity $J^{\mathrm{i}t} J^w = J^w J^{\mathrm{i}t} = J^{w+\mathrm{i}t}$ (for all $t \in \mathbb{R}$ and $w \in \mathbb{C}^+$) and the growth relation (19).*

Bibliography

Akhiezer, N. I. and Glazman, I. M. (1962, 1963). *Theory of Linear Operators in Hilbert Space*, Vols I, II, Ungar, New York.

Ash, R. B. (1972). *Real Analysis and Probability*, Academic Press, New York.

Bachman, G. and Narici, L. (1966). *Functional Analysis*, Academic Press, New York.

Banach, S. (1932). *Theorie des Operations Lineaires*, Hafner, New York.

Berberian, S. K. (1961). *Introduction to Hilbert Space*, Oxford University Press, Fair Lawn, NJ.

Berberian, S. K. (1965). *Measure and Integration*, Macmillan, New York.

Berberian, S. K. (1974). *Lectures in Functional Analysis and Operator Theory*, Springer-Verlag, New York.

Blackadar, B. (2006). *Operator Algebras, Theory of C^*-Algebras and von Neumann Algebras*, Springer-Verlag, Berlin.

Bonsall, F. F. and Duncan, J. (1973). *Complete Normed Algebras*, Springer-Verlag, Berlin.

Bourbaki, N. (1953, 1955). *Espaces Vectoriels Topologiques*, livre V, Hermann, Paris.

Browder, A. (1968). *Introduction to Function Algebras*, Benjamin, New York.

Brown, N. P. and Ozawa, N. (2008). *C^*-Algebras and Finite-Dimensional Approximations*, Providence, Rhode Island.

Brown, A. and Pearcy, C. (1977). *Introduction to Operator Theory*, Springer-Verlag, New York.

Colojoara, I. and Foias, C. (1968). *Theory of Generalized Spectral Operators*, Gordon and Breach, New York.

Davidson, K. R. (1996). *C^*-Algebras by Example*, Providence, RI.

Davies, E. B. (1980). *One-Parameter Semigroups*, Academic Press, London.

Dieudonne, (1960). *Foundations of Modern Analysis*, Academic Press, New York.

Dixmier, J. (1964). *Les C^*-algebres et leurs Representations*, Gauthier-Villar, Paris.

Dixmier, J. (1969). *Les Algebres d'Operateurs dans l'Espace Hilbertien*, 2nd ed., Gauthier-Villar, Paris.

Doob, J. L. (1953). *Stochastic Processes*, Wiley, New York.

Douglas, R. G. (1972). *Banach Algebra Techniques in Operator Theory*, Academic Press, New York.

Dowson, H. R. (1978). *Spectral Theory of Linear Operators*, Academic Press, London.

Dunford, N. and Schwartz, J. T. (1958, 1963, 1972). *Linear Operators*, Parts I, II, III, Interscience, New York.

Edwards, R. E. (1965). *Functional Analysis*, Holt, Rinehart, and Winston, Inc., New York.

Friedman, A. (1963). *Generalized Functions and Partial Differential Equations*, Prentice-Hall, Englewood Cliffs, NJ.

Friedman, A. (1970). *Foundations of Modern Analysis*, Holt, Rinehart, and Winston, New York.

Gamelin, T. W. (1969). *Uniform Algebras*, Prentice-Hall, Englewood Cliffs, NJ.

Goffman, C. and Pedrick, G. (1965). *First Course in Functional Analysis*, Prentice-Hall, Englewood Cliffs, NJ.

Goldberg, S. (1966). *Unbounded Linear Operators*, McGraw-Hill, New York.

Goldstein, J. A. (1985). *Semigroups of Operators and Applications*, Oxford, New York.

Halmos, P. R. (1950). *Measure Theory*, Van Nostrand, New York.

Halmos, P. R. (1951). *Introduction to Hilbert Space and the Theory of Spectral Multiplicity*, Chelsea, New York.

Halmos, P. R. (1967). *A Hilbert Space Problem Book*, Van Nostrand-Reinhold, Princeton, NJ.

Hewitt, E. and Ross, K. A. (1963, 1970). *Abstract Harmonic Analysis*, Vols. I, II, Springer-Verlag, Berlin.

Hewitt, E. and Stromberg, K. (1965). *Real and Abstract Analysis*, Springer-Verlag, New York.

Hille, E. and Phillips, R. S. (1957). *Functional Analysis and Semigroups*, A.M.S. Colloq. Publ. 31, Providence, RI.

Hirsch, F. and Lacombe, G. (1999). *Elements of Functional Analysis*, Springer, New York.

Hoffman, K. (1962). *Banach Spaces of Analytic Functions*, Prentice-Hall, Englewood Cliffs, NJ.

Hormander, L. (1960). Estimates for translation invariant operators, *Acta Math.*, **104**, 93–140.

Hormander, L. (1963). *Linear Partial Differential Equations*, Springer-Verlag, Berlin.

Kadison, R. V. and Ringrose, J. R. (1997). *Fundamentals of the Theory of Operator Algebras*, Vols. I, II, III, A.M.S. Grad. Studies in Math., Providence, RI.

Kantorovitz, S. (1983). *Spectral Theory of Banach Space Operators*, Lecture Notes in Math., Vol. 1012, Springer, Berlin.

Kantorovitz, S. (1995). *Semigroups of Operators and Spectral Theory*, Pitman Research Notes in Math., Vol. 330, Longman, New York.

Kantorovitz, S. (2010). *Topics in Operator Semigroups*, Birkhäuser, Boston, MA.

Kato, T. (1966). *Perturbation Theory for Linear Operators*, Springer-Verlag, New York.

Katznelson, Y. (1968). *An Introduction to Harmonic Analysis*, Wiley, New York.

Kothe, G. (1969). *Topological Vector Spaces*, Springer-Verlag, New York.

Lang, S. (1969). *Analysis II*, Addison-Wesley, Reading, MA.

Larsen, R. (1973). *Banach Algebras*, Marcel Dekker, New York.

Larsen, R. (1973). *Functional Analysis*, Marcel Dekker, New York.

Loeve, M. (1963). *Probability Theory*, Van Nostrand-Reinhold, Princeton, NJ.

Loomis, L. H. (1953). *An Introduction to Abstract Harmonic Analysis*, Van Nostrand, New York.

Malliavin, P. (1995). *Integration and Probability*, Springer-Verlag, New York.

Maurin, K. (1967). *Methods of Hilbert Spaces*, Polish Scientific Publishers, Warsaw.

Megginson, R. E. (1998). *An Introduction to Banach Space Theory*, Springer, New York.

Munroe, M. E. (1971). *Introduction to Measure and Integration*, 2nd ed., Addison-Wesley, Reading, MA.

Murphy, G. J. (1990). *C^*-Algebras and Operator Theory*, Academic Press, Boston, MA.

Naimark, M. A. (1959). *Normed Rings*, Noordhoff, Groningen.

Palmer, T. W. (1994, 2001). *Banach Algebras and the General Theory of $*$-Algebras*, Vols I, II, Cambridge University Press, Cambridge.

Pazy, A. (1983). *Semigroups of Linear Operators and Applications to Partial Differential Equations*, Springer, New York.

Pedersen, G. K. (2018). *C^*-Algebras and their Automorphism Groups*, 2nd ed., Academic Press, London.

Reed, M. and Simon, B. (1975). *Methods of Modern Mathematical Physics*, Vols I, II, Academic Press, New York.

Rickart, C. E. (1960). *General Theory of Banach Algebras*, Van Nostrand-Reinhold, Princeton, NJ.

Riesz, F. and Sz-Nagy, B. (1955). *Functional Analysis*, Ungar, New York.

Royden, H. L. (1968). *Real Analysis*, 2nd ed., Macmillan, New York.

Rudin, W. (1962). *Fourier Analysis on Groups*, Interscience-Wiley, New York.

Rudin, W. (1973). *Functional Analysis*, McGraw-Hill, New York.

Rudin, W. (1974). *Real and Complex Analysis*, 2nd ed., McGraw-Hill, New York.

Sakai, S. (1971). *C^*-Algebras and W^*-Algebras*, Springer-Verlag, New York.

Schmüdgen, K. (2012). *Unbounded Self-Adjoint Operators on Hilbert Space*, Springer, Dordrecht.

Schwartz, L. (1951). *Theorie des Distributions*, Vols I, II, Hermann, Paris.

Stone, M. (1932). *Linear Transformations in Hilbert Space*, A.M.S. Colloq. Publ. 15, Providence, RI.

Stout, E. L. (1971). *The Theory of Uniform Algebras*, Bogden and Quigley, New York.

Strătilă, Ş. (1981). *Modular Theory in Operator Algebras*, Abacus Press, Tunbridge Wells.

Takesaki, M. (2002, 2003, 2003). *Theory of Operator Algebras*, Vols I, II, III, Springer-Verlag, Berlin.

Taylor, A. E. (1958). *Introduction to Functional Analysis*, Wiley, New York.

Tucker, H. G. (1967). *A Graduate Course in Probability*, Academic Press, New York.

Weil, A. (1951). *L'integration dans les Groupes Topologiques et ses Applications*, 2nd ed., Act. Sci. et Ind. 869, 1145, Hermann et Cie, Paris.

Wheeden, R. L. and Zygmund, A. (1977). *Measure and Integral*, Marcel Dekker, New York.

Wilansky, A. (1964). *Functional Analysis*, Blaisdell, New York.

Yosida, K. (1966). *Functional Analysis*, Springer-Verlag, Berlin.

Zaanen, A. C. (1953). *Linear Analysis*, North-Holland, Amsterdam.

Zygmund, A. (1959). *Trigonometric Series*, 2nd ed., Cambridge University Press, Cambridge.

Index